C0-AZG-329

Genetics and Gene Therapy

The International Library of Medicine, Ethics and Law
Series Editor: Michael D. Freeman

Titles in the Series

Death, Dying and the Ending of Life
Margaret P. Battin, Leslie Francis and Bruce M. Landesman

Abortion
Belinda Bennett

Ethics and Medical Decision-Making
Michael D. Freeman

Children, Medicine and the Law
Michael D. Freeman

Mental Illness, Medicine and Law
Martin Lyon Levine

The Elderly
Martin Lyon Levine

Genetics and Gene Therapy
Sheila A.M. McLean

Rights and Resources
Frances H. Miller

Organ Donation and Transplantation
David Price

AIDS: Society, Ethics and Law
Udo Schüklenk

Women, Medicine, Ethics and the Law
Susan Sherwin and Barbara Parish

Legal and Ethical Issues in Human Reproduction
Bonnie Steinbock

Medical Practice and Malpractice
Harvey Teff

Human Experimentation and Research
George F. Tomossy and David N. Weisstub

Medicine and Industry
George F. Tomossy

Genetics and Gene Therapy

Edited by

Sheila A.M. McLean

University of Glasgow, UK

Published by
Dartmouth Publishing Company
Ashgate Publishing Limited
Gower House
Croft Road
Aldershot
Hants GU11 3HR
England

Ashgate Publishing Company
Suite 420
101 Cherry Street
Burlington, VT 05401-4405
USA

Ashgate website: http://www.ashgate.com

British Library Cataloguing in Publication Data
Genetics and gene therapy. – (The international library of medicine, ethics and law)
1. Human gene mapping – Moral and ethical aspects 2. Gene therapy – Moral and ethical aspects
I. McLean, Sheila A.M.
174.2'5

Library of Congress Cataloging-in-Publication Data
Genetics and gene therapy / edited by Sheila A.M. McLean.
p. cm. — (The International library of medicine, ethics and law)
Includes bibliographical references.
ISBN 0-7546-2055-7 (alk. paper)
1. Medical genetics—Moral and ethical aspects. 2. Gene therapy—Moral and ethical aspects. I. McLean, Sheila. II. Series.

RB155.G38745 2005
174.2—dc22

2004054555

ISBN 0 7546 2055 7

Printed in Great Britain by The Cromwell Press, Trowbridge, Wiltshire

Contents

Acknowledgements

The editor and publishers wish to thank the following for permission to use copyright material.

A B Publishers for the essay: Philippa Gannon and Charlotte Villiers (1999), 'Genetic Testing and Employee Protection', *Medical Law International*, **4**, pp. 39–57. Copyright © 1999 A B Academic Publishers.

American Society of Law, Medicine and Ethics for the essays: Susan M. Wolf (1995), 'Beyond "Genetic Discrimination": Toward the Broader Harm of Geneticism', *Journal of Law, Medicine and Ethics*, **23**, pp. 345–53. Copyright © 1995 American Society of Law, Medicine and Ethics; Lawrence O. Gostin (1995), 'Genetic Privacy', *Journal of Law, Medicine and Ethics*, **23**, pp. 320–30. Copyright © 1995 American Society of Law, Medicine and Ethics; George J. Annas, Lori B. Andrews and Rosario M. Isasi (2002), 'Protecting the Endangered Human: Toward an International Treaty Prohibiting Cloning and Inheritable Alterations', *American Journal of Law and Medicine*, **28**, pp. 151–78. Copyright © 2002 American Society of Law, Medicine and Ethics; Deborah Hellman (2003), 'What Makes Genetic Discrimination Exceptional?', *American Journal of Law and Medicine*, **29**, pp. 77–116. Copyright © 2003 American Society of Law, Medicine and Ethics; John A. Robertson (2003), 'Procreative Liberty in the Era of Genomics', *American Journal of Law and Medicine*, **29**, pp. 439–87. Copyright © 2003 American Society of Law, Medicine and Ethics.

Blackwell Publishing for the essays: Julia Black (1998), 'Regulation as Facilitation: Negotiating the Genetic Revolution', *Modern Law Review*, **61**, pp. 621–60. Copyright © 1998 Modern Law Review; Julian Kinderlerer and Diane Longley (1998), 'Human Genetics: The New Panacea?', *Modern Law Review*, **61**, pp. 603–20. Copyright © 1998 Modern Law Review; Roger Brownsword (2002), 'Stem Cells, Superman, and the Report of the Select Committee', *Modern Law Review*, **65**, pp. 568–87. Copyright © 2002 Modern Law Review; Darryl Macer (1991), 'Whose Genome Project?', *Bioethics*, **5**, pp. 183–211; Søren Holm (2002), 'Going to the Roots of the Stem Cell Controversy', *Bioethics*, **16**, pp. 493–507; Stephen Robertson and Julian Savulescu (2002), 'Is there a Case in Favour of Predictive Genetic Testing in Young Children?', *Bioethics*, **15**, pp. 26–49.

British Medical Journal Publishing Group for the essays: Heather Draper and Ruth Chadwick (1999), 'Beware! Preimplantation Genetic Diagnosis May Solve Some Old Problems But It Also Raises New Ones', *Journal of Medical Ethics*, **25**, pp. 114–20; Julian Savulescu (1999), 'Should We Clone Human Beings? Cloning as a Source of Tissue for Transplantation', *Journal of Medical Ethics*, **25**, pp. 87–95; John Harris (1997), '"Goodbye Dolly?" The Ethics of Human Cloning', *Journal of Medical Ethics*, **23**, pp. 353–60.

Cambridge University Press for the essays: Carson Strong (1998), 'Cloning and Infertility', *Cambridge Quarterly of Healthcare Ethics*, **7**, pp. 279–93. Copyright © 1998 Cambridge University Press; Sara Goering (2000), 'Gene Therapies and the Pursuit of a Better Human', *Cambridge Quarterly of Healthcare Ethics*, **9**, pp. 330–41. Copyright © 2000 Cambridge University Press; John Harris (2003), 'Stem Cells, Sex, and Procreation', *Cambridge Quarterly of Healthcare Ethics*, **12**, pp. 353–71. Copyright © 2003 Cambridge University Press.

Elsevier for the essay: Sheila A.M. McLean (2001), 'The Gene Genie: Good Fairy or Wicked Witch?', *Studies in History and Philosophy of Biological and Biomedical Sciences*, **32**, pp. 723–39. Copyright © 2001 Elsevier Science Ltd.

Hastings Center Report for the essay: Mark S. Frankel (2003), 'Inheritable Genetic Modification and a Brave New World: *Did Huxley Have it Wrong*?', *Hastings Center Report*, **33**, pp. 31–36.

Johns Hopkins University Press for the essays: Allen Buchanan, Andrea Califano, Jeffrey Kahn, Elizabeth McPherson, John Robertson and Baruch Brody (2002), 'Pharmacogenetics: Ethical Issues and Policy Options', *Kennedy Institute of Ethics Journal*, **12**, pp. 1–15. Copyright © 2002 Johns Hopkins University Press; Lainie Friedman Ross (2002), 'Predictive Genetic Testing for Conditions that Present in Childhood', *Kennedy Institute of Ethics Journal*, **12**, pp. 225–44. Copyright © 2002 Johns Hopkins University Press.

Taylor & Francis Group for the essay: Graeme T. Laurie (2001), 'Challenging Medical-Legal Norms: The Role of Autonomy, Confidentiality, and Privacy in Protecting Individual and Familial Group Rights in Genetic Information', *Journal of Legal Medicine*, **22**, pp. 1–54. Copyright © 2001 Taylor & Francis.

Every effort has been made to trace all the copyright holders, but if any have been inadvertently overlooked the publishers will be pleased to make the necessary arrangement at the first opportunity.

Series Preface

Few academic disciplines have developed with such pace in recent years as bioethics. And because the subject crosses so many disciplines important writing is to be found in a range of books and journals, access to the whole of which is likely to elude all but the most committed of scholars. The International Library of Medicine, Ethics and Law is designed to assist the scholarly endeavour by providing in accessible volumes a compendium of basic materials drawn from the most significant periodical literature. Each volume contains essays of central theoretical importance in its subject area, and each throws light on important bioethical questions in the world today. The series as a whole – there will be fifteen volumes – makes available an extensive range of valuable material (the standard 'classics' and the not-so-standard) and should prove of inestimable value to those involved in the research, teaching and study of medicine, ethics and law. The fifteen volumes together – each with introductions and bibliographies – are a library in themselves – an indispensable resource in a world in which even the best-stocked library is unlikely to cover the range of materials contained within these volumes.

It remains for me to thank the editors who have pursued their task with commitment, insight and enthusiasm, to thank also the hard-working staff at Ashgate – theirs is a mammoth enterprise – and to thank my secretary, Anita Garfoot for the enormous assistance she has given me in bringing the series from idea to reality.

MICHAEL FREEMAN
Series Editor
Faculty of Laws
University College London

Introduction

Few areas of science and medicine have resulted in the volume of academic and popular literature as has genetics. The so-called revolution in understanding of the causes of disease states, and even behavioural traits, has focused public attention on the influence of genes in making us what we are. Rapidly, however, the potential benefits of such understanding were overtaken, in the public mind at least, by the question of the possible (negative) implications of genetic knowledge and associated technologies. Meanwhile, science developed with considerable speed, holding out new, but by no means uncontentious, avenues of research and therapy.

The essays in this volume show just how wide-ranging concern has become, ranging from regulation to cloning, with the fear of discrimination in between. Such is the breadth and depth of the issues identified as being potentially problematic that structuring the volume has been difficult. Although the volume is split into two distinct Parts, there is no overwhelming theme within each one and there was no easy way of organizing the chapters internally within each Part. Part I begins with a range of general discussions about the genetic enterprise itself, followed by consideration of some specific questions. Part II then addresses cutting-edge debates in genetics.

Genetics – General

In Chapter 1, Julian Kinderlerer and Diane Longley provide a fascinating insight into the history of the science in relation to plants and animals, followed by an overview of the emerging issues, including xenotransplantation, genetic screening, testing and cloning. Their essay is, however, also concerned to make the case for adequate regulation of the science and scientists. Noting that scientists and physicians 'are no better equipped than anyone else to identify the social and moral issues that might result from their use . . .', they conclude that, '[o]nly society can ultimately decide the degree of importance to be attached to the benefits, hazards and impact of these' (p. 20). Thus, they argue, rather than being the enemy of scientific and medical advance – as is sometimes suggested – law has a vital role to play in providing 'the mechanisms that can facilitate effectively the negotiation between fact and value and the taking of necessary decisions about the nature and direction of scarce resources for both research and policy' (ibid.).

This is an important if not uncontroversial, conclusion, which is taken up and expanded on vigorously by Julia Black in Chapter 2. She, too, focuses largely on the role of regulation, highlighting, in particular, the problems of communication between the disciplines involved in the genetics revolution. These problems relate both to assessment of the 'problems' which genetic technology poses as well as to finding the way forward. For Black:

> . . . regulation has a role to play which is not, or not simply, about control, but rather about facilitation. Regulation has an important role to play in connecting the arguments of participants, in facilitating the integration of the wide range of views as to the appropriate course that the technology and its regulation should take. (p. 21)

She further contends that finding the negotiated way forward is unlikely to be an easy task. However, she also argues that the tendency to focus on institutional responses to the challenges posed by genetics should logically be pre-dated by trying to unpick what each of the relevant disciplines are saying and why they are saying it. Thus, it is necessary to establish some appropriate form of communication between the wide range of actors and interested parties. In this sense, she proposes that '. . . regulation needs to facilitate communication by taking on the role of interpreter or translator . . .' (p. 23).

Different perceptions need to be accommodated within regulation – views which recognize that science cannot monopolize the debate, and that the lay perspective is not inevitably irrational or unreasoning. In Black's view, this negotiation is of particular significance in genetics because of the powerful position genetic knowledge now holds in a variety of arenas – not simply the public imagination. This power, she says, is not the only reason why the different voices need to be heard. A further justification is proposed:

> . . . despite many scientists' assertions to the contrary, science is not neutral, just as law is not neutral. It shapes society's expectations, and provides it with choices which it did not otherwise have. (p. 57)

Thus, Black concludes:

> . . . facilitation integration is not simply a question of institutional design. Institutional design is an important aspect, but it is not a panacea for all regulatory ills. Simply introducing the public to the regulatory fora will have no effect if it is the scientific language which remains dominant, and only the scientific voice which is recognised as valid. (pp. 59–60)

In Chapter 3, Darryl Macer asks a series of important questions: for example, who began the project? Whose DNA is being sequenced? Who is funding the project? Who should do the work? How do we achieve coordinated data-sharing? How can we best deal with issues about the ownership of results? Who benefits from the results? What of future generations? And, finally, how do we strategize for the international management of common knowledge?

Macer begins from the perspective that '[t]he international nature of the project and its universally applicable results . . . make it a project of all humanity' (p. 88). For this reason, amongst others, he concludes that 'shared authorship and ownership' become 'legally compelling' (p. 89). This theme dominates Macer's discussion, which argues that research into the genome provides the ideal opportunity for '"scientific altruism" on a global scale' (p. 71) but warns that, if the scientific community does not do this by self-regulation, then external regulation will be necessary.

In fact, the ultimate mapping of the genome was driven both by public and private funding, and the debates about what should or should not be in the public domain continued until the very last minute, culminating in President Clinton and Tony Blair arranging a joint announcement of the mapping of the genome. This outcome would probably disappoint Macer, whose vision was that 'the genome, being common to all people, has shared ownership, is a shared asset, and therefore the maps and sequence should be open to all' (p. 75).

In Chapter 4, Sheila McLean turns to a theme common in much literature on this subject; namely the fear of the 'new eugenics' and the discrimination that might flow from this. Using reproductive choice as a template for evaluation of the impact of genetics, she notes the possible slide from rights to obligations. In the current climate, governments increasingly invest in genetic screening pre- and post-implantation. The linking of more and more disease states to genetic influence will, she argues, very possibly lead to increasing pressure to utilize the screening and testing skills which are becoming available, and may lead to the conclusion that an individual failure to take 'advantage' of them is seen as socially irresponsible – ultimately, perhaps, as something not to be tolerated.

Moreover, McLean argues that we should be cautious about attributing too much weight to the impact that genetics has on the individual. Ceding the capacity to describe us primarily as a combination of our genes, and continuing without serious thought to manipulate genes (in whatever way), pose joint threats which should be carefully critiqued. As she says:

> The DNA which makes us what we physically are is central to our sense of self. When manipulation, even extermination, of certain genetic characteristics is proposed, it is little wonder that people are concerned. When science appears to hold within its grasp the power to reshape the future, we are right to be hesitant before embarking on such a path.

Finally, McLean argues that whether the ultimate conclusions of the genetics revolution will be benefit or disbenefit depends on having 'a clear commitment to articulated and robust principles; a clear set of values, consistently applied, which both accommodate and transcend science' (p. 106).

In Chapter 5, John Robertson identifies three positions which could be used to interrogate the use of genetic information in reproductive liberties and choices. The 'strict traditionalist' stance would hold that:

> . . . reproduction is a gift from God, resulting from the loving intimacy of two persons . . . This view would condemn most uses of technology to control or influence the characteristics of offspring because parental selection necessarily conflicts with the idea of 'unconditional gift' and suggests that the child is a made or chosen 'product'. (p. 112)

Robertson, however, concludes that this approach is an inappropriate foundation for public policy, not least because it has its basis in 'a religiously based or metaphysical view of how reproduction should occur and a breadth that would apparently condemn nearly all forms of technological assistance in reproduction' (p. 114). He also claims that, in any event, the adoption of this perspective is essentially counter-intuitive for most intending parents who wish to have healthy children for their own and their children's sake (ibid.).

The second position he identifies is the 'radical liberty' view. This perspective would permit individuals 'to use any reproductive technique they wish for whatever reason' (ibid.), and would effectively delegitimize any restrictions or limitations on this capacity. This perspective, says Robertson, would leave individuals 'free to select, screen, alter, engineer, or clone offspring as they choose. They are the best judges of what is good for them, including what children they have' (ibid.).

Robertson acknowledges that, although it is likely that few people actually hold to this perspective, it has nonetheless been influential, at least in academic thinking on this area. It is, for him however, 'too extreme in its espousal of personal freedom' (p. 115).

A more moderate approach, in Robertson's view, than either of the first two described is what he terms the 'modern traditionalist'. This perspective 'holds that reproductive choice in a liberal, rights-based society is a basic freedom, including the use of genetic and reproductive technologies that are helpful in having healthy, biologically related offspring' (p. 116).

Of critical significance in his argument is the notion of a protected concept of procreative liberty. Thus, it is a rights-based discourse that will, or should, ultimately determine the uses of genetics and genetic medicine. As he says:

> Ultimately, decisions about how to use or not use genomics in human reproduction will be determined, not by biologic necessity or evolutionary theory, but by how those uses fit into the fabric of rights and interests of individual and social choice and responsibility that particular societies recognize. (p. 122)

In addition to the issue of rights, Robertson also argues that intention is important as a predictor of the legitimacy of prenatal (including pre-implantation) manipulation of the human genome, as will be estimations of the benefits or harms that may flow from them. Clearly, the ethical, social and other concerns surrounding the use of genetic technologies will continue, but Robertson argues for a subtle, rather than a broad-brush, approach to their resolution. In particular, he suggests, we should focus not on the outlandish or extreme possible (ab)uses of these techniques and technologies, but rather on 'ensuring that genetic techniques are used in informed, safe, and productive ways. It is not inherently wrong to achieve traditional reproductive goals in novel, technological ways that use the insights of genomic approaches to human biology' (p. 155).

In Chapter 6 Susan Wolf tackles one of the areas that is, arguably, of most concern to the general public – namely, the issue of genetic discrimination. However, she berates those who analyse this concept from a narrow perspective, arguing instead that we should, in this kind of discussion, not ignore:

> . . . years of commentary on race and gender demonstrating the limits of antidiscrimination analysis as an analytical framework and corrective tool. Too much discussion of genetic disadvantage proceeds as if scholars of race and gender had not spent decades critiquing and developing antidiscrimination theory. (p. 159)

Wolf argues that what is really necessary is an analysis that goes beyond, for example, insurers' use of tests and information; it is rather the '. . . eagerness to draw genetic conclusions, the search for supposedly deviant genes, and the conviction that such genes actually deserve disadvantage . . .' that needs to be tackled (p. 161).

She contends that much could be learned from existing analyses from race and gender studies: categorizing individuals, on the basis of genes or other characteristics, is 'systematic, not just individual; a matter of cognitive mind-set, not just isolated behaviors; and a domain of stereotypes and unfounded beliefs, not just accuracy and rationality' (ibid.).

Just as with race, gender, and disability, the identification of a so-called genetic 'norm' is not value neutral; instead, she argues, it is equally 'socially constructed' (p. 162). Thus, the language of genetic discrimination is inadequate to identify what is really happening, and will serve to 'obscure connections between disadvantaging based on race, gender, and genetics, and it will fail to serve those most burdened by those connections' (p. 163).

In attempting to show that the issue is about unacceptable practices, rather than simply concerned with the uses to which information is, or may be, put, Wolf proposes that a more appropriate concept is that of 'geneticism', which connotes 'an offensive and harmful practice, which remains harmful even when based on accurate rather than exaggerated understanding of the role of genes' (p. 164).

In Chapter 7, Deborah Hellman both responds to Wolf and elaborates her own theoretical position. She says, for example:

> What is not clear from Wolf's critique is how one is to know that certain inequalities in the world, such as inequalities that track genetic predispositions to disease, are morally problematic while others are not. (p. 182)

For Hellman, genetics does require special and unique legislation, because 'the social meaning of treating people differently on the basis of their genetic make-up is different from the social meaning of discrimination on the basis of health or illness . . .' (p. 171). She reviews theories that would support legislation specifically to outlaw genetic discrimination, and what she calls new theories – for example, the promotion of health. If the goal of health promotion can be used to justify specific legislation outlawing genetic discrimination, she claims, several empirical questions need to be addressed. Specifically, she says:

> Whether genetic discrimination wrongfully discriminates depends on whether such discrimination expresses that people with serious genetic conditions are less worthy of concern or respect. (p. 205)

In Chapter 8, Dean Bell and Belinda Bennett consider the issue of 'Genetic Secrets and the Family'. Although this essay is not about discrimination, it does ask questions about whether or not, and if so how, genetic discrimination should be treated differently from other kinds of medical information. Self-evidently, this is of special importance within the family unit, where genetic information about one family member is relevant to others within that same family. The question of the interests or rights to be protected in such situations is a critically important one. Should clinicians adhere to their usual vow of confidentiality or are there other values which might better inform decisions about the use of genetic information in this circumstance?

Bell and Bennett explain their own position thus:

> Our starting point is that the answer to the question of whether genetic information in any particular case should be regarded as 'individual' or 'familial' cannot be resolved scientifically – undoubtedly both an individual patient and genetic relatives of that patient will have an interest in genetic information derived from that patient. The resolution of whether the secrecy of that information should be respected in any particular case is therefore an inquiry of an ethical, and ultimately, legal kind. (pp. 210–11)

Bell and Bennett then review some Australian responses to this question and are critical of the notion that membership of a given family is *in se* sufficient to 'warrant setting aside important legal and ethical protections' (p. 211). Thus, they reject the special treatment of genetic information that some commentators have argued for, and conclude that existing rules of confidentiality are not only important, but may already hold within them sufficient space for doctors to breach confidentiality in exceptional circumstances. The familial issues are, therefore, insufficiently important to mandate dilution of the importance of medical confidentiality.

In Chapter 9, Lawrence Gostin interrogates issues surrounding genetic privacy in a wider sense. He argues that the amount of genetic information being stored at present, and likely to be stored in the future, requires the creation of 'a comprehensive genetic information system. . .' which he believes to be both 'technologically feasible' (p. 245) and valuable. Nonetheless, questions still remain over whether or not it is appropriate (or safe?) to continue to accumulate vast amounts of genetic information. A conclusion on this, he says, means that:

> . . . it is necessary to measure the probable effects on the privacy of these groups [individuals, families, etc.]. The diminution in privacy entailed in genetic information systems depends on the sensitive nature of the data, as well as on the safeguards against unauthorised disclosure of the information. (Ibid.)

Gostin points to the fact that whilst modern systems for storing and handling information have great benefits for research and possibly treatment of genetic conditions, they also pose threats to privacy because of their capacity to 'store and decipher unimaginable quantities of highly sensitive data' (ibid.). Thus, he concludes, legal safeguards would offer one way of providing privacy protection.

However, he also concludes that no system is failsafe. Importantly, the public needs to be assured that 'genomic information will be treated in an orderly and respectful manner and that individual claims of control over those data will be adjudicated fairly' (p. 249). Importantly, he points to the fact that adequate regulation is unlikely to be achieved by focusing on the traditional constraints of the doctor–patient relationship. Unlike Bell and Bennett, then, he argues that, 'data collected in our information age is based only in small part on this relationship . . . Focusing legal protection on a single therapeutic relationship within this environment is an anachronistic vestige of an earlier and simpler time in medicine' (p. 246).

The topic of genetic privacy is also considered by Graeme Laurie in Chapter 10. In common with other commentators, Laurie recognizes the importance to individuals of genetic information, in terms of how it is collected as well as how it is managed. He proposes that 'the nature of the interests in issue must be examined to determine precisely which factors, values, perceived benefits, and harms should be weighed in the balance when deciding how genetic information should be handled' (p. 254).

Recognizing the acute difficulties of negotiating the interests involved – individual or familial – Laurie proposes a model of privacy which he believes may serve as a template for making 'hard' decisions. Thus, he divides privacy into two concepts. First, he identifies an interest in 'spatial privacy', which equates to 'a state of nonaccess to the individual's physical body or psychological person . . .' (p. 279).

For Laurie:

> A spatial privacy analysis underscores a right not to know information. If, in the context of a dilemma about whether someone should be told, the individual has no knowledge at all that familial information exists, then the spatial privacy interest stands as a *prima facie* bar to the person being approached and told the information . . . It requires that we reflect on the act of disclosure and places the onus on us not to disclose unless faced with compelling reasons to do so. (p. 291)

Second, Laurie explains, privacy 'can be viewed as a state in which the individual has control over personal information; what I will term informational privacy' (p. 279). Informational privacy 'concerns the interest of the patient in maintaining . . . information in a state of nonaccess and preventing unauthorized use or disclosure of the information to third parties' (pp. 285–86).

Defining of Laurie's thesis therefore is the question of separateness from others. Importantly, he also conceptualizes privacy as a state rather than a claim or a right, arguing that '. . . to define privacy as a state rather than a right or a claim helps us to describe a concept while at the same time avoids imputing value to it' (p. 282). However, this, he argues, does not in fact *preclude* us from 'seeking to accord (legal) protection to such a state for the good ends that it can further and for the interests it can protect' (p. 284).

In Chapter 11, Philippa Gannon and Charlotte Villiers address an issue of considerable concern – namely, the possible uses of genetic testing in the workplace. Some of the commentators already discussed above have applied their minds to genetics and insurance – an issue of considerable concern, perhaps particularly in healthcare systems which rely on private funding through insurance. However, it is unarguable that the potential for the use of genetics in the workplace is also a matter of considerable interest, both to those already in employment and those seeking it.

The authors of this chapter conclude that:

> Perhaps one of the most controversial issues about genetic testing in the workplace is that such testing invokes issues of control and coercion of the employee by the employer. The employer or potential employer may use the genetic tests to exclude the individual from the workplace. This exclusion may take the guise of either refusing to employ a job candidate or . . . dismissing an employee as a consequence of the test results. (p. 308)

Perhaps inevitably, this raises yet again the issue of discrimination, and whether or not the use of genetic information can or should appropriately be described as discriminatory. Like Wolf, Gannon and Villiers also note that genetic discrimination in the workplace can readily encompass the analysis derived from outlawed forms of discrimination in the United Kingdom – namely, race, sex and disability. Because of the nature of genetic information, like Gostin, they conclude that it is 'suitable for specific privacy laws' (p. 318).

This is not to say that they would ignore the value of such information for employers. Indeed, they argue that it is necessary to recognize the need for balance between these interests and those of the actual or potential employee. They conclude, however, that UK law does not currently achieve such a balance and that there is no likelihood of legislation in the UK in the foreseeable future. In light of this, they recommend that there should be an independent body to supervise and advise on genetic tests in the workplace. To an extent, this proposal may have been met by the work of the Human Genetics Commission, although it seems more likely that Gannon and Villiers would prefer to see a specific body tasked with this sole responsibility.

In Chapter 12 Allen Buchanan, Andrea Califano, Jeffrey Kahn, Elizabeth McPherson, John Robertson and Baruch Brody move us into new territory, exploring the implications of pharmacogenetics, which they describe as 'the study of the effects of genotypic variations on drug-response, including safety and efficacy, and drug-drug interaction' (p. 327).

The development of specific, individual understandings of how medicines work has provoked a new storm of concern over the possibilities of ethical and other dilemmas. Buchanan *et al.*, however, caution that:

> As with other applications of genetic knowledge, the challenge of pharmacogenetics is to take genetic variations among people into account to provide better health care, while treating people with equal respect. The first step towards accomplishing this is to avoid certain misapprehensions that may distort public deliberation. Otherwise, concern about ethical issues may result in ill-conceived constraints that unnecessarily limit the benefits pharmacogenetics could deliver. Two misapprehensions are particularly important to avoid: (1) genetic exceptionalism, the assumption that all things genetic inherently involve exceptionally serious ethical concerns and therefore require novel ethical principles

> or special regulatory responses, and (2) overbroad genetic generalization, the assumption that all 'genetic tests' raise the same ethical issues. (p. 331)

Avoiding these pitfalls will, they argue, allow us then to address the real and important ethical issues which they describe as follows:

> (1) regulatory oversight, (2) confidentiality and privacy, (3) informed consent, (4) availability of drugs, (5) access, and (6) clinicians' changing responsibilities in the era of pharmacogenetic medicine. (Ibid.)

Pharmacogenetics seems set to play an important role in the healthcare of the future, and for this reason it is important that we address the ethical implications clearly, both in terms of research and therapies. For Buchanan *et al.*, this will also require 'careful consideration and wise policy responses' (p. 340).

Gene Therapy/Testing/Cloning

Part II of this book deals broadly with issues concerning even more novel aspects of what might be called 'practical' genetics – that is, the (possibly futuristic, but nonetheless important) applications of genetic knowledge to the creation, manipulation or enhancement of human beings. Moreover, the fact that such interventions are currently under consideration arguably makes the resolution of the issues discussed in Part I even more urgent.

In Chapter 13 Heather Draper and Ruth Chadwick consider the emerging science of pre-implantation genetic diagnosis (PIGD). They first take account of the benefits which may be achieved by the use of this technique. Fundamentally, they say, it allows couples to have an unaffected child; second, it is preferable to abortion; and, third, it can maintain the distinction between actively destroying life and failure to save it (by implanting an embryo). Fourth, 'it gives greater choice to couples because it gives them the scope to make a decision about which of the embryos to implant . . .' (p. 346).

Each of these benefits protects the freedom of parents to secure the best possible future for their children – a value which most would probably acknowledge. Draper and Chadwick then go on to hypothesize three scenarios in which PIGD might be sought by potential parents. First, they consider a scenario in which a couple decide to implant an embryo even knowing that it is affected with a genetic disorder because they want a child which is genetically related to both of them (p. 347). Although they have identified the value of parental decision-making above, they conclude that the question raised is:

> . . . whether being genetically related to the child is an incidental or fundamental good of parenting. Or, put another way, does the right to found a family include the right to be genetically related to one's child? On balance, the answer to this question is no. (Ibid.)

They then postulate two other situations; first, where intending parents want (or are indifferent to) the implantation of a Down's embryo because they don't believe it is a serious condition and, second, where intending parents seek implantation of a deaf child because they think deafness is a good thing, or at least not a disability. Taking the first case, Draper and Chadwick argue that respecting parental choice, as one of the benefits they have discussed, rests on the 'presumption that parents have the best interests of their children at heart, that they of all people can be trusted to do what is best for their children' (p. 348).

However, they argue that parental freedom in this case has to be constrained by the fact that '[i]t is one thing to say that those who have Down's are valuable and quite another to say that the choice between choosing to implant an affected or an unaffected embryo has no moral significance' (p. 349). We need, it is argued to take into account the interests of the future children, not least because: '[b]y offering a choice between possible future persons, PIGD provides not only the opportunity to avoid avoidable harms, but also the possibility of maximising advantage or enhancing quality of life' (ibid.).

The second case, that of selecting a deaf child, has already taken place in the United States since this essay was written. In this case two deaf lesbian women used sperm from a man who was congenitally deaf in order to have a deaf child (see Savulescu, 2002). For Draper and Chadwick, the deliberate choice to have a disabled child:

> . . . highlights one of the disadvantages of giving parental decision making moral authority, namely that it can act as a trump card in those areas where there is no right answer, but where the decision which parents want to make, to choose a state in which a child is born with a hearing system which does not work, is one that most rational bystanders would not take. (p. 350)

However, in an echo of the judgment in the case of *Re B (a minor)*,[1] Draper and Chadwick do seem to concede that the constraints which should be imposed on parental choice in this area are less justifiable where the quality of life for the future child is not 'so disadvantaged that its life would not be worth living' (ibid.).

Finally, the authors consider the question of whether or not availability of PIGD is in fact an enhancement of reproductive freedom. They note that nobody is forced to participate in PIGD, but express concern that, once a decision to do so has been made, the balance of power may shift to others. Thus, they conclude:

> It remains to be seen . . . whether PIGD should be marketed as affording greater autonomy and reproductive freedom to couples when, as things stand, they are effectively putting the decision in the hands of the treating clinician. (p. 351)

A subject of considerable controversy is analysed by Lainie Friedman Ross in Chapter 14 – namely, the use of predictive genetic testing for childhood onset genetic conditions. Ross notes that there is a plethora of guidelines on predictive genetic testing of children, but laments that most such guidelines do not address the testing of 'predictive testing of asymptomatic children for disorders that present later in childhood when presymptomatic treatment cannot influence the course of the disease' (pp. 353–54).

Many consensus statements about genetic testing of children express the concern that they should not be precluded from making autonomous decisions in the future about whether or not to be tested. Such concerns are generally sufficient to mandate delaying the tests until the child becomes an adult. However, this clearly cannot cover the situation where the condition being screened for is likely to manifest itself (if present) *before* adulthood is attained. Testing of this latter type is not subject, therefore, to the concern that future autonomous choice is being frustrated. However, and problematically, predictive testing for childhood onset conditions could have associated disbenefits, which – as Ross points out – are presently unknown and 'probably will be found to vary depending on individual familial circumstances' (p. 357).

Ross then goes on to consider the benefits and disbenefits of such predictive testing in families known to be at high risk of certain conditions. She argues that parents should, as a default position, have 'presumptive medical decision-making authority' in such situations 'unless there is evidence to show that their decisions are abusive or neglectful' (p. 361). This conclusion is, however, subject to two important caveats. First, such testing should be conducted as a research project, so that relevant evidence about benefits and harms can be collected and, second, physicians should have no obligation to offer or encourage uptake of such tests, given that the possible harms are currently unknown.

Ross then turns to the question of predictive population screening where the condition screened for is untreatable or evidence as to the efficacy of presymptomatic treatment is equivocal. Here, she asserts that:

> One of the main dangers with expanding newborn screening programs to include predictive screening programs, even within well-defined research protocols, is that they may be confused with the more traditional diagnostic newborn screening programs in which treatments provide significant clinical benefits to the child . . . consent for predictive screening must be clearly separated from consent for diagnostic screening. (p. 365)

Ross therefore regards the ethics of predictive testing as resting effectively on a kind of risk–benefits analysis, which we, as yet, lack the empirical evidence to resolve.

In Chapter 15, Stephen Robertson and Julian Savulescu also consider the question of predictive testing in young children. They define a predictive test as:

> . . . one in which: (i) an individual has no evidence of pathology (and thus no symptoms or signs of disease) at the time of the testing; and (ii) the result of the test indicates a high chance of developing pathology in the future, either in the person tested or his/her offspring. (p. 376)

They take a different approach from that of Ross. One of the main arguments against such testing, as they see it, is that knowledge of an impending genetic condition would be 'unbearable'. They argue, however, that there is little evidence to support this claim, and instead suggest that '[t]he paediatric oncology literature suggests that children cope better with concrete and frank information if it is available, as opposed to a policy of non-disclosure or suspended uncertainty' (p. 379).

The authors identify three further arguments for not testing in the circumstances under consideration. Broadly, these are:

> 1. Failure to respect the child's future autonomy.
> 2. Breach of confidentiality.
> 3. Harm to the child. (p. 385)

In response to the first argument, they argue that it is not correct to conclude that childhood testing necessarily reduces future choice. Instead, they suggest that:

> The child who is not tested is denied an option of growing up and adapting to the knowledge of their genetic status during their formative years. Thus the choice is not between two courses of action, one of which simply has more choice for the later adult, but between two mutually exclusive futures . . . . (p. 386)

In other words, there may be benefits to testing, even if it appears to deny future autonomous choice. Indeed, Robertson and Savulescu propose that genetic testing may even 'promote the development of autonomous decision-making' (p. 389).

As to concerns about breach of confidentiality, the authors argue that families are often privy to otherwise private information about other members of the family. Thus, they claim:

> Disclosure of such information to parents is hardly a significant breach of confidentiality itself . . . Parents are privy to all sorts of sensitive and personal information about their children. Family relations are built around trust. What is more important is what parents do with that information.

Finally, on the question of harm, the authors seem to agree with Ross that evidence about the possibility of causing harm to the child is lacking. Given that they believe that the other arguments against testing are not sufficiently robust to prevent it, it becomes of significant importance to identify whether or not any such harm would in fact accrue.

In Chapter 16 Mark Frankel moves on to the question of genetic modification – a topic that has generated several volumes of literature, often based on popular fiction such as Aldous Huxley's *Brave New World*. Frankel suggests that the nightmare scenario outlined by Huxley is, in fact, now outdated. Rather, he says:

> . . . the greater danger, I believe, is a highly individualized marketplace fueled by an entrepreneurial spirit and the free choice of large numbers of parents that could lead us down a path, albeit incrementally, that abandons the lottery of evolution in favor of intentional genetic modification. The discoveries of genetics will not be imposed on us. Rather, they will be sold to us by the market as something we cannot live without. (p. 398)

Frankel is concerned with genetic modifications: that is, 'interventions capable of modifying genes that are transmitted to offspring and to generations beyond' (ibid.). He concludes that a matter of particular concern about the use of inherited genetic modifications (IGMs) is that they will become most useful for genetic enhancement, that is: 'improving human traits that without intervention would be within the range of what is commonly regarded as normal, or improving them beyond what is needed to maintain or restore good health' (p. 399).

Frankel identifies the possibility that the rapid advances that now seem possible will seduce us into focusing on the genetic characteristics of a particular condition or trait at the expense of 'solving problems that are really embedded in the structure of our society' (ibid.). This perspective echoes many of the commentators whose work has already been discussed. In other words, a preoccupation with genetic explanation obscures the fundamental, underpinning individual or institutional attitudes which determine how one is dealt with in or by a particular society. And Frankel points to one further important consideration. In his analysis it is difficult to separate out the developments in genetics which are designed to develop new or improved medical interventions from those which may result in the temptation to enhance. The techniques will be the same, leading, he suggests, to 'creeping enhancement applications as well' (ibid.).

Of course, some authors would argue that there is no real difference between striving to enhance people's existence by genetic means and doing so by means of medication, for example. This Frankel dismisses. Rather, he argues, there is a difference, and this difference is that:

> Inheritable genetic enhancement would have long-term effects on persons yet to be born. Thus we have little, if any, precedent for this way of using IGM. We would be venturing into unknown territory, but without any sense of where the boundaries should lie, much less with an understanding of what it means to cross such boundaries. (p. 400)

At this stage, Frankel takes us back to his original assertion: namely, that external (market) forces will ultimately influence the use of IGM. He describes the 'cottage industry' that has, he says, grown up around assisted reproduction and concludes that:

> What began as an effort by fertility clinics to help infertile couples have a child is now a growth industry offering a range of services no longer confined to the infertile. (p. 401)

The concept of enhancement, or improving human beings, is taken up by Sara Goering in Chapter 17. Goering argues that we already see value in 'bettering humanity and/or our own children' (p. 403), but suggests that 'while no one denies the importance of this quite general goal, we are still left with difficult issues about *how* we ought to proceed in addressing that goal' (ibid.). If we accept that improving or bettering people is a good thing, then, says Goering, it might be assumed that if we *can* do so using genetic therapy, then we *should* do so. However, this assumption should, it is argued, not go unchallenged. Two main reasons for this caution are given. First, that we do not know what will be the effect of changing or removing one gene on others and, second, that we might lose something valuable

In any event, Goering also raises the important question '[H]ow are we to decide what is to count as "bettering" children?' (p. 405). Even if the answer to this seems clear – that avoiding disability would amount to 'bettering', for example – and even if we can distinguish between valuing living people with disabilities and preventing the birth of future people so affected, 'in practice, public attitudes toward such individuals are likely to be prejudiced and will likely affect public financial support of the disabled' (ibid.).

So, Goering argues, caution is required in using available or future technology in this way. A robust decision-making process 'that will help to delineate what traits are acceptable candidates for genetic therapy or genetic engineering . . .' is needed (p. 406). Goering concedes that using the standard medical model may assist in making decisions about some 'defects or diseases' (ibid.). This, however, does not answer questions about enhancement and may not even completely resolve the question of disease, because, as Goering suggests 'the label "disease" is not metaphysically pure' (p. 407), and the line between enhancement and disease is, therefore, not always completely clear.

In an interesting move, Goering then suggests that one way of distancing ourselves from 'the particular biases that our society has for traits that are otherwise not genuinely physically desirable' (p. 408) would be to use Rawls's concept of the 'veil of ignorance'. Although conceding that this is not free of the criticisms that have been made of Rawls's work, she argues that 'it may at least help to figure out a way to start the difficult process of distinguishing between legitimate genetic intervention and discriminatory or arbitrary intervention' (p. 409).

If we could use some such conceptual analysis, Goering concludes, then:

> . . . genetic enhancement is not clearly an evil deserving of outright prohibition. Rather, we need to make careful and reasoned decisions about what genetic and/or physical changes would truly constitute human improvement, and what changes would only serve to reproduce our societal biases. (p. 411)

In Chapter 18 George Annas, Lori Andrews and Rosario Isasi propose that inheritable alterations and human cloning should become subject to an international treaty which would prohibit them. In their view:

> . . . cloning and inheritable alterations can be seen as crimes against humanity of a unique sort: they are techniques that can alter the essence of humanity itself (and thus threaten to change the foundation of human rights) by taking human evolution into our own hands and directing it toward the development of a new species, sometimes termed the 'posthuman'. (p. 417)

According to their argument, safety concerns alone are sufficient to justify an outright international ban. However, they also try to deal with two arguments in favour of cloning, should the safety issues be resolved. First, they say, it could be argued that cloning is in reality just another form of reproduction which could help the infertile and, second, that it represents scientific progress and that it would be 'antiscientific' to ban it.

In answer to these arguments, the authors say that the first ignores the rights and interests of some of those involved (for example, children and women) but also that this contention wrongly equates 'duplication or replication' with human reproduction. As they say:

> Asexual replication may or may not be categorized by future courts as a form of human reproduction, but there are strong arguments against it. First, asexual reproduction changes a fundamental characteristic of what it means to be human . . . Second, the 'child' of an asexual replication is also the twin brother of the male 'parent', a relationship that has never existed before in human society . . . Third, the genetic replica of a genetically sterile man would be sterile himself and could only 'reproduce' by cloning. (p. 423)

If, they say, infertility is seen as a bad thing, then – in their scenario – it would be unethical for a doctor deliberately to create someone who will be infertile.

The authors describe the main arguments against cloning and inheritable genetic alterations in this way:

> The primary arguments against cloning and inheritable genetic alterations which we believe make an international treaty the appropriate action, have been summarized in detail elsewhere. In general, the arguments are that these interventions would require massive dangerous and unethical human experimentation, that cloning would inevitably be bad for the resulting children by restricting their right to an 'open future', that cloning would lead to a new eugenics movement for 'designer children' (because if an individual could select the entire genome of their future child, it would seem impossible to prohibit individuals from choosing one or more specific genetic characteristics of their future children), and that it would likely lead to the creation of a new species or subspecies of humans, sometimes called the 'posthuman'. (pp. 425–26)

Thus, they conclude, it is necessary that a universal ban be enacted, by way of their proposed treaty, to prevent doctors and scientists from simply crossing boundaries to defeat the law in any country in which such techniques are banned. Their treaty, they argue, is to be commended because it conceptualizes the events under consideration as human rights issues and reflects the general concerns of most societies. It will also prevent descent down a slippery slope towards a neo-genetic world.

In Chapter 19 John Harris focuses specifically on the ethics of human cloning. Taking a very different approach from that of Annas *et al.*, Harris claims that '[i]f the objection to cloning is to the creation of identical individuals separated in time (because the twin embryos might be implanted in different cycles, perhaps even years apart), it is a weak one at best' (p. 443).

For Harris, the central ethical issue is whether or not it is defensible to deny women the chance to have a desperately wanted child. In terms of cell nuclear replacement (CNR), the technique used to create 'Dolly', Harris notes that a prominent argument used against permitting people to be created in this way relates to the assertion that it is an affront to human dignity. However, he says that '[a]ppeals to human dignity . . . while universally attractive, are comprehensively vague . . .' (p. 444).

It is also often argued that the use of CNR will inevitably allow prospective parents to achieve goals of their own. For Harris, if the child's existence is a benefit, then the reasons for its creation are either morally irrelevant, or at least subordinate, matters. Further, he points to the fact that – at least in some societies – it is permissible to use human embryos in research and also that many of them will necessarily be destroyed. If this is acceptable, he asks, on what basis should we not also use them to create cells lines for therapeutic purposes? In other words, Harris sees no inherent moral objections to beneficial uses of the human embryo, nor does he see the method of its creation as necessarily of moral relevance. He appeals to a moral principle that it is better to do some good than to do none at all – a principle which, he says, carries, at least intuitively, every bit as much weight as others, such as Kant's moral imperative.

Harris then turns to the argument that people have a right to two parents – something which the clone could not have, at least genetically. He doubts the strength of any such assertion and points to the fact that:

> If the right to have two parents is understood to be the right to have two social parents, then it is of course only violated by cloning if the family identified as the one to rear the resulting child is a one-parent family. (p. 446)

In any event, he suggests, this situation is no more likely to arise in cloning than in any other reproductive situation.

On the other hand, if the 'right' to have two parents is understood as an argument based on genetics, he claims that:

> . . . the supposition that this right is violated when the nucleus of the cell of one individual is inserted into the de-nucleated egg of another, is false in the way this claim is usually understood. There is at least one sense in which a right expressed in this form might be violated by cloning, but not in any way which has force as an objection. Firstly, it is false to think that the clone is the genetic child of the nucleus donor. It is not. The clone is the twin brother or sister of the nucleus donor and the offspring of the nucleus donor's own parents. (p. 447)

Finally, Harris turns to the argument that reproductive cloning would encourage people down a dangerous self-gratifying path of creating others in their own image. In response he asserts that:

> . . . there is no way that they could make such an individual a duplicate of themselves. So many years later the environmental influences would be radically different, and since every choice, however insignificant, causes a life-path to branch with unpredictable consequences, the holy grail of duplication would be doomed to remain a fruitless quest. (p. 448)

In conclusion, Harris argues that reproductive freedoms are guaranteed by states, whether or not in writing. If the state wishes to interfere with these freedoms, then it would have to provide compelling arguments; simple intuitive hostility is not enough.

In Chapter 20 Carson Strong analyses the question as to whether or not cloning could be ethically defensible for the infertile. The essay begins, however, by setting the debate in the future, at a time when the safety problems of cloning have been resolved. Thus, he asks the reader 'to consider the possibility that in the future humans could be cloned without a significantly elevated risk of birth defects from the cloning process itself' (p. 451).

Strong cites six reasons why having a child is important. These are as follows:

> . . . having a child involves participation in the creation of a person; it can be an affirmation of a couple's mutual love and acceptance of each other; it can contribute to sexual intimacy; it provides a link to future persons; it involves experiences of pregnancy and childbirth; and it leads to experiences associated with child rearing. (p. 453)

The essay then reviews some of the main arguments against cloning. First, it addresses the alleged psychological effects of not being unique, but rather sharing a genetic make-up with an adult already in existence. The fear is that the adult's life would be set up as a kind of 'standard' for measuring the clone against. Strong argues, however, that a convincing response to this concern can be made. For one thing,

> . . . parents' lives often are held up as standards, even in the absence of cloning . . . Similarly, a clone's being given a role model or standard is not necessarily bad . . . If it is used by parents in a loving and nurturing manner, it can help children develop their autonomy, rather than inhibit it. (p. 455)

Another argument is that a clone would suffer inevitable psychological harm, but Strong maintains that this misunderstands the concept of harm. Persons are, Strong argues, 'harmed only if they are caused to be worse off than they otherwise would have been . . . The claim that cloning harms the children who are brought into being, therefore, amounts to saying that the children are *worse off than they would have been if they had not been created*' (p. 456).

Strong does, however, concede that cloning could sometimes equate to a harm, using the example of deliberately creating a child with an inherited condition (such as cystic fibrosis) when there is no cure. This, Strong says, *would* amount to a harm because so doing would knowingly breach a posited right to be born free of impediments. In such cases, we might reasonably argue that the child is harmed because her existence could be seen to be worse than non-existence. For those born without such problems, he maintains, the assumption that they would suffer psychological harm is based on no foundation in fact.

The author then moves on to the potential difficulties for the parent–child relationship, and the possibility that children would be 'objectified' (p. 458). Strong's argument in favour of some exceptions to a general ban on cloning rests on cases where there is both male and female infertility, and so he posits that because such cases would be so few in number, there would be no general effect on the parent–child relationship that would justify banning this particular use of cloning.

It is also argued that cloning is open to abuse by unscrupulous people. However, in counter-argument, Strong suggests that this possible scenario bears no relationship to the infertility case on which his argument is based, and can therefore be clearly distinguished from it. He argues that:

> . . . cloning does not harm the child, nor is it clear that the right to a decent minimum opportunity for development would be violated. The argument that parent-child relationships generally would be adversely affected continues to be unpersuasive because enhancement is not involved and, although we are now dealing with a larger class of infertile couples, the number is still relatively small compared to the general population. (p. 461)

Strong concludes that cloning may sometimes be ethically defensible in cases of infertility, but not before countering two additional and apparently strong reasons against it. First, is the argument that the possible abuses of cloning warrant banning it, even if it could be argued to be ethical in some cases. Also, it could be difficult to enforce a general policy against cloning if you allow exceptions. Strong argues that, even if there are difficulties in enforcement, there would still in all likelihood be 'widespread compliance' (p. 463), not least because clinicians rarely deliberately break the law. The second argument is that, if it is ethical to permit human cloning for the infertile, then why could it not also be argued to be ethical for the fertile? Here, Strong refers back to the six reasons why it is good to have a child and points out that, by denying cloning to the fertile, we deprive them of none of these possible benefits. On the other hand, we do deny them to the infertile if we place an outright ban on cloning in all circumstances.

In Chapter 21 Julian Savulescu considers a different aspect of cloning – namely, that which is designed to produce therapies by cloning tissue for transplantation. As he notes, cloning could produce a variety of therapies by generating cells, tissue or even organs for therapeutic purposes. As he says, '[a]n embryo, fetus, child or adult could be produced by cloning, and solid organs or differentiated tissue could be extracted from it' (p. 468).

Clearly, even if this were possible, the use of human embryos in this way is not uncontroversial, particularly for those who believe that a human life begins at conception. Although cloning does not involve conception in the traditional understanding of the concept, he notes that '[p]roponents of the persons-begin-to-exist-at-conception view might reply that cloning is like conception' (p. 470). However, Savulescu asks whether or not we need to accept this argument, saying, '[c]onception involves the unification of two different entities, the sperm and the egg, to form a new entity, the totipotent stem cell. In the case of cloning, there is identity between the cell before and after nuclear transfer – it is the same cell' (ibid.).

The moral question, for Savulescu, is whether cells *can* become human beings and whether or not they *should*. To those who use the potential of embryos as an argument against, for example, abortion, he points out that, if all our cells could become people, then there is no moral distinction between any cell and an embryo. As he says:

> If all our cells could be persons, then we cannot appeal to the fact that an embryo could be a person to justify the special treatment we give it. Cloning forces us to abandon the old arguments supporting special treatment of fertilised eggs. (p. 471)

Like Harris, Savulescu also dismisses the importance of the 'yuk factor' that often characterizes discussion about cloning, noting that previous medical advances have also been regarded with distaste and yet have become acceptable – techniques such as artificial insemination and IVF, for example. For him, 'the fact that people find something repulsive does not settle whether it is wrong. The achievement in applied ethics, if there is one, of the last 50 years has been to get people to rise above their gut feelings and examine the reasons for a practice' (p. 473).

Savulescu concludes that '[t]he most justified use of human cloning is arguably to produce stem cells for the treatment of disease' (p. 474). Indeed, he asserts that, rather than destroying millions of surplus embryos, they could be used for the benefit of others.

Søren Holm's essay (Chapter 22) provides a valuable overview of the debate surrounding stem cells. Holm's argument is not concerned with the possibility of creating 'a fully-grown mammal by nuclear replacement' (p. 479). Rather, it is about the use of cloning techniques and embryonic stem cell culture. As he notes, '[w]hen these two techniques are combined it becomes possible to produce embryonic stem cells that are almost genetically identical to any given adult human being' (ibid.).

If stem cells could be developed and grown in significant quantities, then at least in theory, a wide range of therapeutic potential is opened up. Stem cells could be grown in the laboratory and used to cure or alleviate diseases. Moreover, it may even become possible to grow whole organs, alleviating a real and pressing shortfall between supply and demand.

So, the question becomes how one could regulate this whole area: '. . . how the use of embryos for stem cell research and therapy can be fitted into a legislative structure that either relies on a view that embryos have some moral value, or is a direct result of political compromise' (p. 482). Savulescu argues that:

> Giving some moral status to embryos does not automatically rule out embryonic stem cell research, since it can be argued that the likely benefits in terms of reduction of human suffering and death in many cases outweigh the sacrifice of a (small?) number of human embryos. (Ibid.)

Finally, he says:

> The most interesting of these [questions] are the questions surrounding how public policy should be formed in an area where there is 1) agreement about the value of the goal of a particular kind of research (i.e. the creation of effective stem cell therapies), 2) genuine scientific uncertainty about exactly what line of research is most likely to achieve this goal, and 3) disagreement about the ethical evaluation of some of these lines of research but not about others. This question is perhaps more a question of political or legal philosophy than a question of ethics, but it is nevertheless an issue that should be of interest to those bioethicists who want their elegant analyses transformed into public policy. (pp. 490–91)

In a second essay, John Harris (Chapter 23) also turns his attention to the stem cell debate and, like Savulescu, he notes the considerable, theoretical, therapeutic potential of developments in this area. He argues that, although it is right to be cautious about new therapies, 'for the sake of all those awaiting therapy, we should pursue the research that might lead to therapy with all vigor. To fail to do so would be to deny people who might benefit from the possibility of therapy' (p. 495).

For Harris, there are three major reasons why stem cell research is ethically significant:

1. It will for the foreseeable future involve the use and sacrifice of human embryos.
2. Because of the regenerative properties of stem cells, stem cell therapy may always be more than therapeutic – it may involve the enhancement of human functioning and indeed the extension of the human lifespan.

3. So-called therapeutic cloning – the use of cell nuclear replacement to make the stem cell clones of the genome of their intended recipient – involves the creation of cloned pluripotent and possibly totipotent cells, which some people find objectionable. (p. 500)

Considering these, Harris claims that what he calls the 'Principle of Waste Avoidance' (p. 501) can appropriately be used. This would argue that if we can do good, when the alternative is the waste of resources, then 'we have powerful moral reasons to avoid waste and to do good instead' (p. 502). He also argues that 'natural' pregnancies involve the waste of human embryos; thus he says that nature or God already creates 'spare' embryos which are doomed to die so that another may survive. This, he notes:

> . . . may not be intentional sacrifice, and it may not attend every pregnancy, but the loss of many embryos is the inevitable consequence of the vast majority (perhaps all) pregnancies. For everyone who knows the facts, it is a conscious, knowing, and therefore deliberate sacrifice; and for everyone, regardless of 'guilty' knowledge, it is part of the true description of what they do in having or attempting to have children. (Ibid.)

However, Harris is not arguing that just because something happens in nature, we should replicate it deliberately. What he is arguing is that:

> . . . if something happens in nature *and* we find it acceptable in nature given all the circumstances of the case, then if the circumstances are relevantly similar it will for the same reasons be morally permissible to achieve the same result as a consequence of deliberate human choice. (p. 505)

He concludes that:

> Only those who think that it is more important to create new humans than to save existing ones will be attracted to the idea that sexual reproduction is permissible whereas the creation of embryos for therapy is not. (p. 509)

Finally, in Chapter 24 Roger Brownsword considers and reviews the work of the House of Lords Select Committee on Stem Cells, to which he was a legal adviser. He notes that there are four 'recurring objections' to human embryonic stem cell research (hES); 'one scientific, one ethical, one practical, and one legal' (p. 518). The scientific argument relates to the possibility of using adult stem cells rather than hES cells. This would, of course, avoid the problems (of an ethical nature) associated with the use of human embryos. Further, if the potential of adult cells became a reality, then there would be sound scientific reasons for pursing this line of research vigorously. However:

> . . . if therapies derived from adult stem cells involve some reverse engineering (dedifferentiation) before being redirected (differentiated), ES cells provide a crucial point of comparison along the way. At the end of this discussion, the Committee concludes that, whilst recent research on adult stem cells looks promising and should be strongly encouraged by funding bodies and Government, the dual track approach is essential if maximum medical benefit is to be obtained. (p. 519)

The ethical objection, unsurprisingly, relates to the inevitability of the destruction of human embryos involved in this research. Many of the arguments about this are conducted from a religious perspective whereas, according the Brownsword, the House of Lords Select

Committee's Report reflected 'not simply that this is a pluralistic society, but that moralists hold very different views about the status of the embryo and, concomitantly, whether it has rights or protectable interests or whether it represents a protectable good . . .' (p. 520).

Brownsword further draws attention to two recommendations which he says are of considerable significance although they have not received much attention to date. First, the report recommends that where a licence has been given by the Human Fertilisation and Embryology Authority (HFEA) for stem cell research, any cell line generated should be deposited in a national stem cell bank. Second, the HFEA should satisfy itself, before granting any new licences, that there is not already a cell line in existence that would be suitable (ibid.). In this way, Brownsword notes, the Committee was concerned to minimize the need to use large numbers of embryos in research.

The practical objection is based on the slippery slope argument. This argument suggests that, for example, permitting A (which might be a 'good' thing) will inevitably lead us to permitting B (which is a 'bad' thing). Brownsword, however, argues as follows:

> Slippery slopes are only a cause for concern if they create uncertainty or if they carry us towards end states that we judge to be problematic. The momentum of the slope can be relevant to the way in which we address issues of principle or it can be important for its practical consequences; and, whether the bearing is on principle or practice in the first instance, the one can have implications for the other. Although there can be a tendency to appeal rather quickly to slippery slope fears, the objection is far from rhetorical or trivial . . . . (p. 522)

Finally, there was the (legal) concern that the regulations permitting hES research were *ultra vires*. Although this was effectively dismissed as a concern, the Select Committee did consider 'how well the Regulations map onto stem-cell science and its potential development' (p. 524).

Two issues arise. First, two of the new, permissible purposes of stem cell research refer specifically to 'serious disease'. The question, therefore, has to be what is meant by 'serious' for these purposes, and who should decide this?

Second, many of the early research applications will inevitably be for basic research, raising the question as to whether or not this is in fact covered by the new regulations. Although legal advice suggested that this may indeed be permissible, the Committee recommended that this should be put beyond doubt.

Brownsword also points to one interesting omission in the new regulations – namely, increasing knowledge about the *creation* of embryos. As he says:

> It would be ironic if, after the struggle to bring CNR research within the HFEA's regulatory control, it was found that the Regulations had omitted to extend the licensing powers to this strand of research. (p. 527)

Finally, on the legal issues, Brownsword notes that it is now clear that the regulations cover embryos created by fertilization and those produced by cell nuclear replacement. However, he asks, what of embryos created in other ways? He asks:

> What would we make, for example, of eggs or even stem cells that are induced to become embryonic, or of CNR embryos generated from enucleated *animal* eggs, or of embryos that have used an egg engineered to overcome mitochondrial defects (so-called oocyte nucleus transfer)? How far can we continue to stretch the scope of the legislative framework as science devises ever more ways of producing organisms that seem to have embryonic functions and potential? (Ibid.)

Brownsword concludes that the House of Lords report has produced balanced conclusions, which do not seek to prevent or stifle research but at the same time leave open 'the way for new and apparently extremely promising lines of research, but . . . without in any sense seeking to close the door on debate about the necessity, desirability, or acceptability of using human embryos'.

Conclusion

It goes without saying that selecting essays for books of this sort can be invidious. As I said at the beginning of this Introduction, there is a wealth of – often very high quality – academic commentary on the issues raised by the topics covered. Indeed, there are whole topics that have been paid little attention in this collection, largely for reasons of space. For this reason, a short additional bibliography is included below. Nevertheless, the essays selected for inclusion show the wide range of the debate and the very real significance that genetics and its associated developments have for human beings, both individually and collectively.

Note

1 (1981) reported in [1990] 3 All ER 927.

Reference

Savelescu, J. (2002), 'Deaf Lesbians, "Designer Disability", and the Future of Medicine', *British Medical Journal*, **325**, pp. 771–73.

Further Reading

Books

Bodmer, W. and McKie, R. (1994), *The Book of Man*, London: Little Brown.

British Medical Association (1998), *Human Genetics: Choice and Responsibility*, Oxford: Oxford University Press.

Buchanan, A., Brock, D.W., Daniels, N. and Wikler, D. (2000), *From Chance to Choice*, Cambridge: Cambridge University Press.

Kevles, D.J. (1985), *In the Name of Eugenics*, Cambridge, MA: Harvard University Press. Reprinted 2001.

Kitcher, P. (1997), *The Lives to Come: The Genetic Revolution and Human Possibilities*, New York: Touchstone/Simon and Schuster.

Nussbaum, M.C. and Sunstein, C.R. (eds) (1998), *Clones and Clones: Facts and Fantasies About Human Cloning*, New York and London: W.W. Norton & Company.

Rose. S., Lewontin, R.C. and Kamin, L.J. (1984), *Not in Our Genes: Biology, Ideology and Human Nature*, London: Penguin.

Wertz, D.C. and Fletcher, J.C. (2004), *Genetics and Ethics in Global Perspective*, Dordrecht: Kluwer Academic Publishers.

Journal Articles/Chapters

Bonnicksen, A.L. (1997), 'Procreation by Cloning: Crafting Anticipatory Guidelines', *Journal of Law, Medicine and Ethics*, **25** (4), p. 273.

Churchill, L.R. *et al.* (1998), 'Genetic Research as Therapy: Implications of "Gene Therapy" for Informed Consent', *Journal of Law, Medicine and Ethics*, **26**, p. 38.

Colby, J.A. (1998), 'An Analysis of Genetic Discrimination Legislation Proposed by the 105th Congress', *American Journal of Law and Medicine*, **XXIV** (4), p. 443.

Franklin, U. (1995), 'New Threats to Human Rights Through Science and Technology – The Need for Standards', in K.E. Mahoney and P. Mahoney (eds), *Human Rights in the Twenty-First Century*, Dordrecht: Kluwer Academic Publishers, p. 733.

Kand, A.S.F. (1995), 'The New Gene Technology and the Difference Between Getting Rid of Illness and Altering People', *Human Reproduction & Genetic Ethics: An International Journal*, **1** (1), p. 12.

O'Neill, O. (1998), 'Insurance and Genetics: The Current State of Play', in R. Brownsword, W.R. Cornish and M. Llewelyn (eds), *Law and Human Genetics: Regulating a Revolution*, Oxford: Hart Publishing.

Purdy, J.S. (1998), 'Dolly and Madison: The American Prospect Online', Issue 38, May–June, available at http://www.prospect.org/archives/38/38purdfs.html.

Rothstein, M. (1998), 'Protecting Genetic Privacy by Permitting Employer Access Only to Job-Related Employee Medical Information: Analysis of a Unique Minnesota Law', *American Journal of Law and Medicine*, **XXIV** (4), p. 399.

Wellcome Trust, 'Public Perceptions on Human Cloning', available at http://www.wellcome.ac.uk.

Part I
Genetics – General

[1]

Human Genetics: The New Panacea?

Julian Kinderlerer and Diane Longley*

Considerable advances have been made in human genetics in recent years, often outstripping the knowledge and understanding of the medical professions as well as the general public and taking regulators by surprise. In this paper, we seek to give a realistic indication of the many developments in human genetics and of what might or might not be scientifically possible in time, without the clutter and sensationalism of media hype. This, we feel, is an essential exercise to enable certain key elements and concerns to be taken on board, putting developments in context prior to any fresh consideration of the need for and potential effectiveness of regulation of human genetics.

The elucidation of the structure of DNA during the 1950s[1] provided a model for understanding the process for the transfer of genetic information between generations of the same organism. In bacteria and fungi identification of a variety of enzymes capable of modifying this group of large molecules has made possible the science termed 'modern biotechnology', and opened up our understanding of the mechanisms which lead from information molecule to function. Until the identification of these new enzymes scientists had used a variety of mutagenic devices (including, for example, ultra-violet light) to modify the genetic information in bacteria and fungi and observe the change in characteristics (or phenotype). From the late 1970s it became possible both to insert and remove genes in bacteria and to observe the consequences. An understanding of the control mechanisms followed rapidly. It became possible to sequence and extract genes from higher organisms and insert them into bacteria. Numerous quantities of the bacteria could then be grown, which meant that the amount of both the gene and the protein derived from that gene was relatively large, allowing analysis.

In 1990 a fateful decision was made. The entire human genome was to be sequenced. A 'genome' is the complete set of genes and chromosomes of an organism.[2] The intention was to construct a 'high-resolution genetic, physical and transcript map' of the human, with ultimately, a complete sequence. The Human Genome Project is the largest research project ever undertaken with the intention of analysing the structure of human DNA and determining the location of the estimated 100,000 human genes. According to Hieter and Boguski:

> The information generated by the human genome project is expected to be the source book for biomedical science in the 21st century and will be of immense benefit to the field of medicine. It will help us to understand and eventually treat many of the more than 4000 genetic diseases that afflict mankind, as well as the many multi-factorial diseases in which genetic predisposition plays an important role.[3]

About 40,000–50,000 genes have been identified, although for a majority the

---

* Sheffield Institute of Biotechnological Law and Ethics.

1 J.D.Watson and F.H.C. Crick, 'A Structure for Deoxy-ribose Nucleic Acid' (1953) 171 *Nature* 737.
2 P. Hieter and M Boguski, 'Functional Genomics: It's All How You Read It' (1997) 278 *Science* 601.
3 *Human Genome News* (1998)9,1–2. http://www.ornl.gov/TechResources/HumanGenome/project/project.html.

function is still unknown.[4] Whilst more than 95 per cent of the human genome remains to be sequenced, the acceleration in the process as new techniques are introduced means that the 15 year timescale originally envisaged for the completed project is likely to be met.[5]

It should be pointed out that a complete sequence does not provide information that allows an understanding of the mass of data. It can (to some extent) be compared to the possession of a very large encyclopaedia written in an unknown language. The complete sequence will not be 'sufficient to understand its functional organisation, neither for individual units nor at a more integrated level'.[6] The emphasis will quickly shift from the huge databases that store the recorded information to a functional analysis. It is assumed that there are about 100,000 genes with specific functions in the genome. The function of most of these is unknown. 'In the past we have had functions in search of sequence. In the future, pathology and physiology will become "functionators" for the sequences'.[7]

DNA profiling and similar techniques show very clearly that (virtually) no two individuals share the same genome. There will be differences in many of the genes on their chromosomes. Whose genome is, therefore, being sequenced? Sequences are not being determined for an individual, but rather for the genetic information of a large range of persons. This has resulted in an appreciation of the 'polymorphism' in our genetic make-up. Many of the amino-acids found in the linear sequence of a protein cannot be changed, for the change is likely to have a deleterious impact on the function. As proteins are directly coded in the DNA, there must be a similar constraint on the DNA. Many of the proteins found in humans are also found in other organisms. Even though the function is the same or similar, their sequence differs significantly. Hence exact replication is unnecessary. The DNA sequence that makes up many genes will differ from organism to organism, and even from person to person.[8]

Duboule[9] raises the question that lies at the heart of this article. How will it be possible to assimilate the mass of newly available information and translate it into clinical practice 'in a way that fulfils scientific criteria and respects ethical as well as social concerns'?

Genetic and biological advances

For obvious reasons, development of modern biotechnology proceeded apace in bacteria, viruses and fungi much earlier than in higher organisms. Within bacteria, fungi and plants it is now possible to move almost *any* 'gene'[10], from *any* one

4 L. Rowen, G. Mahairas and L. Hood, 'Sequencing the Human Genome' (1997) 278 *Science* 605 and Schuler *et al* 'A Gene Map of the Human Genome' (1996) 274 *Science* 540–546.
5 S.E. Koonin, 'An Independent Perspective on the Human Genome Project' (1998) 279 *Science* 36.
6 D. Duboule, 'The Evolution of Genomics.' (1998) 278 *Science* 555.
7 D. Tosteson, Symposium on 'Genomics and Gene Therapy: Meaning for the Future of Science and Medicine' (1997) Harvard Institute of Human Genetics, Cambridge MA 26 March 1997: cited in P. Hieter and M. Boguski, n 2 above.
8 D. Wang *et al* 'Large Scale Identification, Mapping and Genotyping of Single-Nucleotide Polymorphisms in the Human Genome' (1998) 280 *Science* 1077.
9 See n 6 above.
10 Gene is defined as 'The fundamental physical and functional unit of heredity. A gene is an ordered sequence of nucleotides located in a particular position on a particular chromosome that encodes a specific functional product (i.e., a protein or RNA molecule).' Gene expression is defined as 'The process by which a gene's coded information is converted into the structures present and operating in the cell. Expressed genes include those that are transcribed into mRNA and then translated into protein and those that are transcribed into RNA but not translated into protein (e.g., transfer and ribosomal RNAs).' The definitions are taken from *A Primer on Molecular Genetics* (1992) US Department of Energy, Office of Energy Research, Office of Health and Environmental Research, Washington, 36.

organism to *any* other. The use of 'gene' here includes the coding sequence that provides the information necessary to produce the gene product (protein)[11] and any sequences that identify when, where (in which tissue, within the cell or in interstitial fluids) and how much of the product will be produced. Scientists are now able to choose a known gene product from virtually any source, identify its DNA sequence, manufacture it in the laboratory modifying the sequence so that it will be better expressed in the organism into which it is to be placed — attach to it the various signal sequences needed to identify:

- **when** in the life cycle of the cell or organism that product will be expressed;
- **where** in the cell or organism expression will occur; and
- **how much** will be produced; and

insert this new construct into an organism of choice. Clearly non-human biotechnology genetic modification knows few bounds. As a consequence, efforts have been made to establish international regulatory measures, it being considered inappropriate to leave regulation to the market and industrial initiatives in most countries. There are negotiations currently in progress to produce a protocol to the Convention on Biological Diversity[12] to ensure the safe transport of genetically modified organisms between countries, and the United Nations Environmental Programme (UNEP) has produced a set of guidelines which define minimum standards for the safe manufacture and use of modified organisms.[13]

The transformation of bacteria and viruses is now routine. Viruses carry a small number of genes, and for many viruses the function of most of these is known. Insertion of a gene into a specific position is easily accomplished, and in general, the effect is predictable. Live viruses are often used as vaccines — the pathogenic impact for human and animal viruses having been attenuated by a variety of techniques. However, in many cases, the mechanism of attenuation is still poorly understood, and the impact on the virulence of the organism due to an inserted gene has to be investigated before the modified virus may be used on human or animal tissue.

There remain many limitations to the use of genetic modification in higher organisms. The transformation of some plants remains difficult, and insertion of genes into large animals is problematic as the normally long generation time means that a very high yield of transformed animals is needed if the technique is to be used effectively. In general we cannot choose where the construct is inserted into the genome of the host plant or animal, even if the complete sequence of that genome is known. In many instances the insert may go into the middle of a vital gene, rendering the cell incapable of growing. Techniques presently available for insertion of genes into plant cells make it likely that a number of copies of the gene

11 'A gene product is the biochemical material, either RNA or protein, resulting from expression of a gene. The amount of gene product is used to measure how active a gene is; abnormal amounts can be correlated with disease': *ibid* 36.

12 Article 19 para 2 of the Convention on Biological Diversity (CBD) requires the Parties to the Convention to 'consider the need for and modalities of a protocol setting out appropriate procedures, including in particular, advance informed agreement, in the field of the safe transfer, handling and use of any living modified organism resulting from biotechnology that may have adverse effect on the conservation and sustainable use of biological diversity'. The CBD is a legally binding agreement (Rio de Janeiro, 1992) which has been signed by over 150 countries. It came into force on 29 December 1993.

13 *UNEP International Technical Guidelines for Safety in Biotechnology* (1996). These Guidelines arose from the requirement in Chapter 16 of Agenda 21 (adopted at the United Nations Conference on Environment and Development in Rio de Janeiro in 1992) for the 'Environmentally Sound Management of Biotechnology'.

will be inserted randomly into the genome. For plants, traditional breeding practices (over a period of time) allow choice of a plant carrying both a small number of insertions and with as little loss of other characteristics as possible. As these separate 'constructs' are inserted into different places within the plant genome, selection over time means that they segregate, and it is possible to choose plants that demonstrate a particular, desirable characteristic.

The efficiency of transformation of plant cells is extremely low. Only a small proportion of the targeted cells is usually modified successfully depending on the technique.[14] There are, however, a very large number of cells in the tissue targeted for transformation and in most plants it is possible to regenerate a complete plant from a single cell. By using markers such as antibiotic resistance, or herbicide resistance, it is possible to select those cells which have survived and are not susceptible to (say) the antibiotic — implying the resistance marker is present and is being expressed. Even where the conversion is only one in a million, selection systems are able to find the successfully transformed single cell and grow it into the entire organism, or grow numerous copies.

The insert used is usually a small piece of DNA. Even for a large protein, the number of bases[15] in the 'gene' is likely to be less than 5000. In addition most of the DNA in a chromosome has no known function and is thought of as 'junk'. Insertion of a gene into these regions is unlikely to have a significant impact on the characteristics of the organism.

Table 1 shows the size (in base pairs) of the genome or chromosome for a number of different organisms. A gene may require a few hundred to a few thousand base pairs, and, as we have said, even in humans, there are thought to be less than a hundred thousand genes. If a new gene is inserted into the genome amongst the 'junk' it ought to have no impact on other than the desired characteristics for the modification. Bacteria and viruses have almost no junk DNA, but the modification of their genetic information can be achieved precisely.

Table 1[16]

Comparative Sequence Sizes	**Base pairs**
Largest known continuous DNA sequence (yeast chromosome 3)	350×10^3
Escherichia coli (bacterium) genome	4.6×10^6
Largest yeast chromosome now mapped	5.8×10^6
Entire yeast genome	15×10^6
Smallest human chromosome (Y)	50×10^6
Largest human chromosome (1)	250×10^6
Entire human genome	3×10^9

14 Between 1 in 1,000 and 1 in 10,000,000 cells are successfully transformed.

15 DNA is made up of a set of 4 monomeric units called bases that are linked together in a linear chain to form a very long molecule. Each of the bases is able to associate in a specific manner with one of the others, providing 2 pairs of 'base-pairs'. The DNA molecule is in general actually two, complementary strands wound around one another in a double helix. The four bases are usually named A, C, G and T. A and T pair, as do C and G. If a strand of DNA has the sequence AACGTAAGCTGGG, its complementary strand will have the sequence TTGCATTCGACCC. The two strands will be aligned as

AACGTAAGCTGGG
|||||||||||||
TTGCATTCGACCC.

16 n 10 above, 7.

Plant cells are routinely modified using a variant of *Agrobacterium tumifaciens*. This bacterium possesses an element (termed a plasmid) capable of entering a plant cell and taking over the function of the cell in order to produce a chemical upon which the bacterium lives. Scientists have subverted this plasmid, removing some of the genes that are deleterious to the plant and inserting the 'construct' of choice. There are many plants, however, that are not susceptible to the bacterium, and a ballistic method is used instead. The technique, termed biolistics, literally involves firing fine particles of a metal, usually gold, that have been soaked in the 'construct' at a leaf or similar plant tissue. The DNA is taken up by a very tiny number of cells in the tissue. The selection procedures allow for the destruction of all cells other than those that contain genes resistant to the antibiotic or herbicide used. The remaining cells are those in which insertion has been successful and the inserted genes are working. These may then be cajoled into reforming an entire plant.

It was realised from the inception of the use of this technology in organisms other than humans that there existed a potential for harm. Consequently regulatory structures were put into place that attempted, primarily, to identify the potential hazards and that endeavoured to reduce risk to within acceptable limits.

The regulatory structures for *non-human use* of biotechnology were designed initially to examine risk only, and were largely confined to the identification of risk to human health and safety resulting from the manufacture of new organisms with new properties. At first the risk was considered only to be to those working in the environment in which the technology was being used. However, it was soon realised that there was a significant risk to the environment in the event of escape of organisms modified to be fitter than their corresponding wild-type in particular conditions. Risk to the environment therefore has to be seriously considered. In many instances, however, scientists working with modified organisms may forget that humans are part of the environment and the risk to humans outside the confines of the laboratory, factory or hospital may be notable.

Today, virtually every country in the world has instituted some form of regulatory regime for the use of modern biotechnology in micro-organisms and plants. These are almost exclusively concerned with the minimisation of risk to human health and safety, or with minimisation of risk to the environment. The UNEP guidelines introduced in 1996 constituted an attempt to ensure a minimum baseline for regulatory structures in all countries.

Conversion of animal cells is much more difficult. Although it is possible to select those where transformation has had some success — for example in animals that have extremely short generation times, as is the case with many insects — in general, generation times are too long. It is not possible to regenerate an entire animal from a single cell in the same way as has been achieved for plants. Insertion of DNA sequences into animal cells usually involves micro-injection, whereby the DNA construct is injected into individual cells. The proportion of successful insertions is relatively high — it must be at least 70 per cent for the process to be considered useful. If the fertilised egg is used as the host cell, *all* the cells of the animal will contain the insertion. If the DNA is inserted into one of a group of cells that will eventually form the embryo, the resulting animal is a chimera. Some of the cells contain the construct, others do not, as only those cells which derive from that transformed will contain the inserted genes. If the construct produces 'gene-products' (proteins) which are only important within the cell in which the gene is found, then the chimaeric nature of the animal will be obvious. If the gene is intended to be manufactured within a cell, but used outside the cell, either in the whole body or in a particular organ, the chimaeric nature becomes less important. This particular factor may be important for gene therapy in humans, where a specific gene cannot be inserted into the correct tissue. As long as the gene is present and expressed somewhere, the therapy may be effective.

For reasons identified above, the techniques are not able to be applied directly to humans or, for the most part, animals. Not only are some procedures ethically questionable, but they may not be possible scientifically. We do not, as yet, have the scientific knowledge that would allow the regeneration of an entire human from a single cell. Nor would it be acceptable to use a selectable marker that would 'kill' off the vast majority of those we chose to modify. The insertion of genes, at will, into the genome of individual humans is therefore not yet a viable proposition. On the other hand, the technology is developing rapidly, and it may become possible to insert a particular sequence near to a sequence known to occur in a genome. Although there are only four bases that provide the alphabet for a DNA molecule, there are 1,024 different random sequences each consisting of five bases and over a million different sequences consisting of 10 bases. Recognition of a sequence of bases is a fundamental element to this technology and it is likely that systems that must be in existence already in some organisms will be found which are capable of directed insertion of a construct into a particular position in a DNA molecule.

It is clear that many aspects of biotechnology already impact directly on humans, or will do so in the near future. Modification of plants and animals (by traditional means) so that they provide more effective foods is a centuries old practice. Modification of viruses, bacteria and fungi so that they act as less effective pathogens, or are usable as vaccines, is also a process that has been available for some generations. It is now possible to introduce 'human' genes into a whole range of organisms so that they can be used as factories to produce proteins that are important in food or in medicine; as models to allow a better understanding of the function of particular genes or gene products; as media for the growth (and study) of human pathogens; or for the preparation of cells or organs which will ultimately be used in human medicine.[17]

'Genetic disease'

Advances in human genetics over the last few years — particularly those resulting from the greater understanding that the human genome project is providing — are staggering. It is this aspect and the inherent possibilities, however remote, that have caught the imagination of so many. Technology now provides an opportunity to link many characteristics and 'pathologies' with the product of a gene, even where we are not sure of the actual function of that product. As the human genome project proceeds, it has become apparent that there are many variations in the standard complement of genes that identify each one of us as a unique individual. There are many natural modifications of individual genes in the population, most of which have no (as yet) discernible impact on the individual. What we observe are characteristics, or 'phenotype', not the presence or absence of a particular form of a gene. These characteristics may be the result of a multitude of causes to which many genes may contribute. The presence of a

17 All cells have on their surfaces factors which allow them to be recognised as self, and therefore prevent them being destroyed by the immune system. The humanisation of organs in animals such as the pig requires these factors to be replaced by those found on human cells.

particular 'allele'[18] or gene product may also result in alterations in multiple characteristics.

Significant modification or mutation in a vital part of many genes would result in loss of function of the gene product, and in most cases, a failure to develop into a viable organism. Many mutations, however, result in disease, or increased susceptibility to disease, or simply to 'undesirable' characteristics.

Up to 3,000 diseases so far have been identified as having a genetic element; approximately two per cent of new-born children suffer from a perceptible genetic disorder. All of the characteristics we possess are decided by both the genes we carry (nature) and by the environment in which we live (nurture). Whilst it has been debated over many years as to which is the more important, for the purpose of this paper, it is of no concern — if there is a 'gene' which is implicated in a particular characteristic, then modification of that gene may result in a modification of the characteristic. It is now possible to test for the absence of a normal gene, or the presence of an abnormal one, even where the function of the gene is unknown. If tests are available to detect the gene, or any modification of it that has occurred, in an individual, the results may be used for both the alleviation of a condition, or for discrimination against the carrier. The development of tests thus presents us with a double-edged sword. In addition, for many conditions, we can identify the cause of the 'deviation from the norm' — which we might term 'disease' — but we have no way of mitigating the impact of that deviation or of modifying the person. Intended modification of a gene or its expression by human intervention may modify the expressed characteristic, even though it is not the only gene involved, but other changes may also be observed.

The definition of disease is, of course, problematic: to some, blond hair may be seen as normal, and dark hair would then be seen as pathological. *Perceived disease* (colour of hair, gayness, cystic fibrosis, likelihood of contracting cancer, diabetes, thallasaemias, Lesch Nyhan syndrome, Tay-Sachs disease, Huntington's disease) may be due to: presence or absence of a gene or allele different from that preferred; absence of expression, although the coding sequence would appear to be present; expression of too much or too little of the gene product; or wrong time or timing of the appearance of the gene product.

Many genetic diseases are due to changes in just a single gene (monogenic), such as adenosine deaminase (ADA) and purine nucleoside phosphorylase (PNP) deficiencies. More than two hundred specific enzyme defects cause known human clinical syndromes, and over a hundred other genetic diseases have been biochemically characterised.[19] Cystic fibrosis is the most common genetic disease, affecting some 30,000 individuals in the United States; about 2,000 children suffering from the disease are born each year.[20] Tay-Sachs disease[21] and

18 An 'allele' is an alternative form of a genetic locus; for example, at a locus for eye colour there might be alleles resulting in blue or brown eyes; alleles are inherited separately from each parent. The allele is the actual nucleotide sequence of a gene on a chromosome. Changes in sequence from one allele to another arise as a result of mutation in the germ-line and can be transmitted to the next generation.

19 J.B. Stanbury, J.B. Wyngaarden, D.S. Fredrickson, J.L. Goldstein and M.S. Brown, *The Metabolic Basis of Inherited Disease* 5th ed (New York: McGraw Hill, 1983) ch 1.

20 US Congress, Office of Technology Assessment, *Genetic Counseling and Cystic Fibrosis Carrier Screening: Results of a Survey–Background Paper*, OTA-BP-BA-97 (Washington, DC: US Government Printing Office, September 1992).

21 Tay-Sachs disease is a recessive disorder that affects the central nervous system to cause mental retardation and early death. It is a disease which predominantly occurs among Jews of Eastern and Central European descent and populations in the United States and Canada descended from French Canadian ancestors.

Table 2

Disorder	Number of Patients
Adenosine deaminase deficiency	40–50 worldwide
Purine Nucleoside Phosphorylase deficiency	9 patients in 6 families, worldwide
Lesch-Nyhan Syndrome	1:10,000 males
Arginosuccinate synthetase deficiency	53 cases known
Ornithine carbamoyl transferase deficiency	110 cases known

Huntington's disease[22] are both associated with a single gene defect. However, although there are numerous diseases that may be genetically linked, as Table 2 shows, the number of individuals affected by particular monogenic disorders is quite small.[23]

Familial history and epidemiology used in conjunction with modern gene technology make it relatively simple to identify a gene associated with a particular condition. There is very little that can be done, in terms of mainstream clinical procedures, to alleviate the symptoms of most monogenic diseases.[24] *Theoretically,* treatment of many such monogenic diseases by means of gene therapy is fairly straightforward. If it were possible to replace the gene by one that does not cause disease, in the same place in the genome, then the disease would be removed. But, modern gene technology has not, as yet, provided us with the tools that allow the removal of a gene within a chromosome and its replacement by a corrected form. It is not presently possible to replace genes in the 'right' place with absolute precision. However, if the disease is a result of the absence of a gene product, the provision of that product may alleviate all or most of the symptoms of the disease — insulin for diabetes is an obvious example of one such treatment.

Multigenic disorders have always been seen as much more problematic in terms of treatment. In multigenic disease a number of genes, or their products, interact and very little is known about how changes in one gene or product may affect the interaction with others. However, even though other genes are involved, if a single gene modifies the 'defective' characteristic, then in time it may be possible to 'alleviate' the condition using that gene or gene product.

It is the genes that indicate a predisposition to a disease that are most likely to cause ethical and moral dilemmas when their 'treatment' is considered. If an individual has an increased susceptibility to a particular disease, then genetic information may be used to attempt to ensure that the individual is (so far as is possible) isolated from the vectors that carry the disease. If, however, there is an increased likelihood of developing an auto-immune disease, or suffering from breast cancer later in life, due to the presence of an allele different from that found in the normal population, the question arises as to whether that information should be available to the individual concerned. In the first case there is currently no treatment or prophylaxis. In the latter case the impact of any treatment on the development of the disease is not known.

22 Huntington's disease is a dominant trait encoded in chromosome 4 that causes a debilitating brain disease that usually becomes evident only in a patient's 40s or 50s (after reproductive decisions have been made). All who carry the gene will develop the disease.

23 *Human Gene Therapy — A Background Paper* (Washington, DC: US Congress, Office of Technology Assessment, OTA-BP-BA-32, December 1984).

24 E. Tracy, C.R. Childs and Scriver (1995) 56 *American Journal of Human Genetics* 359, cited in N.A. Holtzman, P.D. Murphy, M.S. Watson and P.A. Barr, 'Predictive Genetic Testing: from Basic Research to Clinical Practice' (1997) 278 *Science* 602.

Probably more at issue than the treatment of disease which has a genetic base will be the identification of such diseases, and of individuals who are likely to be affected. If tests are developed which identify those likely to suffer, or those with a predisposition to a pathological condition, a variety of choices become available.

Advances in gene technology present a number of possibilities:

- to link a particular gene or gene product to a characteristic or to test for the presence or absence of a particular allele, and use the information to predict disease or predilection to disease. It may or may not be possible to offer treatment;
- to replace the gene and thereby alleviate all or some of the unacceptable characteristics;
- to replace the gene product and alleviate some or all of the unacceptable characteristics. The obvious example is diabetes, where treatment with insulin, originally derived from pigs and now manufactured, alleviates some of the disabilities suffered. Phenylketonuria is another example which is 'treated' by ensuring that the diet is deficient in phenylalanine.

Genetic testing and screening

There is a growing number of tests for diseases that might have a heritable element. In some cases there is a chemical present in abnormal quantities within body fluids and testing for the difference from the norm is an indication of a disease state. This method of disease detection or corroboration has long been used in medicine. Where a mutation has occurred within an allele, the identification of the absence of the 'normal' gene may depend on relatively simple genetic tests. If the exact mutation has to be identified, then sequencing of the gene may be required. Genetic testing is defined in a recent report by the National Institutes of Health (NIH) Task Force on Genetic Testing, as

> [t]he analysis of human DNA, RNA, chromosomes, proteins, and certain metabolites in order to detect heritable disease-related genotypes, mutations, phenotypes, or karyotypes for clinical purposes. Such purposes include predicting risk of disease, identifying carriers, and establishing prenatal and clinical diagnosis or prognosis. Prenatal, newborn and carrier screening, as well as testing in high risk families, are included. Tests for metabolites are covered only when they are undertaken with high probability that an excess or deficiency of the metabolite indicates the presence of heritable mutations in single genes. Tests conducted purely for research are excluded from the definition, as are tests for somatic (as opposed to heritable) mutations, and testing for forensic purposes.[25]

Testing can allow for the identification of 'diseases' in individuals before the appearance of any symptoms. If a gene has been implicated in an increased probability of a pathological condition developing, then the test informs the patient of a possible increase in the likelihood of their becoming 'ill' or affected at some time in the future.

Many questions arise about the use of genetic testing which have ethical and subsequently legal implications. If there is no available therapy to alleviate a disease that might be predicted by a test, questions have to be asked about whether

25 Neil A. Holtzman and Michael S. Watson (eds) *Promoting Safe and Effective Genetic Testing in the United States* (1997) — Genetic Testing: Task Force, National Institutes of Health — Department of Energy Working Group on Ethical, Legal, and Social Implications of Human Genome Research (http://nhgri.nik.gov) (NIH Task Force on Genetic Testing); and Holtzman, Murphy, Watson and Barr, n 24 above.

it should be performed. Even for those diseases which have been identified as being essentially monogenic — a single defective gene being responsible for the condition — we have no effective means of preventing the disease, and in most cases have little in the way of treatment to alleviate the worst features. Of course, in some instances we have been able to use therapies which improve the quality of life of those who suffer from a particular disease, and even ensure that the individuals live beyond the age of reproductive competence. Insulin dependent diabetes and cystic fibrosis are both examples. But the prognosis or time of onset of a disease cannot be detected by genetic testing.

The impact of tests is so important that it is crucial that those performing them are competent, and that those tested understand fully the implications of choosing to take them. Cultural attitudes to illness and abnormality 'run deep'.[26] These attitudes may be important in the application and interpretation of genetic testing and genetic screening procedures. Where a test indicates a likelihood of impairment, and is performed before birth or before implantation, should the provision of the test imply abortion? Predictive tests may have severe implications for the applicant or their family to obtain health or life insurance. Positive test results might not mean the disease will inevitably develop, or even anticipate the severity of the disease. For late onset diseases, there can be no reliable prediction of the age at which the disorder will manifest itself, the degree of debilitation, or the response to treatment.

Markel suggests that '[u]nrelenting vigilance is necessary on the validity of tests and the reliability of the laboratories providing them, both as the tests are developed and as they are used on large numbers of people'.[27] This is also highlighted by the NIH Task Force which recommends that 'the genotypes to be detected by a genetic test must be shown by scientifically valid methods to be associated with the occurrence of a disease. The observations must be independently replicated and subject to peer review'.[28]

The symptoms observed in many illnesses might have many causes. Genetic testing may provide an additional diagnostic tool, a method of identifying which of the causes is responsible for the symptoms in an individual, and hence indicate the likely therapy. Can the information derived from genetic tests be limited to determining the therapy (as is done currently with blood tests)? However, in many cases, symptoms of one disease may be linked (genetically or otherwise) with a predisposition to other problems. The question then arises as to whether the results of the test should be used to inform the patient about those other problems.

Tests provide information about a particular individual, but also provide information about their family — not only parents or siblings. This raises issues about individual autonomy, the possibility of making informed choices and the nature of the wider relationship between individuals and society. If a test has been performed on an individual's genetic material, should the information be made available to all those who may also be affected? Should permission for the test be obtained, not only from the person directly affected, but also from those whose lives might be altered following the test? The impact of what are often uncertainties resulting from testing may affect an individual's identity, their privacy and their relationships. These decisions cannot be made by the scientist, but are extremely important in determining what should and should not be done.

Because of the methods by which deleterious mutations are identified, many of the new pathological conditions that have been associated with unusual alleles

26 H. Markel (1997) Appendix 6 of the Report of the NIH Task Force on Genetic Testing (*ibid*).
27 *ibid*.
28 n 25 above.

have been attributed to Ashkenazi Jews. The number of diseases now being described might suggest that Jews from Eastern Europe are the carriers of rather a large number of genetic disorders. That this is an artefact of the methodology may often be forgotten. A recent US Government report, for example, quotes the discovery of a genetic alteration that, in early studies, appears to double a person's risk of colon cancer: 'The genetic alteration, which can be identified with a $200 blood test, is most prevalent among Jews of Eastern European descent. Once identified, people who carry this mutation can use regular colon examinations to detect cancer growth early when it is most easily treated'.[29]

The NIH Task Force on Genetic Testing points out that '[it] is unacceptable to coerce or intimidate individuals or families regarding their decision about predictive genetic testing.'[30] Respect for personal autonomy is regarded as paramount, informed consent must be obtained and it must be made clear that testing is voluntary. Prior to any predictive test in clinical practice, health care providers should describe not only the features of the test but its potential consequences. Whatever decision potential test recipients make, their care is not to be jeopardised.

Testing also carries profound implications for those unable to give consent or fully understand the implications. Should tests be undertaken, for example, on children? Tests are already carried out on almost all children born in the United Kingdom or the United States to determine whether they have phenylketonuria, but this is of direct benefit to the child, as avoidance of foods containing phenylalanine serves almost as a complete 'cure' for the disease. If a test is undertaken on children, should this only be where there is an immediate or direct benefit?

The NIH Task Force has also recommended that: 'No individual should be subjected to unfair discrimination by a third party on the basis of having had a genetic test or receiving an abnormal genetic test result. Third parties include insurers, employers, and educational and other institutions that routinely inquire about the health of applicants for services or positions.'[31] For this recommendation to be effective it would require the introduction of innovative legislation in most countries of a proactive rather than reactive nature. For example, in January 1998, the Vice-President of the United States called for federal legislation to bar employers from discriminating against employees on the basis of their genetic make-up. 'Progress in genetics should not become a new excuse for discrimination,' he said. 'Genetic discrimination is wrong — and it's time we ended it'.[32]

Notwithstanding discrimination, genetic testing is likely to be an important tool in identifying individuals susceptible to disease, not only for attempting to protect them (or their offspring) from the consequences of the disease, but for the future consideration of determining priorities and allocation of resources within health services. It is important that the technology is regulated to ensure that: information is accurate; individuals are provided with information about the identified problem in order to make informed choices; the implications for relatives of those tested are clearly understood by the individual undergoing the test; testing does not result in unfair discrimination at work or for life and health insurance; and priorities and resource allocation decisions are based on as sound evidence as possible taking account of 'the state of the art'.

29 *Genetic Information and the Workplace*, Department of Labor, Department of Health and Human Services, Equal Employment Opportunity Commission & the Department of Justice — 20 January 1998.
30 n 25 above.
31 *ibid.*
32 (1998) 9 *Human Genome News*, 1–2.

Gene therapy

If a gene product is known to be faulty, or absent, it may be possible to treat the condition simply by providing the missing chemical that would have appeared had the gene-product not been absent. This has proved extremely difficult in many cases because it is not easy to place the required chemicals at the correct site (within a cell) at the time or with the timing needed. Instead of providing the protein, or the chemicals that result from the presence of the protein, it may be possible to provide the gene. Thus:

> Gene delivery can be achieved either by direct administration of gene-containing viruses or DNA to blood or tissues, or indirectly through the introduction of cells manipulated in the laboratory to harbour foreign DNA. As a sophisticated extension of conventional medical therapy, gene therapy attempts to treat disease in an individual patient by the administration of DNA rather than a drug.[33]

A major goal of 'gene therapy' is to provide the patient with healthy copies of missing or flawed genes.

It is theoretically possible to replace a gene that is absent or not being expressed, but we have little in the way of mechanisms for the removal of a gene within human tissue. If an abnormal gene product is being expressed, and this is resulting in a disease condition, the presence of a gene expressing the normal gene product may not alleviate the condition. Identification of the presence of genes that are associated with abnormal conditions does not necessarily provide a solution. In the case of the genes BRCA1 and BRCA2 that have been linked with a significant increase in the likelihood of breast or ovarian cancer, possible therapy includes the complete removal of the 'offending' tissue but there is no evidence as yet that this radical procedure would relieve the problem.

There have been a number of attempts to *mask* the effect of genes, in effect *removing* them by using anti-sense oligo-nucleotides[34] that are believed to bind to the RNA before translation or to the DNA and stop the formation of the protein gene-product.

Therapies that involve 'changing the gene' may include:

- gene insertion, in which a new version of a gene is introduced into a cell;
- gene modification, in which a gene already in place is altered; and
- gene surgery, in which a particular gene is excised and may also be replaced by its normal counterpart.

At present, only the first of these alternatives is available.

It is not only the replacement of defective genes that may be important. If a gene is available which can allow the targeting of particular cells, then its insertion into a patient may allow novel therapeutic action. Anti-tumour cytokines is an example where the aim is to enhance the *in vivo* production of therapeutic chemicals. A further use of the technology involves the insertion of *suicide genes* into unwanted cells (such as cancers) the gene-product of which is capable of converting relatively benign drugs circulating in the bloodstream into cyto-toxins.

It is possible to place a 'new' gene in germ-line cells (sperm or egg cells) which

33 Stuart H. Orkin and Arno G. Motulsky, *Report and Recommendations of the Panel to Assess the NIH Investment in Research on Gene Therapy* (1995) National Institutes of Health, Washington DC.

34 An oligo-nucleotide is a short length of DNA or RNA. As DNA is double stranded and consists of two complementary strands, the production of a piece of DNA or RNA which is complementary to that which is translated to form a protein ('anti-sense') may stop translation and little or no gene product is produced.

would result in the inserted information being passed from that generation to all succeeding generations. Germ line gene therapy is not considered acceptable at present because of its permanence, and the lack of precision in the use of the technology. To some, the technique is 'immoral', as exemplified in the Directive on the Legal Protection of Biotechnological Inventions that was agreed earlier this year in the European Parliament. Germ line gene therapy remains unacceptable at the present time, and will remain so until we have much more control and understanding of the processes, both immediate and delayed, that may result from the insertion of a gene into what would be all the cells within that person and their offspring.

The alternative is to place the replacement genes into the genome within 'somatic' cells. Somatic cell gene therapy involves the introduction of a gene into cells with the expectation that the gene will be expressed and that a disease state will be alleviated. The gene is expressed only in those cells and is not passed on to future generations. It is hoped that the gene will remain in the cell and be expressed for a reasonable period of time, but it is not necessarily inserted into the genome within the cell. The technology may be performed *in vitro* on explanted autologous or other cells. The desired gene is inserted into the cell that is then placed within its human host. It is also possible to introduce the gene *in vivo*.

Gene therapy may also be administered simply by using the gene in a way similar to a drug, providing the patient with an encapsulated gene capable of providing, for a short period, the necessary gene products and functions. Such genes would not be incorporated into the genome and would, in general, be lost ultimately from the target cells.

None of these techniques is without problems. Only a small proportion of cells is modified using any of the processes mentioned above. Generally, in humans, modified virus particles would be used to act as vectors for the insertion of genes into the human genome, and the number of cells infected may be small. The amount of gene product expressed may then be significantly less than might have been expected. It is not usually possible to insert the replacement gene into the affected tissue, ie the organ in which the normal gene product would be produced. For example, insulin is produced in the Islets of Langerhans in the pancreas. It may not be necessary to produce the insulin at that same site, as the functional site is different. If it were necessary to introduce the gene into that tissue, the difficulties might well be insurmountable. Many genes appear to function at a particular time or in response to particular environmental stresses or changes. It has not, as yet, proved possible to mimic the time at which a gene is switched on, or the response to stress. In addition, the gene product may not be produced continuously. Until we have a greater understanding of the promoter and signal sequences attached to genes, it will remain difficult to mimic the control.

The insertion of 'foreign' genes into a patient is not new. All transplant patients carry genes which may be different from their own, and which may be expressing a slightly different protein with different properties that impact on the whole patient. Blood transfusion also involves the transfer of genes that may (for a short time) express proteins different from those in the patient. Somatic gene therapy is therefore part of the natural progression in medical science. The technology may enable the alleviation of many disease conditions and even the cure of human disease including cancer, and AIDS.

It was originally envisaged that gene therapy would be used to treat monogenic disorders, but this has not proved to be the case. The number of individuals who are suffering from any single monogenic disease is too small to make the research effort economically viable, and the majority of protocols for gene therapy throughout the world have concentrated on diseases of concern to much larger groups of individuals.[35] For the first 2,300 patients where treatment has been attempted, only 10 per cent have been for a range of monogenic diseases, 60 per cent have been for cancer, and about 18 per cent for infectious diseases (primarily AIDS).[36] At the beginning of 1998, approximately three hundred protocols had been approved, involving 2,293 patients.[37] The vast majority of patients have been in the USA (75 per cent, 1,708 of 2,293); 367 patients have been treated in the European Union and Switzerland, of whom 95 patients have been treated in the UK. The diagram below shows the number of patients treated for identified problems.

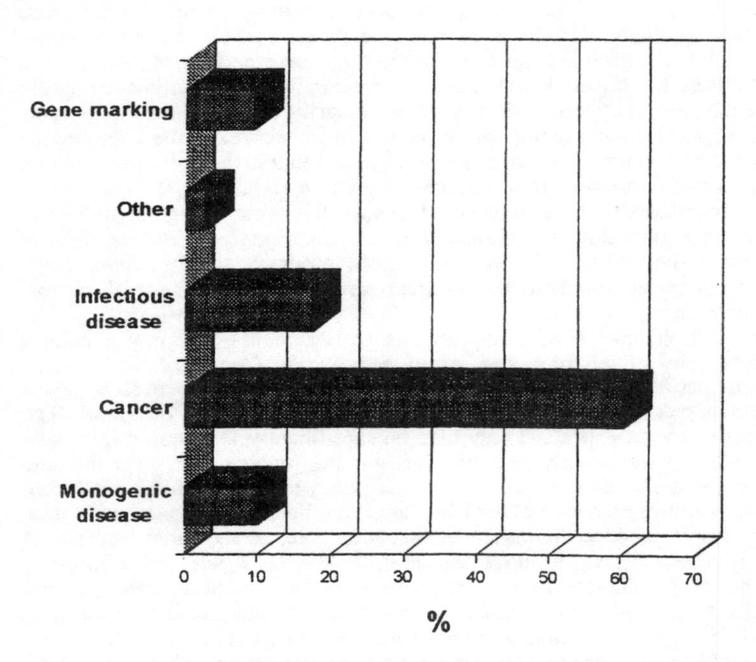

Gene therapy protocols for work with cancer patients have included both gene marking and 'true' gene therapy. Amongst the major areas of work have been: use of suicide genes which provide drug sensitivity to cancer cells; use of cells carrying genes which confer drug resistance into non-malignant blood stem cells to protect them from chemotherapy; insertion of tumour suppression genes into cancers; boosting of the host anti-tumour immune system; transfer of oncogene down-regulating genes to shut off cancer-inducing genes; and use of anti-sense genes to block the expression of cancer-promoting genes.

Getting the gene into the relevant cell in a live human remains problematic. If an attenuated virus is modified so that it is capable of infecting the patient and carrying the required genes into appropriate cells, there is a possibility that the

35 http://www.wiley.co.uk/genetherapy/diseases.htm (January 1998).
36 http://www.wiley.co.uk/genetherapy/ (5 March 1998).
37 http://www.wiley.co.uk/genetherapy/genes.htm (January 1998).

virus may regain virulence, or infect persons other than the target. There are a wide variety of vectors used to insert the DNA into the desired cells. Techniques include micro-injection, where naked DNA is inserted into isolated cells, complexed DNA within liposomes, retroviral RNA vectors and DNA viruses (eg adenovirus, herpes and pox viruses). As Table 3 shows, retroviruses are best understood, and have therefore been used in most protocols, although they have the disadvantage of only infecting dividing cells. The virus is incorporated into the host genome in a random position. Adenoviruses have the advantage of being able to infect non-dividing cells, are said not to integrate within the genome, but are highly immunogenic.

Table 3[38]

Vector	Protocols	Patients
Adeno-associated virus	3	21
Adenovirus	34	265
Electroporation	2	8
Gene gun	4	30
Liposome	56	433
Liposome/Adenovirus	1	3
Naked DNA	9	54
Poxvirus	16	121
Retrovirus	139	1100
Retrovirus/gene gun	1	6
Retrovirus vector producing cells	22	155
RNA transfer	1	0
Transfection	5	94
Totals	*301*	*2293*

The tools are available to allow a very considerable expansion in the use of genes in providing new ways of treating disease but the problems highlighted in terms of the application of the technology must first be overcome.

The NIH Panel that reported on the use of gene therapy in 1995 was extremely critical of that which had been done at that stage. They were excited by the promise of gene therapy but were concerned that 'clinical efficacy had not been definitively demonstrated at this time in any gene therapy protocol, despite anecdotal claims of successful therapy and the initiation of more than one hundred Recombinant DNA Advisory Committee (RAC)-approved protocols'.[39] The Panel identified difficulties in almost every aspect of the technology. They were concerned not only about the vectors used but that the scientific basis for comprehending the manner in which some vectors interact with human tissue was not well understood. There was concern that the basic research required to underpin much of the work in human patients had not yet been done adequately, and 'in the enthusiasm to proceed to clinical trials, basic studies of disease pathophysiology, which are likely to be critical to the eventual success of gene therapy, have not been given adequate attention'.[40] The Panel believed that these studies could have been attempted in animal models before their use in humans, and that such studies could well have offered alternative treatment procedures. On the other hand, the report went on to

38 http://www.wiley.co.uk/genetherapy/vectors.html (5 March 1998).
39 See n 33 above.
40 *ibid.*

identify the need for further clinical studies to evaluate the gene therapy approach. This makes it clear that animal experiments cannot provide complete information. In particular, it is important to use human clinical studies for cystic fibrosis, cancer and AIDS.

The Panel was unsure about the interpretation of the results of many of the gene therapy protocols that had been attempted at the time of their report. Problems of 'very low frequency of gene transfer, reliance on qualitative rather than quantitative assessments of gene transfer and expression, lack of suitable controls, and lack of rigorously defined biochemical or disease endpoints'[41] were cited. The overall impression of the Panel was that only a minority of clinical studies had been designed to yield useful, basic information. A further problem highlighted was that the 'hype' regarding the results achieved implied a greater degree of success than had actually been observed. The Panel emphasised that this could undermine confidence in the integrity of the field and might ultimately hinder progress in development of the successful application of gene therapy to human disease.

Xenotransplantation

Modern biotechnology may also be used to overcome rejection of transplanted animal tissues. This procedure would reduce the problem of many more organs being required for transplantation than are currently available. The transplantee acutely rejects animal organs unless the immune system has been suppressed (or modified). This happens normally when the donor is human, but careful choice of donor and the use of immuno-suppressive drugs slows or stops rejection. Rejection of animal organs is likely to be much more severe than that of human organs. It will be possible to make the animal tissue 'closer' to that of the human into whom it is to be placed by changing a variety of 'recognition sites' which occur on the surface of all cells. This would decrease the amount of suppression of the immune system required to allow acceptance of the new tissue whilst retaining the capacity to resist disease. There are many scientific issues which might be addressed, including the similarity of physiology between the donor and human tissue, sterility of donated tissue, and the possibility of transfer of dormant viruses to the immuno-suppressed human (or animal). Scientists are working on the 'humanisation' of animals, particularly pigs — seen to be the most likely donors of organs — and many of the problems of rejection have been solved. The UK is likely to allow transplantation from pigs to humans, but has ruled out the use of primate organs both for moral and safety reasons.

The risks of disease transmitted from animal to man through xenotransplantation are significant. Even where organs are humanised, the patient will still require suppression of their immune system, and any infectious agents in the animal tissue may be passed on. In order to maintain 'sterility' the animals will have been kept in isolation, and are likely to have been born using caesarean section. It is possible to screen for known pathogens likely to infect humans, whether immuno-suppressed or immuno-competent, and only use animals that appear not to be infected. There are, however, likely to be dormant viruses incorporated into the genome of the animal which, under the new physiological conditions, could be activated to affect the transplanted tissue within the patient, the patient's own tissue, or even be

41 *ibid.*

transferred from the patient to the general public. It is well established that most new emerging human infectious diseases generally have their origins in other species.[42] The risks to the general public are extremely small, are impossible to estimate, and may be too small to justify monitoring patients over an extended period of time. Such risks may well be considered worth taking, but the hazard may be substantial.

An editorial in *Nature* in January 1998 strongly advises a worldwide moratorium on the use of xenotransplantations, arguing that the rejection problem requires much more work, and that the potential risks of disease transfer are currently too great to accept.[43] Given the potential risk to the public, the issue is first and foremost an ethical one. Before introducing a regulatory framework driven by technical considerations, an informed public debate is needed so that the public can decide whether it wishes to consent to clinical xenotransplantation at all and, if so, under what conditions.[44] The United Kingdom currently takes one of the most stringent positions, having introduced a moratorium on clinical trials until further research shows that transplants are safe and that the science is sufficiently developed to offer transplant recipients real benefits.[45] The government has set up a Xenotransplantation Interim Regulatory Authority, and is expected to produce legislation later this year.

Artificial organs and human cells and tissues provide alternatives to xenotransplantation, and would avoid the risk of xenozoonoses. One company is to market an artificial skin, based on cultured human cells, for example, while stem cells extracted from umbilical cord blood offer a promising alternative to bone marrow transplants.[46]

Cloning

The debate about cloning has concentrated on the possibility of individuals attempting to make identical copies of themselves, and the repugnance felt by many that this might be possible. It is likely that the scientific lessons derived from the cloning of animals will lead primarily to the cloning of tissue rather than of whole organisms. The brunt of the discussion on cloning has concentrated on the development of techniques to produce exact genetic copies of an individual (with respect to nuclear DNA). The main advantages of experiments on cloning would appear to be the identification of the biochemical and physiological controls which might allow the reconstitution of an organ from a small number of cells. Thus: 'Nuclear replacement research can improve our knowledge about physiological processes and the genotype. For example, it is hoped that this work will offer a greater insight into the origins of cancer and other cellular development processes such as ageing and cell commitment.'[47]

42 See eg, B. Hjelle *et al* 'A Novel Hantavirus Associated with an Outbreak of Fatal Respiratory Disease in the Southwestern United States: Evolutionary Relationships to Known Hantaviruses' (1994) 68 J Virol 592; S.S. Morse and A. Schluederberg, 'Emerging Viruses: The Evolution of Viruses and Viral Diseases' (1990) 162 J Infect Dis 1; F.A. Murphy, New, emerging, and reemerging infectious diseases' (1994) 4 Adv Virus Res 1.

43 'Halt the Xeno-bandwagon: Xenotransplantation's Risks Make a Moratorium Essential' (1998) 319 *Nature* 309.

44 F. H. Bach and H.V. Fineberg 'Call for Moratorium on Xenotransplants' (1998) 391 *Nature* 326; and R. A. Weiss 'Transgenic Pigs and Virus Adaptation' (1998) 391 *Nature* 327.

45 (1997) 385 *Nature* 285.

46 (1996) 382 *Nature* 99.

47 *Cloning Issues in Reproduction, Science and Medicine. A Consultation Document* Human Genetics Advisory Commission (HGAC) and Human Fertilisation & Embryology Authority (HFEA), (1998) Department of Health.

Development of therapeutic agents

The development of gene products which will be significant as therapeutic agents in other organisms (clotting factors in sheep's milk, or vaccines in plant products) is already important in the pharmaceutical industry. The use of products derived from human tissue is problematic because these may contain factors capable of infecting humans. It has long been held that recombinant blood products (produced in yeasts or bacteria), such as blood clotting factors used for haemophiliacs, are much safer than those derived from human blood as contaminating agents are absent. This is theoretically true, but currently both serum-derived Factor VIII and recombinant factor VIII are stabilised by the addition of serum albumin. In the United Kingdom, the cost of the recombinant factor VIII is more than double that of factors derived from pooled human serum. Because of concerns about the infectivity of new-variant CJD, the Government has recently decided that it will not use pooled serum from within the UK. The production of human gene-products in plants in particular is likely to provide a source of relatively safe products, indistinguishable from those derived from human tissue in large quantities and at reasonable prices.

Conclusion

Clearly there are many avenues open to the scientist to use biotechnology in humans or in other organisms for the benefit of humans. All of these are being explored. Many of these approaches may carry considerable risks, or may prove unacceptable. Human genetics may or may not succeed in becoming the new panacea. In the meantime we may have been presented with a veritable Pandora's Box. Whilst scientists and physicians are capable of identifying the possibilities which may advance therapeutic procedures, they are no better equipped than anyone else to identify the social and moral issues that might result from their use. Only society can ultimately decide the degree of importance to be attached to the benefits, hazards and impact of these. Such fundamental decisions within an area of inherent uncertainty can only be made within a properly constituted regulatory framework. Only law can provide the mechanisms that can facilitate effectively the negotiation between fact and value and the taking of necessary decisions about the nature and direction of scarce resources for both research and policy. Regulation also carries with it a number of other advantages. By ensuring that controversial developments and the practices of scientists and clinicians are properly authorised and monitored a potent regulatory framework can assist public acceptance of 'cutting edge' techniques such as those discussed above. Besides being instrumental in the policy process law thus promotes legitimacy and accountability.

The issue of which regulatory devices are to be given preference is of course a matter of conjecture and debate,[48] but lessons can surely be drawn from the growing provisions and literature on their use in the field of non human biotechnology, in addition to those in the ever widening field of innovative medical technologies.

48 cf the discussion in Julia Black's paper in this volume.

[2]

Regulation as Facilitation: Negotiating the Genetic Revolution

Julia Black*

Regulation of genetic technology, some would claim, is an oxymoron. Genetic technology is simply out of control.[1] If you turn to ask what structures exist to regulate genetic technology, however, then you find a mass of legal regulations, non-legal rules, codes, circulars, practice notes, international conventions, and ethical codes. There exists an enormously complex set of advisory bodies, regulatory bodies, committees, professional bodies, and industry associations, operating at an international, national and sub-national level. In the UK, at the national level alone, there are over eleven different bodies involved in the regulation of some aspect of genetic technology. Surely in this morass of regulation someone, somewhere, must be exerting some form of control?

But the charge is more complex: it is essentially that control is not being exercised over what it should be. That regulation is not finding the right answers, or indeed asking the right questions. Moreover, that science is defining the agenda: in Beck's phrase, the debate about its course occurs as an obituary for activities begun long ago.[2]

Indeed, one of the striking aspects both of the debate about genetic technology and of its regulation is the number of different conceptualisations of the 'problem' which genetic technology poses and thus of the solutions that should be found. It is seen variously to be an issue of risk (to health, the environment), a question of choice (of patients, consumers), a matter of property rights (to patents, an individual's DNA), of confidentiality (against employers, insurance companies, other family members), or a question of ethics. Moreover, the definition of the problem which is adopted by regulation and the solutions proposed are not always those that others would share.

Given these fundamental divergences of views, it is suggested that regulation has a role to play which is not, or not simply, about control, but rather about facilitation. Regulation has an important role to play in connecting the arguments of participants, in facilitating the integration of the wide range of views as to the appropriate course that the technology and its regulation should take. The first task of this article is to explore the extent to which this occurs at present: to examine just what questions the current regulation of genetic technology asks, and what questions others consider it should be asking. What do the regulatory systems and

---

* Law Department, London School of Economics and Political Science.
My thanks go to Rob Baldwin, Anne Barron, Damian Chalmers, Tim Cross and Gunther Teubner for reading and commenting on an earlier draft, and to the participants of the Modern Law Review conference on Law and Genetics for their responses to the paper given at the conference. Responsibility for views, errors and omissions remains my own.

1 A view most forcefully put by Beck: U. Beck, *Risk Society: Towards a New Modernity* (London: Sage, 1992, trans M. Ritter); further U. Beck, *Ecological Politics in an Age of Risk* (London: Polity Press, 1992, trans A. Weisz); *id*, *Ecological Enlightenment: Essays on the Politics of the Risk Society* (New Jersey, Humanities Press, 1995).

2 Beck, *Ecological Politics in an Age of Risk*, *ibid* 203.

their participants consider their purpose and rationale to be, and what gaps exist between that internal perspective and that of those outside the regulatory system as to the central issues which should be addressed.

The main purpose of the article is then to consider how regulation may facilitate the integration of those different perspectives. Integration does not mean the replacement of a multitude of perspectives with the regulatory imposition of just one. Rather integration is the full recognition of different perspectives in the regulatory process. The call for integration in this form is one which is often made. The ways to achieving that integration, it is frequently advocated, are the development of forms and forums of consensus-building and co-operation, and the adoption of communicative, procedural models of regulation. The labels vary across writers and disciplines: 'proceduralisation',[3] 'civic science',[4] 'scientific proceduralism',[5] or simply 'democratisation',[6] but they share a common desire. That is to open up the decision process, to deny any one voice authority in that process, and through the integration of views and perspectives to arrive at accepted solutions to intractable problems. The negotiation of regulation.

To the extent that the facilitation of such negotiative, or integrationist, models requires simply institutional re-design, then finding ways to achieve this is complicated although solutions can probably be found, at least at the level of policy formation. However it is suggested that facilitating integration also requires that attention be paid to other dimensions of regulation which are sometimes overlooked, notably those of cognition and communication. In particular, attention has to be paid to the standing of different parties in that negotiation, to the weight or authority which will be ascribed by participants to each voice, and to the barriers to communication which it is suggested that different cognitive systems give rise.

With respect to the issues of standing and of cognitive authority, then, at present, the scientific voice is that which is granted the status of objectivity, and thus, in the current rules of debate, authority and legitimacy. Lay views are often seen as irrational, based in ignorance, as mere emotions or prejudices. As such they have only such weight as is necessary to afford them in a democratic society. They are either an irritation or something which should be indulged, depending on your view of popular politics. But the scientific language is that which is accepted; it is the official language of debate. For 'lay' views to be accepted, then (to an extent) they have to re-translate themselves into the language of science, and in so doing accept the scientific definition of the problem. For a negotiation to occur which fully recognises and gives standing to other voices, there is a need for a reorientation of the cognitive aspect of regulation. This requires in part a re-conceptualisation of the view that 'expert = objective, lay = irrational'. It is on such an exercise that a number of writers have embarked, and the nature of that exercise will be explored in part here.

3 R. Mayntz, 'The Conditions of Effective Public Policy: A New Challenge for Policy Analysis' (1983) 11 *Policy and Politics* 123; G. Teubner, 'Substantive and Reflexive Elements in Modern Law' (1983) 17 *Law & Soc Rev* 239; G. Teubner, 'After Legal Instrumentalism? Strategic Models of Post-Regulatory Law' in G. Teubner (ed), *Dilemmas of Law in the Welfare State* (Berlin: de Gruyter, 1985); H. Willke, 'Societal Regulation through Law?' in G. Teubner and G. Febbrajo (eds), *State, Law, Economy as Autopoietic Systems* (Milan: Guiffre, 1992).

4 T. O'Riordan, 'Exploring the Role of Civic Science in Risk Management' in C. Hood and D. Jones (eds), *Accident and Design: Contemporary Debates in Risk Management* (London: UCL Press, 1996).

5 K.S. Shrader-Frechette, *Risk and Rationality: Philosophical Foundations for Populist Reforms* (Berkeley: University of California Press, 1991).

6 Beck, *Ecological Politics in an Age of Risk*, n 1 above.

There is a more fundamental problem, however, which it is suggested here faces an approach to regulation which seeks to facilitate integration. It is one which occurs on the communicative dimension, and is one which the idea of 're-translation' emphasises. It is that the languages of science, commerce, ethics, ecology, law are foreign to each other; neither can hardly understand what the other is saying, let alone why they are saying it. Until this problem is addressed, simply focusing on the structural dimension, providing the structures in which different actors can participate, will not lead to a negotiated agreement. This is because the different participants speak different languages. The conditions for real communication do not exist. This 'dialogue of the deaf' is particularly striking in the context of genetics. What is needed, it is suggested, is the means by which this language barrier can be overcome. This cannot be provided by the development of a common language, however, for the cognitive differences are too fundamental. Nor should it be provided by an 'official' language, for that would represent the dominance of one cognitive perspective over all others. Rather, it is suggested, regulation needs to facilitate communication by taking on the role of interpreter or translator: putting the views of each set of participants into a language that the others can understand. It can thereby enable the integration, the negotiation of regulation, which is required.

Regulation: structural, cognitive and communicative dimensions

In analysing the regulation of genetic technology it is proposed to focus on three dimensions of regulation: the structural, the cognitive and the communicative. There are thus other dimensions which are here excluded, notably that of preferences: the article will not explore the different interests which participants have nor how they are pursued.[7]

The structural dimension is essentially the institutional context in which regulation occurs. The ability of actors to enter a particular regulatory or decision making forum is structured by institutional rules and norms, both legal and non-legal.[8] Some may be able to access that forum relatively easily, others may find that they are excluded from it, or can only enter in a limited way. So there may be insiders and outsiders with respect to a particular forum: those who have access to a particular forum and those who do not.

The cognitive dimension refers to the perceptions of participants, their 'world views', their rationalities, their operating logics.[9] The communicative dimension refers to the communicative interaction between participants.[10] The dimensions may interact. Thus the non-legal norms which may define entry to a particular forum or the standing of a participant in that forum may themselves be defined by

7 There are a number of analyses of the relationship between preferences and perceptions: for a discussion of the relationship between the cognitive aspect and that of preferences in the context of institutional analysis see J. Black, 'New Institutionalism and Naturalism in Socio-Legal Analysis: Institutionalist Approaches to Regulatory Decision Making' (1997) 19 *Law and Policy* 51.

8 I am here drawing on sociological institutionalist analyses of decision making: see in particular R. Scott, *Institutions and Organizations* (California: Sage, 1995); W. Powell and P. DiMaggio (eds), *The New Institutionalism in Organisational Analysis* (Chicago: Univ Chicago Press, 1992).

9 For a discussion see for example the references cited in n 8 above, and R. Friedland and R. Alford, 'Bringing Society Back In: Symbols, Practices and Institutional Contradictions' in Powell and DiMaggio, n 8 above.

10 It thus clearly has resonance with the communicative models of Habermas; most recently J. Habermas, *Between Facts and Norms: Contributions to a Discourse Theory of Law and Democracy* (Cambridge: Polity Press, 1996).

the cognitive dimension: the institutional actors' own sense of who should have access and who should not. Institutional structures may facilitate communication. Indeed it is the adjusting of such structures to this end which is advocated by many as the solution to the problem of cognitive fragmentation (fundamental divergences of perspective or world views).

Focusing on these three dimensions of regulation illustrates very clearly the task facing regulators in facilitating negotiation and integration. It is that cognitive fragmentation itself places a considerable barrier to communication. Different cognitive structures simply have different languages. This barrier thus exists even if the appropriate institutional structures are in place. Participants in the debates on the course and regulation of genetic technology simply speak different languages; simply providing the structures in which they can all participate in a decision will not serve to overcome this communication barrier. Focusing on the structural, communicative and cognitive dimensions of regulation indicates the differences in perspective which different actors have as to the definition of and solution to the 'problem' of genetic technology and the implications which this has for the languages in which issues relating to genetic technology are debated, and the role for regulation which this entails.

The next three sections of the article explore three different aspects of genetic technology: genetically modified organisms (GMOs) and genetically engineered products (GEPs), genetic technology in the human and medical context, and that of the rights to the exploitation of genetic technology. With respect to each, the analysis will focus on the structural dimension of the regulation, and ask who has access to the regulatory or decision making fora and who does not ('insiders' and 'outsiders'), on the cognitive dimension, asking what the different conceptualisations are of the problems posed or issues to be addressed, and on the communicative dimension, the languages which are used. Whose definition of the problem is the definitive one? What language dominates? Which discourse do others have to adopt in order to be heard? Is the language of science, or medicine, for example the only one which is seen as valid; do all other arguments have to re-translate themselves into this language in order to be recognised as ones that have to be addressed? To what extent, finally, is there an attempt to negotiate a regulatory decision between different participants and what form does that negotiation take?

Regulating GMOs and GEPs

Techniques of genetically modifying organisms are well advanced, and an increasing number of genetically engineered products are being developed and marketed.[11] Regulation of this aspect of genetic technology is fragmented and there are a number of different regulatory fora in which issues concerning research into genetic engineering and the development and marketing of genetically engineered products are raised. The principal actors are scientists, both in universities and in

11 Genetically engineered maize, soya, cotton, tomatoes and potatoes are being grown in the US and Canada, and a tomato puree made from genetically modified tomatoes was first sold in the UK in 1996. In total 124 consents for the release of GMOs had been granted between 1993 and June 1997; in December 1996 the European Commission decided to give its approval to authorise the marketing of GM maize for unrestricted use; approval has been given for a GM oilseed rape, and several other applications are pending. Details are given in ACRE's newsletters (available at http://www.shef.ac.uk/doe).

industry, industry itself, and a plethora of regulatory bodies whose membership reflects the traditional corporatist relationship of government, business and unions. There are few opportunities for the wider public to enter the regulatory fora, and they are recognised principally in their capacity as consumers. The issues or problems raised by genetic engineering are seen solely in terms of risk. The organising principle of the regulation is thus risk, and moreover, whilst there is some recognition in principle that different perceptions of risk may exist it is not clear to what extent a technical, scientific measurement is discounted in favour of public perceptions of risk.[12]

There are tensions which arise between those who are inside the regulatory fora, the regulators and regulated, but these are as to whether or not the regulation is accurately targeted towards the right risks, not as to whether risk is the most appropriate organising principle. The concerns of those who are outside are often much broader and more fundamental, often going not to issues of risk, although those are raised, but to the activity of genetic engineering itself, and to the wider implications it can have with respect to related issues of for example, intensive farming, the growth and dominance of agri-business, or the disadvantaged position of smaller farmers particularly in developing countries. Such 'ethical' objections are however effectively excluded from the regulatory fora.

Regulatory fora and principal actors

The conduct of genetic research and the development and marketing of GEPs is regulated by laboratories themselves and by governmental bodies who bear a wealth of acronyms. The key regulatory instruments are the Code on Good Laboratory Practice, and the regulations on contained use[13] and deliberate release,[14] both of which implement EU directives.[15] The contained use regulations are issued under the Health and Safety at Work Act 1974 (HSWA) and their operation is principally the responsibility of the Health and Safety Executive (HSE), acting in consultation with the Department of the Environment (DoE). In practice this is undertaken by the Advisory Committee on Genetic Modification (ACGM), a non-statutory body. The ACGM advises on the granting of consents, on policy development and EU negotiations, and produces guidance.[16] Enforcement is the function of the HSE, which has specialist inspectors for the purpose. The deliberate release regulations are issued under the Environmental Protection Act 1990 (EPA), and their administration is principally the responsibility of the DoE, acting in consultation with the HSE. Again, in practice this task is undertaken by

12 On the distinctions between them, see further below.

13 GMO (Contained Use) Regulations 1992, SI 1992/3217; amended by the GMO (Contained Use) (Amendment) Regulations 1996, SI 1996/967 and the GMO (Risk Assessment) (Records and Exemptions) Regulations 1996, SI 1996/1106 (exemption from risk assessment with respect to human health for imported or acquired GMOs).

14 Genetically Modified Organisms (Deliberate Release) Regulations 1992, SI 1992/3280 (as amended), made under Part IV of the EPA 1990.

15 Council Directive on the contained use of genetically modified micro-organisms 90/219/EEC OJ No L 117/1 and Council Directive on the deliberate release into the environment of genetically modified organisms 90/220/EEC OJ No L 117/15. These Directives were based on the OECD's report, *Recombinant DNA Safety Considerations*, OECD, 1986. In addition, the Advisory Committee on Dangerous Pathogens publishes advice on work with dangerous pathogens, and GMOs may also be biological agents under the Control of Substances Hazardous to Health Regulations 1994.

16 The latter task has now been given to a permanent Technical Sub-Committee, established in 1996. See further ACGM Newsletter No 21, February 1997.

an advisory body, the Advisory Committee on Releases to the Environment (ACRE), which is established under the EPA.[17]

The regulation of products which are to be placed on the market is also governed by regulations rooted in EU directives. The regulation of genetically engineered products, or products containing GMOs, initially occurred under the deliberate release regulations but gradually several discrete systems of regulation are being developed for different types of products, or where appropriate the regulation is being assimilated into existing regulation of similar products. So whereas the initial 1992 regulations covered the marketing of all GMOs,[18] subsequent changes have created separate regimes for medicines and veterinary products,[19] animal feed additives[20] and foods and food products which are produced from GMOs or contain GMO products.[21] The change is often described as a move from 'process' to 'product-based' regulation: from scrutinising any act of genetic modification to scrutiny of the product itself.[22] In fact process regulation continues, and moreover GM products are still subject to a risk assessment which is over and above that which is applied to products of the same type.

The principal actors in this regulatory environment are thus scientific researchers in universities and industry, industry itself, and those who sit on the different regulatory quangos. Regulation occurs within laboratories and industry, and those working with GMOs or producing GEPs are required to have local safety committees to advise on risk assessments. The membership of ACGM and ACRE follow a corporatist model, combining government, industry and union representatives. All have a scientific background and the latter are meant to have some kind of broad remit to represent the public, and they include a number of ecologists.[23]

Access

The operation of the regulation is structured essentially as a series of discrete and individualised negotiations between applicant and regulator. Those working with

17 ACRE used to be the sub-committee of the ACGM on releases to the environment.

18 But not non-living GMOs or products derived from GMO products which were not themselves GMOs (bread containing modified wheat, for example).

19 Under the Genetically Modified Organisms (Deliberate Release) Regulations 1995, SI 1995/304, implementing EC Directive 94/15/EC OJ No L 103, the authorisation and supervision of medicinal products for human and veterinary use is to occur on a European basis: Council Regulation EEC No 2309/93 (OJ No L214) lays down an EC procedure for their authorisation and supervision and establishes a European Agency for the Evaluation of Medicinal Products. The Regulation provides for the specific risk assessment of medicinal products containing or consisting of GMOs.

20 The change was again introduced by the 1995 Deliberate Release Regulations, which provide that with respect to the marketing of additives which fall under the Feeding Stuffs Regulations 1991 (SI 1991/2840, as amended by SI 1993/1442, 1994/499 and 1994/2510) incorporated or to be incorporated in any feeding stuff, a specific risk assessment of additives containing GMOs has to be undertaken. The 1995 regulations implement Directives 70/524 OJ No L270, and 93/114/EC OJ No L334.

21 The Genetically Modified Organisms (Deliberate Release and Risk Assessment Amendment) Regulations 1997, SI 1997/1900 add to the exemptions from the requirement for consent novel foods or ingredients containing GMOs which fall under Regulation EC No 258/97 OJ No L 43 concerning novel foods and food ingredients. The regulation provides for specific environmental risk assessment of novel foods or novel food ingredients which contain or consist of GMOs and introduces a labelling regime. It is administered in the UK by the Advisory Committee on Novel Foods and Processes and MAFF.

22 For a discussion of the distinction see J. Tait and L. Levidow, 'Proactive and Reactive Approaches to Risk Regulation: The Case of Biotechnology' (1992) 24(3) *Futures* 219.

23 For an early assessment of the composition of ACRE see L. Levidow and J. Tait, 'Advice on Biotechnology Regulation: the Remit and Composition of Britain's ACRE' (1993) 20 *Science and Public Policy* 193.

GMOs are required to notify HSE of the intention to use premises for genetic modification work for the first time and for certain subsequent individual activities, and in some cases await specific consent from HSE before starting the work. Those intending to release or market GMOs must apply for consent to the appropriate body, usually ACRE. The controls to be adopted are negotiated between the applicant and the regulator.

There is some attempt to broaden the conversation which currently occurs between regulator and applicant to include third parties. This is made through the publication of a register of notifications and approvals for both contained use of GMOs and their release. The HSE maintains a register of notifications of contained use which is open to public inspection.[24] It is required to publish a specified minimum amount of information concerning the identity of the notifier, the type of activity being undertaken, plans for monitoring the GMO and for emergency response, and the valuation of any foreseeable effects of the GMO.[25] The applications for deliberate release are also placed on a public register,[26] and the ACRE Guidance requires applicants to provide a summary conclusion of the risk assessment which would be comprehensible to a layman.[27] Applications to release for non-market purposes must in addition be publicised in one or more newspapers which circulate in the locality of the release, and must contain a statement of the purpose and nature of release to be made and which other specified bodies have been notified.[28] ACRE's advice as to consent will also be entered on the register. Although there is no formal mechanism for public objections to be made or taken into account, the ACRE guidance indicates that ACRE will consider public comments prior to the final decision; to this end its advice is entered at least two weeks before the final decision.[29]

There is a tension, however, between providing transparency of the regulatory system and protecting the confidentiality of the regulated. To this end, beyond this minimum amount the notifiers under the contained use regulations can claim that the information should not be disclosed as it would adversely affect their competitive position or ability to obtain a patent (on the basis that the information has to have been previously unpublished).[30] The onus is however on the notifier to give full justification of this claim.[31] Under the deliberate release regulations information may be witheld from publication (but not from ACRE) if it is commercially confidential.[32] Applicants have tried to make extensive use of these exclusions, marking all information as commercially confidential and on occasion witholding information from the regulators themselves on that basis, practices which the regulators have criticised.[33] Further, following pressure from applicants

24 Reg 16. The register is not complete in that it only contains those notifications for which express consent has to be given by the HSE, that is first uses of Group II organisms, and Type B operations with Group II organisms.

25 Reg 15.

26 Reg 17.

27 Guidance, paras 3.32, 3.44 and 3.90(g).

28 Reg 8.

29 ACRE guidance, para 3.91.

30 See further below.

31 Reg 15(2).

32 EPA s 123(3)(a). The criterion used for deliberate release is thus slightly different than that which applies under the Contained Use Regulations, which applies to information which would adversely affect the applicant's competitive position. Note in any event that the Environmental Information Regulations 1992, SI 1992/3240, apply, governing freedom of information with respect to the environment.

33 ACRE Newsletter, Issue No 7, June 1997.

that the information put on the register was allowing protesters to vandalise the release sites, the regulations were amended in 1995. Information to be included in advertisements for crop plants now does not have to include details either of the dates of release or their location.[34]

The principal way in which the public enters the area of decisions as to the research and development of genetically engineered products is as a consumer. Participation takes the form of either buying or refusing to buy genetically modified products. In order that consumers can participate even in this way, they need information as to whether the product has been genetically engineered or contains engineered ingredients. Parts of the industry have resisted labelling on the basis that consumers will not understand the information which is being given and will be scared off buying the food, or have said that it is not feasible because of the practices adopted by suppliers: in the US, companies have refused to separate out genetically modified soya from normal soya. Nevertheless, in November 1997 an EU regulation came into force requiring the labelling of genetically engineered foods.[35]

Conceptualisation/definition of the problem

The issue at which the regulation is addressed is that of risk. Risk has been the defining conceptualisation of the problems posed by GMOs throughout the development of the regulation. Initially focused on the risk to human health, the regulation subsequently broadened to include risks to the environment.[36] In its Thirteenth Report, which was to a considerable extent the precursor to the regulation, the Royal Commission on Environmental Pollution considered the range of impacts, particularly undesirable impacts, which the release of GMOs could have on the environment.[37] The manipulation of a virus, for example, could unintentionally alter its virulence or widen the range of susceptible organisms; indirectly, it could change the range of insects who carry the virus, so bringing it into contact with previously unaffected plant species. Projects that engineered plants to produce toxins may have the danger that the toxin will appear in a part of the plant that might be eaten by non-target animals or humans; the cultivation of the crop on a wide scale may encourage the development and spread of insects resistant to it. Herbicide resistant genes could spread to weeds, which could lead to greater use of herbicides overall. The release of antibiotic resistant genes could accelerate the dissemination of antibiotic resistant genes in pathogens. Genes inserted into new host organisms may transfer after release to other organisms with undesirable consequences. Non-pathogens could be converted to pathogens. Finally, consideration has to be given to the extent to which the effects of the release of GMOs could be reversed: whether the GMOs could be recovered or eradicated, and how their disposal would be managed, particularly if they had the capacity to become converted into novel pathogenic agents.

The solution proposed and adopted was to adopt a precautionary approach to risk

34 1995 Deliberate Release Regulations, taking into account Commission Decision 94/730/EC OJ No L 292.

35 Regulation (EC) No 258/97; as yet no guidance has been issued by MAFF, or indeed any other member state, as to the implementation of the regulation.

36 However an amendment which would have introduced an ethics commission was rejected during the passage of the EPA: see L. Levidow and J. Tait, 'The Greening of Biotechnology: GMOs as Environment Friendly Products' in V. Shiva and I. Moses (eds), *Biopolitics* (London: Atlantic Highlands, 1995) 134.

37 Royal Commission on Environmental Pollution, Thirteenth Report, *The Release of Genetically Engineered Organisms to the Environment*, Cmnd 720 (London: HMSO, 1989) chapters 4 and 5.

management. A precautionary approach to risk is one in which the risks which are potentially posed are assessed in advance and attempts made to reduce or eliminate them.[38] In Wildavsky's phrase, it attempts to ensure 'trial without error'.[39] Controls are put in place even in the absence of information on the extent of the risks posed. It contrasts with a reactive approach, in which a product is assumed to be safe until a particular hazard is identified and proven. The rationale underlying the regulation was that the act of genetic modification itself posed risks above and beyond those posed by traditional breeding techniques. It is this idea that genetic engineering is inherently risky, coupled with a precautionary approach and a case by case assessment of risks, which was thus adopted by the 1990 EU directives on contained use and deliberate release which form the basis of the current system of regulation.[40]

The regulation implements this precautionary approach by adopting a strategy of individualised assessment and approval, rather than the formulation of rules which attempt to prescribe in advance what is or is not permitted, or to set out particular safety requirements that have to be met. Release of GMOs is permitted only after an assessment has been made of their capacity to cause harm to the environment or to the health of humans or other living organisms, or to human senses or property.[41] For both non-marketing and marketing releases the regulations require a detailed assessment of the risks to the environment including potential impacts on the ecosystem (eg genetic stability and mobility, pathogenicity, ecological and physical traits, antibiotic resistancy, ecological aggressiveness) and the risks to human and animal health. The applicant must also give details of measures to monitor and control the spread of GMOs, means of cleaning up waste and emergency plans to abort the release.[42] The precautionary approach to risk regulation is maintained with respect to GEPs by the deliberate release regulations. Moreover, as noted, even though the regulation is moving to the development of separate product regimes, the risk assessments which are required parallel those required by the deliberate release regulations.

Internal debates

Although those within the regulatory fora agree that the issue to be addressed is that of risk, they disagree on whether or not the regulation is appropriately targeted and as to what degrees of risk are posed. The contained use regulations

38 The precautionary principle was originally enunciated by the West German government in 1976 in the field of pollution control, and has become highly influential in debates both on environmental regulation and on approaches to risk regulation more generally. See further T. O'Riordan and J. Cameron, *Interpreting the Precautionary Principle* (London: Earthscan, 1994).

39 A. Wildavsky, *Trial Without Error: Anticipation vs Resilience as Strategies for Risk Reduction* (London: CIS, 1985).

40 The precautionary principle underlies the Government's sustainable development strategy in environmental regulation: *Sustainable Development. The UK Strategy*, Cm 2426 (London: HMSO, 1994).

41 There are exceptions which recognise that the purpose of releasing some genetically modified organisms may be to further pest control or to break down toxic wastes; in these cases the provisions on prohibition notices and the powers of inspectors where they fear imminent harm are disapplied: 1992 Deliberate Release Regulations, reg 4.

42 Regs 6 and 11; Schedules 1 and 2. The GMO (Deliberate Release) Regulations 1995, SI 1995/304 introduced a simplified system of notification for higher plants. The GMO (Deliberate Release and Risk Assessment) Regulations 1997, SI 1997/1900 added a requirement with respect to applications for marketing that they contain information which could be relevant to the establishment of a possible register of modifications introduced into a species; and information about proposed labelling to indicate that the product contains or consists of GMOs.

have been strongly criticised by scientists, industry, and by the regulators themselves on the ground of lack of 'fit' with scientific understandings and laboratory practice.[43] In its report on the competitiveness of the biotechnology industry, the Select Committee on Science and Technology found that the original categorisation of organisms adopted in the regulation, which was meant to be based on the degree of risk which they pose, was 'risible' in the eyes of scientists, with the consequence that the whole risk assessment system was being brought into disrepute.[44] The need for any system of regulation over and above that required by existing standards of good laboratory practice is itself questioned. Many scientists engaged in genetic engineering take the view that the hazards presented are low or even minimal, and that a precautionary regulation focused specifically on genetic products is unnecessary: genetic engineering poses no more hazards than the introduction of non-indigenous species and is in fact more precise than traditional cross breeding techniques.[45] Indeed even the chairman of the ACGM which administers the regulations stated that there was no evidence that the technique of modification was itself hazardous.[46] Industry argues strongly that the regulations place it at a considerable competitive disadvantage vis à vis the US and Japan, and that it should be relaxed and refocused.[47] In the words of the BioIndustry Association, 'much of the regulation put in place by the European Commission is based on old science and reflects concerns that have not proved justified.'[48]

Outsider perspectives

The regulation thus sees the problem posed by genetic engineering as one of risk, and within those fora the debate is as to the appropriate fit of the regulation to the degree of risk which is in fact posed. The debate about genetic engineering which occurs outside the regulatory fora shares the risk-based conceptualisation of the issue, but adopts a very different perception of the risks which are posed. But in contrast to the regulatory focus, those who at present are outside the regulatory fora do not see the issues to be confined to risk. So not only do the conclusions which are reached differ given the common definition of the problem, but that definition is itself challenged. The conceptualisations of the issues and the language in which they are expressed are thus fundamentally different.

Attitudes to genetic engineering, in particular to its acceptability, tend to be rooted in a wide range of factors. Evidence from a number of public attitude surveys conducted indicates that acceptability is linked, *inter alia*, to the perceived need of the technology or its products, to the interaction of genetic technology with

43 See for example comments in ACGM Newsletter No 19, April 1996.

44 Select Committee on Science and Technology, *Regulation of the UK Biotechnology Industry and Global Competitiveness*, 199–203, HL Paper 80, paras 5.25–5.26. The basis of distinguishing between Group I and II organisms was altered in 1995, and the classification is now based not on the detailed characteristics of the organism, but on a qualitative assessment in each individual case of whether it is likely to cause adverse effects to humans, animals, plants or the environment: EC Directive 94/51/EC; The GMO (Contained Use) (Amendment) Regulations 1996, SI 1996/967. Guidance on the application of the criteria is contained in EC Decision No. 95/1579/EC, and further in ACGM, *A Guide to the GMO (Contained Use) Regulations 1992, as amended in 1996*. Both this distinction and that between Type A and B operations is to be abolished under proposals for a fundamental revision of the Contained Use Directive.

45 Evidence to the SCST, Report on Biotechnology, *ibid* paras 5.8–5.12.

46 *ibid* para 5.11.

47 See further *ibid*.

48 BioIndustry Association, *A Charter for Biotechnology*, 1997, 9.

other practices of which the person may not approve, and to the attitude taken to the wider social impacts it is perceived that the use of genetic technology will have.[49] Thus it is resisted on the basis that it may lead to more intensive farming, the use of pesticides, or exploitation of farmers in developing countries ('genetic imperialism'). The use of the growth hormone rBST, for example, is opposed on the basis that it will simply be to the benefit of large agri-businesses at the expense of smaller farmers; genetically modified seed because of its impact on the issue of farmers' rights[50] or on developing countries' farming practices.

Opposition is manifested in a number of ways. There has been strong resistance to genetic engineering from environmental groups and consumers in both the US and Europe. In Austria, for example, a referendum last year on the continuation of the ban of genetically altered corn received 1.2 million supporting signatures. A referendum in June 1998 in Switzerland on the 'gene protection initiative' proposed a ban on, *inter alia*, the breeding and purchase of transgenic animals and the release of GMOs into the environment (the proposal was actually defeated). In November 1997 Greenpeace filed petition with the US Environmental Protection Agency demanding that it revoke permits for certain GE crops, and English Nature is pressing for a three year moratorium on such crops.[51] Others have taken more direct action, for example digging up crops.[52] In April 1997 a week long 'Global Days of Action Against Biotechnology' was organised, with over 200 groups in 24 countries participating.[53]

Attempts at negotiation

An attempt at negotiating or mediating between the different perspectives is perhaps made in the definition of risk which appears to be adopted by the regulatory bodies. It has been suggested that in its membership ACRE embodies a tension between technical risk assessment and policy issues irreducible to scientific evidence,[54] a tension which is explored further below. In the absence of empirical evidence it is not clear exactly how that tension in practice is resolved. The guidance issued by ACRE does however recognise the essentially qualitative nature of much risk assessment. It states that whilst risk is to be assessed as far as possible on the basis of technical and scientific measurements, given the uncertainties which pervade consideration of the possible impacts of GMOs on human health or the environment much of that assessment is necessarily qualitative rather than quantitative in nature.[55] The requirement is thus that the risk assessment statement should comprise a reasoned appraisal based on the best available quantitative or qualitative measure of risk to the environment and human health.

49 See generally, D.P. Ives, *Public Perceptions of Biotechnology and Novel Foods: A Review of the Evidence and Implications for Risk Communication* (Norwich: Centre for Environmental and Risk Management, UEA, 1995); J. Durant (ed), *Biotechnology in Public: A Review of Recent Research* (London: Science Museum, 1992).

50 The ability to retain a certain proportion of seeds from that year's crop to sow for the following year (which obviously reduces the amount of seed which the farmer will buy from the seed seller/manufacturer).

51 *The Guardian*, 12 April 1998.

52 For example, in Ireland the Gaelic Earth Liberation Front dug up the first field of a genetically engineered crop (sugar beet) to be planted in there: *Nature Biotechnology* vol 15, November 1997, 1229, and crops have been dug up in Scotland: *The Guardian* 12 April 1998.

53 The week was organised by Jeremy Rifkin's Foundation on Economic Trends and the Pure Food Campaign (both in the US). For details see B. Nasto and J. Lehrer, 'Antibiotech week raises tension over transgenic food', *Nature Biotechnology*, vol 15, June 1997, 499.

54 Levidow and Tait, n 23 above.

55 ACRE Guidance, para 4.5.

This view is shared by the HSE and by a recent interdepartmental committee on risk assessment.[56]

There are necessarily limits, however, to the extent that the risk assessment process can be used as a means of mediating the different perceptions which surround this aspect of genetic technology. For whilst the regulation is firmly risk-based, risk is not seen by those outside the regulatory fora to be the only problem which is posed by GMOs and GEPs. There is an opposition to the very activity of genetic engineering, issues of risk aside. There is no formal venue in the regulatory or approval process for these concerns to be expressed, and industry is very opposed to there being one. The BIA's *Charter for BioIndustry* urges the government to 'stand firm against attempts from whatever direction, to attach ethical considerations to safety or indeed, other technical legislation.'[57] However, the result is a closure, structural and cognitive, of the regulatory system to the concerns voiced by those who do not have access, formally or informally, to the regulatory fora in which decisions are taken. The legal remits of the regulatory bodies define the issue in terms of risk, and risk is the only acceptable regulatory criterion for many of those who seek regulatory approval. The result is that those who argue against the release of genetically engineered crops or the sale of genetically engineered seeds or foods on the grounds that they will, for example, contaminate organic crops, or lead to undesirable farming and commercial practices, are simply met with the answer that release is safe, or that the risk can be controlled. That safety is not, or not the only, point, is not a debate in which the regulatory system will currently engage.

Human and medical genetics

Issues of human and medical genetics are also addressed in a number of different regulatory fora, and the main actors are scientific researchers, the medical profession, and a number of regulatory or advisory bodies which are staffed principally by scientific and medical professionals. The principal contrast between the regulation of human and medical genetics and that of GMOs and GEPs is the broader definition of the problems which are raised. The issues which are raised by genetic technology in the human and medical context are seen not solely in terms of the risks posed but of the appropriate adaptation of medical ethics to accommodate that technology. Ethical issues appear in turn to be defined and driven in large part by the medical profession. Non-medical or scientific professionals enter principally in their capacity as patients, although the Advisory Committee on Genetic Testing and the Human Genetics Advisory Commission have a remit to consider the wider social implications of genetic engineering. Again, however, there are differences in cognition, in the conceptualisations of the problem and the perceptions of the issues to be addressed between those who are within the decision making and regulatory fora and those who are outside it. Attempts to negotiate or mediate between these different perspectives have been patchy; an exception is with respect to cloning, where there is currently a concerted effort to address the matters which non-scientists and medical practitioners consider to be important.

56 *Use of Risk Assessment within Government Departments: Report prepared by the Interdepartmental Liaison Group on Risk Assessment*, HSE 1996; and see further, The Deregulation Initiative, *Regulation in the Balance: A Guide to Risk Assessment*, DTI 1993; DTI, *The Use of Scientific Advice in Policy Making*, DTI 1997.

57 n 48 above, 9.

Regulation of genetic technology in the human and medical context is fragmented, and not nearly so structured as that which relates to GMOs and GEPs. For that reason, we will look first at the different regulatory fora,[58] their principal participants and the main access points, but then explore the dimensions of cognition and communication in more depth with respect to three different contexts in which genetic technology is deployed. These are testing (which is increasingly routine), gene therapy (which in scientific and clinical terms is a nascent technology, but in which research trials is occurring) and cloning (which is nascent).

Regulatory fora and principal actors

In the area of human and medical genetics, the fora in which decisions are taken as to the development, application and use of genetic technology are dispersed. They operate at a sub-national level, within laboratories and the medical profession, at a national level, both within the internal bureaucracy of the health service and as statutory and non-statutory executive bodies, and at an international level. The diffuse and often opaque nature of the decision making structures and the partial nature of the regulation mean that access points are dispersed, and thus the categorisation of 'insiders' and 'outsiders' is often fluid and unclear. Again, in the absence of detailed empirical research, only published documents and guidelines can be used as witnesses to the perceptions which are held of the issues to be addressed.

Laboratory research is the subject of the GMO regulation outlined above. In addition, research programmes are reviewed by the relevant funding bodies for their ethical acceptability, and all research involving patients must be approved by local research ethics committees (LRECs) set up in each health authority. The terms of reference of the Human Genetics Advisory Commission (HGAC) include the review of scientific progress in human genetics. There are also separate regulatory fora relating to discrete areas of research. Research into gene therapy must be approved by the Gene Therapy Advisory Commission (GTAC),[59] which was set up to supplement the work of the LRECs. It must also be authorised by the Medicines Control Agency, which must be satisfied that it meets the required levels of safety, efficacy and quality.[60] Research which involves the use of human embryos requires approval from the Human Fertilisation and Embryology Authority (HFEA).[61] Pathology laboratories (which investigate the causes of disease) are accredited by the Clinical Pathology Accreditation (UK) Ltd which audits the standards of professional practice existing in the laboratories.

Further, a number of international conventions apply to genetic research, in particular to research into human cloning. In November 1996 the Council of Europe adopted a Convention for the protection of human rights in biomedicine and has agreed that protocols on, *inter alia*, genetics and medical research will be developed under it.[62] The Convention includes a moratorium for a minimum of

58 Given that there are gaps in the regulatory structure, this will in fact also include fora in which decisions as to the clinical deployment of genetic technology are made.

59 GTAC has issued two sets of guidance: *Guidance on Making Proposals to Conduct Gene Therapy Research on Human Subjects*, September 1994, and *Writing Information Leaflets for Patients Participating in Gene Therapy Research*, August 1995.

60 Under the Medicines Act 1968 and Directive 65/65/EEC, as subsequently modified.

61 Under the HFEA Act 1990.

62 Convention for the Protection of Human Rights and Dignity of the Human Being with regard to the Application of Biology and Medicine, adopted by the Committee of Ministers on 19 November 1996 (Doc DIR/JUR(96)14 — Legal Affairs Directorate of the Council of Europe).

five years on germ line gene therapy in humans. In November 1997 UNESCO adopted the Declaration on the Human Genome and Protection of Human Rights, which seeks to establish universal principles covering all research into the human genome and its application when it appears that it might conflict with human dignity and the rights of the individual.

The use of genetic technology in medical applications is also subject to regulation which derives from a number of different sources. The supply and manufacture of genetic tests is regulated by the Advisory Committee on Genetic Testing (ACGT), which has issued a voluntary code of practice on testing services which are supplied directly to the public rather than via the NHS.[63] A Directive on *in vitro* diagnostic tests, which would include genetic tests, is also being negotiated by the EC Council of Ministers.[64] Any product developed from genetic engineering which is intended for medical use has to be approved by the European Medicines Evaluation Agency before it can be marketed in the EU[65] or by the MCA for market authorisation in the UK,[66] and there are obligations to maintain records relating to medicinal products.[67]

Access

Such an array of regulatory bodies would appear to contradict Beck's contention that 'medicine possesses a *free pass* for the implementation and testing of its "innovations" '.[68] To a significant extent, however, the principal actors within these different regulatory and decision-making fora are medical professionals, and as noted, non-professionals enter principally in their capacity as patients. Their access is defined in terms of the medical norms of informed consent, and those norms determine what information they should receive, in what manner, and what principles of confidentiality should attach to that information.

Conceptualisations/definitions of the problems

The issues which genetic technology raises are to some extent seen as questions of risk, but to a large extent seen in terms of the ethics of patient care. In other words, how the advances in genetic technology should be passed onto the patient, rather than what those advances should be. There is one important exception, and that is with respect to the use of genetic technology in the reproductive context. The main

63 ACGT, *Code of Practice and Guidance on Human Genetic Testing Services Supplied Direct to the Public*, September 1997.

64 See further, ACGT, *Consultation Report on Genetic Testing for Late Onset Disorders* (November, 1997). The Directive would require all *in vitro* diagnostics (IVDs) to be CE marked in declaration of conformity with the essential requirements of the Directive in order to be placed on the market.

65 Established under Council Directive 87/22/EEC (OJ No L 15, 17.1.87, p 38) on the approximation of measures relating to the marketing of biotech medicinal products, following the procedure in Council Regulation 2309/93. Authorisation is given by the Committee for Proprietary Medicinal Products (part of the Agency), which takes into consideration the environmental safety requirements set out in the Deliberate Release Directive. EMEA operates in addition to the National Biological Standards Board, set up under the Biological Standards Act 1975, which manages the National Institute for Biological Standards and Control (NIBSC) which monitors the safety and quality of biological substances used in medicines such as vaccines, hormones, whose purity or potency cannot be adequately tested by chemical or physical means.

66 The MCA is supported by independent advisory committees, chiefly the Committee on the Safety of Medicines and the Medicines Commission (which also consider appeals).

67 EC Directive 91/507/EEC on updated standards and protocols for the testing of medicines for human use.

68 Beck, *Risk Society: Towards a New Modernity*, n 1 above, 207 (emphasis included).

areas of debate at present concern the activities of genetic testing, gene therapy and cloning, which are considered separately below.

Genetic testing

Genetic testing for some diseases is already part of established medical practice. Testing is used either for diagnostic purposes, to test for the causes of particular symptoms, or for predictive purposes. Predictive testing provides information as to whether a person has a genetic disorder which may lead to, or which may mean that they are susceptible to, particular diseases. Genetic testing can be of benefit both to an individual or couple, and to the overall gene pool of a particular group. Genetic testing of Ashkenazic Jews in the US has reduced the incidence of Tay-Sachs disease in the population by 90 per cent.[69] Testing may be carried out on adults or children (neo-natally or pre-natally). Since the 1950s, most babies in the UK have had blood tests to establish whether they have inherited the genetic defect responsible for PKU. More recently, it has become possible to test at various stages of pregnancy for genetic conditions responsible for Tay-Sachs disease, Huntington's disease, cystic fibrosis, Duchenne muscular dystrophy (DMD), and thalassaemia. Tests for genetic susceptibility to common diseases such as cancer, heart disease and diabetes are being developed, and it is reasonable to expect that an increasing range of tests will become available in the next few years.[70]

Most genetic testing in the UK is carried out within the NHS, and is based on regional genetics centres which serve groups of health authorities. Other genetic testing services may be provided by general practitioners, within particular medical specialisms such as oncology and haematology, and within research settings, often in collaboration with regional genetic centres.[71] There is also an increase in the sale of genetic tests developed in pharmaceutical companies directly to the public, bypassing the normal medical structures.[72] Direct services aside, the decisions as to when testing should be offered are made within the relatively diffuse and opaque decision structure of the NHS. Discussion and guidance on when testing should occur has come from a number of medical bodies, including the Royal College of Physicians,[73] the British Medical Association,[74] and from the Nuffield Council on Bioethics.[75] The consensus view of these organisations is that testing should be for therapeutic purposes only, and in particular only for serious diseases, that it should allow carriers for a given abnormal gene to make informed choices regarding reproduction, and/or go towards alleviating the anxieties of families and individuals faced with the prospect of serious genetic disease.[76] Regulation, however, is currently patchy.

Regulation of genetic testing is found principally in professional guidelines and in the single voluntary code which has thus far been issued by the Advisory Committee on Genetic Testing. Established in 1996 on the recommendation of the

69 BMA, *Our Genetic Future: The Science and Ethics of Genetic Technology* (Oxford: OUP, 1992) 192.
70 See ACGT, Consultative Report, n 64 above.
71 *ibid*; Chief Medical Officer and Chief Nursing Officer, *Population Needs and Genetic Services: An Outline Guide*, PL/CMO (95) 5 and PL/CNO (93) 4.
72 The subject of ACGT, *Code of Practice*, n 63 above.
73 Royal College of Physicians of London, *Ethical Issues in Clinical Genetics* (London: Royal College of Physicians, 1991).
74 n 69 above.
75 Nuffield Council on Bioethics, *Genetic Screening: Ethical Issues* (London: Nuffield Council on Bioethics, 1993).
76 *ibid* para 3.9.

Select Committee on Science and Technology,[77] the remit of the ACGT is to consider and advise health ministers on developments in genetic testing, the ethical, social and scientific aspects of testing and establish requirements to be met by suppliers of genetic tests. It considers tests in use in clinical practice and to be supplied to the public. Those which are to be supplied directly to the public have first to seek the approval of the ACGT,[78] although the Committee has no powers to proceed against those who do not seek its approval.

The ACGT has not sought to regulate the availability of tests in the clinical setting, and the decision making fora here are health authorities. It has sought to provide guidance on the procedures which should be followed in conducting certain types of tests, those for late onset disorders,[79] but not for other types of disease. This is otherwise the subject of NHS circulars and the professional ethics of patient care. The process of testing itself is subject to a fair degree of regulation from other sources aimed at ensuring the accuracy of the test results, and much of this will be put on a statutory footing when the IVD Directive is introduced.

Those inside and outside the profession and the regulatory fora appear generally to accept the view that genetic testing raises a wide range of issues. There is doubt as to the predictive reliability of the tests, particularly for late onset disorders and diseases which are not caused by a single gene. Tests are also not wholly accurate, or do not test for all forms of a disease.[80] The absence of treatment for a diagnosed disease and the uncertainty surrounding predictive testing mean that people may receive information the significance of which is uncertain and to which they cannot respond by taking preventative or curative action. Moreover, pre-symptomatic testing may have significant psychological implications.

At present, these issues are addressed in the context of the doctor-patient relationship, and by the doctrine of informed consent. It is considered essential to ensure that full consent to be tested has been given and that the information is imparted in an appropriate manner with full counselling and support. The need for such clinical support (and doubts as to the adequacy of its current provision) has been repeatedly emphasised by the Nuffield Council,[81] the BMA,[82] in the Select Committee's Third Report on Human Genetics,[83] and by the ACGT.[84] It is required by the ACGT's Code of Practice for those who supply tests directly to the public.[85]

Testing is thus seen essentially to be a matter of individual patient choice,[86] although there are particular issues where testing of children is concerned.

77 Select Committee on Science and Technology, Third Report, *Human Genetics: The Science and its Consequences* HC 1994/5, HC Paper 41.
78 ACGT, *Code of Practice*, n 63 above.
79 ACGT, *Consultation Report*, n 64 above.
80 For example, there are over eighty mutations of the CF gene; the current tests will identify only those which occur in around 85 per cent of the population, leaving the other 15 per cent undetected.
81 Nuffield Council, n 75 above, chapter 4, especially para 4.21.
82 BMA, n 69 above, 193–200.
83 SCST, Third Report, n 77 above, paras 72–106.
84 See the ACGT *Code of Practice*, n 63 above, and *Consultation Report*, n 64 above.
85 n 64 above, paras 5 and 6. Indeed, the unlikelihood of such support services being adequately given by such suppliers, as well as concerns about the accuracy of the tests, underlay the ACGT's recommendation that only certain tests should be supplied outside the clinical context. The ACGT's Code of Practice (para 3) provides that only tests for carrier status of recessive disorders, where such status carries no direct health implications for the individual (eg cystic fibrosis), should be provided directly to the public. They should not be provided for inherited dominant and X-linked disorders, chromosomal disorders (eg Downs syndrome), late onset disorders (eg Huntington's), or for the genetic component(s) of multifactorial diseases (eg heart disease or cancer). Genetic tests should not be provided over the counter for children.
86 See ACGT *Consultation Report*, n 64 above; *Population Needs*, n 71 above, Annex 3 para 3.

Concerns have been raised at the wider pressures that may be exerted on the patient affecting the way in which that choice is made, however.[87] For example, it has been suggested that pre-natal screening conveys a signal that a positive result justifies an abortion: why else offer the test pre-natally instead of, for example, neo-natally? There have moreover been reports that screening for Huntington's disease in some areas is only given on the basis that the pregnancy will be terminated if the test is positive.[88] Attitudes to pre-natal testing tend therefore to be linked to the position a person takes on abortion more generally. Experience with genetic testing has shown that pre-natal testing can however lead to societal pressure to abort. There is some evidence that in Sardinia, where there is routine pre-natal testing for thalassaemia, women who fail to abort a foetus diagnosed as having thalassaemia are stigmatized by the local community.[89] Testing is thus not simply a neutral technique and a private issue; it has significant social consequences. Mediating these concerns has not thus far been a significant feature of either regulation or medical decision making on the provision of testing services; there has recently been a citizens' jury held in Wales on the issue of the availability of genetic testing, but it is not clear what impact this will have in the relevant decision making fora.[90]

A somewhat different issue which is seen to be central to genetic testing is the confidentiality of the information which the test produces. Confidentiality issues arise in a number of forms. First, there is the issue of whether the person to whom the information relates has a right to receive that information. Where the person has requested and/or consented to a genetic test, then they should be told the result of that test. Difficulties arise where the test shows other information which was not the specific subject matter of the test (and can be particularly problematic if it relates to paternity), or where subsequent research on a person's DNA shows that he or she has a genetic abnormality the testing for which was not part of the original reason why the DNA was taken. Should the person be told of the results? At present, there is no specific regulation on the issue, although with respect to information obtained in the course of subsequent research the current practice is not to inform that person. On this issue, the ACGT's proposed solution is that samples should be anonymised, so as to prevent the situation arising, and in any event information should not be given to the individual unless a clear and specific arrrangement has been made at the outset.[91]

Secondly, there is the question of whether other family members should be made aware of the test results. As has been well observed, genetic testing may reveal information not only about those who have given their consent to testing, but about members of their families who have not.[92] Again, there is no specific regulation on the matter. After reviewing existing case law and professional guidelines, the Nuffield Council concluded that in certain circumstances that information could be revealed to other family members, without the patient's consent. In all cases individuals should be made aware that the test would reveal information that could

87 See further S. Shiloh, 'Decision-making in the Context of Genetic Risk' in T. Marteau and M. Richards (eds), *The Troubled Helix: Social and Psychological Implications of the New Human Genetics* (Cambridge: CUP, 1996).

88 See SCST Third Report, n 77 above, para 90.

89 Professor Sir David Weatherall, 'Genetic Science — Looking Ahead', paper given at the 21st Century Trust Conference on Genetics, Ethics and Identity, Oxford, 29 March–4 April 1998.

90 Welsh Institute for Health and Social Care, *Report of the Citizens' Jury on Genetic Testing for Common Disorders* (University of Glamorgan: WIHSC, 1998).

91 ACGT, *Consultation Report*, n 64 above, 24–25.

92 Nuffield Council, n 75 above, para 5.1.

be relevant to other family members, and should be encouraged to disclose the information to those people. However, if the individual refused, then if disclosure of that information might avoid grave damage to other family members it could be revealed. Although they recommended that the medical professional have a discretion to reveal that information, the Council was against the imposition of a legal duty to do so.[93] Here there is a noted difference of view between the profession and others. In particular, the Council's conclusion was categorically rejected on policy grounds by the Select Comittee on Science and Technology. In the Select Committee's view, the individual's decision to withold information should be paramount. It would place relatives in no worse position than they were already, and failure to respect individual privacy could discourage people from having tests done.[94]

Thirdly, there is the question of disclosure of genetic information to third parties who are not members of the family group, notably employers and insurance companies. The tension between the profession and those outside it is here even clearer. Insurance companies argue strongly that they should see the information; that disclosure is entirely consonant with the norms and practices of the insurance industry and the principle of *uberrimae fidei* which underlies all insurance contracts. In the employment context, employers could attempt to justify the requirement for the person to have a test or to disclose the results of tests on the basis of their duty to provide a safe system of work under the HSWA. Refusing to employ someone on the basis of the results of those genetic tests would not be unlawful, unless it also constituted sex or race discrimination. Dismissing someone on the basis of a test result would not necessarily constitute unfair dismissal.

Whether information should be disclosed is thus contentious. Some attempt has been made to negotiate a solution by the Nuffield Council and more directly by the HGAC. The Council appeared to conclude that information which bore a direct relation to the employment context could be disclosed. Such information would include that a person suffered from a disorder which would be exacerbated by the conditions in which he or she would be working, or a disorder that posed a particular and serious risk to others. The SCST agreed, although it is notable that its position on this issue is in conflict with its reasoning with respect to the disclosure of information to other family members. In that case, the SCST argued for non-disclosure on the grounds that the family members would be no worse off in their ignorance than they would have been before, even though that information could provide the basis for preventative action on their part. In the employment (and indeed insurance context), this argument was clearly rejected.

In the insurance context, in December 1997, prompted by the threat of statutory regulation, the ABI produced a code of practice on the use of genetic test information in the context of life insurance.[95] The code provides that insurance companies will not require tests or test results for life insurance of up to £100,000. In excess of that amount, they may require genetic tests before giving insurance. The HGAC has stated that a moratorium on the use of genetic test results in any insurance contract should be imposed, as serious doubts remain as to the accurate and valid use of such information.[96]

There are thus concerns expressed within the medical profession as to the context in which genetic testing occurs and the impacts which it can have, but it

93 *ibid* paras 5.7, 5.29.
94 SCST, Third Report, n 77 above, paras 227–228.
95 ABI, *Code of Practice on Genetics and Life Insurance* (London: ABI, 1997).
96 HGAC, *The Implications of Genetic Testing for Insurance* (London: HGAC, December 1997).

would seem testing continues unabated, and the resolution of the issues which are raised can be seen in large part to be decided by patients themselves within the context of the patient-medical professional relationship. On the other key issue of confidentiality, there is a clear conflict between the perceptions of employers and insurers, and those of the medical profession and patients themselves, and there has been some attempt at negotiation made by the HGAC in the insurance context.

Gene therapy

Gene therapy is a far more difficult technique than testing, and this difficulty has consequences for the focus and design of the regulation.[97] The regulation both anticipates and requires that gene therapy be conducted purely in a research context. This is not because the use of gene therapy in a medical context is deemed un-needing of regulation, but because gene therapy has not yet reached the stage in which it can be applied in that context. The nascent state of gene therapy techniques is of course in contrast to genetic testing or screening, where the relative scientific ease of the testing process means that the operational context is either a clinical one or one of direct supply to the public. The state of science thus dictates the regulatory focus. GTAC requires all proposals for gene therapy research to be submitted to it for approval and the regulation operates on an individual, case by case assessment of those protocols. Most of the research in the UK on single gene disorders has so far been directed at cystic fibrosis.[98] Gene therapy is also being investigated as a way of managing other diseases such as AIDS, some forms of cancer, and chronic forms of diabetes. Indeed GTAC has noted a continuing shift from gene therapy for single gene disorders towards strategies aimed particularly at tumour destruction in cancer patients.[99] Genetic modification carries certain risks, however. It might not work: there may just be no effect. Potentially worse, the correcting gene could be inserted into the wrong cell type, or be expressed either in the wrong amount or at the wrong time, or be inserted in such a way as to cause a new mutation, initiating a new genetic disease. It might also be 'infective', moving from the cells to other somatic or germ line cells.

The ethical issues arising out of gene therapy were the subject of a report by the Clothier Committee,[100] and that report has largely defined the approach taken in the regulation of gene therapy. The Committee took the view that gene therapy should be confined to somatic cells. On the grounds of risk and uncertainty, it recommended that gene therapy of germ line cells should not yet be permitted.[101] Too little was yet known about the possible consequences and hazards of gene therapy to permit genetic modification which would be deliberately designed to affect subsequent generations.[102] Moreover, given the uncertainties surrounding the safety and effectiveness of gene therapy, it should be limited to patients in whom the potential for benefit is greatest in relation to possible inadvertent harm. It should thus be restricted to disorders that were life threatening or caused

97 By January 1997 only 18 protocols had been approved, with 13 studies carried out involving 134 patients. GTAC, *Third Annual Report 1996* (London: GTAC, June 1997) para 2.1.

98 *ibid.*

99 *ibid.*

100 *Report of the Committee on the Ethics of Gene Therapy* (chair, Sir Cecil Clothier) Cm 1788 (London: HMSO, 1992).

101 Germ line cells provide sperm or ova, and so the genes carried by them may be transmitted to successive generations. Somatic cells carry genes which operate only in the body of the individual.

102 Clothier Committee, para 2.26, and Part 5.

serious handicap, and for which treatment was either unavailable or unsatisfactory.[103]

Whilst arguing that somatic gene therapy (alteration of the genes in the cells of particular individuals) raised no new ethical issues, the Committee suggested that given the risks arising and the technical competence which was necessary to assess those risks, it should be subject to regulation which went beyond the normal review given by local research ethics committees (LRECs).[104] Further, gene therapy was not sufficiently developed a technique to be considered part of medical practice, and rather should be treated as research on human subjects. It should therefore conform to the slightly higher ethical requirements which applied to such research. These required that the therapy be useful for biomedical knowledge; that it be conducted in a way that maintained ethical standards of practice, protected the subjects of research from harm, and preserved the subject's rights and liberties. Moreover, reassurance had to be provided to the professions, the public and Parliament that these standards were being upheld.[105]

As a consequence of the Report, GTAC was established in 1993 on a non-statutory basis.[106] GTAC's remit is to consider the scientific merits and potential risks of gene therapy, and it has not appeared to see it to be necessary to consider itself the wider social or ethical implications of interfering with genetic make-up. At least, if it has, it has not made its deliberations or their outcomes public. GTAC's Guidance maintains that its primary concern is to ensure that research proposals meet accepted ethical criteria for research on human subjects.[107] Ethical aspects are stated to include scientific merit and safety;[108] there is no requirement, in contrast to the Clothier Committee's recommendation, that a separate ethical assessment of the research proposed be given.[109]

GTAC perhaps takes the view that the wider issues raised by gene therapy were covered by the Clothier Committee. However, Clothier gave scant attention to these arguments. Whilst it noted that genetic modification might cause harmful or unacceptable genetic alterations or lead to social abuses[110] and stated that it was alert to 'the profound ethical issues that would arise were the aim of genetic modification ever to be directed to the enhancement of normal human traits',[111] it did not expand on these views. In particular, the objection to germ line therapy was that it was too risky, rather than being rooted in any deeper philosophical argument (although there was a nod towards human dignity).[112] Two paragraphs sufficed to dismiss the issue.[113]

It may also be that the nascent state of science in this area affects the regulatory agenda. There is perhaps seen to be little need to have a debate on, for example, the reasons why one should confine gene therapy to the treatment of diseases as given the present state of science and the cost of research no-one would wish to fund anything else. Or it may be that there is no point discussing wider ethics of genetic modification for non-human traits as such modification is just not possible. Or that GTAC feels that such debates are occurring elsewhere in the regulatory structures:

103 *ibid* para 4.3.
104 *ibid* para 4.8 and Part 6.
105 *ibid* para 8.3.
106 All trials do however have to gain approval from the statutory Medical Control Agency.
107 Guidance, Part I, para 11.
108 Guidance, Part I, para 6.
109 Clothier Committee, para 6.2.
110 *ibid* para 1.1.
111 *ibid* para 2.16.
112 *ibid* para 4.22.
113 *ibid* paras 5.1–5.2.

in LRECs, for example.[114] There is little public articulation on these matters. The issues raised by somatic cell gene therapy may well be adequately met in GTAC's current approach. Certainly its work has been praised as being 'sensible and effective'.[115] Germ line therapy raises quite fundamental issues, however, which have not been aired by any of the current regulatory bodies, or, as noted, by the Clothier Committee.[116] These include whether germ line manipulation should be permitted as a matter of individual choice, with only manipulation which is deliberately intended to cause harm being prohibited; or whether manipulation should be seen as an act which directly affects another, and so limited to improving the health of that person (with the inevitable problems in both cases of defining 'harm' or drawing the line between 'cure' and 'enhancement'); how, if at all, germ line manipulation affects the dignity of another human being; the psychological effects of germ line manipulation on parents and children; and the broader implications for genetic diversity and variability. There is a danger in adopting the attitude that there is no point in discussing these issues on the basis that such manipulation just would not happen. As has been stated:

> in science as in life it is important to distinguish between chastity and impotence. They both have the same effect and that is why every scientist that I know will put his hand to his heart and say that he will not conduct germ line therapy. He is not being chaste.[117]

Nevertheless, there is little sense that GTAC sees its role to be the discussion of these issues.[118] Moreover, it is not clear what contributions non-scientists are seen to be able to make to such a debate. The membership of GTAC is drawn widely, and includes a psychologist, a genetic counsellor, religious, legal and industry representatives. GTAC is beginning to show greater willingness to open up its proceedings, and is considering holding them in public. However, private hearings were considered to be essential initially as it was felt that lay members would have been inhibited from showing their ignorance of science by asking questions.[119] Further, whilst one does not want to read too much into these things, the title of GTAC's first workshop (held in 1996) shows some discounting of popular views of gene therapy: 'Myth and Reality: Hype and Practicality'. The criticism being levelled is not that GTAC is not careful in its assessment or that its procedures for considering individual cases are flawed, but of the apparent narrowness of its focus. Such narrowness would be partly excusable if there had already been a full consideration of the issues surrounding gene therapy; this was what Clothier was in part meant to provide, but failed to do so.

Cloning

Cloning provides perhaps one of the starkest examples of the fundamental differences in the way that genetic technology can be perceived. For a scientist, cloning is the production of genetically identical unicellular or multicellular

114 The MCA has no remit to consider ethical issues per se, rather its role is to approve products for marketing on the basis of their safety, quality and efficacy.

115 SCST, Third Report, n 77 above, para 110 and further para 108.

116 For discussion see for example, W. French Anderson, 'Human Gene Therapy: Scientific and Ethical Considerations', in R. Chadwick (ed), *Ethics, Reproduction and Genetic Control* (London: Routledge, 1990).

117 Evidence to the SCST, Third Report, n 77 above, para 116.

118 Although in its defence it could be argued that germ line therapy which involves the manipulation of embryonic cells is covered anyway by the HFEA, (though this does not cover genetic manipulation of the gametes or ovum) or that the HGAC is now the body which should air such issues.

119 See the comments of the then Chairman of GTAC, Dame June Lloyd, to the SCST: Third Report, n 77 above, para 110.

organisms by natural or assisted processes. For a non-scientist, cloning is the key to immortality, the 'resurrection' of dead loved or admired ones, the ultimate ego trip, Jurassic Park. These perceptions are fuelled by some scientists: in the US, Richard Seed recently announced that he was proposing to clone human beings.[120] There have even been reports of proposals to take DNA from the Turin shroud and clone it.[121]

The current debate was of course triggered by the birth of Dolly in February 1997. Dolly was the first example of an adult vertebrate cloned from another adult. The nucleus of a cell taken from the udder of a six year old Dorset Finn ewe was introduced into an unfertilised egg from which the nucleus had been removed. That egg was then placed in the uterus of another sheep, and Dolly was born.[122] Dolly is the exact genetic twin of her 'mother', the Dorset Finn. The purpose in cloning Dolly was to develop methods for the genetic improvement of livestock.[123] The technique could also be used for the production of transgenic livestock: animals which contain genes from other species. Instead of taking the nucleus from the cell of one sheep and introducing it as it is to the ennucleated egg of another sheep, a gene from another animal or human is inserted into that nucleus before it is introduced to the egg. It was through this process that Polly and her identical clones Holly, Molly and Olly (short for Olivia, not Oliver) were born in December 1997. Polly and her sisters are transgenic sheep: they carry a human gene for a blood clotting factor which is used for treating haemophilia.[124]

Human cloning is still some distance away in terms of scientific development.[125] Nevertheless, the birth of Dolly prompted a flurry of regulatory activity nationally and internationally.[126] There is no UK legislation specifically directed at human cloning, but current techniques which would be used in human cloning, in that they involve the introduction of DNA into an ennucleated oocyte, probably fall under the HFEA.[127] There are a number of international instruments banning cloning: in

120 *Nature Biotechnology*, vol 16, January 1998, 6. He would not be the first: in 1993 two scientists at George Washington University, Robert Stillman and Jerry Hall, announced they had cloned human embryos by splitting them (which replicates the natural process which occurs when identical twins are formed and is a quite different technique to that used to produce Dolly): *Nature* vol 365, 28 October 1993, 778.

121 'Today', BBC Radio 4, 31 March 1998.

122 In fact, Dolly was the only success from 277 attempts. The procedure was used on 277 eggs, only 29 of which were successfully reconstructed (ie the nucleus was successfully introduced). Of those 29 which were then implanted in surrogate ewes, only Dolly was born. See 'Viable Offspring Derived from Foetal and Adult Mammalian Cells', *Nature*, vol 385, 27 February 1997, 810, 811.

123 For discussion of the potential applications of the technique, see SCST, Fifth Report, *The Cloning of Animals from Adult Cells*, 1996–97 HC 373-I, paras 7–10; *Nature Biotechnology*, vol 15, April 1997, 306.

124 'Transgenic Sheep Expressing Human Factor IX', *Science*, 19 December 1997, 2130–2133.

125 Although the editorial of the issue of *Nature* in which the birth of Dolly was announced stated that cloning humans from adults' tissue is likely to be achievable 'any time from one to ten years from now': *Nature*, vol 385, 27 February 1997, 753.

126 Within days President Clinton called on the US National Bioethics Advisory Committee (NBAC) to investigate the ethics of cloning and gave instructions to the head of executive departments and agencies that no federal funds be given for the cloning of human beings. The NBAC report, published in June 1997, focused primarily on the risks involved and called for a five year moratorium on human cloning until the risks were better understood. It has been criticised for taking too narrow a view to the issue (see for example *Nature Biotechnology*, n 123 above), but following the report President Clinton introduced the Cloning Prohibition Bill into Congress, where it is currently being considered.

127 This is the view taken by the HFEA: HGAC and HFEA, *Cloning Issues in Reproduction Science and Medicine: A Consultation Document*, but the Select Committee on Science and Technology took the view that the position was more ambiguous: SCST, Fifth Report, n 123 above (January 1998) paras 23–33.

November 1997 UNESCO published the Universal Declaration on the Human Genome and Human Rights, clause 11 of which states that 'practices which are contrary to human dignity, such as reproductive cloning of human beings, shall not be permitted.' A protocol forbidding the cloning of human beings has been developed under the Council of Europe's Convention on Bioethics.[128] The recent Biotechnology Patents Directive forbids the issue of a patent on work leading to the deliberate cloning of human beings.[129] Human cloning of various forms was already banned, explicitly or implicitly in several different countries.[130]

The HGAC and HFEA have taken steps to negotiate a position on human cloning by producing a consultation paper which seeks to address a number of the ethical issues which this area of science raises.[131] The paper distinguished two different types of cloning. First, reproductive cloning, where the intention is to produce identical fetuses and babies, and second, what it termed 'therapeutic' cloning, which involved other applications of nuclear replacement technology. Those could include studying cell development or developing stem cell lines with a view to developing medical applications. The strategy would be to take cells donated by one patient, transfer the nucleus to an ennucleated oocyte,[132] then grow it in culture to generate stem cells, and transfer the stem cells to a patient.[133] Possible applications could be for Parkinson's disease or the treatment of blood diseases.

The paper deliberately seeks to address the far wider questions that human cloning raises. It asks whether the use of nuclear replacement techniques or embryo splitting to create embryos would raise any new issues with respect to the special status of the human embryo or what may ethically be done within the first fourteen days of the embryo's development. It asks whether there are any medical or scientific areas that might benefit from research involving the creation of a cloned embryo, and whether any of the potential applications of nuclear replacement that would not result in cloned fetuses or babies raise any new ethical concerns.[134] With respect to reproductive cloning, the consultation paper asks some fundamental questions. To what extent can a person be said to have a right to an individual genetic identity, particularly given that the experience of genetically identical twins suggests that a unique genetic identity is not essential for a human being to feel and to be an individual? Would the creation of a clone of a human being always be an ethically unacceptable act? Would it be beyond the limit of what is ethically acceptable to resolve a couple's infertility problem? Do the large 'wastage' rates and uncertainties about malformations of embryos which would be involved, certainly in the nascent stages of the technology, make experimentation in humans involving the implantation of cloned embryos ethically impossible? And finally, what ethical importance might be attached to the distinction between artificial technologies which have a counterpart in natural processes (IVF), and those which do not (nuclear replacement technology)?[135]

128 Council of Europe, Convention for the Protection of Human Rights and Dignity of the Human Being with Regard to the Application of Biology and Medicine 1996 (Strasbourg: ETS 164).

129 European Parliament and Council Directive on the Legal Protection of Biotechnological Inventions, COM (97) 446 final. Approved by the European Parliament on 12 May 1998.

130 For summaries of different countries' positions see HGAC and HFEA, *Consultation Document*, n 127 above, Annex D; Union for Europe Group, *Report on Cloning*, May 1997.

131 HGAC and HFEA, *Consultation Document*, n 127 above.

132 An egg mother cell.

133 Alan Colman, 'Dolly — the implications of cloning', paper presented to the 21st Century Trust seminars on Genetics, Ethics and Identity, 27 March–4 April 1998.

134 *Consultation Document*, n 127 above, Section 7.

135 *ibid* Section 8.

The approach taken by the HGAC and HFEA to the issue of human cloning is one which seeks to marry scientific knowledge with ethical concerns; to use the one to inform the other. It contrasts with the approach taken by the Clothier Committee which, as noted above, simply dismissed the consideration of germ line therapy in two paragraphs, and then principally on the basis of safety, and which did not seek to explore the wider social and ethical impacts of somatic cell therapy. The consultation paper is an attempt to pre-empt science by discussing the acceptability of human cloning before all aspects of it become scientifically possible. It is also an attempt to forestall a blanket ban on human cloning which could potentially lead to regulation which in fact inhibits the development of research which might not offend public ethics either at all or to the same degree as full reproductive cloning. Achieving a position in which science and ethics can develop a mutually reinforcing relationship is not an easy task; the consultation paper suggests that at least with respect to this issue the regulatory bodies are willing to try.

Rights to exploit genetic material

The final aspect of the regulation of genetics to be explored is the issue of the rights to exploit genetic material. Defined broadly, this embraces questions of the ownership of genetic material, in particular whether human tissue should be treated as property, and the ability to gain intellectual property rights over it. The first issue, although important, will not be explored here.[136] Instead discussion will focus on the second: the issue of intellectual property rights.

Regulatory fora and principal actors

At first glance, including intellectual property rights in a discussion of the regulation of genetic technology seems misplaced. The Patent Office is not a regulator of such technology. It does not prescribe what research should be done or how, nor is its approval necessary for the development and sale of genetic products or services. However the controversy surrounding the role of patents in genetic research and product development and the significant impact which patents can have on those activities mean that patent law is a significant forum in the genetic debate.

A patent is a monopoly right granted to a person to exploit an invention, and thus to exclude others from its exploitation.[137] Patents can be sought from national bodies (in the UK the Patent Office) or in Europe from the European Patent Office, set up under the European Patent Convention — a regional arrangement limited to European countries, although not solely to members of the EU. The EPO has the capacity to grant applicants patent rights in any number of signatory states nominated by the applicant. The requirements for the granting of a patent under the UK Patents Act 1977 and the EPC are similar.[138] They are novelty, inventiveness, industrial applicability and sufficiency of description. The EPC and the UK Act

136 The issue was the subject of an extensive report by the Nuffield Council on Bioethics, *Human Tissue: Ethical and Legal Issues* (London: Nuffield Council, 1995). On the question of ownership of human tissue see G. Dworkin and I. Kennedy, 'Human Tissue: Rights in the Body and its Parts' [1993] 1 *Medical Law Review* 291.

137 See generally, W.R. Cornish, *Intellectual Property* (London: Sweet and Maxwell, 3rd ed, 1996).

138 s 130(7) states that certain provisions of the 1977 Act, in particular those dealing with criteria for patentability, revocation and infringement, are 'so framed as to have, as nearly as practicable, the same effect in the UK as corresponding provisions in the EPC... have in the territories to which it applies.'

exclude from patentability mere discoveries, biological processes, animal or plant varieties and those things which are contrary to public order or morality.[139] A Directive on the patenting of biotechnological inventions has also recently been agreed which aims to harmonise the criteria for the patentability of genetic material across the EU Member States and to facilitate a uniform application of the immorality exclusion in all Member States.

The principal actors are thus the members of the patent offices, patent agents, patent lawyers, and those who are seeking patents: in this context biotechnology companies, universities, and other research institutions. They are thus lawyers, scientists and industry.

The internal conceptualisation of the issues

As noted, a patent is a monopoly right to exploit an invention, and thus to exclude others from its exploitation. It is the notion of a patent as an exclusionary right which tends to be most frequently emphasised by patent lawyers (and patent holders), particularly when defending the patent system.[140] Patents give no obligation to use: they simply allow the patentee to stop others using and making the invention. The classic justifications for granting that monopoly are that it thereby promotes dissemination of information about an invention, information which would otherwise not be disclosed and which would remain as a trade secret, and that it provides an incentive to invent and to exploit that invention.[141] It is the latter which has become the basis for the more recent justification for patents: viz that a strong patent system is a core aspect of commercial development and thus economic policy. The patents that a firm holds are seen by investors as the most important factor in deciding whether or not to invest in a company,[142] and a strong system of patent protection is frequently argued to be essential for a country's economic development and international competitiveness.[143]

Internal debates

The internal debates reflect the rationalities of the three principal groups of participants: those of law, science and economics. They are thus principally as to the appropriate application and development of legal terms to scientific practices, and as to whether the practices of the courts and patent offices adequately meet the economic goal of providing incentives for invention and commercial development.[144]

139 Article 53(a) EPC; the UK counterpart, s 1(3)(a) states that a patent will not be granted for 'an invention the publication or exploitation of which would be generally expected to encourage offensive, immoral or anti-social behaviour.'

140 See for example, BioIndustry Association, *Innovation from Nature: The Protection of Inventions in Biology* (London: BIA, undated).

141 See generally Cornish, n 137 above.

142 Ernst and Young, *European Biotech '97: A New Economy* (Frankfurt: Ernst and Young International, 1997) 36, figure 15.

143 Reflected in the recitals of the recent Directive on Biotechnology; E. Armitage, 'EU Industrial Property Policy: Priority for Patents' [1996] EIPR 555; J. Lerner, 'Patenting in the Shadow of Competitors' (1995) 35 *J Law & Econ* 463; A. McInerny, 'Biotechnology: *Biogen* v *Medeva* in the House of Lords' [1998] EIPR 14; these arguments are also loudly voiced in the context of TRIPS (the Agreement on Trade-related Intellectual Property Rights): see S.K. Verma, 'TRIPs and Plant Variety Protection in Developing Countries' [1995] EIPR 281; M. Blakeney, 'The Impact of the TRIPs Agreement in the Asian Pacific Region' [1996] EIPR 544.

144 Thus particular decisions have been criticised on the basis that they do not provide sufficient reward and incentives to patentees in biotechnology: for example R. Ebbnik, 'The Performance of Biotech Patents in the National Courts of Europe' [1995] *Patent World* 25; N. Jones, 'The New Biotechnology Directive' [1996] EIPR 363.

The advent of biotechnology has posed significant challenges for patent law, and many argue that patent law is struggling to meet that challenge.[145] The requirements of novelty and inventiveness are those around which much of the contention arises in deciding whether and when genes and gene technology are patentable. The novelty requirement demands that the invention claimed has not been made available anywhere in the world prior to the filing date.[146] The question, therefore, is not whether what is claimed already exists, but whether its existence is known. Thus where a product claim is in issue, it is arguable that even if an identical substance to that which is claimed occurs in nature, it may be regarded as novel when isolated and identified, or 'characterised' (eg by means of a DNA sequence) for the first time. Alternatively, a claimed substance may satisfy the novelty test if it is produced in a refined or purified form, ie a form that does not occur naturally.[147] The inventiveness requirement, however, demands in addition that the invention claimed be not obvious to a person ordinarily skilled in the relevant art. Gene technology is not of itself regarded as inventive, and nor, without more, are its products. Thus it will generally be regarded as obvious to attempt to sequence a gene or to produce the substance for which it codes in a pure form using standard recombinant DNA techniques.[148] The applicant will need to point to a method which overcomes particular difficulties attendant upon these techniques.[149] Further, the sufficiency requirement will ensure that broad claims, extending to other recombinant methods than those actually invented by the claimant, will be disallowed.[150]

As in any legal system, there are areas of contention and uncertainty as to the application of legal provisions to particular fact situations. In the context of gene related patents, disputes arise as to the interpretation of the provision which excludes plant and animal varieties from patentability,[151] as to the meaning of non-obviousness and inventiveness in the biotechnological context,[152] and as to whether patents are being granted which are too wide in their scope.[153] Concern is also being expressed in the US over the recent decision of the US Patent Office to grant patents for expressed sequence tags of no known utility on the basis that patents are being granted to reward very little commercial or scientific outlay, and in a way which will severely inhibit future research.[154]

145 See B. Sherman, 'Patent Law in a time of Change: Non-obviousness and Biotechnology' (1990) 10 OJLS 278. For discussion of the application of patent law to biotechnology see further Nuffield Council, n 136 above; Cornish, n 137 above; G.T. Laurie, 'Biotechnology and Intellectual Property: A Marriage of Inconvenience?' in S. McLean (ed), *Contemporary Issues in Law, Medicine and Ethics* (Aldershot: Dartmouth, 1996). I am grateful to Anne Barron for her advice on this section.

146 In contrast to the US, where it is the date of publication rather than of filing which is critical.

147 Decision of the EPO in the Opposition to Patent No 112 149 in the name of the Howard Florey Institute of Experimental Physiology and Medicine, 18 January 1995 (the Relaxin Opposition) (1995) OJEPO 388, and in the UK see *Genentech*, which suggests that both genes and the substances for which they code are mere discoveries, and as such not patentable: [1989] RPC 147 (CA).

148 *Genentech Inc's Patent* [1989] RPC 147.

149 *Biogen Inc* v *Medeva plc* [1997] RPC 1 (HL).

150 *ibid.*

151 Article 53(b) EPC, discussed (despite the title) by D. Beyleveld and R. Brownsword, *Mice, Morality and Patents: The Onco-mouse Application and Article 53(a) of the European Patent Convention* (London: Common Law Institute of Intellectual Property, 1993).

152 For the most recent discussion of the issue in the UK context see *Biogen* v *Medeva*, n 149 above.

153 S. Crespi, 'Biotechnology, Broad Claims and the EPC' [1995] EIPR 371; T. Roberts, 'Broad Claims for Biotechnological Inventions' [1994] EIPR 371; *Biogen* v *Medeva*, n 149 above.

154 Human Genome Sciences has also recently filed an application with the EPO for patents on the genetic sequence of one of the bacteria which cause meningitis: *The Guardian*, 7 May 1998.

Outsider perspectives

These issues are of undoubted importance to the operation of the patent system, and they are sources of controversy both within and to an extent outside that system. However, the granting of patents to gene sequences in humans, plants or animals is opposed by many on more fundamental grounds. These are essentially that patents should not be granted over such material not because existing legal definitions are being wrongly applied (although those arguments are also made)[155] but because the granting of monopoly rights for its exploitation is simply inappropriate.

The inappropriateness argument takes a number of forms, not always distinguished or explicit. For these purposes three main versions can be identified. The first is that granting patents over genetic sequences amounts to the patenting and thus commodification of life.[156] The objection is rooted in concepts of property and ownership, and the appropriateness of the application of those concepts to the DNA particularly, although not uniquely, of humans. So in its opposition in the *Relaxin* case, which was an application for a patent of the DNA sequence which codes for the hormone relaxin, opponents argued that to patent human genes was to patent life, and that it amounted to slavery contrary to the fundamental human right to self determination.[157] The objection that patenting genes is tantamount to patenting life has also been put by a number of non-governmental organisations.[158] Further, the revised EU draft directive on patents in biotechnology provides that 'the human body, at various stages of its development, and the simple discovery of one of its elements including the sequence or partial sequence of a gene, cannot constitute patentable inventions.'[159]

The second type of objection to the granting of gene related patents is essentially directed not so much at the patent itself as at the process or substance which is being patented. Patents are being objected to on the grounds not that only *one* person should be able to conduct the activity for which the patent is sought, but that *no-one* should. Thus the recent EU directive on patents and biotechnology expressly provides that the following shall be considered unpatentable: the procedures for human reproductive cloning; processes for modifying the germ-line genetic identity of human beings; methods in which human embryos are used; and processes for modifying the genetic identity of animals which are likely to cause them suffering without any substantial medical benefit to man or animal and animals resulting from such processes.[160] This is not because such activities should be open to all to exploit; quite the opposite. It is because it is felt that these activities are morally insupportable. Opposing a patent on essentially moral

155 eg Beyleveld and Brownsword, n 151 above.

156 For a discussion see A. Wells, 'Patenting New Life Forms: An Ecological Perspective' [1994] EIPR 111.

157 *Relaxin* opposition, n 147 above.

158 See for example the campaigns of Jeremy Rifkin in the US; further, *The Case Against Patents in Genetic Engineering: A Special Report by the Genetics Forum* (London: The Genetics Forum, 1996), the 'No Patents on Life' campaign which the Forum co-ordinates in the UK, and the 'Blue Mountain Declaration' issued by NGOs in the US opposing patents on living organisms and their component parts.

159 Article 5, para 1. Note however that the Article also provides in para 2 that 'An element isolated from the human body or otherwise produced by means of a technical process including the sequence or partial sequence of a gene may constitute a patentable invention, even if the structure of that element is identical to the natural element.'

160 Amended Proposal for a European Parliament and Council Directive on the Legal Protection of Biotechnological Inventions, COM(97) 446 final, article 6, approved by the Parliament on 12 May 1998.

grounds is an attempt to prevent the activity altogether. It is an attempt which is not made through a direct ban on the activity however, but obliquely, by a withdrawal of the incentives to perform it.

The third form of objection is not so much to the activity per se, but to the commercial exploitation which patents facilitate, or indeed enable. It is an objection to the introduction of commercial norms in areas where many regard them as inappropriate. At its most extreme, it is an objection to the biotechnology industry itself. Thus the application for patents over expressed sequence tags generated in the course of the Human Genome Project which was filed by the US National Institutes of Health was widely condemned as violating the spirit of academic co-operation in which the Project had begun.[161] The claiming of patents over gene sequences obtained from indigenous peoples or plants has been strongly criticised as a form of 'bio-piracy', or as 'gene prospecting'.[162] The example of the agreement reached in 1991 between Merck and an agency of the Costa Rican government is often cited in this context. Merck paid US $1.2m for the right to inventory, test and commercialise plants, micro-organisms and insects from the rainforest. The agency would then receive five per cent of any royalties.[163] In the agricultural context, the granting of patents over genetically modified crops is resisted on the grounds that it will adversely affect farmers, plant and animal breeders and biodiversity.[164] The ethical code of the Human Genome Diversity Project now contains a provision that no patents will be sought on the basis of genetic material obtained as part of the project.

Access

Patent law tries hard to remain insulated from these arguments. However, they enter the patent arena directly through the gateway of article 53(a) of the EPC. Essentially dormant until the arrival of gene related patents,[165] article 53(a) provides that patents may not be granted for inventions the publication or exploitation of which would be contrary to *ordre public* or morality. Through this entry point have come arguments which are totally alien to the closed world of patents and the EPO has tried to narrow the gateway as far as possible.[166] The EPO Guidelines state that the test is whether the public in general would consider the invention to be 'so abhorrent that the grant of patent rights would be inconceiveable.' It has made it clear that its assessment of morality is essentially utilitarian in form. So creating a transgenic mouse which contained an activated gene for cancer, the Harvard onco-mouse, was found to be acceptable (although the case is still under consideration), but creating a transgenic mouse for the purposes of attempting to alleviate baldness (the Upjohn case) was not. Morality also seems

161 For an outline of the moves and countermoves involved see SCST, Third Report, n 77 above, paras 35–45; for an excellent discussion of the background to the HGP see T. Wilkie, *Perilous Knowledge* (Harmondsworth: Penguin, 1993).

162 See for example, D. Dickson, 'Whose Genes are they Anyway?' *Nature*, vol 381, 2 May 1996, 11; V. Shiva and R. Hollar-Bhar, 'Intellectual Privacy and the Neem Tree', *The Ecologist*, vol 23 no 6, November/December 1993.

163 See further Rural Advancement Foundation International (RAFI), *Communique*, Sept/Oct 1995, Ottawa, Canada.

164 See for example, the Genetics Forum, 1996.

165 L. Bentley and B. Sherman, 'The Ethics of Patenting: Towards a Transgenic Patent System' (1995) 3 *Medical Law Review* 275.

166 For discussions of, variously, the Onco-mouse case, *Greenpeace UK* v *Plant Genetic Systems NV*, the *Relaxin* case and the *Upjohn* case, see for example Beyleveld and Brownsword, n 151 above, Bentley and Sherman, n 165 above, Laurie, n 145 above.

to be an essentially risk-based assessment: Greenpeace's opposition to the Genetech patent for a herbicide resistant crop was rejected on the basis that the environmental risks had not been made out. Finally, the morality arguments are ultimately countered by technical ones in what often appears to be a dialogue of the deaf. Thus the claim in the *Relaxin* case that allowing patents of genes or gene fragments is 'patenting life' was rejected with the explanation, scientifically and legally correct, that what is being patented is a chemical substance which carries a genetic code to produce medically useful proteins.[167]

Attempts at negotiation

The EPO, practising patent lawyers, and biotechnology companies, are visibly frustrated at the attempts which are being made to use the morality clause to introduce arguments which they see to be completely inappropriate in the world of patents.[168] Again, the issues which outsiders see to be relevant to the administration of the regulatory regime, this time of intellectual property, are not those which that regime considers relevant to its operation. There is moreover a manifest tension between the perceptions of such 'insiders' of the patent system as to what patents are and the views of outsiders.

These arguments are articulated to varying degrees in a number of places, but can be analysed as follows. Essentially, insiders see patents as objective, technical and legal, and so in express contrast to ethics which are malleable, subjective and emotive. Outsiders implicitly or explicitly see patents as social privileges conferred on inventors which could be witheld or to the exercise of which certain conditions could be attached. The insiders retort that they have already paid for the privilege. In return for the patent they grant disclosure of the invention: the classic view of the exchange relationship which underlies patent law. However, the implicit rejoinder is either that this is not enough, or this is not the point. In certain instances society does not see any merit in granting the patent at all, regardless of the disclosure to which it may lead. This is not because it does not want that disclosure, but because it does not think that social privileges should be granted for its exploitation. In turn, to echo the arguments given above, this could either be because it does not think the invention should be exploited at all (as with germ line therapy or human cloning) or because it does not think that a monopoly should be given over that exploitation.

Some have suggested that these alternative perspectives should be introduced into the patent system through a change in the institutional structure of that system. A proposal put forward by the Nuffield Council and other commentators[169] is that the patent system should better equip itself to assess ethical issues by altering the membership of the Patent Office. Alternatively, or in addition, it should be given better guidance by the appropriate legislative body as to what types of things are considered to be contrary to morality, which it should then apply.

However, although altering the institutional structure of the regulation or its remit could perhaps go some way to ensuring that the patent system takes into account ethical issues in granting patents, it is not clear that this would really

167 *Relaxin* case, n 147 above.
168 See for example, S. Crespi, 'Biotechnology Patenting: The Wicked Animal Must Defend Itself' [1995] 9 EIPR 431; J. Woodley and G. Smith, 'Conflicts in Ethics/Patents in Gene R&D', *les Nouvelles*, September 1997, 119.
169 Nuffield Council on Bioethics, n 136 above; Beyleveld and Brownsword, n 155 above; Bentley and Sherman, n 151 above; Wells, n 156 above.

address the core issues. Withholding a patent on the grounds of morality may send a signal that wider societal approval for the activity will not be conferred and may remove the incentives for anyone to undertake the activity, but it does not prohibit it. As Laurie has commented, '[a]t best, the "morality" exception can draw attention to matters requiring attention and action. In reality, it demonstrates the inappropriateness and inadequacy of using such a system to regulate matters which are clearly outside its scope.'[170] The attempt to use patent law to prevent unwanted commercial exploitation is again simply to use an instrument which is badly fashioned for the task. If what is wanted is to regulate the biotechnology industry, then this has to be done through regulation specifically designed for that end.[171]

Nonetheless, patent law has been one of the central areas in which ethical and moral issues have been raised. It is suggested that one of the principal reasons why such a wide range of opponents are trying to use the immorality door to enter this forum is not (or not just) because they want to deny the implicit conferral of society's approval which the patent manifests. Rather it is because it is the only door available. It is the only point in the different fora in which decisions are made relating to the development and application of genetic technology at which widely based ethical and social objections to genetic engineering itself are given the opportunity to make a formal, and highly visible, appearance.

Facilitating regulation: towards negotiation and integration

There are thus a multitude of regulatory fora in which questions of the use and application of genetic regulation are determined. In each, there are different participants, different rationalities, and different criteria on which regulatory decisions are based. There are, however, tensions between the rationalities, the cognitions, of the different systems of science, academia, industry, government and between the rationalities of those systems and other sub-systems of society.[172] At a risk of caricature, scientists tend to see issues simply in terms of scientific feasibility and risk. Regulation is justified only when directed at particular risks, and debate as to what regulation should be is seen as irrelevant unless and until the scientific means exist for actually doing what it is which is discussed. Industry seeks regulation which is based in scientific assessments of risk, and which facilitates and protects its ability commercially to exploit genetic technology and which does not put it at a disadvantage with respect to international competitors. Academia's position is more complicated: on the one hand it seeks to engage in pure research and values collaboration and co-operation; on the other, it is fiercely competitive, with researchers competing nationally and internationally for reputation and for finance. It also seeks increasingly to exploit its research commercially. For government, the bio-industry's voice speaks loudly and there are a number of initiatives to encourage its development: bio-technology is arguably one of the new 'national champions'.[173] In 1992 the Biotechnology Industry Regulatory Advisory

170 Laurie, n 145 above, 255.

171 For one suggestion as to how this could be achieved see M. Llewelyn, 'The Legal Protection of Biotechnological Inventions: An Alternative Approach' [1997] 3 EIPR 115 (regulate the exercise of the right, not its grant).

172 Again, the focus is still on the way in which issues are viewed, rather than the interests which are pursued; the two are related, but this is not the occasion to explore the interaction between them. See further Black, n 7 above.

173 Industries deliberately supported and promoted by government as symbols of national success, traditionally those of heavy industry.

Group was set up by government to discuss the regulation of biotechnology,[174] and in 1995 a £17 million awareness programme, 'Biotechnology Means Business', was launched.[175] The programme includes funding for small companies and for joint ventures between academia and industry,[176] funding for academia to encourage it to develop patentable products,[177] the creation of business 'mentoring and incubation' services, and a finance advisory service.

The concerns which may loosely be attributed to the wider public, or those outside the individual regulatory fora, tend, as we have seen, to be broader than those which are recognised by those within the fora. Moreover, each forum has a different sense of its own purpose and rationale. The patent system in particular is striking in its closure and its resistance to external interventions. The other regulatory structures also operate with their own logics, however. Risk dominates regulation of laboratory work, and the research, development and marketing of genetically modified products. Risk is also a key aspect of the regulation of the direct supply of genetic testing and gene therapy. In both, risk is considered essentially to be a scientific and technical assessment. Considerations of medical ethics play a strong role in human and medical genetics, and wider social concerns are evident in the development of the policy with respect to cloning.

At present, non-scientists have limited access to the regulatory fora, and tend to be recognised by those fora in the capacity of either consumers or patients. As consumers, the regulatory emphasis is on their ability to exercise informed choice in their purchasing decisions, to which end product information is seen to be the key. In their capacity as patients, a different logic has prevailed. The context is that not of the market place but of the hospital; the norms of the medical profession have informed the way in which issues arising from genetic technology are defined and treated. The overriding concerns are for informed consent, counselling, communication and confidentiality. In both contexts the individual's participation is restricted, however, to the ability to say yes or no to what is offered, not to shape the choice.

There are, however, moves to open up the regulatory decision process. Various attempts have been made to broaden the views to which different systems have to respond: the wider membership of the HGAC, for example, or the role of the European Group on the Ethics in Science and New Technologies, which advises the EU Commission.[178] In addition, some industry associations (although not in the UK) have drawn up their own ethical codes,[179] and there have been some initiatives

174 The industrial representatives come from a range of biotechnology users, including pharmaceuticals, agriculture and food, and biotechnology-specific SMEs (small and medium enterprises). The regulatory departments represented are the DTI, the DOE, the HSE, MAFF, the Patent Office and the MCA.

175 For details see the DTI's Bioguide at http://www.dti.gov.uk/bioguide.htm#contents.

176 The SMART scheme helps SMEs and individuals to 'research, design and develop technologically innovative products and processes for the national benefit': Bioguide. It provides funding of 75 per cent of project costs, up to a maximum of £45,000 to assist in technical and commercial feasibility studies and 30 per cent to help the development of technological products. The LINK scheme aims to 'accelerate the exploitation of technology and to bridge the gap between science and the market place.' Government Departments and Research Councils provide up to 50 per cent of the eligible costs of a project with the balance coming from industry. So far over 120 projects have been supported.

177 Under the banner 'Biotechnology Exploitation Platforms', grants of up to £250,000 are available to encourage higher education institutions to develop exploitable patent portfolios, and identify commercial opportunities.

178 This was established in December 1997 and replaces the Group of Advisers on the Ethical Implications of Biotechnology, established in 1992.

179 In the US, the Bio Industry Organisation has issued a *Statement of Principles* (http://www.bio.org), and in Europe, EuropaBio has issued for consultation *Draft Core Ethical Values* (June 1997) (http://www.europa-bio.be).

in improving public understanding of genetics.[180] These include a Consensus Conference on plant biotechnology[181] and a Citizens' Jury on genetic testing.[182] Such initiatives appear to provide the structures in which the integration which is sought by the participatory or procedural models could be achieved. It would seem that they begin to meet the theoretical demands made from sociology, science, and law for the proceduralisation of regulation in the face of competing rationalities. But these initiatives seem often not so much to integrate different views as simply aggregate them, and in so aggregating them afford science a voice which is regarded as more authoritative, and indeed more legitimate, than that of others.[183]

The call that is being made here is not simply for the broadening of participation in regulation, however, but for regulators to adopt a different role: that of facilitators, of negotiators. For that, it is suggested, they need to take on the role of interpreters: of re-translating the views of different groups and putting them into a language that the others can understand. In order to develop regulation which truly aims to facilitate the integration of contending views, it is thus suggested, there has to be a focus on more than institutional design. There has to be a focus on status, on language and on understanding.

The warning, in other words, is that focusing on institutional design will not necessarily alter much. The mere existence of lay participants in the regulatory process may not in fact result in regulation which is very different from that which would have been formed by scientists alone. For example, in a context where the problem is defined solely in terms of risk, and in which scientific assessments are seen as rational and their risk assessments as assessments of 'real' risk, the inclusion of ethics committees or non-scientists in the decision making process will not necessarily achieve the integration sought simply because it is only the scientific view which is afforded legitimacy. Unless lay views are seen as equally valid, they will be marginalised and will therefore not be afforded full standing in the debate.[184]

In this context, regulation should seek to enable contending groups to arrive at an accepted decision or set of decisions. For this to occur, however, a number of pre-conditions have to be met. It has to be recognised that scientific approaches are not necessarily 'correct', nor can they claim a monopoly over rationality. In turn, it has to be accepted that lay views of science, and in particular of risk, are not necessarily incorrect nor irrational. The argument is thus that to facilitate the integration of scientific and non-scientific views, we need to remove the idea of scientific objectivity *versus* lay irrationality and to replace it with one which recognises different rationalities.

The task is a significant one, and its nature can be illustrated briefly by looking at the question of risk. As we have seen, risk is an important aspect of the debate surrounding genetic technology, and although not the sole aspect of that debate it is nonetheless one of the principal justifications that science would accept for regulation, analogous to the role that welfare economics plays for economists in

180 For example the establishment of the Committee on the Public Understanding of Science.
181 UK National Consensus Conference on Plant Biotechnology, Final Report (November 1994), available at ftp://ftp.open.gov.uk/pub/docs/sci_museum/consensus.txt.
182 WIHSC, n 90 above.
183 On integration and aggregation see further S. Krimsky and D. Golding, 'Reflections' in S. Krimsky and D. Golding (eds), *Social Theories of Risk* (New York: Praeger, 1992).
184 This is a point made by Beck, *Risk Society: Towards a New Modernity* n 1 above, and has been picked up and developed in some of the 'socio-scientific' literature: for a review see S. Eden, 'Public Participation in Environmental Policy: Considering Scientific, Counter-Scientific and Non-Scientific Contributions' (1996) 5 *Public Understanding of Science* 183.

justifying economic regulation.[185] By examining the debate which surrounds the definition and regulation of risk, we can begin to see the nature of the task which a facilitative, integrationist approach to regulation entails.

In the area of risk, the question of the relative roles of scientists and non-scientists in defining societal risk is a matter of considerable debate, both in academia and in government circles.[186] Few would deny individual patients or consumers the opportunity to determine the extent to which they expose themselves to risk through, for example, the principles of informed consent to medical treatment, or the provision of information to enable consumers to make purchasing decisions,[187] but the extent to which non-experts should be making decisions as to the degree of risk to which society should be exposed is a more contested question.

The shades and nuances in that debate are multifarious, but for our purposes we can distinguish between three broad propositions.[188] First, that scientific identification of hazards and determinations of risks should determine the regulatory response. This may be termed the technocratic view. The dominant language should be that of science. To the extent that it is seen to be a problem, the lack of public understanding of the scientific language, or of public acceptance of the scientific view (including the adequacy of its regulation), is to be met through public education. Until then the public is simply ill-equipped to participate in regulatory decision making, and indeed their participation could be positively harmful both to science and to the commercial exploitation of its activities.

This attitude is evident in the approach of government, science and industry to the issue of public understanding. Encouraging public understanding is essentially seen as encouraging public *appreciation* of science. It is an attitude to public understanding which is essentially one-way, based on the view that the public is currently ignorant, and that ignorance needs to be dispelled.[189] If it were, then there would be a greater chance of the public accepting genetic research and its outcomes. This would not only improve the public image of science, and so help secure public funding, it would also mean that the public would adopt a proper attitude towards science and its findings, by for example, adopting proper risk calculations.[190]

185 On which see generally, A. Ogus, *Regulation: Legal Form and Economic Theory* (Oxford: Clarendon Press, 1994).

186 eg O. Renn and D. Levine, 'Trust and Credibility' in H. Jungermann, R.E. Kasperson, P.M. Wiedemann (eds), *Risk Communication* (Julich: KPA, 1992); R.H. Phildes and C.R. Sunstein, 'Reinventing the Regulatory State' (1995) 62 *Univ Chicago LR* 1; A. Hutchinson, *Dwelling on the Threshold: Critical Essays on Modern Legal Thought* (London: Sweet and Maxwell, 1988); A. Ogus, 'Risk Management and "Rational" Social Regulation' in R. Baldwin (ed), *Law and Uncertainty: Risks and Legal Processes* (Berlin: Kluwer, 1997); D.J. Fiorino, 'Technical and Democratic Values in Risk Analysis' (1989) 9 *Risk Analysis* 293.

187 Although exactly what these principles require in any one circumstance can of course still be contested; witness, for example, the debate on the labelling of novel foods and food ingredients.

188 This characterisation of positions focuses simply on the assessment of risks posed by particular hazards; it does not include aspects of risk management, for example risk or cost benefit analysis. A number of other classifications are possible: for discussion see O. Renn, 'Concepts of Risk: A Classification' in Krimsky and Golding, n 183 above; R. Baldwin, 'Introduction – Risk: The Legal Contribution' in R. Baldwin (ed), *Law and Uncertainty: Risks and Legal Processes* (Berlin: Kluwer, 1997).

189 For a discussion of the different approaches to public understanding of science more generally see B. Wynne, 'Public Uptake of Science: A Case for Institutional Reflexivity' (1993) 3 *Public Understanding of Science* 321; 'Public Understanding of Science Research: New Horizons or Hall of Mirrors?' (1991) 1 *Public Understanding of Science* 37.

190 See for example the report of the Committee to Review the Contribution of Scientists and Engineers to the Public Understanding of Science, Engineering and Technology, November 1995 (the Wolfendale Report).

The need for public understanding initiatives to improve public appreciation of science is an explicit aim of the Committee on the Public Understanding of Science (COPUS), of the Wolfendale Committee,[191] and underlay the Select Committee's report on biotechnology.[192] Thus Walter Bodmer, chairman of COPUS, deplores the low levels of 'genetic literacy',[193] with individuals being unable, for example, to provide correct definitions of DNA.[194] The question is seen to be how science can go out to meet the public, not how the public comes in to science.[195] The assumption is that there is a knowledge deficit on the part of the public which it is the role of science to fill. Once filled, the public will see, and accept, the scientific light.

The scientific light in the field of genetics, at least as shone by the proponents of gene technology, is one which encourages support for that technology on a number of grounds. Those involved in genetic research, and indeed its commercial exploitation, emphasise that it has the potential to confer considerable benefits. These include medical applications (diagnosis and treatment), benefits for the environment (less use of herbicides and pesticides), for the farmer (greater yields, less expenditure on chemicals), and for the consumer (lower fat pork, tastier tomatoes).

Fears of genetic engineering are addressed by stressing that its development is evolutionary rather than revolutionary: it is simply a further step building on existing techniques. Analogies are drawn between genetic engineering and 'traditional' biotechnological techniques of fermentation in the production of wine, beer, or bread, or the traditional cross-breeding of plants and animals.[196] It is argued that genetic engineering poses no greater hazards than those which society already tolerates; that in fact it could pose fewer hazards as it is more precise than traditional methods.[197] Genetic engineering is sometimes termed 'precision genetics',[198] allowing specific traits to be introduced through the introduction of a single gene rather than a random mixing of thousands of genes from two parents as occurs in traditional cross-breeding techniques.[199] Such precision, it is stressed, means that it confers greater benefits for the plant breeder and for the animal[200] as well as in medicine.

191 Wolfendale Report, para 1.8.

192 n 44 above.

193 W. Bodmer and R. McKie, *The Book of Man* (London: Little Brown, 1994).

194 For further examples see J. Durant, A. Hansen and M. Bauer, 'Public Understanding of the New Genetics' in *The Troubled Helix*, n 87 above, 239.

195 For example the Wolfendale Committee recommended the linking of research grants to initiatives to disseminate results to the public, encouraging universities to give training in communication skills, linking success in promoting public understanding to an individual's appointment and promotion prospects, and through continuing education initiatives. It proposed that initiatives be assessed by counting the number of hours spent on doing 'popular lectures, interviews, popular articles etc' (para 4.2.1), with individuals keeping logs of activities and portfolios of the outcome, and using questionnaires or audience interviews to assess quality (presumably to test how much the person had 'learned').

196 See for example the SCST Report on Biotechnology, n 44 above, ch 2.

197 See for example, evidence of the then head of the ACGM to the SCST *ibid* (para 5.11), and of Sir Walter Bodmer, head of COPUS (para 5.14) concerning genetic engineering and animals; or that given with respect to genetically modified foods (para 5.24); see further the OECD 1992 report on food safety.

198 eg M. Cantley, 'On LMOs, Catch 220 and DNA, Editorial Overview' (1996) 7 *Current Opinion in Biotechnology* 259

199 *Genetically Modified Crops and Their Wild Relatives — a UK Perspective*, GMO Research Report No 1, DOE 1994, (i). See generally the scientific advice cited in the SCST Report on Biotechnology, n 44 above, paras 5.4–5.8; 5.15–5.18, and the Committee's endorsement of this view, at para 6.10.

200 G. Bulfield, 'Genetic Manipulation of Farm and Laboratory Animals', in P. Wheale and R. McNally (eds), *The Bio-Revolution: Cornucopia or Pandora's Box?* (London: Pluto Press, 1990).

Finally it is argued that the risks that do exist are controllable. The particular hazard posed can be identified, an assessment of the likelihood of harm occurring made, and the nature of the consequences posited. The aim of regulation should be to ensure the adequate control of risks; not necessarily to eliminate them, but to reduce them to an acceptable level.

On this view, therefore, regulation should be 'science-based, proportional to real risk and unclouded by other issues'.[201] Science should be the benchmark. Lay views are irrational deviations from that benchmark, which should be corrected by education and which, if they insist on remaining, should be first, negotiable items in determining policy and second, extrinsic to the real business of regulation. What is necessary is that the public needs to be educated in the language and methods of science. Once people are made aware of science and of the benefits that genetic engineering can bring, and of the relatively low risks involved, they will be more prepared to accept the technology itself and the fruits it can bear.

The second proposition is that there is a role for public participation, but that the public's role should be confined. This view is thus one which may be termed modified technocracy. The scientific view is still that which is seen as the true 'objective' view, but to the extent that public concern relating to particular risks is high, or the public would derive particular utility from a regulatory approach which recognised such concerns, some deviations from this objective view may be permitted. Such deviations would be justified either on economic grounds,[202] or for the more pragmatic reasons of political expediency and the need to maintain the acceptability of the regulation in the eyes of the public.

The third proposition is more radical. It involves the rejection of the authoritative position of science in regulatory decision making, giving it no stronger voice than any other. Science is not infallible and it is not neutral; lay perceptions, moreover, are not arbitrary or irrational but based on identifiable criteria, exhibit a systematic pattern, and are institutionally embedded. It is the acceptance of this proposition which has to underlie any system of regulation which seeks to facilitate full integration. Whether this acceptance can be achieved, both theoretically and in practice, has been a significant topic of debate in social science writings on risk and of parts of the scientific literature on the broader issue of the objectivity of science and on public understanding.

In this proposition, contesting the idea of the objectivity of science is a central part of the discussion. Indeed, that discussion has reached such a point that the idea that scientific risk analysis is not objective is the accepted point of departure, and not something which could seriously be questioned. There are two core operating premises: first, that objectivity is particularly absent in areas characterised by a high degree of uncertainty and so risk analysis can never be free from judgement; and secondly, that in making those judgements science and scientists are not immune from bias or from the impact of institutional and social norms which affect all other aspects of society.[203] Scientists' assessments cannot claim to be 'rational' in any universal sense, but only according to the norms which operate within the scientific system. Other views can be equally 'rational', in the sense of flowing logically from different sets of premises. There are simply competing constructions of risks, those of scientists and those of anybody else. As Shrader-Frechette argues,

201 BIA, 'The BIA Leads the Way in the UK', *Nature Biotechnology Supplement*, June 1997, 29.
202 A. Ogus, 'Risk Management and "Rational" Social Regulation' in Baldwin, n188 above.
203 See in particular, K.S. Shrader-Frechette, *Risk and Rationality: Philosophical Foundations for Populist Reforms* (Berkeley: UCLA Press, 1991).

'there is no distinction between perceived risks and actual risks because there are no risks except perceived risks',[204] experts have no 'magic window on reality'.[205]

With respect to the second, it has been repeatedly demonstrated that lay ideas or understandings of risk are based on fundamentally different criteria than those of scientists. Attitudes towards risk are not dependent on the narrow criteria of the rate of expected fatalities which scientists use in risk estimation or measurement. Scientists are frustrated that people will undertake higher risk activities but not lower risk ones and that they continually mistake the level of risks of activities. That frustration, however, fails to recognise the complex matrix of variables which comprise the risk decisions that individuals make. These include: the voluntary or involuntary nature of exposure; the degree of personal control over the outcomes; the degree of uncertainty over the probabilities or consequence of exposure; the familiarity of the risk; the immediacy or delay of exposure to the risk; the potential for catastrophic consequences; the distribution and visibility of benefits; and whether the risk is individual or societal.[206] Moreover, and critically for regulation, public trust in the institutional arrangements which exist to manage risk plays a key role in attitudes towards risks.[207]

Public attitudes towards genetic technology bear out a number of these findings. UK survey findings show that risk perceptions tend to focus on safety issues and on a perceived lack of control of the technology. They show a generally low estimation of the competence of regulatory agencies and their ability to effectively control the risks of technology.[208] They also show a lack of trust, with people feeling that the public was being kept in the dark about genetic engineering (for example, although 95 per cent felt the public should be told about releases of genetically engineered products only 25 per cent felt it would be). This low level of trust in government is echoed in the perceptions which are held of the relative trustworthiness of different sources of information about genetic engineering. Work by Martin and Tait on public perceptions of trust in different sources of information shows that the public on the whole rate special interest groups as the most trustworthy source, but government, together with industry and tabloid newspapers as the lowest.[209]

Psychometric studies show that perceptions are systematic: they can be mapped according to a range of variables. Lay perceptions of both the magnitude and acceptability of risk may thus be explicable, but are they 'rational'? There are a number of responses. Cultural theorists, for example, would argue that they are indeed rational, in that a person's perception of and attitude to risk is consistent with the rest of their understanding and approach to the world around them.[210] Others do not need to enter the world of grid/group analysis to suggest that

204 Shrader-Frechette, *ibid* 84.

205 *ibid*. See also J. Hunt, 'The Social Construction of Precaution' in O'Riordan and Cameron, n 38 above.

206 B. Fischoff, S. Lichtenstein, P. Slovic, D. Derby and R. Keeney, *Acceptable Risk* (Cambridge: CUP, 1981); H.J. Otway and D. von Winterfeldt, 'Beyond Acceptable Risk: On the Social Acceptability of Technologies' (1982) 14 *Policy Sciences* 247; Royal Society, *Risk: Analysis, Perception, Management* (London: Royal Society, 1992).

207 Royal Society, *ibid*; Wynne, n 189 above.

208 Ives, n 49 above.

209 S. Martin and J. Tait, 'Attitudes to Selected Public Groups in the UK to Biotechnology' in Durant (ed), n 49 above.

210 M. Douglas and A. Wildavsky, *Risk and Culture* (California: UCLA Press, 1982); M. Thompson, R. Ellis, A. Wildavsky, *Cultural Theory* (Boulder, Colorado: Westview Press, 1990); M. Schwartz and M. Thompson, *Divided we Stand: Redefining Politics, Technology and Social Choice* (Philadelphia: University of Pennsylvania Press, 1990).

attitudes to risk, and indeed to science, are indeed rational. Thus, drawing on a number of empirical studies conducted with workers in nuclear plants, Wynne argues that the technical ignorance which scientists lament as an intellectual vacuum and social defect is instead 'a complex "active" social construction': people decide what they need to know, based on their trust in regulations, and in operating processes.[211] Responses to risk and 'irrational' approaches are not 'irrational', but are rather understandings which are institutionally embedded.

Risk is only one example of how different approaches to the role of science in regulatory decision making can play out. As the above discussion shows, a conceptualisation of genetic technology as simply posing the problem of risk is itself too narrow to embrace the range of perceptions which exist. Nevertheless, it illustrates the point that an approach to regulation which aims to facilitate integration would require that these different conceptualisations, which are based on and expressed in very different languages to that of science, be included in any regulatory and decision making forum. For that inclusion to occur, regulators have to be able to understand the different cognitive structures which give rise to those different perceptions, and to explain to others what those perceptions are and why they are held in a language that others can understand.

Conclusions: facilitating integration and its implications

For technocrats, the integrationist model is a recipe for chaos: the replacement of regulation based on rigorous analysis with regulation based on public whim. A whim, moreover, which is shaped by the media portrayals of genetic technology, portrayals which prefer sensationalism to caution, black and white to shades of grey.[212] That attitude, although it pervades many of the current regulatory fora, has to be seriously challenged. For the technical, scientific definition of genetics simply misses the enormous emotive power of genetics: as Nelken and Lidden observe, the gene is more than a biological structure, 'it has become a cultural icon, a symbol, an almost magical force'.[213] This power alone means that debates over the appropriate course of genetic research, its uses and applications cannot remain a closed issue, defined in ostensibly technical language by experts. The myth of the gene is not the only reason why non-scientists demand a voice, however. There is a much more fundamental reason, and justification for that voice. It is that despite many scientists' assertions to the contrary, science is not neutral, just as law is not neutral. It shapes society's expectations, and provides it with choices which it did not otherwise have.[214] Indeed, it is the dislocation of the 'chance/choice' boundary which genetic technology provides, the opportunities to exercise choices where previously matters were left to chance, which Dworkin rightly identifies as being the explanation for the current, confused reactions to that technology.[215] In so shaping choices and expectations, science has a structuring role which extends far beyond the confines of a laboratory.

211 Wynne, n 189 above (1991), (1993) and 'Risk and social learning: reification to engagement' in Krimsky and Golding, n 183 above.

212 On the role of the media in portraying genetic technology see in particular F. Neidhart, 'The Public as a Communication System' (1993) 2 *Public Understanding of Science* 339; Durant, Hansen and Bauer, n 194 above; and more generally D. Nelkin and S. Lidden, *The DNA Mystique: The Gene as a Cultural Icon* (New York: W.H. Freeman & Co, 1995).

213 D. Nelkin and S. Lidden, *ibid* 2.

214 See for example Beck, n 1 above.

215 R. Dworkin, 'Justice and Fate', paper presented to the 21st Century Trust seminar on Genetics, Identity and Justice, Oxford, 27 March–4 April 1998.

The facilitation of the integrationist approach which is being suggested here may itself have significant implications for science, and they are ones which technocrats fiercely resist. Nevertheless, the constructivist approach to science on which it draws finds a chord in some scientists' own reflections on their activity and rationality, reflections which have strong resonance in the genetics context.[216] The problems which science faces, it has been suggested, are more fundamental than those which it has traditionally faced. In order to address them, science has to shift from the traditional scientific methods to those of what Funtowicz and Ravetz term 'post normal' science.[217] The move to post normal science, they suggest, is necessary for two reasons. First, the conditions of extreme uncertainty in which science now operates pose a critical challenge for science. It has to recognise that in these conditions the traditional, Kuhnian approach to science as problem solving is impossible. Second, science-based technology has created moral complexities. Science simply cannot maintain the stance that it is neutral: that it simply provides information or capacities to society, and it is for society to decide what to do with them (and, moreover, that it is not the fault of science if society chooses badly). Science can thus no longer be a process of routine puzzle-solving conducted in ignorance of the wider methodological, societal and ethical issues which are raised by scientific activity and its products. It can no longer ignore the debate about itself. It cannot in effect borrow and adapt the slogan, 'science doesn't kill, people do'. Despite the continual assertion of this closure of science, it has to be abandoned, for where the decision stakes are high, no scientific argument can be logically conclusive.[218]

The move from the traditional model of scientific objectivity to one which recognises the limits of that objectivity does not however necessarily entail an ineluctable slide into subjectivity. It is maintained by these writers that it is still possible for there to be objectivity in science, but that it has to be one which is defined not simply by the current scientific methodology. What they seek is not de-objectification but re-objectification. Critical assessment of scientific claims is essential, but has to be undertaken by an extended peer community, whose roots and affiliations lie outside those who create the problems which it is sought to address.[219]

Those who advocate such a re-conceptualisation of science's own rationality urge science to adopt this reflexivity, this questioning of itself, if it is to maintain its legitimacy and credibility amongst non-scientists.[220] The public is no longer

216 See in particular A. Chalmers, *Science and its Fabrication* (Milton Keynes: Open University Press, 1990).

217 S.O. Funtowicz and J.R. Ravetz, 'Three Types of Risk Assessment and the Emergence of Post-Normal Science' in Golding and Krimsky, n 183 above, and 'Risk Assessment, Post-Normal Science, and Extended Peer Communities' in C. Hood and D. Jones (eds), *Accident and Design: Contemporary Debates in Risk Management* (London: UCL Press, 1996). For similar arguments on the role of science in risk management see K. Lee, *Compass and the Gyroscope: Integrating Science and Politics for the Environment* (New York: Island Press, 1993); O'Riordan, n 4 above.

218 Funtowicz and Ravetz, n 217 above.

219 *ibid*; see also Shrader-Frechette's model of 'scientific proceduralism', n 203 above, especially 169–196; H. Nowotny, 'Socially Distributed Knowledge: Five Spaces for Science to Meet the Public' (1993) 2 *Public Understanding of Science* 307; Wynne, n 189 above. Nevertheless it is still feared that there is a danger of moving from a position in which the core set of 'experts' is too narrowly defined, to one in which it is too extended, where 'inexperienced and untrained outsiders assume and are granted the right to communicate authoritatively on scientific matters': Nowotny, *ibid* 317; H. Collins, 'Public experiments and displays of virtuosity: the core set revisited' (1988) 18 *Social Studies of Science* 725.

220 See for example, Wynne, n 189 above (1993), who disputes the argument that science is the quintessential reflexive system and argues that '[i]n eschewing an automatic assumption of authority

content to observe science and accept its products, or indeed its definitions. Rather, it has been suggested, the boundaries separating science and the public are becoming more fluid.[221] Science has to recognise the role that non-scientific factors play in shaping responses to it. Otherwise it will continue to be bemused and frustrated by those responses, and unable to engage in debate with them simply because the language in which they occur is one which is alien. If it continues simply to discount those concerns, however, and not to accept its own fallibility or its own limitations, it is liable to lose its authority and trustworthiness.[222]

What are the implications of a facilitative, integrationist approach to regulation for the role of science in regulatory decisions? Is science, for example, simply to be discounted as just another social construct? To an extent, this is a debate for science; in so far as regulation relies on science to determine its content, however, it becomes one for regulation. If regulation sees science as simply one voice amongst many, which in part is what is sought, then the debate is not so central to regulation's own validity. With respect to many issues science is the dominant language, however, frequently manifested in the adoption of risk as the organising principle of regulation. The debate about science then has a correspondingly greater significance. But science and scientific definitions of the issues should not be the only ones which are relevant for regulation, no more than should those of medicine, law or business. As the above discussion suggests, however, facilitating the wider negotiation of regulatory norms is far from a simple task.

To start, diluting the connection between regulation and science is difficult, largely because the voices of science and of industry are those which tend to be heard loudest. It may be that the first step is to move away from the dominant regulatory model of scientific-bureaucratic decision making to one in which opposing views are publicly debated; to develop a forum in which a wide range of groups can participate simultaneously, debating directly and in public.[223] Making changes along the structural dimension of regulation is a proposal which has already been advanced by many commentators in many disciplines. It is not the occasion to pursue this proposal here, but it is perhaps worth emphasising that the debate need not become stuck in the rut of statutory vs non-statutory, legal vs non-legal: many different institutional structures are possible, with many different combinations possible of regulatory instruments which derive from a range of sources and which can or can not have formal legal validity. Moreover, although the structural aspect of this model may have some feasibility in the context of policy formation, difficulties of ensuring participation may well remain in the day to day conversations which will inevitably occur between regulator and regulated.[224]

But as has been stressed throughout, facilitating integration is not simply a question of institutional design. Institutional design is an important aspect, but it is not a panacea for all regulatory ills. Simply introducing the public to the regulatory

to a canonical model of science, and allowing greater problematization of its own founding commitments, science would trade in its presumptions of control for greater public identification and uptake, hence "understanding" ': *ibid*, at 335.

221 See further Nowotny, n 219 above.

222 A point forcefully made in much of the socio-scientific literature; see for example, Wynne, n 189 above; Funtowicz and Ravetz, n 217 above.

223 What Hood calls the 'shark' approach: C. Hood, 'Where Extremes Meet: "SPRAT" versus "SHARK" in Public Risk Management' in Hood and Jones, n 217 above, drawing on Dunsire's model of collibration: A. Dunsire, 'Holistic Governance' (1990) 5(1) *Public Policy and Administration* 4; 'Modes of Governance' in J. Kooiman (ed), *Modern Governance* (London: Sage, 1992).

224 J. Black, 'Talking about Regulation' [1998] *Public Law* 77.

fora will have no effect if it is the scientific language which remains dominant, and only the scientific voice which is recognised as valid. The more interesting challenges thus lie, it is suggested, in the other two dimensions: the cognitive and the communicative. In cognitive terms, there has to be a change in the terms of the debate: a re-definition of the 'problem' posed based on a recognition of others' understandings and identification of the issues which are raised and how they need to be addressed.

On the communicative dimension, then as the discussion above demonstrates, there still remains a more fundamental problem which the facilitation of integration faces. That is that participants in the debate simply speak different languages. The languages of science, of commerce, of ethics, of ecology, of law are foreign to each other; each can hardly understand what the other is saying, let alone why they should be saying it. As we have seen, for each, the others' view simply misses the point. Addressing this problem of mutual incomprehension is fundamental to facilitating integration in any form. In so doing, it has been suggested, there is a need not for a common language, which is unattainable. Nor is there a need for an 'official' language, for that would be a denial of the philosophy underlying the integrationist model. Rather, it has been suggested, there is a need for translators, for interpreters. By acting as interpreters, regulators can then facilitate the negotiation of regulation and the integration which is sought. It is how to perform this interpretive and facilitative function which the controversies surrounding genetic technology show to be one of the central challenges facing its regulation.

[3]

WHOSE GENOME PROJECT?

DARRYL MACER

The human genome project is a multinational project aimed at obtaining a detailed map and a complete DNA sequence of the human genome. It will have many scientific, medical, economic, ethical, legal and social implications. A fundamental question to ask is "Whose genome project is it?" We can answer this question from different perspectives, and this aids our thinking about the issues that arise from the project. We can think of who proposed the idea, who should fund the research, who should perform the research, whose genome is mapped and sequenced, who should own the data, who should benefit from the results, and who should make these decisions. We can also compare the answers to these ethical questions with what is occurring in practice.

By September 1990 we possessed the gene sequences of over 5,000 human genes, and the location of 1,900 genes to areas of specific chromosomes.[1] The number is growing exponentially, but the total number of human genes is thought to be about 100,000. This compromises only 5-10% of the total DNA in the human genome. The rest of the DNA is of unknown function, and much is thought to be nonfunctional. The total sequence is about 2.8 billion linear bases on 23 chromosomes. The total length of human gene sequences known today is about 40 million base pairs.[2]

WHO BEGAN THE GENOME PROJECT?

The genome project is often compared to the Apollo project. The analogy highlights the glamour of the project, and may represent the importance of both projects to human pride. The genome project should have many more practical benefits, because not only will it lead to the development of useful new technology, but unlike the Apollo project, the goal itself is also of immense direct practical use. The link between initiating the project and reaching the goal is also

[1] Victor McKusick, *Mendelian Inheritance in Man* (on line Computer) Baltimore: John Hopkins University Press 1990.

[2] James D. Watson, 'The human genome project: past, present and future', *Science* 248, 6th April 1990, 44–49.

different; people would not have gone to the moon if a positive decision to set up the Apollo project had not been made, but the human genome map will be obtained, with or without a positive effort, though over a longer time scale if undirected. Mapping of the human genome has been progressing for decades.[3] The beginnings of the genome project can be at least traced back to Mendel's genetics on peas, the mapping of the trait for colour blindness to the X-chromosome of *Drosophila* by T.H. Morgan and workers, to Avery and colleagues who found DNA was the physical substance of genes, to Crick, Franklin, Watson and Wilkins who determined the structure of DNA, to those who discovered the genetic code, to Sanger and others who developed DNA sequencing, and to many others who contributed to our knowledge of genetics and molecular biology. In this respect, no single group of persons can claim to have initiated the goals of the genome project.

There were several people in the USA who saw the goals of the genome project as ideal for initiating the first large scale biological research project with a definite end-point.[4] The Human Genome Project or Human Genome Initiative is the collective name for several projects begun in the late 1980's in several countries, following the United States Department of Energy (DOE) decision to create an ordered set of DNA segments from known chromosomal locations, to develop new computational methods for analysing genetic map and DNA sequence data, and to develop new techniques and instruments for detecting and analysing DNA.[5] Whether the motive was to fill vacant DOE Laboratories, to provide renewed emphasis for science or to put US biotechnology companies in a better international position,[6] the idea itself was sure to catch the imagination of politicians. Some biologists commented that they do not think that physicists are able biologists, so the project should not be left to the DOE. The NIH joined the project as it is the major funder of U.S. biomedical research.

The goal of discovering our genetic makeup will remain one of the pinnacles of human endeavour. We may pursue the exploration of space infinitely, but once we have obtained the complete DNA

[3] Barbara J. Culliton, 'Mapping terra incognita (humani corporis)', *Science* 250, 12th Oct. 1990, 210–212.

[4] Charles R. Cantor, 'Orchestrating the Human Genome Project', *Science* 248, 6th April 1990, 49–51.

[5] U.S. Congress, Office of Technology Assessment, *Mapping Our Genes – The Genome Projects: How Big, How Fast?* Washington D.C.: U.S.G.P.O., April 1988, OTA–BA–373.

[6] Roger Lewin, 'In the beginning was the genome', *New Scientist*, 21st July 1990, 34–38.

sequence of ourselves we have in one sense reached a pinnacle. Understanding ourselves, one of the dreams or nightmares that many people share, will have come closer.

There have been numerous scientists who have contributed to our knowledge, and it will also be fitting that the completion of this project will require the collaboration of many international scientists. When we ask who initiated the project, who does the work and whose knowledge is needed, the answer is clearly that many people are, and will be, directly responsible for the mapping and sequencing, and the later interpretation of the data.

WHOSE DNA IS BEING SEQUENCED?

The actual DNA that will be sequenced will be a composite of different human tissue cell lines; it will not be the DNA of a particular person, but of the species in general. Geneticists estimate that any two people are about 99% similar in their genetic makeup. To put it a different way, about 0.3-0.5% of the nucleotides in our DNA vary between different people. These differences vary from person to person, therefore it does not matter whose genome is actually sequenced.[7] Different laboratories often use different human tissue culture cell lines, which are derived from different people. Therefore, the DNA sequences will be different. The characterisation of standardised marker regions will make it easier to compare the DNA between different individuals and to produce a single general map and an eventual model of sequence.[8] Most of the cell lines derived from patients are given for the benefit of research, and a large number of patients will have direct claims to parts of the final sequence. However, all can say that the sequence is 99% similar to their own. The answer to the question: "Whose genome is being sequenced?" is: "Everyone's genome." This is a key point. The project is of direct relevance to all human beings.

We can ask whose DNA should be sequenced? Should a DNA close to those who are funding the project, or one close to the most universal in the world, be used? In practice, because the amount of individual variation in DNA sequence is greater than the inter-racial DNA differences, this is unimportant. The mapping and sequence information will be, on average, of similar use to all peoples. There are

[7] With regard to this variability, it also means that the degree of accuracy required for sequencing can be lower, which significantly lowers the cost of sequencing. An error rate of 1 in a 1000 bases is adequate for most purposes, though if sequencing becomes automated lower rates may be possible. See L. Roberts, 'Large-scale sequencing trials begin', *Science* 250, 7th Dec. 1990, 1336–1338.

[8] Leslie Fink, 'Whose genome is it, anyway?, *Human Genome News* 2, July 1990, 5.

some difficult questions, however; for example, if one nation decided that the DNA of its citizens should not be sequenced, could it stop the sequencing? The same sequence is shared by others who want to know. Which is the stronger claim, the right to know or the right not to know? This question has been avoided, and for practical purposes we can expect it to be ignored. This avoidance, however, may be unethical, and those who command the project should ask the general population whether they want the sequence known or not, rather than just assume that the sequence should be known.

While those who object to the sequence being known may be unable to prevent other people from characterising their own DNA, it is a different question whether those who do the sequencing have the right of control over the use of such information. In the United Nations Declaration of Human Rights, Article 27, there are two basic commitments that many countries in the world have agreed to observe (in their regional versions of this declaration). These are (italics added for emphasis) (1) Everyone has the right freely to participate in the cultural life of the community, to enjoy the arts and *to share in scientific advancement and its benefits*. (2) Everyone has the right to the *protection of the moral and material interests resulting from any scientific*, literary or artistic *production of which he is the author*.[9] An important question arising from section (2) is whether all people are the authors of information that is shared by all people. The writers of this Declaration may not have considered DNA, but it would certainly be in the spirit of the Declaration to interpret the DNA sequence as something of shared ownership. On a more general note in section (1), it is said that everyone has the right freely to share in scientific advancement. Article 27 expresses two important and relevant guiding statements of law that reflect a strong body of philosophical support from the principles of justice and beneficence, and from the idea of legal property rights. I shall discuss property rights further in a later section on data-sharing. The common claims for authorship of the genome should be considered in all aspects of the genome project, especially in the questions of who should make the decisions in the project and how the data should be used.

WHO IS FUNDING THE GENOME PROJECT?

Biomedical research is performed and funded in many countries. There are several reasons why research funding should be shared. The basic ethical reason is a principle of justice: that we should all

[9] Paul Sieghart, *The Lawful Rights of Mankind*, Oxford: Oxford University Press 1985, 176.

contribute to shared knowledge that benefits all people. People from every country will be able to benefit from the information, though to different degrees. Many people also believe that the pursuit of knowledge itself is good. There is also the connection between basic research and the standard of university teaching. If we consider education to be a basic moral good we can claim that to perform research has a positive effect on the teaching standard which may be a more significant benefit in small countries than the direct results of research. Another reason in favour of research is national prestige. Countries that participate in high level research gain prestige over those unable to participate in such activities, and this is widely regarded as a good thing by many in those countries.

Economic profit is another important reason given to justify research. For many years science has not only been the individual pursuit of people with unusual ideas, but has involved funding by public or private money, by taxes, charities and business investments. We should not be surprised to see the justifications for funding the project in terms of the business opportunities it will offer. The US Congress was partly convinced of the usefulness of funding the project by the opportunity it provides to boost US biotechnology.

The U.S. portion of the project (possibly 50% of the total) is estimated to cost US$3 billion over the next 15 years, to the intended completion date in 2005 A.D. The total cost is unknown because the project is being broadened to include other organisms as models, and many countries are contributing additional money to it. Data handling will be an important portion of the total. The 1991 government funding in the USA specifically for the human genome project (not including other indirect research on genetic mapping) is US$136 million (NIH US$90 million/DOE US$46 million),[10] and this figure is increasing. However, when one compares this with the cost of the development of a single drug, at US$50-100 million, or the annual U.S. health care expenditure of over US$600 billion, it is a small price to pay for such a large amount of information. It is less than one week's costs for the recent Gulf war. Biotechnology is very big business, and the projected average US$200 million annual contribution for the human genome project in the USA[11] is minor. The scientific methodology for sequencing DNA is routine, but the present cost of US$3-5 per nucleotide must be reduced by nine-tenths before the major sequencing effort begins. This should occur as automatic DNA sequencing methods are improved.

[10] L. Roberts, 'A meeting of the minds on the genome project?' *Science* 250 9th Nov. 1990, 756-7.

[11] U.S. National Research Council, *Mapping and Sequencing the Human Genome*, Washington D.C.: National Academy Press 1989.

In the USA James Watson leads the NIH Genome Center's initiative. His powerful advocacy of the project has obtained funding, but the style in which it has been done has resulted in some criticism.[12] The initial fear of scientists was that the money would come from other biological research, which traditionally involves many small projects encouraging many individual scientists.[13] Critics would prefer the project to stop after generating a general genetic map of the chromosomes, so that the DNA in which more interesting genes are located could be isolated and sequenced. Identifying a particular disease-causing gene can take several years of intense investigation, as seen in the tracing of the cystic fibrosis gene. A more detailed map would decrease the amount of DNA that must be searched through for each gene. Various leading scientists have called for an evaluation of the project priorities after the map is completed so as to use the money in the best way to encourage research.[14] In the initial phase some of the funds for the DOE and NIH projects came from existing research funding. The NIH is spending 2-3% of its total budget on the genome project, which, in view of the importance of a coordinated mapping project, is worth the cost.[15] During 1990 the NIH announced the recipients of special genome research centre grants, which rather than creating new big institutions, will give accountability to expansions of existing research teams. They will be required to produce definite results, such as maps of the mouse genome, as well as maps of specific human chromosomes and the sequence of the yeast genome.[16] Since the original idea was conceived, the project has expanded to include other organisms, and is even closer to what people would be doing anyway, except for the centralised planning and funding. The project now aims to understand how the human genome functions and its relationships with genomes of other organisms. In late 1990, the US DOE announced that it would first attempt to sequence all the expressed human genes (cDNA) before sequencing all the human DNA.[17]

[12] R. Wright, 'Mad scientist? James D. Watson and the human genome project', *New Republic*, 9th & 16th July 1990, 21–31.

[13] Bernard D. Davis et al. 'The human genome and other initiatives', *Science* 249 27th July 1990, 342–343.

[14] Philip Leder, 'Can the human genome project be saved from its critics . . . and itself? *Cell* 63 5th Oct. 1990, 1–3.

[15] Robert M. Cook-Deegan, 'Human Genome Mapping: Policymaking aspects', in *Genetics, Ethics and Human Values: Human Genome Mapping, Genetic Screening and Therapy*, Proceedings of the 24th CIOMS Round Table Conference, Tokyo 22–27th July, 1990, Geneva: CIOMS 1990.

[16] Leslie Roberts, 'Genome center grants chosen', *Science* 249, 28 Sept. 1990, 1497.

[17] See Roberts, note 7.

Given the potential direct medical benefits of the project, the research money that is being spent on this project can be ethically justified from the principle of beneficence. Perhaps even more should be spent. From the ethical principle of justice, other countries that can afford to pay and will benefit should share the cost of the project. This is because all countries will benefit from the knowledge, and in the spirit of the Declaration of Human Rights, article 27 (1), all people should benefit. There are multi-million dollar projects in Europe, Japan and Australia. The 1990 funding in Japan was approximately US$10 million, but is expanding. France is spending US$40 million in 1991, and US$50 million in 1992.[18] The U.K. government is providing £11 million over the next three years, and has stated that it hopes this is enough to buy its stake in the use of the information. Other countries are now joining the project, not entirely because of their recognition of the principle of justice, but because of the potential economic benefit, and the fear of being denied access to the USA-based databases. The answer to the question of who should fund the project on a national level, within the international community, is: "Every country that can afford to pay". This is close to the current situation and trends. This is based on the assumption that all will benefit, which in practice will be more easily and rapidly assured if countries have contributed money and developed their genetic research capabilities over the course of the project.

WHO SHOULD DO THE WORK?

The distance unit used in gene mapping is called a centi-Morgan (cM); one cM is equivalent to two markers being separated from each other in chromosome crossing over in normal reproduction 1% of the time. The actual physical length of 1cM varies, being approximately 1 million base pairs. The use of restriction fragment linkage patterns (RFLPs) in combination with genetic linkage analysis allowed the construction of linkage maps for each chromosome with an average spacing of 10-15cM.[19] Other techniques are being developed to give finer resolution such as radiation hybrid mapping, and the current map has an average of about 6 cM spacing. Different researchers have published maps of the same chromosomes using different markers, so that a combined total map is impossible to draw. The initial target is to construct a linkage map with an average spacing of 2cM,

[18] M. Durand, 'France launches Human Genome Program', *Human Genome News* 2, Jan. 1991, 12.

[19] Helen Doris-Keller et al. 'A genetic linkage map of the human genome', *Cell* 51, 23rd Oct. 1987, 319–337.

and maximum gaps of 5cM. Some chromosomes are already covered by markers at 1-5cM intervals.[20]

The current mapping paradigm is based on a proposal in 1989 to use physical sequence-tagged sites (STS) as the map labels.[21] Different researchers use different cloning vectors for gene analysis, so the exchange of DNA pieces in these different vectors, or DNA clones, is not possible. It is possible using the DNA polymerase chain reaction (PCR), to generate DNA sequences for any DNA if short sequences are known, from which primers for the PCR can be made. Therefore, the ends of large DNA fragments should be sequenced, and the data combined to make an STS map of each chromosome. This approach means that researchers can continue to use different methods and develop better procedures, while the information obtained can be integrated to develop the actual physical map. This will avoid the need to exchange different clones of DNA between laboratories, because each laboratory can use the marker sequence as a starting point.[22]

In two major U.S. Reports on this project, one by the Office of Technology Assessment (note 5), and another by the National Research Council of the National Academy of Sciences (note 11), it was recommended that a map of the human genome be made prior to full scale sequencing. While mapping will benefit from improved methods, sequencing requires much improved and cheaper technology. A map is also essential to efficient sequencing so that a library of DNA fragments can be systematically sequenced. Different research groups have begun to concentrate on different chromosomes in order that they can all have the complete map in a shorter time. There are actually 24 chromosomes to be sequenced, 22 autosomes and the X and Y chromosomes. The five year goal of the NIH program is to construct a map with STS markers spaced at about 100,000 base pairs, and to assemble overlapping contiguous cloned sequences (called contigs) of about 2 million base pairs length. The sequencing can be started from this physical and informational library system. For chromosomes 16, 19 and 21 the contig maps are over 60% complete.[23] Data management technology, such as using advanced computing technology in programmes to search the DNA sequence

[20] J. Clairborne Stephens et al. 'Mapping the human genome: current status', *Science* 250, 12th October 1990, 237–244.

[21] M. Olsen et al. 'A common language for physical mapping of the human genome', *Science* 245, 29th Sept. 1989, 1434–1435.

[22] Leslie Roberts, 'New game plan for genome mapping', *Science* 245, 29th Sept. 1989, 1438–1440.

[23] Leslie Fink, 'DOE-NIH retreat improves understanding', *Human Genome News* 2, Nov. 1990, 1–3.

libraries, must also improve.[24] Even such seemingly basic tasks as selecting the correct tube from the DNA clone library is being automated by robotics.[25]

Using the STS approach allows small teams of researchers to contribute results. There are worldwide efforts, although the major international effort is centred around the USA, Europe and Japan. There are many people from Latin America who are involved in work, though they may still be lacking access to international DNA databanks.[26] There are also projects in Australia, Canada, and there will be contributions from many other countries. Most countries fund research in their own country, so the question of who funds research and who performs the research are overlapping. The acknowledged shortage of trained biologists suggests that it would be an advantage to the project to provide funds for use in places where there are personnel but no funds. This would also aid the internationalisation of the project. However, until now most of the funds are distributed in accord with national boundaries. There are clearly some countries that lack sufficient funds in view of more pressing economic problems, but support of UN or other internationally based research centres would aid the internationalisation of the project.

There does not need to be any repository of DNA pieces; what is required is a computer data bank of the sequence. The project requires the establishment and constant improvement of databases containing the sequences of genes, and their location. There are several international databases, and the information should be openly shared among them to make the best and most up-to-date database possible.

We must ask whether private companies should do the work. If we are only interested in the goal, the sequence, then it does not matter who does it. If a company can develop a cheap method of sequencing then it should be used, and some suggest that we use a free market approach to select the cheapest approach for sequencing. Some of the mapping and sequencing could be performed by contract research. This has the advantage that more diverse but risky approaches may be tested than are likely to be pursued in government funded laboratories. Methods such as direct reading of the sequence by advanced microscopy are possible alternatives to

[24] Susan Watts, 'Making sense of the genome's secrets', *New Scientist*, 4th August 1990, 37–41.

[25] W.J. Martin & R.M. Walmsley, 'Vision assisted robotics and tape technology in the life-science laboratory: applications to genome analysis', *Biotechnology* 8, Dec. 1990, 1258–1262.

[26] Bettie J. Graham, 'Latin American symposium on molecular genetics and the Human Genome Project', *Human Genome News*, Sept. 1990, 14.

current methods.[27] We must address the data-sharing and ownership questions before answering this question.

COORDINATED DATA-SHARING IS REQUIRED

The international mapping and sequencing is being coordinated by the Human Genome Organisation (HUGO). Coordination of the international effort is needed to avoid duplicity of effort which could slow overall progress. The European Commission is also coordinating research in Western Europe. There had been talk of giving different countries the tasks of sequencing different chromosomes to avoid duplication. However, such a system was considered impractical given the way scientists work. It may be possible for different researchers to take responsibility for coordinating the maps for each chromosome. However, many researchers remain more interested in pursuing specific disease-causing genes. Therefore there are only a few chromosomes that are being extensively mapped at the moment. These include chromosomes 21, 7, and X. Other chromosomes, such as number 8, which have few known genetic diseases linked to it, are poorly known. In this way HUGO can play a very useful coordinating role, pointing out the number of existing projects on each chromosome to those who submit research proposals mapping such chromosomes. National medical funding bodies can reject funding applications that are overlapping. To duplicate work is important for verification, but certain regions may have a dozen teams working toward the same goal, which is a waste of effort.

There are special needs for the information to be freely shared, though the director of the U.S. NIH genome project, Dr. James Watson, earlier threatened that countries that do not contribute funds may not get the information immediately.[28] This remark was aimed at encouraging the Japanese Government to provide funds to HUGO, and has been widely criticised by those who believe that the information resource belongs to no country, but to the world, for its use in medicine. The idea of introducing secrecy would defeat the purposes of coordinating international efforts. The idea is that governments will lose control of the decisions about how to use the information unless they contribute money. There are already some examples of US-based databases restricting access to outside users.

[27] William Bains, 'Alternative routes through the genome', *Biotechnology* 8, Dec. 1990, 1251–1256.

[28] Christopher Joyce, 'US faces demands for secrecy on genome programme' *New Scientist*, 18 Nov. 1989, 5; Editorial, 'Diplomacy please', *Nature* 342, 2nd Nov. 1989, 1–2.

Chemical Abstracts obtains data worldwide, but restricts full availability to certain US computers. The National Library of Medicine runs Medline and has taken over Genbank (the DNA data bank) and PIR (the protein sequence database). During the Afghanistan conflict it denied access to Soviet users. In view of this, Europe is combining public and commercial databases to improve the European based EMBL DNA database.[29] Recently the US moved the Genbank database from Yale, which required completely free access for researchers, to Baltimore, which may not, and some scientists are suspicious of the motives for this. The first round of threats has resulted in wider international funding, but the real test will come with the actual results. The most rapid progress will be obtained if data is shared between all researchers. The full value of one part of the sequence is only known when compared to the rest. Even if one government declines to support such a project, the information still belongs to all people of that country and for this reason other countries ought to share it with them.

The problem of data-sharing from the viewpoint of individual competing scientists is one that is already with us. The new technology, for example automated DNA synthesisers make replication of results very rapid, could encourage researchers to delay publication while they get more of a head start in the next stage of the research. In a system where academic jobs depend on the number of publications one has, everyone wants to get papers published. Researchers may not reply to letters requesting data, or may just reply within a selected peer group. The U.S. DOE has drafted guidelines that stipulate that data and materials must be made publicly available within 6 months of generation, but there is considerable pressure among scientists to lower this to 3 months. The NIH is not in favour of rules, but encourages researchers to share information to avoid bureaucracy. The results of discovering a disease-causing gene, or a detailed map which advances the discovery of such genes by many years must outweigh the short-term interests of individual scientists. A recent example of collaboration is between the 30 research teams working on chromosome 21 (it contains the Down's syndrome and Alzheimer genes).[30] In this respect the genome map may be the ultimate collaborative research project. The most rapid progress will come from immediate data-sharing, and it is a chance for "scientific altruism" on a global scale. There is an ethical obligation on researchers, especially those using public money, to share data as

[29] John Hodgson, 'Europeans feel US squeeze', *Biotechnology* 8, Jan. 1990, 15.

[30] Leslie Roberts, 'Genome project: an experiment in sharing', *Science,* 248 25th May 1990, 953.

soon as it is available. If the scientists cannot do this on their own initiative, which is by far the best option, then regulations need to be enforced. It will be possible to make all users of the database first submit their sequence before using the other sequences. Secrecy will undermine the enthusiasm of scientists to participate in the project if they think that other researchers will hide information. Perhaps there is even a chance for scientists to restore confidence in the section of the population that has lost faith in science.

WHO SHOULD OWN THE RESULTS?

The question of who legally owns the data is very topical because some of the work will be funded by businesses. The question of patenting of genetic material is a potentially contentious issue. The U.S. Congress wants publicly funded science to be commercialised, and during the 1980's intellectual property rights were decentralised from government to research institutions to create commercial incentives (note 15). The usefulness of map and sequence databases will be determined by how much data they have. If privately run databases contain more or important information, which is withheld from public databases, researchers will need to use them.

The information arising from the human genome project could be classified as biotechnology. There are two basic approaches to applying patent law to biotechnology inventions. In the USA, Australia, and many other countries, the normal patentability criteria shall apply, that is, the invention must have the attributes of novelty, non-obviousness and utility, and the invention should be deposited in a recognised depository. While a country may accept the first type of criteria, some countries have specifically excluded certain types of invention. What is ethical is not the same as what is legal, though we can attempt to reduce the difference. Concerns over ethics have affected patent laws, for example European countries who joined the European Patent Convention have barred the patenting of plants or animals. Denmark has an even more strongly worded exclusion in its national law. There is public rejection of the idea of patenting animals in some countries, and the patenting of human genetic material is potentially more contentious.

The public attitudes to the patenting of different types of things, including living organisms was measured in New Zealand in mid 1990. Fully 90% of the public had heard of inventors being able to obtain a financial reward through patents or copyright. Those who had heard of patents or copyrights were asked if they agreed that patents should be obtainable for different subject matter. In response, 93% thought that the patenting of new inventions is acceptable, 85%

thought information could be patented, but there was less acceptance of patenting new plant or animal varieties, 71% and 60% respectively. Only 51% agreed with patenting of "genetic material extracted from plants and animals".[31] There was more acceptance of the patenting of genetic material among those who thought that there were benefits to New Zealand from genetic engineering, and among farmers. There was less acceptance of patenting among young people (15-24 years old) and among scientists and high school science teachers (in separate surveys). The negative reaction reflects the general feeling that genetic material is special, and should be different to other types of information. If the question had been asked regarding the acceptance of patents on "genetic material extracted from humans" we can safely assume that there would have been an even lower level of public agreement.

Some of the ethical arguments that are commonly expressed when talking about patenting of animals are also relevant to the question of the patentability of genetic material. The major arguments for patenting genetic material include:

- patent law regulates inventiveness, not commercial uses of inventions;
- patenting promises useful consequences (e.g. new products/research);
- other countries support patents, so our country needs to if the biotechnology industry is to compete;
- if patenting is not permitted, useful information will become trade secrets;
- patenting rewards innovation.

The arguments against patenting include:

- metaphysical concerns about promoting a materialistic conception of life;
- patenting promotes inappropriate human control over information that is common heritage;
- some countries do not permit similar patents;
- patenting produces excessive burdens on medicine (increased costs to consumers, payment of royalties for succeeding generations).

Most of these issues will not be affected by permitting patents as the issues are similar to those existing prior to the patenting debate; for

[31] Paul K. Couchman & Kenneth Fink-Jensen, *Public Attitudes to Genetic Engineering in New Zealand*, DSIR Crop Research Report 138. Christchurch: Department of Scientific and Industrial Research, Crop Research Division, New Zealand 1990.

example the distribution of wealth and international competitiveness.[32] This issue remains contentious and the fact that different countries have conflicting policy reflects this. The issue is closely related to the commercialisation of genetic engineering, but some sort of information protection is already accepted as an incentive to invest in research of benefit to society. Ethically, we can apply the principle of beneficence. Does commercialisation of the genome project give more benefits than a ban? The benefits should be in terms of general medical or agricultural development, rather than the economic prosperity of one company or country over another. Until now there have been very few medical procedures that have been patented, despite their suitability. Only instruments and products used in diagnosis or treatment have been patented, actual procedures have been left open for all to use free of patent liability in accord with the principle of beneficence.

We can rephrase this question by asking how the means we choose will affect the end result of the project. The end is the map or sequence, which is open to all. However, pursuing different means of achieving this end may lead to distinct outcomes. If we allow more privately funded research, there may be more restrictions on the end information in order to encourage private funding. If restrictions are applied to the general access to information then it is clear that we will have a different short-term end if we use a shared ownership approach rather than a private market approach allowing private ownership. After the period of patent exclusion then the direct result is the same, i.e. the full sequence is available to all. The indirect results will be affected by factors such as whether during the period of patent exclusion certain companies have been well established and are able to provide beneficial services or monopolies, and the relative advantages of the open knowledge after patenting compared to industrial secrecy that could occur if patenting was difficult to obtain. More significantly, there may be a greater amount of total knowledge and an earlier completion date using the approach involving private companies. The means by which these approaches are pursued differs. In one approach the government laboratories spend their resources on the project, at the expense of other projects, but with the cumulative results being openly available to all. In the other approach, the private companies do the research, which would create more total biomedical research knowledge, but certain parts of this would be tied up in patents, though the knowledge would also be available with a small delay.

[32] U.S. Congress Office of Technology Assessment, *New Developments in Biotechnology, 4: Patenting Life*, Washington: U.S.G.P.O., March 1989.

Other arguments used to support patenting are not so compelling. The claim that the function of patents is to regulate inventiveness rather than to regulate commercial uses of inventions is minor in practice. There have been some recent controversies regarding the commercial monopoly held by the company which was able to patent AZT, the current HIV/AIDS treatment, which has been reaping large profits in view of its monopoly. It is all the more questionable whether this should be allowed because of the key roles that government funded research played in developing AZT and showing it was active against AIDS. There are other examples where the commercial monopolies obtained cannot be said to be in the best public interest, and the existence of patent laws is certainly relevant to the later commercial uses of inventions. Patenting is said to reward innovation, which is a basis of the successful modern democratic and Asian economic systems. Patents do recognise property rights in inventions. However, there is an existing difference in the protection of property rights compared with other rights in international law and declarations of human rights. Property rights are not absolutely protected in any society. For the sake of justice, "public interest", "social need", and "public utility", societies can confiscate property.[33] Therefore there is an existing precedent for exemption from property ownership, which is the point of the exclusiveness of patents, when some property is of great benefit to the public. As for the argument that we should support patents because other countries do, there are certainly many countries that have exemptions for patent protection.[34] The exemption because of social need would apply if there would be more benefit from patent exclusion, which is not necessarily so. It depends on the way information is used. Therefore one answer to the question as to who should own the results of the genome project is that no one in particular should own these results.

Using a more positive argument, the knowledge gained should be considered as the common property of humanity. There is an existing legal concept that things which are of international interest on such a scale should become the cultural property of all humanity. It can be argued that the genome, being common to all people, has shared ownership, is a shared asset, and therefore the maps and sequence should be open to all. Some of the common factors that derive from the shared ownership are that the utilisation must be peaceful, access should be equally open to all while respecting the rights of

[33] See note 9, pp. 130–2.

[34] William H. Lesser, *Equitable Patent Protection in the Developing World: Issues and Approaches*, Christchurch: Eubios Ethics Institute 1991.

others, and the common welfare should be promoted.[35] As discussed earlier there are existing legal principles from the Universal Declaration of Human Rights (see note 9). The authorship of the genome can be answered in two ways. The DNA could be viewed as a random sequence of bases, and the author as the sequencer, but we would normally talk the sequencer as a discoverer rather than an author or inventor. In the days of colonial rule discoverers could claim a land as their property, but later it was recognised that the pre-existing people had claims to the property no matter how it was developed by the colonisers. The sequencers of DNA are not sequencing unowned land but are sequencing un-characterised land: the name of mappers is suitable for this analogy. Some critics of ownership could go as far as to call those who seek to profit and to control the decisions concerning the human genome project, without general consultation, a type of "genomic imperialist". The DNA is not random, it is merely unknown. This is an important difference. The common possession of the DNA sequence by every member of humanity means that the sequencers are not authors. While it may be unconventional to call the possessor of information the author, they have more claims to that title than the sequencers. In addition they can be called the owners (although in this general way the human sequencers are shared owners because they also possess DNA). The method for sequencing, or mapping, can be invented and patented, but whether that side of the project can be ethically patented lies more with the question of benefit and utility as discussed above.

If these arguments are insufficient to refute those who support the private ownership of genome data, public opinion could force a policy change regarding the patenting of such genetic material. It could be excluded, as patents appear to be excluded for animal and plant varieties in Europe. In the broader context, in the USA the commercialisation of human cells and tissues is generally permissible *unless it represents a strong offence to public sensitivity*.[36] The sale of the human genome map and sequence data may be a strong offence to many and may incite adverse public reaction forcing legislators to exclude it from patenting. As education about the project and whose property it is grows, we may hope for the exclusion of major profit making from this project. The law should exist to benefit humanity. The

[35] Bartha N. Knoppers, 'Human Genome Mapping: Ethics and human-value aspects', in *Genetics, Ethics and Human Values: Human Genome Mapping, Genetic Screening and Therapy*, Proceedings of the 24th CIOMS Round Table Conference, Tokyo 22–27th July, 1990, Geneva: CIOMS 1990.

[36] U.S. Congress Office of Technology Assessment, *New Developments in Biotechnology, 1: Ownership of Human Tissue and Cells*, Washington: U.S.G.P.O., March 1987, 9.

debate will continue, as companies will naturally desire to obtain some information protection for their investment, but they will have to be sensitive to strong public feelings that could easily be aroused. As I have argued, these feelings have an ethical backdrop. The idea that the human genome sequence should be a public trust and therefore not subjected to copyright was also the conclusion of the U.S. National Research Council (note 11), and of the American Society of Human Genetics.[37] The European Parliament "Human Genome Analysis" program limits contracting parties' commercial gains with the phrase "there shall be no right to exploit on an exclusive basis any property rights in respect of human DNA".[38] This idea would also include the option that the donor of genetic information, in terms of a cell line, should be able to make that information publicly available. The motive to aid humanity in general rather than a commercial interest is usually a reasonable interpretation of the motives for patients to provide material for medical research.

There are further legal issues and not just ethical problems with the patenting of genetic material. To qualify for a patent an invention must be novel, non-obvious and useful. If the claimed invention is the next, most logical step which is clear to workers in that field, then it cannot be inventive in the patent sense. If a protein sequence is known, then the DNA sequences that code for it will not in general be patentable, unless there is a sequence which is particularly advantageous and there is no obvious reason to have selected this sequence from the other sequences that code for the protein.[39] In the case of natural products there are often difficulties because many groups may have published progressive details of a molecule or sequence, that may have lost its novelty and become obvious. These are essentially short pieces of the human genome. There are also patents on protein molecules which have medical uses. In this case the protein structure is patentable if it, or the useful activity, was novel when the patent was applied for. The invention must also be commercially useful. There are patents on short oligonucleotide probes used in genetic screening. If someone can demonstrate a use for a larger piece of DNA then they can theoretically obtain a patent on it. An example

[37] E. Short, 'Proposed American Society of Human Genetics position on mapping/sequencing the human genome', *American J. Human Genetics* 43, 1988, 101–102.

[38] European Parliament, 'Human Genome Analysis program', COM, section 3.2, 13th Nov. 1989, 532.

[39] N.H. Carey & P.E. Crowley, 'Commercial exploitation of the human genome: what are the problems?, pp. 133–147 in *Human Genetic Information: Science, Law and Ethics*, Ciba Foundation Symposium 149, Amsterdam: Elsevier North Holland 1990.

of a larger patentable section of genetic material would be a series of genetic markers spread at convenient locations along a chromosome.[40] Another set of genetic markers on the same chromosome can be separately patented if they also meet those criteria. The direct use of products, such as therapeutic proteins, is well established. The information may also be used in the study of a particular disease, for example, by the introduction of a gene into an animal to make a model of a particular human disease, and it was for this reason that the much-discussed cancer-developing "Oncomouse" was patented. The genetic information can also be used to cure a disease, for example using the technique of gene therapy with a specific gene vector.

With the completion of the genome sequence of many organisms, including humans, any new genetic material will no longer be novel as it will be available in a database. If researchers decide to apply for patents on every new protein sequence prior to making it public, they may also fail the usefulness criterion. It is therefore likely that in the near future patents will be difficult to obtain on gene products, though it is expected that prior to the sequence determination there will be many applications for these different types of patent. The completion of the genome maps and sequences of many organisms will have several implications for the future of biotechnology patents.

There are private companies embarking on the project,[41] and they may be able to patent ways of expressing the genome, or genetic maps, but not the genes themselves. The company Genome Cooperation has been created by Walter Gilbert, with the intention of selling databases that contain sequences of key segments of the genome.[42] Companies may undertake contracts for research and development with the respect to the technical aspects, but the final product, the sequence, should not be used for profit. Others will be able to obtain the same genes. It may be cheaper to buy the genes from companies that have found them first, or to use commercially patented genome maps. Perhaps such companies could be rewarded with some funds to reimburse their costs, but it may end up being another case of commercial companies making profits out of human disease. In fact, one commonly voiced aim of the U.S. genome project is to promote the U.S. biotechnology industry: for example it can sell genetic probes

[40] R. Saltus, 'Biotech firms compete in genetic diagnosis', *Science* 234, 12th Dec. 1986, 1318–1320.

[41] Robert Kanigel, 'The genome project', *New York Times Magazine*, 13th Dec. 1987, 43–44, 98–101, 106.

[42] Marc Lappe, 'The limits of genetic inquiry', *Hastings Center Report* 15, Aug. 1987, 5–10.

that are made from the gene sequences, and other new technology. The political aim is to try to put the U.S. biotechnology industry above that of other countries, especially Japan.

In answer to the ethical question of whether private companies should perform such research we can say yes, but not in order to make large profits. They should, however, be able to recover costs if they are more economical than government research, providing the results of mapping and sequencing are openly accessible. In practice much of the project will be publicly funded, but contracts to do the work may be awarded to the most competitive. In France two private non-profit groups (Centre d'Etude des Polymorphismes Humain and Généthon) are spending more money than the French Government on genome mapping, in pursuit of genes for humanitarian not economic reasons. The role of non-profit private organisations is also very important in biomedical research in other countries leaving less room for economically motivated private companies.

WHO BENEFITS FROM THE RESULTS?

The genome project will be a huge resource of information for medicine in the next century.[43] There will be much useful information arising prior to the completion of the project, as growing numbers of disease causing and susceptibility genes are sequenced and the mutations characterised. Most of the major single gene disorders and some of the genes involved in complex diseases should be known within the decade.[44] It will be possible to develop DNA probes to diagnose any known genetic disorder, and it also will be easier to characterise new disorders. The project will expand the number of human proteins that can be made by genetically-modified organisms, which would allow conventional symptomatic therapy for many more diseases, which could be supplemented by somatic cell gene therapy when appropriate. It would also expand our basic knowledge of human biology, which allows medical treatments to be developed. We may not be able to predict when therapies will emerge after the genes are discovered, because there can still be a long delay in clinical applications following biochemical understanding.[45] It is obvious that

[43] Victor A. McKusick, 'Mapping and sequencing the human genome', *New England J. of Medicine* 320, 6th April 1989, 910–915.

[44] 'Statement of the Working Group on Human Genome Mapping', in *Genetics, Ethics and Human Values: Human Genome Mapping, Genetic Screening and Therapy*, Proceedings of the 24th CIOMS Round Table Conference, Tokyo 22–27th July, 1990, Geneva: CIOMS 1990.

[45] Theodore Freidmann, 'The Human Genome Project – Some implications of extensive reverse genetic medicine', *American J. Human Genetics* 46, 1990, 407–414.

within the next few decades medicine will undergo a major change. This is the beneficial side of the extra knowledge. The amount of new knowledge is hard for us to comprehend. It will take decades to process it all, but it offers the potential for understanding all genetic diseases sometime during the next century.

The time is right for discussion regarding how we use the information and who should use it. Town meetings have begun to be held in the USA to inform the general public about the human genome initiative, and to solicit opinions on the ethical, social and legal issues that it raises.[46] The human genome project has even found its way into French school books. It is essential for widespread education to be available in a way that the public can understand it and become involved in decisions about their project. An adequately prepared lay community is the best way to ensure that misuse of genetics does not re-occur.

There should also be education to show that, despite all the information, we should not expect disease to be cured within twenty years, and it will not be a panacea for the world's woes. The relatively low cost of ethical and legal studies of the implications of the project compared to the biological research should encourage funding bodies to provide some funding (at least 1%) to ensure society is more prepared for the data.[47] In the USA the NIH and DOE have awarded research funds for study of these issues, and have established a joint working group.[48] The NIH and the DOE genome project budgets have set aside 3% for ethical, social and legal studies, and the preparation of educational material for the public. The European Commission emphasised more consideration of these issues before it will fund much scientific research.[49]

The ethical debate must focus on how to use the new information, rather than on whether to discover it, if for no other reason than that the knowledge will come. Most religious approaches support the rationale for obtaining better genetic information which can be used to alleviate human suffering.[50] The question is how to use it pro-

[46] Christine McGourty, 'Public debates on ethics', *Nature* 342, 7th Dec. 1989, 603.

[47] Karen Dawson & Peter Singer, 'The human genome project: for better or for worse?', *Medical J. of Australia* 152, 7th May 1990, 484–486.

[48] Robert M. Cook-Deegan, 'NIH-DOE joint working group on ethical, legal, and social issues established', *Human Genome News 2*, May 1990, 5–8.

[49] Debora MacKenzie, 'European Commission tables new proposals on genome research', *New Scientist*, 25th Nov. 1989, 6.

[50] Rihito Kimura 'Religious aspects of human genetic information', pp. 148–166 in *Human Genetic Information: Science, Law and Ethics*, Ciba Foundation Symposium 149, Amsterdam: Elsevier North Holland 1990.

perly. There are dangers in any large scientific projects. In becoming the sole ideal for progress, they tend to take control of people. We have seen this in the past with the Manhattan project and the Apollo project. From the initial response to the human genome project, this is also happening here.[51] The possibility of mastery and control over the human DNA raises the issue of genetic selection. Ideas of eugenics could be explored. We need to maintain a distinction between diagnosis and treatment of disease, and selection for desirability.

Most importantly we need to elevate the importance of individual autonomy, especially in reproduction. The human genome project raises similar ethical and legal issues to those in current genetic screening, such as confidentiality of the results. However, it will lead to screening on a huge scale, for many disease traits and susceptibility to disease. It is important that we deal satisfactorily with the test cases, before we are faced with all the new information. The technology may affect the way we think. The amount of information obtained will overwhelm existing genetics services, and geneticists. The ownership and control of genetic information, and the issue of consent to use such information, must be addressed. More training in genetics (as well as ethics) will be required for health care workers, scientists, and the general public.[52]

The question of fairness in the use of genetic information with respect to insurance, employment, criminal law, adoptions, the educational system and other areas must be addressed. In those countries with private medical insurance, some people may be put into high risk or uninsurable groups because of genetic factors (for example, high blood choloesterol, or family history of diseases). Some insurance companies and employers perform screening in the U.S.A.[53]

Some legislation has been passed in the USA, such as the Americans with Disabilities Act, in 1990, or the proposed Human Genome Privacy Act. The only adequate solution, however, is a national health insurance system with equal access to all. Such a system is not only desirable, but inevitable, and the sooner governments realise this the fewer problems will have accumulated when the time comes to switch to nationalised health schemes. The injustice of private health care schemes will be accentuated once additional genetic information becomes available. We must constantly focus on the question of whose

[51] George J. Annas, 'Who's afraid of the human genome?', *Hastings Center Report*, Jul/Aug. 1989, 19–21.

[52] Darryl R.J. Macer, *Shaping Genes: Ethics, Law and Science of using Genetic Technology in Medicine and Agriculture*, Christchurch, New Zealand: Eubios Ethics Institute, 1990.

[53] U.S. Congress Office of Technology Assessment, *Genetic Monitoring and Screening in the Workplace*, Washington: U.S.G.P.O., October 1990, OTA-BA-455.

project it is. The neglect of the principle of justice with regard to health care is bad enough, but the neglect is even worse when we regard the common ownership of the information in the human genome in the context of the shared taxes that continue to fund biomedical research.

We can say that all people should benefit from the results of the genome project, but we must ask whether in fact they will. Some people hope that the knowledge that we are all equal in our genetic differences might end discrimination (note 35). This, however, will require much education and the passage of specific laws to ensure that equality is respected. There will also be a change in attitudes to ourselves, and genetic determinism might become popular. A danger with simple-minded adherence to genetic hypotheses for behaviour is that it oversimplifies the complex interaction of genetics and environment. In the extreme, determinism eliminates the idea of genuine choice, leaving no room for the belief that we can create, or modify ourselves, or that we can make moral choices. Whether higher human attributes are reducible to molecular sequences is a controversy in the philosophy of biology. The knowledge of human genetics will make scientific understanding of human life much more sophisticated. There may be alteration in social customs, especially if the genetic information is misunderstood by the public as occurred earlier this century.

How can we ensure that all benefit from the results? In order to achieve this there will need to be control of the use of the results, not just the discovery of them. If we include the use of the results in the broad meaning of the "genome project", then it is appropriate to discuss briefly options on the use of results. Let us assume that the full gene sequence is available (as it probably will be within ten years). Let us also assume that technology has allowed the production of cheap and very simple genetic screening, for example colorimetric testkits. Should these be available to the public, as do-it-yourself pregnancy tests are today in the USA? There has to be serious concern about such an approach despite the common ownership claims that people can make to the sequence data. Personal reproductive decisions will need more serious consideration in the future, making life more complicated while hopefully improving its quality. While the people who can make decisions regarding the availability of such kits cannot claim to understand the social consequences of such a move, the general public also cannot understand the broader consequences of their combined individual actions. There is a case for control of public property in order to avoid doing harm. While it may be possible to regulate the use of such kits via intermediatory control, by health care workers, there could be particular testkits which may not even be made because of the fear of misuse. A contemporary example of the regulation of a technique with the potential for abuse is the use of sex as a marker for muscular dystrophy genes; in Japan the fears of misuse of this test have meant that sex-linked gene selection is officially unavailable even for medical reasons. Some other countries allow the use of sex selection itself, and in most countries it is condemned but not illegal. There is a case for making it illegal but many physicians want to resist the imposition of laws in medicine. When a technique requires a specialised kit, as in genetic diagnosis for the general physician, the use of the selection can be controlled by production of the kits, though it will be stronger with the outlawing of the selection.

There does need to be control over the use of screening and therapy that lacks a compelling medical reason. How will this control be effected? In the 1990 German Embryo Protection Law there is specific mention of Duchenne muscular dystrophy as a serious genetic disease for which genetic screening (for pre-implantation diagnosis) can be performed.[54] There has been criticism of this approach because of the possibility of increasing discrimination towards the handicapped who suffer from the legally designated "serious" diseases in which embryos suffering from such diseases can be discarded. It is certainly a sensitive issue, but some control will be required to prevent future abuses, and a slide down the slippery slope to screening of offspring on "cosmetic grounds". The production of a list of diseases is one aspect that needs to be sensitively investigated as a possible means of control. It may be better not to mention the disease in the law but to have a list to be used by the regulatory committee. Life will get complicated, but that is the price of such powerful information and technology. It will be further complicated by combinations of various diseases which may be "judged" permissible for parental selection after genetic screening or for treatment using gene therapy. The extremes of a free market approach or a total ban on genetic testing are both strongly undesirable, but attention must be directed toward the method of control. The question of who decides the application of technology in individual cases must be addressed, whether it be individual genetic counsellors, codes of practise, legally established regulatory committees or parents, and whether it is freely available to all or only to those who can pay, or only to those judged to be at "significant" risk.

Who will really benefit from the project? Will the next generation benefit from being genetically selected? Is a life suffering from serious

[54] German Bundestag, 'Embryo Protection Act 1990'. For a partial English translation see *Bulletin of Medical Ethics* 64, Dec 1990, 9–11.

disease better than no life? As is the case today, these questions need to be answered by individual parents, but they will become much more apparent with the number of conditions and ease of screening. It may be easier to ask the question whether the parents who use such screening on their gametes, embryos and fetuses will benefit from it. They do have benefits of avoiding problems of the extra time that they may need to spend with children, and the extra costs, but they also change themselves by using such screening, by making presumed health a condition of acceptance of children. It is also quite debateable whether the extra time parents spend with children and the potentially greater opportunity to give love, is time better spent than pursuing previous life goals such as time with other children or careers. When people debate these issues there are arguments on both sides. What society really does need is some harder data on the real effects. Rather than presuming outcomes based on incomplete psychological and sociological knowledge, there needs to be detailed study of the effects of genetic screening. While we should not be afraid for society to change, we should be wary of change when there may be adverse social attitude changes. There is a case for limiting the introduction of any population scale use of genetic screening until the effects of the current genetic screening policies are characterised. We will have much data over the course of the next decade, if people are prepared to fund research on this subject. Rather than speculating about the outcomes, there needs to be rigorous study of what data we have before embracing all that new technology can provide. Such study could follow up cases where parents sought genetic counselling, and cases where they didn't, and also those who accept or reject selective abortion, for example. The major issues to which philosophy can contribute are the need to consider the autonomy of the parents who are all different people, the status of the fetus, and the enforcement of genetic screening by society. However, we need data to measure the effects on personal, family and social attitudes.

There is less experience with pre-symptomatic genetic testing of disease risk, such as for Huntington disease. This is another area where the data needs to accumulate before we will be able to make reasonable predictions about the more widespread use of such testing. The other questions arising from the screening of children or adults for disease susceptibility is somewhat easier to address. As mentioned above, from the principle of justice we should work against genetic discrimination and establish national health schemes and equal access to employment (except when there is an actual, current, risk of third party harm). There should be a right to privacy of genetic information. Some employment performance based testing can be used, when there is a reasonable and potential risk of harm, but not mandatory genetic tests. There needs to be guidance over the storage of genetic information for legal purposes, in immigration and in crime. There are already some committees which oversee police records with regard to protecting privacy, such as the Supervisory Board of Interpol's computer records.[55] We know in general what is required; what is needed is translation of ideas into practise.

There are important questions about who has the right to know our individual genetic makeup.[56] Of course, our general genetic makeup will be common knowledge. It could be argued that because we all share in the information to be made public, we all have a say in the discovery and presentation of it. The sequence must be protected from abuse. It will be the most detailed common knowledge about every individual and will provide many opportunities for abuse. It is not unique in this; psychologists have understood common complicated features of the human mind for many years and such knowledge is of similar risk to that being unravelled by geneticists.

In existing genetic services we recognise a right not to know our genes. Could this right be extended to general genome sequencing? We may have a right not to know that we will develop Huntington disease, but what about the right not to know that we are at high risk for schizophrenia, alcoholism, or a life in academia? Do we have a right not to know that the common human DNA sequence "programs" us to die at 85 years of age? Is the right not to know limited to features that distinguish us from the norm in our society? Common knowledge such as general life expectancy are not hidden, though death may be a taboo topic in most countries. Are people still afraid of being different, even in the supposedly individualistic Western societies? This is another question that needs answering before we can work out appropriate means of regulation.

We do not know the effects of these different options, and research is required. Health care should be equally available to all people, and there should be more medical care given to those with greater medical need. Society does discriminate against the apparently "unusual" including handicapped persons, and laws have been enacted to attempt to lessen this discrimination. More laws will be required, but changes in social attitude and education are necessary to change this behaviour. The words "unusual", "abnormal", "disabled" and

[55] Budimir Babovic, 'Interpol and human rights', *International Criminal Police Review*, July/Aug. 1990, 2–8.

[56] Walthur Ch. Zimmerli, 'Who has the right to know the genetic constitution of a particular person?', pp. 93–110 in *Human Genetic Information: Science, Law and Ethics*, Ciba Foundation Symposium 149, Amsterdam: Elsevier North Holland 1990.

"handicapped" can all have negative connotations, a more appropriate word may be required. In Japan, recent law has required that the words used in the media to describe a blind or deaf person cannot be "blind" or "deaf", but rather "people that have difficulty hearing, or seeing", so that if you listen to an old movie the former words are censored. While this is an attempt to address some of the discrimination that such words can carry, society requires more than word changes to end discrimination. Most people have a poor knowledge of genetics, which must be improved before they will be able to understand the new knowledge. Incomplete knowledge can be very dangerous when combined with existing discrimination, as seen with eugenic programmes earlier this century. We should all realise that we are genetically different, and normality is culturally defined, perhaps in terms of the ability to live comfortably or anonymously in a given society.

Article 23 of the International Covenant on Civil and Political Rights,[57] states that "the right of men and women of marriageable age to marry and found a family shall be recognised". This covenant has been signed by over 75 countries, and should guarantee that compulsory eugenics is not introduced. It is a very strong statement based on the ethical principle of respect for human autonomy. Social pressures, however, are very difficult to control. Such a law needs to be supported by equal access to social and health services in order to make it effective. In the same covenant there is also supposed recognition of equal access to health care, but what is required is wording such as "equal access to *equal* health care". This is one avenue that action could be taken from well accepted ethical principles, but education about technology and how to make decisions in an ethical way is also necessary.

FOR OUR GENERATION OR FUTURE GENERATIONS?

Will society allow individuals to have free choice over the use of genetic manipulation and screening when there is no medical reason for it? Ironically such screening can be used both to reduce individual differences, and to highlight differences in ability. There are various arguments used against genetic intervention that has no therapeutic value. It would be a waste of resources, may present risks to offspring, it could promote a bad family attitude, could be a harmful support of society's prejudices and may reduce social variability. It will probably not have any significant effect on genetic variability as there will be plenty of alternative healthy alleles. Some may also defend

[57] p. 186, note 9.

the idea of a natural genetic autonomy, arguing that we should let the genes come together naturally, and allow individuals to develop their genetic potential without unnecessary interference by parents or society.[58] A criteria for transgenerational ethics is that not only must a gene alteration be safe, but there must be unquestionable objectives and benefits for many generations.

A common feature of many issues raised by the human genome project data is that we need to consider the effects of knowledge and technology on future generations. We have a responsibility to future generations. The beneficiaries and those at risk may not yet exist. In the sense of benefits and risks, it is their genome project more than ours. We have an obligation to the future based on the principle of justice.[59] Our traditional view of morality only involves short term consequences. Human action is seen as only having a small effective action range. If another agent intervenes, or something unexpected happens, it is not considered our fault. Genetic engineering changes our moral horizon. For that we should be very grateful, as for too long we have only examined short range effects.[60]

An ethic of long range responsibility is needed. It implies that there is a moral imperative to obtain predictive knowledge and data about the wide-ranging possibilities of some action. Secondary consequences may be sufficient to prevent the primary action, even when the primary action may be good. This imposes a restraint on the use of technology. This is important in making public policy decisions which are beyond the physicians' concerns with each patient, or the scientists concerns with increasing knowledge of genetics. It means that researchers may be held accountable for secondary consequences of their research. Of course it may be very difficult to predict what will happen in the future. Social pressures and thinking are already very distinct between different countries. If social ideas change, then so may the pressures, such as the desire to use genetic enhancement. We need to ensure future generations retain the same power over their destiny as we do, while benefiting from the culture and technology we have developed.

INTERNATIONAL MANAGEMENT OF COMMON KNOWLEDGE

It would be unethical to withhold information that could provide medical therapy if released. The genetic information of the human

[58] pp. 258–260, note 52.
[59] John Rawls, *A Theory of Justice*, Oxford: Oxford University Press 1971.
[60] pp. 308–313, 323–324, 345–347, note 52.

being belongs to all humanity. It should be available to all at an affordable price, and without discrimination. The benefits that come from its discovery and use should show us that all humanity is one. We will see how the genetic constitution of all humans from all races is the same. We will see how all of us have mutations, no one is perfect in his or her genetic structure, (or should we say perfectly normal?). Decisions on the use of genetic manipulation in one country will affect other countries because people migrate. It is therefore imperative that the decisions about any future germline genetic manipulation, especially of humans, take into account people's opinions worldwide. This may be best handled by an international forum, with which national committees should interact. The coordination needed for the genome project may aid this process, but the developing countries need to be adequately represented, especially because they represent such a large proportion of the world's population.

The answer to the question "whose genome project?" is "Everyone's". This implies that everyone should make these decisions. Practically this means that the legal representatives of the people of the world, and the representatives of the various viewpoints of people's of the world (these may be different) need to join together to make decisions. This is the ethical answer. What will happen in practice has so far differed from this approach. It has been assumed that the financial provider for the research can control the project, and it has been suggested that the financial provider for the research may also be able to control the information arising. There should be increasing awareness that the international nature of human DNA necessitates an international common interest in all decisions. Even if people do not accept international ownership, they should be convinced by the "international interest" and "utility" argument that such information should be openly and equally available.

The international nature of the project and its universally applicable results, as discussed in this paper, make it a project of all humanity. It is essential to have international organisations such as HUGO and the UN bodies taking an active part in the work and considerations of the ethical, legal and social issues and solutions. Many countries are unable to significantly contribute material resources to the scientific project, but they share in the material that is being sequenced, their genes, and must be involved in the project's benefits and decisions. By intensifying international coordination at this stage when we are shifting from asking questions to working out solutions, we will be better able to ensure that more people can benefit from the project. People in developing countries will also indirectly benefit from the technology which will be applicable to many pathogenic diseases which are currently more important diseases in those countries than genetic diseases. There are fundamental ethical questions to be answered from an international perspective, which should certainly be broader than local groups such as the Ethical Committees within the USA, Denmark, France, or even the Council of Europe. Rather organisations such as the International Association of Bioethics, and UN-bodies like Council for International Organization of Medical Sciences should attempt to develop international approaches to managing the way the project is performed and the results distributed and applied. From these ethical approaches the appropriate interpretations of international law should be applied, and if necessary new articles added. Society's interests should transcend proprietary rights. Moreover, the special nature of the genome project and the claims that we can all make upon the genome should make the shared authorship and ownership legally compelling. Humanity does have a chance to build on the supposedly improved international climate in a very fitting way. We should take this appropriate opportunity to move beyond the influence of our "selfish genes" in our combined efforts to sequence them.

ACKNOWLEDGEMENTS

I would like to thank the *Bioethics* editors and reviewers for their thoughtful comments.

Institute of Biological Sciences
University of Tsukuba
Ibaraki
Japan

[4]

The Gene Genie: Good Fairy or Wicked Witch?

*Sheila A. M. McLean**

The so-called genetics revolution rests on a history which at its least can be described as controversial. Modern genetics needs to bear this history in mind. In particular, as with the past, the area of reproductive choice seems particularly vulnerable to potential abuse. Courts in the UK and elsewhere have already shown themselves willing to interfere with the choices of women in the management of their pregnancies. Medical advance, perhaps particularly the capacity to visualise the developing foetus, has added complexity to the question of whether the health care provider has one patient (the woman) or two patients (the woman and the foetus). Additionally, pregnancy is thoroughly monitored in modern medical practice and genetics may provide a further impetus or incentive to mandate increased policing of pregnancy. Gene therapy, once offered, will add further to the desire to ensure that women make the 'right' choice, especially when the invasion required is relatively minimal. Further, genetic information is at best predictive, but may, because of its scientific nature, appear to those receiving it to be certain. Thus, the provision of genetic information may reduce rather than enhance choice, unless carefully and sensitively provided. A mature and sophisticated debate about the role of genetics in reproduction is required—engaging rather than bypassing the public—if the real potential of genetics is to be vindicated.

Keywords: Genetics; Reproduction; Procreative Choice.

Science often confounds with its seemingly unstoppable advances, and clinical medicine follows not far behind, in the wake of the leviathan of progress. The media compete to trumpet the phenomenal strides made by the combination of these two great enterprises. Whether it is creating life in a petri dish or the cloning of Dolly, the frontiers of knowledge are constantly pushed further and further back. What we *can* do is amazing; what we *will* be able to do, almost unimaginable—almost, but not quite, that is. The science fiction of only a few years ago is our present-day reality. The obvious question seems to be; where will we go next?

An even more important question, however, demands attention before the answer

* School of Law, The Stair Building, 5–8 The Square, University of Glasgow, Glasgow G12 8QQ, U.K. (*e-mail*: s.mclean@law.gla.ac.uk)

to the first one can be sensibly reached. Where we *will* go is a question which arguably is predicated on answering the question where we *should* go. Equally, where we *should* go may to an extent depend on considerations which are both ethical and temporal. In other words, the decision as to whether or not to proceed with any particular scientific technique may be based either on underpinning ethical conclusions, or more pragmatically on the basis of *when* progress should be initiated. In the often rather negative academic response to the genetics revolution, it is sometimes apparently overlooked that whether or not something is 'right' may change depending on the ethical and intellectual infrastructure on which it builds, or on the time at which it is proposed to undertake it.

Nor should we underestimate the power of science and medicine to control themselves. While the public (or, more accurately, the media) generated vociferous and often scaremongering debate after the birth of the world's first cloned mammal, scientists waited patiently to discover the truly important aspect of Dolly's birth—that is, how old she actually is and whether she would be fertile. In fact, it appears that she is ageing normally and has delivered a number of lambs. This, of course, will not necessarily affect the debate about whether or not to endorse human cloning, but it does show that we should be as guarded in our critique of science as we are in its ready acceptance.[1]

It is important, therefore, that comment on genetics and its potential is couched in terms which both respect the science and explore the ethics. As Schmidtke has said: 'Scientists are at least no better and no worse than the society of which they are members, and criticism of science is part of legitimate social critique.'[2] Few scientific advances come without a down side, but seldom has the negative been the object of so much attention as has been the case with the genetics revolution. Some commentators find such a response problematic and encourage a rigorous defence. Maddox, for example, has said that '. . . the widespread fear of genetics cannot be justified. On the contrary, the research community should speak out strongly to defend the good sense of what it is about.'[3]

However, there are very good reasons for people to be concerned. The DNA which makes us what we physically are is central to our sense of self. When manipulation, even extermination, of certain genetic characteristics is proposed, it is little wonder that people are concerned. When science appears to hold within its grasp the power to reshape the future, we are right to be hesitant before embarking on such a path.

If we are to harvest the good and avoid the bad of genetics, it is essential that an informed, intelligent and sophisticated debate is undertaken. As the British Medical Association put it:

[1]For a full discussion, see McLean (1999).
[2]Schmidtke (1992), p. 205.
[3]Maddox (1993), p. 97.

> People's lack of knowledge about genetic modification has in the past given rise to fear and to opposition to new developments. We believe that these fears are largely unfounded, but that combating them by adopting a paternalistic or secretive approach is not the answer. Instead, the scientific community, both in academia and commerce, has a duty to inform the general public of new developments in the applications of genetic modification in a manner comprehensible to lay people.[4]

More than this is also required. The public must not only be informed, they must be engaged in the debate. Perhaps inevitably, however, this often results in concentration on what might go wrong with genetics. And there are good reasons for this. Recent history in the Western world has shown that, armed with genetic information, states have been all too willing to manipulate it to satisfy social and political agendas.[5] Seldom is this more obvious than in terms of human reproduction—perhaps the single most private and important life event.

The Nazi abuse of 'inferior' people is well documented, but some years before their policy of genocide became a hideous reality, other countries—perhaps most notably the United States—had been taking advantage of genetic information to carry out their own attack on vulnerable members of the community. In the early part of this century in the United States, laws were put in place which permitted the proxy authorisation of compulsory sterilisation of those who were thought unfit to parent. This 'unfitness' was based on a variety of characteristics from ethnic origin, through skin colour, to intelligence levels and criminality.[6] People's reproductive liberties were savagely abused, immigration policies tightened to avoid the influx of those thought to be genetically inferior, and—even against the background of a written Bill of Rights—courts were unprepared for many years to declare such laws to be unconstitutional.[7] The potential for science to fulfil the aspirations of existing agendas, and to give them a patina of respectability, was seldom more obvious than in these early years of the century.

However, it must not unthinkingly be assumed that the world which we presently inhabit is so much more sophisticated that similar problems could not arise again, and it is for this reason that the potential negatives of genetics must be explored. Certainly, it is plausible to imagine that the form of any potential misuse in current times would be different. It is unlikely that wholesale sterilisation would find present-day favour on any ideological agenda, or at least it is unlikely that the political will would be found to translate it from rhetoric to reality. However, the fact that the agenda may be less structured, less coherent or less overt does not mean that societies can afford to relax their vigilance. For good reason or not, modern societies still make, and contemporary judges still endorse, decisions which are essentially about fitness for parenting. Cases involving the non-consensual sterilisation of women with learning disabilities still reach our courts, and very often—

[4]BMA (1992), pp. 227–8.
[5]For further discussion, see McLean (1986).
[6]For discussion, see Meyers (1971).
[7]*Buck v Bell* 274 US 200 (1927).

although not inevitably—these same courts are willing to comply with the request to authorise sterilisation.[8]

More broadly, of course, our genetic makeup has now effectively been exposed as scientists declare themselves to have mapped the majority of the human genome. The aim of the Human Genome Project was to acquire 'complete knowledge of the organization, structure and function of the human genome—the master blueprint of each of us'.[9] The completing of this project will raise more and more dilemmas which present societies seem unwilling or unable to face. The use which is made of genetic information, and the understanding of its implications, will be critical in determining how to maximise the positive and minimise the negative. Our genetic inheritance both celebrates our differences and identifies our similarities. Perhaps most significantly, it is the extent to which diversity is seen as a problem or a blessing that will inform and shape our response to the plethora of information currently emanating from the genetics laboratories.

As has been said, the field of human reproduction is one area in which our resilience to the quick fix and our ethical sophistication will be most challenged. Of course, few, if any of us, would choose to have a child who is in some way disabled or unwell. What genetics offers nowadays is the capacity to make the dream of the healthy child a reality. Or at least; this is how the promise of genetics is often interpreted. In reality, what genetic information can do is to predict, with degrees of certainty, the likelihood of certain (not all) conditions becoming manifest. As Hubbard and Wald note, '[m]ost prenatal tests offer little precise information. They can suggest problems, but cannot say how significant these problems may be. Genetic predictions, like all medical tests, involve setting arbitrary norms.'[10]

There are several stages at which such information can be identified and may be used. Each poses its own dilemmas, under the overarching umbrella concern about the potential slippery slope that separates disease or disorder prevention from the spectre of genetic enhancement and the birth of 'designer babies'. Like the uninvited and unwanted presence of the wicked fairy at the birth celebrations of Sleeping Beauty, the ghost of harm hovers even as celebration of the birth of the new genetics continues.

1. Genetics and Reproduction

The well documented abuse of early genetic knowledge does not require restatement. As was mentioned above, societies have in the past shown themselves

[8]Cases include *Re B (A Minor)(Wardship: Sterilisation)* [1988] 2 FLR 997; *Re P (A Minor)(Wardship: Sterilisation)* [1989] 1 FLR 182; *F v Berkshire Health Authority* [1989] 2 All ER 545; *Lawrence, Petitioner* 32 BMLR 87 (1996). Exceptions include *Re Eve* 31 DLR (4th) 1, 1987 (Canada) and *Re D (a minor)* [1976] 1 All ER 326.

[9]*Human Genome 1991–92 Program Report* (1992), foreword, p. iii.

[10]Hubbard and Wald (1993), p. 30.

all too willing to use such information to achieve political or social goals, worthy or unworthy. However, genetic information has continued to play a major role in reproductive choice, albeit in the absence of the direct and overt coercion that tainted its early use. The intelligible desire of couples (or individuals) to achieve a healthy pregnancy and birth, coupled with increasing knowledge of teratogenesis and genetic predispositions, has led to what has been called as 'evangelistic fervour'[11] for pre-natal screening.

The modern pregnancy is monitored, screened and policed with a thoroughness hitherto impossible.[12] Of course, this should mean that pregnancy is safe both for women and foetuses, and this, where true, is unequivocally a benefit. However, there are also consequences of the intense interest in the progress of pregnancy which both expose part of the rationale for scrutiny and pose ethical dilemmas of their own.

Modern screening techniques permit, indeed depend in some cases, upon visual identification of the foetus *in utero*. The ability to see the developing foetus can encourage bonding, with women and their partners identifying the humanity of the contents of the womb. Equally, however, visualisation, when coupled with medical capacity, can encourage doctors to believe that they are treating not one patient (the pregnant woman) but rather two patients (the woman and the foetus).[13] Although it is clear that foetuses have no legal rights,[14] it would be unrealistic to suggest that we have no moral or personal interest in them. Thus, doctors may be tempted—as indeed they have been—to view the foetus as a separate entity with interests (if not rights) which it is the obligation of the doctor to protect.[15]

By and large, this may prove to be both unproblematic and uncontroversial. However, this will not always be the case. Viewing the foetus as a patient can also result in the generation of conflict between women's decisions about their pregnancy and the professional obligation felt by clinicians. This can lead to a situation where doctors believe that women should be coerced into accepting treatment which is in the interests of their foetuses.[16] Medical invasion may be either minimal or substantial, but there is no doubt that the capacity, for example, to treat foetuses in the womb, or the instinct to treat them as if they were already born, may pose stark problems for some pregnant women. This so-called maternal/foetal conflict is a reality of the contemporary management of some pregnancies, and poses at the theoretical level a serious risk to respect for women's rights.[17]

Doubtless, we might feel that a woman who refuses intervention which could

[11]Stone and Stewart (1994), p. 45.

[12]McLean and Petersen (1996).

[13]For discussion, see Mattingley (1992).

[14]*De Martell v Mertin and Sutton Health Authority* [1991] 2 Med LR 209; *B v Islington Health Authority* [1991] 2 Med LR 133; *Montreal Tramways v Leveille* [1993] 4 DLR 337; *X and Y v Pal and Others* [1992] 3 Med LR 195; *Hamilton v Fife Health Board* [1993] 4 Med LR 201.

[15]Johnsen (1986).

[16]Kolder *et al.* (1987) and Gregg (1993).

[17]Annas (1986).

save the life or the health of her foetus is worthy of moral condemnation, but this does not imply either that such condemnation is appropriate, or that choice should be removed from the woman. As Robertson and Schulman have said:

> Developments in obstetrics, genetics, foetal medicine and infections diseases will continue to provide knowledge and technologies that will enable many disabled births to be prevented. While most women will welcome this knowledge and gladly act on it, others will not. The ethical, legal and policy aspects of this situation require a careful balancing of the offspring's welfare and the pregnant woman's interest in liberty and bodily integrity.[18]

Perhaps most significantly, it does not logically follow that moral condemnation should be translated into legal control. As Swartz has said,

> Although it may be morally and ethically appropriate in most cases in which 1) a woman's own health would not be adversely affected and 2) the foetus is viable for the woman to make decisions that would enhance the foetus' chance for good health, legislating for morality in these cases raises more questions than it answers.[19]

Yet courts in the United Kingdom and the United States, for example, have already shown themselves willing and able to intervene in the choices of otherwise competent women in the progress of their pregnancy.[20]

Perhaps most poignantly this arose in the US case of Angela Carder.[21] Terminally ill and pregnant, Mrs Carder refused to undergo a caesarean section which had a small chance of permitting live birth. This she did with the endorsement of her husband and parents (although it should be noted that legally their agreement is irrelevant), and to the dismay of her doctors. The capacity to envisage the foetus as a separate patient to whom duties are owed, as has been mentioned above, may have played a part in their subsequent actions. An emergency court was established at the hospital and over Mrs Carder's repeated protestations an emergency caesarean was performed. Neither she nor the child survived. Although this decision was eventually overturned,[22] it was too late for Mrs Carder, whose last moments were desecrated by enforced surgery. This may be an extreme example, but it nonetheless points to the potential problems associated with forcing women to act in the best interests (however defined) of their foetus.

The possibilities of such conflict arising will, it is suggested, increase with developments in gene therapy. It has been suggested, for example, that, as the gap between diagnostic and therapeutic capacity narrows, gene therapy will become

[18] Robertson and Shulman (1987), p. 32.

[19] Swartz (1992), p. 56.

[20] Cf. *Re S (Adult: Refusal of Medical Treatment)* [1992] 4 All ER 671; *Norfolk and Norwich Healthcare (NHS) Trust v W* [1996] 2 FLR 613; *Jefferson v Griffin Spaulding County Hospital Auth.* 247 Ga 274; but see also, for a different conclusion, *Re MB* 38 BMLR 175 (1997) and *St George's Healthcare National Health Service Trust v S* [1998] 3 All ER 673.

[21] In *Re AC* 533 A 2d 611 (1987)

[22] *Re AC* 573 A 2d 1235 (D.C. 1990)

the treatment of choice for certain conditions.[23] For many, this will of course be true. The possibility of curing the affected foetus in the womb, and the prospect of a healthy live birth, will doubtless encourage pregnant women and their doctors to opt for such therapy. However, it must be borne in mind that this form of therapy involves both the foetus and the pregnant woman. No invasion of her body is permissible without her agreement, and this may not always be forthcoming. As Annas says:

> Fetuses are not independent persons and cannot be treated without invading the mother's body . . . Treating the fetus against the will of the mother degrades and dehumanizes the mother and treats her as an inert container.[24]

Just as phobias, such as needle phobia, have in the past been sufficient to deter a woman from agreeing to a clinically indicated caesarean section,[25] so too the implications of foetal surgery may be too difficult for some women to accept, even if the outcome of their reluctance or refusal is the birth of a genetically compromised child. As Ruddick and Wilcox put it, '[p]erhaps a woman ought to choose abortion or fetal surgery, but it might be wrong to prevent her from choosing the morally impermissible option of a continued, untreated pregnancy'.[26]

The prospect of foetal gene therapy, therefore, seems likely to engender more examples of maternal/foetal conflict and to continue to pressurise women into adopting or accepting behavioural and decision-making patterns which are essentially against their own judgement. The woman who does not succumb to the pressure to accept foetal therapy is as likely to be vilified as the woman who chooses to have an affected child. This is so despite the fact that no woman (or man, for that matter) could be legally forced to undergo surgery (major or minor) to save an already existing child:

> It is certain that a court could not order a parent to donate bone marrow in order to save a child's life. Yet courts order women to submit to surgery for the sake of a fetus. This is illogical, for the born child has constitutional rights, which the fetus does not, and the pregnant woman is subjected to a more dangerous, intrusive procedure than the donor would experience.[27]

Again, we might argue that they *should* do this, but this is a long way from forcing them to do so.

Equally, screening fulfils the function of permitting termination of pregnancies which are in some way compromised. The capacity to screen for, and screen out, certain defects or abnormalities has become an accepted facet of the contemporary management of pregnancy. Where screening is offered and accepted there is an unspoken acceptance that a 'bad' result will result in the termination of the affected

[23] Davies and Williamson (1993).
[24] Annas (1988), p. 24.
[25] *Re L (Patient: Non-Consensual Treatment)* Family Division 13 December 1996 (transcript).
[26] Ruddick and Wilcox (1982), p. 12.
[27] Lew (1990), p. 642.

pregnancy. In some cases, it appears, this expectation has moves from advisory to mandatory. The British Medical Association, for example, notes that:

> In the past, some health professionals restricted access to prenatal diagnosis to those individuals who planned to terminate an affected pregnancy. Although this approach is now widely regarded as paternalistic and unacceptable, a 1993 survey of obstetricians found that one-third still generally required an undertaking an affected pregnancy before proceeding with prenatal diagnosis. The BMA considers this approach to be unacceptable . . .[28]

In addition, abortion laws in the United Kingdom make severe foetal disability one of the three grounds which allow for pregnancy termination up to full term.[29] Thus, in the United Kingdom at least, it has already been accepted that certain births can be avoided—indeed that it is lawful (perhaps even moral) so to do. These so-called foetal grounds for pregnancy termination are often perceived as among the least contentious ground for abortion, avoiding—as they may—the birth of a child destined to suffer.

However, this is not to say that terminating these pregnancies is entirely uncontroversial. Even for those who do not always oppose abortion on any grounds, there is reason to be cautious, at least in some cases. While termination on severe grounds (which is what UK law permits) may be accepted by many people, this begs the question of what degree of severity must be demonstrated before abortion is permissible. Where the source of the problem is genetic, the issue carries with it even more reasons for concern. With or without the assistance of geneticists, there is a temptation to give absolute credibility to science. This is especially true when, as is undoubtedly the case with genetics, the science is far beyond the understanding of those who will live with its implications and outcome. However, as has been said:

> Neither the fact that a discipline is scientific nor the professionalism and technical expertise of those with the relevant knowledge justifies a lack of scrutiny from non-science perspectives. Like all human endeavours, technology and its application are value-laden enterprises: they must be tested against values such as compassion which emerge from culture, or principles like justice which as societies we claim to endorse.[30]

Nonetheless, the mere existence of genetic abnormality may be sufficient to provoke a 'choice' for termination, by apparently rendering the abortion issues less morally laden, cloaked in the patina of certainty and respectability which often characterises our view of science, making us 'participants in a cultural process, a fundamental mentality, overvaluing the contributions of medical science and tech-

[28]BMA (1998), p. 52.
[29]Abortion Act 1967 (as amended by the Human Fertilisation and Embryology Act 1990), s. 1(1)(d).
[30]McLean (1999), p. 7.

nology to the pursuit of human happiness and wellbeing, and believing medicine's promises to eliminate human suffering and mortality'.[31]

Leaving aside the question as to whether or not this view of science is realistic in general, it is one which can plausibly be argued to be less realistic in terms of genetics. The reality is that genetic screening is not a 100% accurate technique. Even if it could be shown that the tests themselves contain no false negatives or false positives, screening can tell us only something, but by no means everything, about the likely course of future events. Where the condition is genetically predisposed but is not inevitably going to arise, factors such as environment and so on will play a significant role in determining whether or not the child ever suffers from a particular condition. Equally, genetic screening cannot indicate the time, the gravity or even the inevitability of disease onset in most cases. Even where onset is certain, it cannot predict when, or how severely, a person will suffer.[32] And, where the disease is late onset (that is not arising until, say, after 30 or 40 years of life), the mere knowledge that the gene is present is insufficient morally to answer the question as to whether or not termination is the appropriate or the best answer.

Comforted, however, by the apparent certainty of science, potential parents may feel reassured as to the wisdom of their choices, and relieved of some of the moral concerns surrounding their decision (either to abort or not). However, confronted with the revelation that the science is by no means as certain or as value free as at first appears, choice in these circumstances becomes fraught with difficulty. In a world in which geneticists struggle to explain the science and implications of what they are doing to professional colleagues, how likely is it that the decision of a lay person will be as informed as it could be—perhaps as it should be?

The sad but unavoidable conclusion here must be that the mere provision of additional information is insufficient to guarantee enhanced capacity to choose. Yet the mindset which informs screening will also likely inform the decision ultimately made, and the fear of having afflicted children may push women (and their partners) into choices which sit uncomfortably with their own or society's values.[33] It is reported, for example, that some women in the United States have chosen to abort foetuses which have been found to have chromosomal abnormalities, even though neither they nor their clinicians can say whether or not these would actually have any effect on the child if born.[34] Genetics may not actively seek to achieve 'designer babies' but its consequences may be to ensure that only those declared genetically 'fit' will survive. Of course, to an extent this does represent choice—

[31] ten Have (1993), p. 43.

[32] *Ibid.*, p. 29: 'A prenatal diagnosis of sickle-cell anemia or cystic fibrosis does not predict at what age the condition will become manifest, how disabling it will be or to what extent it will shorten the life of the affected individual.'

[33] *Ibid.*, p. 28: 'The Mind-set behind genetic testing rests on societal view of disabilities that should not go unchallenged.'

[34] *Ibid.*, p. 30.

a value generally subscribed to. On the other hand, however, it also connotes the dissonance between what is possible and what is understood, limiting the availability and the reality of any purported choice.

People are, it is accepted, free to make choices based on ignorance or misunderstanding. In the general rule, a decision—even when life-threatening—need not be rational or intelligible to anyone but the person making it.[35] However, this rule also presupposes that the decision has been taken after the intelligible and honest provision of information (save, of course, in those cases where people have refused to receive it). While doctors and others may argue about the capacity of patients to understand general medical information (a view which has been used to reject the notion that full information should be given to patients[36]) it is arguable that the provision of *genetic* information can sometimes be less than autonomy-enhancing.

The general rule, therefore, does not happily accommodate a situation where information obscures rather than elucidates, as may well be the case with genetic information. This is not to say that genetic information should not be sought or provided, but rather relates back to the temporal issues raised earlier. Virtually every report and academic comment on genetics has stressed the need for public education in order that the best opportunity is presented for quality decision-making.[37] Yet this programme of education has not yet been in evidence, leading to the question whether or not it should be put in place before rather than after people are offered such ethically weighty options. However, some comfort can be drawn from the recent report from the Wellcome Trust which found that people were in fact able to understand even complicated genetic information, if adequately explained to them.[38]

Of course, it could also be argued that the issue of autonomy should follow rather than precede decisions in this area. In other words, whether or not genetic information is autonomy-enhancing might take second place to the question of whether or not such decisions should be made at all. To echo the thoughts raised at the very beginning of this chapter, it is important that we decide what *should* be done before we consider actually doing it. For some, if not many, people the capacity to create designer babies is anathema—the biological lottery is essentially value-free and 'natural', and should not be interfered with. For others, as has already been noted, the desire to have a child either free from disease (which is a form of genetic enhancement) or displaying certain characteristics is both normal and value-neutral. In analysing this debate, it is important to bear in mind the fact

[35]Cf. Butler-Sloss, LJ in *Re MB*, note 20 above, p. 186: 'A competent woman, who has the capacity to decide, may, for religious reasons, other reason, for rational or irrational reasons or for no reason at all, choose not to have medical intervention, even although the consequence may be the death or serious handicap of the child she bears, or her own death.'

[36]For further discussion, see McLean (1989).

[37]See, for example, reports by the BMA (1992), the Royal College of Physicians of London (1991) and the Nuffield Council on Bioethics (1993).

[38]Wellcome Trust, Public Perspectives on Human Cloning, p. 200; available at *www.wellcome.ac.uk*.

that it cannot be simply unravelled. Those who object to genetic enhancement must also concede that any intervention designed to 'improve' the embryo or foetus is a form of this—it is no less enhancement to try to improve health than it is to improve, for example, physical or mental characteristics. If one form of enhancement is to be permitted, and another outlawed, there will have to be a robust set of ethical principles to guide us and explain the distinctions drawn in coherent terms.

It is clear, therefore, that some consequences of screening are acutely difficult. Whether they set women in some kind of conflict with their foetus (and perhaps their clinician) or seem set to encourage the termination of compromised pregnancies, the dilemmas are clear.

2. Reproductive Responsibility

There is one further outcome of screening which has traditionally received less attention. Although it has been suggested that the expected outcome of a poor prenatal diagnosis is the termination of the pregnancy, not all women will choose this route. The decision to continue with a pregnancy, while apparently less value-laden than the abortion decision, may not always be perceived as such. Increased awareness of the avoidability of the birth of a genetically compromised child may well lead to moving reproductive choices back into the public arena. One major success of the struggles of the early part of this century which have been described above was the ultimate recognition that reproduction is a central part of the private sphere, and that privacy (however defined) requires that individuals are free to make their own reproductive decisions.[39] Whether it is the choice to use contraception or as to the number and spacing of babies, reproductive choice is a matter for the individual herself and is not, by and large, a legitimate interest of the state. It seems likely that the implementation of the UK Human Rights Act 1998 will enhance the focus on privacy in matters of reproductive choices.

However, where it is possible to know, with reasonable accuracy, that continuing with a pregnancy will result in the birth of a child with a handicap or illness—and therefore one who will likely be a cost on the state, or who may continue to pass the deleterious gene through the generations—it might be argued that a choice to proceed with such a pregnancy is morally reprehensible or wrong. Whittaker, for example, makes the point that, '[w]ith the availability of genetic tests, bringing an affected child into the world could be construed by some as reproductive irresponsibility'.[40]

Two notions are encapsulated in this perspective. First, it envisages that the state may claim a legitimate interest in controlling reproductive choices where they inherently affect its resources. Thus, in a community where health, education and

[39] *Skinner v Oklahoma* 316 US 535 (1942); *Griswold v Connecticut* 381 US 479 (1965).
[40] Whittaker (1992), p. 296.

other social services are partly, largely or wholly funded by the state, it may claim a legitimate remit in the decision-making process. Or as Kevles says:

> Private decision-making in the realm of genetic disorder and disease may ultimately lead to public consequences, and thus to demands for public regulation of reproductive behavior. A sizable number of people may argue that the right to have genetically diseased children, or even to transmit deleterious genes to future generations, must be limited or denied.[41]

Second, the growing weight given to concepts such as intergenerational justice[42] gives credence to the development of notions of 'reproductive responsibility'. Not only might the decision to proceed with an affected pregnancy have cost implications, but it may also be accused of being irresponsible, either in respect of that particular child or for future generations.[43] It is here that the concern expressed earlier about moving reproductive choice from the private to the public arena is at its most acute.

3. Genetics and Assisted Reproduction

Perhaps unsurprisingly, the combination of two revolutions—the genetic and the reproductive—poses additional and novel challenges. Although some dilemmas will be similar to those which can arise in the course of a pregnancy established in the traditional way, there are issues which are unique to assisted reproductive—more specifically to the technique of *in vitro* fertilisation (IVF).

When eggs are fertilised outside of a woman's body (as in IVF) the resulting embryo (or pre-embryo) may also be the subject of screening—in this case pre-implantation rather than pre-natal. This raises a number of special matters worthy of consideration. First, it must be said that if the status of the embryo or foetus *in utero* is the subject of debate and dispute, this debate is likely to be even more intense where what is under consideration is a fertilised egg with no prospect of life unless implantation is successful. To an extent, this may affect our perception of the moral worth of the entity and change the framework within which decisions about its future are taken.

For example, the fact that a pregnancy has not been established may make the decision to discard the pre-embryo at this stage seemingly less worthy of moral concern or condemnation. Potential parents, and their doctors, may be less troubled by the decision not to establish a pregnancy than they would by terminating one. In avoiding abortion, it may be argued that pre-implantation diagnosis is ethically

[41]Kevles (1985), p. 300.

[42]Cf. Fletcher and Wertz (1991), p. 103: 'the completion of the human genome project will provide a basis for acting on a moral obligation for *future* generations, a claim that has appeared weak in the past. A generation *with* such knowledge who neglected to use it in reproduction could hardly be said to respect the requirements of intergenerational justice.'

[43]Cf. UNESCO Declaration, Paris 1997, Declaration of Responsibilities of the Present Generation Towards Future Generations.

to be preferred—where possible—to pre-natal identification of a problem. However as always, there is a potential down side here too.

Concern has already been expressed about how to achieve clarity about the kind and severity of problem which may trigger a lawful abortion on so-called foetal grounds, and although abortion law may be silent on this, it is a decision not taken lightly by women, their partners or their doctors. However, where diagnosis is reached pre-implantation, no pregnancy has been established and no abortion laws need to be satisfied. Given that the pre-embryo is less likely to be the object of an emotional bond than is the embryo already implanted, the temptation may exist to deal with the pre-embryo in a manner less careful—perhaps even more cavalier—than its implanted equivalent.

If this were to be the case, then it is also plausible that pre-implantation diagnosis might be used to look for conditions less severe than those which are sought in the course of a pregnancy. If the pre-embryo is perceived as being of lesser moral standing than one which has been implanted, then arguably it would be less harrowing to discard embryos with trivial or minor problems. In addition, it might seem easier to reach a disposal decision about pre-embryos which carry genes for late-onset conditions. Even if the implanted embryo is viewed as a potential person, the unimplanted embryo is likely to be seen as having less potential and therefore less moral worth.

These issues, of course, may in reality be incapable of resolution. The decision of one person will not be the decision of another, and in societies (and legal systems) which claim to hold libertarian values, it may be that they too should be the result of private rather than public scrutiny. However, the fact that one thrust of this article is that reproductive choice should be a private matter does not conflict with the suggestion that matters of this sort also demand sensitive and sophisticated analysis—by those who are faced with the decisions. The ethical dilemmas posed in the circumstances are similar, but even more complicated, when the entity exists outside of the womb.

Further, screening out of certain conditions in an established pregnancy may already lead those alive with the same condition to feel marginalised, even where they accept that the abortion decision is not an easy one. However, the easier the decision, as arguably is the case with the pre-embryo, the more devastating the potential impact on these individuals. The effect on the well-being and the moral tone of communities cannot be underestimated when decisions which are essentially about the quality of life can be taken, in some cases at least, without the full weight of the ethics involved being experienced and addressed. Naturally, this will not always be the case—many people will feel as reluctant to lose the pre-embryo as they would be to lose an established pregnancy, but for some the decision is easier. The drive in assisted reproductive clinics to increase the availability of pre-implantation diagnosis and to extend the range of conditions for which it may be carried out, makes this an area worthy of urgent moral scrutiny.

4. Lifting the Veil on the Future

To an extent, all of the problems described above may vary with time. As has already been said, our response to progress may depend not just on the sophistication of the intellectual processes which lead to our conclusions, but also—and much more pragmatically—on when such decisions are offered and taken. For the moment, genetics has a vast array of diagnostic skills but few therapeutic ones.[44] The shortfall between diagnosis and therapy has implications which go far beyond the merely practical. The consequences also inform the ethics of decision-making. For example, when genetic disorders can be both identified *and* treated, then the continuation or establishment of a wanted, but compromised, pregnancy will be a simpler matter. Equally, it will not be tainted with allegations of reproductive or other irresponsibility.

When genetic manipulation can be used to correct the problems in the pre-embryo, without affecting the likely success of implantation, then only good can come of this. Agonising decisions for those who have been through a great deal of physical and emotional stress in order to circumvent their infertility will be obviated.

5. Conclusion

At the beginning of this article, it was contended that before we do what we can do, it would be wise to reflect on the implications of progress and its application. There are many examples of why this is a valuable question to ask, but, as has been suggested, reproduction is one area in which the pros and cons of genetic information are at their most clear. Not only is the new genetics opening the door to the possibility of reproductive choice becoming yet again an interest in the public sphere, but what is already feasible in terms of the use of genetic information can be shown to present complex ethical and personal dilemmas.

The aim of this discussion is not to take sides in private decision-making, but rather to expose the tension which exists between the 'good' and the 'bad' that genetics can offer. In addition, it seeks to expose and explore some of the controversies that can result from the use of genetic information. The consequences of the new genetics will be nothing less than to change the face of medicine forever. Wilkie, for example, claims that 'the new genetical anatomy will transform medicine and mitigate suffering in the twenty-first century'.[45] In addition, genetic knowledge will provide individuals with an increased range of options and with a different sense of self. Brenner suggests that '[w]hat is most important about the enterprise [the Human Genome Project] is the scientific knowledge that it will

[44]Cf. Freidmann (1990), p. 411: 'there remains a serious gap between disease characterization and treatment'.

[45]Wilkie (1993), p. 1.

generate, and the insights it will give us into our structure, function and origins'.[46] Relationships within the community will be modified, since genes are shared;[47] fundamental aspects of the doctor/patient relationship will require reappraisal, and values will need to be iterated clearly and consistently.[48] Failure to recognise the dramatic effect of genetics on our present and our future may lead to the insidious absorption of objectionable attitudes and dubious practices.

Only serious and thoughtful consideration of these matters will permit societies to reap the beneficial harvest of genetic progress without suffocating on the ethical chaff that comes with it. There is no obligation—ethical, legal or otherwise—to find out everything, or to use what has been discovered. And certainly, there may be reasons to pause for reflection before walking too far down certain paths.

One most obvious such path which appears to be generating considerable concern is, of course, the possibility of reproductive cloning. The furore of public and legislative debate surrounding this topic makes it one worth mentioning before finally concluding this chapter. In the aftermath of the cloning of Dolly, therapeutic cloning was outlawed in federal institutions in the US,[49] the UK Government made it clear that this is banned in Britain,[50] and the Council of Europe confirmed its ban.[51] In drawing the distinction between reproductive and therapeutic cloning the joint report from the Human Genetics Advisory Commission and the Human Fertilisation and Embryology Authority[52] reaffirmed both their own, and what they held to be the public's, antipathy to the cloning of people. Indeed, it would appear that most are opposed to such cloning, at least based on the 'natural' equivalent which occurred in the case of the Nash family who recently chose to have a child who was both of a particular sex and also most likely to be able to have a DNA suitable for saving the life of his sister.

For this author, the debate on reproductive cloning has generated more heat than light. Arguably, there are few robust arguments against human cloning, beyond the inevitable slippery slope argument, which, as Dworkin has pointed out, is fundamentally weak.[53] However, it is clear that this view is unfashionable. What is also certain is that we have already started down this path in the United Kingdom, with

[46]Brenner (1990), p. 6.

[47]Danish Council of Ethics (1993), p. 57: 'One of the peculiar aspects of genetic screening is that the things being screened for—gene and chromosome changes—can be passed down from one generation to the next . . . Therefore, genetic screening can very easily affect relatives of the people in question, genes being shared with relatives.'

[48]For discussion see McLean (1996), p. 41.

[49]Cf. National Bioethics Advisory Commission (1997); *Human Genome News* 8 (3 and 4), January–June 1997.

[50]Government Response to the Fifth Report of the House of Commons Select Committee on Science and Technology 1996–97 Session (Cm 3815).

[51]The Council of Europe in 1995 added an additional protocol to the Convention on Biomedicine prohibiting 'any intervention seeking to create a human being genetically identical to another human being, whether living or dead'. See also UNESCO Declaration on the Human Genome and Human Rights, Article 11.

[52]Cloning Issues in Reproduction, Science and Medicine, HGAC and HFEA, December 1998.

[53]Dworkin (1994).

the recent endorsement of stem cell research, involving embryo cloning.[54] Although this is endorsed for 'therapeutic' purposes only, it is one tentative step towards developing the technique which could ultimately result in reproductive cloning, unless, of course, strong arguments, rather than knee-jerk responses or intuitive distaste, can be provided which point inexorably to the 'wrongness' of reproductive cloning.

Whether the ultimate assessment of the genetics revolution will clothe it in the black robes of the wicked witch or the sparkling tiara of the good fairy will depend on the extent to which societies and individuals can show themselves capable of a mature and measured response to the new genetics. The genie is out of the bottle—there is no turning back. The challenge is to carve a way forward which combines respect for dignity and diversity with concern for individual freedom. That education in its widest sense may help to shape this mature response is unquestionable. That scientists must carry some responsibility for this is also clear.[55] Alone, however, these will only provide the backdrop against which communities can reach decisions for the future of genetics and genetic research. Underpinning the way forward there will need to be a clear commitment to articulated and robust principles; a clear set of values, consistently applied, which both accommodate and transcend science. These have yet to be formulated, but in this author's opinion, they have seldom been more urgently required.

References

Annas, G. (1986) 'Pregnant Women as Fetal Containers', *Hastings Center Report* **16(6),** 13–14.

Annas, G. (1988) 'She's Going to Die: The Case of Angela C', *Hastings Center Report* **18(1),** 23–25.

Brenner, S. (1990) 'The Human Genome: The Nature of the Enterprise', in *Human Genetic Information: Science, Law and Ethics*, Ciba Foundation Symposium 149, (Chichester: John Wiley and Sons), pp. 6–12.

British Medical Association (BMA) (1992) *Our Genetic Future: The Science and Ethics of Genetic Technology* (Oxford: Oxford University Press).

British Medical Association (BMA) (1998) *Human Genetics: Choice and Responsibility* (Oxford: Oxford University Press).

Danish Council of Ethics (1993) *Ethics and Mapping of the Human Genome* (Copenhagen: Danish Council of Ethics).

Davies, K. and Williamson, B. (1993) 'Gene Therapy Begins', *British Medical Journal* **306,** 1625–1626.

Dworkin, R. (1994) 'When is it Right to Die?' *New York Times* (May 5).

Fletcher, J. C. and Wertz, D. C. (1991) 'An International Code of Ethics in Medical Genetics Before the Human Genome is Mapped', in Z. Bankowski and A. Capron (eds), *Genetics, Ethics and Human Values: Human Genome Mapping, Genetic Screening and Therapy*, xxiv CIOMS Round Table Conference.

[54] *The Times*, 24 January 2001.

[55] Cf. House of Commons Committee on Science and Technology, Third Report, *Human Genetics: The Science and Its Consequences*, 41–I (London: HMSO), at paras 255–6.

Freidmann, T. (1990) 'Opinion: The Human Genome Project: Some Implications of Extensive "Reverse Genetic" Medicine', *American Journal of Human Genetics* **46,** 407–414.

Gregg, R. (1993) '"Choice" as a Double-Edged Sword: Information, Guilt and Mother-Blaming in a High-Tech Age', *Women and Health* **20(3),** 53–73.

ten Have, H. (1993) 'Physician's Priorities—Patients' Expectations', in Z. Szawarski and D. Evans (eds), *Solidarity, Justice and Health Care Priorities* (Linkoping, LCC: Linkoping University), pp. 42–52.

Hubbard, R. and Wald, E. (1993) *Exploding the Gene Myth* (Boston: Beacon Press).

Human Genome 1991–92 Program Report (1992) (Washington, DC: United States Department of Energy, Office of Energy Research Office of Environmental Research).

Johnsen, D. (1986) 'The Creation of Fetal Rights: Conflicts with Women's Constitutional Rights to Liberty, Privacy and Equal Protection', *Yale Law Journal* **95,** 599–625.

Kevles, D. (1985) *In the Name of Eugenics: Genetics and the Uses of Human Heredity* (Harmondsworth: Penguin).

Kolder, V. *et al.* (1987) 'Court-ordered Obstetrical Interventions', *New England Journal of Medicine* **316,** 1192–1196.

Lew, J. B. (1990) 'Terminally Ill and Pregnant: State Denial of a Woman's Right to Refuse a Caesarian Section', *Buffalo Law Review* **38,** 619–645.

McLean, S. A. M. (1986) 'The Right to Reproduce', in T. Campbell, D. Goldberg, S. McLean and T. Mullen (eds), *Human Rights: From Rhetoric to Reality* (Oxford: Basil Blackwell).

McLean, S. A. M. (1989) *A Patient's Right to Know: Information Disclosure, the Doctor and the Law* (Aldershot: Dartmouth).

McLean, S. A. M. (1996) 'The New Genetics: A Challenge to Clinical Values', *Proceedings of the Royal College of Physicians of Edinburgh* **26,** 41–50.

McLean, S. A. M. (1999) *Old Law, New Medicine* (London: Rivers Oram/Pandora).

McLean, S. A. M. and Petersen, K. (1996) 'Patient Status: The Foetus and the Pregnant Woman', *Australian Journal of Human Rights* **2(2),** 229–241.

Maddox, J. (1993) 'New Genetics Means No New Ethics', *Nature* **364(8 July),** 97.

Mattingley, S. S. (1992) 'The Maternal–Fetal Dyad: Exploring the Two-Patient Model', *Hastings Center Report* **22(1),** 13–18.

Meyers, D. (1971) *The Human Body and the Law* (Edinburgh: Edinburgh University Press).

National Bioethics Advisory Commission (1997) 'Cloning Human Beings' , Report and Recommendations, June 1997 (Rockville, MY: National Bioethics Advisory Commission).

Nuffield Council on Bioethics (1993) *Genetic Screening* (London: Nuffield Council on Bioethics).

Robertson, J. and Shulman, J. (1987) 'Pregnancy and Prenatal Harm to Offspring: The Case of Mothers with PKU', *Hastings Center Report* **17(4),** 23–33.

Royal College of Physicians of London (1991) *Ethical Issues in Clinical Genetics* (London: Royal College of Physicians of London).

Ruddick, W. and Wilcox, W. (1982) 'Operating on the Fetus', *Hastings Center Report* **12,** 10–14.

Schmidtke, J. (1992) 'Who Owns the Human Genome? Ethical and Legal Aspects', *Journal of Pharmacy and Pharmacology* **44(Suppl. 1),** 205–210.

Stone, D. and Stewart, S. (eds) (1994) *Towards a Screening Strategy for Scotland* (Glasgow: Scottish Forum for Public Health Medicine).

Swartz, M. (1992) 'Pregnant Women vs Fetus: A Dilemma for Hospital Ethics Committees', *Cambridge Quarterly of Healthcare Ethics* **1,** 51–62.

Whittaker, L. A. (1992) 'The Implications of the Human Genome Project for Family Practice', *Journal of Family Practice* **35(3),** 294–301.

Wilkie, T. (1993) *Perilous Knowledge: The Human Genome Project and Its Implications* (London: Faber and Faber).

[5]

Procreative Liberty in the Era of Genomics

John A. Robertson[†]

I. INTRODUCTION: ETHICAL AND POLICY CHALLENGES OF THE GENOME

Twentieth century biology began with the rediscovery of Gregor Mendel's work on peas and ended with the sequencing of the human genome. In between came Thomas Morgan's studies of fruit flies, the grand synthesis between genetics and evolutionary biology in the 1930s, and Watson and Crick's publication in 1953 of the double helix structure of deoxyribose nucleic acid ("DNA"), the substance in the nucleus of cells, which carries the genetic code of all eukaryotic life.[1]

The genetics of the second half of the century focused on learning how DNA coded for proteins, how to splice, clone, and recombine pieces of DNA, and how genetic mutations caused disease. In the late 1980s, a project to identify the actual sequence of all 3.2 billion base pairs of the human genome began.[2] In June 1999, President Clinton and Prime Minister Blair announced that a working draft of the human genome was complete, with the final completed draft to 99.9% accuracy expected in May 2003, fifty years after the publication of Watson and Crick's landmark paper.[3]

This remarkable achievement now provides scientists with the tools for understanding the molecular details of how living cells function and evolve, and thus the means for diagnosing, treating, and preventing many diseases. It will also lead to applications in reproduction, in personal and kinship identity, and possibly in social control. Increasing "geneticization" of medical, reproductive, and social spheres of life will bring many benefits, but may also lead to public and private

---

† Vinson & Elkins Chair, University of Texas School of Law. The author is grateful to participants in the Yale Legal Theory Workshop, to Jane Cohen, Rebecca Dresser, and Richard Markovits for helpful comments on earlier drafts, and to the National Human Genome Research Institute for support (#RO3 HG002509-02).

1 James Watson & Francis Crick, *A Structure for Deoxyribose Nucleic Acid*, 171 NATURE 737 (1953). For the best history of these events, see HORACE FREELAND JUDSON, THE EIGHTH DAY OF CREATION (1996). *See also* KEVIN DAVIES, CRACKING THE GENOME: INSIDE THE RACE TO UNLOCK HUMAN DNA (2001).

2 Nicholas Wade, *Reading the Book of Life: The Overview; Genetic Code of Human Life Is Cracked by Scientists*, N.Y. TIMES, June 27, 2000, at A1.

3 Nicholas Wade, *Scientists Complete Rough Draft of Human Genome*, N.Y. TIMES, June 27, 2000, at A1.

misuse, and new forms of power over individuals.[4] A special threat is genetic reductionism and its effect on how humans view themselves and their place in the universe.[5]

Coming to grips with the social implications of human genomic knowledge presents a series of diverse, but related, challenges about how to make use of genetic information in human affairs in ethically, legally, and socially acceptable ways. One major group of issues concerns ownership and control rights in the genome. Researchers need access to human DNA and patient medical records to identify genes, and to develop drugs and treatments based on them. Intellectual property rights in genes or gene products may also be necessary to spur investment in genomic research. At the same time, the privacy rights of individuals in their bodily tissue, DNA, and medical records demand strong protection.

A different set of issues arises from the many potential medical uses of genomic information. Genomics will play a major role in understanding the mechanics of disease, and in designing drugs and treatments to prevent and treat disease. Screening individuals or populations for genetic susceptibility or late-onset conditions, so that prevention may occur, will become much more common. Pharmacogenomics—genomic factors influencing drug metabolism[6]—may enable physicians to prescribe drugs tailored to a patient's genotype.[7] But unraveling the body's genetic secrets and turning them into effective therapies poses a major scientific challenge. As Dr. Francis Collins, the director of the National Human Genome Research Institute, has put it, genomics "will occupy science and medicine for the next 50 or 100 years."[8]

A third set of issues—the use of genomic knowledge in reproduction—shows how socially and morally complicated genomic applications can be. Because genes are inherited systems of information passed on to progeny,[9] genomic knowledge increases the ability to predict or even control the genes of offspring. Persons planning reproduction might want to know something about their genetic makeup or that of their embryos or fetuses before they conceive, bear, or give birth to offspring. While much of the resulting control will operate by excluding undesirable genomes, at some point, attempts to rewrite or engineer sections of the genetic code of prospective offspring may also occur.

---

4 The term "geneticization," coined pejoratively by Abby Lippman, is meant as a criticism or warning of the grave risks that come with increased reliance on genetics in human activities. Abby Lippman, *Prenatal Genetic Testing and Screening: Constructing Needs and Reinforcing Inequities*, 17 AM. J.L. & MED. 15, 18-19 (1991). But "geneticization" will not be sought unless it also brings benefits.

5 Alex Mauron, *Is the Genome the Secular Equivalent of the Soul?*, 291 SCIENCE 831 (2001); GREGORY STOCK, REDESIGNING HUMANS: OUR INEVITABLE GENETIC FUTURE (2002); ROGER GOSDEN, DESIGNING BABIES: THE BRAVE NEW WORLD OF REPRODUCTIVE TECHNOLOGY (1999).

6 "Pharmacogenomics is the study of how an individual's genetic inheritance affects the body's response to drugs. The term comes from the words pharmacology and genomics, and is thus the intersection of pharmaceuticals and genetics." U.S. DEP'T. OF ENERGY, HUMAN GENOME PROJECT INFORMATION, PHARMACOGENOMICS, *at* http://www.ornl.gov/TechResources/Human_Genome/medicine/pharma.html (last modified Oct. 29, 2003).

7 David W. Feigal & Steven I. Gutman, *Drug Development, Regulation, and Genetically Guided Therapy*, *in* PHARMACOGENOMICS: SOCIAL, ETHICAL, & CLINICAL DIMENSIONS 99, 100-01 (Mark A. Rothstein ed., 2003).

8 Francis S. Collins, *Shattuck Lecture—Medical and Societal Consequences of the Human Genome Project*, 341 NEW ENG. J. MED. 28 (1999).

9 PAT SPALLONE, GENERATION GAMES: GENETIC ENGINEERING AND THE FUTURE FOR OUR LIVES 28-29 (1992).

Although each set of genomic issues involves some application of genetic knowledge to human activities, each area has its own particular set of normative challenges and conflicts, and is thus best considered independently. This article will focus on the ethical, legal, and social challenges presented by the use of genetics in reproduction in the genomics era. Those uses raise morally complex and politically charged issues, where slogans and shibboleths often replace the careful analysis needed to resolve them.

Part II of this Article describes the controversies that arise with the use of genetics in reproduction and three approaches to resolving them. Part III explores the concept and meaning of procreative liberty, explains why it is valued, and describes its constitutional status. Part IV then applies a procreative liberty analysis to four key areas of debate and controversy over the use of reprogenetic techniques in reproduction. Section III.A. addresses the use of genetic knowledge in screening of prospective children for health reasons. Section IV.B. looks at non-medical selection, with a focus on sexual orientation and gender. Moving then to positive techniques of genetic choice, Section IV.C. discusses the case for reproductive cloning. Section IV.D then addresses positive alteration of embryo genomes for both medical and non-medical purposes. Finally, Part V discusses the problems of making policy in this area.

II. TECHNOLOGY AND REPRODUCTION: THE CONTROVERSY

Reproductive uses of genetic knowledge have been especially controversial for several reasons. First, they come with a bad pedigree. Attempts at the beginning of the twentieth century to improve the gene pool led to a repressive system of involuntary eugenic sterilization in thirty American states.[10] These laws provided a role model for Hitler's eugenic sterilization program,[11] which preceded Nazi efforts to annihilate Jews and other groups with disfavored genes. The shadow of eugenics hangs over all attempts, whether medical or non-medical, to select offspring genes.

Secondly, the use of genes to select offspring is quickly embroiled in social and political battles involving prenatal life, the status of women, and disability rights. Battles over the status of the fetus and abortion arise with the use of techniques to screen embryos or fetuses for genes of interest. Women assume greater burdens than men in the use of most reproductive technologies, thus implicating concerns about the equality and autonomy of women. Persons with disabilities are concerned about biases in genetic screening programs that disfavor persons with disabilities.

Third, the use of genetic information in reproduction inevitably raises questions about the permissibility of any selection of offspring traits, as well as about the particular grounds of selection. The 1996 birth of Dolly, the sheep cloned from the mammary gland of an adult ewe,[12] has upped the ante of concern by stimulating fears that people will attempt to engineer offspring traits, turning children into commodities or objects to serve parental needs. Much of the current concern reflects fears that technologies to silence genes or insert DNA into the genomes of

---

10 PHILLIP P. REILLY, THE SURGICAL SOLUTION: A HISTORY OF INVOLUNTARY STERILIZATION IN THE UNITED STATES 94 (1991). As the noted geneticist J.B.S. Haldane observed, ". . . many of the deeds done in America in the name of eugenics are about as much justified by science as were the proceedings of the Inquisition by the gospels." J.B.S. HALDANE, POSSIBLE WORLDS AND OTHER ESSAYS 144 (1930).

11 REILLY, *supra* note 10, at 106.

12 Gina Kolata, *With Cloning of Sheep, the Ethical Ground Shifts*, N.Y. TIMES, Feb. 24, 1997, at A1.

prospective children will become available and will pose serious threats to the well-being of children, society, and the very meaning of reproduction.

How then are we to reconcile the conflicts between reproductive choices and respect for prenatal life, offspring, families, women, other groups, and societal values that arise in using genetic knowledge in reproduction? A central dilemma is that accepting any instance of genetic selection in principle implies accepting most other instances of selection as well. But some uses seem much more questionable and less beneficial than the one initially accepted. Can acceptable lines be drawn, or is it better, as some would argue, to permit little, if any, genetic selection to occur?

To draw sound lines one needs a realistic sense of what those techniques involve, how they might help people in realizing their reproductive plans, and how they might harm them, their offspring, or society. But answers to those questions will be heavily influenced by more basic attitudes or normative stances that one takes toward the use of technology in reproduction. One's understanding of the meaning and significance of reproduction, parenting, the status of offspring, and a variety of other interests will be a key determinant in resolving these issues.

Three different stances (*strict traditionalism*, *modern traditionalism*, and *radical liberty*) have vied for recognition in ethical, legal, and policy discourse about these issues. I describe each of them, and the reasons why this article uses the *modern traditionalist* perspective to evaluate the many uses of genetic knowledge in reproduction.

A. STRICT TRADITIONALISM.

A *strict traditionalist* holds that reproduction is a gift from God, resulting from the loving intimacy of two persons.[13] They receive the gift of an embryo, fetus, and then child who is to be unconditionally cherished for its own sake. This view would condemn most uses of technology to control or influence the characteristics of offspring because parental selection necessarily conflicts with the idea of "unconditional gift" and suggests that the child is a made or chosen "product."

The leading contemporary articulator of this tradition is Leon Kass, Chair of the President's Council on Bioethics, who has expressed that view in articles and books since the 1970s.[14] Because his views underpin the President's Bioethics Council's 2002 report, *Human Cloning and Human Dignity*,[15] I will take that report as representative of the *strict traditionalist* position.

In referring to children born of technological assistance, the report notes:

> we do not, in normal procreation, command their conception, control their makeup, or rule over their development and birth. They are, in an important sense, 'given' to us. Though they are *our* children, they are not our *property*. . . . Though we may seek to have them for our own self-fulfillment, they exist also and especially for their own sakes. Though we seek to educate them, they are not like our other projects, determined strictly according to our plans and serving only our desires.[16]

---

13 PRESIDENT'S COUNCIL ON BIOETHICS, HUMAN CLONING AND HUMAN DIGNITY 110-11 (2002) [hereinafter CLONING REPORT].

14 Leon R. Kass, *"Making Babies" Revisited*, *in* THE PUBLIC INTEREST 32 (1979); Leon R. Kass, *Making Babies–the New Biology and the "Old" Morality*, *in* THE PUBLIC INTEREST 18 (1972).

15 CLONING REPORT, *supra* note 13, at XIII.

16 *Id.* at 9.

The report goes on to note that:

> If these observations are correct, certain things follow regarding the attitudes we should have toward our children. We treat them rightly when we treat them as gifts rather than as products, and when we treat them as independent beings whom we are duty bound to protect and nurture rather than as extensions of ourselves subject only to our wills and whims.[17]

An important implication of the view that offspring are "gifts" and not "products" is that humans should have no say in the outcome or makeup of a child. They must simply accept the "gift" that is provided and make no attempt to change, direct, control, design, or exclude it. Reproductive cloning is the sin extraordinaire because "cloned children would thus be the first human beings whose entire genetic makeup is selected in advance."[18] Although the report does not address other modes of selection, the gift ethic would seem to condemn all other forms of prenatal selection as well, whether positive or negative, medical or non-medical.[19] All such uses would make children "like other human products, brought into being in accordance with some pre-selected genetic pattern or design, and therefore in some sense 'made to order' by their producers or progenitors."[20] Acceptance of such actions would:

> provide at best only a partial understanding of the meanings and entailments of human procreation and child-rearing . . . [and undermine] the unconditional acceptance of one's offspring that is so central to parenthood.[21]
>
> ...
>
> In short, the right to decide '*whether* to bear or beget a child' does not include a right have a child *by whatever means*. Nor can this right be said to imply a corollary—the right to decide what kind of child one is going to have.[22]

---

17 *Id.*

18 *Id.* at 117.

19 The report's mechanistic view of reproductive technology comes out most clearly in its discussion of cloning. It describes cloning as "control of the entire genotype and the production of children to selected specifications." *Id.* at 118. Cloning is different from *in vitro* fertilization ("IVF") because "the process begins with a very specific final product in mind and would be tailored to produce that product." *Id.* at 119. It goes on to note that "the resulting children would be products of a designed manufacturing process, products over whom we might think it proper to exercise 'quality control.'" *Id.* Using such techniques would teach us "to receive the next generation less with gratitude and surprise than with control and mastery." *Id.* One possible result would be "the industrialization and commercialization of human reproduction," a clearly dehumanizing force. *Id.* Much of this article is a refutation of those claims.

20 The report never addresses the extent to which parents through education, religion, medical treatments, and other interventions shape the child. This is deemed acceptable, and within parental freedom while any "shaping" of the child beforehand is inconsistent with the "dignity" of human reproduction. For further discussion of this point, see *infra* notes 184-85 and accompanying text.

21 CLONING REPORT, *supra* note 13, at 92.

22 *Id.* at 93. It is unclear, of course, whether the report is actually as extreme as it appears to be, since one could be against "any" method without being against "all" methods. The sentences following that quote are also open to a more narrow understanding, but the report gives no indication or basis for choosing a narrow or broad view. Indeed, it suggests that any use of selection risks violating the gift perspective, which it asserts is essential to human reproduction.

The main problem with this view as a guide to public policy is its roots in a religiously based or metaphysical view of how reproduction should occur and a breadth that would apparently condemn nearly all forms of technological assistance in reproduction. As a religiously based view with which many persons would disagree, it has no claim to special respect in a liberal secular democracy, where individuals define within a broad range their own sense of the good. Its condemnation of "unnatural" ways of reproduction is not required to protect the well-being of offspring, because in nearly all cases resulting offspring appear to benefit from the technologies used.[23] The fact that *techne*[24] is used should not itself disqualify techniques that help parents fulfill goals of having healthy children to rear. Disagreements about the ethics of particular cases do not justify having the government impose one "correct" view of how reproduction should occur in all cases.

In addition, this view conflicts with the natural instinct of parents to have healthy children for their own sake and that of the children. In appealing to the natural, the *strict traditionalist* overlooks the most natural fact of all—that people have strong interests in passing on their genes and in having healthy offspring who will do the same. Unless Kassians are adopting the untenable view that any interference with nature is wrong, they must recognize that *techne* can help humans deal with the limits which nature has placed on them, as it does with other limits.[25] Within broad bounds, using technology to accomplish that task is no more objectionable than using technology after birth to enable survival to continue. There are limits, such as harm to children or others, but those harms must be serious when a substantial reproductive interest is aided by a technique.

B. Radical Liberty

The *radical liberty* view is the polar opposite to *strict traditionalism*. It holds that individuals are free to use any reproductive technique they wish for whatever reason, and no limits can appropriately be placed on what they do before the birth of a child.[26] Individuals are thus free to select, screen, alter, engineer, or clone offspring as they choose.[27] They are the best judges of what is good for them, including what children they have.

The justification of this position appears to be general libertarian principles of freedom without government interference, though one strand of the position draws on utopian notions of humans perfecting themselves by engineering their very nature. Libertarianism in reproduction means that a person has the right to select for specific genes or do anything she chooses in the course of reproduction. For *radical*

---

23 The question of what results count as harm to offspring raises a complicated philosophical problem that is discussed at greater length in the Appendix.

24 "Techne" means "[t]he principles or methods employed in making something." Webster's Third New International Dictionary 2348 (Philip Babcock Gove ed., 3d ed. 1986).

25 The view that any interference with nature is unnatural is ultimately untenable because it would deny use of medicine, heating, transportation, agriculture, etc. They all interfere with nature in some respect. At the same time, they may all be viewed as part of nature, because nature selected for the cognitive capacities that allow human control and manipulation of nature to occur.

26 *See, e.g.*, John B. Attanasio, *The Constitutionality of Regulating Human Genetic Engineering: Where Procreative Liberty and Equal Opportunity Collide*, 53 U. CHI. L. REV. 1274, 1285-86 (1986).

27 *See* JOHN A. ROBERTSON, CHILDREN OF CHOICE: FREEDOM AND THE NEW REPRODUCTIVE TECHNOLOGIES (1994).

libertarians, the technical ability to "rewrite" or "edit" the genome of offspring is cause for huzzas not homilies.

Radical liberty proponents are probably few in number. The Raelian sect, Randolph Wicker, and others promoting reproductive cloning have made such a claim.[28] Dr. Brigitte Boisselier claimed at the National Academy of Sciences hearing on cloning that she "has the right to have any kind of child that she wants."[29] Gregory Stock, and to some extent, Lee Silver, predict that parents will want to use any technique that will enhance the well-being of offspring, and that such will come to be accepted because of the strong parental interest and commitment to their offspring.[30] But their views are descriptive, and do not necessarily present arguments for why unlimited choice is desirable.

Although held by relatively few, the *radical liberty* view hovers in the background and casts a shadow over many official, scholarly, and popular accounts of reproductive issues. *Strict traditionalists* often assume that anyone who does not share their view is in favor of *radical liberty* (rather than the *modern traditionalist* view, which does recognize some constraints).[31] Many popular and policy discussions about reproductive technology often assume that if cloning, genetic enhancement, and other selection or alteration techniques were safe, many people would seek them as a matter of right, thus radically altering reproduction and relations with children to the detriment of all.[32]

The problem with the *radical liberty* view is that it is too extreme in its espousal of personal freedom. Just as *strict traditionalism* admits no nuance in assessing possible benefits in reproduction technology, *radical libertarians* see no reason ever to place limits on any choice related to conception, screening, alteration, or the production of children.[33] But such an extreme view denies the validity of the harm principle[34]—personal liberty is justifiably limited when it causes direct harm to others—applied to reproductive choices just as it does to other exercises of autonomy. Even assuming that reproduction deserves special protection, however,

---

28 *Raelian Leader Says Cloning First Step to Immortality, at* http://www.cnn.com/2002/HEALTH/12/27/human.cloning (Dec. 28, 2002).

29 NAT'L ACAD. OF SCIENCES, *Scientific and Medical Aspects of Human Reproductive Cloning, in* THE IMPACT OF FEDERAL POLICY ON REALIZING THE POTENTIAL OF STEM CELL RESEARCH: INFORMATIONAL HEARING OF THE SENATE HEALTH AND HUMAN SERVICES COMMITTEE, 35 (2002).

30 STOCK, *supra* note 5; LEE M. SILVER, REMAKING EDEN: CLONING AND BEYOND IN A BRAVE NEW WORLD (1997).

31 Other bioethicists also often understand claims about procreative liberty to be those of the *radical libertarian*, rather than those of the more nuanced *modern traditionalist* view. *See* Thomas H. Murray, *What Are Families For?: Getting to an Ethics of Reproductive Technology*, 32 HASTINGS CENTER REP. 41 (2002); Dena Towner & Roberta Springer Loewy, *Ethics of Preimplantation Diagnosis for a Woman Destined to Develop Early-Onset Alzheimer Disease*, 287 JAMA 1038, 1038-40 (2002) (challenging perceived discursive prominence of radical libertarian viewpoints and advocating for a nuanced approach).

32 *See* Judith F. Daar, *The Prospect of Human Cloning: Improving Nature or Dooming the Species?*, 33 Seton Hall L. Rev. 511, 526-30 (2003). *See also* Skylar A. Sherwood, Note, *Don't Hate Me Because I'm Beautiful . . . and Intelligent . . . and Athletic: Constitutional Issues in Genetic Enhancement and the Appropriate Legal Analysis*, 11 HEALTH MATRIX 633, 638-46 (2001).

33 The opposing views are mirror images of each other. Curiously, each accepts a strongly reductionist or deterministic role for genes and human manipulation of genomes. One view finds this to be good, even utopian, while the other sees a dystopian evil.

34 JOHN STUART MILL, ON LIBERTY 30 (John Gray & G.W. Smith eds., 1991) (stating that the harm principle holds "[t]hat the only purpose for which power can be rightfully exercised over any member of a civilized community, against his will, is to prevent harm to others.").

the *radical liberty* view provides no way to distinguish which activities surrounding reproduction are truly reproductive and which ones are not. The best or ultimate judge of whether an activity is reproductively important and whether it violates the harm principle cannot be the actor being judged, as the *radical liberty* view asserts.

C. Modern Traditionalism

Modern traditionalism is midway between the other two positions. Although a heterogeneity of views parade under the *modern traditionalist* banner, that approach has much to offer as a methodology for dealing with genetics in reproduction. It holds that reproductive choice in a liberal, rights-based society is a basic freedom, including the use of genetic and reproductive technologies that are helpful in having healthy, biologically related offspring.[35] This view is *modern* in its acceptance of new technologies, but *traditional* in demanding that those techniques ordinarily serve traditional reproductive goals of having biologically related offspring to rear. Its acceptance of reproductive and genetic technologies, however, exists only insofar as they aid the task of successful reproduction, and do not directly harm offspring, families, women, society, or others.

As a result, many uses of reproductive technology will be protected, but not automatically. The connection with reproduction is key, as is the absence of direct harm to others. Some techniques will not be acceptable precisely because their connection to gene transmission and rational investment strategy in offspring will be lacking or unclear. In other cases, harm might result, though what counts as "harm" from reproduction is itself hotly contested.[36]

The problem for the *modern traditionalist* is to give a persuasive account of why some uses of reproductive technology are acceptable, but others are not. To do this she must provide a convincing method or set of criteria for determining which uses are "reproductive" and what counts as "harm." Her challenge is to show how her approach gives reasonable answers to the conflicts that genetic uses inevitably raise. Rather than appeal to Procrustean principles that neatly give answers for all cases, *modern traditionalism* adopts instead a pragmatic, context-specific approach that looks at how proposed techniques are likely actually to be used, and the problems, if any, which might then arise. Although less definite, this approach is best suited for handling issues of reproductive technology in the era of genomics, as analysis of several genetic uses in reproduction will show.

III. WHAT IS PROCREATIVE LIBERTY?

The *modern traditionalist* view translates easily into the language of individual rights. Although not the only relevant perspective to take on these issues, a rights-based perspective focuses attention on key aspects of the individual and societal concerns at issue with these techniques. For example, it reminds us of the importance of the more fundamental decision of whether to assign decisions to use reproductive technologies to individuals and their professional advisors or to

---

35 *See* National Bioethics Advisory Commission, Cloning Human Beings: Report and Recommendations of the National Bioethics Advisory Commission 95 (1997) [hereinafter NBAC Report]; Maura A. Ryan, *Cloning, Genetic Engineering, and the Limits of Procreative Liberty*, 32 Val. U. L. Rev. 753, 754 (1998).

36 For example, there is a question whether harm can occur to offspring from the use of technologies that make their birth possible. *See* discussion *infra* Appendix.

legislative majorities. It also focuses analysis on the context of likely use by assessing the reproductive interests that a disputed technology serves, and the severity and probability of the harm or objection that it generates. A coherent account of what procreative liberty is and why it is protected can provide a workable set of principles for a *modern traditionalist* to use in resolving the normative and policy conflicts that arise when genetic knowledge is used in reproductive decision-making.

Procreative liberty is best understood as a liberty or claim-right to decide whether or not to reproduce.[37] As such, it has two independently justified aspects: the liberty to avoid having offspring and the liberty to have offspring. Because each aspect has an independent justification, each may be conceived as a different right, connected by their common concern with reproduction.[38]

The liberty to avoid having offspring involves the freedom to act to avoid the birth of biologic (genetically related) offspring, such as avoiding intercourse, using contraceptives, refusing the transfer of embryos to the uterus, discarding embryos, terminating pregnancies, and being sterilized. In contrast, the liberty or freedom to have offspring involves the freedom to take steps or make choices that result in the birth of biologic offspring, such as having intercourse, providing gametes for artificial or *in vitro* conception, placing embryos in the uterus, preserving gametes or embryos for later use, and avoiding the use of contraception, abortion, or sterilization.

As with other liberties in a rights-based society, an actor is not obligated to exercise a particular liberty right. He or she may or may not choose to reproduce, or to use or not use genetic or reproductive technologies in making those decisions. An actor may have no need to use a technology or lack the means to do so; or he or she may reject uses of particular technologies for a wide range of personal reasons, including moral or ethical concerns about the effect of particular techniques on children, on society, or on deeply held personal values, including values of how reproduction should occur.[39] The technological imperative—that if something can be done, it will be done—is not nearly as powerful as often claimed. No one is

---

37 A person has a liberty-right if she would violate no moral duty by engaging in an action or omission. A claim-right adds the additional ingredient that other persons have a moral duty not to interfere with her exercise of liberty. *See* RICHARD FLATHMAN, THE PRACTICE OF RIGHTS 33-63 (1976). It is generally understood that the great personal importance of procreative liberty makes it a claim-right against state or private interference with its exercise. JOHN A. ROBERTSON, CHILDREN OF CHOICE: FREEDOM AND THE NEW REPRODUCTIVE TECHNOLOGIES 35-38 (1994). As will be discussed below, procreative liberty may also have constitutional status as a fundamental right. If so, procreative liberty would be presumptively protected against state action unless there were compelling reasons for restricting it.

38 Whether described as one or two rights, each is independently justified and "stands on its own bottom," to paraphrase Justice Harlan's comments about privacy in *Griswold v. Connecticut*, 381 U. S. 479, 499 (1965) (Harlan, J., concurring). Although each comes into play by foregoing or waiving the other, the reasons for protecting each aspect are separate—the one in avoiding the burdens of reproduction, the other in avoiding the burdens of not being able to reproduce.

39 Persons are thus free to reject genetic screening or modification, even if doing so leads to a child born with a congenital handicap. While physicians have ethical and legal duties to inform women of the availability of carrier and prenatal tests, no state has placed a legal duty on parents to be tested or to avoid the birth of such children. Such requirements would presumably be unconstitutional despite the fact that *Buck v. Bell* has not yet been officially overruled. *See* John A. Robertson, *Genetic Selection of Offspring Characteristics*, 76 B.U. L. REV. 421, 468–74 (1996).

obligated to reproduce or to use particular reproductive and genetic technologies in avoiding reproduction or in reproducing.[40]

Like most moral and legal rights in liberal society, procreative liberty is primarily a negative claim-right—a right against interference by the state or others with reproductive decisions—not a positive right to have the state provide resources or other persons provide the gametes, conception, gestation, or medical services necessary to have or not have offspring. Some persons, however, would argue that it should have positive status as well, with the state or public health system providing reproductive health services, including infertility treatment, genetic screening, or abortion.

As should be clear from this discussion, recognizing procreative liberty as a moral or legal right or important freedom does not mean that it is absolute, but rather that there is a strong presumption in its favor, with the burden on opponents to show that there is a good case for limiting it. Many critics, however, assume that claims of procreative liberty are claims of an inalienable or absolute right.[41] But a right can be inalienable—not transferable to others—without also being absolute. And no serious proponents of procreative liberty argue that it is absolute and can never be limited. Rather, the debate is (or should be) about whether particular exercises or classes of exercise of the right pose risks of such harm to others that they might justly be limited.[42]

An important set of related issues concerns the scope of procreative liberty—what activities related to avoiding or engaging in reproduction a coherent conception of procreative liberty includes. This can be determined only by assessing the role that those other activities play in avoiding or engaging in reproduction. Some activities seem so closely associated with, or essential to, reproductive decisions that they should be considered part of it and judged by the same standards. An example is a woman's need to acquire and then use genomic information about herself, her partner, her gametes, her embryos, or her fetus before deciding whether or not to reproduce. Because such information will often be determinative of whether a person or couple would or would not reproduce, freedom to acquire and use it would seem to be part of procreative liberty, unless its use posed substantial risks to others.

In contrast, other activities in and around reproduction might not be part of procreative liberty and thus not deserve the same protection (though some of them might deserve strong protection on other grounds). Thus, actions occurring in the course of reproduction, such as home-birthing, having the father present in the delivery room, using drugs during pregnancy, and the like are not part of procreative liberty *per se*; nor is adopting a child or rearing children not related by genetic kinship, because those activities arise only after reproduction has occurred and are not themselves determinative of whether reproduction will occur. As I argue below, some uses of reproductive and genetic technology, such as reproductive cloning when fertile and intentional diminishment of offspring characteristics, may also fall outside the protective canopy of reproductive liberty.

---

40 Thus, allowing one category of uses does not mean that every person in a similar reproductive situation will use that technology.

41 Towner & Loewy, *supra* note 31, at 1038-40.

42 Severe overpopulation might justify restrictions on the number of children, while severe underpopulation might make limiting access to contraceptives acceptable. However, much more would need to be known about societal circumstances and the efficacy of alternative measures to adequately assess the legitimacy of such restrictions.

Unsurprisingly, there may be intense debate about whether something is central or material to reproduction and thus properly regarded as part of, or an aspect of, procreative liberty, just as there is sharp debate about the seriousness and risk of resulting harms. Before reaching questions of harm, however, it is useful first to ask whether a particular use of reproductive technology, such as embryo screening for medical and non-medical traits, genetic engineering of prospective offspring, or reproductive cloning, is itself an exercise of procreative liberty. All such arguments, it seems, relate to how essential or material those activities are to the values that underlay the importance to individuals of their decision to avoid or engage in reproduction.[43] While people may disagree over the precise limits, the argument, if properly focused, should be about the closeness of the activity in question to the values that support freedom in reproductive decision-making and whether the effects on others of exercising that freedom justify limiting it.

As argued below, however, while a material connection to the reproductive decision is a necessary condition to qualify the choice as one of procreative freedom, it may not always be a sufficient one.[44] The connection may not be sufficient if the materiality to the individual is not in keeping with ordinary understandings of why having offspring is so important to individuals.[45] A certain degree of conformity with common understandings of why reproduction is important is thus necessary for inclusion in procreative liberty. But such conventionalism is limited to clarifying reproductive goals and not to determining the acceptability of means to those goals.[46] From this perspective, the use of new technologies to overcome infertility or to avoid the birth of children with disease are ways to reach the traditional goal of having healthy children and should not be rejected, as the conventionalism of *strict traditionalism* would, merely because the means are novel or new types of rearing relationships result.

At the same time, not all uses of new technologies should be acceptable just because, as the *radical libertarian* would argue, they are an instance of reproductive choice. If they are used to achieve goals not clearly grounded in our ordinary understandings of why reproduction matters to individuals, for example, they employed means that seemed unnecessary, such as reproductive cloning when fertile, or that did not advance the interests of offspring, such as intentional diminishment of capacities of otherwise healthy offspring, they would not serve the values that make having offspring of such key importance to persons.[47]

---

43 A more robust theory of procreative liberty might argue that any preference concerning how reproduction occurs or the characteristics of its results should be protected. In this article, I address only the situation of when use of a technique is material or essential to the reproductive choice and not a mere preference that is not itself determinative of whether a choice to reproduce or not would otherwise occur. For further discussion of this methodology, see Robertson, *supra* note 39, at 429-32.

44 *Id.* at 432-40.

45 Although reproduction is often sought for the experience of rearing, it is not essential in all cases that rearing also be sought or occur. The liberty-right not to reproduce protects the right not to have children and the rearing duties which they ordinarily entail.

46 Karen Lebacqz confuses my support for community conceptions of the importance of having offspring as support for communitarian views of the acceptability of means to achieve that goal. Nor, as she asserts, are community conceptions of when a reproductive activity reasonably advances or ill-serves those goals "symbolic." *See* Karen Lebacqz, *Choosing Our Children: The Uneasy Alliance of Law and Ethics in John Robertson's Thought*, *in* PRINCETON FORUM ON BIOETHICS (2003) (on file with AM. J.L. & MED.).

47 *See* discussion *infra* notes 149-52 and accompanying text, notes 189-95 and accompanying text, and Appendix.

Answering the question of what procreative liberty includes requires us to determine how centrally implicated are the underlying reasons for valuing reproductive choice with the technology under discussion. Rather than adopt *strict traditionalism* that rejects almost all selection technologies or *radical libertarianism* that rejects none, I adopt a *modern traditionalist* approach, which looks closely at the reasons why choice about reproduction is so important for individuals. The more closely an application of genetic or reproductive technology serves the basic reproductive project of haploid gene transmission—or its avoidance—and the rearing experiences that usually follow, the more likely it is to fall within a coherent conception of procreative liberty deserving of special protection. At a certain point, however, answers to questions about the scope or outer limits of procreative liberty will depend upon socially constitutive choices of whether reprogenetic procedures are viewed as plausible ways to help individuals and couples transmit genes to and rear a new generation.

A. WHY PROCREATIVE LIBERTY IS VALUED

Why should procreative liberty have moral or legal rights status? The answer might be so obvious that one wonders why the question is even asked. But asking the question will help us understand the interests and values that undergird the scope of procreative liberty and, by implication, help resolve conflicts that arise from its exercise. Quite simply, reproduction is an experience full of meaning and importance for the identity of an individual and her physical and social flourishing because it produces a new individual from her haploid chromosomes.[48] If undesired, reproduction imposes great physical burdens on women, and social and psychological burdens on both men and women. If desired and frustrated, one loses the "defence 'gainst Time's scythe" that "increase" or replication of one's haploid genome provides, as well as the physical and social experiences of gestation, childrearing, and parenting of one's offspring.[49] Those activities are highly valued because of their connection with reproduction and its role in human flourishing.

Good health in offspring is also greatly prized. Past cultures have sometimes exposed weaker or handicapped newborns to the elements, thus concentrating resources on those who are healthy. We serve some of the same interests by a strong commitment to the health of all children, such as elaborate neonatal intensive care units that go to great expense to save all newborns, and norms for treating all newborns no matter the cost or scope of their handicaps.[50] Even though parental behavior, and social and legal norms are strongly committed to the well-being of children once they are born, parents strongly prefer having healthy offspring and may use mate or gamete selection, and screening of fetuses and embryos to serve that goal.

It is not surprising that an interlocking set of laws, norms, and practices exist that support reproduction. Deeply engrained social attitudes and practices celebrate

---

48 In sexual reproduction, each procreator contributes a haploid chromosome that combine in humans to form 46 chromosomes.

49 William Shakespeare, *Sonnet 12*, *in* THE OXFORD SHAKESPEARE: THE COMPLETE WORKS OF WILLIAM SHAKESPEARE (W.J. Craig ed., 1922).

50 For example, see the controversy that erupted around the issuance of "Baby Doe Rules" that aimed to protect all newborns regardless of their handicaps. Bowen v. Am. Hosp. Ass'n, 476 U.S. 610 (1986); Frank I. Clark, *Withdrawal of Life-Support in the Newborn: Whose Baby Is It?*, 23 SW. U. L. REV. 1 (1993); Nancy Rhoden, *Treatment Dilemmas for Imperiled Newborns: Why Quality of Life Counts*, 58 S. CAL. L. REV. 1283 (1985).

the importance of family and children. Laws, ethical norms, and institutions protect and support human desires to have or avoid having offspring, and the rearing that follows. The deep psychological commitment one has to the well-being of one's offspring is reflected in the strong family and constitutional law protections for rearing rights and duties in biologic offspring, in special tort damages for loss of children and parents, in the law of rape, in the rise of an infertility industry, and in the wide acceptance of prenatal screening programs for the health of offspring. Many other social institutions and practices also support individual and social interests in producing healthy offspring who are fit to reproduce in turn. Strong protection of procreative liberty and family autonomy in rearing offspring is yet another way that social recognition of the importance of reproduction is shown.

Although the importance of reproduction for individuals and society is intuitively accepted, the search for a deeper or more ultimate explanation of its importance might turn to evolutionary biology and psychology. A biologic perspective on human behavior suggests that reproductive success is as important an issue for humans as it is for other organisms. Whether gene, organism, group, or species is the unit of selection, natural selection selects those entities that are best suited to reproduce in the environments in which they exist. Although genes encoding sexuality and sexual attractiveness have not yet been identified, it is likely that many aspects of sexual reproduction reflect physical and perhaps even behavioral tendencies for reproductive success selected at earlier stages of human development.[51]

Given the importance of culture and environment in shaping human behavior, one should be leery of attempting to explain all aspects of human reproductive decisions in evolutionary terms. Yet there are enough similarities between the reproductive challenges that humans and other organisms face to make further inquiry into the biologic basis of human reproduction worthwhile. Humans, like other sexually reproducing organisms, face specific challenges that differ for each sex.[52] Typically, each sex faces the challenge of finding healthy members of the opposite sex with whom to mate and produce progeny. Because females typically have larger and fewer gametes that require internal fertilization, they use different strategies than males for identifying good mates and controlling their reproductive capacity.[53] In either case, some selection of the gametes or reproductive partners may be necessary to maximize the chance of successful reproduction.[54] Similarly, each sex must solve the problem of adequate nurture and protection of offspring, so that they may reproduce in turn.

As a result, it should be no surprise that many human reproductive choices and practices reflect efforts to have healthy offspring to carry genes into future generations.[55] An evolutionary perspective on reproduction cannot itself define the limits or scope of procreative liberty. As the naturalistic fallacy teaches, no "ought"

---

51 RICHARD DAWKINS, THE SELFISH GENE (1976); GEORGE C. WILLIAMS, ADAPTATION AND NATURAL SELECTION: A CRITIQUE OF SOME CURRENT EVOLUTIONARY THOUGHT (1966).

52 JOHN TYLER BONNER, THE EVOLUTION OF CULTURE IN ANIMALS 77-92 (1980).

53 MARLENE ZUK, SEXUAL SELECTIONS: WHAT WE CAN AND CAN'T LEARN ABOUT SEX FROM ANIMALS (2002). Hence the importance of abortion and contraceptive rights for women, who face much greater physical burdens than men in reproduction. *See* SARAH BLAFFER HRDY, MOTHER NATURE: MATERNAL INSTINCTS AND HOW THEY SHAPE THE HUMAN SPECIES (2000).

54 *See* Robert L. Trivers, *Parental Investment and Sexual Selection*, *in* SEXUAL SELECTION AND THE DESCENT OF MAN (Bernard G. Campbell ed., 1972); ZUK, *supra* note 53.

55 Larry Arnhart, *Human Nature is Here to Stay*, NEW ATLANTIS (Summer 2003), *available at* http://www.thenewatlantis.com/archive/2/arnhart.htm.

follows logically or inexorably from any "is" about the world. But a biologic perspective helps explain why reproductive urges are so powerful and widely respected, and why so many secondary norms, practices, and institutions have grown up around them.

Ultimately, decisions about how to use or not use genomics in human reproduction will be determined, not by biologic necessity or evolutionary theory, but by how those uses fit into the fabric of rights and interests of individual and social choice and responsibility that particular societies recognize. Still, understanding how assisted reproductive and genetic technologies serve issues of reproductive fitness is relevant to the ethical, legal, and social debates that surround use of those techniques. The biological concept of reproductive fitness can help at an ultimate level explain what is intuitively felt and culturally protected, even though more proximate analyses are needed to resolve the ethical, legal, and social conflicts that use of reproductive technologies may pose. At the very least, an evolutionary perspective, if not directly supportive, makes comprehensible the *modern traditionalist* intuition that procreative liberty deserves respect because of the individual importance of having and rearing offspring in order to transmit genes to the next and later generations.

B. Is Procreative Liberty Constitutionally Protected?

Given the immense importance of reproduction to individuals and societies, it is not surprising that many cultures have elaborate rule systems for how and when reproduction should occur, and who may reproduce with whom. It is unclear whether those rules serve the reproductive fitness of a particular culture or group, as can be shown for the division of reproductive roles that exist in ant or bee colonies, or whether they simply serve the interests of those who have gained power and wish to preserve it.

In any event, in recent years much social and political conflict has arisen over reproductive behavior. In the United States, much of that conflict has focused on the freedom of women to avoid reproduction through contraception and abortion.[56] As technologies for contraception and abortion have improved, steady progress in expanding a woman's right to methods of avoiding reproduction has occurred both in the United States and in Europe.[57]

Recognition in the United States of a woman's right to avoid reproduction through contraception and abortion has occurred mainly through constitutional decisions by the U.S. Supreme Court. In a series of celebrated but still contested cases, starting with *Griswold v. Connecticut* (1965),[58] continuing through *Eisenstadt v. Baird* (1972)[59] and *Roe v. Wade* (1973),[60] and then in *Casey v. Planned Parenthood* (1992),[61] the Court established a Fourteenth Amendment fundamental liberty-right to avoid conception when having sex, and if pregnancy has occurred, the right to terminate the pregnancy up until viability.

---

56 *See* Bradley E. Cunningham, *Implications of FDA Approval of RU-486: Regulating Mifepristone Within the Bonds of the Constitution*, 90 KY. L.J. 229, 238-39 (2001-2002) (discussing the historical debate in the United States over contraception and abortion).

57 *See* Stephanie B. Goldberg, *The Second Woman Justice: Ruth Bader Ginsburg Talks Candidly About a Changing Society*, 79 A.B.A. J. 40, 40, 42 (1993).

58 381 U.S. 479 (1965).

59 405 U.S. 438 (1972).

60 410 U.S. 113 (1973).

61 505 U.S. 833 (1992).

The liberty interest in engaging in reproduction has received much less attention, no doubt due in part to the infrequent attempts by the state to limit coital reproduction. Although fornication and adultery laws existed in many states,[62] one of their main purposes was to keep reproduction within marriage.[63] In most cases, they have been repealed or are simply not enforced.[64] Within marriage, there have been few attempts to limit coital conception and reproduction. In *Buck v. Bell* (1927), the Supreme Court upheld a state law mandating sterilization of a mental defective, thus validating the eugenic sterilization laws then on the books in many states.[65] But in *Skinner v. Oklahoma* (1943), the Court recognized reproduction as one of the basic civil rights of man, which could not be removed by sterilization, at least if not done equally.[66] Although federal Courts of Appeals have upheld bans on reproduction by prisoners, many Supreme Court cases have discussed the fundamental right to marry and raise a family, which assumes that conceiving and having a child is a protected right.[67] Indeed, in *Bragdon v. Abbott*, the Court found that reproduction is "a major life activity" in holding that a person with HIV who was not able to reproduce without risking an infected child fell within the protection of the Americans with Disabilities Act.[68]

Neither the Supreme Court nor lower courts, however, have provided guidance on how far the explicit protection of decisions to avoid reproduction and the implicit protection of decisions to engage in coital reproduction takes us in resolving conflicts over assisted reproductive and genetic technologies. One could reasonably view the Court's decisions as having established a broad principle of negative reproductive freedom, both to avoid reproduction and to engage in it without state interference, at least until those who would restrict that freedom have shown that important interests would be harmed by the choice in question.

If so, use of a wide range of assisted reproductive and genetic technologies would fall within an individual's discretion. A person would then have a presumptive right not to transfer embryos or gametes, to selectively abort, and to abort to get tissue for transplant, as well as to have carrier, embryo, and fetal genetic screening to decide whether to conceive, transfer embryos, or continue a pregnancy. The use of noncoital means of conception, such as artificial insemination and *in vitro* fertilization ("IVF"), might also be protected, as would egg donation and gestational

---

62 *See* Richard Green, *Griswold's Legacy: Fornication and Adultery as Crimes*, 16 OHIO N.U. L. REV. 545 (1989).

63 *See* People v. Bright, 238 P. 71, 73 (Colo. 1925); Robert A. Brazener, Annotation, *Validity of Statute Making Adultery and Fornication Criminal Offenses*, 41 A.L.R. 3d 1338 (1972); Martin J. Siegel, *For Better or for Worse: Adultery, Crime & the Constitution*, 30 J. FAM. L. 45 (1991/1992).

64 Lawrence v. Texas, 123 S. Ct. 2472, 2483 (2003).

65 274 U.S. 200 (1927). *Buck v. Bell* has never been overruled.

66 316 U.S. 535, 541 (1942). Technically, the Court found an equal protection violation in the categories drawn by the state to require sterilization of some property offenders but not others, thus leaving open the possibility that sterilizing all property offenders would be acceptable. Given that *Skinner* is now cited for its dicta about the right to procreate as a "basic civil right of man," requiring sterilization of convicted offenders under any circumstances is not likely to be upheld.

67 ROBERTSON, *supra* note 27, at 35-38. *See* Gerber v. Hickman, 291 F.3d 617 (9th Cir. 2002) (en banc) (holding that prison inmate has no right to provide sperm to his wife for artificial insemination outside the prison). *See also* Goodwin v. Turner, 908 F.2d 1395 (8th Cir. 1990).

68 524 U.S. 624 (1998). Although not a constitutional decision, *Bragdon* recognizes that the ability to have healthy offspring is a major life activity, the loss of which constitutes a disability within the meaning of the Americans with Disabilities Act.

surrogacy.[69] One could even argue for a right to engage in reproductive or therapeutic cloning, or the right to alter the genes of prospective children, with all the issues of enhancement and engineering which that raises. Whether all those actions would fall under the rubric of constitutionally protected procreative liberty, however, would depend upon whether they were centrally or intimately connected with reproductive decision-making. If so, those choices would be presumptively protected under the principles that underlay the Court's decisions and dicta to date, and be subject to limitation only if their use posed great harm to others.

It would be naive, however, to expect the current Supreme Court to accept the full implications of the principles of procreative freedom that are embedded in the Court's reproductive liberty cases. Although past cases and dicta might plausibly be read to adopt a broad principle of procreative freedom in both its aspects, one suspects that the Court would be quite hesitant to do so. The originalist bias of the Court, and its reluctance to find new fundamental rights make it unlikely that five justices would find most specific uses of assisted reproduction or genetics constitutionally protected, even if direct connection with more general principles of reproductive choice could be shown.[70]

The 2003 decision in *Lawrence v. Texas*, striking down laws against sodomy because of their impact on the intimate personal choices of homosexuals, suggests that the Supreme Court might recognize some rights to use technological assistance and genetics in reproduction.[71] Indeed, the *Lawrence* majority drew on the importance of reproductive rights as the basis for finding an unenumerated right to homosexual sex. It would be surprising if cases directly raising questions of technological choice in reproduction did not receive some protection as well.

After all, the justices do agree that Fifth and Fourteenth Amendment liberties include the right to marry and presumably to have biologic offspring.[72] If unenumerated basic rights are protected, marriage and reproduction are strong contenders for protection. But coital reproduction often is not possible, and the technical means to overcome coital infertility and genetic disease are available. Given these connections, it would be surprising if the Court did not grant protection to some reproductive and genetic technologies if cases involving them arose.[73]

---

69 A closer analysis of each reproductive variation will be necessary to determine whether it falls within a constitutionally protected right to procreate. Strictly speaking, the gametically infertile person who consents to gamete donation to his or her spouse is not reproducing, though their spouse is reproducing, as is the donor. By contrast, in gestational surrogacy the couple who has provided the gametes for embryos would be reproducing, but the gestating woman who has provided no genetic contribution will not be. If the surrogate provides the egg as well, she also would be reproducing, but the wife of the engaging couple will not.

70 For the reluctance of many justices to find new specific rights from general principles of liberty, see *Bowers v. Hardwick*, 478 U.S. 186 (1986), in which Justice White wrote for the majority that a claim that the right to engage in sodomy was "'deeply rooted in this Nation's history and tradition' or 'implicit in the concept of ordered liberty' is, at best, facetious." Justice Blackmun, in dissent, argued that past cases stood for a larger principle–the right of personal intimacy, which would include autonomy in sexual matters. For further discussion of tradition and the level of generality of substantive due process rights, see the dueling opinions of Justices Scalia and Brennan in *Michael H. v. Gerald D.*, 491 U.S. 110 (1989), and Laurence Tribe & Michael Dorf, *Levels of Generality in the Definition of Rights*, 57 U. CHI. L. REV. 1057 (1990).

71 123 S.Ct. 2472, 2481-83 (2003).

72 *E.g.*, Loving v. Virginia, 388 U.S. 1 (1967); Turner v. Safley, 482 U.S. 78 (1987).

73 Such a case might arise if a state banned IVF treatment of infertility, artificial insemination, egg donation, or gestational surrogacy. *See* John A. Robertson, *Embryos, Families, and Procreative Liberty: The Legal Structure of the New Reproduction*, 59 S. CAL. L. REV. 939, 959-61 (1986). A plausible argument for extending constitutional protection to reproductive cloning for gametic

Whether such protection would extend to most non-coital or genetic selection techniques must await further scientific development and social and legal engagement with those issues.

But one need not wait for Supreme Court guidance to determine genetic and reproductive policy or practice. Indeed, Court decisions holding that novel reproductive and genetic technologies are or are not constitutionally protected would designate whether individuals or the state had final say over whether particular uses can occur, and not provide a definitive assessment of their ethical acceptability for individuals and providers. Rather than count on the Supreme Court to provide answers, policymakers and providers should ask whether use of a technique is centrally connected with reproductive choice, and whether its use is likely to cause harm to others, even if it does not fit within the Supreme Court's willingness to define fundamental rights. Those inquiries would then turn on how closely related the activity in question is to prevailing understandings about why reproduction is valued, and whether a contested use reasonably serves that interest without causing undue harm. The less connected a use is to those values, the less likely it is to be respected.

IV. PROCREATIVE LIBERTY AND GENETIC APPLICATIONS

The above discussion analyzes procreative liberty as the freedom to engage in or avoid reproduction because of the great importance to individuals of having (or avoiding) offspring. It has argued for a *modern traditionalist* approach to these issues, rather than *strict traditionalism* or *radical liberty*, and will apply that view to the discussion of particular technologies that follows. Disagreements will arise, of course, as to whether an action is tied closely enough to reproduction to deserve the presumptive protection accorded to procreative liberty. Even if it is, there may also be debate about whether the exercise of procreative choice poses such a risk of harm to the tangible or legitimate interests of others that it can justly be limited or morally condemned in the context at issue.

To assess the role of procreative liberty in the era of genomics, I address four uses of genetic knowledge to choose the genes or genome of offspring. The first two techniques—(1) screening of prospective offspring for susceptibility or late-onset medical conditions and (2) screening for gender and other non-medical characteristics—involve selecting or choosing certain aspects of the genetic makeup of offspring by exclusionary or negative means. The last two techniques—(3) reproductive cloning and (4) positive genetic alteration of offspring genomes—involve positive selection or alteration of genes of offspring.[74]

As noted earlier, the *radical liberty* view would find all of these uses within an individual's freedom. A *strict traditionalist*, on the other hand, would be against most, if not all, of them. The reality, however, of how individuals are likely to use these techniques is much more complicated and contextually based than either pole recognizes. A better approach—the *modern traditionalist* view—is to evaluate each

---

infertility can also be articulated. *See* John A. Robertson, *Liberty, Identity, and Human Cloning*, 76 TEX. L. REV. 1371, 1387-1403 (1998) [hereinafter Robertson, *Liberty, Identity, and Human Cloning*].

74 Other applications of current interest and debate, such as using reproductive technology to expand reproductive age, to enable reproduction to occur after death, or to obtain cells or tissue for research or therapy, are not discussed here because they are less directly implicated in choices about offspring genes.

set of uses in terms of how it serves basic reproductive interests and whether it harms others, as the following analysis will show.

A. MEDICAL SCREENING OF PROSPECTIVE OFFSPRING

Most uses of genomic knowledge in reproduction will involve preconception, preimplantation, or prenatal screening to prevent the birth of offspring with genetic disease or predisposition to disease. Screening for genetic disease is now standard practice for couples with a family history of that disease or when population screening is justified. The main controversies concern extension of screening to other Mendelian diseases,[75] and to late-onset and susceptibility conditions.

It is now routine to screen populations or persons with family histories for a variety of autosomal diseases, such as cystic fibrosis, sickle cell anemia, and Tay Sachs.[76] Carriers of autosomal mutations may also learn whether their reproductive partners are also carriers. If so, they can take the one in four chance that their child will have the disease, adopt, go childless, use donor gametes, or conceive and screen at the embryonic or fetal stage, and then decide not to start or not to continue a pregnancy.[77] Embryo or prenatal screening might also occur for dominant or X-linked diseases, such as Huntington's disease, hemophilia, or Duchenne's muscular dystrophy.[78]

Growing knowledge of the human genome will increase carrier and prenatal screening by increasing the number of indications for screening. As more genetic mutations for susceptibility to diseases are identified, such as mutations in the P53 tumor suppressor gene or the BRCA1&2 genes, carrier or prenatal screening could extend to them.[79] Screening might also occur for late-onset conditions, such as early onset Alzheimer's disease or Huntington's disease.[80] Pre-birth or carrier screening, however, is not presently available for complex polygenic disorders which affect millions of persons, such as diabetes, heart disease, stroke, and autoimmune disorders, though screening may become available at some future time.[81]

Parental interest in screening prospective offspring for disease-causing or susceptibility genes is likely to continue and grow as more genes are discovered, and the ease of sampling DNA from embryos and fetuses improves. Such information may often be material or determinative of parental choice whether or not to reproduce. Knowledge of positive status for a disease-causing mutation often results in excruciating dilemmas about whether to proceed with reproduction.[82] While some persons are content to accept whatever "nature" or God provides, others would

---

75 Mendelian diseases are "the result of a single mutant gene that has a large effect on phenotype and that are inherited in simple patterns similar to or identical with those described by Mendel for certain discrete characteristics in garden peas." THOMAS D. GELEHRTER ET AL., PRINCIPLES OF MEDICAL GENETICS 4 (2d ed. 1998).

76 Robertson, *Liberty, Identity, and Human Cloning*, *supra* note 73, at 1407.

77 *See* Lois J. Elsas, II, *Medical Genetics: Present and Future Benefits*, 49 EMORY L.J. 801, 814-15 (2000).

78 Lori B. Andrews, *Prenatal Screening and the Culture of Motherhood*, 47 HASTINGS L.J. 967, 969-72 (1996).

79 *See* Frederick P. Li et al., *Recommendations on Predictive Testing for Germ Line p53 Mutations Among Cancer-Prone Individuals*, 84 J. NAT'L CANCER INST. 1156 (1992), *available at* http://www.nci.nih.gov/cancerinfo/genetics/predictive-testing-p53-mutations.

80 Robertson, *supra* note 39, at 433.

81 Wayne W. Grody, *Molecular Genetic Risk Screening*, 54 ANN. REV. MED. 473, 486 (2003).

82 There is an extensive literature on this topic. For a recent popular presentation, see Bill Keller, *Charlie's Ghost: Perfect Babies and Imperfect Choices*, N.Y. TIMES, June 29, 2002, at A15.

neither want to have a child if it will have a serious disease, nor would they want to take the risk that the disease might be expressed more mildly. Still others would prefer not to have a child if it will have early onset of a neurodegenerative disease such as Alzheimer's or Huntington's, or if it will face a life of monitoring, worry, or preventive surgery or medications, such as people with BRCA1&2 mutations face.[83]

In expressing such choices, prospective parents may say that they are concerned about the best interest of their children (though healthy children would be different children). But the choice is also understandable in terms of the burdens and concerns they would experience, such as increased child care and child rearing costs, more worry, etc. In evolutionary biology terms, they do not want to invest in offspring who themselves will have little chance of successfully reproducing or who will detract them from serving the needs of other healthy offspring.[84]

Because wanting information about the genetic makeup of prospective offspring and then acting on it fits squarely within conventional understandings of procreative liberty, the relevant legal and policy question is whether acquiring and acting on such knowledge causes harms that would justify not allowing persons to do so. Carrier screening prior to conception would seem to pose the fewest risks of harm. When carriers or others then request that embryos or fetuses be screened and then excluded from transfer or birth, however, four concerns arise: the impact on prenatal life, the impact on those who are disabled and dispreferred, the impact on resulting children, and the promotion of private eugenics. In addition, there may be objections based on the fact of selection itself. A brief discussion will show that none of those concerns is sufficient to limit use of genetic screening technology for disease or susceptibility conditions.

Impact on Prenatal Life. Concerns with impact on prenatal life largely track positions on abortion and the status of the fertilized egg and embryo. Persons who believe that fertilized eggs, embryos, and fetuses are already persons or entities with interests would argue that screening and exclusion on the basis of genes is the equivalent of eugenic murder and may support public policies banning or discouraging it. On the other hand, persons who believe that fertilized eggs, embryos, and fetuses are themselves too rudimentary in development to have rights or interests and thus are not themselves the subject of moral duties would have no principled objection to such practices based on the moral status of embryos and fetuses.

At the same time, however, they might argue that although lacking rights or inherent moral status, embryos and fetuses are not like any other human tissue and deserve special respect, e.g., one must have good reasons for manipulating and destroying them. Although they might not support public policies blocking such actions, they might be reluctant to terminate a pregnancy because of a susceptibility gene alone, though they would be willing to do so for a more serious genetic disease, such as Tay Sachs or sickle cell anemia.[85] Screening embryos prior to transfer to the uterus, however, is more acceptable because the embryo is still a clump of undifferentiated cells outside the body, and no abortion is necessary.

Impact on Disabled and Dispreferred. Some persons have argued against current and expanded use of genetic screening of the health of prospective offspring

83 Robertson, *supra* note 39, at 432-33.

84 However, once such children are born, evolutionary or culturally instilled notions of doing everything possible for one's children kick-in, despite the drain on private or public resources.

85 *See* Robertson, *supra* note 39, at 444-46.

on the grounds that it sends a message to persons with those conditions or disabilities that their lives are not valued or that it would be preferable that they had not been born.[86] But preferring children without serious medical conditions does not itself mean that existing persons with those conditions do not have worthwhile lives, nor that their interests and needs should not be respected; nor does allowing private individuals that choice constitute such state involvement that it gives the appearance of the state encouraging or requiring that people take steps to avoid such births.[87] Society can demonstrate respect and concern for persons with congenital disabilities, for example, by protecting them against discrimination in public accommodations and the workplace without also depriving other persons of the means to avoid having children with those conditions.

Impact on Offspring. Expanded genetic screening of prospective offspring will also not harm the offspring. The purpose is to help the parents have a child whose genes do not condemn him to a "nasty, brutish, and short" life.[88] Desiring a healthy child, or protections against well-known genetic diseases or susceptibility conditions, would not seem to implicate concerns about "designer children," or commodifying or treating children as objects to please the parental fancy—charges more plausibly leveled at cloning or positive actions to rewrite the child's genetic code in order to enhance or diminish its characteristics. But as we shall see, even there, the question of harm to offspring is controversial, because in most instances the child would not have been born if the technique in question had not been used.[89]

Private Eugenics. The history of "eugenics" in the United States is so freighted with abuse and misuse that the charge that a scheme or practice is "eugenic" carries great negative weight. But the abuses of the eugenic era at the beginning of the twentieth century came largely from efforts of state-imposed sterilization to prevent people with "bad genes" from reproduction that consumed societal resources and polluted the gene pool.

The resulting involuntary sterilization of 60,000 persons is now uniformly regarded as an unjustified abuse of reproductive rights that could not be supported either by genetic science or by the social costs that reproduction by "mental defectives" was thought to cause. Private use of genetic screening techniques to ensure a healthy child has none of the abusive features of the earlier eugenic era. It is voluntarily chosen, aims at individual and family rather than social well-being, and leaves persons free to choose to screen or not screen as they choose. The mere fact that many people may choose to screen, resulting in many fewer births of children with genetic handicaps, is not itself evil or undesirable.[90]

---

86 Martha A. Field, *Killing "The Handicapped"—Before and After Birth*, 16 HARV. WOMEN'S L.J. 79, 123-24 (1993); Erik Parens & Adrienne Asch, *The Disabilities Rights Critique of Prenatal Genetic Testing: Reflections and Recommendations*, 29 HASTINGS CENTER REP. S1 (1999).

87 Field, *supra* note 86, at 123. Whether insurance companies should be free to require parents to avoid such births or pay higher premiums is another matter.

88 *See* THOMAS HOBBES, LEVIATHAN 89 (Richard Tuck ed., 1991). *See also* Robertson, *supra* note 39, at 471 ("The purpose of genetic screening is to identify at-risk couples so that they may either avoid reproduction or have offspring only after they have screened embryos and pregnancies to prevent the birth of children with disabilities.").

89 *See* discussion *infra* Appendix.

90 A Canadian study showed that increases in prenatal diagnosis and pregnancy termination for congenital anomalies are related to decreases in overall infant mortality. Shiliang Liu et al., *Relationship of Prenatal Diagnosis and Pregnancy Termination to Overall Infant Mortality in Canada*, 287 JAMA 1561 (2002). A charge of eugenics would carry more weight if there were legal duties to be tested and then to exclude embryos and fetuses when tests were positive. A person's right to reproduce and right to bodily integrity would protect against requiring persons to use genetic tests

Selection Itself. The *modern traditionalist* view applied here assumes that there is nothing inherently wrong with selection of offspring characteristics if the purpose is otherwise justified. Having healthy offspring, who are themselves reproductively fit and happy, is so central to the values of the human reproductive enterprise that choices over whether to reproduce should fall within a person's or couple's freedom.

The *strict traditionalist* view, however, apparently would condemn selection for any purpose, including health, as violating the moral duty to accept unconditionally whatever child the "gift" of reproduction brings. Such a view makes any preconception or prenatal form of selection morally unacceptable (and presumably to be banned), because it involves choosing, controlling, selecting, or designing that child. Yet humans have often selected mates to ensure good health or family connections, and have strong desires to have healthy progeny who will in turn reproduce. Traditional indicia of sexual attractiveness, such as beauty and symmetry, appear to be indicators of genetic health, just as mate selection in the animal world has often depended on external traits, such as the size of the peacock's tail or the loudness or pitch of the frog's mating call, that serve as surrogates for good health and reproductive fitness.[91]

Moving selection to the gamete or embryo stage to identify for transfer to the uterus those embryos that are likely to be healthy and reproductively fit performs a related function. The proximate interest of the parent is to have healthy children for both the child's and the parents' sake. Ultimately, however the cause or explanation may be explained by natural selection working witlessly to enable some genes to survive longer than others. An evolutionary explanation does not in itself justify or condemn any particular practice. It provides, however, a further dimension for understanding why using gamete and embryo selection technologies to ensure healthy offspring might be of great importance to individuals.

If screening and selection techniques are accepted as serving important reproductive interests, the question then would be whether selection causes unacceptable harm. Creating and destroying embryos or fetuses on genetic grounds would not, in a world in which those entities lack interests or rights, count as serious harm. Nor does it appear likely that parental efforts to have children free of disease would pose special problems for them or for respect for human dignity more generally. Indeed, the opposite claim—that parents have a moral obligation to take such steps—is likely to be more strongly urged.[92]

Although the *strict traditionalist's* objection to selection often appears to be purely deontological, it also has a consequentialist aspect. *Strict traditionalists* assert that acceptance of any selection, particularly non-medical selection, will open the door to cloning or other forms of alteration. Their assumption appears to be that if any selection is permitted, then all must be as well. But this is a non sequitur. Rejection of *strict traditionalism* does not mean that *radical liberty* holds sway. Permitting one form of selection does not mean that all forms or situations of

---

or otherwise avoid the birth of children with genetic disease or anomalies. *See* Field, *supra* note 86, at 123-24.

91 "It's all about signaling, of course—the antic/Blue of the booby's feet; the lacewing's knock;/Deep in the reeds, the lowdown bullfrog's steady/*Present, present*, even at the risk that the call/Will materialize not a mate but an owl;/The coded fireflies' cool-burning *Ready, ready*;/The trailing plumes of the angelfish and the peacock. . . ." BRAD LEITHAUSER, DARLINGTON'S FALL: A NOVEL IN VERSE 85 (2002).

92 *See* ALLEN BUCHANAN ET AL., FROM CHANCE TO CHOICE: GENETICS AND JUSTICE 222-57 (2000); Carl H. Coleman, *Conceiving Harm: Disability Discrimination in Assisted Reproductive Technologies*, 50 UCLA L. REV. 17 (2002).

selection must also be permitted. As discussed below, a *modern traditionalist* would have great difficulty including in procreative liberty the right to clone when one is fertile or to enhance or diminish genetically abilities of otherwise healthy offspring.[93] Speculation about such future effects should not stop the use of otherwise acceptable technologies now.[94]

In Sum. The earlier the screening occurs and the less intrusive it is, the more likely is it that prospective parents will seek it to ensure that they have healthy offspring who will live satisfying lives and be able to produce and care for offspring themselves. Given the closeness of those desires with conventional (and evolutionary) understandings of why reproduction is important and the lack of direct harm to important interests of others, it would be surprising if law and social policy did not permit a wide range of such practices. In societies that do not accord inherent moral or legal status to embryos or pre-viable fetuses, legal prohibitions on genetic screening of the health of prospective offspring would appear to be an unjustified violation of an individual's procreative liberty.

B. Non-Medical Selection: Gender, Perfect Pitch, and Sexual Orientation

A much harder set of questions arises with non-medical selection, such as for gender, sexual orientation, hearing, perfect pitch, hair or eye color, intelligence, size, strength, memory, beauty, or other traits, which parents might find desirable. The *strict traditionalist* position would strongly object to any non-medical selection of future children, particularly by screening that causes the death of embryos or fetuses. It views any selection as morally wrong because selection makes the child into a product or object, denying its status as a gift and weakening the attitude of unconditional acceptance that *strict traditionalists*, such as Kass, view as comprising the essence of human reproductive dignity.

The *modern traditionalist*, however, is not prepared to exclude non-medical uses without further inquiry into whether they serve important reproductive or other familial interests. Rather than being shocked at the prospect of non-medical selection, the *modern traditionalist* embraces the possibilities that technology offers if they can be shown to help reproduction occur without undue harm. In reality, technology offers few options here, certainly not enough to justify the enormous heat that contemplating those possibilities generates. But debate about highly speculative kinds of non-medical selection drive much of current policy and ethical concern and thus deserve discussion here.

Among non-medical traits, only selection of gender, which is detectable by looking at the embryo's chromosomes without further analysis of DNA, is now possible. The genes for many "desirable" traits are unknown and are likely to remain unknown for the foreseeable future. Few of those characteristics appear to be inherited in a Mendelian fashion with detectable mutations that could be

---

93 *See* discussion *infra* notes 149-52, 189-95 and accompanying text. In those cases, the reproductive connections are much more attenuated than with selection for health and the risk of untoward social effects is much greater.

94 For problems with appeals to slippery slopes or the precautionary principle as reasons for not using new technologies, see ROBERTSON, *supra* note 27, at 163-64, and John Harris & Soren Holm, *Extending Human Lifespan and Precautionary Paradox*, 27 J. MED. & PHIL. 355 (2002).

identified in advance.[95] Small interactions of many genes appear to be involved, which makes unraveling those connections in the near future highly unlikely. Indeed, for many of those traits, environment may be a more powerful determinant than genes, so that looking for a strong genetic basis for them is bound to fail.[96] Given the lack of success in deciphering the genetic basis of major diseases such as diabetes, Alzheimer's, stroke, and heart attack, it will be quite some time before the genetic basis of complex behavioral or physical traits will be known.[97]

Despite the speculative nature of most non-medical genetic selection, I will look at non-medical selection for gender, perfect pitch, and sexual orientation to see how they would fare under the *modern traditionalist* approach. The key questions are: Does selection for such traits serve plausible reproductive needs? If so, do they use methods that harm the child, other persons, or society? Similar questions would arise if genetic tests for other traits became available.

In discussing these uses, one must have a realistic assessment of who might request such procedures and why they would if they were available. A couple contemplating reproduction would have to ask whether the benefits of non-medical selection outweigh the costs of making the selection. In most instances, it would be rare that the benefits of non-medical selection would outweigh the costs if screening of fetuses and selective abortion were the only viable selection technique.[98] Embryo screening for non-medical selection, however, may be more acceptable because of the rudimentary development of the embryo and its location in the laboratory, despite the need to undergo a cycle of IVF. A key question in each case is whether there is enough reproductive or parental benefit to induce couples to undergo IVF and screen embryos to reproduce so that embryo selection can occur. A second question is whether that reproductive interest is substantial enough to justify creating and selecting embryos for transfer. I address non-medical selection for gender, perfect pitch, and sexual orientation.

1. Non-medical Gender Selection

A good test of the analytic methodology proposed here is non-medical selection of the gender of offspring.[99] The topic is controversial for many reasons, including the sexism that it often reflects or fosters and the attitude toward embryos or fetuses that it might convey.[100] Preconception separation of male and female bearing sperm would be least offensive in terms of technique and more easily accessible, but its efficacy has not yet been established.[101] Screening of embryos to determine sex is more accurate than sperm separation, but requires an expensive and intrusive cycle

95 *See* STOCK, *supra* note 5, at 62-63; Jeffrey R. Botkin, *Prenatal Diagnosis and the Selection of Children*, 30 FLA. ST. U. L. REV. 265, 282-83 (2003) (suggesting that complex, polygenic traits are further influenced by other genes, environment, development, and random variations).

96 Sandra Blakeslee, *A Pregnant Mother's Diet May Turn the Genes Around*, N.Y. TIMES, Oct. 7, 2003, at D1 (reporting that diet affects the genes in mice that determine coat color).

97 Investment in research to uncover genes for non-medical traits is unlikely to be forthcoming from public sources because of the lack of medical payoff. Private investors may or may not support such research.

98 However, such decisions are often made; sex selection abortions in India are an example.

99 Nearly all commentators agree that selection against males carrying an X-linked disease is ethically justified even if only 50% of the males will be affected.

100 It is noteworthy that the American Civil Liberties Union, in challenging the Pennsylvania abortion law at issue in *Casey* did not challenge the State's ban on sex selection abortions, although that provision would, under the premises of *Roe* and *Casey*, also likely be found unconstitutional.

101 John A. Robertson, *Preconception Gender Selection*, 1 AM. J. BIOETHICS 2 (2001).

of IVF and the willingness to discard embryos. To assess the ethical, legal, and social issues presented by non-medical gender selection, I will focus on sex selection by preimplantation genetic diagnosis ("PGD").[102]

Requests for PGD for non-medical gender selection have come from two different groups. One is from persons who wish to select the sex of their first-born child (and possibly other children of the same sex). In almost all cases, the preference here is for a male child due to cultural mores that value males more than females, and that assign performance of cultural rituals to males, or from rank sexism.[103] The second group is persons who already have a child of one gender and wish to have a child of the opposite gender. In many cases, the requests are made after a family has had two or more children of the same gender, with no greater preference for males than for females.[104]

A major concern with any form of sex selection is the effect it will have on women, since sex selection practices are likely to strongly favor males.[105] If carried out on a large scale, it could lead to great disparities in the sex ratio of the population. As a result of easy access to sonograms to visualize the fetus and the pressures of a one child per family norm, some parts of rural China have seen 144 boys have been born for every 100 girls, which is far beyond the norm of 106:100 males to females.[106] Because costs and technical requirements will limit access to PGD, its use is only marginally likely to contribute to those disparities, at least by comparison with easier and cheaper methods, such as preconception sperm sorting or more onerous but widely practiced abortion.[107] Its use for first children, however, is likely to reflect culturally founded sexist notions. As middle and upper classes in those cultures grow and have the means to obtain PGD, such demand could increase.

Under the scheme of procreative liberty developed here, bans on gender selection of the first child may not be acceptable, despite the prejudice that they may evince toward women. As the discussion of homosexuality and musical pitch shows, allowing private prejudice is characteristic of individual freedom in the private sphere and may be recognized without causing public discrimination. The state, however, might adopt a policy to balance-off selection of boys and girls. A prohibition on gender selection of the first child would be more tolerable if the parents could choose the gender of the second child.[108]

---

102 Embryo screening by preimplantation genetic diagnosis ("PGD") requires a woman to undergo a cycle of ovarian stimulation and retrieval to obtain eggs for *in vitro* fertilization. A cell is removed from a 4-8 cell embryo, and its chromosomes or DNA analyzed. Based on the analysis, the embryo would be transferred to the uterus or discarded. At some point, selection of gametes prior to fertilization could replace the need to screen embryos.

103 *See* Dorothy C. Wertz, *International Perspectives on Ethics and Human Genetics*, 27 SUFFOLK U. L. REV. 1411, 1430-32 (1993).

104 Ethics Comm. of the Am. Soc'y of Reproductive Med., *Preconception Gender Selection for Nonmedical Reasons*, 75 FERTILITY & STERILITY 861, 862 (2002).

105 *Id.*

106 Erik Eckholm, *Desire for Sons Drives Use of Prenatal Scans in China*, N.Y. TIMES, June 21, 2002, at A3. *See also* Amartrya Sen, *More Than 100 Million Women Are Missing*, N.Y. REV. BOOKS, Dec. 20, 1990, at 61.

107 Rachel E. Remaley, *The Original Sexist Sin: Regulating Preconception Sex Selection Technology*, 10 HEALTH MATRIX 249, 264 (2000) (finding that PGD costs at least $12,000 compared to preconception sperm sorting which costs $2,500).

108 Depending on the numbers choosing this technique, some sex-ratio disparities could still occur. For example, if couples were content with one child if it were male, they might use PGD for the second child only if the first were a girl. A greater number of males would then result. (I am grateful to Neil Netanel for this point).

The use of PGD or other methods to select the gender of second or subsequent children is much less susceptible to a charge of sexism. Here, a couple seeks variety or "balance" in the gender of offspring, because of the different rearing experiences that come with rearing children of different genders.[109] Biologically based differences between male and female children are now well-recognized. They exhibit different spatial and learning rates and produce different hormones.[110] It is not *per se* sexist to wish to have a child or children of either gender, particularly if one has two or more children of the same sex. Although some feminists would argue that any attention to the gender of offspring is inherently sexist, particularly when social attitudes play such an important role in constructing parental and societal sex-role expectations and behaviors, one can recognize difference and celebrate it. U.S. Supreme Court Justice Ruth Bader Ginsburg, a noted feminist lawyer before being appointed to the Court, remarked in an important sex discrimination case that "[i]nherent differences between men and women, we have come to appreciate, remain cause for celebration."[111] Desiring the different rearing experiences that one has with boys and girls does not mean that the parents are sexist or likely to devalue one or the other sex.

Because legal bans on gender selection of the second child would rest on even weaker grounds than bans on selection of first children, clinics able to provide safe and effective methods of gender selection will be free to decide which patients they wish to treat. It certainly would be reasonable for a program to bar provision of sex selection for the first child, but provide it for the second. With regard to the first child, one may be promoting or entrenching sexist social mores. A clinic might also take the view that choosing the gender of the first child is not a strong enough reason to meet the special respect owed to embryos. A proponent, however, might argue that couples desiring gender variety in the family are the best judges of the importance of that need. This is particularly true in cultures where having a male heir is highly prized.[112] If PGD for the second child is not permitted, pregnancy and abortion, if not infanticide, might occur instead.[113] Other circumstances of gender selection might also arise to meet religious demands or to protect the privacy of a couple seeking a sperm donor, as a recent case in Israel illustrates.[114] In western

109 The choice could also be based on religious beliefs. *See* ELLIOT N. DORFF, MATTERS OF LIFE AND DEATH: A JEWISH APPROACH TO MODERN MEDICAL ETHICS (1998).

110 Robertson, *supra* note 101, at 2-9.

111 United States v. Virginia, 518 U.S. 515, 533 (1996).

112 Ethics Comm. of the Am. Society of Reproductive Med., *supra* note 104, at 862.

113 An IVF program in India is now providing PGD to select male offspring as the second child of couples who have already had a daughter. Because of the importance of a male heir in India, those couples might well consider having an abortion if pregnant with a female fetus (even though illegal for that purpose). In that setting, PGD for gender selection for family balancing may well be justified, and be left to the market to provide. *See* A. Malpani & D. Modi, *Preimplantation Sex Selection for Family Balancing in India*, 17 HUMAN REPROD. 11 (2002).

114 An Orthodox Jewish couple with severe male infertility had to resort to a sperm donor but did not want others to know. Because the husband was a Cohen, a male born as a result would not be able to truthfully recite passages at his Bar Mitzvah that only a Cohen could recite. Then, others attending the ceremony would know that the child was not that of the husband. To avoid the unavoidable disclosure of the donor sperm origins of the child that would result, the couple agreed to have a child only if they could be sure that it would be female. Israeli health authorities approved this couple's use of PGD to select female embryos for transfer. Tamara Traubmann & Haim Shadmi, *Couple Allowed to Choose Baby's Gender to Avoid Halakhic Dilemma*, Haaretz, Oct. 17, 2002 (on file with author). *See also* DORFF, *supra* note 109, at 72-79.

464 AMERICAN JOURNAL OF LAW & MEDICINE VOL. 29 NO. 4 2003

societies, providers might be willing to fulfill the couple's request out of respect for their right to make such decisions.

Acceptance of PGD for gender selection, whether for first-born children or only for gender variety, assumes that use for that purpose is sufficiently important to justify the symbolic costs of creating, screening, and discarding embryos on the basis of sex. Persons who believe that gender selection serves no important individual need, even in families with several children of one gender, might then object to postconception methods, such as PGD, as insufficiently respectful of embryos and choose not to seek or provide it.[115]

The President's Council on Bioethics has recently issued recommendations opposing non-medical gender selection.[116] While a conservative approach to gender selection is rational, the Council does not fully confront the case for such a choice or the complexity of determining whether net harm would ensue from permitting such choices. For example, it makes no attempt to assess the importance that gender variety in offspring has to couples who strongly desire the experience of raising both girls and boys, e.g., who would not reproduce again unless they had that choice. They also assume that the desire for a particular trait in offspring would lead to excessive demands or expectations that ultimately harm offspring. Yet it is just as reasonable to view such preferences as less determinative of rearing behavior than the Council fears. A couple might want to have a girl rather than a boy because of the different experiences that rearing her might bring without having a fixed idea of what they expect that child to be or the flexibility to respond to its developing needs.

2. Selecting for Perfect Pitch

Perfect or "absolute" pitch is the ability to identify and recall musical notes from memory.[117] Although not all great or successful musicians have perfect pitch, a large number of them do. Experts disagree over whether perfect pitch is solely inborn or may also be developed by early training, though most agree that a person either has it or does not. It also runs in families, apparently in an autosomal dominant pattern.[118] The gene or genes coding for this capacity, however, have not been mapped, much less sequenced. Because genes for perfect pitch may also relate to the genetic basis for language or other cognitive abilities, research to find that gene is likely.

Once the gene for perfect pitch or its linked markers are identified, it would be feasible to screen embryos for those alleles and transfer to the uterus only those embryos that test positive. The prevalence of those genes is quite low (perhaps 3 in 100) in the population but higher in certain families.[119] Thus, only persons from those families who have a strong interest in the musical ability of their children would be potential candidates for PGD for perfect pitch. Many of them are likely to take their chances with coital conception and exposure of the child to music at an

---

115 *See* John A. Robertson, *Sex Selection for Gender Variety by Preimplantation Genetic Diagnosis*, 78 FERTILITY & STERILITY 463 (2002).

116 PRESIDENT'S COUNCIL ON BIOETHICS, BEYOND THERAPY: BIOTECHNOLOGY AND THE PURSUIT OF HAPPINESS 66-71 (Oct. 2003) (pre-publication version), *available at* http://www.bioethics.gov/reports/beyondtherapy/fulldoc.html.

117 Sandra Blakeslee, *Perfect Pitch: The Key May Lie in the Genes*, N.Y. TIMES, Nov. 20, 1990, at C1; Dennis Drayna et al., *Genetic Correlates of Musical Pitch Recognition in Humans*, 291 SCIENCE 1969, 1969-72 (2001).

118 *Id.*

119 *Id.*

early age. Some couples, however, may be willing to undergo IVF and PGD to ensure this foundation for musical ability in their child. Should their request be accepted or denied?

As noted, the answer to this question for the *modern traditionalist* depends on the importance of the reproductive choice being asserted, the burdens of the selection procedure, its impact on offspring, and its implications for de-selected groups and society generally. The strongest case for the parents is if they would not reproduce unless they could select that trait, and they have a plausible explanation for that position.[120] Although the preference might appear odd to some, it might be understandable in highly musical families, particularly ones in which some members already have perfect pitch. Parents clearly have the right to instill or develop a child's musical ability after birth. If so, they might then plausibly argue that they should have that right before birth as well.

If so, then creating and destroying embryos for this purpose should also be acceptable. If embryos are too rudimentary in development to have inherent rights or interests, then no moral duty is violated by creating and destroying them.[121] Some persons might think that doing so for trivial or unimportant reasons debases the inherent dignity of all human life, but having a child with perfect pitch will not appear debasing to parents seeking this technique. Ultimately, the judgment of triviality or importance of the choice rests within a broad spectrum with the couple. If they have a strong enough preference to seek PGD for this purpose and that preference rationally relates to reproductive goals that deserve respect, then they have demonstrated its great importance to them. Only in the clearest cases, for example, perhaps creating embryos to picking eye or hair color, might a person's individual assessment of the importance of creating embryos be rejected.[122]

A third relevant factor is whether musical trait selection is consistent with respect for the resulting child. Parents who are willing to undergo the costs and burdens of IVF and PGD to have a child with perfect pitch may be so overly invested in the child having a musical career that they will prevent it from developing its own personality and identity. Parents, however, are free to instill and develop musical ability once the child is born, just as they are entitled to instill particular religious views. It is difficult to say that they cross an impermissible line of moral risk to the welfare of their prospective child in screening embryos for this purpose.[123] Parents are still obligated to provide their child with the basic education and care necessary for any life-plan. Wanting a child to have perfect pitch is not inconsistent with parents also wanting their child to be well-rounded and equipped

---

120 That selection of the trait is essential to the parents' reproductive decision is relevant because if they would otherwise reproduce regardless of the ability to select the trait, a ban on selection would not interfere with their ability to have offspring. A more robust theory of procreative liberty might protect less essential preferences as well. However, I will confine my analysis to cases where parents can plausibly establish that they would not reproduce unless they could use the selection technique at issue. *See* Robertson, *supra* note 39.

121 Ethics Comm. of the Am. Fertility Soc'y, *Ethical Considerations of Assisted Reproductive Technologies,* 62 FERTILITY & STERILITY 32S (Supp. No. 1 1994).

122 A private provider may refuse to screen embryos for those purposes. Whether a state ban on such selection would survive scrutiny would turn on whether selection of hair or eye color was deemed so important to the couple's procreative freedom as to fall within their procreative liberty. If so, respect for embryos or fears of slippery slopes to genetic engineering of offspring genes would probably not justify limiting that choice.

123 If the child would not otherwise have been born, they may not have harmed it at all. *See infra* Appendix.

for life in other contexts. If parents seem likely to be over-invested in the child, physicians should consider not offering them the requested selection services.

A fourth factor, impact on de-selected groups, is much less likely to be an issue in the case of perfect pitch, because there is no stigma or negative association tied to persons without that trait. Persons without perfect pitch suffer no stigma or opprobrium by the couple's choice or public acceptance of it, as is arguably the case with embryo selection on grounds of gender, sexual orientation, intelligence, strength, size, or other traits; nor is PGD for perfect pitch likely to perpetuate unfair class advantages, as selection for intelligence, strength, size, or beauty might.[124]

A final factor is the larger societal impact of permitting embryo screening for a non-medical condition such as perfect pitch. A *strict traditionalist* would argue that accepting non-medical selection for this trait will be a precedent for selecting other traits and eventually enhancing or modifying offspring genomes. Acceptance of any non-medical selection moves us toward a future in which children are primarily valued by the attractiveness of their expected characteristics and not as unconditionally accepted "gifts" from God. This will "coarsen" the dignity of reproduction and those engaged in it.[125]

But that threat is too hypothetical to justify limiting what may otherwise be valid exercises of parental choice. It is highly unlikely that many traits would be controlled by genes that could be easily tested in embryos. Gender is determined by the chromosome, and the gene for perfect pitch, if ever found, would be a rare exception to the multi-factorial complexity of such traits. Screening embryos for perfect pitch, if otherwise acceptable, should not be stopped because of speculation about what might be possible several decades from now.

In sum, musician parents are entitled to instill a love of music and skill in playing in their children just as they are entitled to instill their particular religious views.[126] Their willingness to resort to genetic technology to enhance those possibilities would not itself show that they are less than fully committed to their child within the world-view and cultural context in which they live. The *modern traditionalist* would find a plausible case, and perhaps even a right of procreative liberty, to make such a selection.

---

124 Similar issues would arise in parental selection for hearing or deafness. If couples with a history of family deafness want to have a hearing child, they could screen out embryos with a mutation in the gene for connexion, which appears to cause 60% of inherited deafness. However, allowing parents this choice in the privacy of an IVF clinic would not denigrate existing persons with deafness. By the same token, deaf parents who selected *for* children with the mutation would not be denigrating hearing persons. A different objection would be that deaf parents would be harming a deaf child by intentionally making its birth possible, when hearing embryos could have been transferred. John C. Fletcher, *Deaf Like Us: The Duchesneau-McCollough Case*, 5 L'OBSERVATOIRE DE LA GÉNÉTIQUE (2002), *at* http://www.ircm.qc.ca/bioethique/obsgenetique/cadrages/cadr2002/c_no5_02/ca_no5_02_1.html. For analysis of harm to offspring in such cases, see *infra* Appendix. A different outcome might result if the deaf parents sought to silence the gene for connexion in order to have a deaf child. *See* discussion of intentional diminishment of offspring traits *infra* notes 189-95 and accompanying text.

125 Leon Kass and other *strict traditionalists* might also argue that selection practices could alter the societal status of all children by making them appear to be products or commodities. For a similar argument about the effect of surrogate motherhood and paid sex on children generally, see MARGARET JANE RADIN, CONTESTED COMMODITIES (1996).

126 *See* Wisconsin v. Yoder, 406 U.S. 205 (1972) (holding that Amish parents' interest in instilling certain religious values in their children outweighed State's interest in requiring mandatory public education through age sixteen).

3. Sexual Orientation

A popular play several years ago portrayed the conflict confronting a father whose wife is pregnant with a male fetus with the genetic marker that he will be homosexual.[127] The drama focused on the protagonist's struggles over whether to abort or not. Because there is not yet a genetic test for sexual orientation, there is no pre-birth test of gametes, embryos, or fetuses for this condition, nor is it clear that one will be developed in the future. If one were available, however, we would face the question of whether parents would be free to abort fetuses, or more likely, to select or to exclude embryos that have a particular sexual orientation.

As with a gene for perfect pitch, such a gene is likely to be manifested in families and thus be of primary interest to those with some family history of that orientation, rather than to the population at large. Persons who are homosexual might seek it out in order to have a child who also will be homosexual. More likely, some heterosexual couples with family members who are homosexual may care deeply enough about it to prefer not to have a child with genes that strongly correlate with homosexuality.

In either case, a couple's or individual's claim to choose their child's sexual orientation would be reproductive if it would strongly and plausibly affect their willingness to reproduce. Under the analytic scheme of this article, the key question would be how important such selection would be for the parental project of successful gene transmission to the next generation. For some parents, the idea of raising a gay child poses a number of problems, including the difficulties that such a child would face in a prejudiced society, the reduced likelihood that such a child would have progeny that would continue the parents' genes, and the parents' own prejudices. Although few people might seek to screen on grounds of sexual orientation, particularly if the screening were costly or physically intrusive, it would be difficult to argue that parents would not be exercising procreative liberty in seeking to screen and exclude on that basis.[128]

It is true that they may be exercising a bias or prejudice against homosexuality (or against heterosexuality by homosexuals who seek a gay child), but freedom of association permits persons in the private sphere to discriminate as they choose. One could strongly support equal rights for gays in all public and institutional spheres, yet still find that this choice is within their procreative and associational discretion. Nor could one easily show that allowing such choices would be a continued public demeaning of homosexuals, who are still publicly discriminated against in many ways. We may hope that the genetics of sexual orientation never lends itself to simple tests to screen children for sexual orientation. But if that knowledge develops, it may be hard to show that it does not fit within the rights of parents to decide about those characteristics of offspring.

Nor would a child, chosen in part to have a particular sexual orientation, be a product or commodity of manufacture any more than a child chosen for gender might be. One has no particular design for the child beyond being healthy and having the sexual orientation chosen. The child would still be free to be his own person in other regards.

---

127 JONATHAN TOLINS, THE TWILIGHT OF THE GOLDS (1994).

128 The choice of gay parents to have gay children is not inconsistent with a reproductive agenda of gene transmission because those gay offspring might also reproduce, just as their gay parents did. In any event, in selecting for a child with gay genes, gay parents are engaged in the culturally defined project of reproduction as gene transmission and parenting in the next generation.

Finally, permitting parents to use genetic technology to avoid having a child with a homosexual orientation is distinct from the separate question of whether homosexual individuals or couples have the right to reproduce. Indeed, norms of equal respect for all persons would protect the right of homosexual persons to reproduce to the same extent as heterosexuals do, including the use of assisted reproductive and genetic screening techniques.[129] They would be free then to use genetic screening to attempt to have homosexual offspring and might even be free to use haploidization techniques to enable each partner to contribute haploid genes to a new individual.[130]

C. REPRODUCTIVE CLONING

In addition to testing gametes, embryos, or fetuses before birth to exclude (or include) offspring with particular medical or non-medical traits, reproductive technology may also make possible more active selection of offspring genomes. This section discusses nuclear transfer cloning, which chooses a whole genome rather than gametes or embryos. The next section will discuss alteration of particular sections of the genome.

Reproductive cloning would occur by somatic cell nuclear transfer. A somatic cell is de-differentiated at an earlier state, its nucleus is removed, and then is transplanted into an enucleated egg. After activation, the resulting embryo is placed in the uterus with the hope that it will implant, develop, and come to term.

Since the birth in 1996 of Dolly, the sheep cloned from the mammary glands of an adult ewe, mice, rabbits, cats, cows, and pigs have been cloned.[131] Yet mammalian cloning is not easily achieved and remains unpredictable. Success rates have been quite low. More than 200 embryos were created for every successful pregnancy in sheep, with many failures at every stage in the process.[132] Many questions remain about whether re-starting the cellular clock of transplanted nuclei inevitably interferes with imprinting and the methylation necessary for proper epigenetic development.[133] In addition, very little is known about how well clones will do. Although Dolly had offspring, she died after contracting a lethal sheep virus and, thus, before it could be determined whether she would have suffered from a shortened life-span due to the shortening of her telomeres from previous cell divisions, or whether there would have been other epigenetic effects that impair health.

Given the still rudimentary state of cloning science, it would be highly premature to attempt human cloning now. As we will see, the best case for human cloning is quite limited and would not appear in itself to justify the great amount of

---

129 John A. Robertson, *Two Models of Human Cloning*, 27 HOFSTRA L. REV. 609, 633-37 (1999).

130 *See* Zsolt Peter Nagy et al., *Development of an Efficient Method to Obtain Artificially Produced Haploid Mammalian Oocytes by Transfer of G2/M Phase Somatic Cells to GV Ooplasts*, 78 FERTILITY & STERILITY S1 (Supp. 2002); Antonio Regalado, *Could a Skin Cell Someday Replace Sperm or Egg?*, WALL ST. J., Oct. 17, 2002, at B1.

131 Lowell Ben Krahn, *Cloning, Public Policy and the Constitution*, 21 J. MARSHALL J. COMPUTER & INFO. L. 271, 274 (2003).

132 Ian Wilmut et al., *Somatic Cell Nuclear Transfer*, 419 NATURE 583, 583-86 (2002).

133 Rudolph Jaenisch & Ian Wilmut, *Don't Clone Humans!*, 291 SCIENCE 2552 (2001).

embryo research, miscarriages, and possible early deaths from human cloning;[134] nor is it very likely that many people would actually seek cloning if it were safe. Yet enormous public attention has been paid to the possibility of human reproductive cloning. Some *radical libertarians* and reproductive physicians working with them have asserted a right to clone and announced efforts to do so.[135]

Coming at a time when the safety of mammalian cloning has not been established, those statements appeared to be especially irresponsible. The statements suggested to the public that there was in fact a serious danger that scientists would start to clone before its safety had been established or sufficient public discussion of its dangers had occurred. These events led to bipartisan legislation that would ban nuclear transfer reproductive cloning for any purpose.[136] Because some proponents of the ban also want to ban therapeutic cloning, which does not involve transfer of cloned embryos to the uterus and the chance of birth, the legislation has not yet passed.[137]

Although some of the support for a permanent criminal ban is based on safety reasons, others support a ban even if safety and efficacy were established. The widespread opposition to reproductive cloning comes in part from misunderstandings about whether cloning would produce an exact copy of a child and the strength of the demand that is likely for cloning.[138] Even where environmental effects on phenotype are recognized, opponents seem to think that people will have strongly narcissistic urges to replicate themselves rather than reproduce sexually.[139] Some opposition also arises from the perception that any transfer of cloned embryos to the uterus would be an unethical experiment.[140] In any event, most opponents think it would always be immoral to permit cloning, no matter how it is used, and that a criminal ban with strong penalties is needed to stop it.[141]

The most complete case against cloning has been made by Leon Kass, now fleshed out in the President's Bioethics Council's 2002 report, *Human Cloning and*

---

134 The loss of embryos and fetuses from attempts at human reproductive cloning may not violate moral duties owed to those entities, but it could implicate symbolic and expressive values about not cavalierly creating and destroying prenatal human life.

135 *See* Leon R. Kass, *The Wisdom of Repugnance: Why We Should Ban the Cloning of Humans*, 32 VAL. U. L. REV. 679, 688 (1998).

136 Human Cloning Prohibition Act of 2003, H.R. 534, 108th Cong. § 302 (2003).

137 *See* discussion *infra* Part V.

138 The opponents' perception that a clone will be an exact copy of the DNA source assumes a crude genetic reductionism that overlooks the phenotypic effects of uterine and rearing environments. In many versions, opponents appear to think that the cloned child will simply spring full-born into an identical version of the person cloned, as in the Greek myth of Athena springing full-born from the head of Zeus or the several clones of a busy building contractor that resulted from a surgical procedure in the film *Multiplicty*. Another popular-culture version of cloning, *The Boys from Brazil*, is much more accurate in recognizing that to clone effectively Hitler, one would have to recreate the experiences he had as a child.

139 *See* NBAC REPORT, *supra* note 35, at 69.

140 In arguing that any transfer of a cloned embryo to the uterus would be an unethical experiment on an unconsenting child, Leon Kass and the President's Bioethics Council assume that knowingly risking the birth of a child with handicaps is always unethical, even if the child could not otherwise have been born and the parents are committed to loving and rearing it. *See* discussion *infra* Appendix.

141 Paul Tully, Comment, *Dollywood is Not Just a Theme Park in Tennessee Anymore: Unwarranted Prohibitory Human Cloning Legislation and Policy Guidelines for a Regulatory Approach to Cloning*, 31 J. MARSHALL L. REV. 1385, 1405 (1998).

Human Dignity.[142] One argument is that any experimentation with reproductive cloning would necessarily be unethical because there could never be a guarantee that the resulting child would not suffer some physical or other injury.[143] But even if no direct physical or psychological injury were shown, the report finds that cloning would be unacceptable because of its "challenge to the nature of human procreation and child-rearing."[144] Reflecting the *strict traditionalist* view, cloning is wrong because it involves making, rather than "begetting" a child, with the child a "product of wills," chosen for particular characteristics. Because the genome of cloned children is chosen, they are objects or products made to serve the parents' interests and will, and not valued for themselves. Indeed, they are not treated equally because they are denied the individual identity and uniqueness that other humans have, and that view is central to the human condition. The report then proceeds to discuss issues of the clone's identity, its status as a manufactured product, eugenics, family, and society.

The report displays several misconceptions about how close the resulting phenotype would be to that of the nuclear DNA source. Even if corrected, however, the *strict traditionalist* premise of the report would still lead to rejection of cloning in all cases. For example, the report misdescribes choosing nuclear DNA for cloning as "choosing" the entire genome or even "designing" the child. In nuclear transfer cloning, a total package of DNA is chosen for replication but not a complete menu of genes, because genes for most traits are unknown. Moreover, most phenotypic traits depend heavily on environment.[145] A cloned child will not simply replicate the phenotype of the DNA source, for it will gestate in a different uterus, be reared by different persons in a different environment, and will be subject to different mitochondrial DNA influences.[146] For the *strict traditionalist*, however, any degree of choice is anathema, and choice of the entire genome, as phenotypically expressed, is a degree of choice beyond that of excluding or including single genes in healthy offspring.[147]

In contrast, the *modern traditionalist* would not condemn reproductive cloning in all cases without further inquiry into how it is likely to be used and the effects it is likely to have.[148] Since cloning would involve having and rearing a child and would present special, psychological challenges, people would be unlikely to seek to clone unless they had very good reason to do so. The most plausible demand for reproductive cloning is likely to be from people who are at high risk of having offspring with severe genetic disease or who cannot themselves reproduce sexually,

---

142 CLONING REPORT, *supra* note 13.

143 *See* discussion *infra* Appendix.

144 CLONING REPORT, *supra* note 13, at 110.

145 Erin Rentz, *Estimating Additive Genetic Variation and Heritability of Phenotypic Traits*, *at* http://online.sfsu.edu/~efc/classes/biol710/heritability/heritability.htm (last updated May 30, 2002).

146 Nuclear transfer cloning requires transfer of the nucleus to an enucleated oocyte, with mitochondrial DNA in the cytoplasm that will be different than that of the DNA source.

147 The other major misconception is that a cloned child is necessarily harmed by being cloned (as would be cloned children who die early or are born with defects). But cloned children would not otherwise have been born and from their perspective, once born, they have an interest in living. *See* discussion *infra* Appendix. Rather than focus on harm to offspring, one might more usefully ask whether parents would be violating norms of good parenting if they were committed to rearing the resulting child, whatever its impairments.

148 The *radical libertarian* would accept cloning even if a person is fertile as part of the right to select whatever genes of offspring one wishes. *See infra* notes 152-53 and accompanying text.

for example, azoospermic males for whom intracytoplasmic insertion of sperm, a standard treatment for male infertility, is the only possible solution.[149]

In such cases, rather than go childless, adopt, or use donor sperm from a stranger or brother (a 25% rather than 50% sharing of genes with offspring), they might want to have a child with whom they have a close kinship genetic connection, which they could achieve by cloning themselves. In this case, they are interested in a reproductive genetic connection of at least 50% and, if they cannot achieve that, are willing to settle for 99.9% through nuclear transfer cloning. They would not be seeking to "design" a child so much as have a child with a genetic connection to their family. If their spouse provided the oocyte and gestation necessary for reproductive cloning to occur, both partners would have a biologic connection with their child.[150]

A plausible case exists in these circumstances for recognizing reproductive cloning for the gametically infertile as an exercise of procreative liberty. It serves the basic reproductive goal of getting the infertile patient's genes into the next generation—indeed, it is their only way of doing so. The couple is committed to rearing the child and treating it as an individual. In evolutionary terms, cloning may not be a long-run successful reproductive strategy, but without this short-run remedy, their genes will die with him. This suggests that the infertile person seeking to clone is seeking the conventional reproductive goal of rearing genetically related children by the only means, given his medical condition, open to him. As such, it should be treated like attempts at coital reproduction and banned or restricted only if a strong showing of harm can be made.[151]

The greatest claim of harm is that the life of a child produced by cloning is likely to be so psychologically painful or confusing that its interests justify preventing cloning even in cases of true infertility.[152] If parents are truly interested in having and rearing a genetically related child rather than an identical copy, there may be good reason to think that the cloned child will psychologically fare well. There may be a strong physical resemblance to the DNA source, but parents committed to the well-being of their child, as a person in his own right, can avoid treating the child as a mere copy of the DNA source. More experience will be necessary to determine how infertile couples can use cloning to have genetically related children while minimizing psychological or social problems in resulting children. It is plausible to think that a positive rearing experience for parents and child would occur even in these unique circumstances.

A *modern traditionalist*, however, would have more difficulty accepting reproductive cloning by persons who are sexually fertile. Their claim to be

---

149 Gerald Schatten et al., *Cell and Molecular Biological Challenges of ISCI: Art Before Science?*, 26 J.L. MED. & ETHICS 29, 34 (1998).

150 The husband would have a 99.9% genetic connection, while the wife would have a gestational and mtDNA connection. Even then, only a small number of that small group of gametically infertile persons is likely to opt for reproductive cloning.

151 The small number of persons seeking reproductive cloning when infertile does not lessen or change the ethical analysis. Rights matter even if only one person exercises them. As the Supreme Court reminds us in *Casey*, it is the degree of interference, not the number interfered with, that determines whether a fundamental right has been infringed. 505 U.S. at 894 (discussing whether a spousal notification requirement infringes a woman's right to terminate pregnancy if it would result in only a few women being denied an abortion).

152 In a person-affecting system of harm, the cloned child who would not otherwise have been born would not be harmed by being brought into the world in that condition. *See* discussion *infra* Appendix.

exercising reproductive freedom has no basis if they are cloning a third party whom they think has a desirable genome, for in that case they would not be providing any genes and thus would not be reproducing.[153] If they cloned themselves, they would be passing on 99.9% of their genes. In choosing to clone themselves rather than sexually reproduce, they are claiming more than the ordinary interest in passing on a haploid set of chromosomes by sexual reproduction. The claim of a right to pass on more goes beyond what ordinarily occurs in reproduction, and thus would seem much less deserving of special protection on that score. It is not helpful to get into an essentialist argument about whether an interest in getting a diploid set of chromosomes into the next generation is truly "reproductive." It is reproductive *plus*, but there is no reason why an individual's desire to pass on diploid, rather than haploid, chromosomes should be respected when they are able to reproduce sexually.[154]

The claim of the sexually fertile to clone is a claim either to choose whatever genome in children that one wishes or a claim to maximize the number of its genes, which gets into the next generation and presumably generations after. But there is no commonly accepted right to choose whatever genome one wishes for one's offspring, even if there is a right, as this article argues, to select some aspects of the genome of offspring when that choice advances conventional understandings of reproductive interest without harming others.[155] A close analysis of the cases in which selection occurs is needed to show its legitimacy, as the previous discussion has shown. Unless one takes a *radical liberty* approach to these questions or adopts a more robust version of procreative liberty than has been argued for here, it will be difficult for the sexually fertile to sustain that claim. Cloning when fertile does replicate a larger portion of genes, but it does so in a way that changes ordinary understandings of offspring and of why reproducing is so valued by individuals.[156]

Because of doubts that cloning when fertile is truly "reproductive," one need not ask whether there is a strong case for thinking that permitting it would cause direct, substantial harm. If not protected as part of procreative liberty, a state that chose to limit reproductive cloning by the fertile would need show only a rational basis for thinking it posed harm. One ground of concern would be whether parents would be as interested in the well-being of offspring when cloning is chosen in lieu of sexual

---

153 Persons who have obtained the nuclear DNA of another for somatic cell nuclear transfer cloning would not themselves be reproducing. The source of the DNA would be replicating 99.9% of their genes, but they are not claiming a right to do so nor in the scenario described here intending to rear resulting offspring.

154 The claim of such a right would give those with wealth and power additional advantages over others, in much the same way that wealth-based access to non-therapeutic enhancement might. For a discussion of "genetic domination," see BRUCE A. ACKERMAN, SOCIAL JUSTICE IN THE LIBERAL STATE 113-38 (1980). In the long-run, cloning by the sexually fertile is unlikely to be a successful reproductive strategy. Without the genetic diversity generated by meiotic recombination and sexual merging of different chromosomes, cloned individuals who clone in turn will eventually be weakened versus others. Consistent inbreeding would make it much less likely to survive in the long-run. *See* Mark Derr, *Florida Panther's Great Leap Hits a Wall*, N.Y. TIMES, Oct. 15, 2002, at F3. Yet the individual faced with the short-run prospect of no genetic transfer to a new generation or transfer by cloning might rationally choose cloning despite the many long-run disadvantages.

155 This position is a far cry from the perception of a general right to select offspring characteristics in all cases that Karen Lebacqz mistakenly ascribes to her earlier cited critique. *See* Lebacqz, *supra* note 46.

156 This statement is contestable and may appear to contradict earlier statements. *See supra* note 46 and accompanying text. In any event, yielding to community conceptions of the importance of reproduction as intimately tied to sexual reproduction does not also entail yielding to its conceptions of acceptable means of achieving sexual reproduction. *See* Lebacqz, *supra* note 46.

reproduction. But even if some parents would respect such an offspring for its own identity, there is enough of a risk that others would not do so to satisfy the rational basis standard that would justify limitation when reproductive liberty is not at stake.

Finally, a judgment that cloning when fertile goes beyond most plausible accounts of procreative freedom may call into question whether cloning when infertile would not also be deemed non-reproductive. It is true that it is the only way to get genes into the next generation, but in doing so, one loses the advantages of meiotic recombination and interchange of chromosomes that constitutes sexual reproduction. In addition, the resulting child may also have to resort to cloning to transmit its genes, thus further limiting recombination.

These issues will have to be revisited once additional experience with mammalian cloning and cloning human embryos occurs. If the formidable scientific and medical obstacles to safe cloning are overcome, the key ethical inquiry will be whether an individual's choice of reproductive cloning serves important reproductive interests without causing harm to others (as understood at the time the science has improved). I have argued that cloning by an infertile person might plausibly be viewed as "reproductive," because it is the only way that he or she could have a genetically related child to rear. In gene transmission terms, the person cannot get any genes into the next generation unless they transmit a diploid genome containing their nuclear genome.[157] In contrast, preventing the sexually fertile person from cloning themselves does not prevent them from the core reproductive event of transmitting a haploid set of chromosomes.[158] Reproductive disadvantages in the long-run are not a sufficient basis for denying short-run use by persons who have no other way to pass a haploid genome to the next generation. The problem of how the cloned child will then reproduce, and if cloned, how his clones in turn would reproduce would also have to be faced.

D. Rewriting the Genetic Code of Offspring

Frequently mentioned in ethical and policy debates of reproduction in the genomic era is the prospect of active engineering or altering of the genome of prospective offspring.[159] Such a prospect, however, is highly speculative, even more so than negative non-medical selection or nuclear transfer cloning. A major barrier is that the genes associated with desired traits are unknown and, because of their polygenic nature, will be difficult to identify. Once the genes are known, gametes or embryos could be screened for them and then used or not used in reproduction.

Positive alteration, however, requires further steps beyond identifying genes, screening gametes or embryos, and not using or transferring affected ones.[160] Knowing the relevant gene simply sets the stage for the further step of silencing genes or adding new segments of DNA.[161] Few would expect those techniques to be soon available for pre-birth editing of a prospective child's genome.

---

157 *See* NBAC Report, *supra* note 35, at 15.

158 *See id.*

159 *See* Robertson, *supra* note 39, at 436; John A. Robertson, *Oocyte Cytoplasm Transfers and the Ethics of Germ-Line Intervention*, 26 J.L. Med. & Ethics 211, 211 (1998).

160 *See* Kristie Sosnowski, *Genetic Research: Are More Limitations Needed in the Field?*, 15 J.L. & Health 121, 127-30 (2001).

161 Genes may be silenced by adding interfering RNAs to stop DNA transcription. Or genes may be inserted through homologous recombination or in artificial chromosomes.

Yet the prospect is realistic enough to merit discussion now. Scientists are quite skilled at "knocking out" or silencing genes in mice or other organisms.[162] The 2001 discovery of new classes of ribo-nucleic acids ("RNAs") that selectively silence genes after transcription will greatly increase the ability to knock-out genes.[163] Presumably all the techniques done in mice could occur in human embryos, though genetic manipulation experiments in human embryos are now quite rare, if they occur at all. At some point, using gene-editing techniques with human gametes or embryos will seem reasonable, and issues of whether the moral acceptability of such uses justifies allowing research to proceed will have to be faced.[164]

The *radical liberty* view easily accepts "editing" or "rewriting" the genomes of prospective offspring.[165] *Strict traditionalists*, on the other hand, would staunchly oppose any genetic alteration, with the possible exception of therapeutic ones, but they are unlikely to accept the research on embryos that would be necessary to establish the safety and efficacy of therapeutic germline interventions.[166] Even *modern traditionalists* are likely to be leery of non-therapeutic alterations at this time. Ever pragmatic, however, they are willing to examine the facts and see if plausible reproductive uses of gene alterations could be made in the future. Indeed, a *modern traditionalist* risks inconsistency if she dismisses gene alteration out of hand. She should be open to a right to "edit" or alter offspring genomes when necessary as a plausible reproductive strategy, just as she is with non-medical negative selection or reproductive cloning for gametic infertility.

A key issue is whether positive alteration does serve important reproductive goals. Parents demanding the right to alter genes before birth would have to show that alteration is not a mere preference, all other things being equal, but is essential to whether they will reproduce all.[167] But if they are otherwise fertile and likely to have a healthy child, it may be difficult to see why their "need" to alter genes is so key that it should be respected. Even if it would confer fitness advantages on offspring, the resulting distribution of desirable genes would raise serious justice issues and risk genetic domination by the few over the many.[168]

Also important will be the impact on the child. Because alteration in most cases will generally aim at improving the life-prospects of a child, it will be hard to show

---

162 Scientists can transfer genetic material between organisms, for example, transfecting pigs with the silk-producing genes of spiders so that pigs produce silk in their milk. Lawrence Osborne, *Got Silk*, N.Y. TIMES MAG., June 16, 2002, at 49. At present, DNA is simply injected into the nucleus, counting on principles of homologous recombination for it to be taken up at the intended location. Those cells that have taken up the genes can be identified and then cloned to produce many copies. Improved vectors for inserting DNA into cells or artificial chromosomes may be a more efficient way to transfer genes between cells.

163 *Science* magazine, the leading science publication in the world, named this discovery the most important scientific event of 2002. Jennifer Couzin, *Breakthrough of the Year*, 298 SCIENCE 2296, 2296 (2002).

164 Henry E. Malter et al., *Gene Silencing in Mouse Embryos Using Short Interfering Oligoribonucleotide-Based Double-Stranded Constructs*, 78 FERTILITY & STERILITY S75 (Supp. 2002). Such research would raise issues of embryo and human subjects research, research funding policy, and patents, all of which are beyond the scope of this article.

165 If DNA is the "code" for proteins and cellular functioning, then instructions in the code for disease-causing proteins could be edited out or rewritten to ensure a healthy child. *Strict traditionalists* are likely to be offended by this articulation for they would perceive it as assuming that the child is an object or product which the parents may legitimately design or fashion as they wish.

166 *See* CLONING REPORT, *supra* note 13, at 96-110.

167 For discussion of the importance of use of the technique being "essential" and not a "mere preference," see Robertson, *supra* note 39.

168 See ACKERMAN, *supra* note 154, at 113-38.

that the child is harmed as a result.[169] True, the parents might have hopes and expectations for the child based on the engineered trait, but parents could still be loving and respectful of a child whose genes they have altered. Assessing the impact on the child is doubly difficult because it may be difficult to show that the child has been harmed as a result of the alteration. But for the technique in question, the child claimed to be harmed might never have been born.[170]

At the same time the *modern traditionalist* can sympathize with many of the concerns that animate the *strict traditionalist*. Many persons would find genetic manipulation of offspring to engineer traits as the epitome of "designing" or "manufacturing" a baby, of turning the "gift" of a child into a product acceptable only with those designed traits. The strongly negative connotation of the term "designer babies" reflects the fear that parents will use genetic technology to turn their children into objects or commodities that undermine their freedom and dignity.[171] Although only a few parents will do so at first, others will join the race if early movers into non-medical enhancement gain an advantage. A "positional arms race" could ensue that leaves the competitors at roughly the same relative position as they were before expending many resources and changing the social tone by doing so.[172] Those without the resources to compete may be left even further behind. Policies that discourage such inequities might well be desirable.[173]

In the end, the acceptability of positive alteration of human genes before birth will depend heavily on the reasons motivating parents and the benefits and harms of the alterations sought. To assess the arguments and competing interests, I address three situations (therapeutic alteration, non-therapeutic enhancement, and intentional diminishment) in which "editing" or modification of a prospective child's genome code might occur.

---

169 The *strict traditionalist* might argue that any alteration harms the child by robbing it of its qualities and experiences it would have had if it had been "begotten" and not made.

170 Indeed, some persons would argue that the very process of experimentation that would be necessary to perfect these techniques risks harming the children who are born as a result. They call for changes in human subject research regulations to prevent such work from occurring. *See* Rebecca Dresser, Designing Babies: Research Ethics Issues (2003) (unpublished manuscript, on file with American Journal of Law & Medicine). While this position at first blush is appealing, it has not considered the fact that the children in question would not have been born if the experimental technique in question had not been used. Even if those children, strictly speaking, have not been harmed, it does not follow that such research or other gene alteration activity is otherwise acceptable or part of a person's reproductive liberty. *See* discussion *infra* Appendix.

171 Francis Fukuyama gives three arguments against such choices, but the ones he lists (violating the dignity of a human, the right of a human to be human, and the preservation of human nature) are too vague and undeveloped to perform serious work in determining which genetic technological innovations would be acceptable. FRANCIS FUKUYAMA, OUR POSTHUMAN FUTURE: CONSEQUENCES OF THE BIOTECHNOLOGY REVOLUTION (2002).

172 *See* ROBERT H. FRANK & PHILIP J. COOK, THE WINNER-TAKE-ALL SOCIETY: WHY THE FEW AT THE TOP GET SO MUCH MORE THAN THE REST OF US 167-87 (1995) [hereinafter WINNER TAKE-ALL SOCIETY]; ROBERT H. FRANK, LUXURY FEVER: WHY MONEY FAILS TO SATISFY IN AN ERA OF EXCESS (1999); BUCHANAN, *supra* note 92, at 222-56.

173 Whether restrictions aimed at limiting positional arms races would be constitutional will depend upon whether non-medical enhancement falls within procreative liberty, and the importance of communal efforts to prevent genetic segmentation in society.

1. Therapeutic Alterations

A plausible case could be made for genetic alteration on therapeutic grounds in a few circumstances, but they are likely to be rare.[174] The most likely candidate for germline gene therapy would be couples who face the 1 in 4 risk of a child with serious genetic disease, such as sickle cell anemia, Tay Sachs disease, or cystic fibrosis.[175] They could screen embryos by PGD for that condition and have a child by transferring only those which lack the mutation. Given that option, there is little reason why a couple with healthy embryos to transfer (the 75% that are heterozygous) would have an interest in genetic alteration of the affected embryos.[176]

The most plausible situation for a parent requesting therapeutic alteration of affected embryos would be if the only viable embryos they could produce had the disease-causing mutation. Unless they silenced certain genes or inserted new DNA, they would have no healthy offspring. Such cases are also likely to be rare. If all embryos in a given IVF cycle were positive for the disease, most couples could simply go through another IVF cycle to create embryos that were mutation free. While some couples may not ever be able to produce healthy embryos, for example, when one partner has two copies of a dominant gene or both partners have two copies of a gene for a recessive trait, that number is likely to be quite small.

But the importance to individuals, not their numbers, should determine whether a right exists.[177] Assuming that genetic modification of mammalian and human embryos has been shown to be safe and effective, a couple unable to produce healthy embryos would have a plausible claim to use germline gene therapy of affected embryos so that they might have healthy offspring. Unless the genetic alteration occurred, they would have no viable means of producing healthy offspring. In those circumstances, a ban on germline intervention would effectively limit their ability to have offspring.

A major ethical concern with germline alterations would be the risk of a deleterious impact on offspring.[178] The alteration here, however, is designed to benefit the child by permitting it to be born without a disease that would greatly limit its opportunities. Indeed, unless the alteration occurs, the parents might not reproduce at all, thus depriving the child, which opponents of germline interventions seek to protect, of the life that it would have had if the alteration had occurred.[179] The main risk to offspring then is not genetic engineering as such, but rather whether that process itself is safe and effective enough that the intended therapeutic benefits

---

174 The distinction between "therapeutic" and "non-therapeutic" alterations is a rough cut of the issues. If pushed further, the distinction might collapse in many circumstances. *See* BUCHANAN, *supra* note 92, at 107-54.

175 This analysis would apply also to dominant conditions (50% chance of the child being born with the mutation).

176 A strict right-to-lifer might choose to do so to save embryos with the mutation, but it is unlikely that persons with those views would be requesting PGD to screen embryos in the first place. Perhaps a family that wanted siblings for a child born after IVF and PGD and was not able to undergo another IVF cycle might request to gene therapy on remaining embryos with the mutation. But such cases are likely to be infrequent.

177 *See Casey*, 505 U.S. at 894 (holding that husband notification requirement that would likely bar very few women from abortions is unconstitutional despite the small number affected).

178 The argument has also been made that such germline genetic engineering could remove desirable genes from the gene pool. Given the small number of cases in which such deletions would occur, this fear is not realistic. If important genes were lost, the missing genes could be inserted in later generations.

179 For further discussion of this point, see *infra* Appendix.

will be achieved. If safety is established, it is hard to see how the child has been harmed and the parents and provider subject to moral condemnation for their efforts to remove disease genes from it.[180]

Such certainty, however, is possible only if enough cases of human germline gene therapy have already occurred to establish its safety and efficacy. To achieve that level of clinical certainty, one would have to show that the first, experimental transfers to the uterus of altered embryos were themselves justified on the basis of facts concerning apparent risks, including for example, extensive experience with germline engineering of other mammals, including primates, and blastocyst-stage studies of the effects of genetically altering earlier human embryos. When there is a reasonable basis for thinking that actual harm will be minimal, and the parents are committed to rearing and loving the resulting child, proceeding with the first embryo transfers would be ethically justified. The child will not have been harmed if it has no other way to be born healthy.[181]

2. Non-therapeutic Alteration—Enhancement

Discussions of non-medical genetic enhancement are highly theoretical, because altering human embryos for non-medical purposes is even further off in the future than germline therapy. Most desirable traits are likely to be polygenic in origin and thus not subject to easy manipulation, even if the relevant family of genes were known.[182] Without a close connection to treating or preventing disease, however, the resources for finding those genes may never be forthcoming.

Once the genes controlling or affecting those characteristics were known, techniques for inserting them into gametes or embryos would have to be perfected. Only after extensive experience with animal models and human embryos would transfer of genetically altered embryos to the uterus be reasonably considered. Further study would be needed to determine whether children born after alterations have the phenotypic traits sought. As noted, some persons would argue that any experimental embryo transfer is unethical, because the future child is not available to give consent to the research.[183] That child, however, does not exist at the time of the

180 Even if unsafe, the child itself, strictly speaking, is not harmed in a person-affecting system of harm. However, the authenticity of the parents' reproductive project can then be questioned. *See* discussion *infra* Appendix.

181 The ethics of research here is quite complex. Strictly speaking, no child would have existed unless the experimental gene therapy had been done. Even if children are born with injuries or anomalies, they would have had no alternative way to have been born without them. If they suffer inordinately, they may have no interest in continued living, and maintaining their lives would arguably violate their right to be free of inordinate suffering. For discussion of this point, see *infra* Appendix.

182 *See* Michael J. Reiss, *What Sort of People Do We Want? The Ethics of Changing People Through Genetic Engineering*, 13 NOTRE DAME J.L. ETHICS & PUB. POL'Y 63, 75 (1999); Robertson, *supra* note 39, at 436.

183 Leon Kass and the President's Bioethics Council make this move to oppose reproductive cloning. They argue that it would never be ethical to transfer a cloned embryo to the uterus because of the resulting child's lack of consent. Indeed, they suggest that the first IVF embryo transfers were unethical because of the child's lack of consent. As this paragraph shows, however, they are mistaken that the child is harmed by being born after research and thus that its advance consent is necessary. To make that claim, they have to adopt a person-affecting theory of harm. Since the person comes into existence only as a result of the embryo transfer, he or she cannot have been harmed by it (unless truly a wrongful life, in which case cessation of all life support would be morally required). *See infra* Appendix. To make their case, they must appeal to non-person-affecting principles of harm which they have failed to discuss. A general objection to children as "products" does not provide such a theory. *See* CLONING REPORT, *supra* note 13, at 109.

transfer and would only come into being if the transfer occurred. If parents requesting the procedure are committed to loving and rearing the child born after the experimental transfer, e.g., they are consenting to the experiment in order to have that child, then they are not harming the child nor violating social norms of parental commitment to the well-being of offspring.[184] If the subject of research is not able to consent, but its interests are protected or advanced, then the research may still ethically proceed.

Suppose, however, that animal research has demonstrated the safety and efficacy of non-therapeutic genetic enhancement and that genetically engineered children have been safely born and flourish. May a parent claim a right to use such techniques on their own prospective children as an exercise of procreative or familial liberty? The argument for procreative liberty would rest on the parents' claim that genetic enhancement was essential to their decision to reproduce—that they would not reproduce unless they could be assured through genetic enhancement that their child would be well-equipped for the competitions and vicissitudes of life. They would not be claiming a mere preference that the child be enhanced, but that it was a *sine qua non* of their decision to reproduce at all.[185]

Even if such a claim satisfies a necessary condition for procreative liberty, it may not be sufficient. Because the couple is sexually fertile and the expected child is healthy, their reproduction—and their progeny's reproduction—could easily occur without resort to genetic enhancement. The parents might believe that only genetic enhancement will equip the child sufficiently for life's challenges, but they may not be rational in holding such beliefs. The situation might be different if other parents routinely practiced genetic enhancement, thus placing their child at a disadvantage. Until such wider use occurs, however, parents desiring to use genetic enhancement might not have a convincing case that non-medical enhancement is essential to their child's well-being. Although building on the understandable parental commitment to provide for the well-being of their child, they have gone far beyond what is reasonably necessary for why reproduction is valued.[186]

The claim, however, that enhancement is not sufficiently "reproductive" or central to parental rearing concerns must contend with two counterfacts. The first is that parents engage in many sorts of non-medical enhancement of their children's attributes after birth and, indeed, may even have a constitutional right of familial autonomy in rearing offspring to do so.[187] If so, it should not matter, if physical safety is ensured, that the enhancement efforts occur prior to birth. Yet others would argue that genetic changes are different in quality and kind, not merely degree[188]. They are likely to be more permanent, will affect the germline, and are potentially much more dangerous. A key problem for the *modern traditionalist* is to justify the moral difference or lack thereof between pre-birth and post-birth enhancement.

---

184 For further elaboration of this point, see *infra* Appendix.

185 *See* Robertson, *supra* note 39, at 429-32.

186 For example, the claim of a right to alter the genes of prospective offspring to enhance its longevity would also fail. A long lifespan is not necessary for the successful reproduction of that progeny. Requests to extend the life of future children might then reasonably be viewed as not part of reproductive choice. Although a longer life might serve the interests of their children, not everything that serves offspring interests falls within reproductive liberty.

187 Meyer v. Nebraska, 262 U.S. 390, 400 (1923); Pierce v. Soc'y of Sisters, 268 U.S. 510, 534-35 (1925); Wisconsin v. Yoder, 406 U.S. 205, 232 (1972).

188 *See* Maxwell J. Mehlman, *The Law of Above Averages: Leveling the New Genetic Enhancement Playing Field*, 85 IOWA L. REV. 517, 537-39 (2000).

A second obstacle is the status of non-medical selection of traits other than gender. If negative selection of non-medical traits is not permitted, then *a fortiori* there is no case for positive alteration. But if non-medical selection is allowed, it will be difficult on grounds other than safety to argue against inserting genes for the same trait. If safety is established, then the same arguments against germline or inherited modifications as against pre-birth versus post-birth enhancement apply. Genetic changes are likely to be more permanent, affect future generations, and are potentially more dangerous. Again, the persuasiveness of the distinction between non-medical negative selection and non-medical positive alteration will be a key issue in the moral and social acceptance of these techniques.

If non-medical genetic enhancement is not a clear expression of procreative or familial liberty, then a rational case against it would suffice as grounds for restricting it and could easily be established. Government concerns that it might actually harm children, either in the actual engineering or in the expectations that parents then have of them, and the risk of creating classes of differentially endowed citizens provides a sufficient, rational basis for preventing such uses of genetic technology.[189]

A more difficult question arises if genetic enhancement is found to be essential for parental reproduction and rearing.[190] A governmental ban on positive alteration would require stronger justification. One possible justification would be to prevent positional arms races for the genes of children that will be hard for parents, concerned about the well-being of their children, to resist. The danger is that parents will feel obligated to engage in pre-birth genetic engineering, because other parents are doing so, just as SAT prep courses have become routine and athletes feel obligated to use steroids if other athletes are gaining an advantage from them.[191] Although some first movers may gain some advantages for their children, eventually other parents with resources will catch up. As a result, relative positions may not have changed, but everyone would have spent more money to stay where they are, some children may have been injured in the process, and the gap with "have-nots" will have greatly increased. Unless the government acted, there would be no incentives for parents to refuse such techniques.[192]

---

189 The fact that the child is not itself harmed would not mean that government could not rationally try to discourage the activity. *See infra* Appendix.

190 For a discussion of the importance of the use of the technique being essential to reproduction and not a mere preference, see Robertson, *supra* note 39.

191 The phenomenon of wasteful and self-defeating positional arms races where each tries to improve his relative position versus others is now familiar. *See* WINNER-TAKE-ALL SOCIETY, *supra* note 172, at 167-87. If genetic enhancements become possible, agreements to discourage positional arms races will likely emerge through social norms, contracts, or governmental regulation. If so, the argument that restrictions on positional arms races in genetic enhancement of offspring would violate procreative liberty would be weak.

192 *See id.* at 186-87. The case for a ban on enhancement would be strongest if there were physical risks from the procedure, and the child would have been born regardless of the use of the harmful enhancement technique. Frank and Cook posit a situation in which a genetic enhancement technique has a 99% chance of a 15% improvement on SAT-type tests but a 1% risk of no improvement and severe emotional disability as a result. If there is such a high risk, a genetic enhancement could be banned on those grounds, e.g., as not safe and effective, for existing children. If the enhancement occurred prenatally, it might not be harmful to the child who would not otherwise have been born but for use of the technique. However, one might still reasonably question whether parents are exercising procreative liberty when they take a small risk of a great loss in order to make an otherwise healthy child somewhat better off.

480 AMERICAN JOURNAL OF LAW & MEDICINE VOL. 29 NO. 4 2003

3. Intentional Diminishment

The least persuasive case for parental freedom to use non-medical genetic alteration techniques is for intentional diminishment of prospective offspring—genetic alteration that aims to reduce or remove capabilities that would otherwise have made the child normal and healthy. A paradigm case of intentional diminishment occurs in the film *Bladerunner*.[193] An evil scientist genetically engineers human "replicants" with a limited life span to "off-planet" in menial positions.[194] Other frequently noted cases concern deaf or dwarf parents electing to have genes for hearing or height removed from prospective children so that they will have phenotypes similar to their parents.[195] The literature on this topic also occasionally discusses creating limited humans or human-animal chimeras to do menial tasks, to be "meat puppets," or to provide organs for other humans.[196]

The case for including intentional diminishment within the protective canopy of procreative liberty is even weaker than is the case for non-medical enhancement. Although human individuals would be born from the haploid chromosomes of two other individuals, the alteration is neither done for the well-being of the resulting child (except possibly in the case of the deaf), nor to increase its own prospects for successful reproduction. Rather, reproductive resources are being used to produce a human entity to serve parental preferences, not to benefit the child or accomplish transmission of genes to succeeding generations. Only the deaf or dwarf couple might have a plausible claim of reproductive interest. They are seeking a child to raise and love and want it to have size and hearing phenotypes closer to their own.[197]

Given the difficulty of showing that intentional diminishment falls within procreative or family liberty, concerns about harm to the child, who could have been born healthy but the parents have chosen to diminish it, would easily provide a rational basis for action ending such prohibition. Strictly speaking, the child would not itself be harmed if the parents had not brought it into the world whole and undiminished. Yet the government is not obligated to allow all possible children to be born, just because enabling them to be born would not "harm" them.[198] If procreative liberty is not infringed, society might reasonably prefer a different set of children and stop uses of techniques for intentional diminishment when the parents could otherwise have normal, healthy children.[199] The right to diminish offspring is simply not coherent as an expression of procreative or familial liberty, for it does not seek to produce healthy offspring who themselves will be fit to reproduce.

---

193 *See* Katheryn D. Katz, *The Clonal Child: Procreative Liberty and Asexual Reproduction*, 8 ALB. L.J. SCI. & TECH. 1, 17 (1997).

194 The plot is driven by the attempts of some "replicants" who have escaped to earth to find the genetic key that can unlock a normal human lifespan, and thus enable them to live out the love they have for each other.

195 Robertson, *supra* note 39, at 438-39.

196 JAMES BOYLE, SHAMANS, SOFTWARE, AND SPLEENS: LAW AND THE CONSTRUCTION OF THE INFORMATION SOCIETY 150-53 (1996).

197 But even that claim could be contested, even if ultimately accepted. In the end, one would expect to find very few legitimate reproductive cases of diminishment.

198 *See* Robertson, *supra* note 39, at 440-41.

199 *See infra* Appendix.

V. POLICY MAKING FOR REPROGENETICS

The prospect of expanded use of genetics in reproduction has produced demands for a more rational and focused set of public policies in this area. Many people assume that these techniques pose major individual and social risks, and need much closer monitoring than they have received until now. While some countries have central agencies that license and monitor assisted reproductive clinics, most notably the Human Fertilisation and Embryology Authority in the United Kingdom, there is no federal or state agency that examines, reviews, and licenses the provision of reproductive and genetic services;[200] nor is there a well-developed body of state law that defines rights, duties, and resulting legal relations in use of these techniques.[201] As a result, questions of the acceptability of genomic information in reproduction are left largely to the market created by patients and providers of these services. Some decry market domination of technological reproduction, while others see the market as a useful device for accommodating the diverse demands of new technology.[202]

Debates over the need for regulation of reproductive and genetic techniques often conflate two kinds of issues that need to be carefully separated to arrive at defensible solutions. One set of issues concerns the safety and reliability of the reproductive and genetic services provided. In addition to their other complaints about genomics in reproduction, *strict traditionalists*, for example, view the lack of regulation as a serious gap that allows commercially driven operators to create demand for untested and often unsafe products, and eventually for a "Brave New World" of "designer children."[203]

The call for regulation to protect safety and transparency, however, overlooks both the regulatory systems that affect provision of these services and the general hands-off approach that government has typically taken toward medical services. Physicians and hospitals must be licensed, meet tort law standards of good practice, and use only Food and Drug Administration-approved drugs and devices, regardless of the area of practice. They must also comply with institutionally imposed reviews of research with human subjects and, in some cases, may have to seek approval of the Recombinant DNA Advisory Committee before conducting human gene therapy research.[204] Beyond those restrictions, most physicians are free to practice as they choose. Although some highly publicized cases of theft of embryos, use of own sperm, and misleading advertisements have occurred in infertility clinics, cases of

---

200 *See* Human Fertilisation and Embryology Act, c. 37 (1990) (Eng.); Maurice A.M. de Wachter, *The European Convention on Bioethics*, 27 HASTINGS CENTER REP. 13, 13 (1997); UNITED NATIONS EDUCATION, SCIENTIFIC AND CULTURAL ORGANIZATION, *at* http://www.unesco.org (last visited Nov. 13, 2003).

201 *See* Erik Parens & Laurie P. Knowles, *Reprogenetics and Public Policy Reflections and Recommendations*, 33 HASTINGS CENTER REP. S3 (Supp. 2003).

202 *Compare* Kely M. Plummer, Comment, *Ending Parents' Unlimited Power to Choose: Legislation is Necessary to Prohibit Parents' Selection of Their Children's Sex and Characteristics*, 47 ST. LOUIS U. L.J. 517, 518 (2003), *with* David Orentlicher, *Beyond Cloning: Expanding Reproductive Options for Same-Sex Couples*, 66 BROOK. L. REV. 651, 677 (2000/2001).

203 Legally trained commentators emphasize other issues. *See, e.g.*, Judith F. Daar, *Regulating Reproductive Technologies: Panacea or Paper Tiger?*, 34 HOUS. L. REV. 609 (1997); Lori B. Andrews, *Regulating Reproductive Technologies*, 21 J. LEGAL MED. 35 (2000); LORI B. ANDREWS, THE CLONE AGE: ADVENTURES IN THE NEW WORLD OF REPRODUCTIVE TECHNOLOGY (1999).

204 NATIONAL INSTITUTES OF HEALTH, DEP'T OF HEALTH & HUMAN SERVICES, NIH GUIDELINES FOR RESEARCH INVOLVING RECOMBINANT DNA MOLECULES (Apr. 2002), *available at* http://www4.od.nih.gov/oba/rac/guidelines_02/NIH_Guidelines_Apr_02.htm.

dishonesty, fraud, slanted advertisements, and rampant commercialization are not unique to assisted reproduction.[205]

Although some providers have exploited consumers, it is by no means clear that there are any greater problems in this area than in various other areas of medical practice, nor any reason to think that the healthy competition that exists among infertility providers will not doom most of those providing poor quality or dishonest services to short, professional half-lives. Patients are increasingly knowledgeable about clinic and provider success rates, and physicians compete hard for patients.[206] If abuses or egregious mistakes occur in genetic diagnosis or assisted reproduction, tort suits (of which there have been few[207]) and the consumer grapevine should provide adequate correction.

Given the differences between the British and U.S. legal systems, a central licensing or regulatory agency at the state or federal level in the United States may not be necessary. Indeed, creation of a new regulatory agency to govern reprogenetics is unlikely given the general reluctance to "throw" a new agency at social problems, particularly if doing so would reignite the bitter conflicts over embryo status and abortion that have accompanied most legislative forays into this area.[208]

The second motivation for regulation has been the ethical, legal, and social concerns raised by use of these techniques even when they are safe and effective, and consumers are fully informed of their risks, benefits, and proven or unproven status. To date, most regulation in this regard has occurred through state law defining the parental rights and duties arising from use of gamete donors and surrogates.[209] There have also been recurring debates at the federal level about the use of federal funds for embryo research.[210] Indeed, it is the intransigence of those conflicts that make the emergence of a comprehensive regulatory structure for these techniques at the federal or even state level unlikely, as recurring controversies over embryo research and, most recently, over therapeutic and reproductive cloning have shown.

A brief review of recent federal activity will show the slim chances for a national consensus on how to handle the reproductive issues likely to arise in the era of genomics. Since 1980, administrative inaction spurred by right-to-life sentiments had blocked the use of federal funds for embryo research.[211] The Clinton administration was prepared to remove the ban administratively and appointed the

---

205 *E.g.*, John A. Robertson, *The Case of the Switched* Embryos, 25 HASTINGS CENTER REP. 13 (1995); Karen T. Rogers, *Embryo Theft: The Misappropriation of Human Eggs at an Irvine Fertility Clinic Has Raised a Host of New Legal Concerns for Infertile Couples Using New Reproductive Technologies*, 26 SW. U. L. REV. 1133 (1997).

206 The one exception is the 1992 Fertility Clinic Success Act, which fosters a voluntary reporting system through the Centers for Disease Control and the relevant professional groups.

207 Matthew Browne, Note, *Preconception Tort Law in an Era of Assisted Reproduction: Applying a Nexus Test for Duty*, 69 FORDHAM L. REV. 2555, 2559-60 (2001).

208 The creation of the new Department of Homeland Security to deal with terrorist threats may be an exception.

209 *E.g.*, N.H. REV. STAT. ANN. § 168-B:4 (2002); VA. CODE ANN. § 20-158 (2000).

210 Erin P. George, Comment, *The Stem Cell Debate: The Legal, Political and Ethical Issues Surrounding Federal Funding of Scientific Research on Human Embryos*, 12 ALB. L.J. SCI. & TECH. 747 (2002).

211 Shannon Brownlee, *Designer Babies: Human Cloning Is a Long Way Off, But Bioengineered Kids Are Already Here*, WASH. MONTHLY, Mar. 2002, at 25, 27.

Human Embryo Research Panel for guidance on the matter.[212] That panel strongly supported federal funding of embryo research.[213] Before any research was funded, however, Congress passed a ban on any use of federal funds for "embryo research," thus leaving the matter in the hands of the private sector.[214]

The right-to-life versus scientific research fault-line surfaced again in debates over federal funding of embryonic stem cell research. That debate began in 1998 when researchers at Johns Hopkins University and the University of Wisconsin developed ways to culture human embryonic stem cells indefinitely in the laboratory, opening the door to directing them to produce replacement tissue to treat disease.[215] At the behest of President Clinton, the National Bioethics Advisory Commission examined the issue and recommended federal funding both of derivation of and research with embryonic stem cells.[216] The Bush administration, however, pulled back from this recommendation before the National Institutes of Health could make any awards.[217] On August 9, 2000, President Bush announced that he would permit federal funding of research with embryonic stem cells derived before that date.[218] Although this "compromise" gave federal support to some embryonic stem research, fewer cell lines than first imagined were available, thus relegating most research again to privately funded actors.

The latest reprise of the battle has occurred in the fight over a ban on therapeutic cloning. In 2001, the House of Representatives had overwhelmingly voted in favor of a criminal ban on therapeutic as well as reproductive cloning.[219] After intense lobbying by both right-to-life, patient, and biotech industry groups, the issue died in the Senate.[220] With the Republican victory in the 2002 elections, however, it is now back on the public agenda. Once again, we see pro-life forces battling with scientists and patient groups for whether scientific research using embryos may occur. Given the sharp split, reproductive matters remain once again in the private sector with no direct regulatory oversight.

Given how abortion politics entangle every turn of the policy-making road, it is unlikely in the near term that, other than Food and Drug Administration oversight of safety, extensive federal or state oversight of the uses of genetics in reproduction will become part of the regulatory landscape in the United States. Most regulation will occur informally through the market interactions of willing consumers and providers of these services against a background of common law norms, some

---

212 R. Alta Charo, *Bush's Stem-Cell Decision May Have Unexpected–and Unintended–Consequences*, CHRON. HIGHER EDUC., Sept. 7, 2001, at B14.

213 President Clinton, however, did reject the Panel's recommendations on federal support in some cases of creating embryos for research.

214 H.R. 2127, 104th Cong. § 511 (1996).

215 James A. Thomson et al., *Embryonic Stem Cell Lines Derived from Human Blastocysts*, 282 SCIENCE 1145, 1145-47 (1998); Michael J. Shamblott et al., *Derivation of Pluripotent Stem Cells from Cultured Human Primordial Germ Cells*, 95 PROC. NAT'L ACAD. SCI. 13726 (1998).

216 General Counsel Harriet Raab had previously issued an opinion stating that since embryonic stem cells were not themselves embryos, federal funding of research with them did not violate the Congressional ban on federal funding of research with embryos.

217 ZENIT, *Report Fueling Fears About Stem Cell Research: Embryos Created and Destroyed on Purpose*, *at* http://users.colloquium.co.uk/~BARRETT/stems.htm (last visited Nov. 13, 2003).

218 Katherine Q. Seelye & Frank Bruni, *A Long Process that Led Bush to His Decision*, N.Y. TIMES, Aug. 11, 2001, at A1.

219 *See* Human Cloning Prohibition Act of 2001, H.R. 2505, 107th Cong. (2001).

220 Nicholas Wade, *Scientists Make 2 Stem Cell Advances*, N.Y. TIMES, June 21, 2002, at A18.

professional self-regulation, and occasional state legislative intrusions.[221] Unless some major problems develop, some version of the current decentralized system of review, with many different centers of power and influence offsetting each other, is likely to continue to characterize the regulatory landscape of reproductive and genetic technologies. If this mixed system proves grossly incompetent for the task, more direct regulation will come into being.

VI. CONCLUSION

I have looked at the ethical, legal, and social conflicts that are likely to arise from increased use of genetics in reproduction and the approaches to resolving those conflicts that are now in play. A central focus has been to show how a *modern traditionalist*, rather than *strict traditionalist* or *radical libertarian*, perspective on these issues is the most fruitful approach to reconciling the competing interests that arise from the growing use of genomic knowledge in reproduction.

Modern traditionalism strongly supports a liberty claim-right to use genetic knowledge and techniques to have healthy offspring to nurture and rear. Genetic techniques that directly aim to serve those goals are usually ethically acceptable and should be legally available, for their use fits neatly into traditional understandings of why reproduction is valued. Access to them, however, could be limited if they imposed serious harms on the persons most directly affected by them. As genetic techniques grow in importance, providing access to persons without the means to obtain them will also be important.[222]

Applying this perspective to four areas of current or future controversy, this article has shown that the most likely use of these techniques serve standard reproductive goals without causing undue harm to values of respect for prenatal life, the welfare of offspring, the status of women, or social equality. Use of some techniques, however, such as reproductive cloning when fertile, intentional diminishment, and possibly non-medical enhancement do not clearly advance conventionally understood reproductive agendas and deserve less respect than other uses.

The analysis has also shown that many fears of abusive use of reproductive technologies are highly speculative and may not occur even when techniques are perfected. If genomic knowledge is increasingly used in reproduction, it is likely to be used primarily in those cases where it reasonably serves goals of having healthy children to nurture and rear. Techniques that inspire the worst fears—genetic enhancement, intentional diminishment, reproductive cloning when fertile—are least

---

221 Such public bodies as the National Institutes of Health Human Embryo Research Panel (1996), the National Bioethics Advisory Commission (1998-2001), and the President's Council on Bioethics (2002-2004) have provided a useful service in analyzing issues and organizing discussion. So have bodies from the National Academy of Sciences, the Institute of Medicine, the American Association for the Advancement of Science, and interested professional bodies, such as the Ethics Committees of the American College of Gynecology and Obstetrics, and the American Society of Reproductive Medicine. Although only advisory, their reports and recommendations have helped clarify and shape the social and political debate by pointing out the major normative fault lines and the tradeoffs that different public policies could have, with recommendations for how best to balance them. They have also played a role in advising courts, legislatures, institutions, and providers about ethical practices.

222 John Deigh has argued that if genetic enhancement becomes feasible, there will be an obligation to provide it to the poor as well (personal communication with author).

likely to be of interest to people interested in the age-old project having healthy offspring.[223]

As the genomic revolution continues to unfold, ethical, legal, and social controversy over how to use and regulate these techniques will continue. Rather than limit broad categories of use in order to stop the most unlikely and least defensible uses, policymakers should shift their attention to ensuring that genetic techniques are used in informed, safe, and productive ways. It is not inherently wrong to achieve traditional reproductive goals in novel, technological ways that use the insights of genomic approaches to human biology. A strong commitment to these principles should help ensure that the use of genetic technologies in reproduction will increase, rather than diminish, human flourishing.

223 This statement does not hold as strongly for genetic enhancement but there are other features of that practice that could limit its use. *See* discussion *supra* Part IV.D.2.

APPENDIX: HARM TO OFFSPRING WHO WOULD NOT OTHERWISE BE BORN

The question of harm to offspring born as a result of genetic and assisted reproductive technologies raises perplexing issues. The problem arises because most of our ethical reasoning assumes a person-regarding conception of harm. If an action or omission does not set-back the interests of a person, then it does not harm that person.

This question arises repeatedly in situations involving assisted reproduction and genetic technology, as many references in the text indicate. A common concern is that children born as a result of those techniques are "harmed" because of the physical, mental, social, or psychological conditions which attend their births. These techniques include cloning, genetic enhancement or diminishment, surrogacy, egg and embryo donation, and other manipulations of gametes and embryos.

The problem is that but for the use of (or failure to use) these technologies, the child whose welfare is at issue would never have existed and thus, under person-affecting theories of harm, has not herself been harmed. The technique or manipulation that causes the condition of concern is also the technique or manipulation that brings or causes the child to be brought into existence.[224] Viewed from the perspective of the now-existing child, she is better off than the alternative of not existing at all.[225]

Yet many persons are troubled by decisions that knowingly bring a handicapped child into the world when actions to avoid its birth were within reach. Taking the risk that the child with a cloned or altered genome will be born handicapped or damaged, or in a novel parenting situation, seems insufficiently attentive to the welfare of the resulting child.[226] Can we not condemn their actions as wrong and impose sanctions to prevent them from occurring?

Dan Brock and Derek Parfit have explored a non-person-affecting theory of harm to deal with those situations.[227] They argue that the world is better off if a class of 100 persons is all "normal" rather than a world with a class of 100 persons, 99 of whom are "normal" and one who has a disability. If that is so, a person would have an obligation to use reasonably available means to substitute a "normal" for a

---

224 The case is different if the parents have done something affecting the child's present condition which could have been avoided and yet the child be born, such as refraining from using drugs or alcohol in a pregnancy going to term.

225 I put aside the rare cases of truly wrongful life, in which every postpartum moment of life is full of excruciating pain. In such a case, one would have a moral obligation to cause the cessation of that child's life. However, David Heyd would disagree. He thinks that, until there is a person in being, there is no being to whom moral duties are owed, and thus no rights holder until birth occurs. Therefore, there has been no wrong until after the child is born, and efforts are not made to cease its excruciating existence. DAVID HEYD, GENETHICS: MORAL ISSUES IN THE CREATION OF PEOPLE 59-62 (1992). *See also* MELINDA A. ROBERTS, CHILD VERSUS CHILDMAKER: FUTURE PERSONS AND PRESENT DUTIES IN ETHICS AND THE LAW (1998).

226 Derek Parfit's example of the woman who could wait a month to conceive or take a pill and thereby avoid the birth of a child with a withered arm captures this sense perfectly. Derek Parfit, *On Doing the Best for Our Children*, *in* ETHICS AND POPULATION 100 (Michael D. Bayles ed., 1976).

227 DEREK PARFIT, REASON AND PERSONS 487-90 (1984); Dan Brock, *The Non-Identity Problem and Genetic Harms—The Case of Wrongful Handicaps*, 9 BIOETHICS 269 (1995). *See also* BUCHANAN, *supra* note 92, at 222-57. *See* ROBERTS, *supra* note 225; Melinda Roberts, *A New Way of Doing the Best That We Can: Person-Based Consequentialism and the Equality Problem*, 112 ETHICS 315, 315-50 (2002).

disabled child. In making this argument, Brock and Parfit recognize that it applies only if the number of children born is the same and excessive efforts are not required to substitute the "normal" child for the disabled one.

Let us assume that this analysis is correct. If so, their theory provides no basis for judging the vast majority of cases in which the numbers could not have been the same or there is no clear basis for determining that the efforts required to have the "normal" child are "excessive." Due to infertility, medical uncertainty, or strongly held personal beliefs, a couple may not be able to substitute a healthy child for the disabled one. Also, substituting the "normal" for the disabled child could require invasive prenatal diagnostic procedures and destruction of embryos or fetuses, which Brock and colleagues recognize may be too much to ask. Because of these constraints, their approach would also appear to exempt a person who would reproduce only if they could use the genetic enhancement or diminishment technique in question.

Another way to approach this problem, suggested at a few points in the text, is to ask whether parents who are willing to use (or not use) genetic modification and reproductive techniques are acting in ways that serve reproductive needs as commonly valued and understood. A relevant question would be whether the contested use makes sense as a way to satisfy an individual's goals of producing viable offspring in the next generation.

A relevant factor in answering that question is whether parents will rear and care for the resulting children just as other parents do. If so, it will be difficult to claim that they are not involved in achieving the usual goals of reproduction. While cases of extreme or bizarre views will challenge such an approach, a commitment or intent to rear makes the situation different from that of a person producing cloned or altered embryos, or children for others to rear with whom there is no genetic connection.[228] As long as parents who use these techniques are committed to rearing their child, they should be considered to be exercising or engaging in legitimate procreative activity. Questions in implementing this approach, as well as the legitimacy of imposing resulting rearing costs and burdens on others, will also need to be addressed.

---

228 Although that person may not be harming the child in pursuing reproductive goals, she may be harming other interests, such as a collective or communal interest in a certain moral tone. But using such judgments to constrain reproduction in a liberal society where people are generally free to choose their own ends and vision of the good is questionable.

[6]

Beyond "Genetic Discrimination": Toward the Broader Harm of Geneticism

Susan M. Wolf

The current explosion of genetic knowledge and the rapid proliferation of genetic tests has rightly provoked concern that we are approaching a future in which people will be labeled and disadvantaged based on genetic information. Indeed, some have already suffered harm, including denial of health insurance.[1] This concern has prompted an outpouring of analysis.[2] Yet almost all of it approaches the problem of genetic disadvantage under the rubric of "genetic discrimination."[3]

This rubric is woefully inadequate to the task at hand. It ignores years of commentary on race and gender demonstrating the limits of antidiscrimination analysis as an analytic framework and corrective tool.[4] Too much discussion of genetic disadvantage proceeds as if scholars of race and gender had not spent decades critiquing and developing antidiscrimination theory.

Indeed, there are multiple links among race, gender, and genetics. Dorothy Roberts has discussed the historical links between racism and genetics,[5] while she and others have begun to map connections between gender and genetics.[6] Thus genetics has sometimes been used in the past to disadvantage people by race or gender or both. Whatever is wrong with disadvantaging people by race or gender would therefore be wrong with using genetics to do so. In addition, however, harm based on race or sex is a useful model for analyzing harm based on genetics. Both racial and gender differences have historically been construed as biological and "real," thereby justifying differences in treatment.[7] Genetic differences, too, are typically regarded as biological and "real," justifying differences in treatment, with too little attention to the social choices involved.[8]

Much work on race and gender thus speaks to genetics as well. And critique of antidiscrimination theory has been pivotal in that work. Indeed, Martha Minow argues that scholarship on gender can be divided into three stages, depending on the approach taken to notions of equality and discrimination.[9] In the first, women sought to be treated the same as men. In the second, women sought respect for their differences. In the third, the sameness/difference debate has been transcended by focusing on the limits of equality theory and by developing new approaches. These include careful attention to the way that the categories of race and gender are both used to disadvantage individuals.

Most writing on genetic disadvantage thus far seems mired in a first-stage understanding of equality theory and use of antidiscrimination. This writing clings to the Aristotelian notion that equality is simply treating likes alike, and then seeks to render genetic information beyond the reach of insurers, employers, and so on, to force identical treatment regardless of genetics. When this fails, "discrimination" is claimed, relying on antidiscrimination statutes and constitutional equal protection doctrine.

This ignores serious problems posed by the standard antidiscrimination approach, and disregards the lessons of the literature on race and gender. Some lessons are doctrinal; certainly race- and gender-attentive writing on existing statutory and constitutional law can school efforts to use law in the genetics sphere.[10] But this article focuses on deeper lessons: not on strategies for using current law, but on a theoretical critique that should fundamentally revamp the way we frame the genetic problem and devise our legal, moral, and political responses.

I argue below that current state statutes and federal bills prohibiting genetic discrimination by health insurers demonstrate the inadequacy of an antidiscrimination approach to genetics. I then turn to the literature on race and gender to learn why antidiscrimination analysis fails. That

Volume 23:4, Winter 1995

literature shows that an antidiscrimination approach instantiates a norm that does not exist in genetics and merely entrenches genetic bias. Moreover, using antidiscrimination analysis to ensure uniform treatment within the existing system fails to question the system itself and, thus, thwarts structural change. Third, that analysis pretends that the problem is deliberate use of genetic information, ignoring the deeper psychological dimensions—the pervasiveness of stereotypes, unfounded beliefs, and prejudices fueled by genetic notions. And fourth, an antidiscrimination approach obfuscates the connections between problems of genetics, race, and gender, and so helps reinforce, not overturn, the status quo.

Finally, I turn to work on race and gender for alternatives to antidiscrimination analysis. Prominent alternatives suggest reunderstanding the harm as something broader than discrimination, as rather the use of genetic notions to privilege some individuals and subordinate others. Thus, the problem is more fundamental than antidiscrimination analysis indicates. It is seeing individuals as their genes, and struggling to glean genetic information from their medical records as if it were the hidden truth.[11] It is subdividing communities by their genetic characteristics, and promoting the idea that genetic differences are real, biological, and neutral grounds for different treatment.[12] This is something more akin to the pervasive harm of racism and sexism. Indeed, as I have noted, it is historically linked to both. I thus propose a new rubric for the harm at issue: not "genetic discrimination," but *geneticism*.

The inadequacy of discrimination theory: genetics and health insurance

The poverty of current antidiscrimination analysis is illustrated by state statutes now being enacted and federal bills now being considered to prevent the use of genetic information and "genetic discrimination" by health insurers.[13] The standard discrimination analysis embodied in these statutes and bills does not solve the problem it is meant to address.

A spate of recent state statutes aims to prohibit genetic discrimination by health insurers. The states bear the primary responsibility for regulating insurance, though the federal government retains jurisdiction over many features of self-insured health benefits plans through the Employee Retirement Income Security Act.[14] Starting in the 1970s, state legislatures began to show concern about health insurers' use of genetic information, by passing legislation forbidding insurers from denying insurance or charging higher premiums simply because a person had a certain genetic disorder or was a carrier for a certain genetic trait.[15] Throughout the 1970s and 1980s, a few states passed legislation that restricted insurers' use of knowledge that a person had sickle-cell trait, Tay-Sachs trait, or another trait on a limited list, though two states more broadly covered all genetic traits without ill effects on the carrier.[16] However, the protection afforded even by these broader statutes was partial: they only covered traits without ill effects and one statute still permitted use of the genetic information when supported by "actuarial justification."[17] In other words, insurers *could* use genetic information in their usual practice of risk-rating; they were simply forbidden from additional exclusionary practices against carriers. From in the protected category of carriers, this is akin to saying that one cannot discriminate by race or gender unless reasonably related to a legitimate and usual business goal. Only a narrow form of discrimination is forbidden by such an approach.

Since 1990, with the start of the Human Genome Project, a new sort of state legislation has emerged, giving somewhat broader protection.[18] Wisconsin passed a statute forbidding health insurers from requiring or requesting that a person or family member take a genetic test, reveal whether they had previously taken one, or share the results.[19] Moreover, the insurer could not consider whether a test had been taken or test results in determining coverage, benefits, or rates. A "genetic test" was defined as a test using DNA to determine the presence of a genetic disease, disorder, or predisposition.

A number of states have now followed Wisconsin's lead.[20] This second generation of antidiscrimination statutes broadens the protections accorded by the first generation in two major ways. Rather than protecting a limited set of traits, nearly all of these newer statutes cover the category of "genetic tests" or "testing." They also no longer permit actuarial practices to justify insurers' use of these genetic tests.

At first glance, this broader approach may seem adequate to prevent health insurers from disadvantaging people on genetic grounds. But, as recent analyses have shown,[21] these statutes, too, are problematic. "Limiting the scope of protection to results of genetic tests means that insurers are only prohibited from using the results of a chemical test of DNA, or in some cases, the protein product of a gene. But insurers can use other phenotypic indicators, [or] patterns of inheritance...."[22] In other words, the approach of the current statutes is akin to saying that insurers cannot use a certain set of tests or category of information to ascertain a person's race or gender. The insurer can still rely on other indicators to try to glean those characteristics and then disadvantage the person accordingly. The approach of these statutes is thus not akin to ruling out altogether consideration of race or gender, no matter how ascertained.

This problem is further demonstrated by federal bills now under consideration. The Genetic Privacy and Nondiscrimination Act of 1995, introduced by Senators Hatfield and Mack, clearly pursues an antidiscrimination approach. One of its three stated purposes is to "protect against discrimination by an insurer or employer based upon an individual's genetic information."[23] It thus prohibits health insurers from using "genetic information."[24] "Genetic information" is defined more broadly than "genetic tests,"[25] but is still limited to "information about genes, gene products or inherited characteristics that may derive from an individual or a family member."[26] Yet all sorts of health information may suggest genetic propensities. Thus, the bill fails to rule out all genetic considerations. Representative Stearns's House Bill is substantively identical.[27] And Representative Slaughter's bill, the Genetic Information Nondiscrimination in Health Insurance Act, suffers from the same problem by similarly prohibiting health insurers from penalizing an individual or family member based on "genetic information" or the request or receipt of "genetic services."[28]

The concern that insurers will hunt for information on which to base genetic judgments despite these prohibitions is not just theoretical. The Task Force on Genetic Information and Insurance has expressed skepticism that genetic information can be segregated effectively from the rest of a person's medical record, especially as we come to appreciate the complex interaction between genetic and nongenetic factors underlying many diseases.[29] And even if genetic information could be segregated, "much of the information in the record about risk factors, diseases, diagnostic tests and treatments will in fact reveal genetic information to the astute reader of that record."[30] Moreover, insurers may draw conclusions about genetics and may disadvantage people based on misunderstandings about genetics and what the medical record, an individual's history, or a family's history indicates. The 1992 study by Paul Billings et al. demonstrates this, as in the case of an individual "denied a job because a health record noted that the applicant's mother was 'schizophrenic'."[31]

Thus the social practice that needs to be changed is broader than health insurers' accurate use of genetic tests and information. It is the eagerness to draw genetic conclusions, the search for supposedly deviant genes, and the conviction that such genes actually deserve disadvantage. This sort of attitude and practice is not adequately addressed by the notion of "genetic discrimination." That notion, as it has been defined, condemns treating an individual or family differently from another "solely because of real or perceived differences from the 'normal' genome."[32] Yet the social practice involves creating genetic categories, actively looking for any kind of information about people in order to sort them into those categories, and harboring attitudes and prejudices that motivate such behavior. Thus, it is systematic, not just individual; a matter of cognitive mindset, not just isolated behaviors; and a domain of stereotypes and unfounded beliefs, not just accuracy and rationality.

All of this makes the problem more akin to the pervasiveness of racism or sexism than to isolated instances of race or sex discrimination. Combating individual instances in which a person of color was treated differently because of race, or a woman treated differently because of her sex, will not uproot deep and widespread racism and sexism. Similarly, condemning individual instances of genetic discrimination will do little to address systematic genetic categorization, a world view that seeks to sort people by their genetics, and the conviction that supposedly deviant genes merit different treatment.

It will also do little to address what some have correctly argued is a profound structural problem: the shift from community-rating to risk- or experience-rating as the basis of health insurance.[33] As Thomas Murray has written, "[h]ealth insurance in the United States has moved from a system based mostly on community rating where, in a given community, all people pay comparable rates, to a system where the cost to the purchasers ... is based on the expected claims...."[34] This permits insurers to charge higher premiums to those people who will need the most expensive care, or to deny them coverage entirely. In this context, genetic inference becomes particularly important; it "may permit a much more complete and refined classification of people into risk categories, and so move us further away from sharing the financial burdens of illness...."[35] In a system built on risk-rating, simply prohibiting use of genetic testing and information does nothing to reduce insurers' motivation to get other information about a person in order to project risk. Thus, the system still encourages thinking in genetic categories and attempts to glean genetic information from other sources. These problems are unlikely to abate without broader reconsideration of a health insurance system that is built on sorting people into risk categories and then disadvantaging those most likely to need health care services.[36]

Critiques of antidiscrimination in race and gender analysis

These failures of genetic antidiscrimination theory are explained by the work published on race and gender. There, a robust literature has developed criticizing "discrimination" as the best way to understand and attack disadvantage.[37] These writings are diverse, sometimes in conflict, and present distinct but overlapping conversations about race and gender. They nonetheless offer important lessons for the effort to fight disadvantage based on genetics.

First, they teach that formal equality theory, which labels as "discrimination" treatment that unjustifiably deviates from a norm, is based on the view that a norm exists and that all should be treated in conformance.[38] Thus, an antidiscrimination approach counsels that people of color should be treated like whites and that women should be treated like men. It bifurcates the world into those who

Volume 23:4, Winter 1995

nonproblematically fit the norm (whites, men) and those who are problematically different (people of color, women). In genetic terms, this means bifurcating the world into those with nonproblematically "normal" genotypes and those with problematically "abnormal" ones. Discrimination is acting on that difference, rather than treating all the same.

Yet, as Lucinda Finley argues, this approach to equality has serious problems. Its bifurcation of the world produces an "inability to come to terms ... with the reality of human variety."[39] It thus "inherently assumes that the goal is assimilation to an existing standard."[40] In gender analysis, for example, "[i]t leaves unquestioned the notion that life patterns and values that are stereotypically male are the norm, such as the idea that ... focus on work to the exclusion of other concerns is necessary to ... the workplace."[41] This kind of antidiscrimination analysis actually does harm, by supporting the idea that there is a single norm, that those not suffering disadvantage constitute this enviable norm, and that the goal is simply to ensure all are treated the same as the normative group. In racial terms, it supports the idea that race is biological and real rather than a social construction for the purposes of subordination, that being nonwhite is bad and unfortunate, and that people of color merely want to live and be treated as if they were white.[42]

This antidiscrimination approach is profoundly problematic as applied to genetics. It supports the fiction that there is such a thing as a "normal" genotype, and that the goal is to change the treatment of people who deviate.[43] The reality, though, is enormous genetic heterogeneity, with all of us harboring genes capable of expressing themselves as deleterious diseases and disorders.[44] Given this reality, conceptualizing the harm as discrimination is itself harmful. We see this when articles discuss a "biological underclass";[45] they suggest that genetic disadvantage is a harm that the majority "we" may inflict on the vulnerable "them." Yet we are all part of "them." The category of people whose genetic burdens can be labeled today is simply a function of what genetic mapping has occurred so far. The category will grow tomorrow. Certainly some people, such as children with Lesch-Nyhan disease or adults who develop Huntington's chorea, face particularly heavy burdens relating to the expression of their genetic makeup. Yet the health history of each of us is affected by his or her genes, and advances in genetic knowledge are steadily increasing the roster of health problems with some sort of genetic underpinning.

Thus, to preach that the goal of antidiscrimination analysis in genetics is to ensure that all are treated as if they had a "normal" genotype is to create a radically indeterminate standard. There is nothing neutral or scientifically "real" about identifying a genetic norm; the notion is inevitably socially constructed. And if no one actually possesses this fictive "normal" genotype, it is completely unclear what it means to treat someone as if they did have it. To return to the health insurance example, if insurers can draw genetic conclusions from everyone's medical records, then antidiscrimination analysis merely leaves insurers free to do this to everyone, even if using genetic tests on information to harm genetically suspect individuals is ruled out. While some commentary depicts this as a drafting problem, arguing that genetic discrimination statutes should be drafted more broadly,[46] in fact this is a profound problem with antidiscrimination analysis itself.

The second lesson of writings on race and gender is that traditional antidiscrimination analysis actually thwarts systemic and structural reform. Such analysis permits individuals to complain of exclusion from existing structures, or of treatment different from people already included. But it does not provide tools for challenging those structures themselves. Thus, Alan Freeman argues that antidiscrimination analysis has allowed individuals to insist that race not serve as a basis for excluding them, but has not permitted remedies that would have actually required redistribution of wealth.[47] On gender, Mary Becker maintains that "formal equality ... cannot be expected to transform society by equalizing the status of women and men. In the context of employment, it only opens men's jobs to women on the terms and conditions worked out for men."[48] The limitations of antidiscrimination doctrine can therefore actually reinforce disadvantage by merely disallowing consideration of race or gender, and so in every other way legitimizing structures built on a history of disadvantaging certain groups and molded to serve those dominant.[49]

The application to genetics is direct. In health insurance, merely prohibiting negative treatment based on a genetic test or genetic information leaves untouched an entire system of risk- or experience-rating that cannot help but disadvantage people whose genetics renders them likely to manifest health problems. Thus, their supposedly nongenetic cholesterol test may yield "bad" results in part due to an underlying genetic predisposition. A careful reading of their medical record may reveal genetic problems, as suggested above. And their genetic propensities may eventually express as diseases and conditions that insurers may then penalize in the usual course of experience-rating. Merely prohibiting insurers' use of genetic tests or information legitimizes their use of genetic inference in risk- or experience-rating in a host of other ways. At the same time, it blunts any challenge to the rating system itself by suggesting that the problem is adequately handled by prohibiting disadvantage based on genetic tests or genetic information.

The third lesson of work on race and gender is that antidiscrimination doctrine will lead us astray, to the extent it attacks only deliberate acts of disadvantage by race or sex. The problem is much deeper. A person may, of course, deliberately disadvantage someone because of the latter's race or gender. But more often, stereotypes and unfounded beliefs about people of color or women will influence attitudes and behavior.[50] Indeed, Charles Lawrence has argued that much racism is unconscious:

> Americans share a common ... heritage in which racism has played and still plays a dominant role. Because of this ..., we also inevitably share many ideas, attitudes, and beliefs that attach significance to ... race and induce negative feelings ... about non-whites.... [M]ost of us are unaware of our racism.... [A] large part of the behavior that produces racial discrimination is influenced by unconscious racial motivation.[51]

Thus Lawrence complained that the Supreme Court's requirement of "proof of conscious or intentional motivation" in order to find unconstitutional discrimination "create[d] an imaginary world where discrimination does not exist unless it was consciously intended."[52] This, he said, "advances the disease [of racism] rather than combatting it."[53] Minow has similarly warned that behavior disadvantaging women may turn on "unstated assumptions" and "unconscious attachment to stereotypes."[54] "We especially attach ourselves to categories like male/female because of our own psychological development in a culture that has made gender matter."[55]

The history of eugenics in this century and in the United States certainly suggests the power and virulence of genetic stereotypes as well. Over and over we see those in power exercising their dominance by claiming others to be genetically inferior, demonstrating the lure of genetic categories and even mistaken genetic beliefs.[56] The example of Nazi Germany is obvious. But even in the United States, when a Supreme Court Justice as distinguished as Oliver Wendell Holmes justifies involuntary sterilization by exclaiming, "[t]hree generations of imbeciles are enough," and without concern about what is known about actual heritability or patterns of phenotypic expression, the force of genetic misconceptions is clear.[57] And in our own time, Dorothy Nelkin and M. Susan Lindee decry "the rise of a new eugenics," a product now not of state policy, but of "social and institutional pressure."[58] This new eugenics is fueled in part by the old idea that "[s]uperior individuals have a responsibility to replicate their genes.... But those who are flawed ... have a different responsibility, to prevent perpetuation of 'bad' genes."[59] Any attempt, then, to combat disadvantage based on deeply rooted genetic stereotypes and prejudices by focusing merely on deliberate use of genetic tests or information will miss most of the problem. This shallow kind of antidiscrimination effort will lead us astray.

The final lesson is that antidiscrimination doctrine will obscure connections between disadvantaging based on race, gender, and genetics, and it will fail to serve those most burdened by those connections. Kimberlé Crenshaw writes that "dominant conceptions of discrimination condition us to think about subordination as disadvantage occurring along a single categorical axis." Thus people discuss race discrimination separately from sex discrimination, "eras[ing] Black women." Only the sex-privileged members of the race category are then considered, and only the race-privileged members of the sex category. This "marginalizes those who are multiply-burdened."[60] It also obscures the extent to which racial and gender disadvantage are interlocking systems that reinforce one another. Thus antidiscrimination doctrine actually "reinforces the status quo" by blocking alliances across groups suffering supposedly separate harms, and by serving only the least disadvantaged in each group.[61]

Because genetic ideas have so often been linked historically to racism, anti-Semitism, and other forms of subordination and disadvantage, this is an especially powerful point. Roberts argues that "[t]he genetic tie's prominence in defining personal identity arose in the context of a racial caste system that preserved white supremacy through a rule of racial purity."[62] She gives an example: "the institution of slavery made the genetic tie to a slave mother critical to determining a child's social status, yet legally insignificant to the relationship between male slaveowners and their mulatto children."[63] To this day, "[i]n America, perhaps the most socially significant product of the genetic link between parents and children continues to be race.... which determines one's social status."[64]

Thus, the appeal of genetic categories is deeply related to a long history of racism and, Roberts shows, sexism. It is also intertwined with a history of stereotyping and disadvantaging people based on disability. Both the Nazi efforts to exterminate the disabled[65] and the *Buck v. Bell* example demonstrate that. These are interlocking systems of harm. To isolate genetic disadvantage and to craft an antidiscrimination approach that ignores all we have learned about how to conceptualize and combat these other interconnected forms of disadvantage is a dangerous mistake. As Crenshaw argues, it entrenches the status quo and makes the real problem more difficult to solve.[66]

The broader harm of geneticism

To begin to address and combat systems of genetic disadvantage effectively, in health insurance and beyond, we have to transcend the narrow language of "genetic discrimination." Antidiscrimination analysis cannot deal adequately with the pressing issues of genetic labeling and disadvantage. Thus it is no surprise to see complaints that state statutes being enacted to combat genetic discrimination will fail;[67] when a health insurer penalizes an individual based on a "genetic test" or "genetic information," we are

Volume 23:4, Winter 1995

only seeing one way in which genetic disadvantage is imposed. To cling to a simplistic antidiscrimination approach short-circuits a necessary analysis that places this in the broader context of genetic harms and links it to the harms of racism and sexism. But, as I have argued above, clinging to "genetic discrimination" does more than thwart further analysis; it actually does damage by creating a false genetic "norm," frustrating structural reform, obscuring the deep psychological roots of genetic stereotyping and prejudice, and isolating genetic from other harms.

In moving beyond antidiscrimination analysis, writers on race and gender have proposed a number of alternatives. Ruth Colker, for example, has proposed antisubordination analysis instead.

> Under ... [this] perspective, it is inappropriate for certain groups in society to have subordinated status because of their lack of power.... This approach seeks to eliminate the power disparities between men and women, and between whites and non-whites, through the development of laws and policies that directly redress those disparities.[68]

Thus, Colker reconceptualizes the relevant harm not as treating an individual differently from the norm of men and whites, but as perpetuating the subordination of certain groups. "[B]oth facially differentiating and facially neutral policies are invidious only if they perpetuate racial or sexual hierarchy."[69]

Catharine MacKinnon, in a somewhat similar vein, argues for a dominance approach, in which "an equality question is a question of the distribution of power."[70] But she reanalyzes the very idea of dividing the world by gender. That itself is an expression of dominance. Gender is not "merely a question of difference."[71] Instead, it has been invested with importance to preserve a certain hierarchical distribution of power. "[I]f gender ... [has been] constructed as a socially relevant differentiation in order to keep that inequality in place, then sex inequality questions are questions of systematic dominance...."[72]

One can debate the relative virtues of these broader alternatives to a naive reliance on antidiscrimination notions. What such alternatives counsel at the very least, though, is the reinterpretation of "genetic discrimination" as something bigger, the use of genetic notions to create and reinforce power relationships in which some dominate and others are subordinated. This use of genetics might properly be called *geneticism*.[73] Like racism and sexism, it is a long-standing and deeply entrenched system for disadvantaging some and advantaging others.[74] It can be seen in the pervasive individual and institutional use of genetic information and concepts to disadvantage people whether singly or by creating groups. It predates any accurate understanding of genetics, and now refers to social structures, practices, beliefs, and predispositions that together support disadvantaging based on a mixture of accurate and inaccurate genetic ideas.

This use of "geneticism" goes beyond Abby Lippman's "geneticization" by connoting an offensive and harmful practice, which remains harmful even when based on accurate rather than exaggerated understanding of the role of genes.[75] It goes beyond the "genetic essentialism" that Rochelle Dreyfuss and Dorothy Nelkin decry, by suggesting harm even when the role of environment and of other nongenetic factors is appreciated.[76]

Moreover, "geneticism" suggests something whose virulence may approach that of racism and sexism. In fact, geneticism is intimately connected with them. The relationship between geneticism, racism, and sexism is not just one of analogy. Historically, they have been closely linked, certainly in our own century.[77] "Geneticism" is meant to reflect that, as is my entire analysis. Harms to people based on notions of their genetic makeup, their race, and their gender (not to mention their disability and their religion) have been interrelated. It is fully appropriate, indeed essential, to apply the lessons already learned in combating these harms.[78] Here, I apply those lessons to reconceptualize the problem of genetic disadvantage. But we shall have to go further, of course, to devise ways to attack the bigger problem this reconceptualization reveals.

The language of "genetic discrimination" may continue to have a superficial appeal. It suggests, at a glance, something wrong, something akin to race or sex discrimination. But it also suggests an approach that scholarship on race and gender has found flawed and now transcends. To cling to "genetic discrimination" analysis while ignoring decades of relevant work on race and gender only entrenches the genetic problem. The deeper harm is geneticism.

Acknowledgments

Thanks to Dianne Bartels, Gene Borgida, Mary Dietz, Daniel Farber, Philip Frickey, Eric Juengst, Tom Murray, Dorothy Roberts, Karen Rothenberg, and Michael Yesley for help at various stages, and to Bridget McKeon of the University of Minnesota Law School for research assistance. This article grew out of my lecture and a panel discussion for the Institute for Applied Health Care Ethics at the City of Hope Medical Center.

References

1. See P.R. Billings et al., "Discrimination as a Consequence of Genetic Testing," *American Journal of Human Genetics*, 50 (1992): 476–82. See generally NIH-DOE Working Group on Ethical, Legal, and Social Implications of Human Genome Research, *Genetic Information and Health Insurance: Report of the Task Force on Genetic Information and Insurance* (NIH Pub. No. 93-3686, May 10, 1993).

2. In the insurance domain alone, analysis includes K.H. Rothenberg, "Genetic Information and Health Insurance: State Legislative Approaches," *Journal of Law, Medicine & Ethics*, 23 (1995): 312–19; K.L. Hudson et al., "Genetic Discrimination and Health Insurance: An Urgent Need for Reform," *Science*, 270 (1995): 391–93; Ad Hoc Committee on Genetic Testing/Insurance Issues, "Background Statement: Genetic Testing and Insurance," *American Journal of Human Genetics*, 56 (1995): 327–31; T.H. Cushing, "Should There be Genetic Testing in Insurance Risk Classification?," *Defense Counsel Journal*, Apr. (1993): 249–63; NIH-DOE Working Group, *supra* note 1; H. Ostrer et al., "Insurance and Genetic Testing: Where Are We Now?," *American Journal of Human Genetics*, 52 (1993): 565–77; S. O'Hara, "The Use of Genetic Testing in the Health Insurance Industry: The Creation of a 'Biological Underclass'," *Southwestern University Law Review*, 22 (1993): 1211–28; T.H. Murray, "Genetics and the Moral Mission of Health Insurance," *Hastings Center Report*, 22, no. 6 (1992): 12–17; N. Obinata, "Genetic Screening and Insurance: Too Valuable an Underwriting Tool to be Banned from the System," *Computer & High Technology Law Journal*, 8 (1992): 145–67; L. Gostin, "Genetic Discrimination: The Use of Genetically Based Diagnostic and Prognostic Tests by Employers and Insurers," *American Journal of Law & Medicine*, XVII (1991): 109–44; R. Lowe, "Genetic Testing and Insurance: Apocalypse Now?," *Drake Law Review*, 40 (1991): 507–32; R. Pokorski, "Genetic Screening and the Insurance Industry," *Yale Journal of Biology and Medicine*, 64 (1991): 53–57; and J.M. Miller, Comment, "Genetic Testing and Insurance Classification: National Action Can Prevent Discrimination Based on the 'Luck of the Genetic Draw'," *Dickinson Law Review*, 93 (1989): 729–57.

3. See, for example, R.A. Epstein, "The Legal Regulation of Genetic Discrimination: Old Responses to New Technology," *Boston University Law Review*, 74 (1994): 1–23; G.P. Smith II and T.J. Burns, "Genetic Determinism or Genetic Discrimination?," *Journal of Contemporary Health Law and Policy*, 11 (1994): 23–61; NIH-DOE Working Group, *supra* note 1, at 9; Billings et al., *supra* note 1; M.A. Rothstein, "Genetic Discrimination in Employment and the Americans with Disabilities Act," *Houston Law Review*, 29 (1992): 23–84; and Gostin, *supra* note 2. Dan Brock acknowledges that there are different understandings of "equality," but he focuses on a different question than I: "whether equality of opportunity can be fully realized despite ... [differences] in ... genetically based cognitive ability." D.W. Brock, "The Human Genome Project and Human Identity," *Houston Law Review*, 29 (1992): at 8, 11.

4. See, for example, N. Gotanda, "A Critique of 'Our Constitution Is Color-Blind'," *Stanford Law Review*, 44 (1991): 1–68; K. Crenshaw, "Demarginalizing the Intersection of Race and Sex: A Black Feminist Critique of Antidiscrimination Doctrine, Feminist Theory and Antiracist Politics," *University of Chicago Legal Forum* (1989): 139–67; D.L. Rhode, *Justice and Gender: Sex Discrimination and the Law* (Cambridge: Harvard University Press, 1989): 81–107; K. Crenshaw, "Race, Reform, and Retrenchment: Transformation and Legitimation in Antidiscrimination Law," *Harvard Law Review*, 101 (1988): 1331–87; C.R. Lawrence III, "The Id, the Ego, and Equal Protection: Reckoning with Unconscious Racism," *Stanford Law Review*, 39 (1987): 317–88; C.A. MacKinnon, *Feminism Unmodified: Discourses on Life and Law* (Cambridge: Harvard University Press, 1987); R. Colker, "Anti-Subordination Above All: Sex, Race, and Equal Protection," *New York University Law Review*, 61 (1986): 1003–66; L.M. Finley, "Transcending Equality Theory: A Way Out of the Maternity and the Workplace Debate," *Columbia Law Review*, 86 (1986): 1118–82; and A.D. Freeman, "Legitimizing Racial Discrimination through Antidiscrimination Law: A Critical Review of Supreme Court Doctrine," *Minnesota Law Review*, 62 (1978): 1049–119.

For analysis of why bioethics has tended to ignore race- and gender-attentive work, see S.M. Wolf, "Introduction: Gender and Feminism in Bioethics," in S.M. Wolf, ed., *Feminism & Bioethics: Beyond Reproduction* (New York: Oxford University Press, 1996): 3–43.

I focus on writing on race and gender rather than on disability because of the far more developed discussion of antidiscrimination theory in work on race and gender. However, for attention to equality theory and disability, see, for example, W.E. Parmet, "Discrimination and Disability: The Challenges of the ADA," *Law, Medicine & Health Care*, 18 (1990): 331–44.

5. See D.E. Roberts, "The Genetic Tie," *University of Chicago Law Review*, 62 (1995): 209–73.

6. See, for example, *id.*; A. Asch and G. Geller, "Feminism, Bioethics, and Genetics," in S.M. Wolf, ed., *Feminism & Bioethics: Beyond Reproduction* (New York: Oxford University Press, 1996): 318–50.

7. See, for example, Roberts, *supra* note 5, at 211 n.8 ("The concept of race—like the meaning of the genetic tie—is a cultural artifact."). See also A. Fausto-Sterling, *Myths of Gender: Biological Theories about Women and Men* (New York: Basic Books, 1985).

8. See R. Dreyfuss and D. Nelkin, "The Jurisprudence of Genetics," *Vanderbilt Law Review*, 45 (1992): at 315–17 (comparing to race and gender); A. Lippman, "Prenatal Genetic Testing and Screening: Constructing Needs and Reinforcing Inequities," *American Journal of Law & Medicine*, XVII (1991): 15–50; and R. Hubbard, *The Politics of Women's Biology* (New Brunswick: Rutgers University Press, 1990): 70–86.

9. See M. Minow, "Introduction: Finding Our Paradoxes, Affirming Our Beyond," *Harvard Civil Rights-Civil Liberties Law Review*, 24 (1989): at 2–4, citing M.J. Frug, *Feminist Histories* (1987, unpublished manuscript).

10. Indeed, the doctrinal effort to develop law to restrict the use of genetic information by insurers can benefit from comparison to the law curbing the use of race and sex in insurance. On the latter body of law, see, for example, R. Austin, "The Insurance Classification Controversy," *University of Pennsylvania Law Review*, 131 (1983): 517–83; and L. Brilmayer et al., "The Efficient Use of Group Averages as Nondiscrimination: A Rejoinder to Professor Benston," *University of Chicago Law Review*, 50 (1983): 222–49.

11. For abundant illustration of this, see D. Nelkin and M.S. Lindee, *The DNA Mystique: The Gene as a Cultural Icon* (New York: W.H. Freeman, 1995).

In the domain of health insurance, there is growing agreement (as I discuss below) that antidiscrimination measures merely prohibiting insurers from using genetic test results will not prevent them from gleaning genetic information from the rest of the medical record. See Rothenberg, *supra* note 2; and NIH-DOE Working Group, *supra* note 1, at 8 ("much of the information in the [medical] record about risk factors, diseases, diagnostic tests and treatments will in fact reveal genetic information to the astute reader of that record").

12. Compare Nelkin and Lindee, *supra* note 11, at 2 ("In one sense the gene is a biological structure.... But it has also become a cultural icon.... The biological gene ... has a cultural meaning independent of its precise biological properties." (footnote omitted)).

13. See Cal. Health & Safety Code § 1374.7 (West Supp. 1995), and Cal. Ins. Code §§ 10123.3, 11512.95 (West Supp. 1995), all as amended by 1995 Cal. Leg. Serv. ch. 695 (West);

Volume 23:4, Winter 1995

Colo. Rev. Stat. § 10-3-1104.7 (1994); Ga. Code Ann. §§ 33-54-1 to -8 (Supp. 1995); Minn. Stat. § 72A.139 (Supp. 1995); N.H. Rev. Stat. Ann. ch. 141-H (1995); Ohio Rev. Code Ann. §§ 1742.42-.43, 3901.49, 3901.491, 3901.50, 3901.501 (Baldwin 1995); 1995 Or. Laws 680; and Wis. Stat. Ann. § 631.89 (West 1995). Bills have been introduced in other states. See, for example, 1995 Mass. H. 4485, 179th Gen. Ct., 1st Ann. Sess.; and 1995 Tex. H. 343, 74th Reg. Sess.

14. There is a copious literature on the development of state responsibility for insurance and on the persistence of federal jurisdiction over self-insured health benefit plans through the Employee Retirement Income Security Act (ERISA), 29 U.S.C.A. §§ 1001-461 (West 1985 & Supp. 1995). See, for example, W.E. Parmet, "Regulation and Federalism: Legal Impediments to State Health Care Reform," *American Journal of Law & Medicine*, XIX (1993): at 132–40. For analysis of how ERISA affects the use of genetic information by self-insured plans, see, for instance, Ad Hoc Committee, *supra* note 2, at 330; Ostrer et al., *supra* note 2, at 571–75; and NIH-DOE Working Group, *supra* note 1, at 17.

The Americans with Disabilities Act (ADA), 42 U.S.C.A. §§ 12101-213 (West 1995), has also been analyzed by these latter sources for its potential effects on insurers' use of genetic information. The ADA states that it does not prohibit underwriting or risk classification, but that underwriting cannot be used as a subterfuge for prohibited employment discrimination. See 42 U.S.C.A. § 12201(c). This has been interpreted to mean that an employer may not single out an individual for discriminatory treatment, but may structure an entire benefits plan using risk classifications. See, for example, Ostrer et al., *supra* note 2, at 573. However, the ADA says nothing specific about genetics. It clearly covers manifested disabilities, whether or not genetics has played an underlying role. As to ADA coverage of asymptomatic genetic susceptibilities, a March 1995 pronouncement by the Equal Employment Opportunity Commission clarified that they were included in the category of protected disabilities. Equal Employment Opportunity Commission, *Compliance Manual, Vol. II: Definition of the Term "Disability"* (1995): at § 902.8.

15. On the history, see Rothenberg, *supra* note 2; Hudson et al., *supra* note 2, at 392; and NIH-DOE Working Group, *supra* note 1, at 18.

16. See Md. Ann. Code art. 48A, § 223(b)(4) (1994); Cal. Health & Safety Code § 1374.7 (West 1990); and Cal. Ins. Code § 10123.3(a) (West 1993), § 11512.95(a) (West 1988).

17. See statutes *supra* note 16.

18. See Rothenberg, *supra* note 2; and Hudson et al., *supra* note 2.

19. Wis. Stat. § 631.89.

20. See *supra* note 13.

21. See Rothenberg, *supra* note 2; Hudson et al., *supra* note 2; and NIH-DOE Working Group, *supra* note 1.

22. Hudson et al., *supra* note 2, at 392. Some of the statutes indicate as much. See, for example, Colo. Rev. Stat. § 10-3-1104.7(2)(b) ("'Genetic testing' includes only such tests as are direct measures of ... [genetic] alterations rather than indirect manifestations thereof.").

23. S. 1416, 104th Cong., 1st Sess. § 2(b)(3) (1995).

24. *Id.* § 6(a).

25. *Id.* § 3(5).

26. *Id.* § 3(4).

27. H.R. 2690, 104th Cong., 1st Sess. (1995).

28. H.R. 2748, 104th Cong., 1st Sess. § 2(a) (1995). As is similar to the prior two bills, "genetic information" is defined as "information about genes, gene products, or inherited characteristics." *Id.* § 2(e)(3). "Genetic services" are then defined as "health services to obtain, assess, and interpret genetic information for diagnostic and therapeutic purposes, and for genetic education and counseling." *Id.* § 2(e)(4).

29. NIH-DOE Working Group, *supra* note 1, at 8.

30. *Id.*

31. Billings et al., *supra* note 1, at 477.

32. *Id.*

33. See, for example, Murray, *supra* note 2, at 12; and N.E. Kass, "Insurance for the Insurers: The Use of Genetic Tests," *Hastings Center Report*, 22, no. 6 (1992): 6–11.

34. Murray, *supra* note 2, at 12.

35. *Id.*

36. As Nancy Kass argues, "The greatest protection for those with a genetic predisposition ... would be to return to a community-rated system." Kass, *supra* note 33, at 10.

37. See *supra* note 4.

38. See, for example, M. Minow, "The Supreme Court 1986 Term, Foreword: Justice Engendered," *Harvard Law Review*, 101 (1987): at 38–45 (on "The Unstated Norm").

39. Finley, *supra* note 4, at 1148.

40. *Id.* at 1143.

41. *Id.* at 1120.

42. See, for example, Crenshaw (1988), *supra* note 4, at 1370–74; compare Roberts, *supra* note 5, at 211 n.8 (race as "a cultural artifact").

43. Compare Brock, *supra* note 3, at 18-22 (speculating on the challenges that genetic therapy for enhancement may pose to our concept of the "normal").

44. See President's Commission for the Study of Ethical Problems in Medicine and Biomedical and Behavioral Research, *Screening and Counseling for Genetic Conditions* (Washington, D.C.: U.S. Government Printing Office, 1983): App. B, at 111 ("Every human being inherits about six or seven deleterious mutations that under certain circumstances can cause serious illness.").

45. See O'Hara, *supra* note 2; and Billings et al., *supra* note 1, at 479 ("a new social class").

46. See Rothenberg, *supra* note 2.

47. See A. Freeman, "Antidiscrimination Law: The View from 1989," in D. Kairys, ed., *The Politics of Law: A Progressive Critique* (New York: Pantheon, rev. ed., 1982): 121–50.

48. M.E. Becker, "Prince Charming: Abstract Equality," *Supreme Court Review* (1987): at 212.

49. See, for example, *id.* at 214–24.

50. See L.H. Krieger, "The Content of Our Categories: A Cognitive Bias Approach to Discrimination and Equal Employment Opportunity," *Stanford Law Review*, 47 (1995): 1161–248.

51. See Lawrence, *supra* note 4, at 322 (footnotes omitted).

52. *Id.* at 323–25.

53. *Id.* at 324.

54. Minow, *supra* note 38, at 15, 64.

55. *Id.* (footnote omitted).

56. On the history of eugenics, see, for example, D.J. Kevles, *In the Name of Eugenics: Genetics and the Uses of Human Heredity* (New York: Knopf, 1985).

57. See *Buck v. Bell*, 274 U.S. 200, 207 (1927).

58. Nelkin and Lindee, *supra* note 11, at 171.

59. *Id.* at 174.

60. Crenshaw (1989), *supra* note 4, at 140.

61. See *id.* at 166–67.

62. Roberts, *supra* note 5, at 223.

63. *Id.* at 210.

64. *Id.* at 223.

65. See, for example, R.N. Proctor, *Racial Hygiene: Medicine Under the Nazis* (Cambridge: Harvard University Press, 1988); and R.J. Lifton, *The Nazi Doctors: Medical Killing and the Psychology of Genocide* (New York: Basic Books, 1986).

66. See *supra* note 61 and accompanying text.

67. See Rothenberg, *supra* note 2; Hudson et al., *supra* note 1, at 392; and NIH-DOE Working Group, *supra* note 1, at 8, 18.

68. Colker, *supra* note 4, at 1007 (footnote omitted).

69. *Id.* at 1007–08 (footnote omitted).

70. MacKinnon, *supra* note 4, at 40.

71. *Id.* at 42.

72. *Id.*

73. "Geneticism" is defined by the dictionary as "a theory explaining the perceptions, attitudes, and behavior of an individual primarily in terms of his heredity and development." *Webster's Third New International Dictionary of the English Language Unabridged* (Springfield: G. & C. Merriam, 1976): at 946. I use the term more broadly. Comparison to the definition of "racism" ("the assumption that psychocultural traits and capacities are determined by biological race and that races differ decisively from one another which is usu. [*sic*] coupled with a belief in the inherent superiority of a particular race and its right to domination over others") reveals missing elements: questioning of the underlying biologism, recognition of harm, and disapproval. Of course, even this definition of "racism" does not go far enough in questioning the concept of "race" and whether it is biological. Compare Roberts, *supra* note 5, at 211 n.8 ("The concept of race ... is a cultural artifact.").

One author states that "geneticism" was used in 1969 by Sir Peter Medawar "to describe the inappropriate genetic labeling of variations between people." Lippman, *supra* note 8, at 18–19 n.16, citing P.B. Medawar, "The Genetic Improvements of Man," *Australasian Annals of Medicine*, 18 (1969): at 319. However, I have been unable to locate his use of the term in that article. As described, though, his use would accord with the dictionary's definition above, but again would be narrower than my use of the term.

Abby Lippman also cites past uses of "geneticizing" ("to refer to the tendency to label as 'genetic' diseases and disorders" with "'scant or no genetic evidence'") and "'construction' of genetic disease." Lippman, *supra* note 8, at 18 n.16, citing G.J. Edlin, "Inappropriate Use of Genetic Terminology in Medical Research: A Public Health Issue," *Perspectives in Biology and Medicine*, 31 (1987): at 48; and E.J. Yoxen, "Constructing Genetic Diseases," in P. Wright and A. Treacher, eds., *The Problem of Medical Knowledge: Examining the Social Construction of Medicine* (Edinburgh: Edinburgh University Press, 1982): 144–61. All of this is in defense of her choice of "geneticization" to describe a process that goes further. But none of these terms either embraces the range of phenomena I capture under "geneticism" or explicitly establishes the relationship to "racism" and "sexism."

74. MacKinnon's critique of the term "sexism" might be construed to cast doubt on my suggestion of geneticism. However, her complaint is that "sexism" suggests the problem is merely bad treatment of women, rather than a system of dominance that seizes on gender as a means of subordination. See *id.* My article meets that complaint by using "geneticism" more comprehensively to embrace a system of genetic disadvantage based on supposed deviation from a fictive genetic norm.

75. See Lippman, *supra* note 8, at 18–19 and n.16.

76. See Dreyfuss and Nelkin, *supra* note 8, at 316–21. "Genetic essentialism posits that personal traits are predictable and permanent, determined at conception, 'hard-wired' into the human constitution.... [T]his ideology minimizes the importance of social context." *Id.* at 320–21. Nelkin and Lindee also develop the idea of "genetic essentialism" in the *DNA Mystique*, *supra* note 11, at 2, 149–68. "Genetic essentialism reduces the self to a molecular entity, equating human beings, in all their social, historical, and moral complexity, with their genes." *Id.* at 2.

77. See, for example, D.J. Kevles, "Out of Eugenics: The Historical Politics of the Human Genome," in D.J. Kevles and L. Hood, eds., *The Code of Codes: Scientific and Social Issues in the Human Genome Project* (Cambridge: Harvard University Press, 1992): 3–36. For an analysis of the link between geneticism and racism, see Roberts, *supra* note 5.

78. I am grateful to Eric Juengst for raising the terminological concern that the analogue to "racists" and "sexists" cannot be "geneticists" (because we already use that to refer to certain professional specialists), and so requires a different term, perhaps "geneticizers."

[7]

What Makes Genetic Discrimination Exceptional?

Deborah Hellman†

I. INTRODUCTION

Recent advances in understanding the genetic basis of disease has inspired hope but also fear. While establishing a link between a person's genetic makeup and a propensity to disease may lead to better treatment, many scientists, physicians and genetic counselors also worry that it may lead to discrimination. Although access to health insurance is the primary concern, people also fear discrimination in life and disability insurance, employment and other contexts, such as child custody decisions or adoption. In response to this concern, many state legislatures have passed laws forbidding genetic discrimination.[1] While most of these laws focus on health insurance,[2] some also prohibit genetic discrimination in employment[3] or in life or disability insurance coverage.[4]

---

† Associate Professor, University of Maryland School of Law. B.A., Dartmouth College; M.A. in Philosophy, Columbia University; J.D., Harvard Law School. I want to thank Ryan Lemmerbrock for his excellent research assistance.

1 Most of the fifty states have passed laws prohibiting genetic discrimination in health insurance and employment. For charts depicting states and their respective legislation, see both National Conference of State Legislatures (NCSL), *State Genetic Discrimination in Health Insurance Laws*, *available at* http://www.ncsl.org/programs/health/genetics/ndishlth.htm (Aug. 7, 2002), and NCSL, *State Genetics Employment Laws*, *available at* http://www.ncsl.org/programs/health/genetics/ndiscrim.htm (Aug. 8, 2002).

2 Forty-six of the fifty states have enacted laws prohibiting such genetic discrimination by insurers. NCSL, *State Genetic Discrimination in Health Insurance Laws*, *supra* note 1. Examples of insurer practices which are prohibited include: the use of genetic information for risk selection or risk classification purposes, the establishment of eligibility rules based on genetic information and the requirement of genetic tests. *Id.* Many states prohibit all of the above practices in individual and group policies; however, a small number of states prohibit certain practices in group policies only. *Id.*

3 Thirty-one of the fifty states have enacted laws prohibiting genetic discrimination in employment. NCSL, *State Genetic Discrimination in Health Insurance Laws*, *supra* note 1. As with the prohibitions in health insurance, the scope of what is impermissible in the employment context varies widely. *See id.* (noting that examples of prohibitions include requesting or requiring genetic information, performing genetic tests and obtaining genetic test results). All states that have such legislation prohibit discrimination based on the results of genetic tests, and most states prohibit employers from accessing such information or conducting their own genetic tests. *Id.*

4 Seventeen states curtail discrimination in life or disability insurance coverage. NCSL, *State Genetic Nondiscrimination Laws in Life, Disability, and Long-term Care Insurance*, *available at* http://www.ncsl.org/programs/health/genetics/ndislife.htm (Oct. 14, 2002). These states' laws do not expressly prohibit, but rather *restrict* insurer use of genetic information in some manner. *See id.* For example, four states prohibit discrimination in life and disability insurance without actuarial justification. *Id.*

These laws have been the subject of both praise and criticism. Defenders of the laws see them as important and necessary, though arguably incomplete.[5] Critics view them as unjustified and unwarranted.[6] However, the question that dominates current literature is whether genetic discrimination is meaningfully different from discrimination on the basis of general health status; or as the debate is often framed, whether anti-discrimination laws ought to be *genetic* or *generic*.[7] In fact, this is one point on which there is some convergence between the critics and supporters of anti-discrimination laws. Critics of the current laws that target genetic discrimination argue that genetic discrimination is no different from discrimination on the basis of health status and, additionally, that such discrimination is necessarily at the heart of the proper administration of insurance.[8] Therefore, these critics argue, the current laws are both unwise and unjust. Some supporters of anti-discrimination protection acknowledge that there is no good reason to differentiate between those with a genetic predisposition to disease and those who already suffer from disease. However, rather than concluding that no anti-discrimination laws are necessary, these commentators see the current laws as merely a first step in the right direction. The second step, then, would be a generic law that protects people from discrimination on the basis of health status.

This Article also addresses the question of whether special genetics legislation is warranted. In other words, why treat discrimination on the basis of genetic variation differently from discrimination on the basis of health? Although scholars offer a myriad of arguments in support of laws prohibiting genetic discrimination, most of these arguments fail to justify the need for genetic (as opposed to generic) legislation.

---

5 *See, e.g.*, Trudo Lemmens, *Selective Justice, Genetic Discrimination, and Insurance: Should We Single Out Genes in Our Laws?*, 45 MCGILL L.J. 347, 383 (2000) (arguing that access to health insurance should not depend on health status, whether of genetic or non-genetic origin).

6 *See, e.g.*, Colin S. Diver & Jane Maslow Cohen, *Genophobia: What is Wrong with Genetic Discrimination?*, 149 U. PA. L. REV. 1439, 1445-446 (2001) (arguing that laws banning genetic discrimination are unjust both because they will "cause significant welfare losses due to the distortion of allocative efficiency" and because they "selectively favor[] a single type of moral 'bad luck,' while concealing both the extent and the form of its intended cross-subsidies"); Richard A. Epstein, *The Legal Regulation of Genetic Discrimination: Old Responses to New Technology*, 74 B.U. L. REV. 1, 18 (1994) (stating that "the prohibition against genetic discrimination should be seen for what it is—an elaborate set of cross-subsidies that reduces the total level of social wealth as it transfers wealth between parties"); Robert J. Pokorski, *Use of Genetic Information by Private Insurers*, *in* JUSTICE AND THE HUMAN GENOME PROJECT 91, 97 (Timothy F. Murphy & Marc A. Lappé eds., 1994) (arguing that prohibiting genetic discrimination by insurers is unjust because "[t]his mandated subsidization of unfavorable risks by good risks would be tantamount to an indirect governmental tax levied solely against insurance policyholders and stockholders").

7 Mark A. Rothstein & Mary R. Anderlik, *What Is Genetic Discrimination, and When and How Can It Be Prevented?*, 3 GENETICS MED. 354, 357 (2001) (arguing that laws specially targeting genetic discrimination as distinct from health status-based discrimination are both unjust and impratical); Sonia M. Suter, *The Allure and Peril of Genetic Exceptionalism: Do We Need Special Genetics Legislation?*, 79 WASH. U. L.Q. 669 (2001) (claiming that genetic information is not so qualitatively different from other medical information that it warrants special legislation and that law makers should take a more comprehensive approach to health-status discrimination); Susan M. Wolf, *Beyond "Genetic Discrimination": Toward the Broader Harm of Geneticism*, 23 J.L. MED. & ETHICS 345, 347 (1995) (arguing that laws forbidding genetic discrimination are overly narrow because the "social practice that needs to be changed is broader than health insurers' accurate use of genetic tests and information").

8 *See* T.H. Cushing, *Should There be Genetic Testing in Insurance Risk Classification?*, 60 DEF. COUNS. J. 249, 253 (1993) (noting that insurers underwrite policies in accordance with factors known to effect life expectancy and "[f]rom the health and life insurers' point of view, genetic testing [is] useful in assessing the risks they will underwrite.").

There are, however, two reasons for specifically targeting genetic discrimination that warrant further exploration. The first of these—that legislation prohibiting genetic discrimination is needed in order to safeguard the health benefits that the new genetic science offers—is powerful and important. But, it is more complex than has been thus far recognized. This Article will examine that argument in depth, and argue that the strength of the argument depends on the answers to several empirical questions that require further study.

The second argument that warrants a closer look—which implicitly underlies many of the arguments in the literature on genetic discrimination—has neither been teased out nor examined explicitly. Genetic discrimination may be meaningfully different (and worse) than health status discrimination because of what it *expresses*. This claim rests on the more general claim that the expressive dimension of action matters morally and legally.[9] In arguing for protection from genetic discrimination, commentators commonly refer to the history of eugenics both in this country and in Europe, particularly Nazi Germany. However, this reference is made in a cursory fashion without any articulation of why that history matters or what part it plays in building an argument for legislation prohibiting genetic discrimination. This Article will fill that gap by developing the argument that because the social meaning of treating people differently on the basis of their genetic make-up is different from the social meaning of discrimination on the basis of health or illness, special legislation is warranted to prohibit genetic discrimination.

Interestingly, in the related context of prenatal genetic testing, disability rights activists offer a critique that focuses on the meaning expressed by the routine nature of genetic testing in prenatal care.[10] Because this argument rests squarely on the expressive dimension of these actions and because both its strengths and weaknesses have been well-developed,[11] albeit outside of the legal literature, it provides an illustration that will be useful to the examination of the related claim that genetic discrimination by insurers, employers and others is especially wrongful because of what it expresses.

This Article begins by addressing an important preliminary issue regarding the problem of definition. Can one define "genetic discrimination" in a way that adequately differentiates it from health status discrimination? This leads to the central question of whether genetic discrimination is meaningfully different from discrimination on the basis of health. And, if so, do these differences provide good reasons to specially prohibit genetic discrimination? This Article briefly summarizes the familiar arguments for forbidding genetic discrimination as well as the familiar replies. Next, it examines the claim that laws forbidding genetic discrimination are warranted to ensure that fears of genetic discrimination do not thwart the promise of genetic science. While this argument is not new, its complexities have not as yet been adequately explored. Finally, the Article focuses on the claim that genetic discrimination is different because it expresses a morally

---

9 Deborah Hellman, *The Expressive Dimension of Equal Protection*, 85 MINN. L. REV. 1 (2000) [hereinafter Hellman, *Expressive Dimension*] (arguing that a state law or policy violates the Equal Protection clause if it expresses a meaning that conflicts with the government's obligation to treat each person with equal concern and respect); Deborah Hellman, *Judging by Appearances: Professional Ethics, Expressive Government and the Moral Significance of How Things Seem*, 60 MD. L. REV. 653 (2001) (examining whether there are non-consequentialist reasons to be concerned about the appearance of impropriety).

10 *See* PRENATAL TESTING AND DISABILITY RIGHTS (Erik Parens & Adrienne Ashe eds., 2000) [hereinafter PRENATAL TESTING].

11 *Id.*

problematic meaning—an argument that underlies critiques found in the literature, but which has not yet been clearly articulated or evaluated.

II. DEFINITIONAL PRELIMINARIES

Some critics of prohibitions on genetic discrimination favor a generic antidiscrimination approach (in the form of general protections of medical privacy, for example) because they believe it is theoretically or practically impossible to distinguish genetic discrimination from discrimination on the basis of general health.[12] This is a common and well-articulated critique that will be summarized only briefly here. If laws define genetic discrimination as discrimination on the basis of the results of a test of a person's genetic material, such as an examination of DNA, the laws will fail to capture many instances of discrimination on the basis of genetic predisposition to disease. For example, a family medical history, which is the obvious starting point of any medical record or doctor's visit, contains a wealth of information about a person's genetic predisposition to disease. Yet, this information is not the product of any new or sophisticated DNA test. As a result, those state laws that define "genetic information" as information resulting from a test of DNA have been criticized as overly narrow.[13]

This critique has led to a second wave of legislation prohibiting genetic discrimination.[14] These laws define "genetic information" more broadly. For example, in Maryland, "[a]n insurer, nonprofit health service plan, or health maintenance organization may not: use a genetic test, or the results of a genetic test, genetic information, or a request for genetic services to reject, deny, limit, cancel, refuse to renew, increase the rates of, affect the terms or conditions of, or otherwise affect a health insurance policy or contract."[15] This approach has its own problems however. The definition has become so broad that it seemingly prohibits almost all forms of discrimination on the basis of health, except perhaps illness or disability caused by accident. As we learn more about the genetic components of common illnesses, scientists are discovering that most diseases are, at least in part, influenced by our genes. As a result, common tests ordinarily not labeled "genetic tests," such as a blood pressure reading, arguably fall within the purview of these new laws. As Dr. Thomas Murray reports, the lack of a clear line between genetic and non-genetic diseases was one of the central reasons why the Task Force on Genetic Information and Insurance for the Human Genome Project rejected what he terms "genetic exceptionalism"—the view that genetic discrimination is different and therefore

---

12 *See, e.g.*, Diver & Cohen, *supra* note 6, at 1451; Lemmens, *supra* note 5, at 368-69; Thomas H. Murray, *Genetic Exceptionalism and "Future Diaries": Is Genetic Information Different from other Medical Information?*, *in* GENETIC SECRETS 60, 68-9 (1997); Rothstein & Anderlik, *supra* note 7, at 357.

13 *See, e.g.*, Henry T. Greely, *Genotype Discrimination: The Complex Case For Some Legislative Protection*, 149 U. PA. L. REV. 1483, 1495-497 (2001), Lemmens, *supra* note 5, at 368; Karen H. Rothenberg, *Genetic Information and Health Insurance: State Legislative Approaches*, 23 J.L. MED. & ETHICS 312, 317 (1995).

14 *See, e.g.*, CAL. INS. CODE ANN. § 10123.3 (West 1993); COLO. REV. STAT. § 10-3-1104.7 (2002); GA. CODE ANN. § 33-54-1 (1996); 410 ILL. COMP. STAT. 513/10–513/45 (1997); MO. ANN. STAT. §§ 375.1300–375.1312 (West 2002); N.Y. INS. LAW § 2612 (McKinney 2000 & Supp. 2003); OR. REV. STAT. § 746.135 (2001); R.I. GEN. LAWS § 27-18-52 (2002); TENN. CODE ANN. § 56-7-2701 (2000).

15 MD. CODE ANN., INS. § 27-909(c)(1) (1997 & Supp. 2002).

warrants special prohibitory legislation.[16] Consequently, many scholars argue against laws that specially target genetic discrimination for the practical reason that their goal is futile.

Professor Henry Greely provides a helpful resolution to this definitional problem. Greely proposes that legislation should draw a distinction between discrimination on the basis of genotype and discrimination on the basis of phenotype.[17] In his measured support for limited federal legislation prohibiting genetic discrimination in health insurance and employment, Greely suggests that such a law forbid health insurers from charging higher rates on the basis of information about unexpressed genetic traits, regardless of the source of the information. As he explains, "[g]enetic information should thus be defined broadly to encompass any . . . information that provides probabilitic information about a person's genotype . . . from genetic tests, other medical tests, family history, diagnoses of traits or conditions, or the taking of (or even making inquires about) a genetic test."[18] In order to avoid also forbidding discrimination on the basis of health status generally, Greely proposes allowing discrimination when the trait has already manifest in the form of illness or disability.[19] Drawing the line in this way avoids what Greely calls the "feedback effect," when broadly drawn legislation could allow someone with a genetic disease like Huntington's disease to sue for genetic discrimination if he is denied coverage on the basis of having that illness.[20] For Greely, once the genetic predisposition is manifest as illness or in the form of a medically relevant symptom, it would not be forbidden.[21]

Greely's resolution of the definitional dilemma adopts the meaning originally proposed by Paul Billings and his colleagues in an early piece on genetic discrimination.[22] In Greely's view, the definition is "conceptually straightforward though perhaps complicated to implement."[23] At least at first blush, his distinction seems workable, but is it defensible? Whether this definition is satisfactory depends on what moral justification is offered for enacting legislation specifically forbidding "genetic discrimination." The definition of "genetic discrimination" and the moral argument for forbidding genetic discrimination are, obviously, intimately

16 Murray, *supra* note 12, at 68 (explaining that the task force could not find a clear distinction between genetic and non-genetic information because in most cases "the two-bucket theory [to wit, putting specific diseases or conditions in either a genetic or non-genetic 'bucket'] was hopelessly inadequate").

17 Greely, *supra* note 13, at 1502-503.

18 *Id.* at 1502.

19 *Id.* at 1502-503.

20 *Id.* at 1497.

21 Greely admits that the distinction may be complicated to implement as many genetic predispositions are manifest in the body in ways that are not harmful before they cause disease. *Id.* at 1503-504. To delineate which conditions count as discrimination on the basis of phenotype, and are therefore permitted, and which are discrimination on the basis of genotype, and are therefore prohibited, Greely proposes the concept of the "medically significant symptoms." *Id.* at 1504. As I understand his proposal, if the phenotypic difference is medically significant as a symptom of the illness to come, then discrimination on the basis of that difference is permitted.

22 Paul R. Billings defines "genetic discrimination" as "discrimination against an individual or against members of that individual's family solely because of real or perceived differences from the normal genome of that individual. Genetic discrimination is distinguished from discrimination based on disabilities caused by altered genes by excluding, from the former category, those instances of discrimination against an individual who at the time of the discriminatory act was affected by the genetic disease." Paul R. Billings et al., *Discrimination as a Consequence of Genetic Testing*, 50 AM. J. HUM. GENETICS 476, 477 (1992).

23 Greely, *supra* note 13, at 1503.

intertwined, or at least they ought to be. Since Greely's resolution draws a distinction between discrimination on the basis of unexpressed genetic traits and discrimination on the basis of manifested illness, whether of genetic or non-genetic etiology, the reasons for distinguishing between genetic and non-genetic discrimination must make such a definition both sensible and warranted. This issue will be addressed after an examination of the moral grounds for distinguishing genetic discrimination from other forms of health status discrimination.

III. IS GENETIC DISCRIMINATION DIFFERENT?

At the heart of the question of whether genetic discrimination ought to be prohibited lies the issue of whether discrimination on the basis of genetic information is morally different from discrimination on the basis of health or illness. At present, a person who is sick may be charged higher insurance rates or denied coverage. Why then forbid similar treatment of someone who has a genetic predisposition to the same illness? Most Americans with health insurance are covered through group rated plans in which no individual risk assessment is carried out. However, the question whether to prohibit genetic discrimination is germane to assessing whether protective legislation is warranted for those who buy *individual* health insurance policies as well as to the life, disability and long term care insurance markets in which individual underwriting is commonplace.

One might think that distinguishing between the sick and healthy in insurance is "fair," while distinguishing between two healthy persons, only one of whom carries a genetic mutation predisposing him to disease, is not. While the sick person is already sick and will therefore surely need to make a claim against the insurance carrier, it is uncertain whether the person with the genetic predisposition will need to make a similar claim. But of course the person who is already sick will not necessarily need healthcare services. She could be hit by a bus and killed on the first day of the policy period. Moreover, in the case of some genetic diseases—albeit a small number of them—the person with the specific genetic mutation is certain to develop the predicted illness so long as she does not die of something else first.[24] Whenever individual underwriting is employed, the insurer makes a prediction about the likelihood that a given insured will make a claim for reimbursement during the policy period. The difference between the person who is already sick and the person who may become sick is a difference in degree, not a difference in kind.

In this sense, genetic information is like other information that insurers routinely use in setting individually-based rates. It predicts the likelihood that a given person will draw from the insurance pool during the policy period. While information provided by the new genetic science may turn out to be a better predictor of future health, it will never be perfectly predictive. Intervening events, such as accidents, are always possible. Moreover, the impact of environmental factors upon genetic predispositions is likely to leave a substantial amount of uncertainty. How much predictive capacity genetic understanding will provide is a matter of dispute, but the point to stress here is this: genetics offers better predicitive ability but not perfect predictive ability. Although the above discussion focuses on insurance, the question

---

24 Geneticists use the term "penetrance" to describe this characteristic of a genetic disease. Richard R. Sharp, *The Evolution of Predictive Genetic Testing: Deciphering Gene-Environment Interactions*, 41 JURIMETRICS J. 145, 148 (2001). If a genetic mutation is 100 percent penetrant, then all who carry that mutation will develop the disease if they live long enough to do so. NORMAN V. ROTHWELL, UNDERSTANDING GENETICS: A MOLECULAR APPROACH 63 (1993). Huntington's chorea is an example of a genetic disease that is nearly 100 percent penetrant. *Id.*

of whether genetic discrimination is different, morally speaking, from discrimination on the basis of health or illness can be raised in other contexts as well. Thus far, employment discrimination is the other area that has attracted significant attention.[25] Here too, one must ask whether genetic discrimination in employment ought to be specifically forbidden.

Although the insurance and employment contexts are not the only ones in which genetic discrimination may occur, they are likely to remain the most important. The family context may become significant as well. In a child custody dispute, for example, a judge could conclude that placing the child with one parent is in the child's best interest because the other parent carries a genetic predisposition to early onset Alzheimer's disease. Perhaps prospective adoptive parents will be rejected because one parent carries the gene that causes Huntington's disease. Alternatively, prospective adoptive parents may demand that their potential child undergo genetic testing before finalizing an adoption.[26] Lastly, parents who know they carry a mutation for a serious genetic disease may choose to use genetic testing to avoid carrying a child with that disease to term.[27] Should discrimination on the basis of genetic factors be prohibited in these contexts?

A. Familiar Arguments

This section will briefly review the arguments for genetic exceptionalism[28] that justify legislation specifically targeting genetic discrimination. In most instances, these arguments are unconvincing, often for reasons that have been well-articulated by others. In order to add to this dialogue in a way that picks up where others have left off, the arguments will be discussed only briefly. The reason to discuss them is twofold. First, this review will give the reader a feel for the arguments which have already been advanced in support of legislation prohibiting genetic discrimination.

---

25 *See supra* note 3 and accompanying text.

26 The American Society of Human Genetics (ASHG), together with the American College of Medical Genetics (ACMG), report that there is anecdotal evidence that prospective adoptive parents and adoption agencies are requesting "a wider range of genetic tests before, during, or immediately after the adoption process." ASHG/ACMG Statement, *Genetic Testing in Adoption*, 66 AM. J. HUM. GENETICS 761, 761 (2000). In response, the two groups have issued proposed guidelines to govern when genetic tests are performed on newborns and children during the adoption process. *Id.* In brief, the ASHG and ACMG:

> support genetic testing in the adoption process if it is (1) consistent with preventive and diagnostic tests performed on all children of a similar age, (2) generally limited to testing for medical conditions that manifest themselves during childhood or for which preventive measures or therapies may be undertaken during childhood, and (3) not used to detect genetic variations within the normal range.

Id. at 766.

27 Parents can avoid the birth of a child with certain genetic abnormalities in two ways. The first method combines genetic testing with selective abortion. *See* Cynthia Powell, *The Current State of Prenatal Genetic Testing in the United States*, *in* PRENATAL TESTING, *supra* note 10, at 47-48. The pregnant woman would undergo amniocentesis at approximately sixteen to eighteen weeks gestation to test the developing fetus for specific genetic abnormalities. *Id.* The second method, called "preimplantation genetic diagnosis," uses in-vitro fertilization. *See* Jason Christopher Roberts, *Customizing Conception: A Survey of Preimplantation Genetic Diagnosis and the Resulting Social, Ethical, and Legal Dilemmas*, 2002 DUKE L & TECH. REV. 12, 12 (2002). The fertilized embryos are then tested for the genetic abnormality and only ones that are free of the relevant mutation are transferred to the womb of the mother. *Id.* There are variations on these two methods but these two procedures capture most of the ways that genetic testing is used in the prenatal context.

28 Coined by Thomas H. Murray, the term "genetic exceptionalism" refers to the view that genetic information is different from other health information in ways that warrant different treatment. Murray, *supra* note 12, at 61.

Second, this Article builds on these arguments by revealing an argument that lies under the surface of much commentary and by developing and drawing out the complexities and difficulties of one of these familiar arguments. To articulate what is new requires briefly reviewing the points that have been made before.

1. Genetic Discrimination Is Irrational

Some supporters of special legislation argue that prohibiting genetic discrimination is necessary because such discrimination is irrational.[29] First, it is important to clearly define "irrational" in this context. The concept of irrational discrimination is most often used in connection with insurance. Loosely, irrational discrimination in insurance refers to instances when insurers charge higher than normal rates to a group of persons who are not in fact more likely than average to make claims for reimbursement from the insurer. Rational discrimination, by contrast, distinguishes between groups in a way that reflects the real risk of loss each group poses.[30] More precisely, discrimination is irrational if the prices charged by the insurer do not reflect the actual risk of loss each group poses as well as the cost to the insurer of distinguishing between the groups. Clearly defined, it is unclear what is morally important about discrimination being rational. Irrational discrimination is simply synonymous with bad business. To see this point, consider an example from outside of the insurance context. Law firms often select which law graduates to hire based on their grades in the hopes that grades are predictive of legal ability. If grades are not a good predictor of legal ability, then this form of discrimination by law firms is not rational. Would this irrationality make it morally troubling, however? It is entirely plausible that reviewing writing samples of students would provide a more accurate measure of which students are likely to become good lawyers. Law firms may be reluctant to use this screening method until they have narrowed the field somewhat because it is costly to review samples of student writing. The fact that firms may reject the student writing method (on the grounds of cost) emphasizes the fact that rational discrimination is simply a matter of assessing the efficiency of a proxy *from the perspective of the entity employing it.* It is hard to see the moral pull of this criterion.

Even so, is genetic discrimination irrational? While it is true that many people who carry a gene which predisposes them to a particular illness will not in fact become sick, it is also true that such a person is more likely than average to develop that illness, thereby making discrimination, were it to occur, rational. Moreover, the probabilistic nature of genetic information is no different from other information about a person's future health used by insurers to set rates. Not all smokers develop lung cancer. But because smokers are more likely than average to develop lung cancer, charging smokers higher health insurance rates is rational.

Supporters of protective legislation may argue that genetic information can be more complex and less predictive of future health or illness than smoking is of cancer. While this is accurate in the case of some genetic information, other genetic

---

29 Larry Gostin, *Genetic Discrimination: The Use of Genetically Based Diagnostic and Prognostic Tests by Employers and Insurers*, 17 AM. J.L. & MED. 109, 113-15 (1991) (arguing that discrimination on the basis of genetic information is likely to be irrational because the degree of uncertainty surrounding how predictive of future illness a genetic mutation may be); Greely, *supra* note 13, at 1500 (offering the opportunity "to protect those people who are at risk for irrational and ill-informed genetic discrimination" as one reason to support legislation prohibiting genetic discrimination in health insurance).

30 KENNETH S. ABRAHAM, DISTRIBUTING RISK: INSURANCE, LEGAL THEORY, AND PUBLIC POLICY 93 (1986).

information is highly predictive. The term used by geneticists is "penetrance." The "penetrance" of a gene mutation tells us the likelihood that a person with the mutation will develop the disease over the course of his life. If a gene is seventy-five percent penetrant, then seventy-five percent of those who carry the gene mutation will develop the illness.[31] Sometimes then genetic discrimination may be rational when the genetic mutation at issue is highly penetrant and sometimes it will be irrational when the genetic mutation is not highly penetrant. So far, the percentage of highly penetrant mutations among those scientists have identified is fairly significant because the fact that these diseases are strongly genetic has been fairly obvious for a long time. As scientists learn about the genetic bases for more diseases, however, these highly penetrant mutations are likely to become rarer.[32]

Where genetic mutations are not highly penetrant, genetic discrimination may be irrational. But will it occur? In other words, is it a problem that requires a legislative solution? First, it is important to note that the insurance statutes of all states already require that insurance rates be grounded in actuarial data; state law generally requires that insurance rates be rational.[33] In employment law, by contrast, there is no general requirement of rational behavior. If an employer wants to discriminate irrationally by hiring only brown-eyed applicants for example, such

---

31 As Henry Greely and others have pointed out, while a small number of genetic mutations are highly penetrant, the scientific community knew of their penetrance well before the recent boom in genetic knowledge. Their high penetrance made them readily observable in family medical histories before scientists were to identify the specific gene mutation responsible. Therefore, what scientists are likely to discover in the future will be cases where the genetic component of the disease is much less significant. Greely, *supra* note 13, at 1487 (explaining that "the logic of the discovery process means that the strong associations [between gene mutations and disease] are likely to be rare; the less-rare associations are likely to be weak").

32 *Id.*

33 The most common means of state regulation of actuarially supported insurance rates are "standard valuation laws," under which all insurers must submit to the state insurance commissioner actuarial data supporting their insurance premiums. For example, the State of Alabama's Insurance Code requires that:

> Every life insurer doing business in this state shall annually submit the opinion of a qualified actuary as to whether the reserves and related actuarial items held in support of the policies and contracts specified by the commissioner by regulation are computed appropriately, are based on assumptions which satisfy contractual provisions, are consistent with prior reported amounts, and comply with applicable laws of this state. The commissioner, by regulation, shall define the specifics of this opinion and add any other items deemed to be necessary to its scope.

ALA. CODE § 27-36-7 (1986); *see also* ALASKA STAT. § 21.18.110 (2002); ARK. CODE ANN. §§ 23-84-101 to 23-84-113 (Lexis 1999); CAL. INS. CODE § 10489.1-10489.10 (West 1988 & Supp. 2003); CONN. GEN. STAT. ANN. § 38a-78 (West 2000); DEL. CODE ANN. tit. 18, § 1113 (1999); D.C. CODE ANN. § 31-4701 (2001); FLA. STAT. ANN. § 625.121 (West 1996 & Supp. 2003); GA. CODE ANN. § 33-10-13 (2000); HAW. REV. STAT. ANN. § 431:5-307 (Michie 2001); IDAHO CODE § 41-612 (Michie 1998 & Supp. 2002); 215 ILL. COMP. STAT. ANN. 5/223 (2000); IOWA CODE ANN. § 508.36 (West 1998 & Supp. 2002); KAN. STAT. ANN. § 40-409 (2001); KY. REV. STAT. ANN. § 304.6-120 (Michie 1997); ME. REV. STAT. ANN. tit. 24-A, §§ 951-958-A (2000); MD. CODE ANN., INS. §§ 5-301-5-312 (1997 & Supp. 2002); MINN. STAT. ANN. § 61A.25 (West 1996 & Supp. 2003); MISS. CODE ANN. § 83-7-23 (1999); MONT. CODE ANN. §§ 33-2-521-33-2-529 (2000); NEV. REV. STAT. ANN. § 688A.325 (2001); N.H. REV. STAT. ANN. § 410 (1998); N.J. STAT. ANN. § 17:19-8 (West 2001 & Supp. 2002); N.M. STAT. ANN. § 59A-8-5 (Michie 2000); N.C. GEN. STAT. § 58-58-50 (2001); N.D. CENT. CODE §§ 26.1-35-01 to 26.1-35-10 (2001); OKLA. STAT. ANN. tit. 36, § 1510 (West 2002); OR. REV. STAT. §§ 733.300 to 733.322 (2001); R.I. GEN. LAWS §§ 27-4.5-1 to 27-4.5-12 (2002); TENN. CODE ANN. § 56-1-403 (2000); TEX. INS. CODE ANN. § 3.28 (Vernon 2002); UTAH CODE ANN. §§ 31A-17-501 to 31A-17-513 (2002); VT. STAT. ANN. tit. 8, §§ 3781-3789 (2001); WASH. REV. CODE ANN. § 48.74 (West 1999); W. VA. CODE ANN. § 33-7-9 (Michie 2000 & Supp. 2002); WIS. STAT. ANN. § 623.06 (West 1995 & Supp. 2002).

irrational discrimination by itself is not prohibited. It is only prohibited when the particular irrationality is of a special sort, and then not *because* it is irrational. Race, sex and disability discrimination, for example, are largely prohibited whether they are rational or irrational.[34]

Second, why think that irrational discrimination will occur at all? Irrational discrimination is, by definition, irrational and therefore bad business. Those who believe the market is mostly rational should be confident that irrational discrimination will not be a long-term problem.[35] But perhaps the insurance and employment markets are not as rational as the economist likes to suppose. So far, this worry does not appear to be warranted. Despite the fears of genetic scientists and people with genetic diseases, there is very little evidence of genetic discrimination in insurance or employment.[36] But that may change. Genetics is still not well understood by insurers and employers. As genetics makes its way into common discourse, discrimination could become more common. Moreover, because insurers and employers may have only an incomplete and unsophisticated understanding of genetics, irrational discrimination may occur. While this is conjecture, the important question to ask is, if irrational genetic discrimination were to become a bigger problem than it is now, would that be a reason to prohibit it?

Three reasons argue against prohibition. First, the law will be over-inclusive if its aim is to ban only irrational genetic discrimination because some genetic discrimination is rational. Second, current law already bans irrational discrimination in insurance, though these laws are surely less potent than a law specifically banning genetic discrimination in insurance. Third, and most importantly, it is not at all clear why being subject to irrational discrimination is a significant moral harm that requires remediation. Rational discrimination is simply the making of distinctions

34 *Compare* Frontiero v. Richardson, 411 U.S. 677, 690-91 (1973) (striking down a federal law that provided male married armed service members an automatic dependency allowance, but which required female married members to prove the dependency of their spouses) *with* Reed v. Reed, 404 U.S. 71, 76 (1971) (striking down an Idaho law that gave a preference to men over women as estate administrators on the ground that it was "arbitrary").

35 Richard Epstein is the most forceful advocate of this position. *See* RICHARD EPSTEIN, FORBIDDEN GROUNDS: THE CASE AGAINST EMPLOYMENT DISCRIMINATION LAWS 42 (1992) (arguing that "there are natural curbs against irrational contracting behavior" and thus that "the legal system normally has no need to superintend the wisdom of bargains"); *see also* Epstein, *supra* note 6.

36 Greely, *supra* note 13, at 1483 (contending that "[g]enetic discrimination is a much greater threat in people's fears than it is in reality, today or in the foreseeable future, for both scientific and social reasons"); Rothstein & Anderlik, *supra* note 7, at 357 (pointing out that "[t]o date, the evidence of genetic discrimination has been anecdotal (Billings et al. 1992; Geller et al. 1996) or derived from studies with methodological weaknesses such as reliance on self-report (Lapham et al. 1996)" and that, therefore, "a recent study combining in-person interviews with health insurers and a direct market test has attracted considerable attention (Hall and Rich 2000)" because it found that "a person with a serious genetic condition but asymptomatic for disease would have little or no difficulty obtaining individual health insurance under current market conditions"). *But see* E. Virginia Lapham et al., *Genetic Discrimination: Perspectives of Consumers*, 274 SCIENCE 621, 623 (1996) (finding that "43% of the respondents reported that they or members of their family have experienced genetic discrimination in one or more of the three areas" of health insurance, life insurance and employment). This study is based on the type of statistical self-reporting that Rothstein and Anderlik critique. *See* Rothstein & Anderlik, *supra* note 7, at 357. Lapham and her co-authors define genetic discrimination to include cases in which the individual discriminated against is already symptomatic for the disease, whereas Hall and Rich looked for evidence of cases in which discrimination occurred when the person has a genetic predisposition but no manifest disease. *Compare* E. Virginia Lapham et al., *Genetic Discrimination: Perspectives of Consumers*, 274 SCIENCE 621 (1996) *with* Mark A. Hall & Stephen S. Rich, *The Impact on Genetic Discrimination of Laws Restricting Health Insurers' Use of Genetic Information*, 66 AM. J. HEALTH GENETICS 293 (2000); *see also infra* note 70-75. This definitional difference may, in part, explain the wide disparity of results.

that are economically sensible to make from the perspective of the insurer, employer or other actor who draws such distinctions. A bad business judgment, without more, does not constitute a moral wrong to the person disadvantaged by that judgment. The fact that some genetic discrimination is irrational, therefore, does not provide a reason to ban genetic discrimination.

2. Genes Are Beyond Individual Control

A common argument for singling out some attributes for protection from discrimination is that the traits in question are "immutable"[37] or beyond individual control.[38] The moral intuition underlying this argument is that a person ought to be granted or denied a benefit on the basis of what she *does*, not who she *is*. This conception of merit is subject to contention in its own right.[39] Its virtues and complexities will not be examined here, however. For now, this Article argues that immutability fails as a reason to prohibit genetic discrimination for two reasons. First, this principle is not one that is adopted in our current laws. In fact, most goods are distributed according to quite different principles. The basketball player who earns millions of dollars for his performance earns that money only partially in recognition of his effort. His height—a trait beyond his control—and his natural talent also play a role in making him a skilled player. Moreover, while many of the traits on which discrimination is prohibited are immutable (or mutable only at great effort), traits such as race or sex for example, others are accorded special protection, even if they are in fact highly mutable, such as religion.

Second, while the idea that one ought to be held responsible only for those traits over which one has control has important moral appeal and is surely at the heart of much of our criminal law, it is far more complex conceptually than it initially appears. The idea that a smoker ought to pay more for health insurance than a non-smoker seems morally appealing because the smoker, in choosing to smoke, is thereby at least partially responsible for the fact that she is more likely than the non-smoker to become sick. However, even that example is problematic because of the additive quality of nicotine and the fact that scientists are beginning to learn that qualities outside of individual control influence the degree to which each smoker in fact endangers his or her health by smoking. As scientists learn more about genetic predispositions to disease, they discover that our bodies make demands on us unevenly. For example, some people can eat high fat diets with little health risk while others must maintain extremely prudent low fat diets in order to have similar risks of disease. While dietary choices are within individual control, the genetic predispositions that make them necessary are not. If we say that the second person is responsible for her poor health if she does not follow the prudent diet, but that the first person is free to indulge, then we in fact hold the second person responsible for

---

37 *See, e.g.*, ERWIN CHEMERINSKY. CONSTITUTIONAL LAW: PRINCIPLES & POLICIES 551 (1997) (explaining that one of the three central reasons that the Supreme Court scrutinizes the use of some legal classifications closely is the fact that the trait at issue is immutable). As Chemerinsky articulates the moral intuition behind this strand of law, "[i]t is unfair to discriminate against people for a characteristic that is acquired at birth and cannot be changed." *Id.*

38 *See* Gostin, *supra* note 29, at 110-11 (arguing that because genetic conditions are "neither subject to the person's control, nor the result of willful behavior," genetic discrimination ought to be prohibited).

39 For an interesting examination of the complexities of the merit principle, see Judith Lichtenberg & David Luban, *The Merits of Merit*, 17 REP. FROM INST. FOR PHIL. & PUB. POL'Y 21 (1997).

factors beyond his control—the fact that his genetic make-up demands an especially low fat diet.[40]

3. A Small Number of People Are Especially Burdened by Genetic Disease

A related argument for protection from genetic discrimination is based on the fact that genetic predisposition to disease is distributed quite unequally. While most of us may turn out to have roughly equivalent risk profiles—somewhat higher than average risk of X, combined with lower than average risk of Y—there are likely to be smaller groups of people who are extremely lucky or extremely unlucky in terms of their genetic make-up. For the few extremely unlucky, a social policy that permits discrimination on that basis would seem to cruelly compound their misfortune. Professor Greely makes this point in support of limited protection from genetic discrimination in health insurance and employment.[41]

While this argument may seem persuasive, it does not explain why those whose genetic make-up predisposes them to illness deserve special solicitation that those whose poor health has already manifested do not. After all, the argument that it would be especially cruel to compound the misery of the unlucky is stronger still with regard to those who have already become sick. Whatever responsibility our community has to provide basic healthcare to those who are especially likely to become ill surely also extends to those who are already sick.

Greely offers three arguments to support his view without explicitly addressing the question of how these arguments differentiate genetic discrimination from health status discrimination. He reasons (1) that "employment and health insurance are important parts of life in contemporary America;" (2) that "we can afford [it];" and (3) that providing such protection "should have only minor costs to the economy but major benefits for those few people whose jobs or health insurance would be saved."[42] The second two points relate to why special protection for genetic discrimination is warranted. Protecting those with manifest illness from the same sorts of employment and health insurance discrimination would presumably be more expensive, perhaps putting pressure on our ability to afford such protection and disrupting the economy in greater ways. However, the benefits to those whose jobs or health insurance would be saved would also be greater. Why this is the point at which the cost simply becomes too great is a claim that requires more discussion to be persuasive. Without it, there is no clear reason to include some who are unlucky with regard to health in the important goods of society (for Greely is surely right that health insurance and employment are extremely important) but not others.

4. Genetic Traits Are Shared with Racial or Ethnic Groups

One reason the genetically unlucky may deserve special protections that those who are merely sick do not is that particular disease-causing genetic mutations may especially affect identifiable racial or ethnic groups. For example, sickle-cell trait generally affects Africans and African-Americans,[43] and two mutations associated

---

40 *Cf.* Deborah S. Hellman, *Is Actuarially Fair Insurance Rating Actually Fair?: A Case Study In Insuring Battered Women*, 32 HARV. C.R.-C.L. L. REV. 355, 364-69 (1997) (arguing the ability to control one's own health is not in itself enough to justify responsibility for it).

41 Greely, *supra* note 13, at 1500.

42 *Id.*

43 Lemmens, *supra* note 5, at 373 (reporting that "in the U.S., from 8% to 10% of African-Americans are carriers of the sickle-cell trait, and 1 in 400 to 600 has sickle-cell anaemia").

with breast cancer are more common among Ashkenazi Jews.[44] When the group affected is one which has already been stigmatized in our society, there is a risk of further entrenching negative attitudes and of over-reacting to the significance of the association.[45]

Though important, these associations alone do not justify a ban on genetic discrimination. First, while the associations between some genetic diseases and stigmatized racial or ethics groups seem significant today, this significance will probably wane. The nature of genetic research makes it easier to identify genetic mutations among relatively homogeneous and relatively small ethnic groups.[46] But, as scientists' ability to identify the function of genes grows, the relevance of these small populations for study will diminish. It is probably not the case that Ashkenazi Jews, for instance, have more disease predisposing genes than other ethnic groups. It just seems so now because their group is easier to study.

Moreover, there are many disease-causing genetic mutations that are not in fact more prevalent among already stigmatized racial or ethnic populations. Huntington's disease and early onset Alzheimer's disease are two prominent and particularly devastating examples. In addition, as Professor Trudo Lemmens argues, there are similar associations between non-genetic diseases and racial or ethnic groups. For example, there is a "lower incidence of high cholesterol levels among certain ethnic groups" and "HIV/AIDS is more prevalent among gays and intravenous drug users and specific ethnic communities."[47] To ban genetic discrimination because of the risk of further stigmatizing racial and ethnic groups that are already subordinated in our society would be to enact a law that is quite dramatically both over- and under-inclusive.

5. Stigma

Some supporters of protective legislation argue that it is necessary because the genetic predisposition to disease is itself stigmatizing. Like the concern that racial prejudice creates a color hierarchy in our society, one might worry about a genetic hierarchy. For example, those whose genes especially predispose them to disease may be viewed as less desirable as customers for insurance, as employees or as marriage partners. The argument contends that because genetic predisposition to disease is stigmatizing, discrimination on this basis ought to be prohibited.

While superficially appealing, this argument is underdeveloped. First, in order to work as an argument for special legislation forbidding genetic discrimination as distinct from discrimination on the basis of health status, one must claim not just that having a genetic predisposition to disease is stigmatizing, but that it is *more*

---

44 *Id.* (citing S.V. Hodgson et al., *Risk Factors for Detecting Germline BRCA1 and BRCA2 Founder Mutations in Ashkenazi Jewish Women with Breast or Ovarian Cancer*, 36 J. MED. GENETICS 369 (1999)).

45 *See* Janet L. Dolgin, *Personhood, Discrimination and the New Genetics*, 66 BROOK. L. REV. 755, 786-87 (2000-01) (arguing that "[e]mployers, insurers, and others may arrive at assumptions about an individual's genome from information regarding the genome of the individual's ethnic group" and that "[s]uch discrimination may be especially harmful when aimed at individuals belonging to groups that have historically been singled out for racist treatment on the basis of somatic characteristics"); Gostin, *supra* note 29, at 111 (claiming that the "fact that genetic diseases are sometimes closely associated with discrete ethnic or racial groups such as African Americans, Ashkenazi Jews or Armenians compounds the potential for invidious discrimination").

46 For that reason, Ashkenazi Jews, for example, provide a fertile population for study. Dolgin, *supra* note 45, at 789-90.

47 Lemmens, *supra* note 5, at 374.

stigmatizing than having an already manifested illness. Second, and more importantly, the meaning of "stigma" is not clear in the way this argument is generally presented.[48] It is thus hard to evaluate the argument's strength. There are two reasonable reconstructions of what those commentators who object to genetic discrimination on the basis of stigma may have in mind. First, stigma may refer to the *effect* on the persons with the genetic condition. That is, genetic discrimination is wrong because it causes harm. Second, stigma may refer to what the policy of genetic discrimination *expresses*. According to this interpretation, genetic discrimination is wrong because of the meaning expressed in distinguishing people on this basis. This expressivist argument is important and is implicit in much of the critique of genetic discrimination. It has not, however, been carefully articulated and evaluated. I will take up that task below. If stigma refers to the *effect* on the person, one still must ask what harm in particular is at issue and why this harm makes such discrimination wrong. If a person is denied a job or insurance coverage because of the genetic traits she carries, the harm suffered by the loss of the job or insurance is not a stigmatic harm. Moreover, it is surely not a harm that distinguishes genetic discrimination from discrimination on the basis of health. Perhaps stigmatic harm refers to the psychological effect on the person denied the good because of her genetic traits. This way of understanding stigma is familar from arguments made about what makes race discrimination wrong. *Brown v. Board of Education*[49] argued, after all, that racial segregation of black school children was unconstitutional because of the effect on the "hearts and minds" of the children.[50]

As I have argued elsewhere, this conception of what makes discrimination wrong is flawed.[51] If harm to the person subject to discrimination were a necessary component of wrongful discrimination, then racial segregation of facilities for those in a permanent vegetative state would not be wrongful because such persons are incapable of suffering psychological or emotional hurt. As this conclusion seems untenable, the claim that wrongful discrimination requires that those affected feel stigmatized fails.

Moreover, stigma seems to connote something bigger or more systemic than simply the psychological effect on the individuals at issue. The idea of stigma carries with it the idea of a class or caste-like distinction between groups of people. Perhaps what is wrong with genetic discrimination, and distinguishes it from discrimination on the basis of health, is that it threatens to create a genetic underclass. This fear and accompanying argument for legislation prohibiting genetic discrimination appears in several places in the literature, but is most forcefully articulated by Susan Wolf.[52]

6. Geneticism

Professor Susan Wolf argues that the tendency to focus on genetics and to use information about genetics to subordinate people is best termed "geneticism"—a

---

48 *See, e.g.*, Lemmens, *supra* note 5, at 411 (arguing for some regulation of genetic discrimination in Canada on the grounds that "[t]here can be circumstances in which genetic discrimination has such a symbolic, stigmatizing character, that allowing it to be used for insurance purposes would be considered inappropriate *per se*" (emphasis added)). Lemmens' treatment of symbolic as synonymous with stigmatizing implies that he is characterizing the idea of stigma as expressive harm. *See id.* This implication, however, is not clear.

49 347 U.S. 483 (1954).

50 *Id.* at 494.

51 Hellman, *Expressive Dimension*, *supra* note 9, at 10.

52 *See infra* notes 53-58 and accompanying text.

label she uses to call attention to a deeply ingrained mindset and set of structural practices.[53] Wolf does not in fact support the enactment of laws prohibiting genetic discrimination because she believes that the anti-discrimination paradigm is inapt and will result in more harm than good. Her objection is grounded in critiques of the anti-discrimination approach as applied to problems of racism and sexism. She believes that "clinging to 'genetic discrimination' . . . does damage by creating a false genetic 'norm,' frustrating structural reform, obscuring the deep psychological roots of genetic stereotyping and prejudice, and isolating genetic from other harms."[54] Consequently, she argues for a conception of the issue as something deeper, more entrenched, and more problematic, such that change of a systemic nature is required.[55]

Thus, while Wolf does not support laws prohibiting genetic discrimination, her approach does support the idea of genetic exceptionalism. Although she does not argue that genetic discrimination is meaningfully different than discrimination on the basis of health status, Wolf's whole approach seems to endorse the view that *geneticism* is importantly different from *healthism*, to use her terminology. For that reason, I consider her approach here.

How does Wolf argue that geneticism is different? She focuses on demonstrating the harms of an anti-discrimination approach rather than on demonstrating that geneticism pervades society. She takes this to be obvious, and merely calls our attention to "the eagerness to draw genetic conclusions, the search for supposedly deviant genes, and the conviction that such genes actually deserve disadvantage"[56] that she terms *geneticism*. What is not clear from Wolf's critique is how one is to know that certain inequalities in the world, such as inequalities that track genetic predispositions to disease, are morally problematic while others are not. Surely, the mere fact that there is equality along some dimensions in a society does not make that society unjust. It is fair to assume that we are all committed to the idea that each person is entitled to equal concern and respect.[57] Beyond that, there is much debate. For example, traditionally those on the political left stress the importance of lessening social and economic inequalities, while those on the political right emphasize the importance of equality of negative liberties like speech, assembly and economic activity. The important question that Wolf fails to address is how one is to know that the particular inequalities she identifies by her term *geneticism* are morally problematic. A theoretical link is missing from Wolf's argument. She needs to explain why the inequality she highlights is one that matters morally. In order to do so, Wolf needs to make use of a general theory of when discrimination is wrong, a task she avoids because she shuns the anti-discrimination paradigm. Without it, her approach is unable to do more than call our attention to the fact that genetic differences are salient in our society in a way that can work to subordinate people. As Professor Amartya Sen insightfully emphasizes, since

53 Wolf, *supra* note 7, at 350.

54 *Id.* at 350.

55 *Id.* (arguing for a "reinterpretation of 'genetic discrimination' as something bigger, the use of genetic notions to create and reinforce power relationships in which some dominate and others are subordinated").

56 *Id.* at 347.

57 Ronald Dworkin offers this principle of equal concern as the moral principle underlying the Equal Protection Clause. Ronald Dworkin, *In Defense of Equality*, 1 SOC. PHIL. & POL'Y 24, 24 (1983) (positing that equal protection requires that "the interests of the members of the community matter, and matter equally"). I build on Dworkin's principle in my article on the expressive dimension of Equal Protection. Hellman, *Expressive Dimension*, *supra* note 9, at 7-8.

equality among people across all dimensions of life is impossible, moral theories differentiate themselves by articulating what sort of equality is morally significant.[58] To merely note a particular inequality—people with trait X have less of Y—is not enough. One must also explain why this sort of inequality is one that is morally problematic.

In my view, one can only determine whether a given inequality violates the equal concern mandate by looking at the social meaning of that inequality in the particular society. In other words, a policy that discriminates is wrong if the policy *expresses* that people are not of equal moral worth. Clearly this is a theory that requires some elaboration. While a detailed argument for the expressive conception of wrongful discrimination will not be presented in this Article,[59] below I will examine how the argument can be developed in the context of genetic discrimination. It is important to do so because this argument is implicit in much of the critique of genetic discrimination. Before examining this expressivist argument in some detail, however, I will discuss one more familiar argument for legislation prohibiting genetic discrimination.

7. The Health Benefits of the Human Genome Project

The last argument in support of legislation prohibiting genetic discrimination that I will consider here emphasizes the tremendous health-promoting potential of advances in understanding human genetics and the genetic basis for disease. Supporters of protective legislation argue that fears of genetic discrimination may thwart the promise of genetic science in two ways. First, people who fear discrimination may be reluctant to take genetic tests that could identify their risk for particular diseases. If there are therapies or other interventions that may benefit them, this fear may frustrate efforts to improve health through genetic understanding.[60] Second, if people are fearful of discrimination, they may be reluctant to participate in research that includes genetic testing.[61] Because scientists are just beginning to understand how genetic traits influence disease, this research is critically important to achieving substantial health benefits from the Human Genome Project.

These are important issues to consider. An understanding of genetic diseases stands to benefit us all. In addition, this argument provides an answer to the question with which we began this inquiry: in what way is genetic discrimination morally different from discrimination on the basis of health such that it warrants special protection? If fears of genetic discrimination will frustrate the health-enhancing potential of advances in genetic science, then this provides a reason to

---

58 *See generally* AMARTYA SEN, INEQUALITY REEXAMINED (1992).

59 *Cf.* Hellman, *Expressive Dimension*, *supra* note 9.

60 Several commentators offer this argument in support of legislation prohibiting genetic discrimination. *See, e.g.*, Gostin, *supra* note 29, at 113 (emphasizing that "if fear of discrimination deters people from genetic diagnosis and prognosis, renders them less willing to confide in physicians and genetic counselors, and makes them more concerned with the loss of a job or insurance than with care and treatment, the benefits of genetic data collection will not be fully achieved"); Greely, *supra* note 13, at 1501 (arguing that legislation is warranted in part because "people who are afraid of genetic discrimination are afraid to take genetic tests that offer the possibility of improving their health").

61 Greely, *supra* note 13, at 1501 (stressing the fact that reducing people's fears of discrimination is important because "research in human genetics has enormous potential for alleviating human suffering"); Lemmens, *supra* note 5, at 364 (arguing that studies "have indicated that women at risk for breast cancer because of family history often refuse to undergo testing out of fear of the impact of testing on insurability").

specially protect people from genetic discrimination. However, this simple argument ought not to end the inquiry. There are several questions, both empirical and philosophical, that require attention before one can conclude that enhancing health provides a sufficient reason to support protective legislation. This more complete analysis, which has thus far been missing from the literature, will be examined below.[62]

B. NEW ARGUMENTS

This brief review of the most familiar arguments for legislation prohibiting genetic discrimination lays the groundwork for the contributions of this Article. This section focuses on two arguments for genetic exceptionalism. First, I will look closely at the argument that genetic discrimination is different because fear of genetic discrimination will thwart the tremendous health-enhancing potential of the Human Genome Project. Second, I will examine the argument that genetic discrimination is meaningfully different from discrimination on the basis of health and therefore necessitates protective legislation because of what the practice of genetic discrimination expresses. Each of these arguments is promising. Below, I present a detailed analysis of the strengths of each in a way that will, I hope, advance the discussion of whether genetic discrimination really is different.

1. The Policy Argument: Promoting Health

The first argument to carefully consider emphasizes the consequences of policy choices and argues that legislation prohibiting genetic discrimination ought to be adopted in order to ensure the realization of the tremendous health enhancing potential of the Human Genome Project. Rather than claiming that the rights of people with genetic conditions will be violated if legislation is not adopted, this argument contends that laws forbidding genetic discrimination will serve the general social welfare in the form of medical advances. As such, the persuasiveness of this argument depends on the accuracy of significant empirical claims.

The argument is not new and was therefore outlined briefly above. It is, however, an important argument, the complexities and implications of which have not yet been adequately addressed. In order to do that here, let me first restate the claim. The scientific advances in understanding the genetic basis for disease have enormous potential to improve health. Understanding individual predispositions to disease may allow medical advice to be individually tailored, both for prevention and for treatment.[63] For example, some people may be advised to change their diets or to be screened for disease more frequently than is ordinarily recommended. If a person develops the disease nonetheless, more individualized treatment recommendations may be available. Most exciting of all, a greater understanding of genetics may allow doctors to treat people with strong predispositions to serious illness prophylactically so that the illness itself never develops.

Genetic discrimination is a problem, then, because it may get in the way of each of these beneficial developments. If people fear genetic discrimination, they may be reluctant to be tested for genetic conditions. If so, achievement of the health

---

62 *See infra* notes 63-98 and accompanying text.

63 Whether individuals will in fact change their behavior in response to genetic information is uncertain. *See* Theresa M. Marteau & Caryn Lerman, *Genetic Risk and Behavioural Change*, 322 BRIT. MED. J. 1056, 1058 (2001) (finding that "current evidence suggests that providing people with DNA derived information about risks to their health does not increase motivation to change behaviour beyond that achieved with non-genetic information").

benefits described above may be thwarted in two ways. First, if people avoid testing, they may fail to partake in the therapeutic benefits that are currently available. Second, and perhaps more importantly, people who fear discrimination may decline to participate in research involving testing—research that could lead to discoveries that ultimately make pre-symptomatic treatment of genetic conditions possible. This second effect is particularly important because, at present, scientists have learned a lot about which gene mutations predispose one to certain diseases, but they have not yet been able to intervene that productively to help. One hopes that by building on current knowledge scientists will be able to discover how to advise individuals with particular genetic conditions to alter their diets or habits so as to improve health. New knowledge may even make it possible to treat incipient conditions before they develop. However, this is still a hope and promise of genetic science, not the reality. If people are unwilling to participate in research involving genetic testing because they fear genetic discrimination, scientific advances in this area may be dramatically slowed.

This argument surely sounds both plausible and significant. The argument rests on several assumptions, however—some empirical and some ethical—that require further consideration.

a. *Empirical Questions*

Many people considering genetic testing report that they fear genetic discrimination.[64] In order for legislation forbidding genetic discrimination to be successful in removing this barrier to testing or participation in research, proposed legislation must forbid the sorts of genetic discrimination that people actually fear. For example, Greely believes that "genetics has enormous potential for alleviating human suffering" and, thus, that "reducing that fear is important" because "people who are afraid of genetic discrimination are afraid to take genetic tests that offer the possibility of improving their health" and "may be afraid to take part in genetic research that could provide important answers to major diseases."[65] For this reason, among others, Greely supports modest federal legislation forbidding genetic discrimination in health insurance and employment.[66] To be effective, however, people must fear discrimination in health insurance and employment. If people also fear discrimination in life insurance or disability and long-term care insurance, for example, legislation that is limited to the healthcare and employment contexts is

64 *See* Lapham et al., *supra* note 36, at 622 (reporting that in a study surveying the experiences of people with family histories of genetic disease that "[t]he large majority (83%) of respondents said they would not want their insurers to know if they were tested and found to be at high risk for a genetic disorder"). In addition, Lapham and her co-authors found that "fear of genetic discrimination . . . resulted in 9% of the respondents or a family member refusing to be tested for a genetic condition." *Id.*; *but see* P.B. Jacobson et al., *Decision-Making about Genetic Testing Among Women at Familial Risk for Breast Cancer*, 59 PSYCHOSOMATIC MED. 459, 465 (1997) (finding that "the only disadvantages perceived by a majority of women were that learning their genetic carrier status would increase their concerns about developing breast cancer and cause them to worry more about other family members who could be carriers"). Interestingly, when people with a family medical history of a particular disease are considering genetic testing, genetic discrimination may not be of foremost concern because some information about genetic risk (albeit less precise) is already present in the form of that family medical history. Therefore, other issues take precedence in the decision-making.

65 Greely, *supra* note 13, at 1501.

66 *Id.* at 1504 (describing the scope of legislation forbidding genetic discrimination that he favors as "limited to employment questions and to health coverage (whether through traditional health insurance, HMOs, or a self-funded employee health benefit plan)").

unlikely to promote health in the way that Greely hopes. Consider the case of someone contemplating entering a research study dealing with the genetic predisposition to early onset Alzheimer's disease. For this person, the availability of long-term care and disability insurance are critical issues. Because most Americans who have healthcare coverage get that coverage through large group plans that do not individually assess the risk of each covered person, protection from discrimination in health insurance is likely to be the least of this person's concerns. If so, limited legislation may do little to affect people's decision-making about whether to participate in testing or research.

People's fears of genetic discrimination may extend further still. Perhaps their central worry is that they will be unable to adopt children, or that in a custody battle they will be denied custody of the children because of their genetic condition. Also, these fears may extend to genetic discrimination in contexts that no one is likely to advocate regulating. People may fear that no one will want to marry them if they test positive for some conditions, or that mates or family members will abandon them.[67] In order to argue that states or the federal government ought to adopt legislation prohibiting genetic discrimination to ensure that the health-enhancing potential of genetic science is realized, one must ascertain that the legislation will allay precisely the fears that are liable to affect decisions about testing.

It is not enough to cite studies documenting the fact that people fear genetic discrimination. One must determine in what contexts people fear genetic discrimination as well as to what degree. One must then assess whether the goal of promoting health through advances in genetic knowledge warrants legislation forbidding genetic discrimination in those areas, be it life and disability insurance, family law or personal life. Finally, if people fear the sorts of discrimination described above, but society is only willing to forbid discrimination in health insurance and employment, one must determine whether this limited protection will be enough to tip the balance for people toward a decision to participate in research or to undergo testing when it is therapeutically valuable. Otherwise, legislation forbidding discrimination in health insurance and employment will not achieve the goal of promoting participation in genetic research.

This discussion suggests that more empirical research is needed about precisely what sorts of protective legislation would indeed allay people's fears and, more importantly, affect their decisions about whether to be tested or to participate in research involving testing. Additionally, while fear of genetic discrimination in any context may indeed be part of the reason that people shun genetic testing, there are a plethora of other possible reasons. For example, there has been very little interest in testing for the genetic mutation responsible for Huntington's disease,[68] a fatal, degenerative disease for which there is currently no treatment or therapy. While fears of discrimination may explain in part why some people at risk for the disease avoid testing, it is easy to imagine other explanations. Here Sophocles's oft-cited admonition taps a powerful intuition: "what misery to be wise when wisdom profits

---

67 *See* Lori B. Andrews, *A Conceptual Framework for Genetic Policy: Comparing the Medical, Public Health, and Fundamental Rights Models*, 79 WASH. U. L. Q. 221, 245–49 (2001) (discussing data about the actual impact of genetic testing on relationships with spouses and potential spouses).

68 *Id.* at 250 (explaining that contrary to scientists' initial expectation that many people at risk for Huntington's disease would choose to be tested, in reality "fewer than fifteen percent of at-risk individuals chose to undergo the testing").

nothing."[69] Until we know how important fears of genetic discrimination are, relative to other considerations, we cannot say that legislation prohibiting genetic discrimination will in fact promote the fulfillment of the promise of genetic science.

Mark Hall and Stephen Rich have begun this important research in a study in which they conducted interviews of genetic counselors and medical geneticists at major medical centers. The study assessed whether recent state laws that prohibit genetic discrimination have been successful in reducing fear of health insurance genetic discrimination and whether the laws have thereby resulted in more testing.[70] Briefly, they reached the following conclusions on the basis of their study:

1. Fear of losing health insurance is the primary concern of people concerned about genetic discrimination.[71]
2. These fears "do not have very much actual impact on patients' final decisions about testing."[72]
3. Patients may cite fear of discrimination as a reason to decline testing because "it gives them a more socially acceptable reason to decline."[73]

Notably, these conclusions did not differ between states that had adopted laws forbidding genetic discrimination and states that had not.[74]

These conclusions suggest that the argument that one ought to prohibit genetic discrimination in order to promote health by encouraging participation in genetic testing and research may rest on shaky empirical ground. However, though the Hall and Rich study is important and provocative, it is limited. First, the study based its conclusions on the views of genetic counselors. As Hall and Rich acknowledge, some people who are deterred from testing may never make it as far as a session with a genetic counselor.[75] Second, as Hall and Rich acknowledge, the study focused on genetic testing for Huntington's disease and cancer, conditions in which the test accuracy is high and there is often little treatment that has therapeutic benefit. In a different context, potential testers may well react differently. Third, the Hall and Rich study methodology centers on the counselors' and geneticists' impressions of the clients' motivations. While surely useful, these impressions may be inaccurate. Finally, it is also important to remember that this is only one study and more empirical research is necessary.

One should also examine whether the desire to promote health provides a reason to forbid genetic discrimination in a way that distinguishes genetic discrimination from discrimination on the basis of general health status. The health promotion argument in favor of legislation forbidding genetic discrimination rests on the assertion that fear of genetic discrimination keeps people from testing or research that, in turn, thwarts the health-promoting potential of genetic science. There is nothing morally special, however, about promoting health through genetics. The argument examined in this section is appealing because health is important. In order for health promotion to provide a good reason to support legislation specifically forbidding genetic discrimination, one must determine that forbidding genetic

69 SOPHOCLES, OEDIPUS THE KING 28 (Anthony Burgess trans., Univ. Minn. Press 1972).

70 Mark A. Hall & Stephen S. Rich, *Genetic Privacy Laws and Patients' Fear of Discrimination by Health Insurers: The View from Genetic Counselors*, 28 J. L. MED. & ETHICS 245 (2000).

71 *Id.* at 249.

72 *Id.*

73 *Id.* at 251.

74 *Id.* at 252-53.

75 *Id.* at 253.

discrimination would promote health more than would prohibiting discrimination on the basis of health status more generally. The discrimination on the basis of health status that is currently permitted may also undermine health because people may refrain from other non-genetic, therapeutically useful medical tests in order to avoid discrimination.

Finally, if promoting health is important, and legislation forbidding genetic discrimination is warranted as a means to that end, one must also ask whether enactment of this legislation may simultaneously work to thwart that goal. If enactment will sap political will for the passage of some form of universal healthcare or universal health insurance coverage, then legislation forbidding genetic discrimination could undermine, rather than enhance, health. While one can only speculate about this question, access to healthcare for all Americans may be so important to the goal of promoting health that this concern is worth careful consideration.

I argue above that in order for the goal of health promotion to provide a reason to enact legislation forbidding genetic discrimination, one must examine several empirical questions to determine whether enactment of such legislation will indeed achieve that result. Before moving on, it is worth considering an objection to the health promotion argument that would make these empirical questions unimportant. Professors Colin Diver and Jane Maslow Cohen argue that protecting people from the economic consequences of a testing decision by passing legislation forbidding genetic discrimination is a mistake because it will encourage more testing than ought to occur. This is a strange yet familiar "law and economics" type argument:

> When a person decides not to undergo genetic testing, she is presumably deciding that her net utility would be reduced: that is, that the costs (including not only the direct costs of undergoing the testing, but also the expected adverse impact that the resulting knowledge would produce on both her economic prospects and her psychological state) outweigh the benefits (in terms of the improvement in her, and perhaps her offspring's, health that could result from ameliorative actions). By protecting individuals from adverse employment and insurance consequences, the antidiscrimination strategy eliminates two components of the economic costs of testing from the individual's cost-benefit calculus. In doing so, the government in effect encourages overconsumption of genetic testing.[76]

The flaw in this argument is that it assumes the very point it is attempting to show. The authors assume that the correct amount of testing is the amount that would occur if insurers and employers could discriminate on the basis of genetic information. If so, forbidding such discrimination and thereby shielding potential testers from these economic consequences will lead to "overconsumption" of testing. However, the amount of testing that would take place if discrimination were forbidden is only overconsumption if the correct amount is precisely the amount that would take place absent such protection. Missing from their argument is an explanation for why they take this amount of testing to be the baseline from which any deviation demands explanation.

A claim that a certain amount of testing is the right amount requires justification. The claim that the right amount is that amount that will promote health

76 Diver & Cohen, *supra* note 6, at 1469.

in the long run by providing enough people willing to undergo testing to allow productive research to continue is such a reason. Alternatively, one might argue that the right amount of testing is the amount that would take place were people to decide for reasons related to their medical and psychological well-being. The second reason makes individual health and well-being the desired end and the right amount of testing the amount that results when each individual is free to pursue that end. The first reason makes the overall promotion of health the goal and the "right" amount that which best achieves this outcome.

One would clearly need to say more to argue convincingly for either approach. The important point is that one must present an argument for why certain reasons ought to affect testing decisions and not others—which Diver and Cohen fail to do. Greely and others do just that by arguing for antidiscrimination protection (which has the effect of shielding individuals deciding whether to undergo genetic tests from consideration of some reasons) on the grounds that this legislation will promote health. While this argument rests on important empirical assumptions that must be examined (discussed above) and important ethical issues that require consideration (to be discussed below), this argument for protection from genetic discrimination has the virtue of providing a reason for the claim that only some considerations ought to affect individual testing decisions.

b. *Ethical Complications*

The argument that the government ought to enact legislation forbidding genetic discrimination, particularly in the context of health insurance, raises two important ethical concerns. First, one must ask whether it is right to pass laws that ensure that only some of those who are entitled to receive a benefit receive it. To those who believe that everyone ought to have access to healthcare, the fact that many people do not have health insurance is unjust. The interesting point that the current debate about legislation targeting genetic discrimination poses is whether remedying this injustice for some compounds the moral wrong to others by adding unfairness to injustice.

An anecdote—the truth of which I cannot verify—explains well the distinction between justice and fairness that I employ here. As I have heard the story told, Professor Sidney Morgenbesser of the Philosophy department of Columbia University was called for jury duty in New York City. In *voir dire*, he was asked whether the police had ever treated him unjustly or unfairly. The case before the court was, apparently, one dealing with alleged police brutality. Professor Morgenbesser replied, "unjustly yes, but not unfairly." Confused, the judge asked the philosopher to explain. Morgenbesser related that during the anti-war protests on the Columbia campus during the Vietnam War, he participated in a sit-in of the school's administrative offices. The police arrived and arrested the protesters, beating Morgenbesser in the process. This was unjust, according to the Professor, as the police ought not to have beaten him during the arrest. He was not resisting arrest and would have come with the police willingly. But it was not unfair, Morgenbesser went on, since the other protesters were also beaten. Following Morgenbesser's distinction between justice and fairness, then, legislation forbidding genetic discrimination may remedy injustice to some while adding unfairness to the injustice of being without health insurance to others.

Second, and perhaps more problematic, laws forbidding genetic discrimination in health insurance may work to thwart the development of the political constituency necessary to enact legislation providing healthcare coverage to all. Advances in genetic science could help foster the requisite sense of community needed to make it politically possible to enact some form of universal access to healthcare. As scientists learn more about genetic diseases, few of us are likely to escape some genetic flaw. Perhaps this sense of common fragility will help us to feel connected to and responsible for those who lack access to health insurance. If so, legislation forbidding genetic discrimination could work to undermine this momentum. In addition, because it is the relatively privileged who currently have access to genetic testing and who thus could stand to lose their insurance if genetic discrimination was permitted, those who would benefit from antidiscrimination protection are those with the most political power to bring about change.[77]

Assuming, for the sake of argument, that legislation forbidding genetic discrimination will indeed promote health by ensuring that the potential of genetic science is realized, who will benefit? Those individuals who would otherwise be deterred from testing and gaining information that could enhance their own health will benefit. In addition, if further research is not stymied by lack of participation and that research yields new discoveries, people who might otherwise become sick with genetic diseases in the future could be treated or cured, which is a significant benefit. But, if passage of antidiscrimination protection would indeed dampen the political pressure for universal healthcare coverage, then one must look closely at who loses out so that the benefits of genetic science may be realized. While it is not the poorest among us who lack healthcare or health insurance, as Medicaid insures healthcare for the poorest, those Americans without healthcare generally come from the ranks of the near poor or working poor.[78] Accordingly, legislation forbidding genetic discrimination may turn out to exacerbate inequality.

Moreover, the gains made possible by advances in genetic science are likely to be reaped by those who are already relatively privileged. Trudo Lemmens stresses this argument about deepening unfairness, pointing out that "[p]eople who will seek information on their genetic susceptibility will be those who have the financial means to undergo testing or those who have a good insurance plan (and thus already benefit from better healthcare)."[79] One ought to consider, then, whether society acts justly if it enacts legislation that will enhance health, but not evenhandedly. While one cannot know who will develop genetic diseases in the future, and those diseases are likely to affect the rich and poor alike, not everyone who stands to benefit from these health advances will actually do so. Those without access to healthcare are also likely to be without access to new genetic therapies. If legislation forbidding genetic discrimination also lessens the chance of achieving universal access to healthcare, then these laws may well cluster their benefits among the relatively privileged and their burdens among the less well-off.

Like the empirical questions identified above, these ethical concerns also require further study. First, I identified the tension between remedying injustice for some by ensuring access to health insurance, while adding unfairness to injustice for those left behind. In order to make the argument that special legislation forbidding genetic

77 Chetan Gulati emphasizes this point. Chetan Gulati, *Genetic Antidiscrimination Laws in Health Insurance: A Misguided Solution*, 4 QUINNIPIAC HEALTH L.J. 149, 168 (2001) (arguing that laws prohibiting genetic discrimination work to protect the health insurance structure from more systemic revision). Sonia Suter makes an interesting related observation by pointing out that the poor are disproportionately affected by non-genetic factors that affect health, like pollution and violence, and that therefore, special genetic legislation may exacerbate inequality. Suter, *supra* note 7, at 719.

78 Walter L. Stiehm, *Poverty Law: Access to Healthcare and Barriers to the Poor*, 4 QUINNIPIAC HEALTH L.J. 279, 285 (2001).

79 Lemmens, *supra* note 5, at 380.

discrimination is warranted, one must assess whether the unfairness of helping only some of those entitled to a benefit ought to defeat the policy. Second, the claim that forbidding genetic discrimination may stand in the way of universal access to healthcare is quite speculative and difficult to weigh. Thus, how much should we discount the exacerbation of inequality in light of the speculative nature of this supposition? This is a difficult question that requires further consideration.

2. The Expressivist Argument

Much of the commentary about genetic discrimination refers to the history of eugenics both in this country and in Europe. For example, Professor Aviam Soifer and Miriam Wugmeister warn that "[t]he tragic history of eugenics also casts a long shadow over contemporary claims regarding new knowledge about human genetics."[80] Professor Lori Andrews argues that the justification offered by insurers for their use of genetic information in classifying risks "is similar to that used in the earlier eugenics movement—that healthy people (that is, people with 'good genes') should not have to support people who have or may develop genetic diseases (people with 'bad genes')."[81] It seems obvious that this history is relevant to the question of whether genetic discrimination is wrong as well as to whether genetic discrimination is different from discrimination on the basis of health in a manner that warrants special legislation. Exactly why this history matters, however, is not clear from the arguments currently found in the literature. The expressivist argument presented in this Article makes it clear why this history is relevant; it is relevant because it changes the social meaning of current practices. Genetic discrimination expresses something different because of our experience with illegitimate uses of genetics. According to an expressivist conception of what makes discrimination wrong, this is important because it is the expressive dimension of policies that draw distinctions between people which determines whether these policies discriminate wrongfully.

The following argument is a close cousin of the argument that genetic discrimination is wrong because it is stigmatizing. Although "stigma" is a somewhat elusive term, it arguably calls attention to the effect of a law or policy. An expressivist approach, in contrast, focuses on what is expressed by a law or policy, regardless of whether this expression actually harms an identifiable group in a particular way. The expressivist argument is also related to Wolf's argument about "geneticism."[82] Both make the social dimension in which a particular discriminatory practice occurs highly relevant in determining its character. However, Wolf's argument is different in that it does not explain how or why the fact that genetics has been used to subordinate people matters in assessing whether genetic discrimination in insurance or employment is morally problematic.

The argument below explores the intuition that the history of eugenics matters in assessing whether and when genetic discrimination is wrong. It is a sympathetic reconstruction of an argument that is implicit in the literature on genetic discrimination but which has not been clearly articulated and evaluated.

---

80 Aviam Soifer & Miriam Wugmeister, *Mapping and Matching DNA: Several Legal Complications of "Accurate" Classifications*, 22 HASTINGS CONST. L.Q. 1, 25 (1994).

81 Lori B. Andrews, *Past as Prologue: Sobering Thoughts on Genetic Enthusiasm*, 27 SETON HALL L. REV. 893, 904 (1997).

82 *See* Wolf, *supra* note 7 and accompanying text.

a. *What Is an "Expressive" Theory?*

An expressive theory is one in which what an action expresses—its meaning—is relevant in determining its moral permissibility.[83] In this sense, it differs from the more familiar way one judges the moral permissibility of actions, in which it is the *intent* of the actor, the *effect* of the action, or both, that are central. For example, in criminal law, both intent and effect are relevant in judging the moral acceptability of an action. If someone accidentally kills another, the person will not likely be guilty of murder, which requires that the actor have a particular intent. In addition, if someone intends to kill another but fails, she may only be guilty of attempted murder and will thereby be punished more lightly because her action did not cause significant harm. To claim that the expressive character of the action is what matters in a particular instance is to call attention to a third dimension of an action. For example, if I spit on a homeless person, my action is wrong because it *expresses* disrespect. The effect of that action is clearly far less important than what it expresses. The spit may be unpleasant, but its unpleasantness is not what makes the action wrong.

In the above example, perhaps it is my intent to be disrespectful that makes the action wrong. A theory in which the expressive character of action is important is one in which actions can be morally wrong even when done accidentally or without the specific intent to convey the meaning the action expresses. Consider the following example. The Governor of South Carolina decides to fly the Confederate flag over the State House with the intent of building a sense of loyalty to the state and thereby increasing social cohesiveness.[84] This action would be morally problematic despite the innocuous intent of the Governor, and not merely because of the negative effect it is likely to have on social harmony. The action is morally troublesome because of what flying the Confederate flag expresses, regardless of whether the Governor intends to convey that meaning. Given the history of slavery and discrimination in the South, as well as the historical association of the Confederate flag with that history, flying the flag has a meaning that is beyond the control of the Governor's intent. To say that this meaning is relevant in assessing whether flying the flag is wrong is to emphasize our responsibility for how our actions are likely to be understood. So even if I do not intend disrespect by spitting, I cannot spit blamelessly, as the meaning of spitting on someone in our culture is as an act of disrespect. A theory that emphasizes the expressive dimension of action contends that the fact that spitting expresses disrespect matters.

One caveat: to say that the expressive dimension of action matters is not to say that it is all that matters. Intent and effect may also be relevant. However, the expressive dimension is relevant in a way that cannot be reduced to intent and effect. The expressive character of an action is relevant even when the meaning of the action is not what the actor intends to express (intent) and even if that meaning is not what is in fact communicated (effect).

Recently, attention to the expressive dimension of action has become important in constitutional law, particularly in interpreting the Establishment Clause of the

---

[83] For a discussion of the relevance of the expressive dimension of action, see Symposium, *The Expressive Dimension of Governmental Action: Philosophical and Legal Perspectives*, 60 MD. L. REV. 465 (2001).

[84] For a similar discussion regarding flying the Confederate flag over the State House in South Carolina, see James Lindemann Nelson, *The Meaning of the Act: Reflections on the Expressive Force of Reproductive Decision Making and Policies*, IN PRENATAL TESTING, *supra* note 10, at 196.

First Amendment and the Equal Protection Clause of the Fourteenth Amendment. In some cases, the U.S. Supreme Court has held that the correct manner by which to judge whether holiday displays unconstitutionally establish religion is to look at how those displays are likely to be interpreted by a hypothetical reasonable observer in our culture.[85] In the area of Equal Protection, the development has been less a phenomenon of the Court and more a small movement among academic commentators.[86] Here, the claim is that the right way to interpret whether laws or state policies violate the Equal Protection Clause is to focus on the meaning or expressive dimension of the law or policy. Current doctrine makes intent the touchstone of an Equal Protection violation.[87] The dominant critique of that doctrine would focus instead on the effect of laws and policies.[88] The expressivist approach, in contrast, looks at the law's objective meaning.

But why focus on meaning and how can such meaning be described as objective? While I cannot here develop the entire argument for understanding the Equal Protection Clause in expressive terms—as that would (and did) require an article on its own—let me just briefly restate my view. Laws and policies draw distinctions between people in a myriad of ways, some of which our intuitions tell us are morally troubling and some of which are not. Pre-*Brown* racial segregation of public schools seems paradigmatically wrong and in violation of Equal Protection.[89] Single-sex admissions at the Virginia Military Institute seems more ambiguous, though it was ruled unconstitutional by the Supreme Court.[90] Single-sex bathrooms at state schools, such as the University of Maryland School of Law, seem unproblematic. If we use the term "discrimination" to describe each of these cases, and others where people are treated differently on the basis of some trait or characteristic, it is clear that not all discrimination is morally troubling or unconstitutional. A law firm's decision to hire only law graduates in the top ten percent of their law school class—thereby discriminating against the remaining ninety percent of each class—seems fine, but a decision to hire only women seems wrong. Although state-sponsored single-race education seems wrong, single-sex education seems more ambiguous.

---

85 *See* Lynch v. Donnelly, 465 U.S. 668, 692 (1984) (O'Connor, J., concurring) (arguing that "[w]hat is crucial is that a government practice not have the effect of communicating a message of government endorsement or disapproval of religion"); County of Allegheny v. ACLU, 492 U.S. 573, 573 (1989) (adopting the endorsement test of O'Connor's concurring opinion in *Lynch* when finding that a crèche in a city's holiday display violated the First Amendment's Establishment Clause while a menorah did not); *see also* Hellman, *Expressive Dimension*, *supra* note 9, at 24-26 (discussing *Lynch* and *Allegheny* and the scholarship they engendered).

86 *See, e.g.*, Elizabeth S. Anderson & Richard H. Pildes, *Expressive Theories of Law: A General Restatement*, 148 U. PA. L. REV. 1503 (2000) (offering a general defense of why the expressive dimension of state action matters); Hellman, *Expressive Dimension*, *supra* note 9 (arguing that the only way to determine whether classifications by state actors wrongly discriminate is to focus on the what the state action expresses); *but see* Matthew D. Adler, *Expressive Theories of Law: A Skeptical Overview*, 148 U. PA. L. REV. 1363 (2000) (challenging whether any of the so-called expressive theories really are theories which make the permissibility of law depend on its expressive dimension).

87 *See* Hellman, *Expressive Dimension*, *supra* note 9, at 1-2 (discussing Washington v. Davis, 426 U.S. 229 (1976), as the case that marked the point at which the intent-based understanding of the Equal Protection Clause became the dominant understanding of the Supreme Court).

88 *See* Hellman, *Expressive Dimension*, *supra* note 9, at 1 (discussing the work of Kenneth L. Karst and Andrew Koppelman).

89 *See* Brown v. Bd. of Educ., 347 U.S. 481 (1954).

90 U.S. v. Virginia, 518 U.S. 515 (1996).

Articulating a general theory of when and why discrimination is wrong is not an easy task. If the Supreme Court's Equal Protection case law is an attempt to do so, it is a confusing and often conflicting articulation. Taking the dominant strands of that doctrine, the Court's answer appears to be that discrimination is wrong when the law or policy is enacted for the wrong sort of reason, that is, with *invidious intent*. While it is notoriously difficult to clearly articulate which intentions are in fact invidious, the Court is fairly clear that the actual subjective mental state of the people responsible for the law or policy at issue is what matters.[91] This theory has been roundly criticized for many reasons:

> Briefly, the intent doctrine has been criticized as incoherent because determining the intent of a group like a legislative body is both philosophically as well as practically problematic. Second, even at the individual level, it is often difficult to know or assess the precise reasons for an individual's action. Unconscious or subtle motives may guide us without our recognition of their influence. Third, moral responsibility for actions extends beyond those actions one specifically intends. Surely the failure to take the interests of a particular group into account—indifference rather than animus—is to deny those affected the law's equal protection. Fourth and finally, while intent is relevant to assessing the moral culpability of legislative actors, courts ought to interpret the Equal Protection Clause to police how people *are treated* by their government. We ought to be interested in the permissibility of laws, not in the purity of legislative motives.[92]

But neither is focusing on effect, the dominant alternative vision, a helpful way to sort the acceptable from the unacceptable instances of discrimination. First, one faces the difficult task of articulating what effects are problematic. The standard answer is that those effects that entrench caste-like distinctions in our society violate Equal Protection. The problem with this understanding of Equal Protection, however, is that the clause then becomes a general guarantee of substantive social justice. As I have argued previously, this understanding of Equal Protection is misguided:

> Equal Protection does speak to issues of social justice, but it is best understood as a specific protection. While it is difficult to specify precisely what Equal Protection protects—indeed, that is the project of this Article as well as a surfeit of others—we weaken its power and dilute its special appeal to deep and shared moral intuitions if we interpret it as a general guarantee of distributional fairness. For example, the current flat tax proposal would violate an anti-subordination conception of Equal Protection if it turned out that the law worked to entrench racial inequality. While a flat tax may be

---

91 *See, e.g.*, Personnel Adm'r v. Feeney, 442 U.S. 256 (1976) (citing the Court's explanation that "'discriminatory purpose . . . implies more than intent as volition or intent as awareness of consequences, [i]t implies that the decisionmaker . . . selected or reaffirmed a particular course of action at least in part 'because of,' not merely 'in spite of,' its adverse effects" (footnotes and citation omitted)); *see also* Hellman, *Expressive Dimension*, *supra* note 9, at 30-34 (discussing the claim that subjective intent is what has mattered to the Court in assessing violations of Equal Protection in most of the case law); *but see* C. Edwin Baker, *Outcome Equality or Equality of Respect: The Substantive Content of Equal Protection*, 131 U. PA. L. REV. 933 (1983) (arguing that the Court's requirement of invidious intent is used and ought to be used to denote objective rather than subjective intent).

92 Hellman, *Expressive Dimension*, *supra* note 9, at 4 (footnotes omitted).

> inimical to the demands of justice in that justice may well require redistribution, it seems a stretch to claim that the flat tax denies people the Equal Protection of the laws. To read the Clause as equivalent to a general requirement of social justice is to sap the Equal Protection Clause of its unique potency.[93]

The alternative to these two familiar ways of understanding the Equal Protection Clause shifts the focus of the courts' attention to the expressive dimension of laws. Single-sex bathrooms in state universities do not violate Equal Protection because the meaning in our culture of segregating people by sex for bathroom use is not denigrating. Single-race bathrooms violate Equal Protection because that policy carries a different meaning in our culture, a meaning that conflicts with the state's obligation to treat everyone with equal respect. Similarly, a decision by a District Attorney's office to hire only men on the theory that men make better prosecutors because they are more aggressive than women violates the Clause. Yet, the decision to hire only law graduates in the top ten percent of their graduating class on the theory that they are more likely to be successful prosecutors does not. Despite the current doctrine's focus on the degree of fit between the classification and its target, it does not matter whether men really are more aggressive then women. In fact, as feminist scholars have rightly pointed out, stereotypes often fit quite well because of the history of ubiquitous gender stereotyping or because of real differences between men and women.[94] The use of sex as a proxy for lawyerly skill is not wrong because it is irrational or because it harms the group left out. After all, the use of grades as a proxy for legal ability may turn out to be equally irrational and it equally harms those not hired. The difference between these two instances of discrimination is that the first expresses that some people do not equally deserve concern and respect while the second policy does not similarly denigrate those with lower grades. The first policy expresses denigration while the second does not because it is interpreted in the context of American history, a history replete with denials of the intellectual capacity of women.

This summary of the expressive conception of what makes discrimination wrong is necessarily brief, and questions clearly remain. Rather than attempting to fully articulate the argument in favor of an expressive conception of wrongful discrimination, I will assume its reasonableness here. Instead, I examine the strength of the argument that genetic discrimination is wrong because of what it expresses. Much of the critique of genetic discrimination seems to implicitly reach towards this argument. Some of the questions that remain about the expressive theory of discrimination generally will be explored further by looking at it in the context of genetic discrimination. In particular, one may wonder how to determine

---

93 *Id.* at 5.

94 CATHERINE A. MACKINNON, SEXUAL HARASSMENT OF WORKING WOMEN 101-06 (1979). Contrasting two approaches to sex discrimination which MacKinnon terms the "difference approach" and the "inequality approach," she emphasizes that

> [w]hat the second approach grasps, and the first does not, is that it is not only lies and blindness that have kept women down. It is as much the social creation of differences, and the transformation of differences into social advantages and disadvantages, upon which inequality can *rationally* be predicated. Discrimination is often irrational. But under the inequality approach, that is not all, nor even primarily, what is unjust about it. What is unjust about sex discrimination is that it supports a system of second-class status for half of humanity.

Id. at 105.

the meaning of a particular law or policy and whether that meaning is likely to be contested. These important questions will be addressed below.[95]

b. *The Claim: Genetic Discrimination Is Wrong Because of What It Expresses*

According to the expressive theory of wrongful discrimination, whether discrimination is wrong depends on the expressive dimension of the law or policy at issue. Applying this theory to genetic discrimination requires that one ask what a policy of genetic discrimination expresses. Clearly, it is not simple or straightforward to determine the social meaning of genetic discrimination. Fixing social meaning is an interpretive and inherently contestable task. However, drawing on the literature criticizing genetic discrimination, we can glean meanings that others have found in genetic discrimination, two of which recur enough to be worthy of attention. First, one might find that genetic discrimination expresses a kind of genetic determinism, that genes fully and completely determine one's fate.[96] Second, one might find that the practice of making insurance, employment or other decisions on the basis of a person's genetic traits expresses that some people have intrinsic flaws that render them less good, less worthy and less fit to be part of our community.[97] Both of these meanings seem troubling.

The fact that a discriminatory policy's meaning is troubling is not enough, however, to render it wrongful. Consider first the understanding of genetic discrimination as expressing genetic determinism. Genetic determinism may offend our sense of personal identity and dignity because it suggests a limited ability to control one's destiny and it may be misinformed as it inaccurately overstates the role of genetics in determining our futures. However, the mere fact that genetic discrimination expresses genetic determinism does not make genetic discrimination wrong. According to my theory of when classifications constitute wrongful discrimination, only those meanings that conflict with the principle that each person is equally worthy of concern and respect render a policy or practice wrongfully discriminatory. Genetic determinism tends to negate the importance of individual behavior and uncertainty in how our lives play out, but it does so across the board. To assert that genes are "our future diaries," to borrow the metaphor offered by Thomas Murray,[98] may be reductionist as well as inaccurate. It may ignore the creative individual contribution of free will and the importance of serendipitous chance encounters, but it negates the individuality and randomness in everyone's life equally. As such, the fact that genetic discrimination may express a kind of genetic determinism does not render such discrimination wrongful because this meaning does not offend equality.

---

95 *See infra* notes 96-98 and accompanying text.

96 Leon Kass finds this to be the meaning of our society's enthusiasm for genetic science. Leon R. Kass, *Triumph or Tragedy? The Moral Meaning of Genetic Technology*, 45 AM. J. JURIS. 1 (2000); Murray, *supra* note 12, at 68 (finding that the concern with "genetic prophesy"—a phrase borrowed from Nancy Wexler—to be one of the central reasons that some commentators worry about genetic discrimination); Dorothy C. Wertz, *Society and the Not-So-New Genetics: What Are We Afraid of? Some Future Predictions from a Social Scientist*, 13 J. CONTEMP. HEALTH L. & POL'Y 299, 308 (1997) (arguing that a genetic essentialism has dominated our history except during a few periods, and that this view of genetics and its power shapes how we view issues related to genetics).

97 This conception of genetic discrimination seems implicit in the critiques that emphasize the importance of the history of America's eugenic practices. *See* Soifer & Wugmeister, *supra* note 80; Andrews, *supra* note 81.

98 Murray, *supra* note 12.

Genetic discrimination also expresses a second and more problematic meaning. Several commentators emphasize the history of eugenics in order to imply that genetic discrimination is different from discrimination on the basis of health because the history of the misuse of genetic information colors the way that our society understands the practice of genetic discrimination. Read with this history as a backdrop, the meaning of genetic discrimination may be that those with genetic flaws, or more flaws than average, are less worthy or important for that reason.

If this understanding of the meaning of genetic discrimination is correct, then genetic discrimination may be wrong for that reason. This argument distinguishes genetic discrimination from discrimination on the basis of health and thus can support the claim that genetic discrimination is exceptional. According to this view, genetic discrimination is different because the social meaning of genetic discrimination is different. The strength of this argument depends on the strength of the claim that genetic discrimination does indeed express disrespect for people with genetic mutations that especially predispose them to disease or disability. To assess the soundness of this interpretive assessment, it will be helpful to briefly review our society's previous encounters with genetics.

IV. THE HISTORY OF EUGENICS AND ITS ROLE IN FIXING THE MEANING OF GENETIC DISCRIMINATION

This section examines the claim that that the meaning of genetic discrimination is powerfully affected by the history of the misuse of genetics. This section will discuss aspects of that history which may inform the way genetic discrimination is likely to be perceived. Although my treatment of the subject is brief,[99] it is sufficient because a claim about the meaning of a practice refers to how that practice—genetic discrimination—is likely to be perceived or understood by people with an ordinary amount of knowledge about their society's history.

The beginning of the systematic, scientific understanding of genetics and its effect on health stretches back approximately 150 years.[100] It was not until the twentieth century, however, that these scientific developments began to have significant social consequences. Interestingly, the attraction to genetics and its social possibilities came, both in the United States and in Great Britain, from both the right and the left of the political spectrum.[101] Scientists and social critics saw genetics as offering tremendous potential. Rather than focusing on therapeutic efforts for people with genetic diseases, however, scientists and social reformers were mainly interested in the promise of genetics to affect reproductive choices—eugenics.

The social policies that new genetic knowledge generated were eugenic in both the positive and negative sense. First, eugenicists were interested in encouraging reproduction by those perceived as having "better" genetic endowments. For example, in the 1920s, state fairs around the country sponsored "Fitter Families" contests. Along with prizes for the tomatoes and hogs, there were prizes for "Grade

---

99 For a thorough discussion of the history of the eugenics movement of the United States and Great Britain, see DANIEL J. KEVLES, IN THE NAME OF EUGENICS: GENETICS AND THE USES OF HUMAN HEREDITY (1985).

100 *Id.* at 3-19 (detailing Francis Galton's work in the mid nineteenth century).

101 *Id.* at 63 (describing that in the early part of the twentieth century, genetics was popular with both radicals and conservatives in the United States and England).

A individuals" in the "human stock" category.[102] This interest in encouraging the "better" to reproduce was coupled with a fear that the "defective" were over-producing. As a result, the feminist cause of birth control became more broadly acceptable because it offered the promise that the birthrates of the lower classes could be contained. This negative eugenic purpose is well-illustrated in the disturbing quotation from Havelock Ellis in Dr. Daniel Kevles's history of eugenics movements in the United States and Great Britain: "The superficially sympathetic man flings a coin to the beggar; the more deeply sympathetic man builds an almshouse for him so that he need no longer beg; but perhaps the most radically sympathetic of all is the man who arranges that the beggar not be born."[103]

In the United States, the eugenics movement had racial overtones as well. The "better" stock was presumed to be "white Protestants of Northern European Stock," while the lesser stock was made up of "blacks and Jewish and Catholic immigrants."[104] The first widespread use of intelligence testing, which occurred in the armed forces, offered supposedly objective data for these conclusions. This data consisted of information showing that blacks consistently tested lower than whites. The fact that the test measured education as much as intelligence was largely ignored.[105]

It was not until the second World War that the use of genetic science for eugenic purposes was discredited. Revelation of Nazi atrocities did much to halt the popularity of eugenics in the United States as well as to affect later perception of eugenics. For example, when the genetic counseling profession was just beginning to develop in 1947, Dr. Sheldon Reed—one of its founders—chose the term "genetic counseling" to replace "genetic hygiene" and "genetic advice" which had come to be seen as tainted by eugenics.[106] Moreover, the profession adopted the credo of "nondirectiveness"—a commitment to provide information, but not advice to patients seeking genetic services—in response to the problematic history of eugenics, especially in the context of reproduction.[107]

Considering this brief description, one can perhaps conclude that our collective consciousness regarding the misuse of genetic science in the United States, as well as the powerful example of Nazi Germany, has influenced how genetics is currently perceived. The question we are considering is whether genetic discrimination *expresses* disrespect or unequal concern. The social meaning of genetic discrimination, understood against this historical background, may well denigrate the equal value of people with genetic disease. Given the misuse of genetics in our past, this claim is surely plausible.

---

102 *Id.* at 62.

103 *Id.* at 90.

104 *Id.* at 75.

105 *Id* at 83 (explaining that "[c]learly a variety of causes, including the cultural bias of the Army tests themselves and the poor education of many of the test takers, might have accounted for the results[,] . . . [y]et the supposedly objective test data further convinced eugenically minded Americans not only that mental deficiency was genetically determined but that so was intelligence").

106 Wertz, *supra* note 96, at 306.

107 *Id.* (explaining that "[i]t is unlikely that there would have been such emphasis on nondirectiveness had genetics begun as a pediatric or adult specialty outside the context of reproduction").

A. HOW STRONG IS THE ARGUMENT?

Interestingly, some of the critics of special legislation prohibiting genetic discrimination object to these laws by arguing that the social meaning of genetic discrimination is not denigrating, though they do not label the argument in this manner. For example, Diver and Cohen point out that understanding a condition as a genetic disease can in fact make having it less stigmatizing. To illustrate this point, they ask us to "[c]onsider the characterization of alcoholism as a disease, the relabeling of 'senility' as Alzheimer's disease, or the emerging consensus that obesity has a strong genetic component."[108] In addition, they point out that "[t]he most common genetically influenced diseases, such as heart disease or cancer, elicit reactions of sympathy and solicitude far more than fear and aversion."[109]

Although this is an important criticism, it does not respond directly to the expressivist argument for prohibiting genetic discrimination. The reason it is inapposite has much to do with the ambiguity regarding what one means by the term "stigma." Diver and Cohen's point refers to how the particular people with genetic conditions are treated. The expressivist argument, however, focuses on the meaning expressed to everyone by the practice of genetic discrimination: does it convey the message that people with genetic conditions are less worthy of our concern and respect than others? While it may be true that an individual is pitied rather than blamed for being obese or alcoholic in that the genetic link tends to absolve individual responsibility for the condition, it may remain true that social practices which continue to treat the obese or alcoholic and others with genetic conditions less well express that the lives of these people are less valuable. Diver and Cohen's other examples, heart disease and cancer, illustrate this point even more clearly.[110] While the individual may elicit sympathy for having the condition, the meaning expressed in the practice of genetic discrimination may still convey that the lives of these "flawed" individuals is less important and less valuable.

The fact that Diver and Cohen's critique is not persuasive does not settle the question whether genetic discrimination expresses that people with genetic conditions are less worthy of concern and respect. Just as determining the meaning of a novel or play is an interpretive exercise about which people can disagree, the determination of the meaning of a social practice can also result in disagreement. Indeed, the same thing can be said about determining what a statute or prior case law requires. In each of these cases, the interpretation is open to dispute. But this contestability surely does not render the thing interpreted (a social practice, a novel or a law) meaningless. As Professor Ronald Dworkin explains about the nature of law, some interpretations of current law are ruled out by prior law, while a range of others fit reasonably well with current law, making them plausible readings of the law.[111] So, how should one choose among plausible readings of the meaning of genetic discrimination? To answer this question, it may be helpful to examine prenatal genetic testing and its social meaning. This example is likely to be helpful

---

108 Diver & Cohen, *supra* note 6, at 1478.

109 *Id.*

110 *Id.*

111 RONALD DWORKIN, LAW'S EMPIRE 228-38 (1986). Dworkin uses the example of a chain novel to explain what a judge does in deciding a case. The new decision is like the next chapter in an on-going novel. As such, this new chapter must fit reasonably well with what has come before. According to Dworkin, there will be several reasonable interpretations of prior law that make possible a range of ways the new case can be decided. The judge must then choose among them, according to Dworkin, guided by the principle that he or she should make our law "the best it can be." *Id.* at 229.

because the critique of that practice by disability advocates focuses precisely on the expressive dimension of prenatal genetic testing.

B. PRENATAL GENETIC TESTING: WHAT DOES THAT PRACTICE EXPRESS?

Prenatal genetic testing is a procedure by which cells from a developing fetus are tested for genetic mutations associated with particular diseases or disabilities.[112] It is generally performed so that the woman or couple can abort the fetus if a serious anomaly is found. Since amniocentisis, the most common form of prenatal genetic testing, carries some risk of inducing a spontaneous abortion, it is unlikely that the procedure would be offered in the routine fashion that is common today if it were generally undertaken for its informational value alone.[113]

Prenatal genetic testing for disease and disability and the selective abortion that generally follows a finding of an abnormality, or more accurately the routine way in which prenatal testing is offered,[114] has become the subject of criticism by disability rights advocates. In particular, many critics argue that prenatal testing and selective abortion for disability are morally problematic because of what that practice says or expresses about the value of the lives of living people with disabilities.[115] These critics do not oppose abortion per se. In fact, the critics of testing explicitly and ardently support a woman's right to reproductive freedom. Rather, they oppose the fact that prenatal testing has become a part of standard prenatal care. These critics contend that the normalization of prenatal screening for disability expresses that these disabilities are incompatible with a meaningful and worthwhile life. The expressive theory of discrimination captures the intuition that there may be something morally troubling about selective abortion for disability that has nothing to do with the rights of the disabled fetus.

Disability activists emphasize that the routine nature of testing carries two related messages. First, the routine availability of testing changes the landscape for women or couples who seek prenatal care. As Bruce Jennings points out, offering prenatal screening changes the terrain for couples whether they accept the offer and screen or reject it.[116] He provides the following example: "Suppose a couple is offered a test for Tay-Sachs and decides to forego it. If the child eventually becomes symptomatic with the disease, that cannot be considered a tragic surprise or shock to the parents, but rather stands as an example of willful and deliberate choice."[117]

---

112 *See* Powell, *supra* note 27, at 44-48 (discussing the history and current status of available prenatal testing).

113 *Id.* at 44-45.

114 It is more productive to view the routine practice of offering prenatal tests and selective abortions as expressing a problematic meaning rather than focusing on the expressive dimension of the individual woman's choice to abort a disabled fetus, as that decision is usually carried out privately. As Eva Feder Kittay notes, "A woman rarely says: 'Listen up, world. I am having an abortion based on a diagnosis of fetal abnormality and I am about to tell you why I choose to abort a fetus with such an abnormality.'" Eva Feder Kittay & Leo Kittay, *On the Expressivity and Ethics of Selective Abortion for Disability: Conversations With My Son*, *in* PRENATAL TESTING, *supra* note 10, at 186.

115 *See* Bruce Jennings, *Technology and the Genetic Imaginary: Prenatal Testing and the Construction of Disability*, *in* PRENATAL TESTING, *supra* note 10, at 124; Marsha Saxton, *Why Members of the Disability Community Oppose Prenatal Diagnosis and Selective Abortion*, *in* PRENATAL TESTING, *supra* note 10, at 147; Kittay & Kittay, *supra* note 114, at 165 (Leo Kittay's contribution is particularly noteworthy); Adrienne Asch, *Why I Haven't Changed My Mind About Prenatal Diagnosis: Reflections and Refinements*, *in* PRENATAL TESTING, *supra* note 10, at 234.

116 Jennings, *supra* note 115, at 135.

117 *Id.* at 134.

Second, the fact that prenatal testing programs exist and are common suggests that the "normal" response to a positive test result for disability is abortion.[118] Disability activists focus on the meaning expressed by the supposition, which girds the routinization of testing, that most women will want to abort for certain disabilities. They argue that this supposition expresses that these disabilities provide good reasons to abort because the lives of people with these disabilities are not worth living or, alternatively, that raising a child with such disabilities is so difficult that the joys that the child occasions do not compensate for the problems the child's needs engender. As Dr. Marsha Saxton describes, "For people with disabilities, 'the message' implicit in the practice of abortion based on genetic characteristics is . . . : 'It is better not to exist than to have a disability.' Your birth was a mistake. Your family and the world would be better off without you alive."[119] Professor Adrienne Asch poignantly describes the meaning she sees in the practice of testing:

> People with just the disabilities that can now be diagnosed have struggled against an inhospitable, often unwelcoming, discriminatory, and cruel society to fashion lives of richness, of social relationships, of economic productivity. For people with disabilities to work each day against societally imposed hardships can be exhausting; learning that the world one lives in considers it better to 'solve' problems of disability by prenatal detection and abortion, rather than by expending those resources in improving society so that everyone—including those people who have disabilities—could participate more easily, is demoralizing. It invalidates the effort to lead a life in an inhospitable world.[120]

This meaning is surely one that is morally problematic as it certainly offends the principle of equal concern and respect. What is less clear is whether disability activists are correct in reading the routine practice of prenatal testing and selective abortion as carrying this message.

In an exemplary model of moral inquiry, the Hastings Center gathered together a group of people to discuss the disability rights critique of prenatal testing at several meetings over a two-year period.[121] The group included participants who found that critique of testing persuasive and others who did not; it included disabled and non-disabled participants as well as some with experience as parents of disabled children. The group was also diverse in professional backgrounds, including philosophers, genetic counselors, physicians, educators, writers, lawyers, participants from public health and psychiatry.[122] At these meetings, members of the discussion group addressed the precise question of what the practice of testing and selective abortion expresses. All did not believe that the meaning of the practice was as its critics found it to be. The disagreement largely centered on the question of whether selective abortion for disability is the kind of act whose symbolism is clear enough to carry a distinct meaning.[123]

---

118 Saxton, *supra* note 115, at 148.

119 *Id.* at 160.

120 Asch, *supra* note 115, at 240.

121 Erik Parens & Adrienne Asch, *The Disability Rights Critique of Prenatal Genetic Testing: Reflections and Recommendations*, *in* PRENATAL TESTING, *supra* note 10, at 3, 5.

122 Erik Parens & Adrienne Asch, *Introduction* to PRENATAL TESTING, *supra* note 10, at ix, x.

123 *Id.*

These counter-critics (*i.e.*, those who were unconvinced that the meaning of prenatal testing and selective abortion denigrates the lives of living people with disabilities) make three important points. First, they argue that there are simply not enough shared reference points to confidently define the meaning of prenatal genetic testing.[124] Professor James Lindemann Nelson compares the practice of prenatal testing with the act of flying a Confederate flag atop a public building (an act which he sees as carrying a distinct meaning) and stresses that the "semantic force of diagnostic tests and pregnancy termination is not well defined within our shared symbol systems, and whether one can be justified in attributing to such practices clear messages, disrespectful or not, needs careful consideration."[125]

Moreover, if the meaning seems clear, the counter-critics argue, this is not because the practice of routine genetic testing in the prenatal context is denigrating, but rather because the background context of unjust treatment of people with disabilities makes it seem denigrating.[126] Context can legitimately participate in fixing meaning (think of the Confederate flag example). Rather, the counter-critics argue that, since the symbolic significance of prenatal testing is so indeterminate, the context is doing too much work. Professor Eva Feder Kittay explains: "Because codes concerning acts of abortion and selective abortion are so underdeveloped and so contested, context is virtually, though not entirely, determinative."[127]

Third, the counter-critics argue, if prenatal genetic testing expresses something, it is something about the value of the disability, not the value of a person with such a disability. As Dr. Bonnie Steinbock emphasizes, "[f]rom the fact that a couple wants to avoid the birth of a child with a disability, it just does not follow that they value less the lives of existing people with disabilities, any more than taking folic acid to avoid spina bifida to avoid having a child with disabilities does not imply that if that outcome should occur, the child will be unwanted, rejected, or loved less."[128] The disability is not equivalent to the person with the disability, a point that, ironically, disability rights activists have themselves been emphasizing for years.

It is difficult to judge which is the stronger argument. Does routine prenatal genetic testing denigrate the value of the lives of people with disabilities or not? In reading the very thoughtful contributions to the volume assembled by the Hasting Center project, I found myself swayed by each view. The critics of testing are surely right that given the history of the misuse of genetic information, it is significant that detecting genetic disease in utero so as to facilitate abortion has become so common and seemingly important. Underlying this practice may be the view that disabilities are so terrible that an otherwise wanted child should be aborted if she will be disabled.[129] While this view of disability and its significance both for the child and for the parent is a comment on the disability more than the child, is it so disconnected from a comment on the value of life with a disability and, thus, of the life of the person with the disability? The fact that prenatal genetic testing has

---

124 James Lindemann Nelson, *The Meaning of the Act: Reflections on the Expressive Force of Reproductive Decision Making and Policies*, *in* PRENATAL TESTING, *supra* note 10, at 196, 211.

125 *Id.* at 196-97.

126 *See, e.g.*, Bonnie Steinbock, *Disability, Prenatal Testing, and Selective Abortion*, in PRENATAL TESTING, *supra* note 10, at 108, 120-21.

127 Kittay & Kittay, *supra* note 114, at 187.

128 Bonnie Steinbock, *supra* note 126, at 121.

129 I have avoided drawing distinctions among disabilities as disability activists who critque prenatal genetic testing avoid making these distinctions.

become so routine and invisible bespeaks a failure of imagination—an inability to envision the lives of disabled persons as rich and meaningful.[130] In reading the Hastings Center volume, as well as Michael Berube's book about his son who has Down's Syndrome (a book which stretches that imagination),[131] I find myself very close to being persuaded by the critics of testing. And yet, not fully so.

C. Fixing the Meaning of a Practice: "Justice Is Conflict"

How does this discussion of whether prenatal genetic testing and selective abortion expresses denigration of people with disabilities help to advance the inquiry into whether genetic discrimination against living people denigrates the worth of people with genetic predispositions to disease? In one sense, the case for a denigrating meaning is clearer in the prenatal testing example in that the practice at issue determines criteria for selective abortions—quite literally, criteria for being born. If it is not clear that this practice denigrates the lives of people with genetic disease, one might think it is unlikely that a practice that instead affects insurance and employment would be denigrating. On the other hand, in the prenatal testing example, it is not clear that selective abortion expresses anything about the worth of living people with disabilities. Rather, to the extent that it says something, it may be something about the value of the disability rather than the person. In contrast, genetic discrimination in insurance, employment or elsewhere does not have this problem because it directly affects the lives of living persons with genetic conditions. To the extent that it says something, it does so about the lives of living people with genetic conditions.

In both examples, the process of determining what a practice expresses, if it expresses a meaning at all, is not an easy task. As the counter-critics described above noted, when there is no well-recognized code to use in determining the expressive significance of action, it is difficult to say if that action expresses anything. But this is not reason to abandon the task. In discussions about literature, movies or art, we routinely disagree about what meanings a book, film or painting expresses. There is also no agreed-upon code for determining such meanings. While a language itself may be such a code, a scene about a particular subject, to which literary critics of the professional and armchair variety routinely attach meanings, are not accompanied by a code that instructs that scenes about X signify Y.

Most importantly, if the expressive theory of discrimination is right about what makes a policy or social practice discriminatory, then we have no alternative but to try to determine the meaning of practices like genetic discrimination and prenatal genetic testing. The expressive account of when discrimination is wrong instructs one to judge whether a policy or law impermissibly discriminates by looking at the expressive dimension of the policy or law. If it expresses denigration—that is, if its meaning conflicts with the claim that each person is equally worthy of concern and respect—then that policy discriminates wrongfully. To make this judgment, one must do the best one can to determine the meaning of the policy, law or practice at issue.

---

[130] This failure of imagination especially applies to persons with Down's Syndrome. *See* Parens & Asche, *supra* note 121, at 7-8 (discussing the attitudes of obstetricians and genetic counselors to patients who have received a prenatal diagnosis of Down's Syndrome).

[131] Michael Bérubé, Life As We Know It (1996).

Does the difficulty in determining the meaning of the policy, law or practice point to a failure of the expressive theory itself, thereby making this particular argument about why genetic discrimination is wrong misguided from the start? I think not. Interpreting social practices, like interpreting art and literature, is an exercise in which our diverse histories, cultures and experiences are critically important. If diversity among people is deep, then there are bound to be significant disagreements when intellectual tasks draw upon that diversity. But perhaps this disagreement is not problematic for the just resolution of social issues. The philosopher Stuart Hampshire believes, for example, that conflict is a persistent feature of both society and the mind precisely because of the profound diversity among people.[132] The only universally acknowledged value, in his view, is a commitment to procedural fairness, which he understands quite minimally.[133] For Hampshire, procedural fairness requires only that a procedure afford each side to a conflict the opportunity to be heard.[134] If Hampshire is right, both about the persistence of conflict and the significance of procedural fairness, then what is most important is that our society provides opportunities to listen and hear the views of many people regarding what the practice of genetic discrimination expresses. This may be the best we can do in assessing whether genetic discrimination is wrong. If Hampshire is right, actually doing just that is an important act of securing social justice. As Hampshire explains:

> Neither in a social order, nor in the experience of an individual, is a state of conflict the sign of a vice, or a defect, or a malfunctioning. It is not a deviation from the normal state of a city or of a nation, and it is not a deviation from the normal course of a person's experience. To follow through the ethical implications of these propositions about the normality of conflict, these Heracleitean truths, a kind of moral conversion is needed, a new way of looking at all the virtues, including the virtue of justice. We need to turn around the mirror of theory, so that we see ourselves both as we are and as we have been.[135]

The argument that genetic discrimination is meaningfully different from discrimination on the basis of health because of what it expresses is thus an argument about which I cannot reach a final conclusion. Whether genetic discrimination wrongfully discriminates depends on whether such discrimination expresses that people with serious genetic conditions are less worthy of concern or respect. The history of eugenics in western society is relevant to that question and surely makes this claim plausible. As Professor Eva Feder Kittay emphasizes in the context of prenatal genetic testing, however, if context seems to do all the work, then it is hard to say that the particular practice under discussion expresses anything at all.[136]

---

132 STUART HAMPSHIRE, JUSTICE IS CONFLICT 37 (2000) (arguing that "the diversity and divisiveness of languages and of cultures and of local loyalties is not a superficial but an essential and deep feature of human nature—both unavoidable and desirable—and rooted in our divergent imaginations and memories").

133 *Id.* at 4 (asserting that "fairness in procedures for resolving conflicts is the fundamental kind of fairness, and that it is acknowledged as a value in most cultures, places, and times: fairness in procedure is an invariable value, a constant in human nature").

134 *Id.* at 8 (identifying a key feature of procedural fairness as "*audi alteram pertem* ('hear the other side')").

135 *Id.* at 33-34.

136 Kittay & Kittay, *supra* note 114, at 187-92.

114 AMERICAN JOURNAL OF LAW & MEDICINE VOL. 29 NO. 1 2003

This examination of the expressivist argument in support of special legislation prohibiting genetic discrimination demonstrates the importance of grappling with the social meaning of genetic discrimination. It instructs that decisions by states and the federal government regarding whether to enact such legislation ought to focus on the question of what the practice of genetic discrimination by insurers, employers and others expresses about the equal worth of persons with a genetic predisposition to significant disease. We should also keep in mind that this is a question to which the answer may vary depending on whether the discrimination occurs in health insurance, life and disability insurance or in family contexts. Moreover, if Hampshire is right that it is the process of listening to the "other side" that is centrally important to acting justly, perhaps the most important step is ensuring that these questions are addressed in fora that provide opportunities for many voices to be heard.

D. Other Concerns About Expression

Some commentators have pointed out that the passage of laws forbidding genetic discrimination may have the unfortunate consequence of reinforcing the public perception that genetic endowments determine destiny. The laws may communicate that genetics is different and significant in ways that are misleading and dangerous.[137] Thomas Murray captures this point well: "The more we repeat that genetic information is fundamentally unlike other kinds of medical information, the more support we implicitly provide for genetic determinism, and for the notion that genetics exerts special power over our lives."[138] This concern points out an important drawback of special genetics legislation that is worth considering. Before considering it, however, it is important to note the ways in which it differs from the argument explored in this Article.

The claim that genetic discrimination is wrong because it expresses denigration builds on the argument that there are limits on what one ought to express. One ought not (and governments especially ought not) act in a way that express that some people are less worthy of concern and respect than others. This argument does not tell us what we ought to express, rather it merely tells us what we may not. Within the realm of permissible actions (*i.e.*, ones that do not express denigration in ways ruled out by the equal concern and respect principle), some of those actions will be better and worse for the people affected for a variety of reasons.

Murray and others who point out that special genetics legislation may reinforce attitudes that are inaccurate and harmful do not claim that this expressive dimension of the state's action (passing the law) conflicts with any principle that limits expressive action. Rather, they argue that the laws are unwise because they express an unfortunate meaning.[139] This is a policy argument about how best to change attitudes, not an argument about what practices violate the principles that ought to guide action.

137 Greely, *supra* note 13, at 1494 (arguing that "[b]y telling people that their genetic variations *are* so important—are the equivalents of race, sex, religion, and other suspect classifications—we encourage them to believe that it is true, in ways that are both inaccurate and pernicious"); Murray, *supra* note 12, at 71; Suter, *supra* note 7, at 740 (concluding that by "responding to the public's fears with special protections for genetic information, genetics legislation may validate and fuel the flames of these fears as well as the underlying perspective that generates those fears").

138 Murray, *supra* note 12, at 71.

139 *See supra* notes 137-138 and accompanying text.

That said, how should we weigh their concern? First, if we believe that rights claims trump policy considerations, then their argument must give way to the argument of this section if we conclude that genetic discrimination does indeed denigrate people with genetic disease. In addition, like most policy arguments, it must be balanced against the other important policies to be considered. The most pressing policy consideration to be weighed on the other side is the one considered in the previous section—that laws forbidding genetic discrimination promote health.

V. LINKING DEFINITION WITH JUSTIFICATION

Finally, we must make sure that the reasons to support legislation forbidding genetic discrimination do in fact justify prohibiting genetic discrimination as earlier defined. Recall that in response to the difficulty delineating genetic discrimination from discrimination on the basis of health, this Article adopted Professor Henry Greely's proposed solution. Greely suggests that genetic discrimination be defined as discrimination on the basis of information about genotype, whatever the source of that information, whether it is sophisticated DNA analysis or the familiar family history solicited by a physician. However, once a person manifests disease whether that disease is of genetic, non-genetic or mixed origin, discrimination on this basis would not be considered genetic discrimination. It is this definition of genetic discrimination that we must examine to ensure that the reasons to prohibit genetic discrimination that have been examined in this Article make sense when genetic discrimination is so defined.

Of the many reasons that have been offered in support of prohibiting genetic discrimination, it is only the two discussed in detail above that remain plausible candidates for reasons to distinguish genetic discrimination from health-status discrimination. To recap, genetic discrimination may be different and thereby warrant special legislation in order to ensure that the health benefits of genetic science are realized. Second, genetic discrimination may be meaningfully different from health-based discrimination because it expresses denigration of people with genetic disease. Next, we must ensure that these reasons support prohibiting genetic discrimination as defined in this Article. In other words, does discrimination on the basis of genotype express denigration that discrimination on the basis of phenotype does not? Will forbidding discrimination on the basis of genotype remove barriers to reaping the health benefits of genetic science and will it do so more than would forbidding discrimination on the basis of phenotype?

The answer to each of these questions is quite plausibly yes. In the case of the expressive significance of discrimination, when the discrimination is based on an asymptomatic genetic condition, it is likely that such discrimination will be seen to be especially genetic, thereby recalling all of the history of eugenics in our collective consciousness. In the case of the argument based on the promotion of health, it is only those individuals without manifest disease for whom genetic testing will provide new information that may be useful in either therapy or research. It makes sense to target this group because it is this group that may be discouraged from testing without protective legislation.

VI. CONCLUSION: AGENDA FOR RESEARCH AND DISCUSSION

In conclusion, it will be helpful to review the questions that warrant further discussion or research. With regard to the expressive argument in favor of prohibiting genetic discrimination, this Article stresses the importance of continuing the discussion, in public fora as well as in private conversation, regarding whether genetic discrimination denigrates the worth of people with genetic disease. In order to ensure that this argument tracks the definition of genetic discrimination adopted here, this discussion must focus on whether discrimination on the basis of unexpressed information about genotype is especially denigrating and whether it is meaningfully different from the social meaning of discrimination on the basis of expressed disease, whether of a genetic or non-genetic etiology.

In examining the argument that protection from genetic discrimination will enhance health, this Article uncovered several empirical questions that require further study. First, if people refrain from genetic testing because they fear genetic discrimination, one must ascertain in what contexts people fear genetic discrimination—in health insurance, life and disability insurance, employment or family law contexts, or family and private life. This determination is important because if prohibitions on discrimination are to succeed in assuaging the concerns that cause people to decline testing, this legislation must prohibit what people actually fear. Second, if people fear discrimination in multiple contexts, researchers must determine whether limited legislation that reaches only health insurance and employment would be sufficient to affect decision-making about testing. The Hall and Rich study described earlier begins to answer these questions.[140] Further research must work both to confirm and build upon their results. Third, if the promotion of health is the main goal, one must try to determine whether legislation prohibiting genetic discrimination would in fact promote health more than legislation prohibiting general health-status discrimination. If not, health promotion would not be a sufficient reason for genetic, as opposed to generic, legislation.

Finally, one must try to determine how legislation prohibiting genetic discrimination in health insurance is likely to affect the receptiveness of the body politic to universal healthcare coverage. This is important because ensuring healthcare for all will surely enhance health and because if genetics legislation will make universal coverage less likely, then the passage of genetics legislation may deepen inequality.

---

140 *See supra* text accompanying notes 70-75.

[8]

GENETIC SECRETS AND THE FAMILY

DEAN BELL* AND BELINDA BENNETT**

I. INTRODUCTION

The issue of how individual patients and their doctors should act in relation to the knowledge that the patient has a genetic condition—specifically, whether the patient and/or the doctor should or must inform relevant members of the patient's family—is a looming area of medico-legal controversy. Over the last fifteen years or so, the issue of confidentiality versus disclosure has been particularly controversial in relation to HIV/AIDS patients.[1] It has been argued that medical information about genetic disease gives rise to special problems *vis-à-vis* blood relatives. Because genetic disease is transmitted only by way of procreation, information about genetic disease is unique in that there is a propensity (which is highly variable and depends upon a variety of factors) for the condition to be shared by members of a family who are biologically related. Thus, genetic information about an individual may reveal information about relatives of that individual which is 'specific (that the person has or will develop a genetic disease); or predictive (that the person has an unspecified risk of developing the disease)'.[2]

There are other factors, specific to genetic medical information, which contribute to its 'shared' character. Firstly, genetic information about individuals may have to be supplemented by information obtained from relatives in order for that information to be meaningful. This is because in relation to many genetic conditions, there is at present no direct test for the gene itself, and a marker or linkage test is necessary. The accuracy and usefulness of such tests are improved by larger sample pools.[3]

---

* Ph.D. Candidate, Faculty of Law, University of Sydney, Australia. ** Senior Lecturer, Faculty of Law, University of Sydney, Australia. Research for this paper was supported by the Legal Scholarship Support Fund, Faculty of Law, University of Sydney.

1 M. Neave, 'AIDS: Confidentiality and the Duty to Warn' (1987) 9 *University of Tasmania Law Review* 1; L.O. Gostin and J.G. Hodge, 'Piercing the Veil of Secrecy in HIV/AIDS and Other Sexually Transmitted Diseases: Theories of Privacy and Disclosure in Partner Notification' (1998) 5 *Duke Journal of Gender Law & Policy* 9; D.G. Caswell, 'Disclosure by a Physician of AIDS-related Patient Information: An Ethical and Legal Dilemma' (1989) 68 *Canadian Bar Review* 225; R.J. Paterson, 'AIDS, HIV Testing and Medical Confidentiality' (1991) 7 *Otago Law Review* 379.

2 L. Skene, 'Patients' Rights or Family Responsibilities? Two Approaches to Genetic Testing' (1998) 6 Med. L. Rev. 1 at 10.

3 Skene, *ibid.* at 5.

While linkage tests are now less frequently used,[4] even where a direct gene test is available it is important to confirm the mutation in at least one other affected family member.[5] This is especially important where, as is the case for most genetic disorders, there is more than one form of genetic mutation[6] which causes the disorder.[7] Moreover, in relation to some conditions, multiple genes may cause the disorder.[8] Because of this, the starting point for genetic testing 'is to identify the mutation responsible for the disorder in an affected family member, who may be a symptomatic individual or a carrier . . . Testing relatives is a much simpler process as the family's specific mutation is now known and a specific test can be developed to identify it'.[9] Secondly, even if family members are not actually tested, in order to verify a diagnosis clinical practice normally entails obtaining information about family members in successive generations leading to the creation of a family pedigree.[10]

In the light of these unique features of genetic information as a species of medical information, the question arises of how this should affect the treatment of genetic information by doctors, and the extent to which traditional approaches to medical information should be followed or departed from. This article will first summarise a number of Australian responses, both governmental and academic, to this problem. It identifies, at least in some contexts, an argument against the traditional individual-centred approach of confidentiality, and in favour of a more 'communitarian' approach which proposes that the inherently shared nature of such information justifies different ethical and legal rules. The article seeks a critical analysis of this literature, especially the argument for a departure from existing ethical and legal protections. Our starting point is that the answer to the question of whether genetic information in any particular case should be regarded as 'individual' or 'familial' cannot be resolved scientifically—undoubtedly both an individual patient and genetic relatives of that patient will have an interest in genetic informa-

---

[4] Skene, *ibid.* at 5.

[5] Skene, *ibid.* at 6. A linkage test 'uses genetic markers with variable sequences, within or close to the mutant gene of interest, to track it within the family': National Health and Medical Research Council, *Ethical Aspects of Human Genetic Testing: An Information Paper* (Australian Government Publishing Service 2000) at 23, available at <www.health.gov.au/nhmrc/ethics/contents.htm>.

[6] That is, deletions, duplications, inversions, base substitutions, and regulatory mutations: National Health and Medical Research Council, *ibid.* at 22.

[7] For example, there are approximately 140 different mutational forms of cystic fibrosis (although one mutation accounts for 70 per cent of cases): V. Harcourt, 'Genetic Screening Tests: Whose Risk Is It?', *Topics for Attention Issues Paper 12* (Australian Institute of Health Law and Ethics 1999) at 6.

[8] National Health and Medical Research Council, *op.cit.*, n. 5 at 21.

[9] National Health and Medical Research Council, *ibid.* at 22.

[10] Skene, *op.cit.*, n. 2 at 6.

tion derived from the patient. The resolution of the question of whether the secrecy of that information should be respected in any particular case is therefore an inquiry of an ethical, and ultimately, legal kind.

Accordingly, we are critical of the role of the concept of 'family' in some of the literature, and argue that genetic filiation is not enough *in itself* to warrant setting aside important legal and ethical protections. The article acknowledges that there may well be some (albeit limited) contexts in which a doctor's disclosure of genetic information to a relative of a patient may well prevent or lessen a serious or imminent threat to the life or health of a relative. And it is argued that existing Australian law (both common law and also the Genetic Privacy and Non-Discrimination Bill 1998 (Cth.) (subsequently lapsed)) would probably allow disclosure in such circumstances. But the nature of genetic medical information, which is more often predictive than determinative of the occurrence of a particular disease, is such that these cases are likely to be much less frequent than the proponents of treating genetic information as 'communal' information would allow. Furthermore, the interest of relatives of patients in *not* knowing genetic information also needs to be respected.

While there are strong reasons for eschewing a thoroughly 'individualistic' approach to genetic information, this has in fact never been the attitude adopted by the law relating to duties of confidentiality, or even by proposed statutory changes in Australia to introduce a right to privacy in relation to genetic information. Moreover, there is nothing in current legal approaches to support the contention that individuals *own* genetic information which is derived from them. In this context, a departure from the current approach, with its attempt to balance the interests and needs of both individual patients and third parties (including family members), is not as necessary as some writers suggest.

II. COMMUNITARIAN MODELS OF GENETIC INFORMATION

This section of the article looks at what it identifies as an emerging consensus in Australian policy debate about the need for genetic information to be treated differently from other kinds of medical information for the purpose of confidentiality. This approach argues that such information is inherently 'familial' in nature, setting aside the traditional presumption of both ethics and law that an individual patient's interest in the confidential status of that information must be respected. Accordingly, the article labels this approach a 'communitarian' model, an appellation which is in fact adopted by some of the proponents of this approach. While it does not entirely disregard individual interests, the communitarian approach involves a significant departure from the basic presumption or starting point that individuals who are the source of information of a medical nature are entitled to have that information kept secret, even from close relatives. It is appropriate therefore to summarise in detail the arguments in favour of such an approach. In Australia the arguments are typified by a report of the Cancer Genetics Ethics Committee of the Anti-Cancer Council of Victoria, *Ethics and Familial Cancers*,[11] and in an article by legal academic Loane Skene, entitled 'Patients' Rights or Family Responsibilities? Two Approaches to Genetic Testing'.[12]

A. Anti-Cancer Council of Victoria, Cancer Genetics Ethics Committee—Ethics and Familial Cancers

In 1997, the Anti-Cancer Council of Victoria published *Ethics and Familial Cancers* to address the ethical implications of developments in familial cancer genetics.[13] The final chapter of the report contains *Proposed Guidelines on Ethical Aspects of Risk Assessment, Genetic Testing and Genetic Registers in Relation to Cancer*.[14] That chapter has been included in a set of evidence-based clinical practice guidelines on familial cancer that were developed by the Australian Cancer Network and endorsed in November 1999 by the National Health and Medical

---

[11] Anti-Cancer Council of Victoria, Cancer Genetics Ethics Committee, *Ethics and Familial Cancers: Including Guidelines on Ethical Aspects of Risk Assessment, Genetic Testing and Genetic Registers* (March 1997).

[12] *Op.cit.*, n. 2.

[13] The report identifies three main forms of hereditary cancer for which there are presently tests available, n. 11 at 7–10:

(1) Familial Adenomatous Polyposis ('FAP'), in which the affected gene is the APC gene. FAP is a dominant condition and the mutation has a very high penetrance (almost 100 per cent of carriers of a mutation of the gene will develop colorectal cancer), however 'FAP accounts for less than one per cent of the approximately 10,000 new cases of colorectal cancer diagnosed each year in Australia': n. 11 at 7.

(2) Hereditary Non-Polyposis Colon Cancer (HNPCC) involves one of at least four genes, with lower penetrance than the APC gene mutation: about 80–90 per cent of individuals with a HNPCC gene mutation develop colorectal cancer. It therefore 'presents a more complex challenge than FAP for cancer prevention and management of affected families': n. 11 at 9.

(3) Breast Cancer: 'Predisposing gene mutations have been identified in two major genes, BRCA1 and BRCA2'. Penetrance is relatively high but variable: between 70 and 90 per cent of carriers will develop breast cancer. There is debate over the optimal management of women with a gene mutation because unlike FAP or even HNPCC, '[t]here is no entirely effective way of detecting early disease in healthy individuals. Early detection, when achieved, may improve survival, but there is not the same promise of cure as with FAP. Prophylactic surgery to remove the breasts (bilateral mastectomy) and ovaries (oophorectomy) is sometimes considered in the expectation of reducing the risk of cancer': n. 11 at 10.

[14] Anti-Cancer Council of Victoria, *ibid.* at 53–63. The chapter is available at <www.dhs.vic.gov.au/phd/hdev/genetics/append6.htm>.

Research Council.[15] According to the report, at least three features of familial cancers give rise to special ethical considerations:

1. 'the extent and range of relatives and relationships potentially uncovered by a consultation'.[16] A patient may bring into the consultation (either directly or indirectly) other members of his or her family 'who may not know anything about the whole matter, and indeed who may not wish to know about it'.[17] Moreover, the relevant family in such cases may extend well beyond the traditional nuclear unit of mother, father, and children.
2. the 'complexity of the consequences for individuals and their families of coming to possess information about their genetic risk of cancer and especially the results of gene tests'.[18]
3. the extended time scale. Tests will become possible which had not been recognised or possible at the time that a sample was taken.[19]

In relation to the confidentiality of information about genetic cancer, the report notes that under current practice there is a strong presumption in favour of confidentiality, so that such information may be disclosed, even to family members, in special circumstances only.[20] However, the report claims, it is important to remember that the 'flipside' of doctors' obligations is that patients have ethical responsibilities. The duty of confidentiality presupposes that patients are 'undertaking responsibility for improving their health or at least managing their illness',[21] and just as the broader social context in which the doctor–patient relationship is situated places limits on the doctor's obligation of confidentiality (by extending his or her responsibilities to others), so patients 'need to appreciate that the communal context of their medical interaction may involve them in considering more than just what will contribute to their own health'.[22]

These issues, including those in relation to consent, raise the broader issue of the relationship between individual and the community. The report states its position on this relationship in the following terms:

> While we recognise the great importance of confidentiality, privacy and consent in medical practice, we are less disposed than much

[15] National Health and Medical Research Council, *Familial Aspects of Cancer: A Guide to Clinical Practice*, available at www.health.gov.au:80/nhmrc/publications/synopses/cp67syn.htm>. The Anti-Cancer Council of Victoria's guidelines are included as ch. 2 of this publication.

[16] Anti-Cancer Council of Victoria, *op.cit.*, n. 11 at 35.

[17] *Ibid.*

[18] *Ibid.* at 36.

[19] *Ibid.* at 36.

[20] *Ibid.* at 37.

[21] *Ibid.* at 38.

[22] *Ibid.* at 38.

> recent ethical discussion of medical practice has been to frame our reflections on them in the language of individual patients' rights . . . [A] different ethical emphasis is increasingly called for in discussion of medical practice and policy, an emphasis on wider responsibility and communal concern.[23]

Under this more communitarian ethic, the report says, there would be a shift away from a focus on 'ownership' of genetic material and towards an approach under which a range of institutions and people 'look after' genetic material, with medical institutions and personnel 'at the centre' of this stewardship arrangement.[24] Such institutions and personnel 'are essentially the wider community's representatives as guardians of information and material which have both personal and social significance'.[25]

The report states that in relation to familial cancer susceptibility, 'patients have no invariable entitlement to block disclosure of medical information to their relations which may sometimes enable their own identity, and even their genetic test status, to be inferred'.[26] The first consideration relevant to the distinctive status of genetic information is the *context* in which patients make contact with health services. Patients are at risk 'as *members of families*'[27] and normally seek out a genetic test in the first place because of a shared family history of disease. A further consideration is the kind of genetic condition, and in particular the propensity and kind of disease or disorder associated with the genetic condition. For example, in FAP the gene mutation will almost certainly lead to the cancer, creating a stronger case for disclosure.[28]

Having stated what may be described as two criteria for non-consensual disclosure, the report goes on to consider reasons against non-consensual disclosure. It makes the obvious point that non-consensual disclosure should only be made after attempts have been made to persuade the patient voluntarily to disclose the information to relevant family members. One possibility is to advise relations of their cancer susceptibility without identifying the original patient. As the Anti-Cancer Council of Victoria's proposed guidelines state:

> If a patient or other inquirer objects to information on his or her genetic risk becoming known to relations, the aim could still be for the relations to be advised that they may be at risk from a

[23] *Ibid.* at 38–9.
[24] *Ibid.* at 39.
[25] *Ibid.* at 39.
[26] *Ibid.* at 41.
[27] *Ibid.* at 41, emphasis in original.
[28] *Ibid.* at 41.

> family cancer susceptibility, but without in the process identifying the reluctant inquirer. For the purpose of their own genetic testing, the aim should also be to acquaint relations with the specification of the gene mutation carried in their family. Objection by inquirers to knowledge of their risk status being used for the benefit of their relations and descendants should not in itself be regarded as sufficient reason for the information not to be used in de-identified form. This remains the case even if the inquirer's identity and even risk status may be inferred.[29]

In addition, the report acknowledges that blood relations may have an interest in *not* knowing their genetic history, including even the possibility that they may be at risk. The report responds to this consideration by saying that one relevant factor in deciding whether or not to disclose is the nature of the genetic condition. Where the risk is high (for example, in relation to FAP) there is a 'strong presumption' that blood relatives will appreciate the opportunity to ascertain their risk.[30] The best person to approach blood relations, says the report, is the original patient or other family members. But if family involvement of some sort is not possible the doctor may have to initiate contact.[31]

The Victorian Cancer Council's Report and its implications have been further analysed by Loane Skene[32] who identifies and contrasts two approaches to genetic information about individuals, 'especially in the intra-familial context':[33]

1. A *legal model* which is based on the patient's 'right to privacy'. This right normally takes the form of legislation protecting the privacy of genetic information by making it an offence to disclose a person's genetic details without that person's consent or other lawful justification.[34] This approach, Skene asserts, has been adopted in some states in the United States with the enactment of privacy legislation, and has also been adopted in various Canadian, British and European statements on genetic testing, as well as in a Bill which was

---

[29] *Ibid.* at 57–8.

[30] This presumption is subject to the receipt by the doctor of information indicating that a particular relation does not want to be apprised: Anti-Cancer Council of Victoria, *ibid.* at 42.

[31] Anti-Cancer Council of Victoria, *ibid.* at 43. A *spouse* of a person with a genetic condition does not have an interest in disclosure of that condition for the purpose of their own health, but the report claims that a spouse does have an interest where there is a possibility of the couple having children, or where the couple have children who have not yet attained majority: Anti-Cancer Council of Victoria, *ibid.* at 44.

[32] Skene, *supra.*, n. 2.

[33] Skene, *ibid.* at 1.

[34] *Ibid.*

before the Australian Parliament in 1998.[35] While a number of these instruments 'acknowledged that divulging information to family members may be different from disclosure to other third parties, that has not, for the most part, been reflected in any legislation or ultimate recommendations'.[36]

2. A *medical model* by which Skene refers to the kind of model proposed in the Anti-Cancer Council of Victoria guidelines discussed above. According to Skene this approach rejects considerations of rights, focussing instead on the care of patients and their families, and envisages a proactive role for doctors in this process.[37]

Skene considers the implications of these two approaches. She firstly applies the dominant principles informing the 'legal' model—privacy and non-discrimination—to a typical case study of a patient. What follows are some of the implications of a 'legal' approach in the context of the health of an individual and his or her genetic relations:

— *Validating the family history from medical records*. A regulatory approach premised on privacy, and which requires formal, written authorisation before a family member's genetic information can be provided, makes the process of validation significantly more onerous.[38]

— *Access to tissue samples*. Consent is required for an inquirer to obtain access to a tissue sample of the inquirer's relatives (alive or deceased). Moreover, where a specific test is proposed on a tissue sample (rather than information about tests already done) *informed consent* will be required, including not only information about the medical nature of the information sought but also its potential 'social' implications (including information about who will have access to the test results, how securely they will be stored and for how long, etc.).[39]

— *Control of personal information about oneself*. The individual alone must decide the people who will know the existence of the genetic mutation.[40]

— *Duty to warn relatives*. In limited circumstances, the common law in some jurisdictions in the United States based on the rule in

[35] See Skene, *ibid.* at 12–17 for a discussion of developments in the United States, Canada, the United Kingdom and the Council of Europe. In 1998 the Genetic Privacy and Non-Discrimination Bill was introduced into the Australian Federal Parliament. The Bill, which has subsequently lapsed, is discussed below.

[36] Skene, *ibid.* at 1.

[37] Skene, *ibid.* at 2.

[38] Skene, *ibid.* at 18–19.

[39] Skene, *ibid.* at 20.

[40] Skene, *ibid.* at 20–1.

Tarasoff v. *Regents of the University of California*[41] might impose a duty on doctors to warn relatives. But the model genetic privacy statutes enacted in some states in the United States appear to override the effect of any such common law requirement by prohibiting disclosure without the consent of the person concerned. Whether the patient has a legal duty to warn his or her blood relatives is another question, which is not addressed in the statutes and which has not been judicially considered.[42]

By way of contrast, Skene considers the likely consequences of the adoption of the 'medical' model as characterised by the guidelines proposed by the Anti-Cancer Council of Victoria. Under this model, people would not have a 'right' to control their genetic information and the use of their tissue taken for genetic testing, and 'doctors will have a special role in providing and imparting genetic information that may appear contrary to their traditional obligation to maintain patient confidentiality'.[43] Although directed to familial cancers, Skene considers there is no reason why this model should not have a wider application in relation to genetic information.

— *Validating the family history from medical records*. Generally, the medical model envisages a very different role for the doctor from that in which the doctor cares only for an individual patient.[44] As part of the development of a management plan for an inquirer, the doctor may sometimes obtain genetic information about the inquirer's relatives without first seeking their consent.[45]
— *Access to tissue samples*. The model treats tissue samples similarly to genetic information: they should be available to all family members who might benefit from such access.[46]
— *Control of personal information about oneself*. Under the medical model, patients would not have an invariable right to block disclosure of genetic information because it is, of its nature, familial information.[47] The inquirer's consent to such a disclosure must always be sought first. In other words, doctors should encourage inquirers to disclose relevant information to family members, but may do so themselves where the inquirer is unwilling or unable to do so. The disclosure must be non-identifying if possible, and if this is impossible the disclosure should be made 'only after very careful

[41] 551 P. 2d 334 (1976).
[42] Skene, *supra*., n. 2 at 21.
[43] Skene, *ibid*. at 24.
[44] Skene, *ibid*. at 26.
[45] Skene, *ibid*. at 26.
[46] Skene, *ibid*. at 27.
[47] Skene, *ibid*. at 28.

consideration'.[48] Skene observes that a doctor who follows these guidelines is probably acting consistently with the existing common law, which recognises that it may be lawful to breach confidentiality where there is a 'serious' risk to others.[49]

— *Duty to warn relatives*. There is no case law on this issue in Australia, but Skene considers that the argument that doctors owe third parties a duty of care in limited circumstances is unlikely to be successful in the context of genetics.[50]

Skene delineates her understanding of the nature of the differences between the two models. The 'legal/rights/privacy' model, she claims, 'is based on autonomy and self-determination'. By way of contrast, the 'medical/family-centred model . . . is based, not on individual rights but on familial and community obligations'.[51] After weighing up the pros and cons of each model,[52] Skene observes that if the medical model was to be adopted, new laws might be needed.[53] Skene considers three alternative approaches: 'to legislate; to do nothing and allow the law to develop case by case as issues come before the courts; and to attempt to regulate by contract'.[54] The legal form of this new arrangement preferred by Skene is for either an express or implied contractual provision to the effect that the patient agrees to share genetic information with, inter alia, family members.[55]

III. OTHER AUSTRALIAN APPROACHES

In the light of these arguments of the Anti-Cancer Council of Victoria and Skene, a number of reports of other Australian entities on the ethics of non-consensual disclosure of genetic information may be considered. We will see that these recommendations and discussions are much more cautious about discarding the current presumption of confidentiality, and the reason for this caution resides principally in the assertion that under both ethics and law non-consensual disclosure is allowed in very limited circumstances only.

---

48 Skene, *ibid*., quoting from the guidelines.

49 There is no Australian case on this point, though as Skene points out, Australian courts would most likely follow the decision of the English Court of Appeal in *W*. v. *Egdell* [1990] 1 Ch. 359: Skene, *ibid*. at 29.

50 Skene, *ibid*. at 29–30.

51 Skene, *ibid*. at 32.

52 Skene, *ibid*. at 32–5.

53 Skene, *ibid*. at 35.

54 Skene, *ibid* . at 36.

55 Skene, *ibid*. at 36–37.

A. *National Health and Medical Research Council—Ethical Aspects of Human Genetic Testing: An Information Paper*

The National Health and Medical Research Council has published an information paper on *Ethical Aspects of Human Genetic Testing*.[56] The Council emphasises that doctors must be careful not to inadvertently release information to family members, especially where several members of a family are patients of a particular doctor. 'It is very easy for health professionals to assume falsely that there has been a free flow of genetic information within families. This is often not the case'.[57] In rare situations, a patient may refuse to consent to the release of information to family members. The Information Paper states:

> Ultimately, the individual's decision should bind the health professional. However, there may be rare circumstances in which a health professional considers that the risk to the health of relatives is sufficiently large, serious, imminent and potentially preventable that consideration should be given to breaching the individual's confidentiality. Before doing so, the health professional should consider the potential for professional censure or legal action if confidentiality is breached.[58]

The Information Paper also discusses the 'right not to know', especially in the context of ascertaining the family history which will normally be a part of good medical practice in testing an individual patient. While generally 'family members who are contacted will appreciate the approach and the underlying concern for their welfare',[59] for whatever reasons an individual may not wish to discuss the family's illness. However, as the Information Paper notes, '[a] more common situation is that the views of a family member are not known, or that he/she appears to dislike discussing health matters in general'.[60] If the relative is not informed of the genetic condition they are denied the opportunity of considering the issue and their response to it. The Information Paper indicates that '[t]here is a strong argument for informing such people in a sensitive way, almost always through a family member with whom they relate well'.[61]

---

[56] National Health and Medical Research Council, *supra*., n. 5.
[57] National Health and Medical Research Council, *ibid.* at 50.
[58] National Health and Medical Research Council, *ibid.* at 51.
[59] National Health and Medical Research Council, *ibid.* at 51.
[60] National Health and Medical Research Council, *ibid.* at 51.
[61] National Health and Medical Research Council, *ibid.* at 51.

B. National Health and Medical Research Council—National Statement on Ethical Conduct in Research Involving Humans

The National Health and Medical Research Council, *National Statement on Ethical Conduct in Research Involving Humans*[62] provides guidance on, inter alia, human genetic research. While the Statement is of primary relevance in the research context, rather than a clinical context, its provisions relating to privacy and confidentiality in genetic research serve to underscore the importance of these concepts. The *National Statement* requires that researchers maintain the confidentiality of information provided by participants about family members.[63] In addition, paragraph 16.7 provides that: 'identifying genetic information must not be released to others, including family members, without the written consent of the individual to whom the information relates, or a person or institution which may legally provide consent for that person'. The *National Statement* requires that consent be obtained from participants for human genetic research unless the requirement for consent has been waived by a Human Research Ethics Committee.[64] Furthermore, when seeking consent from individuals for prospective collection of genetic material and information, individuals should be informed, inter alia:

— that they can refuse consent. The *National Statement* points out that researchers should be aware that individuals may participate in genetic research out of a sense of obligation to family members.[65]

— of the arrangements that exist 'to ensure the privacy and confidentiality of their genetic information both with regard to family members and persons who are not family members'.[66]

— 'that if the research generates information about participants which may be of relevance to the health of other family members, the consent of participants will be sought before offering to disclose such information to the family members concerned'.[67]

C. Privacy Commissioner—The Privacy Implications of Genetic Testing

In 1996, the Australian Privacy Commissioner released a report on the privacy issues arising in relation to the use of genetic information in a variety of contexts, including clinical settings, employment, insurance,

---

[62] The National Statement is available at <www.nhmrc.health.gov.au/issues/researchethics.htm>.

[63] National Statement, *ibid.* at para. 16.4.

[64] National Statement, *ibid.* at para. 16.9.

[65] National Statement, *ibid.* at para. 16.10(a).

[66] National Statement, *ibid.* at para. 16.10(b).

[67] National Statement, *ibid.* at para. 16.10(e).

scientific research and law enforcement.[68] Chapter 4 of the report, on 'Testing in a Clinical Context', includes discussion of the ethical status of genetic information and its disclosure. This chapter begins by emphasising an individual's right *not* to know particular things about his or her health,[69] and goes on to discuss the relationship of confidentiality that exists between doctors and patients.[70] It notes that the Australian Medical Association's *Code of Ethics*, after stating the basic ethical rule that medical information must be kept in confidence, recognises that 'Exceptions may arise where the health of others is at risk . . . '.[71] The Privacy Commissioner's report notes that there has been some concern generated about the potential for this exception to erode traditional standards of confidentiality, and that the Privacy Commissioner has argued that the code should be supplemented with a clear set of guidelines on:

— 'when disclosure without the patient's permission might occur, *i.e.* what threshold of risk must be satisfied;
— the steps that should be followed before making such a disclosure; and
— a set of safeguards applying to any such disclosures'.[72]

The report also notes that there is an important difference between the principle of medical confidentiality and the Information Privacy Principles ('IPPs') in the Privacy Act 1988 (Cth.).[73] Whereas the IPPs 'are based on the principle, embodied in the International Covenant on Civil and Political Rights, that each individual has a right to privacy and a right to have that privacy protected by the law', medical confidentiality, at least insofar as it has been interpreted by the courts, 'rests on the premise that there is a public interest in patients being candid with their doctor and that this requires strong assurances of confidentiality'.[74]

In the context of personal genetic information which has possible implications for blood relatives of the patient, the report points out that IPP 10 'provides that a keeper of personal information may use it for a purpose other than that for which it was collected if there are reasonable grounds to believe that doing so is necessary to prevent or lessen a serious and imminent threat to the life or health of the individual con-

[68] Privacy Commissioner, *The Privacy Implications of Genetic Testing*, Information Paper Number 5 (Human Rights and Equal Opportunity Commission, 1996).
[69] *Ibid.* at 26–7.
[70] *Ibid.* at 29–32.
[71] *Ibid.* at 30.
[72] *Ibid.* at 30–1.
[73] The Information Privacy Principles in the Privacy Act 1988 (Cth.) provide principles relating to the collection, storage and use of personal information.
[74] Privacy Commissioner, *supra.*, n. 68 at 31.

cerned or another person'.[75] The requirement that a threat be 'serious and imminent' is not likely to be met by many presently known genetic conditions. However, the report does note that:

> [s]uch issues will be more pressing as the degree of consanguinity between the relatives increases. The clearest case would be where identical twins disagree about having presymptomatic testing for a genetic condition. The twins' genes will be effectively identical (barring significant mutation from environmental factors . . .) so that information about genetic makeup of one will apply equally to the other.[76]

The report concludes that it is difficult to formulate:

> hard and fast standards to govern this sort of decision [about disclosure of information]. However, where there is the possibility of effective intervention and the consequences of non intervention are serious for affected relatives, it is hard to see that the privacy rights of the affected person should override pressing health interests of relatives. In such cases, public health considerations argue strongly for disclosure, though this should clearly be limited to what is necessary to address the health risk relatives face.[77]

The report does not discuss whether and what different considerations would apply where breach of confidentiality, rather than breach of privacy, has occurred.

D. Australian Medical Association—Code of Ethics and Position Statement on Human Genetic Issues

The current *Code of Ethics* of the Australian Medical Association contains the following statement in relation to patient information:

> Keep in confidence information derived from your patient, or from a colleague regarding your patient, and divulge it only with the patient's permission. Exceptions may arise where the health of others is at risk or you are required by order of a court to breach patient confidentiality.[78]

The Association has also issued a Position Statement on *Human Genetic Issues*, which basically repeats the provision in the *Code of Ethics*, though with no mention of the exceptions:

---

[75] Privacy Commissioner, *ibid.* at 34.
[76] Privacy Commissioner, *ibid.* at 34–5.
[77] Privacy Commissioner, *ibid.* at 35.
[78] Australian Medical Association, *Code of Ethics* (1996) <http://www.ama.com.au>, para. 1.3.4.

> Genetic information acquired in the context of the doctor–patient relationship should not be disclosed to a third party, without the patient's specific, and where possible, written consent (refer to AMA, *Code of Ethics, February 1996*, 1.3(d)).[79]

The combined effect of these provisions is uncertain. On its face, the effect of this provision appears to be that the exceptional circumstances in which non-genetic medical information may be disclosed do not apply to genetic information.[80] On the other hand, it is also arguable that the reference at the end of the provision to the Code of Ethics, which explicitly states the exceptions to the general prohibition on disclosure, is also intended to apply to genetic information. At any rate, what is clear from both these statements is that the AMA contemplates that the confidential status of non-genetic medical information applies equally to genetic medical information.

E. The Genetic Privacy and Non-Discrimination Bill 1998 (Cth.)

In March 1998, the Australian Democrats introduced the Genetic Privacy and Non-Discrimination Bill in the Federal Senate. The Bill is based broadly on the provisions of the American model Genetic Privacy Act which arose out of research into the ethical, legal and social implications ('ELSI') of the Human Genome Project.[81] The Bill sets out 'the mechanisms by which human genetic material may be collected, stored and used'.[82] The Bill was referred to the Senate Legal and Constitutional Legislation Committee, which handed down its report on the Bill in March 1999.[83] The report recommended that the Bill should not proceed pending further examination of the issues by either a national working party or the Senate Legal and Constitutional References Committee, or ongoing monitoring of the issues by relevant Commonwealth

[79] Australian Medical Association, Position Statement on *Human Genetic Issues* (1998) <http://www.ama.com.au>, para. 5.7.

[80] An interesting development if it is intended to be a clarification of circumstances in which disclosures without the patient's permission might occur in relation to genetic information, as advocated by the Privacy Commissioner's report above.

[81] Genetic Privacy Act and Commentary 1995 by G.J. Annas, L.H. Glantz and P.A. Roche <http://www.ornl.gov/TechResources/Human_Genome/resource/privacy/privacy1.html>. On the model law, see generally T. McGleenan, 'Rights to Know and Not to Know: Is There a Need for a Genetic Privacy Law?' in R. Chadwick *et al.*, *The Right to Know and the Right Not to Know* (Avebury 1997) at 44.

[82] Genetic Privacy and Non-Discrimination Bill 1998 (Cth.): Briefing Sheet.

[83] Report of the Senate Legal and Constitutional Legislation Committee, *Provisions of the Genetic Privacy and Non-Discrimination Bill 1998 (as introduced in the 38th Parliament)* 31 March 1999, <www.aph.gov.au/senate/committee/legcon_cttee/genetic/index.htm>.

government departments. The Committee was of the opinion that there are 'significant privacy and ethical issues in relation to access to, and control over, personal medical records containing genetic information. The committee does not consider current protections, nor those proposed by the bill, to be adequate, particularly in relation to the private sector'.[84] However, the Committee also considered that comprehensive legislation to attempt to address these issues would be premature, and that there should be further monitoring of privacy and discrimination issues with a view to appropriate legislation at some time in the future.[85] The Bill has since lapsed and in August 2000 the Australian Law Reform Commission and the Australian Health Ethics Committee were given a joint reference on gene technology.[86] It is expected that that inquiry will consider privacy and discrimination issues relating to genetic testing.

Despite the ultimate fate of the Bill, it has been widely debated in Australia and so the provisions of the Bill relevant to the disclosure of genetic information to family members will be briefly discussed here. The Bill prohibits genetic discrimination, and contains provisions relating to the use of genetic information by employers and insurance providers. Most importantly for our purposes, however, part 2 of the Bill contains provisions relating to the disclosure of genetic information. It is to this part of the Bill that Skene is referring when she describes the framework established by the Bill as a 'rights' approach.[87] However, the accuracy of this characterisation is not entirely clear. The Australian Bill, despite its short title, at no point refers to any 'right' to privacy. Within part 2 of the Bill, section 10 can perhaps be characterised as providing individuals with a right of access to records containing genetic information about themselves,[88] and section 11 provides a right to request an amendment of information in a genetic record.[89] In addition, section 16 of the Bill lays down requirements which must be satisfied for a valid authorisation to collect, store or analyse a DNA sample. The authorisation must include provisions that permit the individual to

---

[84] Senate Legal and Constitutional Legislation Committee, *ibid.* at para. 4.18.

[85] Senate Legal and Constitutional Legislation Committee, *ibid.* at para. 5.18–5.33.

[86] For information see the Australian Law Reform Commission's website: <www.alrc.gov.au>.

[87] See also the report of the Senate Legal and Constitutional Legislation Committee, which describes the purpose of the Bill as, inter alia, the establishment of 'an enforceable right to privacy of genetic information of an individual, by proscribing disclosure of such genetic information except with the authorisation of the individual, or in other limited circumstances', *supra.*, n. 83 at para. 1.6.

[88] Provided a request is made in writing: s. 10(1).

[89] The section provides, inter alia, that: 'The person must make the amendment if such information is not accurate or complete for the purposes for which such information may be used or disclosed by the person', s. 11(1).

consent to, inter alia, 'commercial use of the DNA sample, with a waiver of, or a provision for, economic benefit to the individual'.[90] Thus the Bill recognises that individuals possess a quasi-economic interest in their DNA.

However, in relation to *disclosure* of genetic medical information, it is more debatable whether a 'rights' approach is adopted. Instead of referring to a 'right' to privacy in relation to genetic information, the statute arguably adopts the common law approach of a blanket prohibition on disclosure except in the limited circumstances defined in section 8(1). Section 8(1) provides:

> A person may disclose genetic information in a genetic record characterised from the DNA sample of an individual only if:
> (a) the individual has authorised the disclosure; or
> (b) the disclosure is required by or under law; or
> (c) the person believes on reasonable grounds that the disclosure is necessary to prevent or lessen a serious and imminent threat to the life or health of the individual or of another person.

The grounds for disclosure in section 8(1)(c) are same as those in which disclosure is permitted in IPP 11(1)(c) of the Privacy Act 1988 (Cth.). While this could be used to characterise the Bill as adopting a rights approach to disclosure, the following section will discuss the existing common law and equitable rules relating to disclosure and suggest that, despite a dearth of Australian case law relating to disclosure of medical information to third parties, section 8(1) of the Bill provides for disclosure in circumstances that are largely the same as would be allowed under Australia common law.[91]

IV. CONFIDENTIALITY OF GENETIC INFORMATION AND DISCLOSURE TO FAMILY MEMBERS

What is so interesting about the Victorian Cancer Council's report is that its proposal for a familial concept of medical information potentially represents a major departure from the presumptions of existing law Within modern health care the duty of confidentiality is clear.

---

[90] Genetic Privacy and Non-Discrimination Bill 1998 (Cth.), s. 16(1)(f)(ii).

[91] An argument which is reinforced if s. 8(1) of the Privacy and Non-Discrimination Bill is compared with the much more limited exceptions in the American model Genetic Privacy Act and Commentary 1995, which allows the disclosure of private genetic information only in accordance with written consent (s. 111(a)), pursuant to compulsory process (s. 115), or for the purpose of research in strictly defined circumstances (s. 132): provisions which are similar to the narrow interpretation of the Australian Medical Association's Position Statement on *Human Genetic Issues*, referred to above.

Health professionals have an ethical and legal duty to maintain the confidentiality of the information of individual patients. From the Hippocratic Oath[92] to the Australian Medical Association's current *Code of Ethics*, health professionals have a clear ethical obligation to maintain their patients' confidences, in all but exceptional circumstances.

This duty of confidentiality has also been given legal expression both in statutory provisions and in case law. Roberta Berry has noted that the transformation of confidentiality 'from ethical command into legal duty was part of a larger transformation of Western medicine in the nineteenth and twentieth centuries from a largely self-regulated practice into a social institution subject to extensive legal regulation'.[93] There are many statutory provisions that impose obligations to maintain the confidentiality of medical information.[94] Case law has also provided a basis for the protection of interests in confidential information.[95]

Confidentiality in health care is neither perfect nor absolute. The changing nature of modern health care means that it is often provided by teams of professionals.[96] As a consequence, private health information may need to be available to all members of the clinical team rather than a single doctor, although the duty of confidentiality would obviously extend to the members of that team. As discussed above, there are also reasons of public policy for which medical information may need

---

[92] The Oath states in part: 'Whatever, in connection with my professional service, or not in connection with it, I see or hear, in the life of men, which ought not to be spoken of abroad, I will not divulge, as reckoning that all such should be kept secret.'

[93] R.M. Berry, 'The Genetic Revolution and the Physician's Duty of Confidentiality: The Role of the Old Hippocratic Virtues in the Regulation of the New Genetic Intimacy' (1997) 18 *Journal of Legal Medicine* 401 at 413–14.

[94] See, for example, in New South Wales the Health Administration Act 1982 (NSW), s. 22; Public Health Act 1991 (NSW), s.17.

[95] See for example, *X.* v. *Y.* [1988] 2 All E.R. 648.

[96] The implications of this for confidentiality have been noted by one commentator who sought to find out how many people could access his patient's medical record and had a reason to examine it: 'I was amazed to learn that at least 25 and possibly as many as 100 health professionals and administrative personnel at our university hospital had access to the patient's record and that all of them had a legitimate need, indeed a professional responsibility, to open and use that chart. These persons included 6 attending physicians (the primary physician, the surgeon, the pulmonary consultant, and others); 12 house officers (medical, surgical, intensive-care unit, and 'covering' house staff); 20 nursing personnel (on three shifts); 6 respiratory therapists; 3 nutritionists; 2 clinical pharmacists; 15 students (from medicine, nursing, respiratory therapy, and clinical pharmacy); 4 unit secretaries; 4 hospital financial officers; and 4 chart reviewers (utilization review, quality assurance review, tissue review, and insurance auditor). It is of interest that this patient's problem was straightforward, and he therefore did not require many other technical and support services that the modern hospital provides.' (M. Siegler, 'Confidentiality in Medicine: A Decrepit Concept' in T.L. Beauchamp and L. Walters (eds.), *Contemporary Issues in Bioethics* (4th edn.) (Wadsworth 1994) at 179.)

to be disclosed, and these are reflected in medical ethics[97] as well as law. A number of statutory provisions require disclosure of medical information, particularly in the context of reporting by health professionals of cases of suspected child abuse,[98] and in the statutory requirements for notification of certain diseases.[99] Confidential information may also be disclosed with the consent of the patient.

The question of whether disclosure of confidential medical information can be justified on the basis that it is in the *public interest* has also been considered in a number of English cases.[100] From the decisions in these cases it seems clear that what is at issue is not an individual's private interest in confidentiality versus a public interest in disclosure, but rather two competing public interests: a public interest in maintaining the confidentiality of medical information and a public interest in disclosure of that information.[101] In addition, the courts have only regarded the disclosure of confidential medical information as justified in exceptional circumstances and where the disclosure itself was of a limited nature, such as that made to a responsible authority.

The public interest basis for disclosure is potentially relevant to a consideration of disclosure of genetic information to family members for two reasons. Firstly, the decided cases suggest that the nature of the public interest is such that it could be invoked to justify the disclosure of confidential patient information in a wide range of circumstances—for example, 'where a patient's medical condition presents an infection risk to others, where a patient's ill health renders him or her unfit to continue certain activities because others would be placed at risk, or where inherited genetic disorders should properly be disclosed to other family members'.[102] Secondly, the *method* of assessment recognises that it is competing *public* interests which need to be assessed. This means that in the context of genetic conditions, a relevant factor to consider in any

---

[97] See the Australian Medical Association's *Code of Ethics*, *supra.*, n. 78.

[98] Children and Young Persons (Care and Protection) Act 1998 (NSW), s. 27.

[99] See *e.g.* Public Health Act 1991 (NSW) Part 3.

[100] *X.* v. *Y.* [1988] 2 All E.R. 648; *W.* v. *Egdell* [1990] 1 Ch. 359; *R.* v. *Crozier* (1990) 12 Cr. App. R.(S.) 206.

[101] In *X.* v. *Y.* [1988] 2 All E.R. 648 Rose J. said at 653: 'In the long run, preservation of confidentiality is the only way of securing public health; otherwise doctors will be discredited as a source of education, for future patients "will not come forward if doctors are going to squeal on them". Consequently, confidentiality is vital to secure public as well as private health . . . ' However, the English courts have also suggested that the evaluation of the competing interests is not just a matter of 'balancing': '[T]he process is to consider fairly the strength and value of the interest in preserving confidentiality and the damage which may be caused by breaking it . . . This is a more complex process than merely using scales: it is an exercise in judicial judgment' (*Science Research Council* v. *Nassé* [1980] A.C. 1028, *per* Lord Wilberforce at 1067).

[102] Laws of Australia, vol. 20, *Health and Guardianship*, 20.6 'Confidentiality', para. 31.

specific case would be possible adverse effects of allowing disclosure of genetic medical information on the willingness of individuals to take advantage of genetic tests in the future, and the likely public health outcomes of such a policy.

Nonetheless, the existing case law on the public interest basis for disclosure indicates a number of limitations on the circumstances in which disclosure may legally occur. First, disclosure must be confined to 'exceptional circumstances', where 'another's life is immediately endangered and urgent action is required',[103] or where there is a 'real risk of consequent danger to the public'.[104] Secondly, in these cases the courts have emphasised that disclosure should be to a responsible authority.[105] These limitations suggest that the public interest basis for disclosure has not, to date, been contemplated for a situation such as disclosure to a family member of a genetic condition, both because such a condition will rarely, if ever, present an immediately life-threatening risk, and also because such disclosure would ultimately have to be disclosed to the family member rather than a responsible authority (although if the relative is not a patient of the doctor's, it may be disclosed to that relative's treating doctor).[106] However, there is nothing in the cases themselves which precludes such a factual scenario being considered by the courts, and indeed some cases of this kind have come before US courts in the context of a consideration of the doctor's duty to warn.

V. A DUTY TO WARN OF GENETIC RISK?

In recent years there has been debate over whether a 'duty to warn' exists in Australian law.[107] The duty to warn developed in the United States following *Tarasoff* v. *Regents of the University of California*[108] which held that health professionals could have a duty to warn identifiable third parties of a foreseeable risk to their safety. Consideration of

---

[103] *Duncan* v. *Medical Practitioners Disciplinary Committee* [1986] 1 N.Z.L.R. 513 (H.C.), Jeffries J. at 521.

[104] *W.* v. *Egdell* [1990] 1 Ch. 359, Bingham L.J. at 424.

[105] *Duncan* v. *Medical Practitioners Disciplinary Committee* [1986] 1 N.Z.L.R. 513 (H.C.), Jeffries J. at 521; *W.* v. *Egdell* [1990] 1 Ch. 359, Bingham L.J. at 424.

[106] A consideration which is relevant to the issue of the 'right' of the family member 'not to know' of any genetic condition that they carry, and which is discussed below.

[107] Neave, *supra.*, n. 1; D. Mendelson, '"Mr Cruel" and the Medical Duty of Confidentiality' (1993) 1 *Journal of Law and Medicine* 120; D. Mendelson and G. Mendelson, '*Tarasoff* Down Under: The Psychiatrist's Duty to Warn in Australia' (1991) *Journal of Psychiatry & Law* 33; A. Abadee, 'The Medical Duty of Confidentiality and Prospective Duty of Disclosure: Can They Co-exist?' (1995) 3 *Journal of Law and Medicine* 75; W. Herdy, 'Must the Doctor Tell?' (1996) 3 *Journal of Law and Medicine* 270.

[108] 551 P. 2d 334 (1976).

a duty to warn in the context of information about genetic risk within families highlights the potential dilemma for doctors and other health professionals.[109] On the one hand, the duty which *Tarasoff* established to warn another of a serious danger of violence[110] is arguably quite different from a situation where a family member has a genetic condition. On the other hand, the doctor may potentially find him/herself at risk of being sued by a disgruntled relative who claims that they should have been warned. It will frequently be the case that family members will be patients of the same doctor, and even if that is not the case, at the very least the doctor is likely to have knowledge of the existence and identity of family members based on the doctor's treatment of the patient in relation to the identified genetic disorder. Certainly in the case where the family member is a patient of the same doctor, it is arguable that that person is relying on the doctor to prevent injury to them by providing proper care and treatment, and that in appropriate cases this would include the provision of information obtained from another family member of a genetic condition which is treatable.

These issues have been addressed by the courts in the United States. In *Pate* v. *Threlkel*[111] the Supreme Court of Florida held that a doctor owed a duty of care to the children of a patient. In 1987 the plaintiff's mother, Marianne New, had been treated by Dr Threlkel and others for medullary thyroid carcinoma. The case arose when the plaintiff, Heidi Pate, learned in 1990 that she had the same condition. The plaintiff and her husband filed a complaint against the doctors, alleging that they knew or should have known that the children of their patient could have inherited the condition, that the doctors had a duty to warn the patient that her children should be tested, that had she been warned the patient would have had her children tested in 1987, and if the plaintiff had been tested then she could have taken preventative action and her condition would probably have been curable. The matter came before the Supreme Court of Florida after the trial court granted the motion by the health care providers to dismiss the complaint, and this decision was affirmed by the district court. The Supreme Court noted that 'the court's dismissal requires us to assume that the factual allegations in the complaint are true'.[112]

---

[109] For analysis of the ethical and legal obligations of health professionals in the United States see, Berry, *supra.*, n. 93; C.M. Parker, 'Camping Trips and Family Trees: Must Tennessee Physicians Warn Their Patients' Relatives of Genetic Risks?' (1998) 65 *Tennessee Law Review* 585; L.J. Deftos, 'Genomic Torts: The Law of the Future: The Duty of Physicians to Disclose the Presence of a Genetic Disease to the Relatives of their Patients with the Disease' (1997) 32 *University of San Francisco Law Review* 105.

[110] 551 P. 2d 334 (1976) at 345.

[111] 661 So. 2d 278; 1995 Fla. LEXIS 1156; 20 Fla. Law. W.S. 356.

[112] 661 So. 2d 278 at 281.

The Supreme Court held that the plaintiff's claim was not barred by privity because in the context of this case 'a patient's children fall within the zone of foreseeable risk'.[113] However the Court fell short of deciding that the doctor had a duty to warn the patient's children *directly*, limiting the duty to the provision of a warning to the patient:

> In most instances the physician is prohibited from disclosing the patient's medical condition to others except with the patient's permission. [. . .] Moreover, the patient ordinarily can be expected to pass on the warning. To require the physician to seek out and warn various members of the patient's family would often be difficult or impractical and would place too heavy a burden upon the physician. Thus, we emphasize that in any circumstances in which the physician has a duty to warn of a genetically transferable disease, that duty will be satisfied by warning the patient.[114]

The Supreme Court overturned the decision of the district court which had affirmed the dismissal of the complaint, and remanded the matter for further proceedings.[115]

In *Safer* v. *Pack*[116] the Superior Court of New Jersey, Appellate Division declined to follow the narrow approach adopted in *Pate*. In 1956 the plaintiff's father was admitted to hospital with a preoperative diagnosis of retroperitoneal cancer. The defendant, Dr Pack, performed a 'total colectomy and an ileosigmoidectomy for multiple polyposis of the colon with malignant degeneration in one area'.[117] The plaintiff's father was admitted to hospital again in 1961 and the defendant performed further surgery for the plaintiff's father's colon cancer. The defendant continued to treat the plaintiff's father postoperatively. The plaintiff's father was hospitalised again in 1963 for treatment of his cancer which had metastasised, and died in January 1964. At that time the plaintiff was 10-years-old.

In 1990 when the plaintiff was 36-years-old she was diagnosed with colon cancer and multiple polyposis. She underwent a total abdominal colectomy with ileorectal anastamosis, had her left ovary removed, and also underwent chemotherapy. In 1991 the plaintiff learnt that her father had suffered from polyposis after she obtained a copy of his medical records. She filed a complaint alleging that the defendant had breached his duty of care in failing to warn of the risk to the plaintiff's

[113] 661 So. 2d 278 at 282.
[114] 661 So. 2d 278 at 282.
[115] 661 So. 2d 278 at 282.
[116] 291 N.J. Super. 619; 677 A. 2d 1188; 1996 N.J. Super. LEXIS 284.
[117] 291 N.J. Super. 619 at 621.

health. The trial judge dismissed the plaintiffs' case holding that there was no doctor–patient relationship between the defendant and the plaintiff and that genetic diseases were different from infectious diseases or threats of harm in relation to the duty to warn because 'the harm is already present within the non-patient child, as opposed to being introduced, by a patient who was not warned to stay away. The patient is taking no action in which to cause the child harm'.[118]

However, the Superior Court took a different view. Citing the *Tarasoff* case the Court held that, 'In terms of foreseeability especially, there is no essential difference between the type of genetic threat at issue here and the menace of infection, contagion or a threat of physical harm.'[119] The Court declined to follow *Pate* stating:

> We decline to hold as the Florida Supreme Court did in *Pate* v. *Threlkel*, [. . .] that, in all circumstances, the duty to warn will be satisfied by informing the patient. It may be necessary, at some stage, to resolve a conflict between the physician's broader duty to warn and his fidelity to an expressed preference of the patient that nothing be said to family members about details of the disease. We cannot know presently, however, whether there is any likelihood that such a conflict may be shown to have existed in this matter or, if it did, what its qualities might have been.[120]

The Superior Court reversed the trial court's dismissal of the complaint and remanded the matter to the trial court for further proceedings.[121]

To date, neither Australian nor English courts have recognised a duty to warn. However, both Australian and English courts have recognised a duty to control in defined circumstances.[122] The relationship between a duty of control and a duty to warn is unclear, although the Californian Supreme Court in *Tarasoff* clearly considered that the duty of control and the duty to warn were cognate legal concepts: 'when the avoidance of foreseeable harm requires a defendant to control the conduct of another person, or to warn of such conduct, the common law has traditionally imposed liability only if the defendant bears some special relationship to the dangerous person or to the potential victim'.[123] In Australia, the High Court in *Sutherland Shire Council* v.

---

118 Quoted at 291 N.J. Super. 619 at 623–4.

119 291 N.J. Super. 619 at 625–6.

120 291 N.J. Super. 619 at 627.

121 A choice of law issue was subsequently considered by the Superior Court: *Safer* v. *Estate of Pack* 314 N.J. Super. 496; 715 A. 2d 363; 1998 N.J. Super. LEXIS 379.

122 For example, *Home Office* v. *Dorset Yacht Co. Ltd.* [1970] A.C. 1004, *Hill* v. *Chief Constable of West Yorkshire* [1989] A.C. 53, *Sutherland Shire Council* v. *Heyman* (1985) 157 C.L.R. 424.

123 551 P. 2d 334 (1976) at 342–3.

Heyman[124] reaffirmed the general principle that the common law does not impose a prima facie duty to rescue another, or safeguard from or warn another of a reasonably foreseeable injury. That prima facie presumption may be rebutted where the defendant had a prior duty to do the act or prevent the ensuing injury. Such a duty of positive action could potentially arise in two ways—by statute or by antecedent conduct of the defendant that creates, or increases the risk of, injury to the victim; or where the victim relies upon the defendant to prevent an injury to him or her, and the defendant knows or ought to know of such reliance. Especially where a genetic relative is a patient of the *same* doctor, reliance of this kind may be present.

Furthermore, there is some recognition in Australian law that a duty of care may be owed to someone who is not a patient of the doctor's. In *Lowns* v. *Woods*[125] it was alleged that a medical practitioner had failed to provide medical assistance when requested to a boy who was suffering from an epileptic fit. Although the boy was not a patient of the doctor, a duty of care was nonetheless found to exist on the basis that, in the circumstances, a relationship of proximity existed between the doctor and the boy.[126] In reaching their decision, the majority justices also took into account the statutory obligation imposed on medical practitioners under the then Medical Practitioners Act 1938.[127] More recently, in *BT* v. *Oei*[128] the New South Wales Supreme Court also held that a duty of care could be owed to a third party. The Court held that a medical practitioner owed a duty of care to the plaintiff, who was the sexual partner of a patient of the doctor's. The plaintiff contracted HIV from her partner and Justice Bell held that the medical practitioner had breached his duty of care to the plaintiff by failing to properly advise the patient in relation to the need for an HIV test. It is important to note

[124] (1985) 157 C.L.R. 424.

[125] (1996) Aust. Torts Reports 81–376.

[126] See for example, the judgment of Cole J.A., finding physical, circumstantial and causal proximity between the doctor and the plaintiff: (1996) Aust. Torts Reports 81–376 at 63,176. It should be noted in *Perre* v. *Apand Pty. Ltd.* (1999) 198 C.L.R. 180, a pure economic loss case, the High Court of Australia reduced the centrality of proximity in the determination of whether a duty of care exists.

[127] As it applied at the relevant time, s. 27(2) provided it was 'misconduct in a professional respect' for a medical practitioner to '(c) refuse or fail, without reasonable cause, to attend, within a reasonable time after being requested to do so, upon a person for the purpose of rendering professional services in his capacity as a registered medical practitioner in any case where he has reasonable cause to believe that such person is in need of urgent attention by a registered medical practitioner but shall not be guilty if he causes another registered medical practitioner to attend as aforesaid'.

[128] [1999] N.S.W.S.C. 1082 (5 November 1999). For discussion of the case see D. Hirsch, 'Doctor's Duty of Care to a Patient's Sexual Partners' (1999) 8 *Australian Health Law Bulletin* 53.

that there was no duty to advise the patient's partner directly. Rather the doctor's duty could be discharged through the provision of appropriate advice to the patient. The Court noted section 12(1) of the Public Health Act 1991 (NSW), which requires a medical practitioner who believes on reasonable grounds that his or her patient is suffering from a sexually transmissible condition to provide the patient with such information as is required by the Regulations made under that Act.

These cases may, of course, be distinguished from cases involving a duty *to warn* the plaintiff. Indeed, in *BT* Justice Bell observed: 'There is no suggestion that the obligation on the doctor extends beyond the provision of adequate advice to the patient.'[129] And one of the bases upon which her Honour found that the doctor owed the third party a duty of care was that there 'is no conflict between the duty owed by the defendant to [the patient] and [the third party] as the two are coincident'.[130]

VI. THE NATURE OF GENETIC MEDICAL INFORMATION AND THE 'RIGHT NOT TO KNOW'

In this section, we put to one side the distinction between a right/entitlement of doctors, and a duty imposed on doctors, to disclose confidential information to third parties, and analyse an issue which is relevant to both contexts—considerations arising from the nature of genetic medical information. In *Safer* v. *Pack*, the New Jersey court claimed that 'there is no essential difference between the type of genetic threat at issue here and the menace of infection, contagion or a threat of physical harm',[131] an assertion which needs to be interrogated more closely. Graeme Laurie makes four observations in his consideration of whether health professionals should be subject to a *Tarasoff*-type duty in the context of genetics.[132] First, he observes that it can be difficult to predict the likelihood of harm to relatives from genetic conditions. 'With diseases such as Huntington's Disease there can be a one in two chance that first degree relatives will be affected, but with multifactorial conditions this probability can drop considerably.'[133] Secondly, Laurie

[129] [1999] N.S.W.S.C. 1082 (5 November 1999), para. 97.

[130] [1999] N.S.W.S.C. 1082 (5 November 1999), para. 98.

[131] 291 N.J. Super. 619 at 625–6.

[132] G.T. Laurie, 'The Most Personal Information of All: An Appraisal of Genetic Privacy in the Shadow of the Human Genome Project' (1996) 10 *International Journal of Law, Policy and the Family* 74.

[133] *Ibid.* at 82. In the words of the report of the Senate Legal and Constitutional Legislation Committee, 'Most personal genetic information deals in probabilities, not certainties', *supra.*, n. 84 at para 3.7; see also R. Chadwick, 'The Philosophy of the Right to Know and the Right Not to Know', *supra.*, n. 81 at 14. On the other hand, the majority justices in *Tarasoff* acknowledged that the judgement of a therapist in diag-

argues that it is necessary to consider the kind of harm that will justify imposition of *Tarasoff*-type duties. As he notes, while some genetic conditions will lead to death, 'with many more the most likely way in which relatives will be affected, if at all, is that they will be carriers of the condition'.[134] If relatives have no adverse effect on their health, other than their carrier status, and the only risk of harm is to any future progeny, is such harm sufficient to justify the imposition of a duty of care?[135] Thirdly, it is necessary to consider whether harm can be averted by the breach of confidentiality. 'With genetics, given the very limited number of conditions for which there is a cure, should a duty be imposed if nothing can be done to prevent the onset of disease?'[136] Finally, in *Tarasoff* the therapist's relationship with the patient, and the therapist's knowledge of the risk of harm the patient might pose to the third party, were relevant considerations in finding that a duty of care existed. Yet, as Laurie and others have pointed out, in the context of genetics, one person does not pose a risk to another simply because they are a relative.[137] As Sonia Suter argues:

> There are clear differences . . . between the risks in failing to warn about a contagious disease or a violent person and failing to warn about disease genes. Someone with a contagious disease or with violent propensities can harm others by exposing them to infection or by assaulting them, respectively. Yet individuals with disease genes do not put relatives at risk by carrying the gene. Their relatives have no risk of becoming carriers; they only have the risk of finding out that they are carriers.[138]

If one considers that a public interest in disclosure is to be weighed against a public interest in maintaining confidentiality, what are the public interests that could be jeopardised through a failure to disclose genetic information?[139] As Laurie asks, 'can harm be prevented by disclosure in the context of genetics?'[140] The existence of a cure or treat-

nosing mental disorders and in predicting whether a patient presents a serious danger of violence involves some uncertainty, and that while a 'perfect' or infallible judgement could not be made, the courts would still expect such a professional judgement to be made and for the therapist to take steps to warn the person or persons at risk in the event that a determination is made (or should be made) that the patient poses a serious danger (551 P. 2d 334 (1976) at 345).

[134] Laurie, *supra.*, n. 132 at 82.

[135] *Ibid.* at 82.

[136] *Ibid.* at 82.

[137] *Ibid.* at 82–3.

[138] S.M. Suter, 'Whose Genes Are These Anyway? Familial Conflicts Over Access to Genetic Information' (1993) 91 *Michigan Law Review* 1854 at 1881.

[139] Laurie, *supra.*, n. 132 at 85.

[140] *Ibid.* at 85.

ment for the condition, the likelihood of harm to relatives, and the degree of severity if the condition does occur are all relevant factors which would need to be considered.[141] There are very few genetic conditions for which a cure exists. Laurie notes that one could still argue in favour of disclosure even in the absence of a complete cure because information could provide relatives with the opportunity to change their lifestyle so as to minimise their risk of the disease occurring.[142] However, given that many genetic diseases are multifactorial, test results from relatives may have limited reliability thus limiting the benefits from disclosure.[143]

Alissa Brownrigg argues that, given the lack of consensus about disclosure of genetic information about patients to family members of that patient, a 'balancing' approach should be adopted.[144] She observes that the courts in *Pate* v. *Threlkel* and *Safer* v. *Pack* did not address a number of factors which will be critical in future cases, including the accuracy of the test performed and the ability of the doctor to interpret its results. Accordingly, she advocates a 'balancing' approach which involves weighing five factors:

1. the severity of the disease identified by testing;
2. the availability of preventive or curative options for that disease;
3. the accuracy and reliability of the test performed;
4. the ability of the physician or health care provider to interpret and address issues relevant to the test performed; and lastly
5. the protections afforded to the tested individual against discrimination.[145]

Moreover, some account needs to be taken of a very important aspect of genetic information within families. While the Anti-Cancer Council of Victoria and Skene focus on the 'right' of family members to know about genetic information derived from a patient, the interests of such family members (perhaps giving rise to a right) in *not* knowing such information must also be considered. The status of this concept in ethics and law is currently uncertain.[146] Graeme Laurie has argued that a right 'not to know' can be derived from the concept of privacy. In this context, the wishes of the individual can help to define the scope of the

---

[141] *Ibid.* at 85.
[142] *Ibid.* at 86.
[143] *Ibid.* at 86.
[144] A. Brownrigg, 'Mother Still Knows Best: Cancer-related Gene Mutations, Familial Privacy, and a Physician's Duty to Warn' (1999) 26 *Fordham Urban Law Journal* 247 at 251.
[145] *Ibid.* at 272–3.
[146] For the philosophical basis of a right not to know, see Chadwick, *supra.*, n. 81 at 17–20.

privacy interest. Laurie argues that an individual who is aware of his or her genetic risk, *e.g.* through their family history, but does not want to be informed of the details, should have those wishes respected.[147]

> Unlike a blanket refusal to receive information, the expression of a wish not to know is meaningful because such relatives are likely to be aware of the nature of the disease from which they are at risk. To seek to inform them of genetic information would be an affront to their dignity and an invasion of their privacy.[148]

However, if an individual's wishes are unknown and there is a risk of considerable harm that can be prevented by a disclosure, Laurie argues that in these circumstances 'arguably, privacy should be invaded to prevent the harm'.[149] Reconciling these competing interests will be difficult, but the response of the Anti-Cancer Council of Victoria to this issue—that it 'will be very hard to ascertain when relations would prefer [not to know], since the question can hardly be discussed with them without intimating to them precisely what they do not want to know'[150]—is not really adequate.

VII. CONCLUSION—INDIVIDUALISM AND THE FAMILY

It has been argued in this article that the call by the Victorian Anti-Cancer Council for a 'different ethical emphasis . . . an emphasis on wider responsibility and communal concern',[151] with its family-based information and obligations, represents a significant challenge to the traditional way of handling medical information about patients. However, there are at least four general criticisms which can be made of such an alternative framework. First, the dichotomy drawn between a 'legal' and a 'medical' model, for the purposes of favourably promoting the latter, is over-simplistic. If either model was adopted in Australia, it would ultimately need to be expressed in legal form, notwithstanding that the so-called 'medical' model would arguably involve a departure from existing legal approaches to medical information. Moreover, the ascription of the label 'medical' obscures the fact that this approach is no less bound to a particular set of values and politics (specifically those of communitarianism) than the so-called 'legal' approach. That is, the debate should really be seen in terms of liberalism versus communitarianism, and the various forms these ideologies may take in law and

---

[147] Laurie, *supra.*, n. 132 at 91.
[148] *Ibid.* at 91.
[149] *Ibid.* at 91.
[150] Anti-Cancer Council of Victoria, *supra.*, n. 11 at 42.
[151] *Ibid.* at 38.

ethics, rather than between 'law' and 'medicine'. This paper has argued that current legal doctrines should be labelled 'individualistic' or 'rights'-based very cautiously, and with considerable qualification. The law of confidentiality does indeed presumptively accord confidential status to an individual's medical information. But this does not automatically entail more extensive rights—for example, a right of access to, or ownership of, that information. And neither is the interest in confidentiality absolute (a point recognised in the National Health and Medical Research Council information paper on *Ethical Aspects of Human Genetic Testing*).

Secondly, the Victorian report appears to be motivated by the quite reasonable concern that a model that leaves decision making in the hands of individuals could result in other individuals suffering harm. Yet this assumption appears to rest on highly individualised and atomised notions of autonomy, which fail to take account of the relational aspects of the exercise of autonomy. While autonomy can be, and generally is, conceptualised in individualised terms, autonomy can also be understood in relational terms that explicitly take account of an individual's relationships with others.[152] As Jennifer Nedelsky argues: 'Autonomy is a capacity that exists only in the context of social relations that support it and only in conjunction with the internal sense of being autonomous.'[153] Leslie Bender has made a similar point when she argues that 'Self-governing in an ethic of care does not mean governing alone by abstract reasoning and distant observations, but means choosing options with respect to responsibilities, relationships, conversations, and dialogues with others.'[154] Given this relational dimension to autonomy, there is no reason to think that maintaining individual rights will potentially create problems for others. Indeed, it is precisely because autonomy is so often exercised relationally that most individuals do share information from genetic tests with other family members.

Thirdly, it is not clear that genetic information is sufficiently different from other medical information to justify the development of an alternative legal framework for that information. As this article has shown, the law protects confidentiality with exceptions provided for certain circumstances. Furthermore, where a patient poses a risk to another individual the courts have accepted that limited disclosures of confidential information may be justified in order to avert that harm. In

152 For discussion see J. Nedelsky, 'Reconceiving Autonomy: Sources, Thoughts and Possibilities' (1989) 1 *Yale Journal of Law & Feminism* 7; B. Bennett, 'Posthumous Reproduction and the Meanings of Autonomy' (1999) 23 *Melbourne University Law Review* 286 at 300–1.

153 Nedelsky *ibid.* at 25.

154 L. Bender, 'A Feminist Analysis of Physician-Assisted Dying and Voluntary Active Euthanasia' (1992) 59 *Tennessee Law Review* 519 at 537.

other words, if the concern is that patients might not advise their relatives of their genetic risk, and that relatives may suffer as a result, the existing law of confidentiality arguably already provides a framework for disclosure to be permitted if the health of another is at risk. In the absence of a duty to warn being recognised in Australian law the legal position of a doctor who took it upon him or herself to disclose genetic information to a patient's relatives, without the patient's consent, is uncertain. What this discussion of the law suggests is that the interaction of the law relating to confidentiality, and the common law tort of negligence, combine to allow doctors to disclose limited information in appropriate circumstances (that is, where there is a real risk of harm to a family member who acts in reliance on the doctor). This approach arguably has much to commend it, and whilst it may be very generally characterised as 'individualistic', the discussion above has suggested that this is an oversimplified characterisation.

Finally, we have concerns with the concept of 'family' which underlies the Victorian report. The report appears to rest on a biological model of the family, in which longstanding assumptions about the sensitivity of medical information are overturned on the basis of the purportedly 'shared' nature of genetic material.[155] The use of this biological model to justify differential ethical and legal obligations is dubious, and is arguably a continuation of the stream of Western political thought which has postulated a dichotomy between the 'private' (the family) and 'public' (the realms of the political and the market) in which each was distinct and separate from the other.[156] This tradition has come under sustained attack in recent years from feminist scholarship. As Martha Fineman has observed:

> The metaphor of 'symbiosis' seems more appropriate than the separate spheres imagery: the family is located within the state. In this conceptualization, family and state are interactive; they define one another . . . If this model of family-state relationship is accurate, it has important implications for public policy. First, it indicates that the relationship between family and state is not fixed—it is potentially dynamic. Second, it illustrates that the family is not a natural entity with a form that is constant and essential—it is a societal creation.[157]

---

[155] Anti-Cancer Council of Victoria: 'It is as *members of families* that they are at risk, and because of a family history which they share with many others that they may end up having a genetic test': *supra.*, n. 11 at 41.

[156] S. Moller Okin, *Justice, Gender, and the Family* (Basic Books 1989) at 8; M.A. Fineman, 'What Place for Family Privacy?' (1999) 67 *George Washington Law Review* 1207 at 1207.

[157] Fineman, *ibid.* at 1209, references omitted.

What this means is that the family cannot be *assumed* to be any more natural or any less contested than the public or political realm. But it is precisely such an assumption which appears to be behind arguments that confidentiality of genetic medical information ought to be different in the case of family members. The question must be asked: what makes biological families different from other social contexts? The answer provided is that it is because family members *share* the genetic information that disclosure is compelled. In other words, genetically-related family members have a special relationship with each other *vis-à-vis* their shared genetic information which overrides other considerations (in our case both individual and broader social interests in very sensitive medical information being kept confidential). But what this ignores is that to say that the information is 'shared' is not a statement of biological fact (and therefore self-evident), but rather a social and political judgement. It is a social and political judgment not just about the nature of genetic information, but also about families. And it is a judgement which reiterates the construction of the family as separate from, and subject to different standards of justice from, the wider realms of society and of the market. There may be very good reasons for treating families differently from other social units or associations, but those reasons must at the very least be articulated and justifiable. The communitarian approach postulates only genetic connection as the basis of difference. Accordingly, not only does it risk participating in an ideology which views the families as essentialistic, but it may conjoin this discourse with one which privileges biological, and specifically *genetic*, connections between people—what has been called an ideology of geneticism.[158] In addition, within contemporary society the family itself is a contested institution. Climbing divorce rates, demands for the recognition of gay and lesbian families, and the impact of the new reproductive technologies[159] have all played a part in reshaping the modern family and in ensuring that the biological and social frameworks traditionally associated with the family are increasingly uncertain.

For these reasons, we have reservations about the 'communitarian'

---

[158] For good accounts of the role of the ideology of 'geneticism' or 'genetic determinism' in creating social understanding of genetics, see R.C. Lewontin, *The Doctrine of DNA: Biology as Ideology* (Penguin 1993), and D. Nelkin and M.S. Lindee, *The DNA Mystique: The Gene as a Cultural Icon* (W.H. Freeman 1995).

[159] As Janet Dolgin has argued: 'Questions about the meaning of "mother", "father", and "child", occasioned by reproductive technology, arise in a world already unsure of the meaning and contours of family. Traditional assumptions about the character, longevity, and membership of the family, as well as assumptions about the connection between the family as a social unit and the biological facts long thought to produce and undergird that unit, are in disarray.' (J.L. Dolgin, 'The Law Debates the Family: Reproductive Transformations' (1995) 7 *Yale Journal of Law & Feminism* 37 at 38.)

project in familial genetics. As we have sought to demonstrate, existing legal rules dealing with disclosure of genetic information to genetic relatives of patients provide for a complex and variegated response to a very difficult issue. The law in relation to confidential information currently contemplates a range of situations in which confidential information may be disclosed (both under statute law as well as common law). There may well be, and arguably probably are, situations involving familial genetic information which would satisfy the requirements for disclosure by a doctor. But such a disclosure would not be subject to a blanket rule: it would depend on a range of factors specific to the case at hand, including the nature of the genetic condition and the precise genetic mutation (itself subject to change over time as the Human Genome Project and related research provides more accurate information about the health risks posed by particular mutations), knowledge which the doctor has about the family member, and so on. It is much less certain whether Australian courts would ever impose a duty of disclosure on a doctor, especially given doubts about whether disclosure of a genetic condition would meet the much more stringent requirements that would inevitably be imposed by the courts as the condition for recognising such a duty.

[9]

Genetic Privacy

Lawrence O. Gostin

Human genomic information is invested with enormous power in a scientifically motivated society. Genomic information has the capacity to produce a great deal of good for society. It can help identify and understand the etiology and pathophysiology of disease. In so doing, medicine and science can expand the ability to prevent and ameliorate human malady through genetic testing, treatment, and reproductive counseling.

Genomic information can just as powerfully serve less beneficent ends. Information can be used to discover deeply personal attributes of an individual's life. That information can be used to invade a person's private sphere, to alter a person's sense of self- and family identity, and to affect adversely opportunities in education, employment, and insurance.[1] Genomic information can also affect families and ethnic groups that share genetic similarities.

It is sometimes assumed that significant levels of privacy can coexist with widespread collection of genomic information. Understandably, we want to advance all valid interests—both collective and individual. We want to believe that we can continue to acquire and use voluminous data from the human genome while also protecting individual, family, and group privacy. This article demonstrates that no such easy resolution of the conflict between the need for genomic information and the need for privacy exists. Because absolute privacy cannot realistically be achieved while collecting genetic data, we confront a hard choice: Should we sharply limit the systematic collection of genomic information to achieve reasonable levels of privacy? Or, is the value of genomic information so important to the achievement of societal aspirations for health that the law ought not promise absolute or even significant levels of privacy, but rather that data be collected and used in orderly and just ways, consistent with the values of individuals and communities? As I argue, the law at present neither adequately protects privacy nor ensures fair information practices. Moreover, the substantial variability in the law probably impedes the development of an effective genetic information system.

In earlier articles, I scrutinized the meaning and boundaries of health information privacy.[2] Here, I build on that work by examining a particular aspect of health information—genetic privacy. I acknowledge a debt to those scholars who have aptly identified and wrestled with the difficult ethical and legal issues inherent in genomic information.[3] This is well-tread territory; what I hope to bring to the literature is a conceptual structure relating to the acquisition and use of genomic information. First, the methods of collection and use of genomic data must be understood and its public purposes evaluated. Second, the privacy implications of genomic information must be measured. To what extent are genomic data the same as, or different from, other health information? Third, an examination of the current constitutional and statutory law must be undertaken to determine whether existing safeguards are adequate to protect the privacy and security of genomic data. Finally, proposals for balancing societal needs for genomic information and claims for privacy by individuals and families must be generated.

Genetic information infrastructure

I define the *genetic information infrastructure* as the basic, underlying framework of collection, storage, use, and transmission of genomic information (including human tissue and extracted DNA) to support all essential functions in genetic research, diagnosis, treatment, and reproductive counseling. Despite the technical problems and the cost, several governmental[4] and private[5] committees have proposed automation of health data, including genomic information. Several conceptual and technological innovations are likely to accelerate the automation of health records: patient-based longitudinal clinical records, which include genetic testing and screening information; unique identifiers and the potential to link genomic information to identifiable persons; and genetic data bases for clinical, research, and public health purposes.

Longitudinal clinical records: testing and screening

The health care system is moving toward patient-based longitudinal health records. These records, held in electronic form, contain all data relevant to the individual's health collected over a lifetime. What is foreseen is a single record for every person in the United States, continually expanded from prebirth to death, and accessible to a wide range of individuals and institutions.[6]

Genetic testing and screening are likely to become an important part of longitudinal clinical records. The principal forms include: fetal (prenatal), newborn, carrier, and clinical (primary care) screening.[7] Prenatal screening seeks to identify disease in the fetus. Prenatal diagnosis of birth defects often involves genetic analysis of amniotic fluid, blood, or other tissues. Prenatal diagnostic methods are used for genetic diseases including Down syndrome, Tay-Sachs, sickle cell, and thalassemia major (Cooley's anemia). Newborn screening often focuses on detection of inborn errors of metabolism. Phenylketonuria (PKU) was the first condition subject to newborn screening; other inborn defects often screened at birth are galactosemia, branched-chain ketonuria, and homocystinuria.[8] Carrier screening seeks to identify heterozygotes for genes for recessive disease. Carrier testing has been used for such conditions as Tay-Sachs, cystic fibrosis (CF), and sickle cell.

The Human Genome Initiative has advanced to the point where it is now possible to conceive of an ever-expanding ability to detect genetic causes of diseases in individuals and populations. Testing for predispositions to disease represents one of the most important developments. For example, testing for predispositions to Huntington's disease, colon cancer, heart disease, and Alzheimer's disease are currently possible or expected.[9] Relatively recent discoveries include genes found for ataxia-telangiectasia (a rare hereditary neurological disorder of childhood),[10] Lowe syndrome (a rare X-linked disorder affecting diverse organ systems),[11] melanoma, pancreatic cancer,[12] and breast cancer.[13] Genetic methods to identify elevated risk for multifactorial diseases are also likely. It may be possible, for example, to identify individuals at risk for such conditions as schizophrenia, manic depression, and alcohol or drug dependency.

Clinical records could potentially be linked to many other sources of genomic information: (i) a lucrative commercial market in self-testing, which is growing even before scientists regard test-kits as reliable (for example, testing for genetic predictors of breast cancer);[14] (ii) workplace screening, through which employers can determine an employee's current and future capacity to perform a job or to burden pension or health care benefit plans[15] (such testing may occur despite some legal restrictions under disability discrimination statutes[16]); (iii) screening to determine eligibility for health, life, and disability insurance, which is likely when tests are more cost-effective;[17] (iv) testing in the criminal justice system, which will increase as more courts recognize the probative value of genomic data;[18] and (v) testing for a wide variety of public purposes (for instance, to prevent fraud in collection of welfare or other social benefits, to identify family ties in adoption, and to adjudicate paternity suits).[19] Automated health information systems hold the capacity electronically to link information collected for these and other purposes. Data from several sources can be compared and matched; and different configurations of data can reveal new understandings about the individual.

It is thus possible to conceive of a genetic information system that contains a robust account of the past, present, and future health of each individual, ranging from genetic fetal abnormalities and neonate carrier states, to current and future genetic conditions at different points in one's life. Genetic data can even explain causes of morbidity and mortality after death; for example, genetic technologies were used to determine whether Abraham Lincoln had Marfan's disease.[20] As will become apparent below, such genetic explanations of morbidity and mortality provide an expansive understanding of the attributes not only of the individual, but also of her family (ancestors as well as current and future generations) and possibly of whole populations.

Unique identifiers and potential links to identifiable persons

Health data can be collected and stored in identifiable or nonidentifiable forms. Data raise different levels of privacy concerns, depending on whether they can be linked to a specific person. The most serious privacy concerns are raised where genomic data are directly linked to a known individual. For reasons of efficiency, many health plans in the private and public sector are considering the use of unique identifiers. These identifiers would be used for a variety of health, administrative, financial, statistical, and research purposes. The identifier would facilitate access to care and reimbursement for services rendered. Some envisage using the social security number (SSN) as the unique identifier, which is controversial because the SSN is linked

Volume 23:4, Winter 1995

to data from the Internal Revenue Service, Department of Defense, debt collectors, the Medical Information Bureau, credit care companies, and so forth.

Where data are collected or held in nonidentifiable form, they pose few problems of privacy. Because anonymous data are not personally linked, they cannot reveal intimate information that affect individual privacy rights. Epidemiological data, including health statistics, are frequently collected in this form. This enables investigators or public health personnel to collect a great deal of information, usually without measurable burdens on privacy interests. The obvious question arises whether genomic data can also be collected in nonidentifiable form. Genomic data that are not linked to identified individuals can significantly reduce, but do not eliminate, privacy concerns. Genomic data are qualitatively different from other health data because they are inherently linked to one person. While nongenetic descriptions of any given patient's disease and treatment could apply to many other individuals, genomic data are unique. But, although the ability to identify a named individual in a large population simply from genetic material is unlikely, the capacity of computers to search multiple data bases provides a potential for linking genomic information to that person. It follows that nonlinked genomic data do not assure anonymity and that privacy and security safeguards must attach to any form of genetic material. It is, therefore, a concern that even the strict genetic privacy statutes that have been introduced in Congress exempt "personal genetic records maintained anonymously for research purposes only."[21] Minimally, such statutes must require that privacy and security arrangements ensure that these "anonymous" data are never linked to identified persons.

Genetic data bases

Data bases collect, store, use, and transfer vast amounts of health information, often in electronic or automated form. The technology exists to transfer data among data bases, to match and reconfigure information, and to seek identifying characteristics of individuals and populations. Data bases hold information on numerous subjects including medical cost reimbursements, hospital discharges, health status, research, and specific diseases.[22] A growing number of data bases also contain genetic information.[23] Genetic research usually requires only DNA, sources of which include not only solid tissues, but also blood, saliva, and any other nucleated cells.[24] Reilly defines DNA banking as "the long-term storage of cells, transformed cell lines, or extracted DNA for subsequent retrieval and analysis"; it is "the indefinite storage of information derived from DNA analysis, such as linkage profiles of persons at risk for Huntington Disease or identity profiles based on analysis with a set of probes and enzymes."[25]

Genetic data bases are held in both the private and public sector for clinical, research, and public health purposes. The National Institutes of Health (NIH), for example, maintains a genetic data base for cancer research, while private universities, such as the University of Utah human tissue repository, conduct genetic research. Commercial companies offer genetic banking as a service to researchers or individuals.[26] Genetic data bases are also created to support nonhealth-related functions, such as identification of the remains of soldiers,[27] detection, prosecution, and post-conviction supervision through "DNA fingerprinting" of persons engaging in criminal conduct,[28] and identification of blood lines in paternity and child disputes.[29]

One problematic source of information is previously stored tissue samples. Stored samples may be regarded as inchoate data bases because the technology exists to extract from them considerable current and future health data.[30] The public health and research communities have shown increasing interest in using existing tissue samples for genetic testing and for creating new genetic data bases. From a privacy perspective, this interest raises a serious problem: any consent that was obtained when that tissue was originally extracted would not meet current informed consent standards because the donor could not have envisaged future genetic applications.

The most prominent example of an inchoate genetic data base is the Guthrie spot program, whereby dried blood spots are taken from virtually all newborns throughout the United States. All states screen newborns for PKU, congenital hyperthyroidism, and other genetic defects. The genetic composition of Guthrie spots remains stable for many years and, if frozen, can be held indefinitely. A recent survey found that three-quarters of the states store their Guthrie cards, with thirteen storing them for more than five years. Of them, several store these cards indefinitely; and a number of other states have expressed an intention to do so.[31] Only two require parental consent for the blood spot.

Perhaps the most ambitious public or private effort to create a data base with both genetic and nongenetic applications is the National Health and Nutrition Examination Survey (NHANES) conducted by several federal agencies.[32] NHANES has collected comprehensive health status data in patient-identifiable form on some 40,000 Americans in eighty-one counties in twenty-six states. About 500 pieces of data are collected from each subject, ranging from sociodemographics, diet, bone density, and blood pressure, to risk status, drug use, and sexually transmitted diseases (STDs). Additionally, NHANES tests and stores biological samples for long-term follow-up and statistical research.

NHANES provides a classic illustration of a massive collection of highly personal and sensitive information that has enduring societal importance. These data pose a significant risk of privacy invasion, but they are critical to understanding health problems in the population.

Clinical and public health benefits of genomic information

Americans seem enamored with the power of genomic information. It is often thought capable of explaining much that is human: personality, intelligence, appearance, behavior, and health.[33] Genetic technologies generated from scientific assessment are commonly believed always to be accurate and highly predictive. These beliefs are highly exaggerated; for instance, personal attributes are influenced by social, behavioral, and environmental factors.

A person's genetic diary, moreover, is highly complex, with infinite possibilities of genetic influence. Ample evidence exists that the results of genetic-based diagnosis and prognosis are uncertain. The sensitivity of genetic testing is limited by the known mutations in a target population. For example, screening can detect only 75 percent of CF chromosomes in the U.S. population. Approximately one of every two couples from the general population identified by CF screening as "at-risk" will be falsely labeled.[34] Predicting the nature, severity, and course of disease based on a genetic marker is an additional difficulty. For most genetic diseases, the onset date, severity of symptoms, and efficacy of treatment and management vary greatly.

Nonetheless, the force of genomic information, even if exaggerated, is powerful. Genomic information is highly beneficial for health care decisions regarding prevention, treatment, diet, lifestyle, and reproductive choices. In particular, collection of genomic data can provide the following benefits to individuals and to society.

Enhanced patient choice. Genetic testing can enhance autonomous decision making by providing patients with better information. Genomic data, for example, can provide information about carrier states, enabling couples to make more informed reproductive choices; about disabilities of the fetus, guiding decisions about abortion or fetal treatment; about markers for future disease, informing lifestyle decisions; and about current health status, providing greater options for early treatment. Some may not agree that genetic information used for these purposes is inherently good, for the information could be used to increase selective abortion to "prevent" the births of babies with genetic disabilities.

Clinical benefit. Often a disconnection exists between the ability of science to detect disease and its ability to prevent, treat, or cure it. Scientific achievement in identifying genetic causes of disease must be tempered by a hard look at scientifically possible methods of intervention. As discussed below, if the possible stigma or discrimination associated with the disease is great, and science remains powerless to prevent or treat it, the potential benefits may outweigh harms. Despite this caveat, the Human Genome Initiative holds the current or potential ability to achieve a great deal of good for patients.

Couples can decide to change their plans for reproduction based on information disclosed in genetic counseling, thus reducing the chance of a child born with disease. Detection of metabolic abnormalities can empower a person to control their diet and lifestyle to prevent the onset of symptomatology. Identification of enhanced risk for multifactorial diseases, such as certain cancers or mental illness, could help people avoid exposure to particular occupational or environmental toxins or stresses.[35] Finally, medicine is increasing its ability to treat genetic conditions. Wivel and Walters discuss several categories of human genetic intervention: somatic cell gene therapy involving correction of genetic defects in any human cells except germ or reproductive cells; germ-line modification involving correction or prevention of genetic deficiencies through the transfer of properly functioning genes into reproductive cells; and use of somatic and/or germ-line modifications to effect selected physical and mental characteristics, with the aim of influencing such features as physical appearance or physical abilities (in the patient or in succeeding generations).[36] While use of germ-line therapy, particularly when designed to enhance human capability, is highly charged, most people agree that the ability to prevent and treat genetic disease offers patients a chance for health and well-being that would not be possible absent genetic intervention. Clinical applications of genetic technologies are also possible in other areas; for example, scientists have reported progress in transplanting animal organs into humans. Insertion of human genes into animals could render their organs more suitable for transplantation into humans without substantial tissue rejection.[37]

Improved research. Despite substantial progress in the Human Genome Initiative, a great deal more must be understood about the detection, prevention, and treatment of genetic disease. Genetic research holds the potential for improving diagnosis, counseling, and treatment for persons with genetic conditions or traits. Research can help determine the frequency and distribution of genetic traits in various populations, the interconnections between genotypes and phenotypes, and the safety and efficacy of various genetic interventions.

Genetic data bases, containing DNA and/or stored tissue, could make this kind of research less expensive by reducing the costs of collecting and analyzing data, more trustworthy by increasing the accuracy of the data, and more generalizable to segments of the population by assuring the completeness of the data.

Protection of public health. While traditional genetic diagnosis, treatment, and research is oriented toward the individual patient, genetic applications can also benefit the public health. There is considerable utility in using population-based data to promote community health. Genomic data can help track the incidence, patterns, and trends of genetic carrier states or disease in populations. Carefully planned surveillance or epidemiological activities facilitate

Volume 23:4, Winter 1995

rapid identification of health needs. This permits reproductive counseling, testing, health education, and treatment resources to be better targeted, and points the way for future research. For example, recent epidemiological research of DNA samples from Eastern European Jewish women found that nearly 1 percent contained a specific gene mutation that may predispose them to breast and ovarian cancer. This finding offered the first evidence from a large study that an alteration in the gene, BRCA1, is present at measurable levels not only in families at high risk for disease, but also in a specific group of the general population.[38] Certainly, evidence of enhanced risk of disease in certain populations, such as sickle cell in African Americans or Tay-Sachs in Ashkenazi Jews, may foster discrimination against these groups. At the same time, population-based genetic findings support other clinical studies to evaluate the risk to populations bearing the mutation or to determine whether BRCA1 testing should be offered to particular ethnic groups as part of their routine health care.

Privacy implications of genomic data

The vision of a comprehensive genetic information system described above is technologically feasible, and a well-functioning system would likely achieve significant benefits for individuals, families, and populations. However, to decide whether to continue to accumulate vast amounts of genomic information, it is necessary to measure the probable effects on the privacy of these groups. The diminution in privacy entailed in genetic information systems depends on the sensitive nature of the data, as well as on the safeguards against unauthorized disclosure of the information.

Genomic data and harms of disclosure

Privacy is not simply the almost inexhaustible opportunities for access to data; it is also the intimate nature of those data and the potential harm to persons whose privacy is violated.[39] Health records contain much information with multiple uses: demographic information; financial information; information about disabilities, special needs, and other eligibility criteria for government benefits; and medical information. This information is frequently sufficient to provide a detailed profile of the individual and that person's family. Traditional medical records, moreover, are only a subset of records containing personal information held by social services, immigration, and law enforcement.

Genomic data can personally identify an individual and his/her parents, siblings, and children, and provide a current and future health profile with far more scientific accuracy than other health data. The features of a person revealed by genetic information are fixed—unchanging and unchangeable. Although some genomic data contain information that is presently indecipherable, they may be unlocked by new scientific understanding; but such discoveries could raise questions about improper usage of stored DNA samples.[40] Finally, societies have previously sought to control the gene pool through eugenics. This practice is particularly worrisome because different genetic characteristics occur with different frequencies in racial and ethnic populations.

The combination of emerging computer and genetic technologies poses particularly compelling privacy concerns. Scientists have the capacity to store a million DNA fragments on one silicon microchip.[41] While this technology can markedly facilitate research, screening, and treatment of genetic conditions, it may also permit a significant reduction in privacy through its capacity to store and decipher unimaginable quantities of highly sensitive data.

A variety of underlying harms to patients may result from unwanted disclosures of these sensitive genomic data. A breach of privacy can result in economic harms, such as loss of employment, insurance, or housing. It can also result in social or psychological harms. Disclosure of some conditions can be stigmatizing, and can cause embarrassment, social isolation, and a loss of self-esteem. These risks are especially great when the perceived causes of the health condition include drug or alcohol dependency, mental illness, mental retardation, obesity, or other genetically linked conditions revealed by a person's DNA. Even though genomic information can be unreliable or extraordinarily complicated to decipher, particularly with multifactorial disease or other complicated personal characteristics (for instance, intelligence), public perceptions attribute great weight to genetic findings and simply aggravate the potential stigma and discrimination.

Maintaining reasonable levels of privacy is essential to the effective functioning of the health and public health systems. Patients are less likely to divulge sensitive information to health professionals, such as family histories, if they are not assured that their confidences will be respected. The consequence of incomplete information is that patients may not receive adequate diagnosis and treatment. Persons at risk of genetic disease may not come forward for the testing, counseling, or treatment. Informational privacy, therefore, not only protects patients' social and economic interests, but also their health and the health of their families and discrete populations.

Legal protection of genetic privacy and security of health information

One method of affording some measure of privacy protection is to furnish rigorous legal safeguards. Current legal safeguards are inadequate, fragmented, and inconsistent, and contain major gaps in coverage. Significant theoretical problems also exist.

The Journal of Law, Medicine & Ethics

Constitutional right to privacy

A considerable literature has emerged on the existence and extent of a constitutional right to informational privacy independent of the Fourth Amendment prohibition on unreasonable searches and seizures.[42] To some, judicial recognition of a constitutional right to informational privacy is particularly important because the government is an important collector and disseminator of information. Citizens, it is argued, should not have to rely on government to protect their privacy interests. Rather, individuals need protection from government itself, and an effective constitutional remedy is the surest method to prevent unauthorized government acquisition or disclosure of personal information. The problem with this approach is that the Constitution does not expressly provide a right to privacy, and the Supreme Court has curtailed constitutional protection both for decisional and informational privacy.[43]

Notwithstanding the Court's current retreat, a body of case law does suggest judicial recognition of a limited right to informational privacy as a liberty interest within the Fifth and Fourteenth Amendments to the Constitution. In *Whalen v. Roe*,[44] the Supreme Court squarely faced the question of whether the constitutional right to privacy encompasses the collection, storage, and dissemination of health information in government data banks. In *dicta*, the Court acknowledged "the threat to privacy implicit in the accumulation of vast amounts of personal information in computerized data banks or other massive government files."[45] However, the Court hardly crafted an adequate constitutional remedy to meet this threat. Justice Stevens, writing for a unanimous court, simply recognized that "in some circumstances" the duty to avoid unwarranted disclosures "arguably has its roots in the Constitution."[46] The Court found no violation in *Whalen* because the state had adequate standards and procedures for protecting the privacy of sensitive medical information. Rather, it suggested deferentially that supervision of public health and other important government activities "require[s] the orderly preservation of great quantities of information, much of which is personal in character and potentially embarrassing or harmful if disclosed."[47]

Most lower courts have read *Whalen* as affording a circumscribed right to informational privacy, or have grounded the right on state constitutional provisions.[48] Courts have employed a flexible test balancing the government invasion of privacy and the strength of the government interest. For example, the Third Circuit in *United States v. Westinghouse Electric Corp.*[49] enunciated five factors to be balanced in determining the scope of the constitutional right to informational privacy: (1) the type of record and the information it contains; (2) the potential for harm in any unauthorized disclosure; (3) the injury from disclosure to the relationship in which the record was generated; (4) the adequacy of safeguards to prevent nonconsensual disclosure; and (5) the degree of need for access—that is, a recognizable public interest.

Judicial deference to government's expressed need to acquire and use information is an unmistakable theme in the case law. Provided that government articulates a valid societal purpose and employs reasonable security measures, courts have not interfered with traditional governmental activities of information collection. Unmistakably, government could enunciate a powerful societal purpose in the collection of genomic information such as public health or law enforcement.

The right to privacy under the Constitution is, of course, limited to state action. As long as the federal or a state government itself collects information or requires other entities to collect it, state action will not be a central obstacle. However, collection and use of genomic data by private or quasi-private health data organizations, health plans, researchers, and insurers remains unprotected by the Constitution, particularly in light of an absence of government regulation of genetic data banking.

Legislating health information privacy: theoretical concerns

Legislatures and agencies have designed a number of statutes and regulations to protect privacy. A full description and analysis of the legislation and regulation is undertaken elsewhere.[50] The Department of Health and Human Services described this body of legislation as "a morass of erratic law."[51] The law is fragmented, highly variable, and, at times, weak; the legislation treats some kinds of data as super-confidential, while providing virtually no protection for other kinds.

Health data are frequently protected as part of the physician-patient relationship. However, data collected in our information age is based only in small part on this relationship. Many therapeutic encounters in a managed care context are not with a primary care physician. Patients may see various nonphysician health professionals. Focusing legal protection on a single therapeutic relationship within this information environment is an anachronistic vestige of an earlier and simpler time in medicine. Moreover, the health record, as I pointed out, contains a substantial amount of information gathered from numerous primary and secondary sources. Patients' health records not only are kept in the office of a private physician or of a health plan, but also are kept by government agencies, regional health data base organizations, or information brokers. Data bases maintained in each of these settings will be collected and transmitted electronically, reconfigured, and linked.

Rules enforcing informational privacy in health care place a duty on the entity that possesses the information. Thus, the keeper of the record—whether a private physician's office, a hospital, or a hospital maintenance organi-

Volume 23:4, Winter 1995

zation—holds the primary duty to maintain the confidentiality of the data. The development of electronic health care networks permitting standardized patient-based information to flow nationwide, and perhaps worldwide, means that the current privacy protection system, which focuses on requiring the institution to protect its records, needs to be reconsidered. Our past thinking assumed a paper or automated record created and protected by the provider. We must now envision a patient-based record that anyone in the system can call up on a screen. Because location has less meaning in an electronic world, protecting privacy requires attaching protection to the health record itself, rather than to the institution that generates it.

Genetic privacy legislation

A genetic-specific privacy statute has been introduced in Congress.[52] Several states have adopted genetic-specific privacy laws,[53] and others have bills pending.[54] Eight states have provisions that prohibit obtaining and/or disclosing genomic information about individuals without their informed consent; one of these is limited to information about sickle cell testing.[55] These genetic privacy statutes are highly variable. While a few, such as California's Hereditary Disorders Act, provide privacy protection across a broad range of genomic information, most statutes have limited application. For example, privacy statutes in Maine, Massachusetts, Missouri, Ohio, Tennessee, and Virginia are applicable principally to genetic screening programs conducted by or under the auspices of the state health department. They may leave the private sector virtually unregulated in its collection and use of genomic data. Other states, like Florida, have strong, generally applicable, provisions giving persons "exclusive property" rights over genomic information, but specify broad exemptions for data collected for criminal prosecutions and determinations of paternity.

Additional statutes protect the confidentiality of genomic information, but do so with narrow purpose. Several states regulate the use of genomic data collected for insurance underwriting[56] or determinations of parentage.[57] Among the eight states that proscribe genetic discrimination in insurance, most simply require actuarial fairness and a few require confidentiality; the actuarial provisions have the effect of promoting accuracy, but little more. In Nevada, the genetic privacy statute applies only to the state university system.[58]

The adoption of a genetic-specific privacy statute at the federal or state level has been proposed.[59] A recently drafted model federal act incorporates traditional fair information practices into the collection and use of genomic data.[60] Under this model act, a person who collects human tissue for the purposes of genetic analysis must provide specific information and a notice of rights prior to collection; obtain written authorization; restrict access to DNA samples; and abide by a sample source's instructions regarding the maintenance and destruction of DNA samples.

Existing and proposed genetic-specific privacy statutes are founded on the premise that genetic information is sufficiently different from other health information to justify special treatment. Certainly, genomic data present compelling justifications for privacy protection: the sheer breath of information discoverable; the potential to unlock secrets that are currently unknown about the person; the unique quality of the information enabling certain identification of the individual; the stability of DNA rendering distant future applications possible; and the generalizability of the data to families, genetically related communities, and ethnic and racial populations.

It must also be observed that genetic-specific privacy statutes could create inconsistencies in the rules governing dissemination of health information. Under genetic-specific privacy statutes, different standards would apply to data held by the same entity, depending on whether genetic analysis had been used. The creation of strict genetic-specific standards may significantly restrain the dissemination of genomic data (even to the point of undermining legitimate health goals), while nongenomic data receive insufficient protection. Arguments that genomic data deserve special protection must reckon with the fact that other health conditions raise similar sensitivity issues (for examples, HIV infection, tuberculosis, STDs, and mental illnesses). Indeed, carving out special legal protection for sensitive data may be regarded as inherently faulty, because the desired scope of privacy encompassing a health condition varies from individual to individual. Some patients may be just as sensitive about prevalent nongenetic or multifactorial diseases like cancer and heart disease as they are about diseases with a unique genetic component. Even if it could be argued that most diseases will one day be found to be, at least in part, genetically caused, this will still raise questions about why purely viral or bacterial diseases should receive less, or different, protection.

Finally, adoption of different privacy and security rules for genomic data could pose practical problems in our health information infrastructure. The flow of medical information is rarely restricted to particular diseases or conditions. Transmission of electronic data for purposes of medical consultation, research, or public health is seldom limited to one kind of information. Requiring hospitals, research institutions, health departments, insurers, and others to maintain separate privacy and security standards (and perhaps separate record systems) for genomic data may not be wise or practical. A more thoughtful solution would be to adopt a comprehensive federal statute on health information privacy, with explicit language applying privacy and security standards to genomic information. If genomic data were insufficiently protected by these legal standards, additional safeguards could be enacted.

Uniform standards for acquisition and disclosure of health information

I previously proposed uniform national standards for the acquisition and disclosure of health information.[61] Below, I briefly describe those standards and outline how they would apply equally to genomic data.[62]

Substantive and procedural review. Many see the collection of health data as an inherent good. Even if the social good to be achieved is not immediately apparent, it is always possible that some future benefit could accrue. But despite optimism in the power of future technology, the diminution in privacy attributable to the collection of health data demands that the acquisition of information serve some substantial interest. The burden rests on the collector of information not merely to assert a substantial public interest, but also to demonstrate that it would be achieved. Information should only be collected under the following conditions: (1) the need for the information is substantial; (2) the collection of the data would actually achieve the objective; (3) the purpose could not be achieved without the collection of identifiable information; and (4) the data would be held only for a period necessary to meet the valid objectives. Thus, collectors of genomic information would have to justify the collection and to use of the information, and they would have to show why collection of tissue or DNA is necessary to achieve the purpose.

The collection of large amounts of health information, such as a tissue or a DNA repository, not only requires a substantive justification, but also warrants procedural review. Decisions to create health data bases, whether by government or private sector, ought to require procedural review. Some mechanism for independent review by a dispassionate expert body would provide a forum for examination of the justification for the data collection, the existence of thoughtful consent procedures, and the maintenance of adequate privacy and security.

Autonomy to control personal data. If a central ethical value behind privacy is respect for personal autonomy, then individuals from whom data are collected must be afforded the right to know about and to approve the uses of those data. Traditional informed consent requires that a competent person have adequate information to make a genuinely informed choice. However, few objective standards have been developed to measure the adequacy of consent. To render consent meaningful, the process must incorporate clear content areas:[63] how privacy and security will be maintained; the person's right of ownership of, and control over, the data; specific instructions on means of access, review, and correction of records; the length of time that the information will be stored and the circumstances when it would be expunged; authorized third-party access to the data; and future secondary uses. If secondary uses of those data go beyond the scope of the original consent (for example, use of human tissue to create cell lines or disclosure to employers or insurers) additional consent must be sought.

Right to review and correct personal data. A central tenet of fair information practices is that individuals have the right to review data about themselves and to correct or amend inaccurate or incomplete records. This right respects a person's autonomy, while assuring the integrity of data. Individuals cannot meaningfully control the use of personal data unless they are fully aware of their contents and can assess the integrity of the information. Individuals can also help determine if the record is accurate and complete. Health data can only achieve essential societal purposes if they are correct and reasonably comprehensive. One method, therefore, of ensuring the reliability of health records is to provide a full and fair procedure to challenge the accuracy of records and to make corrections. Thus, persons must be fully aware of the tissue and genetic material that is collected and stored. Moreover, they must be fully informed about the *content* and *meaning* of any genetic analysis—past, current, or future. For instance, if an individual consents to the collection of tissue for epidemiological research on breast cancer, he/she would be entitled to see and correct any information derived from that tissue. If, in the future, the tissue were used to predict, say, dementia in the patient, he/she would have to consent and would also have the right to see and correct any new information derived from that particular genetic analysis.

Use of data for intended purposes. Entities that possess information have obligations that go beyond their own needs and interests. In some sense, they hold the information on behalf of the individual and, more generally, for the benefit of all patients in the health system. A confidence is reposed in a professional who possesses personal information for the benefit of others. They have an obligation to use health information only for limited purposes; to disclose information only for purposes for which the data were obtained; to curtail disclosure to the minimum necessary to accomplish the purpose; and to maintain an accounting of any disclosure.

The idea of seeing holders of information as trustees has special force with genomic data. Because DNA might unlock the most intimate secrets of human beings and holds the potential for unethical uses, those who possess it must meet the highest ethical standards.

Conclusion

The human genome retains enormous appeal in the United States. Americans, enamored with the power of science, often turn to genetic technology for easy answers to perplexing medical and social questions. This exaggerated perception is problematic. Genomic information can wield considerable influence, affecting the decisions of health care professionals, patients and their families, employers, in-

Volume 23:4, Winter 1995

surers, and the justice system. How does society control this information without stifling the real potential for human good that it offers? The answer to this question must be in recognizing that trade-offs are inevitable. Permitting the Human Genome Initiative to proceed unabated will have costs in personal privacy. While careful security safeguards will not provide complete privacy, the public should be assured that genomic information will be treated in an orderly and respectful manner and that individual claims of control over those data will be adjudicated fairly.

Acknowledgments

This analysis here borrows significantly from my article, "Health Information Privacy" (see *supra* note 2). It is part of a project on health information privacy I chair for the U.S. Centers for Disease Control and Prevention (CDC), the Council of State and Territorial Epidemiologists, and the Carter Presidential Center. I am grateful to Michael Yesley, Ethical, Legal, and Social Implications of the Human Genome Project, Office of Energy Research, U.S. Department of Energy, for providing information on genetic privacy statutes. I am also grateful to Megan Troy, Georgetown University Law Center, for research assistance.

References

1. An expansive literature on genetic discrimination exists. See L. Gostin, "Genetic Discrimination: The Use of Genetically Based Diagnostic and Prognostic Tests by Employers and Insurers," *American Journal of Law & Medicine*, XII (1991): 109–44; G.P. Smith and T.J. Burns, "Genetic Determinism or Genetic Discrimination," *Journal of Contemporary Health Law & Policy*, 11 (1994): 23–61; R.A. Epstein, "The Legal Regulation of Genetic Discrimination: Old Responses to New Technology," *Boston University Law Review*, 74 (1994): at 1–23; and M.A. Rothstein, "Discrimination Based on Genetic Information," *Jurimetrics*, 33 (1992): 13–18.

2. L.O. Gostin, "Health Information Privacy," *Cornell Law Review*, 80 (1995): 451–528.

3. See, for example, G.J. Annas, L.H. Glantz, and P.A. Roche, *The Genetic Privacy Act and Commentary* (Boston: Boston University School of Public Health, 1995); G.J. Annas, "Privacy Rules for DNA Databanks: Protecting Coded 'Future Diaries'," *JAMA*, 270 (1993): 2346–50; J.A. Kobrin, "Comment: Medical Privacy Issue: Confidentiality of Genetic Information," *UCLA Law Review*, 30 (1983): 1283–315; M. Powers, "Privacy and the Control of Genetic Information," in M.S. Frankel and A.S. Teich, eds., *The Genetic Frontier: Ethics, Law and Policy* (Washington, D.C.: American Association for the Advancement of Science Press, 1994): 77–100; and D.L. Burk, "DNA Identification Testing: Assessing the Threat to Privacy," *University of Toledo Law Review*, 24 (1992): 87–102.

4. Office of Technology Assessment, *Protecting Privacy in Computerized Medical Information* (Washington, D.C.: Government Printing Office, OTA-TCT-576, 1993); General Accounting Office, *Automated Medical Records: Leadership Needed to Expedite Standards Development* (Washington, D.C.: Government Printing Office, GAO/IMTEC-93-1, 1993); Task Force on Privacy, Department of Health and Human Services, *Health Records: Social Needs and Personal Privacy: Conference Proceedings* (Washington, D.C.: DHHS, 1993); and Final Report of the Task Force on the Privacy of Private-Sector Health Records, Department of Health and Human Services (Washington, D.C.: DHHS, Sept. 1995).

5. Institute of Medicine, M.S. Donaldson and K.N. Lohr, eds., *Health Data in the Information Age: Use, Disclosure, and Privacy* (Washington, D.C.: National Academy of Sciences, 1994).

6. L.O. Gostin et al., "Privacy and Security of Personal Information in a New Health Care System," *JAMA*, 270 (1993): 2487–93.

7. P.T. Rowley, "Genetic Screening: Marvel or Menace?," *Science*, 225 (1984): 138–44.

8. *Id.*

9. B.S. Wilfond and K. Nolan, "National Policy Development for the Clinical Application of Genetic Diagnostic Techniques: Lessons from Cystic Fibrosis," *JAMA*, 270 (1993): 2948–54.

10. K. Savitsky et al., "A Single Ataxia Telangiectasia Gene with a Product Similar to PI-3 Knase," *Science*, 268 (1995): 1749–53.

11. I.M. Olivos-Glander et al., "The Oculocerebrorenal Syndrome Gene Product is a 105kd Protein Localized to the Golgi Complex," *American Journal of Human Genetics*, 57 (1995): 817–23.

12. A.M. Goldstein et al., "Increased Risk of Pancreatic Cancer in Melanoma-Prone Kindreds with $p16^{INK4}$ Mutations," *N. Engl. J. Med.*, 333 (1995): 970–74.

13. J.P. Struewing et al., "The Carrier Frequency of the BRCA1 185delAG Mutation is Approximately 1 Percent in Ashkenazi Jewish Individuals," *Nature Genetics*, 11 (1995): 198–200.

14. G. Kolata, "Tests to Assess Risks for Cancer Raising Questions," *New York Times*, Mar. 27, 1995, at A1; and E. Tanouye, "Gene Testing for Cancer to be Widely Available, Raising Thorny Questions, *Wall Street Journal*, Dec. 14, 1995, at B1.

15. L.B. Andrews and A.S. Jaeger, "The Human Genome Initiative and the Impact of Genetic Testing and Screening Technologies: Confidentiality of Genetic Information in the Workplace," *American Journal of Law & Medicine*, XVII (1991): 75–108; D. Orentlicher, "Genetic Screening by Employers," *JAMA*, 263 (1990): 1105, 1108; E.F. Canter, "Employment Discrimination Implications of Genetic Screening in the Workplace Under Title VII and the Rehabilitation Act," *American Journal of Law & Medicine*, 10 (1984): 323–47; K. Brokaw, "Genetic Screening in the Workplace and Employers' Liability," *Columbia Journal of Law & Social Problems*, 23 (1990): 317–46; and E.E. Schultz, "If You Use Firm's Counselors, Remember Your Secrets Could Be Used Against You," *Wall Street Journal*, May 26, 1994, at 1.

16. Gostin, *supra* note 1.

17. Report on the Task Force on Genetic Information and Insurance, *Genetic Information and Health Insurance* (NIH-DOE Working Group on Ethical, Legal, and Social Implications of Human Genome Research, May 10, 1993); T.H. Murray, "Genetics and the Moral Mission of Health Insurance," *Hastings Center Report*, 22, no. 6 (1992): 12–17; and S. O'Hara, "The Use of Genetic Testing in the Health Insurance Industry: The Creation of a 'Biological Underclass'," *Southwestern University Law Review*, 22 (1993): 1211–28.

18. C. Ezzell, "Panel Oks DNA Fingerprints in Court Cases," *Science News*, 141 (1992): at 261; and G. Kolata, "Chief Says Panel Backs Courts' Use of a Genetic Test," *New York Times*, Apr. 15, 1992, at A1.

19. S.M. Suter, "Whose Genes Are These Anyway? Familial

Conflicts Over Access to Genetic Information," *Michigan Law Review*, 91 (1993): 1854–908.

20. W.E. Leary, "A Search for Lincoln's DNA," *New York Times*, Feb. 10, 1991, at 1.

21. H.R. 5612, Cong. 101, Sess. 2 (Sept. 13, 1990).

22. Gostin, *supra* note 2.

23. D. Brown, "Individual 'Genetic Privacy' Seen as Threatened; Officials Say Explosion of Scientific Knowledge Could Lead to Misuse of Information," *Washington Post*, Oct. 20, 1991, at A6 (quoting J.D. Watson as saying: "The idea that there will be a huge data bank of genetic information on millions of people is repulsive.").

24. E.W. Clayton et al., "Informed Consent for Genetic Research on Stored Tissue Samples," *JAMA*, 274 (1995): 1786–92.

25. P.R. Reilly, letter, "DNA Banking," *American Journal of Human Genetics*, 51 (1992): at 32–33.

26. *Id.*

27. Deputy Secretary of Defense Memorandum No. 47803 (Dec. 16, 1991).

28. E.D. Shapiro and M.L. Weinberg, "DNA Data Banking: The Dangerous Erosion of Privacy," *Cleveland State Law Review*, 38 (1990): 455–86 (many states authorize the banking of DNA usually for convicted sex offenders; the FBI is establishing a computerized DNA data bank); Burk, *supra* note 3; and Note, "The Advent of DNA Databanks: Implications for Information Privacy," *American Journal of Law & Medicine*, XVI (1990): 381–98.

29. Suter, *supra* note 19.

30. J.E. McEwen and P.R. Reilly, "Stored Guthrie Cards as DNA 'Banks'," *American Journal of Human Genetics*, 55 (1994): 196–200.

31. *Id.*

32. Department of Health and Human Services, *National Health and Nutrition Examination Survey III* (1994).

33. D. Nelkin and S. Lindee, *The DNA Mystique: The Gene as a Cultural Icon* (New York: W.H. Freeman, 1995); D. Nelkin, "The Double-Edged Helix," *New York Times*, Feb. 4, 1994, at A23; and R. Weiss, "Are We More Than the Sum of Our Genes?," *Washington Post Health*, Oct. 3, 1995, at 10.

34. N. Fost, "The Cystic Fibrosis Gene: Medical and Social Implication for Heterozygote Detection," *JAMA*, 263 (1990): 2777–83.

35. P. Reilly, "Rights, Privacy, and Genetic Screening," *Yale Journal of Biology & Medicine*, 64 (1991): 43–45.

36. N.A. Wivel and L. Walters, "Germ-Line Gene Modification and Disease Prevention: Some Medical and Ethical Perspectives," *Science*, 262 (1993): 533–38.

37. P.J. Hilts, "Gene Transfers Offer New Hope for Interspecies Organ Transplants," *New York Times*, Oct. 19, 1993, at A1.

38. Struewing et al., *supra* note 13.

39. Privacy Commissioner of Canada, *Genetic Testing and Privacy* (Toronto: Ontario Premier's Commission, 1992); and Shapiro and Weinberg, *supra* note 28. Later, I show why enacting genetic-specific privacy statutes, instead of a general statute applicable to all health information, may be problematic. This is not intended to undercut the observation that genomic data present distinct privacy concerns. Rather, I argue, that robust privacy legislation should cover all kinds of health information without creating "super" privacy protection for any particular kind of data, whether it be genomic data or data relating to STDs, HIV infection, mental health, or substance abuse.

40. Annas, *supra* note 3.

41. R.T. King Jr., "Soon, a Chip Will Test Blood for Diseases," *Wall Street Journal*, Oct. 25, 1994, at B1.

42. S.F. Kreimer, "Sunlight, Secrets, and Scarlet Letters: The Tension Between Privacy and Disclosure in Constitutional Law," *University Pennsylvania Law Review*, 140 (1991): 1–147; R.C. Turkington, "Legacy of the Warren and Brandeis Article: The Emerging Unencumbered Constitutional Right to Informational Privacy," *Northern Illinois University Law Review*, 10 (1990): 479–520; and F.S. Chlapowski, Note, "The Constitutional Protection of Informational Privacy," *Boston University Law Review*, 71 (1991): 133–60.

43. *Webster v. Reproductive Health Servs.*, 492 U.S. 490 (1989); *Bowers v. Hardwick*, 478 U.S. 186 (1986); and *Paul v. Davis*, 424 U.S. 693 (1976).

44. *Whalen v. Roe*, 429 U.S. 589 (1977). See *Nixon v. Administrator of General Servs.*, 433 U.S. 425 (1977).

45. 429 U.S. at 605.

46. *Id.*

47. *Id.*

48. *Rasmussen v. South Fla. Blood Serv., Inc.*, 500 So. 2d 533 (Fla. 1987).

49. *U.S. v. Westinghouse Electric Corp.*, 638 F.2d 570 (3d Cir. 1980).

50. Gostin, *supra* note 2, at 499–508.

51. Workgroup for Electronic Data Interchange, *Obstacles to EDI in the Current Health Care Infrastructure* (Washington, D.C.: DHHS, 1992): app. 4, at iii.

52. H.R. 5612, Cong. 101, Sess. 2, 101 (Sept. 13, 1990) ("a bill to safeguard individual privacy of genetic information"). The bill provides individuals with certain safeguards against the invasion of personal genetic privacy by requiring agencies, *inter alia*, to permit individuals to determine what personal records are collected and stored; to prevent personal records from being used or disclosed with consent; to gain access to personal records; and to ensure accuracy of records.

53. Hereditary Disorders Act, Cal. Health & Safety Code § 151 (West Ann. 1990) (test results and personal information from the hereditary disorders programs are considered confidential medical records and can only be released with informed consent); Fla. Stat. Ann. § 760.40 (West Supp. 1994) (genomic data are the exclusive property of the person tested, are confidential, and may not be disclosed without the person's consent; genomic data collected for purposes of criminal prosecution, determination of paternity, and from persons convicted of certain offenses are exempted from confidentiality requirement); Me. Rev. Stat. Ann. tit. 22, § 42 (Supp. 1995) (personal medical information obtained in state's public health activities, including but not limited to genetic information, is confidential and not open to public inspection); Mass. Gen. Laws Ann. ch. 76, § 15B (West 1982) (data from state voluntary screening program for sickle cell, or other genetically linked diseases determined by the commissioner, are confidential); Mo. Ann. Stat. § 191.323 (Vernon Supp. 1996) (authorizing the health department to maintain a central registry for genomic information, and providing that identifying information is confidential); Oh. Rev. Code Ann. tit. XXXVII, § 3729.46 (Baldwin 1994) (health department and contractors must keep personal information, including genetic information, confidential); Tenn. Code Ann. § 68-5-504 (1992) (requiring health department to develop statewide genetic and metabolic screening programs including PKU and hypothyroidism, and requiring that the program follow state laws governing confidentiality); Va. Code § 32.1-69 (1950) (records maintained as part of genetic screening program are confidential except with informed consent); Ga. Code Ann. § 33-54-3 (Supp. 1995) (use of genomic information is authorized in criminal investigations and prosecutions, and scientific research); Kan. Stat. Ann. § 65-1, 106 (1994) (sickle cell testing information is confidential); 1995 Or. Laws 680 (requires informed consent

Volume 23:4, Winter 1995

for the procurement of genetic information, and provides that an individual's genetic information is the property of the individual); 1995 La. Acts 11299.6; and Pa. S. 1774 (1993).

54. In Pennsylvania, see Genetic Information Confidentiality Act, Pa. S. 1774 (1993). In New York, see A. 5796, N.Y. Reg. Sess. (1995–96); S. 4293, N.Y. Reg. Sess. (1995); and S. 3118, N.Y. Reg. Sess. (1995).

55. Cal. Health & Safety Code § 151 (West Ann. 1990); Colo. Rev. Stat. 10-3-1104.7 (1994); Fla. Stat. Ann. § 760.40 (West Supp. 1994); Ga. Code Ann. § 33-54-3 (Supp. 1995); Kan. Stat. Ann. § 65-1, 106 (1994) (sickle cell only); 1995 Minn. Laws 251; 1995 N.H. Laws 101; and 1995 Or. Laws 680.

56. For example, Colo. Rev. Stat. 10-3-1104.7 (1994); Ga. Code Ann. § 33-54-3 (Supp. 1995); and Cal. Ins. Code § 10148 (West 1994).

57. For example, Colo. Rev. Stat. §§ 19-1-121, 25-1-122.5 (1995).

58. Nev. Rev. Stat. 396.525 (1991).

59. National Society of Genetic Counselors, *Resolutions* (rev. Nov. 1994).

60. Annas, Glantz, and Roche, *supra* note 3.

61. Gostin, *supra* note 2, at 513–27. Other work on public health information privacy is currently being done under the auspices of the CDC and the Carter Presidential Center.

62. The Medical Records Confidentiality Act, S. 1360, Cong. 104, Sess. 1 (1995), is pending. This statute would create a set of fair information practices for a wide range of health information.

63. In deriving these standards, the author appreciates the work of Professor Robert Weir of the National Human Genome Project and Joan Porter of the Office of Protection from Research Risks of the National Institutes of Health.

[10]

Challenging Medical-Legal Norms

The Role of Autonomy, Confidentiality, and Privacy in Protecting Individual and Familial Group Rights in Genetic Information

Graeme T. Laurie, LL.B., Ph.D.*

INTRODUCTION

Much ink has been spilled discussing the ramifications of genetic advances for individuals, communities, and society at large. A central concern has been the problem of regulating access to, and control of, genetic information that has been produced as a result of rapid progress in the fields of genetic research and genetic testing. To date, discussion has rightly focused on the uses to which genetic test results should be put, and, indeed, on the logically prior question of whether genetic information should be sought at all in certain circumstances. Debate has, however, tended to polarize the issues under scrutiny, setting the individual against the state or other interested parties, such as insurers or employers. Moreover, from the perspective of the individual, the interests that have been identified as being at stake have centered on the autonomy of persons and the "right" that they have to control personal genetic information. While these are important starting points, it should be realized that the discourse barely has begun on the appropriateness of social, ethical, and legal responses to the novel challenges that arise by such scientific advances. This article offers an alternative perspective on these challenges. In particular, three aspects of the debate are considered.

First, focus is placed on assessing the range of interests that are at stake when genetic information is generated, and in particular the interests of family members in shared familial genetic information are examined. It is the fact that genetic information relates to a group of persons and not simply to one

* Lecturer in Law, Faculty of Law, University of Edinburgh, Scotland. The ideas contained in this article form the basis of a monograph entitled *Legal and Ethical Aspects of Genetic Privacy*, to be published by Cambridge University Press in 2001. Address correspondence to Dr. Graeme T. Laurie, Faculty of Law, Old College, South Bridge, Edinburgh, EH8 9YL, Scotland or via e-mail at <Graeme.Laurie@ed.ac.uk>.

individual that sets genetic information as a class apart from other forms of medical information. This requires us to consider the group dynamics of managing and controlling shared information, and the possible rights and interests that might flow from a "group" claim to familial data. These, in turn, must be seen in contradistinction to the more traditional atomistic, autonomy-based approach, which focuses on the rights and interests of the individual from whom the genetic information initially has been obtained (the proband).

Second, the nature of the interests in issue must be examined to determine precisely which factors, values, perceived benefits, and harms should be weighed in the balance when deciding how genetic information should be handled. This becomes particularly important if one adds the interests of family members to the equation. At first blush, the most obvious interest focuses on knowing genetic information, and, on this basis, arguments for a "right to know" are frequently founded. However, the potential existence of a "right not to know" genetic information, which may protect both personal and familial interests, also merits analysis. Only then can a proper assessment be made of the appropriateness of any use of the information in question. It is submitted that proper recognition of the interest in not knowing is urgently needed. Furthermore, to protect such an interest by legal or other means, a paradigm shift in medical-legal norms is required. This is revealed by an assessment of the role of more traditional concepts, such as autonomy and confidentiality in providing a suitable basis on which to found the claim not to know. As is argued, these are wanting in the present context, and a viable alternative must be sought.

In the third and final section of this article, a unique view of privacy is offered as a means of recognizing and protecting the full gamut of personal and familial interests surrounding genetic information, and most particularly, the interest in not knowing information in certain circumstances. The benefits of this concept are manifold, and its particular value for legislative purposes in designing ethically appropriate genetic privacy laws is examined in the shadow of legislative proposals to protect genetic privacy to date.

I. GENETIC INFORMATION: WHAT DO WE KNOW, AND WHAT DO WE NEED TO KNOW?

In seeking an appropriate legal response to advances in genetics, it is trite to observe that we require laws that are informed by ethical debate, that are morally sound, and that reflect as largely as possible our common societal values. These parameters must never be forgotten or obscured in the legislative process, and they require, first and foremost, that we proceed in an informed manner whereby we are apprised of the functional utility of genetic information. The starting point then in deciding whether the promised benefits of genetic advances are truly desirable is to understand the limits of what information of this kind can allow us to do, and perhaps more importantly, what it cannot allow us to do.

A. The Uniqueness of Genetic Information (Individual v. Family)

Information is a unique entity. The same information can be used contemporaneously by a large number of persons for a wide range of ends, and yet the essential character of that information may remain unchanged. Furthermore, a particular use by one person or group of persons does not preclude others from engaging in other uses of the information, for no two uses are mutually exclusive. Traditionally, when we have asked where the control of information should be located and how that control should be exercised, the answer has been that the person to whom the information belongs or to whom it relates should exercise that control; that is, the source of the information. In the genetic sphere, however, such an answer is simplistic and unsatisfactory.

As has been pointed out, genetic information differs from other forms of medical information because it pertains to a range of people and not solely to one individual. In this respect, it gives rise to special problems concerning how the information should be gathered, stored, accessed, and used. While one might choose to locate control of a genetic sample with the person from whom it has been taken,[1] one cannot ignore the fact that genetic information derived from the sample also reveals information about the relatives of the sample source. These persons can base a claim to the information on precisely the same grounds as the source, namely: "I have a claim because it is about me." Moreover, the ends to which this information can be put may affect relatives in much the same way as they can affect the life of the person who has been tested. To locate the control of this information solely with the proband, therefore, might seem to many to be an inadequate response to concerns about how information of this sort should be treated.

B. The Distinction Between Family History and Test Results (Specific v. Abstract Knowledge)

It is sometimes claimed that a family history is simply genetic information in a different guise, and therefore that a genetic test result is no different than a known family history. Yet, family history is abstract knowledge that has been tainted by bad or failing memories, lack of accurate data about why someone has become ill or died, and by an absence of understanding about the pattern of disease in a family pedigree. In contrast, genetic test results can offer a high degree of specificity, both in terms of predicting the likelihood of disease in other family members and in terms of putting flesh on the bones of a suspicion that has heretofore been unconfirmed. Specific information brings

---

[1] The Genetic Privacy Act, which is discussed in section VIII(C) below, draws a distinction between genetic samples and genetic information derived from samples. The Act further creates a property right for a sample source in his or her own sample. The Genetic Privacy Act and Commentary (unpublished, D.O.E. No. DE-FG02-93ER61626, Feb. 28, 1995) is available at <www.bumc.bu.edu/Departments/PageMain.asp?Page=789&DepartmentID=95>.

with it a number of realities that can include a degree of "certainty" about future ill health or even the mode and manner of one's own death. These realities can impact an individual's self-perception in ways that family history cannot, for, with the latter, one has the comfort of having lived with an abstract threat that has always manifested to someone else.

Specific knowledge of one's own genetic constitution, especially when it is accompanied by knowledge of future ill health, requires individuals to reassess themselves and their position within a family unit and to look with fresh eyes upon their family history, which will have suddenly become very unfamiliar. Specificity of knowledge can deprive us of the ostrich's "head-in-the-sand," which can sometimes serve as a valuable psychological coping mechanism.[2]

C. The Perceived Utility of Tests (Introducing Other Interested Parties)

It is precisely because genetic testing is thought to offer a high degree of specificity in determining future ill health that genetic information is seen to have a "value" not only for a proband or his or her relatives, but also for parties outside the familial milieu. Insurers, employers, and researchers can find considerable utility in genetic information where test results might impact their own interests. It is apposite, however, to offer a word of caution on the perceived value of genetic information in this context.

Genetic diseases may have a variety of causes. On the one hand, monogenic diseases are caused by mutations in the genome that directly result in disease. These diseases can either be recessive or dominant. Recessive conditions such as cystic fibrosis or sickle cell anemia are caused when an individual inherits two copies of a defective gene from his or her parents. If only one copy is inherited, then disease is not manifested, but the individual becomes an asymptomatic carrier for the condition. In contrast, dominant disorders are inherited when only one copy of a disease gene is passed on. Its influence overrides the effects of its twin "healthy" copy. In both cases, the predictability of disease in future generations is a relatively straightforward exercise. For a recessive disorder where both parents are carriers, there is a 25% risk in the case of each pregnancy that a child will be born affected, a 50% risk that a child will be born as a carrier, and only a 25% chance that the child will inherit two copies of the healthy gene. For dominant disorders, the chances of having an affected child are 50% in each case. It is understandable that interested parties might place considerable store in test results, in light of such figures. However, it is of crucial importance also to consider other

---

[2] *See, e.g.*, Martin Richards, *Families, Kinship and Genetics*, *in* THE TROUBLED HELIX: SOCIAL AND PSYCHOLOGICAL IMPLICATIONS OF THE NEW HUMAN GENETICS 249-73 (Martin Richards & Theresa Marteau eds. 1996).

forms of genetic disease, which, in fact, represent by far the greater category of genetic diseases that affect individuals and families.

Polygenic disorders are caused by the interaction of two or more defective genes, and the chances of being affected are consequently more difficult to predict.[3] Moreover, these disorders are part of the wider class of multifactorial conditions that involve disease processes caused not only by genetic defects but also by the interaction of those genes with environmental factors, all of which can be operative in the onset of disease. Axiomatically, the predictability of the onset of multifactorial conditions is considerably lower than that for monogenic conditions. The value of genetic test results for such conditions is consequently diminished. The obvious conclusion is that it is impossible to attach a uniform value to the practice of genetic testing for an entire range of conditions.[4]

But, even if third parties were to restrict their interests to monogenic disorders, the predictability value of testing is nonetheless affected by other considerations. Testing can only offer probabilities of onset of disease. Importantly, it cannot give any indications of when disease will arise, nor of the degree to which any one individual will be affected (and this can vary considerably as between individuals), nor can tests necessarily detect mutations for particular conditions. Accordingly, there can be a significant risk of false negatives.[5] All of these factors undermine the utility of genetic testing because they demonstrate that what is predictable is only predictable in a limited number of cases, which, in themselves, are further subject to a range of variables. Each of these factors can have a significant bearing on the outcome. Knowledge and certainty must, therefore, be seen as relative concepts.

D. What Is Genetic Information?

It becomes increasingly clear as more work is carried out on the human genome that a genetic component might have a factor to play in many disease processes, and not simply in those that have been classified to date as genetic. This may impact considerably on the subject matter of any legislation that is enacted to protect genetic information. A definition of this term that is too narrow might prove to be useless in protecting any interests at all, while an overly broad definition might, for example, encompass data used in important research, with the resultant risk that such work might be unduly hampered. Indeed, such a definition might include details of family history, if, as has

---

[3] Examples are ischemic heart disease and certain forms of diabetes.

[4] *See* Ruth Hubbard & Richard Lewontin, *Pitfalls of Genetic Testing*, 334 NEW ENG. J. MED. 1192 (1996).

[5] For example, current tests for cystic fibrosis only can detect up to 75% of at-risk individuals in society. As Gostin states: "[A]pproximately one in every two couples from the general population identified by CF screening as 'at-risk' will be falsely labeled." Larry Gostin, *Genetic Privacy*, 23 J. L. MED. & ETHICS 320, 323 (1995).

been suggested above, a family history is thought to reveal familial genetic information.[6]

In recognition of these problems of definition the Task Force on Genetic Testing, a joint working group of the Department of Energy and the National Institutes of Health, has offered a working definition of genetic test information that seeks to strike a balance between protecting legitimate interests in test results while, at the same time, avoiding the conclusion that any kind of medical test is, in fact, a genetic test.[7] The definition restricts genetic testing to "processes which are carried out for the direct analysis of human DNA and other compounds such as RNA, chromosomes, proteins and certain metabolites, with a view to achieving a number of clearly identified end points; namely, the prediction of inherited disease, the detection of carrier status or the diagnosis of actual inherited disease." This can encompass not only the testing of individuals but also the screening of at-risk populations, and will include prenatal and antenatal screening and the testing of families with recognized histories of genetic disease.

By corollary, this also means that the contents of one's medical file do not necessarily contain genetic information and that testing for certain conditions in which a genetic factor is operative, such as diabetes or ischemic heart disease, will not be classified as genetic testing unless there is a high probability that the genetic form of the disease is at work. While it is the case that all human cellular material contains a complete copy of the genome (with the exception of the gametes), tests that do not involve the direct analysis of the DNA, but rather concern other traits of the cells, will not be deemed to be genetic tests. The Task Force specifically excludes certain testing from their definition, for example, tests conducted purely for research, tests for somatic mutations (compare heritable mutations), and tests for forensic purposes. On this basis, the genetic information that the Task Force would seek to protect would be restricted to information that arises from genetic tests falling within the definition, and would not be so broad as to cover abstract data about family history. A definition such as this would be an important starting point in developing specific genetic-related legislation. Of course, whether that exercise in itself is necessarily a good thing is another matter entirely, and is discussed below.

[6] Some have argued that specific laws to protect only genetic privacy, as opposed to medical privacy generally, are misguided because of this problem of definition and the narrow scope of the protection that would be afforded. *Compare, e.g.*, George Annas, *Genetic Privacy: There Ought to Be a Law*, 4 TEX. REV. L. & POL. 9 (1999) *with* Mark Rothstein, *Why Treating Genetic Information Separately Is a Bad Idea*, 4 TEX. REV. L. & POL. 33 (1999).

[7] NEIL HOLTZMAN & MICHAEL WATSON, FINAL REPORT OF THE TASK FORCE ON GENETIC TESTING: PROMOTING SAFE AND EFFECTIVE GENETIC TESTING IN THE UNITED STATES 6 (1998).

E. Lessons We Cannot (Currently) Learn from Genetic Information

An important rider to this discussion concerns the knowledge that genetic advances do not currently give us, except in a few, rare instances. Primarily, this is knowledge about how to treat or cure a genetic disorder for which a test has been developed. As the United Kingdom's Science and Technology Committee has pointed out: "While genetics is likely eventually to transform medicine, it may take some while before treatments based on genetic knowledge become available. . . . In the short term, the most widespread use of medical genetics will be, as now, in diagnosis and screening."[8]

This poses the dilemma of how we should respond to this limited knowledge. If the pursuit of better health is our goal, then we must ask whether, and how, this current knowledge can assist in achieving that goal. In the absence of therapies or cures, preparedness is often cited as a reason to seek out genetic knowledge. Certainly, one can better prepare for reproductive decisions in the light of proper information about genetic risks, and in the case of multifactorial conditions, lifestyle changes might minimize the health implications of carrying a defective gene. The achievement of psychological preparedness for the onset of future disease through the disclosure of risk is, however, by no means certain. Psychological health may be damaged, rather than improved, by such disclosures. This is explored further below. The lesson, however, is that we should not expect too much of this knowledge in our quest to improve health. To do so may mean that we achieve nothing more than the frustration our own efforts.

A final crucial lesson to be learned is that we cannot know all of the ends to which genetic information might be put. Illegitimate uses of this information, which result in harm, discrimination, and stigmatization, must clearly be guarded against. Yet, the question of where the boundary lies between legitimate and illegitimate use, or between legitimate and illegitimate claims with respect to genetic information, is similarly not answered by the new knowledge that genetic science gives us. We must, therefore, determine for ourselves where these limits are to be drawn. This article offers one model by which we may do so.

II. THE INTERESTED PARTIES

It is through an examination of the respective interests that parties have in genetic information that we can understand the nature of potential problems.

[8] HOUSE OF COMMONS SCIENCE AND TECHNOLOGY COMMITTEE, HUMAN GENETICS: THE SCIENCE AND ITS CONSEQUENCES ¶¶ 71-72 (Third Report, HMSO, 1995). The Report makes clear: "*Diagnosis* is aimed at individuals; genetic screening is routine screening of populations, or identifiable subsets of populations (for example, men or women only, or ethnic groups at increased risk for particular diseases)." *Id.*

Such an analysis also serves to bring the issues within the rubric of a common language, which in turn allows us to compare and contrast various, and at times competing, claims with respect to genetic information.

An interest is here defined as a claim that a benefit can come to the party in question by recognizing that the party has a relationship with the subject of the interest; in this case, genetic information. The question of whether a party has an interest in genetic information is, of course, an evaluative matter. Integral to the notion of interest is the idea that it is in the party's interest to recognize the relationship with the genetic information. To do so normally will lead to the conclusion that it is therefore in the party's interest to know, and to have access to, the information in question.[9] However, such an assumption should not go unchallenged in all circumstances, as is explained below.

A. An Individual's Interest in Personal Genomic Information

It is axiomatic that a person who has been tested for one or more genetic conditions has a significant interest in knowing and determining what happens to the resulting information. Arguably, genetic information is "the most personal information of all."[10] While it can be asserted that any form of personal health information is inherently part of the private sphere of an individual's life, genetic information has a unique relationship with the individual in many specific ways. For example, as Suter has noted: "While contracting chicken pox has virtually no effect on identity, the knowledge that one carries a disease gene may influence one's self-perception and definition of 'one's own concept of existence' in a way most infectious diseases do not."[11]

Furthermore, and again unlike conventional health information, genetic information cannot be completely anonymized. It is a unique marker pointing the way to a single individual. As Gostin puts it:

> Genomic data are qualitatively different from other health data because they are inherently linked to one person. While nongenetic descriptions of any given patient's disease and treatment could apply to many other individuals, genomic data are unique. But, although the ability to identify a named individual in a large population simply from genetic material is unlikely, the capacity of computers to search multiple data bases provides a potential for linking genomic information to that person. It follows that non-linked genomic data do not assure anonymity and that privacy and security safeguards must attach to any form of genetic material.[12]

---

[9] *See also* Ann Sommerville & Veronica English, *Genetic Privacy: Orthodoxy or Oxymoron?*, 25 J. MED. ETHICS 144 (1999).

[10] *See* Graeme Laurie, *The Most Personal Information of All: An Appraisal of Genetic Privacy in the Shadow of the Human Genome Project*, 10 INT'L J. L. POL. & FAM. 74 (1996).

[11] *See* Sonia Suter, *Whose Genes Are These Anyway?: Familial Conflicts over Access to Genetic Information*, 91 MICH. L. REV. 1854, 1893 (1993).

[12] Gostin, *supra* note 5, at 322.

Moreover, genetic information does not simply provide information about an individual's medical past, which is the case with most medical records. Genetic information also can furnish knowledge about an individual's medical future. This knowledge can be vague, in that we know only that the person has a certain percentage risk of developing disease, or it can be certain: we know that, given time, disease will develop. Either way, such knowledge permits those who hold it to make judgments about the future life of the individual. Not all such persons will be the individual.

For these reasons, an individual has a very strong claim to control the circumstances in which this information is generated and to determine what happens to the information subsequently. In essence, persons have an interest in this information because it relates to them and can affect their lives in profound ways. As moral agents, their decisions regarding this information are entitled to respect.

B. The Interest of Relatives in a Proband's Genetic Information

In an entirely unique way, exactly the same reasons specified above can be advanced by the blood relatives of a proband to claim an interest in genetic test results because a test result also will reveal information about them.[13] On this basis the "right to know" is frequently founded.[14] Yet, in one important respect, relatives stand in a very different position to a person who has sought out testing, for the latter has made a conscious decision to acquire the information in question, while this might not be true for relatives. We must, therefore, recognize the possibility that family members might be surprised, or even loath, to learn of a relative's predisposition to a particular genetic condition, given the likelihood that they carry a similar risk. Yet, once such information exists, questions of security, access, and control arise. Furthermore, if the individuals to whom the information relates do not agree on such issues, problems of weighing the competing interests in the balance must be addressed.[15]

---

[13] The existence of this interest has been recognized by a variety of bodies, including: COMMITTEE ON HUMAN GENOME DIVERSITY, NATIONAL RESEARCH COUNCIL, EVALUATING HUMAN GENETIC DIVERSITY (1997); National Institutes of Health Office of Protection from Research Risks, *Human Genetic Research*, *in* PROTECTING HUMAN RESEARCH SUBJECTS: INSTITUTIONAL REVIEW BOARD GUIDEBOOK (1993); NUFFIELD COUNCIL ON BIOETHICS, GENETIC SCREENING: ETHICAL ISSUES (London, Dec. 1993); THE ROYAL COLLEGE OF PHYSICIANS OF LONDON, ETHICAL ISSUES IN CLINICAL GENETICS: A REPORT OF THE WORKING GROUP OF THE ROYAL COLLEGE OF PHYSICIANS' COMMITTEES ON ETHICAL ISSUES IN MEDICINE AND CLINICAL GENETICS (1991); and the DANISH COUNCIL OF ETHICS, ETHICS AND MAPPING THE HUMAN GENOME 62 (1993).

[14] *See* Somerville & English, *supra* note 9. *See also* THE RIGHT TO KNOW AND THE RIGHT NOT TO KNOW (Ruth Chadwick et al. eds. 1997).

[15] Because of technological advances in the last 50 years in the field of computers, the means now exist to store and access all forms of information for indefinite periods of time. In this way, genetic information also could prove relevant for future generations of the same genetic line. *See* Barry Barber, *Securing Privacy in Medical Genetics*, Second Symposium of the Council of Europe on Bioethics, Strasbourg, CDBI-SY-SP (93) 2, at 6 (Nov. 30-Dec. 2, 1993) Kåre Berg, *Confidentiality Issues in Medical Genetics:*

The question of whether the interest of relatives is as strong as that of the proband is more difficult to answer. Certainly, the risk of more distant relatives being affected by a particular condition is reduced because of the different genetic influences to which they have been subjected compared with the proband.[16] Those relatives with the strongest interest of all are the first-degree relatives of the person who has been tested. The interests of such relatives include those of the children of a proband who might want to know whether they have any risk of disease that might affect themselves or their progeny. Siblings, too, have a strong interest in each other's test results given their common parentage. A further complicating factor is the potential claims of non-blood-related relatives such as spouses, whose reproductive decisions can be profoundly affected if they are denied access to genetic information that might indicate the presence of disease within their partner's family.

Finally, it is important to stress that, even if test results show no risk of disease, it should not be presumed that individuals will be happy to surrender control of genetic information. Relatives retain an interest in each other's genetic information even if it reveals nothing sinister. The information is intimately connected with their private sphere and their sense of self and therefore to disrespect the information is to disrespect the persons concerned. Moreover, it should not be thought that accuracy of information is in any way a prerequisite to discrimination or stigmatization at the hands of third parties.[17]

From the above, it is clear that conflict will arise when the proband wishes to keep test results secure and family members wish to know them. That is, when the individual wishes to keep the data private and the family wishes to invade that private sphere, or perhaps, become part of that private sphere. The converse is, of course, also true. Conflict can arise when the proband is willing to permit access to genetic information, for example, to third parties outside the family, yet relatives are unwilling to relinquish control of these

---

The Need for Laws, Rules and Good Practices to Secure Optimal Disease Control, Second Symposium, *supra*, at 3.

[16] *See* JEAN WILSON ET AL., PRINCIPLES OF INTERNAL MEDICINE 30 (1991) ("as the degree of relation becomes more distant, the likelihood of a relative inheriting the same combination of genes becomes less. Moreover, the chances of any relative inheriting the right combination of genes decrease as the number of genes required for the expression of a given trait increases").

[17] For survey evidence from the United States and the United Kingdom of discriminatory practices in insurance that had little foundation in scientific fact, see Paul Billings et al., *Discrimination as a Consequence of Genetic Testing*, 50 AM. J. HUM. GEN. 476 (1992); Joseph Alper et al., *Genetic Discrimination and Screening for Hemochromatosis*, 15 J. PUB. HEALTH POL. 345 (1994); Lisa Geller et al., *Individual, Family and Social Dimensions of Genetic Discrimination: A Case Study Analysis*, 2 SCI. ENG. ETHICS 71 (1996); Lawrence Low et al., *Genetic Discrimination in Life Insurance: Empirical Evidence from a Cross-Sectional Survey of Genetic Support Groups in the UK*, 317 BRIT. MED. J. 1632 (1998).

familial data. Here, family privacy might be in jeopardy. Moreover, relatives might be unwilling to receive such data into their own private sphere when they have previously been in ignorance, given the implications this knowledge might have for their future lives. Here, the privacy of the relatives might be invaded by unsolicited disclosures of information to them.

The question arises of whether individuals—either a proband or his or her relatives—have an interest in not knowing test results. For example, a proband might agree to be tested but then change his or her mind. Equally, relatives might be approached by a proband willing to reveal test results but they might refuse to accept the information. On what basis might individuals have an interest in not knowing information?

It is frequently argued that knowledge of genetic information can bring many benefits to individuals. If a cure or therapy is available, then it can be sought and ill health may be averted. Yet, even if a cure or therapy is not available, knowledge can serve several ends. For example, because multifactorial conditions are by definition affected by many influences including the nongenetic, knowledge of a predisposition to such a condition can provide individuals with the opportunity to change aspects of their lifestyle. This can in turn influence the onset of disease.[18] Moreover, it has been argued that, with knowledge, comes preparedness for the risk of developing a disease at a later stage in life.[19] Similarly, the discovery of disease or predisposition to disease means that any reproductive decision that is taken thereafter will be an informed one. Unfortunately, such arguments all suffer from one fundamental weakness: they presume that only benefit can result from knowledge. This is not necessarily so.

The availability of a cure or a therapy carries with it the certainty that disclosure can avert harm uncontrovertibly, or at least minimize it considerably.[20] For a person to whom such a disclosure is made this only can be seen as a good thing. If, however, disclosure is made to avoid an ancillary harm, such as psychological upset, then there is less of a guarantee that the harm in question will, *de facto*, be avoided. Evidence exists from empirical studies that both

[18] *See* Mark Ryan et al., *An Ethical Debate: Genetic Testing for Familial Hypertrophic Cardiomyopathy in Newborn Infants*, 310 BRIT. MED. J. 856 (1995); Philip Reilly, *Rights, Privacy, and Genetic Screening*, 64 YALE J. BIO. & MED. 43 (1991).

[19] *See* David Ball et al., *Predictive Testing of Adults and Children*, *in* GENETIC COUNSELLING: PRACTICE AND PRINCIPLES (Angus Clarke ed. 1994); Mary Pelias, *Duty to Disclose in Medical Genetics: A Legal Perspective*, 39 AM. J. MED. GEN. 347 (1991).

[20] This having been said, in circumstances where a cure is available but an individual would not choose to take it—for example, for religious reasons—it is hard to see how disclosure ever could be justified because the perceived harm could not be avoided. Of course, one could argue that, faced with the reality of the situation, the individual might nevertheless accept treatment, but this is to adopt a strong paternalistic perspective, the ethical propriety of which is doubtful to many.

supports[21] and refutes[22] the benefits of disclosure to facilitate preparedness. Thus, it is entirely possible that individuals might be loath to learn of a relative's genetic status because of the implications this knowledge can have for their own well-being. The Danish Council of Ethics has warned of the risk of "morbidification": the notion of falling victim to some inescapable fate through knowledge about risk of disease.[23]

The possible adverse effects of knowledge of genetic predisposition have been well documented by Hoffman and Wulfsberg.[24] They cite three examples of child screening programs in Sweden, the United States, and Wales involving respectively, $alpha_1$-antitrypsin deficiency,[25] cystic fibrosis,[26] and Duchenne's muscular dystrophy.[27]

In 1972, the Swedish government initiated a nationwide screening program of newborns. As part of the program parents were (1) told whether the child had $alpha_1$-antitrypsin deficiency, (2) counseled to protect the child from environmental factors such as smoking or high dense-particle atmospheres, which could exacerbate the child's problems, and (3) followed to determine the psychological impact of the information. Follow-up studies showed that more than half of the families with affected children suffered adverse psychological consequences, some of which continued for five to seven years. Moreover, there was little evidence of reduction in smoking among parents of affected children. In some cases, an increase was noted. This led

---

[21] *See* Jessica Buxton & Marcus Pembrey, *The New Genetics: What the Public Wants to Know*, 4 EUR. J. HUM. GEN. 153 (Supp. 1996); Marja Hietala et al., *Attitudes Towards Genetic Testing Among the General Population and Relatives of Patients with a Severe Genetic Disease: A Survey from Finland*, 56 AM. J. HUM. GEN. 1493 (1995); Ball et al., *supra* note 19 (quoting several others including, Michael Hayden, *Predictive Testing for Huntington's Disease: Are We Ready for Widespread Community Implementation?*, 40 AM. J. MED. GEN. 515 (1991); Jason Brandt et al., *Presymptomatic Diagnosis of Delayed-Onset with Linked DNA Markers: The Experience of Huntington's Disease*, 261 J.A.M.A. 3108 (1989)).

[22] *See* DANIEL KEVLES, IN THE NAME OF EUGENICS: GENETICS AND THE USES OF HUMAN HEREDITY 298 (1985); Elizabeth Almqvist et al., *Risk Reversal in Predictive Testing for Huntington Disease*, 61 AM. J. HUM. GEN. 945 (1997); Joanna Fanos & John Johnson, *Perception of Carrier Status by Cystic Fibrosis Siblings*, 57 AM. J. HUM. GEN. 431 (1995); Lori Andrews, *Legal Aspects of Genetic Information*, 64 YALE J. BIO. & MED. 29 (1990); David Craufurd et al., *Uptake of Presymptomatic Predictive Testing for Huntington's Disease*, 2 LANCET 603 (1989).

[23] DANISH COUNCIL OF ETHICS, *supra* note 13, at 60. Whereas this is arguably true of all disease, the problem can be particularly acute with genetic disease because individuals can have future ill health predicted. Thus, a person can be affected even when they are perfectly healthy. With nongenetic disease, usually one is actually affected by the disease before suffering psychological sequelae.

[24] Diane Hoffman & Eric Wulfsberg, *Testing Children for Genetic Predispositions: Is It in Their Best Interest?*, 23 J. L. MED. & ETHICS 331 (1995).

[25] This is a genetic enzyme deficiency which is common in persons of Scandanavian descent. Those with the gene have a high risk of developing adult-onset emphysema.

[26] Cystic fibrosis results in thick secretions in the lungs and pancreas that lead to chronic pulmonary and digestive disease.

[27] This condition is typified by chronic muscle wasting. The disease usually manifests itself in children of between two and four years of age. Death normally results by the middle teenage years.

directly to the abandonment of the program by the Swedish government in 1974.[28]

In like manner, Hoffman and Wulfsberg note that cystic fibrosis screening programs in the United States, which commenced as early as 1968, were abandoned because "many people think (even in cases where there is a familial risk for the disease) that early detection has no value and may, in fact, cause the family significant psychological distress prior to the time when the individual might become symptomatic."[29] For these reasons, the authors assert that the United States has not instituted a program of screening newborns for Duchenne's muscular dystrophy, unlike Wales in the United Kingdom, where such a program has run since 1990.[30]

Similar evidence is available for adults. Kevles has noted, citing several studies, that "[t]he revelation of genetic hazard has been observed to result not only in repression but in anxiety, depression, and a sense of stigmatisation."[31] Most recently, Almqvist and colleagues have found in an international study that the suicide rate among persons given a positive genetic test result for Huntington disease was 10 times higher than the United States average.[32] While this rate is no greater than that for the symptomatic Huntington disease population (nor is it vastly greater than the rate for persons with other debilitating and progressive diseases), it is significant that the survey primarily focused on the two years after test results were given. This would tend to indicate that the deaths were more directly related to the disclosure of the genetic information, rather than to some other factors, such as the onset of disease itself.[33]

Finally, it has even been observed that confirmation of one's status as a nonaffected person also can have adverse psychological effects. Huggins and colleagues,[34] as well as Wexler,[35] have carried out studies in families

---

[28] Hoffman and Wulfsberg cite the following articles as authority: T. Thelin et al., *Psychological Consequences of Neo-natal Screening for Alpha$_1$-Antitrypsin Deficiency (ATD)*, 74 ACTA PAED. SCAN. 787 (1985); T. F. McNeil et al., *Psychological Effects of Screening for Somatic Risk: The Swedish Alpha$_1$-Antitrypsin Experience*, 43 THORAX 505 (1988).

[29] Hoffman & Wulfsberg, *supra* note 24, at 333.

[30] *Id.*

[31] KEVLES, *supra* note 22, at 298.

[32] *See* Elizabeth Almqvist et al., *A Worldwide Assessment of the Frequency of Suicide, Suicide Attempts, or Psychiatric Hospitalization after Predictive Testing for Huntington Disease*, 64 AM. J. HUM. GEN. 1293 (1999). The authors surveyed 100 centers in 21 countries and gathered data on 4,527 individuals who had undergone predictive genetic testing for Huntington's disease. Of those reviewed, 1,817 people had received a positive result, of whom five had taken their own lives. This extrapolates to 138/100,000 suicides per year, compared to the United States average of 12-13/100,000 per year. *See* Thomas Bird, *Outrageous Fortune: The Risk of Suicide in Genetic Testing for Huntington Disease*, 64 AM. J. HUM. GEN. 1289 (1999).

[33] *See* Bird, *supra* note 32.

[34] Marlene Huggins et al., *Predictive Testing for Huntington Disease in Canada: Adverse Effects and Unexpected Results in Those Receiving a Decreased Risk*, 42 AM. J. MED. GEN. 508 (1992).

[35] Nancy Wexler, *Genetic Jeopardy and the New Clairvoyance*, 6 PROGRESS MED. GEN. 277 (1985).

affected by genetic disease that show "[m]any may suffer 'survivor guilt,' particularly characteristic of wartime soldiers who live while their buddies are killed."[36]

The possibility that any or all of these forms of harm can result means that individuals can cite a strong interest in not knowing genetic information about themselves.[37] However, one should not imagine that potential harm is the only reason for claiming an interest in not knowing. The question of respect also arises. To disclose genetic information to someone who has not expressed a desire to know may be disrespectful in two ways. First, if the individual has specifically stated they do not wish to know the information, then it is an affront to them as moral chooser to furnish the information. Second, even if no such wish has been expressed, then it can be offensive to provide information in the absence of a justified reason for doing so. While no tangible harm might result from disclosure, the fact that the individual's private sphere is invaded with such information can be problematic. For example, it was stated in the preamble to the World Medical Association Declaration on the Human Genome Project that "[t]his area of scientific progress will profoundly affect the lives of present and future members of society, bringing into question the very identity of the human individual and intruding upon the snail's pace of evolution in a decisive and probably irreversible manner."[38]

The implications these advances have for personal privacy are extremely far-reaching. To discover that one is likely to develop a debilitating condition in later life or that one might pass on such a condition to one's children can be a devastating and profound experience. Exposure to such knowledge can challenge notions of self-identity and alter considerably one's self-perception.[39] It requires individuals to take on board information which then cannot be unknown. The knowledge becomes a factor that will necessarily become part of many future life decisions of the individual. Individuals are coerced into self-reflection and forced to evaluate and reevaluate themselves. While it might be argued that it is in the individual's best interests to know the information, this is to make an evaluative judgment which, to be justified, must surely weigh in the balance the possibility that disclosure might be unwanted or harmful in certain circumstances. To presume that individuals will always and necessarily wish to know familial genetic information is not only to ignore these possible adverse consequences, but it is also to disrespect such persons, for such a presumption disregards the individuality of subjects and subjugates them to a view of life which is not their own.

---

[36] *Id.*

[37] In one study, only 43% of women tested for BRCA1 wanted to know the result. *See* Caryn Lerman et al., *BRCA1 Testing in Families with Hereditary Breast-Ovarian Cancer*, 275 J.A.M.A. 1885, 1888 (1996).

[38] This declaration was adopted at the World Medical Association's 44th assembly in 1992.

[39] This point is made particularly well by the Danish Council of Ethics, *supra* note 13, at 52.

For all of these reasons it is submitted that both the proband and his or her relatives could have an interest in not knowing genetic information. However, the recognition of this interest complicates matters considerably. The various claims require close scrutiny, especially given that the resolution of the matter also will have implications for family members further down the genetic line. In order to conduct this scrutiny adequately, it is necessary to consider the key principles, values, and factors that are of relevance in resolving conflict dynamics in the medical-legal sphere.

III. WESTERN PRINCIPLES AND VALUES: A BRIEF ANTHOLOGY

A. Principles of Ethics

The so-called "four principles of ethics" have significantly influenced much of Western thinking and action, particularly in the medical-legal sphere. These four principles are autonomy, beneficence, nonmaleficence, and justice.[40] Autonomy refers to a state of moral independence, and an autonomous individual is one who is a "moral chooser."[41] The principle of respect for patient autonomy is fundamental to good medical practice and is the cornerstone of many ethical and legal requirements concerning the way in which health care professionals treat their patients. Among other things, the principle requires that patients be consulted about health care provision, that their consent be sought to proceed with medical interventions, and that their wishes concerning treatment be respected, even when such wishes run counter to the advice or wishes of health care professionals. This extends to respect for the patients' wishes about their personal health information. Beneficence and nonmaleficence prescribe, respectively, that one should strive where possible to bring benefit to individuals and that, contemporaneously, one should endeavor at all times to minimize harm to them and others. Justice requires that comparable cases be treated alike and that no unjustifiable decisions be made that prejudice one individual or group over another on irrelevant or unjustified grounds.

B. Confidentiality

Confidentiality is characterized by a relationship involving two or more individuals, one or more of whom has undertaken, explicitly or implicitly, not to reveal to third parties information concerning the other individual in the relationship. It is widely accepted that health care professionals owe a duty

---

[40] These four principles typify the model of bioethics developed by Thomas Beauchamp & James Childress, *Principles of Biomedical Ethics* (4th ed. 1994). This is not the only model of medical ethics in existence, but it is the one preferred by the present writer.

[41] This term is borrowed from Stanley Benn, inter alia, from his work, *A Theory of Freedom* (1988).

of confidence to their patients and that only rarely should disclosure without patient consent be made. While exceptions to the duty exist, in practice no breach is made lightly or without good cause. Confidentiality is the duty of the health care professional and the right of the patient.

C. Privacy

More generally, "privacy" as used herein is an interest that is premised on setting the "private"—which is bound up inherently with the personal—in contradistinction to the "public." The maintenance of a public/private divide, and the location of certain personal attributes within the latter and therefore out of the reach of the former, is commonly taken as a "good." It should be clear that certain interests may be common to privacy, confidentiality, and autonomy, and may be protected to varying degrees by each. This is especially true in the context of personal information.

D. Public Interest

Public interest is an amorphous concept that has a role to play in both ethics and law. It acts as a safeguard equally for individual and collective interests, but suffers from a lack of precise definition. It is open to abuse as a result. Nevertheless, the concept reflects many important values and must be considered in this debate. When public interest enters the equation, it is usually weighed in the balance with another public interest. For example, the public interest in disclosure compared to the public interest in maintaining confidences generally. In the context of public interest, community interests and values are at stake, and, as such, necessarily and frequently subsume private interests within their scope.

E. Additional Factors

In addition to the above, there are several factors that must be considered when trying to resolve complex issues surrounding genetic information. These are not only highly relevant but context specific and can be invoked—alone or in combination—in particular situations to assist in making the strongest argument for the most appropriate outcome. These factors are listed below.

1. The Availability of a Therapy or Cure

If death or disease can be avoided incontrovertibly, or if the effects of disease can be substantially diminished, then it is trite that very strong arguments must be advanced to prevent disclosure of genetic information to those likely to be affected, especially in the absence of some other means of preventing harm. If, however, nothing can be done to prevent the onset of genetic disease or to alleviate physical suffering, then the argument for disclosure is accordingly weakened.

2. The Severity of the Condition and Likelihood of Onset

A fatal condition intuitively calls for action if death can be prevented. In contrast, a mild condition for which nothing can be done makes arguing for disclosure more difficult. In like manner, a 50% risk of developing a genetic condition, which lies with a first-degree relative, is more compelling than a 1% or 2% risk to unidentified third cousins.

3. The Nature of Genetic Disease

The affliction of one individual with genetic disease does not pose any direct threat to any other living human being. In this respect, genetic disease is very different from many other diseases. Also, with recessive disorders that render people asymptomatic carriers, there is additionally no threat to the health of the carrier. Only future progeny might be affected. Facts such as this can have a bearing on how one views particular complex scenarios.[42]

4. The Nature of Genetic Testing

The point already has been made that predictive genetic testing (and family history) are imprecise tools for assessing future risk. Thus, it is important to appreciate that any trade in information is trade in further uncertainty. People may be alerted to a possibility, but they cannot be apprised of a medical certainty in respect to their own health without undertaking further steps, such as additional testing. If there is good reason to suspect that such further steps will not be taken, then there is good reason to reflect seriously on any decision to disclose information at all.

5. The Nature of the Request

If individuals are asked to disclose or receive genetic information, then the specific nature of the request might have an influence on the outcome one would recommend. For example, if an individual is asked simply to take part in linkage tests to determine a relative's particular risk (for procreative purposes) and the tested individual receives guarantees that he or she will not be given the test results, then such an altruistic gesture is unlikely to conflict in any way with that individual's interests. Compare this with an unexpected advance from a health care professional or relative to disclose a 50% chance of developing a late onset condition in the future. In the former example, the

[42] For example, the United Kingdom Advisory Committee on Genetic Testing (ACGT) has issued a Code of Practice that recommends strongly that over-the-counter test kits should only be made available in respect to carrier status, and not for more severe conditions such as late-onset Huntington's disease or X-linked disorders. The rationale is that, while the discovery of carrier status has no direct implications for a proband's health, the discovery of a fatal condition such as Huntington's disease should not occur outside the clinical setting where full and appropriate counseling can be provided. ACGT, CODE OF PRACTICE FOR GENETIC TESTING OFFERED COMMERCIALLY DIRECT TO THE PUBLIC (London, Dep't of Health 1997).

individual is not being asked to take on board any information about himself or herself, while in the latter they are placed in a position where they have no other option but to do so.

6. The Views and Likely Reaction of the Disclosee

Evidence of how individuals might react to information about their genetic makeup can be of considerable assistance in determining whether a disclosure should be made. Clearly, of most value is evidence that the individual has specifically requested to know or not to know the information in question. This is an expression of autonomy and as such should be respected where possible.

IV. SCENARIO

The following scenario is offered to demonstrate the problems that can arise when the factors, principles, and values discussed above are brought to bear on practical situations. The scenario will be considered from the perspective of the three central stalwarts of medical law and ethics: autonomy; confidentiality; and privacy. It will be argued that the first two of these fail to protect adequately the interests at stake, and an argument will be put in defense of a particular conception of privacy as a means to recognize and protect the interests in question.

BRCA1 is the gene responsible for between 5% and 10% of female breast cancers. It was discovered in 1994 and is known to be 10 times longer than most human genes.[43] This fact means that the likelihood of mutations is increased and this, in turn, has implications for the efficacy of test kits designed to identify the gene, for they cannot detect all mutations. There is a high risk of secondary cancers associated with this disease, but early detection and radical intervention in the form of mastectomy can reduce this risk. Preventative measures, also in the form of mastectomies, can reduce the instances of disease.[44] The condition is also thought to be multifactorial, further complicating matters.

> Nicola is aware of a history of breast cancer in her family. Her mother, her great-grandmother, and one of her aunts died from the disease. Nicola has a sister, Nadia, and three female cousins, Norma, Romana, and Elvira. She does not know the extent to which these relatives are aware of the pattern of disease in the family. Recently, Nicola discovered a lump in

[43] The gene contains around 100,000 base pairs of nucleotides. It was discovered on September 15, 1994, by a team of researchers at the University of Utah.

[44] Tamoxifen has been used in the treatment of breast cancer for over two decades, but its preventative efficacy remains unknown. *See* Susan Nayfield et al., *Potential Role of Tamoxifen in Prevention of Breast Cancer*, 83 J. NAT'L CANCER INST. 1450 (1991).

her breast which has been diagnosed as malignant. She is concerned that the BRCA1 gene runs in her family and that her sister and cousins are at risk. Nicola's physician has advised a mastectomy and has strongly urged her to contact her relatives to arrange testing. Should she approach her sister and cousins with the news of her own disease and urge them to seek medical advice? She is aware, for example, that Nadia is phobic about operations and that Elvira is prone to bouts of depression.

V. SHIFTING PARADIGMS: THE EFFICACY OF AUTONOMY, CONFIDENTIALITY, AND PRIVACY IN MEETING CHALLENGES POSED BY GENETIC ADVANCES

In this section, we explore the nature and content of the ethical and legal principles of autonomy and confidentiality to discern what assistance, if any, they offer in the resolution of this dilemma. In the next section, the conclusions of this section are contrasted with the solution proposed by the author's concept of privacy.

A. The Merits of Existing Paradigms: Autonomy

The term autonomy is derived from the Greek words autos ("self") and nomos ("law" or "rule"). While there is no unifying definition of the principle of autonomy from the philosophical or ethical perspective,[45] certain core elements can be identified that provide a workable model of autonomy for use in the health care setting.

First, the idea of choice is central to the principle of respect for autonomy.[46] To be respected as an autonomous person is to have one's choices respected. Second, crucial to this respect is noninterference. To make one's own choices—that is, for those choices to be autonomous—one must be unrestrained by unwarranted interference by others.[47] Finally, bound up with all of this is possession of the capacity to make one's own choices.[48] Although autonomy is concerned with choice and the exercise of that choice in relation to life decisions, realistically it must be accepted that no person can control, at all times, all aspects of his or her life.[49] It is only necessary that a certain degree of autonomy is reached and that capacity to make a choice is present in

[45] This point is cogently made by Gerald Dworkin, *The Theory and Practice of Autonomy* 5-6 (1988).

[46] *See, e.g.*, JOEL FEINBERG, HARM TO SELF 54 (1986); ISAIAH BERLIN, FOUR ESSAYS ON LIBERTY 131 (1969); Dworkin, *supra* note 45, at 20; JOSEPH RAZ, THE MORALITY OF FREEDOM 370-72 (1986); Joseph Raz, *Autonomy, Toleration, and the Harm Principle, in* ISSUES IN LEGAL PHILOSOPHY 313-33 (Ruth Gavison ed. 1987); Beauchamp & Childress, *supra* note 40, at 121.

[47] *See* Beauchamp & Childress, *supra* note 40, at 121-22; BERLIN, *supra* note 46, at 131; Dworkin, *supra* note 45, at 18-19; Raz, *supra* note 46, at 408-11.

[48] On capacity, see Dworkin, *supra* note 45, at 20; Raz, *supra* note 46, at 314 & 408; Beauchamp & Childress, *supra* note 40, at 132-41.

[49] *See* Raz, *supra* note 46, at 314.

relation to the choice that must be made. The standard that is required in practical terms is always a question to be answered with reference to the facts and circumstances of each case.[50] Rather, what is important is that autonomy and autonomous choices be respected. This is embodied in law in most Western jurisdictions in ways that do not require repetition here.[51]

It should not be thought, however, that the principle of respect for autonomy and the other ethical principles discussed above always function harmoniously. It is easy to imagine situations where an individual might wish to exercise his or her autonomy in a manner that might interfere with the autonomy of others and/or cause them harm and/or treat them unfairly. As Beauchamp and Childress point out:

> Respect for autonomy . . . has only prima facie standing and can be overridden by competing moral considerations. Typical examples are the following: If our choices endanger the public health, potentially harm innocent others, or require a scarce resource for which no funds are available, others can justifiably restrict our exercises of autonomy. The justification must, however, rest on some competing and overriding moral principles.[52]

Thus, just as the principles of nonmaleficence, beneficence, and justice can serve to accord respect to individuals and their autonomy, the same principles can be used to impose restrictions on individual action and autonomy if this conflicts with other third-party interests. Ethical principles provide us with a framework of moral reference within which to analyze human behavior and human interaction. Only in a very crude way, however, do they provide us with the means of resolving conflict.

The relevance to genetic information of what has been said above should be obvious. It already has been argued that aspects of the self such as the body and personal information require respect under the principle of autonomy. The principle also dictates that individuals deserve respect concerning the choices they make about what happens to their bodies and their personal information. Thus, the principle prescribes that choices concerning genetic information are equally deserving of respect. Several problems, however, become immediately apparent. First, given that genetic information concerns many individuals in a family, how can the principle of autonomy help us to resolve conflicts that arise about the control and use of the information? For example, if patient A is tested and found to be a carrier of cystic fibrosis but wishes to keep this to himself, does his pregnant sister nonetheless have a right to the information so

[50] This point is made by Beauchamp & Childress, *supra* note 40, at 123.

[51] The classic pronouncement by Justice Cardozo in Schloendorf v. Society of New York Hospital, 105 N.E. 92 (N.Y. 1914) scarcely requires repetition: "Every human being of adult years and sound mind has a right to determine what shall be done with his body. . . ."

[52] Beauchamp & Childress, *supra* note 40, at 126.

that she can make an appropriate and autonomous choice about her pregnancy? In other words, what is to be done when two autonomies conflict?

Second, it was noted earlier that there exist certain fundamental criteria that are necessary to be an autonomous individual. Central to the principle of autonomy is choice. In particular, choices must be made free from interference and by someone who has the capacity to make those choices. Fundamental to such choices is knowledge. One cannot choose in a meaningful sense if one is not informed of the parameters within which one must choose. This is why informed consent is crucial to ethically and legally acceptable health care. However, in the context of genetics this may be problematic. For, in many circumstances, the problems surrounding genetic information are precisely concerned with the absence of knowledge: this is the basis of the claim to respect the interest in not knowing genetic information. The choice, if there is one, is whether to receive or not to receive information about oneself. This is problematic for the concept of autonomy because it is difficult to see how one can exercise meaningfully a choice not to know unless one has a certain degree of knowledge that there is something to know. Of course, an obvious practical solution would be to approach the individual and ask, "do you want to know this information?," but as Wertz and Fletcher have pointed out, "[t]here is no way . . . to exercise the choice of not knowing, because in the very process of asking, 'Do you want to know whether you are at risk . . . ?' the geneticist has already made the essence of the information known."[53]

This is not to say that one cannot simply state "I wish to know no information about my genetic makeup whatsoever," nor is it to suggest that such a wish should not be respected. However, the requirement that autonomous choices be informed choices tends to imply that the credibility of an uninformed choice is more easily questioned. It leaves it open to be argued that actual knowledge about circumstances might nevertheless affect the chooser who might choose differently if furnished with relevant information. Alternatively, the situation might be seen as analogous to the problem of the incapax. Individuals who are incapax cannot choose for themselves and so must have choices made for them. In the same way, individuals who are ignorant of genetic information might be seen as a pseudo-incapax and therefore it might be assumed that it is legitimate to make choices about the genetic information on their behalf, including the choice whether to know. Choices for the incapax are often made in the person's best interests. It is far from clear, however, how one would determine an individual's best interests concerning genetic information, given that the passing on of knowledge itself can be harmful.[54]

---

[53] *See* Dorothy Wertz & John Fletcher, *Privacy and Disclosure in Medical Genetics Examined in an Ethic of Care*, 5 BIOETHICS 212, 221 (1991).

[54] *See* Roberta Berry, *The Genetic Revolution and the Physician's Duty of Confidentiality*, 18 J. LEGAL MED. 401, 435-36 (1997).

Applying an autonomy perspective to our scenario reveals how these limitations prove problematic in the context of deciding what is the best thing to do with genetic information. Nicola must determine whether to approach her relatives with a possible index of genetic risk in the absence of any views about her relatives' wishes. What guidance might be offered by the principle of autonomy?

At first blush, one might assume that, because no views have been expressed by the relatives, the principle of autonomy is unhelpful. In like manner, the principles of nonmaleficence and beneficence also would appear unhelpful because of the nature of the condition and the circumstances of the family. These principles require that harm should be avoided and benefit conferred wherever possible. It is not clear, however, whether this could be achieved by the subject of our scenario by disclosing information about her condition and the risk to relatives. As has been argued above, harm can result from the mere fact of disclosure and the personal circumstances of two of the relatives would tend to indicate that psychological trauma is probable. Also, it is important to consider the nature of the treatment that is offered. Mastectomy is a very traumatic and potentially psychologically devastating operation. The sequelae can include altered perception of self-image and feelings of loss of identity. The preference of some women might be not to have the operation. This is likely to be true of Nadia who is phobic about surgery. Furthermore, even if testing proves to be negative, exposure to the knowledge of increased risk can heighten concerns about future ill health. Testing and counseling might not allay such fears, especially in someone such as Elvira who is depressive. These factors mean that Nicola should consider very seriously whether to disclose the information.

Yet, the perceived utility of autonomy-based arguments extends beyond circumstances in which a meaningful choice is within the grasp of an individual. For, even in cases where no choice has been made or where no meaningful choice is possible for want of information, autonomy is frequently advanced as a reason to put individuals in a position whereby they can choose. Indeed, the facilitation of autonomous choices is generally a given good in contemporary health care.[55] However, it is important to distinguish between cases in which the physician-patient alliance has been established at the behest of the patient, and those in which an individual is approached by a physician, or some other third party, with information that is perceived to be of benefit to the individual's future health. In the former case, an alliance has been established wherein the goals of the union have been agreed by the parties, and when

[55] *See* JAY KATZ, THE SILENT WORLD OF DOCTOR AND PATIENT 141 (1984) who argues that "[t]he inevitable conflict that such insistence [on disclosure and conversation] creates between the values of autonomy and privacy should be resolved in favour of autonomy. Such invasions of privacy must be tolerated in order to enhance patients' psychological autonomy through insight and not allow it to be further undermined by too hopeful promises, blind misconceptions and false certainties."

the promotion of the patient's health (and autonomy) is one of those goals. In the latter case, there is no mutually agreed alliance, and unilateral efforts to "optimize [someone else's] future health"[56] are ontologically and ethically different. Indeed, Malm has gone so far as to argue in the context of screening that, while a recommended treatment for a patient should be justified by the "preponderance of the evidence" as embodying a benefit, in the case of preventive medicine "the evidence must show it to be beyond reasonable doubt that the recommended procedure will benefit the patient on balance."[57]

This is not, however, a well-accepted view. The preferred view is that autonomy should not only be respected, but sought out where possible. Thus, in our scenario, Nicola might be drawn on an autonomy analysis to disclose her family information to her relatives, in spite of the risks, to allow her relatives to choose for themselves what they wish to do.

For all of these reasons, it is submitted that the principle of autonomy is particularly unhelpful in addressing the question of an interest not to know. Nicola cannot simply approach her relatives to ask if they would like to know, because this in itself immediately compromises the interest in not knowing. If she treats them as incapax, she then must consider what is in their best interests, but this is not easily discernible on the facts. Nor does it have anything to do with the autonomy of her relatives, but rather their perceived incapacity. Finally, if she seeks to facilitate their autonomy, then disclosure is likely, because the bias is to allow persons to decide for themselves by knowing the options that are available. This, however, ignores the fact that the interest at stake is one about whether to receive knowledge at all.

All of this would tend to indicate that the basis for a claim not to know information cannot be the principle of autonomy alone.

B. The Merits of Existing Paradigms: Confidentiality

Confidentiality is concerned with security of information. To be precise, it is concerned with the security of confidential information. To be confidential, information must be in a state of limited access from individuals, groups, bodies, and institutions generally.[58] The nature of the confidential relationship has been described above. Most particularly, it is accepted almost unquestionably that health care professionals owe a duty of confidence to their patients and that only exceptionally should disclosure without consent be made.[59]

---

[56] Bruce Charlton, *Screening Ethics and the Law*, 305 BRIT. MED. J. 521 (1992).

[57] Heidi Malm, *Medical Screening and the Value of Early Detection: When Unwarranted Faith Leads to Unethical Recommendations*, 21 HASTINGS CENTER REP. 26, 36 (Jan.-Feb. 1999).

[58] Information that is in the public domain cannot be confidential and therefore cannot be protected by confidentiality. Similarly, once information moves from the private sphere where it is confidential to the public sphere, it loses the necessary quality of confidence.

[59] *See, e.g.*, Alberts v. Devine, 479 N.E.2d 113 (Mass. 1985); Bryson v. Tillinghast, 749 P.2d 110 (Okla. 1988); People v. Doe, 435 N.Y.S.2d 656 (Sup. Ct. 1981).

It is trite, therefore, to confirm that the woman who has been diagnosed with breast cancer in our scenario is owed a duty of confidence by her health care professional. This entitles her to decide whether and how the information, which is the object of the duty, should be disclosed to others. However, in the case of her relatives, it would be possible for the health care professional to justify disclosing the information to them even without the patient's consent, because the duty of confidentiality is not absolute, and certain exceptions are admitted, including actions to prevent harm to third parties.[60] A considerable and far-reaching discretion to disclose on such grounds is afforded to health care professionals. Yet, this is not our problem in the scenario. Nicola is willing to tell her relatives about her condition; our problem is that it is not clear that she should do so. What is clear, however, is that she cannot herself breach a duty that is owed to her by another. If she decides to disclose her condition to her family, then there could be no question of a breach of the duty of confidence. Thus, for her disclosure to amount to a breach of confidence, it must be seen to be an invasion of someone else's right and a breach of her duty to maintain confidentiality. Can the family dynamic envisaged by our scenario fit into such a rubric?

The first matter to determine is the circumstances in which a duty of confidence arises as between two parties. Legally,[61] professionally,[62] and ethically,[63] health care professionals owe duties of confidentiality to their patients. While the sources of this duty in law are many and varied,[64] in each case the duty arises with respect to the specific relationship that the professional has

[60] For example, in Hague v. Williams, 181 A.2d 345, 349 (N.J. 1963) it was stated: "[A]lthough ordinarily a physician receives information relating to a patient's health in a confidential capacity . . . where the public interest or the private interest of the patient so demands . . . disclosure may, under . . . compelling circumstances, be made to a person with a legitimate interest in the patient's health."

[61] *See, e.g.*, Horne v. Paton, 287 So. 2d 824 (Ala. 1973); McDonald v. Clinger, 446 N.Y.S.2d 801 (N.Y. Sup. 1982); *In re* Lasswell, 673 P.2d 855 (Or. 1983). *Cf.* Logan v. District of Columbia, 447 F. Supp. 1328 (D.D.C. 1978); Quarles v. Sutherland, 389 S.W.2d 249 (1965). For the statutory legal position as of 1996, see LARRY GOSTIN ET AL., LEGISLATIVE SURVEY OF STATE CONFIDENTIALITY LAWS, WITH SPECIAL EMPHASIS ON HIV AND IMMIGRATION (1996); JANET SAUNDERS, PATIENT CONFIDENTIALITY 56-67 (1996). For the common-law position, see Judy Zelin, *Annotation: Physician's Tort Liability for Unauthorized Disclosure of Confidential Information*, 48 A.L.R.4TH 668 (1998).

[62] Principle IV of the American Medical Association *Code of Medical Ethics* states: "A physician shall respect the rights of patients, of colleagues, and of other health professionals, and shall safeguard patient confidences within the constraints of the law." *See also* Council on Ethical and Judicial Affairs, *Fundamental Elements of the Patient-Physician Relationship*, 264 J.A.M.A. 3133 (1990): "The patient has the right to confidentiality. The physician should not reveal confidential communications or information without the consent of the patient, unless provided for by the law or by the need to protect the welfare of the individual or the public interest."

[63] Of the numerous versions of the Hippocratic Oath that are available, the following passage is typical in respect of the obligation to maintain confidences: "[A]ll that may come to my knowledge in the exercise of my profession or outside of my profession or in daily commerce with men, which ought not to be spread abroad, I will keep secret and will never reveal." *See* Berry, *supra* note 54, at 408-13 (overview of the history of the Hippocratic Oath).

[64] For an account, see WILLIAM ROACH, JR., MEDICAL RECORDS AND THE LAW 91ff (3d ed. 1998).

with patients, qua patients, and it is not thought to be the case that legal duties arise merely by virtue of the fact that an individual comes into possession of personal information about another.[65] Thus, absent some specific customary or professional, contractual or impliedly contractual relationship,[66] a duty to maintain confidences is unlikely to arise.[67] This would suggest that, in our scenario, no duty of confidentiality is owed to the relatives of Nicola, either by the health care professional or Nicola.[68] Each is therefore free to disclose the information, subject solely to Nicola's wishes.

Moreover, even if a duty of confidentiality were owed, protection of the interest in not knowing information cannot flow from this legal construct. It would be unreasonable to suggest that the duty can be breached by one relative telling the person to whom the duty is owed the information in question. A duty of confidence is breached when confidential information is used or disclosed to those outside the confidential relationship. A breach of duty is constituted by making the information in some way public. Precisely how public any use or disclosure must be is a matter of debate, but it cannot be the case that disclosure of information from one party to a confidential relationship to the other party in any way makes the information public. This then means that even if a duty of confidence is owed in our scenario by Nicola to her female relatives, she could not breach that duty by disclosing the information to the women themselves. Thus confidentiality provides no means to protect the interest that these women have in not knowing information about themselves.

One might suggest it is the confidential relationship that receives protection and not the information conveyed between the parties. If this were not true, then why would it matter who was in possession of the information or how they came by it? If the information were protected, then it would be protected irrespective of the circumstances in which it was imparted or received. The confidential quality of the information would be enough to merit protection. In the authorities cited above, however, it is clear that the decisions

---

[65] *See* Note, *Breach of Confidence: An Emerging Tort*, 82 Colum. L. Rev. 1426 (1982).

[66] In Quarles v. Sutherland, 389 S.W.2d 249 (Tenn. 1965), the provision of free medical treatment meant that no contractual duty arose between physician and patient, and so no duty of confidentiality. In Macdonald v. Clinger, 446 N.Y.S.2d 801 (N.Y. App. 1982) the court stated: "We believe that the relationship contemplates an additional duty springing from but extraneous to the contract that the breach of such duty is actionable as a tort." This is a fiduciary relationship arising from the contractual nature of the relationship. Hammonds v. Aetna Cas. & Sur. Co., 243 F. Supp. 793 (N.D. Ohio 1965).

[67] In Darnell v. Indiana, 674 N.E.2d 19 (Ind. 1996) a physician-patient privilege statute was construed narrowly to exclude nurses from the duty of confidentiality. Similarly, in Evans v. Rite Aid Corp., 478 S.E.2d 846 (S.C. 1996) and Suarez v. Pierard, 663 N.E.2d 1039 (Ill. App. 1996), pharmacists were held to have no duty in the absence of a contractual obligation.

[68] This would be the case in other jurisdictions, such as those of the United Kingdom, where a duty of confidence can arise provided that a reasonable person would realize, or should have realized, that he or she was receiving information in circumstances that imported a duty of confidence. *See* Attorney General v. Guardian Newspapers (No. 2) [1990] AC 109; Lord Advocate v. Scotsman Pub. Ltd., 1989 SLT 705; Stephens v. Avery [1988] Ch. 449.

have little to do with the nature of the information and everything to do with the recognition of a relationship as privileged. The right of confidentiality is, therefore, a right in personam and not in rem.

The consequence of this is that the concept of confidentiality does not accord a right to relatives of a proband to control the flow of familial genetic information toward themselves. If they are to be informed of familial information, then this either will be as an exercise of the right of the proband to control his or her own information, or as a result of a discretion exercised by a health care professional to breach confidentiality without fear of sanction.[69] More importantly for present purposes, it is not clear that, even if a duty of confidence is owed between relatives concerning their common genetic information, such a duty could ever be breached simply by telling relatives themselves about their own personal information. It therefore becomes apparent that the law of confidence cannot address the question of protecting a possible interest in not knowing information, such as might arise in the scenario described above.

C. The Protection of Privacy Interests by Autonomy and Confidentiality

This section draws the first part of this article to a conclusion. We have examined the nature of the interests that individuals have in genetic information and these can be described as being of two sorts. First, there are interests that concern issues of security of existing information, and second, there are interests that relate to the protection of the self from unwarranted intrusion, including intrusion with information about one's own being. We have seen in the last two sections how the existing concepts of autonomy and confidentiality fare in protecting the latter type of interest and we can conclude that the major problem arises in the context of an interest in not knowing. While both confidentiality and autonomy can, to an extent, help to protect interests concerning access and control of known genetic information, these concepts, alone or in combination, cannot furnish a useful, precise, and effective means of articulating all of the interests involved, or of protecting them in an appropriate fashion. The solution that is proposed in the next section is that of a concept of genetic privacy, for it is submitted that the interests at stake are, in essence, privacy interests. Presently, therefore, a definition of privacy

---

[69] The Committee on Genetic Risks of the Institute of Medicine has gone even further and has suggested that an obligation of disclosure to relatives should be imposed in certain circumstances:

> The Committee recommends that confidentiality be breached and relatives informed about genetic risks only when attempts to elicit voluntary disclosure fail, there is a high probability of irreversible or fatal harm to the relative, the disclosure of the information will prevent harm, the disclosure is limited to the information necessary for diagnosis or treatment of the relative, and there is no other reasonable way to avert the harm.

ASSESSING GENETIC RISKS, IMPLICATIONS FOR HEALTH AND SOCIAL POLICY 278 (Lori Andrews et al. eds. 1994). *See also* PRESIDENT'S COMMISSION FOR THE STUDY OF ETHICAL PROBLEMS IN MEDICINE AND BIOMEDICAL AND BEHAVIORAL RESEARCH, SCREENING AND COUNSELING FOR GENETIC CONDITIONS (1983).

is argued for and defended within the context of the wider debate about the value of privacy per se and its current protection both at common law and under the United States Constitution. This novel definition is then applied to the genetic information scenario to show how the privacy interests involved can best be protected by an appeal to privacy itself.

VI. PRIVACY

A. Privacy: A Definition

A valuable concept of privacy should reflect the privacy needs of persons in society. In Western society, these needs are reflected in two views of privacy. First, privacy can be seen as a state of nonaccess to the individual's physical body or psychological person; what I will call spatial privacy. Second, privacy can be viewed as a state in which the individual has control over personal information; what I will term informational privacy. From these two conceptions of privacy, one can deduce a single unifying definition: privacy as a state of separateness from others. Thus, privacy should be taken to refer to a state in which an individual is separate from others, either in a bodily or psychological sense, or by reference to the inaccessibility of certain intimate adjuncts to their individuality and personality, such as personal information. The reasons for this choice of definition are more fully considered and properly justified in the next section.

B. Why Protect Privacy?

A number of arguments can be offered to justify privacy protection. First, it has been posited by several commentators that a state of physical separateness from others is necessary in order to allow personal relationships to begin and to grow. The levels of intimacy that typify the modern personal relationship only can be achieved by ensuring and securing separateness from others. Trust, which is essential to the establishment and maintenance of all relationships, requires not only a degree of intimacy to develop but also a currency in which to deal. An important part of that currency is personal information. Individuals trade private information both as a sign of trust and on the basis of trust. The security of the information is guaranteed by the tacit undertaking that it will not be noised abroad. In this way, personal and professional relationships flourish and an important part of the fabric of society is woven more tightly. As Fried has said:

> Love and Friendship . . . involve the initial respect for the rights of others which morality requires of everyone. They further involve the voluntary and spontaneous relinquishment of something between friend and friend, lover and lover. The title to information about oneself conferred by privacy provides the necessary something. To be friends or lovers persons must be intimate to some degree with each other. Intimacy

is the sharing of information about one's actions, beliefs, or emotions which one does not share with all, and which one has the right not to share with anyone.[70]

Second, a degree of separateness—that is, being alone with no company or merely selected company—allows the individual personality to reflect on experiences and learn from them. Constant company, and so constant interaction, deprives the individual of time to assimilate life experiences and to identify with one's own individuality.[71]

Third, it has been argued that the modern psychological makeup of individuals is such that a degree of separateness is required to ensure that individuals retain a degree of mental stability. Jouard has put a forceful argument that (Western) public life puts considerable strain on individuals who must assume certain personae to integrate with others. These personae, not being full and true reflections of the personality of the individual, cannot be maintained indefinitely without serious psychological consequences. A state of privacy allows the masks to be dropped and a degree of release to be obtained.[72]

Fourth, tangible harm can come to an individual who is not granted a degree of privacy. As regards spatial privacy, invasion on the body, which is unauthorized, is disrespectful of the individual and might, of course, cause physical harm. The criminal and civil laws of assault recognize and protect to a degree the inviolability of the human body. Perhaps less obvious but no less valid, however, is the mental harm that can arise if one's spatial privacy is not respected. For example, clandestine observation can produce profound feelings of violation in individuals even though no actual physical contact occurs and/or no personal information is gathered.[73] Similarly, unauthorized use or disclosure of personal information can lead to harm to individuals. Information about one's personal condition, behavior, or habits, which others find distasteful, can lead to individuals being ostracized from communities or becoming the object of violence and discrimination. As Greenawalt puts it: "One reason why information control seems so important is precisely because society is as intolerant as it is, precisely because there are so many kinds of activity that are subject to overt government regulation or to the informal sanctions of loss of job or reputation."[74]

There is, moreover, one final argument in support of the protection of privacy. While the above arguments concentrate on individual interests, it is important to recognize that there are also public interests in privacy protection.

---

[70] CHARLES FRIED, AN ANATOMY OF VALUES: PROBLEMS OF PERSONAL AND SOCIAL CHOICE 142 (1970).

[71] JULIE INNESS, PRIVACY, INTIMACY AND ISOLATION (1992).

[72] Sidney Jouard, *Some Psychological Aspects of Privacy*, 31 LAW & CONTEMP. PROBS. 307 (1966). *See also* PETER PETSCHAUER, HUMAN SPACE: PERSONAL RIGHTS IN A THREATENING WORLD (1997).

[73] *See* Stanley Benn, *Privacy, Freedom and Respect for Persons*, *in* PHILOSOPHICAL DIMENSIONS OF PRIVACY: AN ANTHOLOGY 230-31 (Ferdinand Schoeman ed. 1984).

[74] Kent Greenawalt, *Privacy and Its Legal Protections*, 2 HASTINGS CENTER STUD. 45 (1974).

For example, it can be argued that it is in the public (societal) interest to have a community inhabited by complete individuals as opposed to two-dimensional characters.[75] For a society that holds the individual in esteem and seeks to accord him or her respect, it is surely in the public interest to reduce to a minimum all potential harm to individuals. Moreover, it should not be overlooked that harm can come to society if privacy is not respected. If the element of trust which is so crucial to the development of relationships is lost, because individuals cannot seek and receive guarantees about the security of information, important and valuable information will not be communicated. This can render important social organs powerless to deal with a variety of social conditions. This is especially true in the health care context, where physician-patient trust is seen to be essential to an effective and beneficial therapeutic relationship.[76] If that trust is compromised because individual privacy is not protected, then public and private health may suffer as a result.

These arguments support the effort to protect privacy as a construct of general social good. The specific definition of privacy advanced in this work is, however, two-pronged: it relates both to spatial and informational privacy. There are strong reasons for recognizing and protecting both kinds of individual privacy—reasons that also are grounded in both private and public interests. These are best discussed in the context of the health care setting for which the definition of privacy that is offered is intended.

C. Spatial and Informational Privacy: A Medical-Legal Definition of Privacy

Privacy was defined above broadly as a state of separateness from others. Such a state encompasses two forms of separateness. Physical or psychological separateness from others (spatial privacy) and separateness of certain intimate adjuncts to one's personality; namely, personal information (informational privacy). The argument for viewing privacy in such terms is as follows.

First, consider informational privacy. Undoubtedly, patients have considerable interests in their own medical information because, inter alia, it can be used against them by others and this can lead to harmful outcomes such as discrimination and prejudice. Informational privacy therefore concerns the interest of the patient in maintaining such information in a state of

---

[75] Benn, *supra* note 73, at 237, observes that

> the children of the kibbutz have been found by some observers defective as persons, precisely because their emotional stability has been purchased at the cost of an incapacity to establish deep personal relations. Perhaps we have to choose between the sensitive, human understanding that we achieve only by the cultivation of our relations within a confined circle and the extrovert assurance and adjustment that a Gemeinschaft can offer. However this may be, to the extent that we value the former, we shall be committed to valuing the right of privacy.

[76] The Supreme Court has expressly recognized the importance of the public and private interests in protecting the confidential nature of the therapeutic alliance (most particularly in the context of the psychotherapeutic relationship). *See* Jaffee v. Redmond, 518 U.S. 1 (1996).

nonaccess and preventing unauthorized use or disclosure of the information to third parties. For purposes of this article, the information in question is genetic information. Thus, a concern about informational privacy is a concern about maintaining a state of nonaccess to personal genetic information. For reasons already articulated, an interest in genetic informational privacy can be claimed both by a proband and his or her blood relatives.

Second, let us examine spatial privacy. It is submitted that, as a caveat to the above, a concept of privacy that is defined solely in informational terms does not adequately reflect the interests patients have in privacy matters and so cannot purport to protect comprehensively such interests. The concept of spatial privacy is therefore offered as a complement to the concept of informational privacy. The concern of spatial privacy is not simply information. Rather, spatial privacy relates to the sphere of the self—a bubble of privateness around the individual that cannot and should not be invaded without due cause. Such a sphere of separateness from others can be invaded either by unwarranted physical contact (such as unauthorized treatment or continued futile medical treatment) or by uninvited intrusion into the sphere of psychological integrity that individuals create for themselves. In the context of genetic information, it is submitted that spatial privacy can be invaded by the revelation of genetic data about an individual to that self-same individual, if there is no indication that the individual would want to know such information. This cannot appropriately be seen as an informational privacy issue because this latter privacy interest concerns the interest in maintaining nonaccess vis-à-vis third parties. In the example under discussion, the concern is revelation of information about oneself to oneself. Informational privacy focuses on the control that an individual can exercise over his or her personal information. Spatial privacy protection cannot focus on control of information because its domain is the maintenance of a state of ignorance, wherein information is unknown and therefore beyond the reach of any meaningful exercise of control. One cannot control that which one does not know to exist.

The justifications for this two-fold conception of privacy are numerous. First, the conception of what is private in lay terms accords to a high degree with the view of privacy advocated in this work.[77] This is important because it goes a long way toward helping us formulate a view of the law that can address actual social needs. Moreover, this definition pinpoints interests that already are recognized by privacy laws in the United States, yet which are currently underprotected, as will be demonstrated presently.

Second, to define privacy as a state rather than a right or a claim helps us to describe the concept while at the same time avoids imputing value to it. As

---

[77] *See, e.g.*, the Electronic Privacy Information Center survey list at <www.epic.org/privacy/survey>. *See also* RAYMOND WACKS, PERSONAL INFORMATION, PRIVACY AND THE LAW (1989).

has been noted, privacy is defined as a state of separateness from others, be that society in general, the family, or other individuals. This is not to say that others cannot enter that sphere, nor that individuals simply can act however they would wish when in such a sphere, nor that such a state necessarily protects undesirable activities. Rather, it is to say that prima facie a state of privacy places the individual apart from others.[78] Yet, merely to say that I am apart from others will not always lead us to conclude that I am in a state of privacy. For example, if I am marooned on an island, then I am certainly apart from others, but few of us would say that I have privacy. This is in part because privacy implies something more than mere isolation, which can be seen as undesirable. To be in a state of privacy, one must be in a context where there are others from whom one can be separate. On a desert island, this is not possible for one is alone. This is isolation, which implies a state of enforced nonaccess to others. Privacy, on the other hand, is a state that easily can be relaxed or maintained because it occurs in a social setting. Isolation concerns the removal of individuals from a social context and therefore cannot accurately be described as privacy.[79]

Third, simply because I am in the presence of others does not necessarily mean that I cannot claim privacy interests. For example, an aspect of spatial privacy is the interest in maintaining bodily integrity. It is not because I am in a crowd that unwarranted interferences with my bodily integrity are not offensive and cannot be classed as invasions of privacy. Intentional contact with my person by another easily can be seen as an invasion of privacy. However, incidental touchings are a necessary and obvious part of entering a crowded public forum and could not reasonably be treated as an invasion of privacy.[80] Similarly, one would not say that one's privacy interest in not being observed is invaded by being in a crowd. Arguably, in such cases, one has consented to a degree of observation or physical contact—that which flows directly and naturally from one's presence in the public sphere.[81] This having been said, if one's movements were to be recorded clandestinely, then a strong argument could be made that this does indeed infringe privacy interests. There is a considerable difference between the anonymity of the crowd and the specific identification of an individual within a crowd. In the former case, any observation that occurs is merely incidental and readily

---

[78] This notion can of course extend to the privacy of groups that are apart from other groups or society in general, but the common denominator is the individual and his or her separateness.

[79] Consider the position of the prisoner condemned to solitary confinement. The prisoner has been removed from a social context (not simply society in general but also the community of the prison population) and has been placed in isolation. Such a person does not have privacy. But, a prisoner who retires to the cell to read does have privacy in that the prisoner is separate from the rest of the prison community.

[80] Of course, mere jostling or accidental contact is subject to the *de minimis* principle and can be explained by the implied consent of the individual who is in a public setting.

[81] One could say precisely the same about bodily contact and spatial privacy interests in a public crowd.

can be anticipated by the individual in question. If, however, one is being clandestinely observed, then one cannot reasonably anticipate being the focus of someone else's attention. Moreover, one becomes a means to someone else's end: a factor that, in itself, is offensive and disrespectful of the individual.[82] The specificity of the information obtained by recorded observation is an additional factor that differentiates the two experiences. In like manner, the specificity of detail that can accompany genetic test results, and the implications the use of that information can have for those identified, should serve to heighten our privacy concerns.

Fourth, to describe privacy as a state and therefore to seek to offer a neutral description of the concept of privacy does not preclude us ultimately from attributing value to such a state. Nor does it prevent us from seeking to accord (legal) protection to such a state for the good ends that it can further and for the interests it can protect. It already has been argued that a state of separateness can protect good ends—both private and public. But, in essence, such a state can be seen as one in which the interests of the individual are paramount. If one chooses to accord respect and protection to such a state, then this is evidence of a degree of commitment to valuing individuals. But, the obvious question that arises from this is, why should we seek to protect such a state of privacy when we already have mechanisms for respecting individuals and protecting their interests? The response is that such existing mechanisms cannot always provide adequate protection. Furthermore, the concept of privacy advanced here allows us to recognize a broad range of interests that might otherwise go unrecognized. To view privacy either as solely concerned with personal information, or to argue that autonomy or confidentiality (or liberty) can adequately protect privacy interests is to fail to protect important interests and to miss many interesting nuances. This having been said, one criticism that might be leveled at the view of privacy presented here is that it confuses privacy with concepts such as autonomy, confidentiality, or even liberty. For example, a state of separateness implies a state of noninterference, which is arguably simply one definition of liberty or freedom. Similarly, it might be argued that the state in question is one that depends largely on the notion of autonomy—the individual as self-ruler. This would be an important criticism and even if it were not raised in respect of the definition of privacy advanced here, the relationship between privacy and these related concepts nevertheless must be examined.

[82] This is not to say that strong counterarguments cannot be made to justify such observation—for example, closed circuit television in shopping malls for security purposes. However, if such tapes are sold to television shows for entertainment purposes, then arguably this becomes an offensive use of the images obtained.

D. Privacy and Related Concepts

Many writers associate the beginning of legal interest in privacy in the United States with the seminal article by Warren and Brandeis, *The Right to Privacy*, published in the 1890-91 volume of the *Harvard Law Review*.[83] From such humble beginnings was born the tort of invasion of privacy.[84] Warren and Brandeis examined cases drawn from areas as diverse as defamation,[85] breach of confidence,[86] and copyright,[87] and concluded that the common law recognized common interests in each of these actions, which could be subsumed under the rubric of a general right to privacy.[88] This they classified as a "right to be alone."[89] For present purposes, it is neither intended to praise[90] nor particularly criticize[91] this work, but rather to offer it as an illustration of a common problem that arises in the field of privacy study; namely, conflation of concepts and confusion of terminology. The association of privacy with the "right to be alone" has been made by many writers since Warren and Brandeis,[92] and all have been subject to the same criticism: by conceiving privacy to be a "right" to be free from intrusion or interference, they have equated privacy with liberty. This is not only confusing generally, but for those who seek to argue positively about privacy, it can have adverse consequences. For example, Fried has recognized that "to present privacy only as an aspect of or an aid to general liberty is to miss some of its most significant differentiating features."[93] Similarly, Posner has observed: "[W]e already have perfectly good words—Liberty, Autonomy, Freedom—to describe the interest in being allowed to do what one wants (or chooses) without interference. We should not define privacy to mean the same thing and thereby obscure its other meanings."[94]

---

[83] Samuel Warren & Louis Brandeis, *The Right to Privacy*, 4 HARV. L. REV. 193 (1890-91).

[84] For a seminal analysis, see Dean Prosser, *Privacy: A Legal Analysis*, 48 CAL. L. REV. 338 (1960).

[85] Warren & Brandeis, *supra* note 83, at 205.

[86] *See, e.g.*, Abernathy v. Hutchinson, 3 L. J. Ch. 209 (1825); Prince Albert v. Strange, 1 McN. & G. 25 (1849).

[87] *E.g.*, Tuck v. Priester, 19 QBD 639 (1887).

[88] It is often pointed out that, ironically, the authors relied heavily on English common-law cases to support their argument and yet, it was not until the closing days of the twentieth century that the English Courts recognized a common-law right to privacy. *See* Douglas v. Hello! Limited (Court of Appeal, Dec. 21, 2000) (unreported).

[89] The "right to be alone" was first expounded in Justice Cooley, *A Treatise on the Law of Torts* 29 (2d ed. 1888).

[90] *See, e.g.*, Ruth Gavison, *Too Early for a Requiem: Warren and Brandeis Were Right on Privacy vs. Free Speech*, 43 S.C. L. REV. 437 (1992).

[91] *E.g.*, Diane Zimmerman, *Requiem for a Heavyweight: A Farewell to Warren and Brandeis' Privacy Tort*, 68 CORNELL L. REV. 291 (1984); Walter Pratt, *The Warren and Brandeis Argument for a Right to Privacy*, 1975 PUB. L. 161; Edward Bloustein, *Privacy, Tort Law and the Constitution: Is Warren and Brandeis' Tort Petty and Unconstitutional as Well?*, 46 TEX. L. REV. 611 (1968); Harry Kalven, *Privacy in Tort Law: Were Warren and Brandeis Wrong?*, 31 LAW & CONTEMP. PROBS. 326 (1966).

[92] *See, e.g.*, Louis Blom-Cooper, *The Right to Be Let Alone*, 10 J. MEDIA L. & PRAC. 53 (1989).

[93] Charles Fried, *Privacy*, 77 YALE L.J. 475, 490 (1968).

[94] Richard Posner, *An Economic Theory of Privacy*, *in* PHILOSOPHICAL DIMENSIONS, *supra* note 73, at 274-75.

E. Conflation of Concepts

Today, privacy is protected in the United States at a number of different levels and by a number of different means.[95] Central among these is the common-law right, of which Warren and Brandeis were the progenitors and to which we shall return presently, and the Supreme Court's creation: the constitutional right to privacy. The latter has been much criticized ever since it was "interpreted out" of the Constitution by the Court in 1965[96] in *Griswold v. Connecticut*.[97] Once again, however, one major criticism, which is frequently voiced, is the alleged confusion of "privacy" with "liberty." Parent, for example, argues:

> The defining idea of liberty is the absence of external restraints or coercion. A person who is behind bars or locked in a room or physically pinned to the ground is unfree to do many things. Similarly, a person who is prohibited by law from making certain choices should be described as having been denied the liberty or freedom to make them. The loss of liberty in these cases takes the form of a deprivation of autonomy. Hence we can meaningfully say that the right to liberty embraces in part the right of persons to make fundamentally important choices about their lives and therewith to exercise significant control of different aspects of their behaviour. It is clearly distinguishable from privacy, which condemns the unwarranted acquisition of undocumented personal knowledge.[98]

Parent is of the opinion that all of the United States constitutional privacy cases "conflate the right to privacy with the right to liberty."[99] While one might not agree with his particular definition of privacy, his point on confusion of concepts is, nevertheless, a valid one. DeCew offers the following explanation: "Given early association of a legal right to privacy as a right to be let alone and the well-known explanation of a concept of negative liberty in terms of freedom from interference, it is hardly surprising that privacy and liberty should often be equated."[100]

There is, however, an additional problem that stems from the fact that, although one may accept wholeheartedly that privacy and liberty, as defined

---

[95] For specific accounts in the health care context, see ROACH, *supra* note 64; JONATHAN TOMES, HEALTHCARE PRIVACY AND CONFIDENTIALITY: THE COMPLETE LEGAL GUIDE (1994).

[96] Louis Henkin, *Privacy and Autonomy*, 74 COLUM. L. REV. 1410 (1974) has argued that *Griswold* and its progeny have given rise to "an additional zone of autonomy of presumptive immunity to governmental regulation." This constitutional right of privacy he considers, "may not add much protection to "traditional value privacy." *Id.* at 1424-25. A similar criticism has been advanced by Herman Gross, *Privacy and Autonomy, in* PHILOSOPHY OF LAW 246-51 (Joel Feinberg & Herman Gross eds. 2d ed. 1980).

[97] 381 U.S. 479 (1965). For an account of the historical precursor to this case and its impact on privacy rights, see Jed Rubenfeld, *The Right of Privacy*, 102 HARV. L. REV. 737 (1989).

[98] William Parent, *Privacy, Morality and the Law*, 12 PHIL. & PUB. AFF. 269, 274-75 (1983). This is Parent's definition of privacy.

[99] *Id.* at 284.

[100] Judith Wagner DeCew, *The Scope of Privacy in Law and Ethics*, 5 LAW & PHIL. 145, 162 (1986).

by Parent, are completely separate, it does not necessarily follow that the two concepts raise issues wholly unconnected to each other. Furthermore, as DeCew points out in relation to the case law, "it is not at all clear that Parent has shown that the constitutional privacy cases involve no 'genuine' privacy interests."[101]

Clearly, however, the two concepts are by no means synonymous. As DeCew states, it is simple to show how one's notion of privacy can be shown to be distinct from that of liberty. The example she gives is where one's privacy is being constantly invaded by surreptitious surveillance, of which one is unaware, thereby having no effect on one's liberty. To this, one could add the example of genetic testing where information is gathered about oneself from family members when one is wholly ignorant of the fact. Both of these examples involve invasion of one's private sphere yet entail no impingement on one's liberty. DeCew comments: "While the word 'privacy' could be used to mean freedom to live one's life without governmental interference, the Supreme Court cannot so use it since such a right is at stake in every case. Our lives are continuously limited, often seriously, by governmental regulation."[102]

In fact, the Supreme Court has expressly rejected this idea.[103] However, we can once again accept that, while this particular conflation of privacy with liberty might be wrong, this does not necessitate that we reject completely the possibility of a relationship between the two concepts. Just as DeCew gives examples of privacy issues that do not involve liberty, and vice versa, she equally talks of autonomy examples that exclude all mention of privacy.[104] She qualifies this immediately, however, by acknowledging that

> a subset of autonomy cases, however, certain personal decisions regarding one's basic lifestyle, can plausibly be said to involve privacy interests as well. They should be viewed as liberty cases in virtue of their concern over decision-making power, whereas privacy is at stake due to the nature of the decision. More needs to be said about which decisions and activities are private ones, but it is no criticism or conflation of concepts to say that an act can be both a theft and a trespass. Similarly, acknowledging that in some cases there is both an invasion of privacy and a violation of liberty need not confuse those concepts.[105]

What a defense of privacy can do, however, is protect some forms of liberty—principally those relating to the personal sphere of individuals' lives. The same is true for autonomy, and, in the case of personal information, this

---

[101] *Id.* at 161.

[102] *Id.* at 162. More recently, and more generally, see JUDITH WAGNER DECEW, IN PURSUIT OF PRIVACY: LAW, ETHICS AND THE RISE OF TECHNOLOGY (1997) (especially chapter 7).

[103] Paris Adult Theatre I v. Slaton, 413 U.S. 49 (1973). *See also* Washington v. Glucksberg, 521 U.S. 702 (1997); Vacco v. Quill, 521 U.S. 793 (1997) (assisted suicide cases).

[104] DeCew, *supra* note 100, at 164-65.

[105] *Id.* at 165.

can be said of confidentiality too. Many commentators who concern themselves with the concepts of liberty or autonomy face problems of conceptual confusion, difficulty of definition, and ambiguities of scope. Beauchamp and Childress, for example, point out that autonomy is terribly conceptually confused and "not an unequivocal concept in either ordinary English or contemporary philosophy."[106] Dworkin similarly considers a plethora of definitions of autonomy offered by writers in that field almost none of which is in conformity with any other.[107] Berlin has noted in the context of liberty: "Almost every moralist in human history has praised freedom. Like happiness and goodness, like nature and reality, the meaning of this term is so porous that there is little interpretation that it seems able to resist."[108]

As a way through this conceptual mire, it is helpful to recognize that notions such as liberty, autonomy, and privacy are interrelated. One could go as far as to say that they are interdependent, each one relying on the other to fulfil its true function in the best possible way.[109] Consider the impossibility of making autonomous choices without a degree of freedom from interference.[110] Consider the residual value of liberty if one's life choices are never respected. Consider whether it is feasible to be truly free or fully autonomous without some sphere of the private? Liberty and autonomy cannot properly fulfil their function or potential in protecting individuals and their interests without a concomitant commitment to a respect for privacy. Each of these concepts performs the same function, albeit in different ways: each represents an expression of the fundamental respect that a liberal society has for its citizens. Yet, each is also open to criticism as ill-defined, anticommunitarian, and conceptually obfuscated. Accordingly, these reasons are insufficient in themselves to deny a healthy respect for privacy.

This having been said, it might be that we see liberty and autonomy as ends in themselves rather than as means to an end, while we may view privacy purely as a device to achieve a certain end. Even so, it is submitted that it is not necessary to show privacy to be a fundamental and ultimate value of itself to argue validly for its protection. Furthermore, as Gavison points out:

---

[106] Beauchamp & Childress, *supra* note 40, at 120-21.

[107] Dworkin, *supra* note 45.

[108] BERLIN, *supra* note 46, at 121.

[109] This is supported by Robert Hallborg, *Principles of Liberty and the Right to Privacy*, 5 LAW & PHIL. 175 (1986) who argues for a view of privacy that is deduced from fundamental principles of liberty. This he does "to obtain a right to privacy which is not easily defeasible, and a right which ought to be a permanent part of our legal system."

[110] Greenawalt has argued, for example, that "[g]iven a society in which many lifestyles and points of view evoke negative reactions if publicly known, a substantial degree of freedom from observation is essential if there is to be any genuine autonomy; and real choice also depends on the ability of persons to enjoy states of privacy without intrusion." RIGHTS OF PRIVACY 199 (John Shattuck ed. 1977). The original version of this material may be found at Greenawalt, *supra* note 74.

> Privacy has as much coherence and attractiveness as other values to which we have made a clear commitment, such as liberty. Arguments for liberty, when examined carefully, are vulnerable to objections similar to the arguments . . . [against] privacy, yet this vulnerability has never been considered a reason not to acknowledge the importance of liberty, or not to express this importance by an explicit commitment so that any loss will be more likely to be noticed and taken into consideration.[111]

Gavison argues that the case for an explicit commitment to privacy is made by pointing out the distinctive functions of privacy in our lives. Are there, then, specific functions for privacy to perform over and above a general support for other concepts such as liberty and autonomy? It has been argued thus far in this article that this is indeed the case in the context of genetic information. We have seen how concepts such as autonomy and confidentiality do not, and cannot, address the concerns and interests that surround the availability of genetic information. Moreover, it is submitted that, while the constitutional right to privacy and the common-law tort of invasion of privacy reflect the interests that are protected by the view of privacy described above, these conceptions of privacy are currently inadequate in their protection of such interests.

Thus, for example, it might be argued that the interest protected by the constitutional right corresponds to a spatial privacy right, in that it affords individuals a zone of personal space into which the state cannot intrude without adequate justification. The development of this right has been hampered, however, by the piecemeal nature of court jurisprudence, and by the inconsistencies of the Supreme Court in recognizing the parameters of the interest it believes it is protecting.[112] Moreover, the right in question is good only against the state, and would not provide any horizontal protection vis-à-vis other individuals. Most importantly, toward the end of the twentieth century, the Supreme Court signaled a rejection of privacy as the key value under which constitutional rights of individuals in the health care context are to be protected, preferring instead liberty under the fourteenth amendment.[113] For these reasons, the notion of a constitutional right of privacy is an inappropriate means to protect the spatial privacy interests in genetic information. Nonetheless, the history of jurisprudence of such a right demonstrates well the recognition of the need for adequate protection of interests of this sort.

Relatedly, one might speculate on the common-law right of privacy as a means of protecting informational and spatial privacy. As a right *in rem*, this

---

[111] Ruth Gavison, *Privacy and the Limits of the Law*, 89 YALE L. J. 421, 428 (1980).

[112] *See, e.g.*, Bowers v. Hardwick, 478 U.S. 186 (1986). For trenchant criticism of the case, see Kendall Thomas, *Beyond the Privacy Principle*, 92 COLUM. L. REV. 1431 (1992); Mark Kohler, *History, Homosexuality and Homophobia: The Judicial Intolerance of* Bowers v. Hardwick, 19 CONN. L. REV. 129 (1986).

[113] *See, e.g.*, Cruzan v. Director, Missouri Dep't of Health, 497 U.S. 917 (1990); Planned Parenthood of SE Pennsylvania v. Casey, 505 U.S. 833 (1992).

privacy right is good against the world at large, and so is unlike the right of confidentiality, which must be owed specifically and voluntarily to one individual or group of individuals.[114] We can test the efficacy of this privacy right by applying it, once again, to our imaginary scenario. The relatives might, for example, claim that an invasion of their privacy had occurred if certain uses of the information were employed, either by the health care professional, or arguably, Nicola herself. But can this vision of privacy protect the interest in not knowing? Of the four subsets of the privacy tort that have been indentified,[115] the most appropriate might be public disclosure of private facts or unreasonable intrusion upon the seclusion of another. The first of these, however, would only assist if the information were revealed, as with confidentiality, to third parties outside the familial relationship, and no interest would be infringed if the women themselves were told the news.

With respect to unreasonable intrusion, it might be thought that the offensive intrusion could be the receipt of burdensome information, but a perusal of the case law does not bear this out as a means of constituting the tort. Rather, an element of intentional invasion of private space is required, which is bound up with the possibility that personal information will be acquired or removed from that space by unacceptable means, rather than, as with our concern, that personal information will be added to that personal space.[116] Thus, the tort is constituted when an illegal search of one's property is carried out,[117] or when one's home is physically invaded,[118] or when eavesdropping or spying occurs,[119] or even when one is the subject of harassing telephone calls.[120] There is no authority to suggest that an actionable tort is committed by adding private information to the private sphere. It would seem that the conceptual underpinnings of the tort do not encompass such an invasion. Thus, the common-law right of privacy cannot help to establish a valid legal basis for a right not to know information. The focus of the tort on the need to extract information from the private sphere (usually with a view toward placing it in the public sphere[121]) renders it ill-equipped to protect against such intrusions into the private sphere as occur when an interest in not knowing is compromised.

---

[114] *Supra* section V(B).

[115] *See* Prosser, *supra* note 84. The four privacy rights are (a) appropriation of an individual's name or likeness, (b) unreasonable intrusion upon the seclusion of another, (c) public disclosure of private facts, and (d) subjecting an individual to publicity that casts them in a false light in the public's eye.

[116] *See* Doe v. Mills, 536 N.W.2d 824 (Mich. App. 1995) in which the Michigan Court of Appeals made it clear that the tort is constituted when secret and private subject matter is obtained by means that would be objectionable to a reasonable person.

[117] Sutherland v. Kroger Co., 110 S.E.2d 716 (W. Va. 1959).

[118] Ford Motor Co. v. Williams, 134 S.E.2d 483 (Ga. 1963).

[119] Rhodes v. Graham, 37 S.W.2d 46 (Ky. 1931); McDaniel v. Atlanta Coca-Cola Bottling Co., 2 S.E.2d 810 (Ga. 1939).

[120] Carey v. Statewide Finance Co., 223 A.2d 405 (Conn. Cir. 1966).

[121] Public disclosure is not, however, a prerequisite. *See* Cort v. Bristol-Myers Co., 431 N.E.2d 908 (Mass. 1982); Themo v. New England Newspaper Pub. Co., 27 N.E.2d 753 (Mass. 1940).

The parallel that might be drawn, then, is that the common-law tort is more akin to an informational privacy right and not a spatial privacy right.

These conclusions about current privacy protection in the United States should not lead us to the belief, however, that pursuing further privacy protection is a fruitless task. Rather, as has been argued above, there can be considerable value in recognizing the importance of protecting privacy. The definition of privacy proposed in this article finds its roots in moral notions about individuality, and presupposes social norms, such as respect for individuals. In this, it is allied with the related concepts of autonomy, confidentiality, and liberty, for all of these concepts perform essentially the same function—to define how individuals are perceived and treated in Western society and to establish and maintain the boundaries between the individual and society. Moreover, these concepts are necessary adjuncts to a view of human dignity and respect that is prevalent in our society. To see privacy in such terms allows one to comprehend better why a state of separateness should be sought. It further allows us to put forward valid and legitimate reasons for arguing that such a state should be protected and that invasion should only be on legitimate grounds and for legitimate reasons. Below, a more specific defense of privacy is presented, which applies the definition here advanced to the genetic information scenario outlined above to show that privacy *in se* can best protect the interests involved.

VII. A RIGHT NOT TO KNOW: A PRIVACY PERSPECTIVE

A spatial privacy analysis underscores a right not to know information. If, in the context of a dilemma about whether someone should be told, the individual has no knowledge at all that familial information exists, then the spatial privacy interest stands as a *prima facie* bar to the person being approached and told the information. Spatial privacy requires that, before such an approach is made, we consider how the individual might be harmed by disclosure and what good, if any, might come from disclosure. It requires that we reflect on the act of disclosure and places the onus on us not to disclose unless faced with compelling reasons to do so. Finally, it goes some way toward ensuring that the decision-maker does "not rest content with assumptions that flow from preconceived value preferences."[122] That is, a privacy analysis reveals the broader and more complex reality of scenarios involving genetic information. This does not happen when we analyze the problem from the perspectives of autonomy or confidentiality. As we have seen, autonomy is susceptible to argument for autonomy enhancement through disclosure of information. On the other hand, confidentiality permits wide exceptions—such as the amorphous

[122] Ruth Macklin, *Privacy and Control of Genetic Information*, *in* GENE MAPPING: USING LAW AND ETHICS AS GUIDES 164 (George Annas & Sherman Elias eds. 1992).

public interest—whereby disclosure easily can be justified at the discretion of those in possession of confidential information and when the value judgments of those persons dictate when information is so disclosed.[123]

Thus, in the circumstances of our scenario, Nicola must consider the privacy interests of her female relatives in not knowing the familial genetic information. It has been argued from the perspective of autonomy that Nicola could approach her relatives with the news of her condition and let them decide for themselves whether they should do something about discovering their genetic composition. From the perspective of privacy, however, Nicola must seriously consider the spatial privacy interests of her relatives. This might lead her to conclude that the information should not be imparted, for example, in the case of the sister who is unlikely to take advantage of the cure available because she is phobic about operations, or the cousin who is likely to react badly to the information given that she is prone to bouts of depression.

Our spatial privacy analysis provides Nicola with a more sophisticated model than is currently available from either of the concepts of autonomy or confidentiality with which to determine how she should proceed with the news about her condition. It is undeniable that this is a paternalistic stance. It cannot be otherwise in the absence of more information about what the relatives would want. Yet, such a paternalistic approach must be accepted for what it is and not be eschewed automatically in favor of an autonomy-enhancing disclosure. While autonomy-based arguments tend to create an imperative to let the individual make the personal choice, too frequently this amounts to an abrogation of responsibility on the part of the discloser of the information; for, with the passing of the information also comes the responsibility for assisting in how the information should be used. But, the transference of the burden of decision does not in itself absolve the first party of his or her moral obligations to the recipient of the information. Because of the susceptibility of autonomy to value-laden enhancement or facilitation arguments, it is often overlooked that decisions to enhance the autonomy of another are just as paternalistic as decisions not to disclose information at all.

What should happen in the case of a refusal based on limited knowledge? For example, if Nicola's cousin, Norma, was known to have expressed a disinclination to know her own health status when she was aware of the family history of disease, should she nevertheless be told? In these circumstances, we have an indication that an individual might not wish to know information. Autonomy indicates that we should respect such a wish and a spatial privacy analysis gives us another good reason to do so. It is accepted that a privacy

[123] Paul Lombardo, *Genetic Confidentiality: What's the Big Secret?*, 3 U. CHI. L. SCH. ROUNDTABLE 589, 593 (1996) (suggesting that so many exceptions have been made to confidentiality over the years that it has now been rendered meaningless and outlived its usefulness).

analysis does not necessarily make it easier for us to respect a wish not to know if that wish seems irrational (for example, if a cure for the condition is available and yet refusal is still made), but it does give us all the more cause to reflect that the refusal should be respected nonetheless. In addition, while autonomy-based arguments can be undermined because the subject in our scenario is not in full possession of all material facts to enable her to make a truly autonomous decision, a privacy paradigm offers a prima facie starting point of noninterference, which places the onus of justifying disclosure firmly on the shoulders of those who would do so.

None of the above should be taken as suggesting that disclosure should never be made. Rather, it is offered as a model for reevaluating the information disclosure decision-making process, and for considering the weight and merit of a range of factors in deciding if, when, and under what circumstances, a disclosure should be made. Furthermore, it should not be forgotten that hypotheticals rarely translate easily into real-life situations. It is acknowledged that, in a family context, it is very difficult to keep matters secret or private. Also, faced with the prospect of death, many would consider that everyone would wish to know of a predisposition to disease, no matter how upsetting the knowledge. This article cannot adequately address such issues. But, the point to be made is that the privacy analysis advanced here can be seen as a reflection of a wider trend in medicine and the care of others. The principle of sanctity of life is no longer seen to be the governing value in health care. Quality of life has taken over that role. Acceptance of this requires many paradigm shifts. If it were thought that the supreme value were to save life at all cost, then subtle privacy issues such as those advanced here would not arise. If, however, one values quality of life and accepts that we might prefer quality to the mere continuation of life, then this requires us to acknowledge that individuals might have an interest in preserving their current quality of life, even if that comes at the cost of life itself. The privacy model suggested here provides us with one way of seeking to respect such an interest.

Further utility for this model can be found when claims to have access to familial genetic information come from outside the family context. The requests of employers or insurers that individuals undergo genetic testing can be seen as an invasion of privacy given that these individuals are required to know information about themselves that they might not otherwise discover or seek out. In the balance of interests that could be undertaken, it would be hard to justify the promotion of the interests of employers and insurers—being primarily financial—over the significant personal spatial privacy interests of individuals that might be compromised by such requests. While the focus of this article is the family unit, this brief example demonstrates the potential extension of our privacy model to other areas. It is a sphere that requires considerably closer examination.

VIII. WHAT KIND OF GENETIC PRIVACY RIGHT SHOULD THERE BE?

If the above arguments are accepted, and the current legal protection of privacy is rejected as inadequate, then the question that arises is, what kind of privacy right should there be? We can approach this question from several different perspectives. On the one hand, we can consider other means within the existing law for recognizing these privacy interests. This could be in one of two ways: (1) by the refusal to impose any duty to disclose through the law of negligence; or (2) by recognizing or creating a common-law duty not to disclose. Alternatively, we could contemplate the introduction of a new statutory right of privacy specifically designed to protect the privacy interests in question.

A. No Duty to Inform

The negligence action has been used widely in tort law to delimit the extent of the duty of care that a physician owes to patients. Occasionally, however, a duty is deemed to be owed to persons outside the therapeutic relationship,[124] and, in such circumstances, the courts rely heavily on policy arguments to shape and temper such extensions of the law. The beginnings of a trend to extend the duty of care to relatives of persons diagnosed with genetic disease can be discerned in a number of states. Thus, in *Pate v. Threlkel*,[125] the Florida Supreme Court specifically addressed the question: "Does a physician owe a duty of care to the children of a patient to warn the patient of the genetically transferable nature of the condition for which the physician is treating the patient?" In answering this question in the affirmative, the court concluded that

> when the prevailing standard of care creates a duty that is obviously for the benefit of certain identified third parties and the physician knows of the existence of those third parties, then the physician's duty runs to those third parties. . . . [A] patient's children fall within the zone of foreseeable risk.[126]

---

[124] *See, e.g.*, Bradshaw v. Daniel, 854 S.W.2d 865 (Tenn. 1993); DiMarco v. Lynch Homes-Chester County, 559 A.2d 530, *aff'd* 583 A.2d 422 (Pa. 1990); Shepard v. Redford Comm. Hosp., 390 N.W.2d 239 (Mich. 1986); Gooden v. Tips, 651 S.W.2d 364 (Tex. 1983); Bradley Ctr. v. Wessner, 287 S.E.2d 716 (Ga. 1982); McIntosh v. Milano, 403 A.2d 500 (N.J. 1979); Renslow v. Mennonite Hosp., 367 N.E. 2d 1250 (Ill. 1977); Tarasoff v. Regents of Univ. of Cal., 551 P.2d 334 (Cal. 1976); Hofmann v. Blackmon, 241 So. 2d 752 (Fla. 1970).

[125] 661 So. 2d 278 (Fla. 1995).

[126] *Id.* at 282. In Schroeder v. Perkel, 432 A.2d 834 (N.J. 1981), the court recognized a duty of a physician to the parents of a child whose cystic fibrosis had not been correctly diagnosed to inform the parents of the child's condition. The court said:

> The foreseeability of injury to members of a family other than one immediately injured by the wrongdoing of another must be viewed in light of the legal relationships among family members. A family is woven of the fibers of life; if one strand is damaged, the whole structure may suffer.

The court stressed, however, that the duty did not require that relatives be approached directly by the physician: "[T]he duty will be satisfied by warning the patient."[127]

This view is unsatisfactory as a matter of policy for two reasons. First, it says nothing about the nature of the physician's duty if the patient refuses to disclose to relatives. Second, it assumes that the interests of patient and relatives necessarily coincide. For example, one can foresee circumstances in which it might not be in a patient's interests to be told that he or she is dying of a genetic condition, yet a failure to do so would be a breach of the physician's duty to the patient's relatives. The two duties of care are not, therefore, always reconcilable.

A 1996 decision of the Superior Court of New Jersey has addressed at least the first of these problems. In *Safer v. Estate of Pack*,[128] the court refused to follow the Florida court's restriction of the duty, and "declin[ed] to hold . . . that, in all circumstances, the duty to warn will be satisfied by informing the patient."[129] The court continued: "It may be necessary, at some stage, to resolve a conflict between the physician's broader duty to warn and his fidelity to an expressed preference of the patient that nothing be said to family members about the details of the disease."[130] Here, the court contemplates preferring a physician's duty of care to third parties to the patient's right to confidentiality. That it does so at least recognizes that the physician's duty of care to those third parties is separate from the relationship the physician has with the patient (albeit that it might have its origins in that relationship). Furthermore, it recognizes that the requisite standard of care should come from the physician and should not be discharged merely by telling the patient about his or her condition.

But, even if this line of authority is thought to be persuasive,[131] the fundamental premise for extensions of this sort in tort law should not be forgotten; namely, that public policy considerations must dictate the future course of the negligence action. It is for the courts to decide this matter, and a number of factors have a direct bearing on whether such an extension should be made. Several considerations should be immediately apparent, such as the burden

---

The filaments of family life, although individually spun, create a web of interconnected legal interests.

Id. at 839.

[127] *Pate*, 661 So. 2d at 282. An analogous decision in the context of HIV infection is Reisner v. The Regents of the Univ. of Cal., 37 Cal. Rptr. 2d 518 (Cal. 1995).

[128] 677 A.2d 1188 (N.J. 1996).

[129] *Id.* at 1192-93.

[130] *Id.*

[131] It is by no means certain that it will be followed. *See, e.g.*, Olson v. Children's Home Soc'y of Cal., 252 Cal. Rptr. 11 (Cal. 1988); Ellis v. Peter, 627 N.Y.S.2d 707 (N.Y. App. 1995); Conboy v. Mogeloff, 78 N.Y.2d 862 (N.Y. App. 1991); Sorgente v. Richmond Mem. Hosp., 539 N.Y.S.2d 269 (N.Y. Sup. 1989).

that a duty would place on health care professionals,[132] the difficulty in knowing who should be contacted and how,[133] and the possible detrimental effect that such a duty would have on the physician-patient relationship if confidentiality can be disregarded in favor of the duty to disclose.[134] But, of most importance in the present context, the courts should not rely unquestioningly on an assumption that nondisclosure is necessarily a (legal) harm.

As has been argued above, the interest in not knowing can be very important, and it will not be served by imposing a duty on health care professionals to make disclosures without first considering the consequences, both for the patient and the patient's relatives to whom disclosure will be made. One way, therefore, to recognize and protect the interest in not knowing would be to refuse to endorse the extension of tort law to impose a duty to disclose. The problem with such an approach is that it leaves the matter of the recognition of spatial privacy interests to the judiciary, which can only recognize such interests as and when relevant disputes come to court. Also, and more importantly from the individual's perspective, such an approach does not accord any right of compensation to those who have had their privacy interests invaded. It merely acts to pay abstract lip service to the interests in question.

B. A Duty Not to Inform

An alternative means to enshrine a right not to know in law would be to make an unauthorized disclosure a cause of action leading to the payment of damages, either for consequential harm or, simply, because the privacy of the individual had not been respected. As far as the common law is concerned, however, the analysis carried out above has shown that the primary concern of the privacy tort is with informational privacy interests and not spatial privacy interests. It is, thus, ill-equipped to be developed along such lines, absent some act of unprecedented judicial activism. Thus, in the absence of a viable alternative at common law,[135] any right to compensation would henceforth require to be introduced by statute at the state or federal level.

---

[132] For an interesting attitudinal survey about the responsibility of patients and genetic services providers to remain in contact, see Jennifer Fitzpatrick et al., *The Duty to Recontact: Attitudes of Genetics Service Providers*, 64 AM. J. HUM. GEN. 852 (1999).

[133] Lori Andrews, *Torts and the Double Helix: Malpractice Liability for Failure to Warn of Genetic Risks*, 29 HOUS. L. REV. 149, 181 (1992) (recognition of a duty of disclosure to relatives with whom a physician has no direct professional relationship should, logically, also give rise to a duty for physicians to tell strangers of the health risks that they run).

[134] *See also* Graeme Laurie, *Obligations Arising from Genetic Information—Negligence and the Protection of Familial Interests*, 11 CHILD & FAM. L.Q. 109 (1999).

[135] While the tort of intentional infliction of emotional distress might initially appear helpful, the requirement that there be evidence of intention to cause harm through extreme and outrageous conduct, as judged from the perspective of the reasonable person is unlikely ever to be met in circumstances such as those under discussion, which deal with nuanced matters of professional and individual judgment. *See, e.g.*, Harris v. Jones, 380 A.2d 611 (Md. 1977).

During the 105th Session of Congress, 110 bills seeking to protect genetic privacy were introduced. None was debated beyond the subcommittee stage.[136] While a number of states have taken the initiative to introduce protective measures,[137] no federal initiative has ever been successful. Yet, as Starr has pointed out: "Almost everyone agrees that the absence of stronger protections for the privacy of health data is a national problem and that this problem has become more urgent in recent decades."[138] The Health Insurance Portability and Accountability Act of 1996 set a deadline of August 21, 1999, for Congress to enact federal health privacy legislation, but this deadline passed without action. In default, the Act empowers the Secretary of Health and Human Services to introduce privacy regulations, and this duly happened in December of 2000. However, these measures are not underpinned by a coherent concept of privacy. Accordingly, they are not helpful to the present debate. Instead, it is apposite to examine model federal legislation designed specifically for genetic privacy concerns.[139]

C. The Genetic Privacy Act

The Genetic Privacy Act (GPA) was produced for the Human Genome Project's Ethical, Legal and Social Issues division by George Annas, Leonard Glantz, and Patricia Roche of Boston University's School of Public Health.[140] This draft model law is in the format of a federal statute and has already been a source of inspiration for several state legislatures.[141]

The introduction to the Act states:

> [T]he overarching premise of the Act is that no stranger should have or control identifiable DNA samples or genetic information about an individual unless that individual specifically authorizes the collection of DNA samples for the purpose of genetic analysis, authorizes the creation of that private information, and has access to and control over the dissemination of that information.

---

[136] *See* Dorothy Wertz, *Legislative Update: Genetic Privacy Bills*, 3 THE GENE LETTER (Feb. 1999) (<www.geneletter.org/0299/legislativeupdate>). For comment, see Jeremy Colby, *An Analysis of Genetic Discrimination Legislation Proposed by the 105th Congress*, 24 AM. J. L. & MED. 443 (1998).

[137] By January of 1999, 44 states had introduced laws concerning genetic privacy or discrimination, see William Mulholland & Ami Jaeger, *Genetic Privacy and Discrimination: A Survey of State Legislation*, 40 JURIMETRICS 317 (1999).

[138] Paul Starr, *Health and the Right to Privacy*, 25 AM. J. L. & MED. 193 (1999).

[139] *Cf.* Tony McGleenan, *Rights to Know and Not to Know: Is There a Need for a Genetic Privacy Act?*, *in* THE RIGHT TO KNOW, *supra* note 14.

[140] For information on this, see *Human Genome News*, 6 Mar.-Apr. 1995, at 4.

[141] The text of the Act can be found at <www.ornl.gov/TechResources/Human_Genome/resource/privacy/privacy1.html>. It would seem that the most far-reaching legislation has been passed in New Jersey. For comment, see Fred Charatan, *New Jersey Passes Genetic Privacy Bill*, 313 BRIT. MED. J. 71 (1996). In Maryland, a bill based on the GPA was defeated in the state senate. For comment, see Neil Holtzman, *Panel Comment: The Attempt to Pass the Genetic Privacy Act in Maryland*, 23 J. L. MED. & ETHICS 367 (1995).

Thus, the GPA envisages a highly individualistic approach to the question of control of genetic samples and information. It should be noted, however, that the GPA defines the term "private genetic information" to mean

> any information about an identifiable individual that is derived from the presence, absence, alteration, or mutation of a gene or genes, or the presence or absence of a specific DNA marker or markers, and which has been obtained: (1) from an analysis of the individual's DNA; or, (2) from an analysis of the DNA of a person to whom the individual is related.[142]

This clearly seeks to take account of the interests that relatives of a "sample source" can have in genetic information. Yet, the GPA gives a property right in the DNA sample to the sample source.[143] Moreover, it is not clear how well the distinction is drawn between a DNA sample and private genetic information derived from a sample.[144] Axiomatically, the first is unique and personal to the person from whom the sample was taken. The same is not true of the information, but the GPA nevertheless provides that the exclusive right over such information (as with samples) is retained by the sample source.[145] The provisions of section 101(b)(8) stipulate, however, that prior to the collection of a DNA sample from individuals they should be informed, among other things, that "the genetic analysis may result in information about the sample source's genetic relatives which may not be known to such relatives but could be important, and if so the sample source will have to decide whether or not to share that information with relatives."[146]

It is fortunate that the text of the GPA is accompanied by a commentary prepared by its authors in which they seek to clarify their general aims and to expand upon the specific terms contained therein. Of the above provision, they say the following:

> Creating either a contractual or statutory obligation for individuals to share [genetic] information with their family members would not only be unprecedented, but inadvisable. The creation of new substantive rights or duties of family members is not our intention and is beyond the scope of this Act. However, because the Act creates rules that govern the use and disclosure of information, it is imperative that individuals be informed of the fact that by seeking genetic information about themselves through genetic analysis, they may also become privy to information about

[142] Genetic Privacy Act (GPA), § 3. For the text of this section, see <www.bumc.bu.edu/www/sph/lw/gpa/GPA_cmid.htm>.

[143] *Id.* § 104(a). This move clearly has very far-reaching implications for a great number of areas within the disciplines of law, medicine, and science generally. These are, however, outside the scope of this article. For comment, see Michael Lin, *Conferring a Federal Property Right in Genetic Material: Stepping into the Future with the Genetic Privacy Act*, 22 AM. J. L. & MED. 109 (1996).

[144] Part A of the Act deals with collection and analysis of DNA samples; Part B concerns disclosure of private genetic information.

[145] See Part B of the Act which is concerned with matters of consent and disclosure.

[146] GPA, § 101(b)(8).

> other family members who would also want and/or need such information. . . . While it will be an individual choice as to whether or not to share that information with others, this disclosure should instigate discussion between the sample source and the collector of the sample.[147]

Thus, the GPA allows sample sources to decide for themselves whether to disclose genetic information to relatives. Many would argue that this is not necessarily a bad thing because often such a person will be better (or even best) placed to establish how relatives might feel about receiving such information. However, the GPA does not give any guidance to a sample source on how to decide whether or not disclosure should be made. In particular, there is no recognition of the possible spatial privacy interests that relatives who are the potential recipients of such information might have in not knowing. If it is accepted that individuals can have valid interests in such notions, then it is submitted that an Act that purports to deal with genetic privacy should include provisions aimed at recognizing and protecting such interests.

The only part of the GPA to recognize such interests is that concerned with minors. Section 141 of the Act provides as follows:

> (a) INDIVIDUALS UNDER 16 — . . . the individually identifiable DNA sample of a sample source who is under 16 years of age shall not be collected or analyzed to determine the existence of a gene that does not in reasonable medical judgment produce signs or symptoms of disease before the age of 16, unless:
>
> (1) there is an effective intervention that will prevent or delay the onset or ameliorate the severity of the disease; and
> (2) the intervention must be initiated before the age of 16 to be effective; and
> (3) the sample source's representative has received the disclosures required by section 101 of this Act and has executed a written authorization which meets the requirements of section 103 of this Act and which also limits the uses of such analysis to those permitted by this section.

The authors justify these provisions as follows:

> There are two reasons for this prohibition on the exercise of parental discretion. First, if someone learns that the child is a carrier of a gene that disposes the child to some condition later in life, this finding may subject the child to discrimination and stigmatization by both the parents and others who may learn of this fact. Second, a child's genetic status is the child's private genetic information and should not be determined or disclosed unless there is some compelling reason to do so.[148]

---

[147] *See* George Annas et al., *Drafting the Genetic Privacy Act: Science, Policy, and Practical Considerations*, 23 J.L. MED. & ETHICS 360 (1995). For a critique, see Edwin Troy, *The Genetic Privacy Act: An Analysis of Privacy and Research Concerns*, 25 J.L. MED. & ETHICS 256 (1997).

[148] The Genetic Privacy Act and Commentary (1995) is available on request from the Health Law Department, Boston University School of Public Health, 80 East Concord Street, Boston, MA 02118 and also on the Department's Web site at <www.bumc.bu.edu/sph/internet.htm>.

This corresponds to arguments that have been made above concerning the spatial privacy interests of individuals. Arguably, the GPA here recognizes the spatial privacy interests of children, and further, it recognizes that these should not be invaded without due cause.[149] The GPA is remiss, however, in not recognizing the spatial privacy interests of all persons about whom genetic information is known but who have not sought it out themselves.

Of course, the situation of the minor is not in all respects the same as that of the adult relative of a proband. One clear point of difference concerns the initial generation of information. In the case of the minor, the legal prohibition concerns the initial collection or analysis of genetic material. In the case of an adult relative of a proband, this is not the point at issue because no one can (or should) prevent others from having their genetic material analyzed. This is, however, a distinction without a difference for present purposes. For, the essential issue in both cases is the same; namely, the unwarranted intrusion of personal genetic information into the private sphere of the individual in question. Thus, the interest of the adult relative is not in seeking to control the proband's access to the information, but rather it is in having his or her own spatial privacy interests of nonintrusion respected. For the minor, precisely the same interest is at stake. The means to protect the child's interest may lie in securing control over the minor's own sample, but simply because the same means are not available to relatives of a proband should not lead to the conclusion that the spatial privacy interests of the proband's relatives are any less deserving of protection.

Relatedly, because the minor is the proband in such cases, the minor has the primary right to decide what happens to his or her genetic sample and to any genetic information derived from that sample. Given this, one might argue that the above provisions simply ensure that the choice of accessing genetic information be left until the child is capable of making independent choices. So, one might conclude that no specific provision is necessary in the case of an adult because it is axiomatic that the adult may choose to know or not to know his or her own information. However, the question here is not simply one of access, but also one of nonaccess. The interest is not merely one of control but of maintaining a state of ignorance: a state of nonaccess to the person. Yet, the focus of the GPA on control of samples (and so on autonomy and choice) means that the child is only protected from attempts to gain specific access to personal genetic information. The minor is not protected from unwarranted disclosure of genetic information from relatives.

Adults are in an equally vulnerable position. In fact, one can draw a parallel between the child and the unknowing adult in that, in many senses,

[149] The GPA does not go so far as to recognize any spatial privacy interests for fetuses. Sections 151 and 152 provide that a competent pregnant woman has the sole right to determine both when DNA samples shall be taken from her fetus and how genetic information about the fetus shall be used.

they are both incapax with respect to the genetic information. While the child is generally incapax, the adult can certainly make a choice to know, in that the adult has the capacity to choose to know. But, to offer the individual the opportunity to choose might be to offend the very interests with which one is concerned. Thus, it is submitted that it is acceptable in the case of genetic information to adopt the position that both adult and child are incapax. The consequence of this is equally the same; namely, that neither should be approached with unsolicited disclosures of genetic information without due cause and justification.

The conclusion to be drawn is that it would not be inappropriate to extend the form of protection offered by the GPA to relatives of a proband. The prohibition on disclosure could not only cover requests for direct testing, but also could extend to unwarranted approaches to family members with genetic information about which they are unaware. The determination of a warrantable approach would need to be settled by more debate on the legitimate nature of competing interests and a proper assessment of genetic risks and consequences within the family and the wider community setting. In this way, a spatial privacy right could be established that would require proper justification before a legally acceptable approach to a person could be made.

D. Privacy Problems

A well-defined spatial privacy right could embody a clear account of the kind of factors that would make disclosure in different circumstances acceptable or unacceptable. These factors have been described above as: (1) the availability of a cure or therapy; (2) the severity of the condition and likelihood of onset; (3) the nature of the genetic disease; (4) the nature of the genetic testing; (5) the nature of the request; (6) the question of how the individual might be affected if subjected to unwarranted information, and whether the individual has expressed any views on receiving information of this kind.

An obvious problem with this approach, however, is that the existence of a right not to know implies that a duty not to disclose information should exist in certain cases. Yet, an important factor in determining whether such a duty exists is the question of how the individual to whom the information is to be disclosed might react. This is a very subjective matter that can be especially difficult for any third party to assess. It leads to the possibility that individual A might determine that individual B should not be informed of information, when in fact individual B actually would want to know, had he or she been given the opportunity. As we have seen, the privacy argument in favor of nondisclosure is based primarily on a desire to respect and not to harm the individual, but in such a case the very fact of nondisclosure might cause harm and might be an act of disrespect in itself. In recognition of this, a number of additional factors could be brought to bear on the problem.

First, an objective assessment of circumstances could serve to delimit the parameters of any duty not to disclose. That is, the person in possession of the information would assess factors such as likelihood of onset and availability of cure, together with an objective consideration of what a person in the subject's position would or would not want to know. Such a reasonable subject could assume the particular characteristics of the actual subject. At the end of the day, provided that the assessment was a reasonable one, one could conclude that no legal redress should lie against someone who had decided (not) to disclose information. The relevance of any views of the subject clearly would be significant in the assessment of reasonableness, as would the extent of the effort made by the duty-holder to seek out evidence of those views.

None of this detracts from the fact that the assessment of the factors to be considered is in itself a difficult exercise. On the one hand, the clinical data concerning the extent of risk or the likely success of therapy or cure are best assessed by health care professionals, while the question of which characteristics should be taken into account to determine if this subject should be told, is better determined by those close to the person, such as the subject's relatives. While a health care professional might be in a position to gather a range of data to assist in the assessment of the situation, it is far less clear whether family members are in a position to make meaningful assessments of such factors, let alone whether they should be the subject of a legal action if they disclose information in unjustified circumstances.

It is therefore submitted that it is more permissible to impose a duty not to disclose or seek information on parties outside the family milieu, or at least to require that they do so only in the most justified of circumstances. Primarily, this would affect employers, insurers, and the state. Thus, for example, requests for genetic testing by such parties would be seen to be a clear invasion of spatial privacy. This does not mean, however, that the range of interests under consideration, or their importance, should not be discussed with individuals who seek genetic testing and who might contemplate disclosure to their relatives. It is simply to admit that the law does not always have a role to play in determining what should be done with genetic information, especially within the family setting. Nonetheless, it could become a duty for health care professionals to discuss such interests with probands. The GPA requires, for example, that a number of matters be discussed with a sample source.[150] To this list could be added a specific requirement with respect to the interest of relatives in not knowing.

[150] GPA, § 101(b). The GPA requires, among other things, that the sample source be told: (1) that consent to the collection and taking of the DNA sample, and its analysis, is voluntary; (2) about the information that reasonably can be derived from the analysis; (3) that the genetic analysis may result in information about the sample source's genetic relatives, which may not be known to such relatives but could be important, and if so the sample source will have to decide whether or not to share that information with relatives; and (4) about the existence of genetic counseling.

An additional defining factor for a duty not to disclose could be the need to show the reasonable prospect that a tangible benefit would come to the person to whom disclosure would be made. The benefit should be more than the facilitation of preparedness or the promotion of autonomy, and should represent some clinical benefit to the subject. Thus, for example, employers and insurers would not be able to rely on the argument that individuals can choose whether to undergo testing at their behest, but rather would have to show some real medical benefit to those persons to justify their requests for testing and/or access to genetic data. Also, state screening initiatives would be justified only if such a benefit could be shown.[151] Health care professionals similarly could be obliged to discuss with probands the likely real benefits of disclosure to relatives.

Even if all of the above is accepted, a concern may remain: is this approach not simply a paternalistic assessment of spatial privacy interests? To an extent it is, but perhaps this can never be avoided in circumstances where one cannot approach an individual directly to determine how to proceed. Rather, it is submitted that the worth of this approach is found in its responsiveness to broader, less atomistic interests. It is an approach that does not view the beginning and the end of ethical discourse as lying with the autonomy of individuals, but rather responds to the wide range of interests from the perspective of an ethic of care, wherein autonomy has a significant role to play, but where privacy is also required to complete the model.

Finally, as Powers has rightly pointed out, "a commitment to privacy rights does not entail a commitment to absolute rights."[152] Indeed, in the context of genetic privacy there are many reasons why this cannot be so, not least of which is the fact that we are dealing with a plethora of privacy (and other) interests stemming from the familial nature of the information in question. What, then, might the limits and exceptions to privacy protection be? Some of these already have been discussed above. In addition, it is difficult to reject the argument that public interest is a valid exception to such a right, just as it is a determining factor in the law of confidentiality. The classic paradigmatic tension is that between the public interest in protecting private interests (such as privacy and confidentiality) and the protection and promotion of other public interests (such as protection of the community from harm, or freedom of the press). However, the devil is in the detail of determining what is meant by public interest in each case.

A number of well-accepted public interests are self-evident and certainly would be included in any genetic privacy legislation. These include the prevention and detection of crime, scientifically valid and ethically justified

---

[151] *See* Malm, *supra* note 57.

[152] Madison Powers, *Privacy and the Control of Genetic Information*, *in* THE GENETIC FRONTIER: ETHICS, LAW AND POLICY 82 (Mark Frankel & Albert Teich eds. 1994).

research, and court-ordered disclosures.[153] Public health initiatives aimed at particular populations similarly might be justified where, for example, tangible harm can be avoided by effective screening and treatment. Screening of newborns for phenylketonuria and hypothyroidism is acceptable on such grounds.[154] In each of these examples, the public in question is the community at large. But does this definition of public necessarily exhaust the concept? More specifically, when we ask who is the public in the genetic privacy public interest exception, should this include the families of persons who have been tested for genetic conditions?

A public is a collective defined, like society, by reference to the individual. Relatively speaking, the individuals in a family unit might constitute a public by virtue of the fact that, as a common collective with a common interest in familial information, they have claim to the information in question. However, this is not to suggest that a familial public necessarily has the same common interest in the information,[155] for as we have seen, a number of potentially competing interests may be in play. Nor is it to suggest that a familial public, by virtue of its strength of numbers alone, should have an automatic or strong(er) claim to the information in question, as compared to the person who is the original source of the information. The balance, were there to be one, would need to be between the familial public interest weighed against another public interest, such as the interest in respecting individual privacy generally, or the public interest in protecting individuals from potentially harmful uses of their personal information.

This view of a collective familial claim presupposes, of course, that a familial public interest could be formulated. No comment is offered here on this point. While such a communitarian approach to privacy and genetic information scarcely has been contemplated,[156] it is a self-evident and natural

---

[153] The Secretary of Health and Human Services recommended five guidelines in 1997 to shape the future of health privacy laws. Among these was the principle of "public responsibility" which states that "individuals' claims to privacy must be balanced by their public responsibility to contribute to the public good through use of their information for important socially useful purposes, with the understanding that their information will be used with respect and care and will be legally protected." *See Confidentiality of Individually-Identifiable Health Information: Recommendation of the Secretary of Health and Human Services, Pursuant to Section 264 of the Health Insurance Portability and Accountability Act 1996*, (Sept. 1997) available at <*http://aspe.os.dhhs.gov/admnsimp/PVCREC1.htm*>.

[154] But, the security of the samples taken from such infants deserve no less stringent security measures for having been the source of valuable medical interventions. *See generally* STORED TISSUE SAMPLES: ETHICAL, LEGAL AND PUBLIC POLICY IMPLICATIONS (Robert Weir ed. 1998) (especially chapter 1).

[155] Note, however, that Ruth Chadwick has pointed out that the concept of solidarity can be invoked to justify a claim that families do have a collective claim concerning the use and control of their genetic information. *See* Ruth Chadwick, *The Philosophy of the Right to Know and the Right Not to Know*, *in* THE RIGHT TO KNOW, *supra* note 14, at 20.

[156] *Cf.* AMITAI ETZIONI, THE LIMITS OF PRIVACY (1999); Morris Foster et al., *The Role of Community Review in Evaluating the Risks of Human Genetic Variation Research*, 64 AM. J. HUM. GEN. 1719 (1999); Helena Rubinstein, *If I Am Only for Myself, What Am I? A Communitarian Look at the Privacy Stalemate*, 25 AM. J.L. & MED. 203 (1999).

corollary to the recognition of the range of claims surrounding this sort of information. If the family is to come to be seen as community in microcosm, then the collective claims and interests of that community also must be determined and weighed in any balance of values when assessing the appropriateness of any dealings with familial genetic information.

This article has argued for recognition of one small area of this new field; namely, the interest not to know. But, if this thesis is accepted, and a paradigm shift is undertaken, which refocuses attention away from purely monoindividualistic, autonomy-based concerns, then the relevance of other similar claims also falls to be considered. While it has not been appropriate to do so in this article, the interconnectedness of the disparate elements of this discourse must ultimately be fully explored to determine the optimal role for the law within such a dynamic. At no point, however, should we dismiss the notion that the role of the law might, in fact, be limited in a complex domain such as this.

CONCLUSION

The need to address privacy issues in the field of genetics has been appreciated by a number of international bodies in a number of international documents. For example, the Bilbao Declaration highlights the main problem areas likely to arise from the work of the Human Genome Project and pinpoints matters considered to be worthy of immediate attention by the legal systems of the world. Included in this Declaration is "[p]rotection of the personal privacy or confidentiality of genetic information, and determination of cases in which it could feasibly be altered or overstepped."[157] Moreover, the interest in *not* knowing has been recognized. For example, the Council of Europe in its Convention for the Protection of Human Rights and Dignity of the Human Being with Regard to the Application of Biology and Medicine, states in Article 10(2): "Everyone is entitled to know any information collected about his or her health. *However, the wishes of individuals not to be so informed shall be observed*."[158] Similarly, the UNESCO Universal Declaration on the Human Genome and Human Rights states in Article 5c "[that t]he right of every individual to decide whether *or not* to be informed of the results of genetic examination and the resulting consequences should be respected."[159] These instruments embody the best and the worst features

---

[157] The Bilbao Declaration on the Human Genome was drafted at the International Workshop on Legal Aspects of the Human Genome Project, which took place in Bilbao, Spain, in May of 1993.

[158] COUNCIL OF EUROPE, CONVENTION FOR THE PROTECTION OF HUMAN RIGHTS AND DIGNITY OF THE HUMAN BEING WITH REGARD TO THE APPLICATION OF BIOLOGY AND MEDICINE: CONVENTION ON HUMAN RIGHTS AND MEDICINE (Oviedo, Apr. 1997) (emphasis added).

[159] Adopted unanimously on November 11, 1997 in Paris at the Organization's 29th General Conference (emphasis added).

of the dilemma that we currently face. They recognize the value of an interest which has hitherto received short shrift, but they offer mere aspirational means to protect this interest, which is without substance in the absence of specific national interventions. Furthermore, while these instruments recognize the value of rights-based discourse, they subsume the protection of the interest in not knowing within a rubric of rights of autonomy and choice, when these constructs are ill-suited to the task at hand. Thus, while these documents offer a new way of looking at genetic information, they represent only one means of addressing the problem—one that is typical of the current focus on autonomy-based argument.

This article has argued for an original concept of privacy that would provide recognition and protection of the interest that individuals might have in not knowing genetic information about themselves. Furthermore, it has offered a view as to how such an interest could be protected by legal means within a domestic system. The argument has been as much about identifying the problem and the most appropriate tools to use to solve the problem, as it has been about offering concrete means to address all of the nuanced issues that arise. It is as much an appeal to view the problem from an alternative perspective, as it is an offer of a solution to the dilemmas at hand. Just as the solution offered is not without its problems, so the reader is invited to reflect that the current approach yields little by way of solution.

[11]

GENETIC TESTING AND EMPLOYEE PROTECTION

PHILIPPA GANNON AND CHARLOTTE VILLIERS

*Law School, Centre for Regulatory Studies, and Institute of Law and Ethics in Medicine, University of Glasgow**

ABSTRACT

Given advances in the science of genetics it is increasingly possible for individuals to acquire an increased understanding about their DNA. Employers may wish to access such information or may request that employees participate in genetic testing. An examination of the UK legislative framework to accommodate or to prevent such demands raises concern about the need to balance the employer's economic interests and the autonomy of the employee

INTRODUCTION

In the UK genetic tests are at present usually administered through traditional state-provided health care, for example, consultant-led NHS genetics services. Yet, it is increasingly feasible that genetic tests will in time, be supplied upon a commercial basis to the public, thereby augmenting and facilitating the circumstances in which individuals learn of their genetic health.[1] As a result of such developments, in the near future, more individuals will know the state of their genetic health. Whilst this knowledge may have profound implications for many life decisions it may also be misinterpreted or used for inappropriate purposes by third parties.

Currently employment related medical examinations do not encompass genetic diagnosis on a large scale.[2] There are numerous reasons for this position. First, genetic diagnosis is still in the research phase and is relatively expensive and time-consuming to provide.[3] Secondly, the present and future validity of genetic tests is debatable since there remain doubts about reliability, accuracy and sensitivity.[4] For example, there is the possibility of maladministration and human error in performing and interpreting the tests. Thirdly, the nature of genetic conditions also makes testing difficult; some conditions are multi-factorial. Some genetic dis-

*Thanks are due to Professors Roger Brownsword, Hugh Collins, Keith Ewing, Sheila McLean and Tony Prosser for their helpful comments on earlier drafts of this article.

orders such as Huntingtons Chorea have a late onset, which may not affect the immediate employment of the individual. These aspects of genetic tests lead to concern about how and to what extent genetic tests should be utilised and relied on, especially in the employment context. Yet, given the speed at which advances in genetics change the existing status quo, as witnessed for example in the insurance industry it may only be a matter of time before genetic testing is used for employment purposes.[5] Although genetic tests will not be used uniformly in all employment contexts they will undoubtedly be useful for certain employment positions, for example if the employment involves contact with hazardous chemicals which may, lead to onset of a genetic condition.[6]

Perhaps one of the most controversial issues about genetic testing in the workplace is that such testing invokes issues of control and coercion of the employee by the employer. The employer or potential employer may use the genetic tests to exclude the individual from the workplace. This exclusion may take the guise of either refusing to employ a job candidate or by dismissing an employee as a consequence of the test results. The employer may regard exclusion or dismissal as a more appropriate response to that of altering the work processes in order to eliminate the occupational risks. Alternatively the employee could be moved to a different job within the same organisation which could itself limit opportunities, for example promotion or the ability to obtain higher pay.

There are also other legal issues that are not unique to the employment contract. For example, the use of information obtained by genetic tests may affect the employee's privacy because his or her personal health details may be used inappropriately or disclosed to another party. In addition, the employee could also suffer emotional distress and psychological damage if he or she feels compelled to have a genetic test, the results of which indicate problems with his or her genetic profile. In this article we shall explore the existing legal provisions that could be used to protect the employee in any of these possible circumstances before going on to consider possible reforms.

THE LEGAL IMPLICATIONS OF WORKPLACE GENETIC TESTING

UK law is under-developed with regard to workplace genetic testing. Some existing rules on discrimination may apply (for it may be considered discriminatory to afford different treatment to individuals who have a genetic condition) but these rules are not specific to genetics. The main issues to consider will be exclusion, health and safety and discrimination as well as privacy.

Exclusion or dismissal

The use of genetic tests could lead an employer to exclude or dismiss an individual from the workplace. This response has been described as "victim-blaming", emphasising the individual's susceptibility rather than the employer's use of a dangerous substance or process.[7] This response does not encourage an employer to alter the working system or to remove dangerous substances or materials. The dangerous conditions are thus allowed to continue but the higher risk individuals are prevented from being involved in the process. This approach may ultimately only be delaying the potential costs of litigation rather than avoiding the onset of a condition since those employees deemed to have a lower susceptibility are not necessarily guaranteed immunity from such harms. In the UK, an individual who has been dismissed may seek a remedy for unfair dismissal, but this remedy is of course dependent upon employment.

Unfair dismissal protection

A major problem with unfair dismissal is its limited availability.[8] It offers no scope for those who are not given employment. If the employee is already working for the employer and is removed from the workplace due to the results of a genetic test (that has been required or disclosed to the employer) the unfair dismissal provisions contained in the Employment Rights Act 1996 might apply. Under these provisions an employer must offer a reason for the dismissal based on one of the grounds specified in s.98 and in addition, the employer is required to adopt fair dismissal procedures. For example, where an employee is dismissed for reasons of ill health the employer ought to consider alternative employment for the employee before dismissing them.

One of the grounds in s.98 that could be used to justify dismissal is capability. Capability may include health and safety reasons since capability is assessed by reference to skill, aptitude, health or any other physical or mental quality.[9] The employer's health and safety obligations could therefore be used as a reason to exclude the employee who has an unfavourable genetic test result. Yet, a dismissal on such grounds would not necessarily be fair. This is because when the reason for the dismissal is incapability on grounds of competence or ill health the employer should "treat each case individually where there is genuine illness with sympathy, understanding and compassion".[10] Factors that might be relevant to this assessment include: the nature of the illness; the likelihood of its recurring; the length of absences in relation to the periods of good health; the employer's dependence on the work done by the employee; the effect of the absence on fellow workers; and the ease with which work could be redeployed during the employee's absence.[11] Thus, the specific

42

features of the genetic condition would need to be assessed by employers, which may be difficult in relation to late-onset or progressive genetic conditions. Genetic tests are usually predictive and may be revealing a condition that might not develop in the near future. However, future illness could be prompted by the workplace activities of the employee. If those processes are extremely difficult to change the dismissal might then be justifiable. In any event, even if there is a statutory provision against continued working by a sick employee, the employer may be found to have acted unreasonably in failing to allow an employee time to take remedial or cautionary measures or in failing to look into the possibility of suitable alternative employment.[12]

Two other potentially fair reasons for dismissal are first, that the employee could not continue to work in the position which he or she held without contravention (either on his part or on that of his employer) of a duty or restriction imposed by or under an enactment[13] and secondly, for "some other substantial reason".[14] The first of these two potentially fair reasons could arguably include the duties imposed on the employer by the Health and Safety at Work etc Act 1974, which is discussed in greater detail below. The second potentially fair reason has been criticised widely for the broad discretion it grants to the employer. Frequently, this ground opens the door to business efficiency considerations, thus favouring the employer's rather than the employee's position.[15]

Generally, the approach of the courts is to pay regard to the so-called "managerial prerogative".[16] The range of reasonable responses test[17] within this prerogative is well known for giving to employers a broad margin of discretion in their treatment of an employee whom they wish to dismiss. The judicial approach to this test focuses on the employers' actions rather than on the protection intended to the employee and therefore fails to take into account substantial justice issues.[18] This test would therefore not take into account an employee's concerns regarding genetic testing, but would focus upon the requirements of the employer. Employees are expected to fit into the work practices that have been established by an employer. The employer is supported by the fact that an employee's refusal to cooperate in the operation of genetic tests might itself give grounds for disciplinary action.[19]

Occupational and Health and Safety Liabilities

The employer has a general health and safety duty under Section 2 of the Health and Safety at Work etc Act 1974. Section 2(1) requires the employer, generally, to ensure, so far as is reasonably practicable, the health, safety and welfare at work of all his or her employees. Sub-section (2)(a)–(e) provides a more specific list of requirements. Although the sub-section does not impose any requirements on employers that are directly related to genetic testing the provisions are relevant in their requirement that the employer takes reasonable care to ensure that the plant and systems of work are safe and without risks to health and that employee contact with articles and substances is risk-free. Sub-section (2) does not explicitly prohibit or seek to prevent contact by employees with dangerous articles or substances but requires that where such contact is made the employer will arrange, as far as is reasonably practicable, that such contact will not incur risks for the employee.

These provisions emphasise the employer's control of the situation. For example, the employer is required to arrange for storing and handling and transport of articles and substances. The wording of subsection (2)(d) expressly states: "so far as is reasonably practicable as regards any place of work under the *employer's control*, (emphasis added) the maintenance of it in a condition that is safe and without risks to health and the provisions and maintenance of means of access to and egress from it that are safe and without such risks".

However, there is opportunity in section 2 for developing the employee's participation in the decisions to be made. This opportunity arises from sub-section (2)(c) which requires: "the provision of such information, instruction, training and supervision as is necessary to ensure, so far as is reasonably practicable, the health and safety at work of his employees". This provision emphasises the importance of the employee's participation in the health and safety aspects of the relationship. The provision of information is recognised as a necessary element in health and safety protection. Again, this is prescribed by sub-section (3) which states: "Except in such cases as may be prescribed, it shall be the duty of every employer to prepare and as often as may be appropriate revise a written statement of his general policy with respect to the health and safety at work of his employees and the organisation and arrangements for the time being in force for carrying out that policy, and to bring the statement and any revision of it to the notice of all his employees." Further support for employee participation is offered by sub-section (6) which requires consultation with safety representatives in making and maintaining health and safety arrangements.

The employer could adopt the view that it is reasonably practicable to reduce the risks to health and safety of employees by removing the risk altogether. The employer might do this, not by altering the working procedures, but by removing the employees perceived, as a result of the tests, as being more susceptible to the potential harm. The Health and Safety at Work etc Act 1974 itself does not offer guidance on such a response by the employer. For example, the Act does not expressly mandate genetic tests nor does it require that employees be removed where risks arise from certain work processes. This makes uncertain any judgment about unfair dismissal. However, some Regulations, which have been created secondary to the Health and Safety at Work etc Act, do

44

provide for removal of employees on medical grounds. For example, the Management of Health and Safety at Work (Amendment) Regulations 1994[20] require employers to assess the risks to the health and safety of new and expectant mothers and ensure that they are not exposed to those risks. The responses deemed appropriate by these Regulations include the possibility of the employer temporarily adjusting the worker's hours or conditions of work to avoid the risk or offering her alternative suitable and appropriate work, or suspending her from work for as long as necessary to protect her health and safety. In these Regulations, the emphasis on safety seems to override the individual choices of the employee and the employer is required to make the decision.

In general, health and safety provisions emphasize an employer's responsibility, leaving the strategic health and safety decisions firmly in the employer's hands. Whilst we appreciate that the employer has to take into account the interests of all employees, perhaps in relation to genetic testing, a greater account should be taken of the effects of any proposed practices upon individual employees.

Protection from Discrimination

Perhaps the main concern of the use of genetic testing in the workplace is discrimination. Genetic testing at work potentially involves all three forms of unlawful discrimination in the UK: racial, sexual and disability discrimination. Published research reports that the sickle cell trait appears in approximately 7% of the African population whereas it appears in less than 1% of the white population[21] and pulmonary disease appears more frequently in the North European population.[22] These established facts give rise to the possibility of combined medical and racially-focused employment decisions. In turn, such decisions could be responded to by allegations of unlawful racial discrimination. Gender may also be a relevant factor for genetic issues. For example, some genetic conditions affect one sex such as genetic breast cancer or haemophilia. If the employer responds to these potential problems by refusing to employ such applicants or by dismissing employees who fall into that category, sex discrimination claims might arise. Finally, if a genetic condition is regarded as a disease or is likely to develop into a disease or illness, leading the employer to treat such persons differently, this might give rise to disability discrimination claims.

Race Discrimination and Sex Discrimination

The Race Relations Act 1976 and the Sex Discrimination Act 1975 include similar definitions of discrimination. According to each Act a person discriminates against another if he treats him or her less favourably on racial or sexual grounds. Indirect discrimination occurs if the employer (a) applies a requirement or condition equally which is such that the proportion of persons of the same racial or sexual group who can comply with it is considerably smaller than the proportion of persons not of that racial or sexual group who can comply with it, and (b) the employer cannot justify the requirement or condition objectively, and (c) the person who cannot comply with the requirement or condition suffers a detriment.[23] Both Acts contain anti-discrimination provisions in employment which cover arrangements for determining who should be offered, refused or omitted employment; the terms of employment; promotion, transfer or training, or other 24 benefits, facilities or services; and dismissal.[24]

In the context of indirect discrimination an employer might choose to show that medical grounds justify the discriminatory conduct. For example, it is conceivable that an employer may use the sickle cell trait to exclude job applicants. This strategy could have an indirect discriminatory effect against members of the afro-caribbean population. In some circumstances there would, of course, be justification for such exclusion, such as public safety. Thus to exclude a person with sickle cell from a job as an air pilot would manifest a recognition of the public safety issue as well as protecting the applicant from personal harm. However, the justification for indirect discrimination is unpredictable. It is not clear to what extent a medical justification may be accepted by a court or tribunal. Originally the approach of the courts was stringent but it has gradually been relaxed particularly in race discrimination cases. For example, the Court of Appeal in *Ojutiku v Manpower Services Commission* required simply that the reasons given for a discriminatory policy would be "acceptable to right-thinking people as sound and tolerable reasons".[25] On balance, the problem lies in the lack of information that is outwith the capabilities of both parties to obtain. Yet, this works against the interests of the employee with whom as a complainant, the burden of proof of discrimination lies. Thus, the lack of information available to the employee, especially in a pre-hiring context[26] and the uncertainty of the accuracy of tests are a problem as much for the employee as for the employer, making any challenge on the basis of discrimination by the employee more difficult still.

The combined effect of the legislative discrimination and health and safety provisions is illustrated by the case of *Page v Freight Hire (Tank Haulage) Ltd.*[27] Here, the applicant, Mrs Page, had been removed from a job which was considered dangerous to women of child-bearing age. The Industrial Tribunal held that she had been discriminated against but not unlawfully since it was an answer for the employer to show that what had been done was done in the interests of safety. On appeal the EAT held that, despite the fact that the employee, a divorcee, had stated that she did

46

not intend to have children and was willing to take the risk of exposure to a dangerous chemical, the employer had acted in the interests of safety. However, that was held not to be an answer to the case once discrimination had been found. Yet the EAT went on to say that Section 51 (1) of the Sex Discrimination Act 1975 provided that if an action was necessary in order to comply with a requirement of an Act passed before the SDA, it would not be unlawful discrimination. This was the case here since the employer was complying with section 2 (2)(b) of the Health and Safety at Work Act 1974. The EAT said that the woman's wishes were relevant where there was a risk of sterility or to a present or future foetus, but these wishes could not be a conclusive factor.

From the point of view of the present discussion the important aspect of *Page v Freight* was that the woman's intentions were not enough to influence the EAT and the employer's extra cautious approach was considered to be appropriate. The result was that the applicant was denied the opportunity to choose what to do in her work. The court felt that she had suffered no detriment. This case shows a combined application of discrimination and health and safety laws which emphasises the employer's, rather than the employee's, views. For example, the employer was not expected to seek to change the work practices in order to accommodate the woman. One may ask whether this case is confined to its particular facts (for example, it is unclear whether the handling of the substances could not have been avoided) or whether more generally the combined interpretation of discrimination and health and safety provisions will always be paternalistic. One response to the *Page v Freight* case was that,

> "it is only in the rarest possible cases that consent to danger should ever be relevant; it is correct statutory policy to place on the employer by far the greatest responsibility for ensuring compliance with safety legislation. Further, there can no guarantee that her future plans would remain unaltered. Finally, and perhaps strongest of all, it is one thing to permit employees to consent to possible danger to themselves and quite another when the danger relates to as yet unborn children."[28]

This response may be justified because of the particular facts of *Page v Freight*. However, some may object to the paternalistic element of this view because, ironically, one implication is that this view might actually lead to a situation that completely disregards the needs of the employee. This possibility is relevant to genetics as it is not unrealistic to predict the rejection of a candidate or the dismissal of an employee whose genetic make-up does not meet with the requirements of the job.

Disability discrimination

The Disability Discrimination Act 1995 was introduced to protect disabled persons from discriminatory treatment. Section 1 of the Act defines disability as, "a physical or mental impairment which has a substantial and long term adverse effect on a person's ability to carry out normal day to day activities." Schedule 1 elaborates this definition stating *inter alia* that: mental impairment includes an impairment resulting from or consisting of a mental illness only if the illness is a clinically well-recognised illness; the effect of an impairment is a long term effect if it has or is likely to last at least 12 months; mobility; ability to lift, carry or otherwise move everyday objects; ability to concentrate; and where (a) a person has a progressive condition (such as cancer, multiple sclerosis or muscular dystrophy or infection by the human immunodeficiency virus) he shall be taken to have an impairment which has such a substantial adverse effect if the condition is likely to result in his having such an impairment.

Thus, under s.1 of the 1995 Act, a person with genetic defects but without symptoms of disease or illness might not be considered as disabled and so may be denied protection from discrimination. However, paragraph 8 of Schedule 1 states that a person suffers a substantial adverse effect where a progressive condition impairs or is likely to impair his ability to perform daily activities. This provision suggests that the person must already be having or have encountered some difficulty with his or her work. It does not appear to cover a person with a predisposition to a disease but which has not been manifested by any symptoms. This raises doubt as to whether potential future incapacity is covered by the Act. Added to this problem is the lack of reliability of the genetic tests so that the likelihood of illness at a later stage is not guaranteed.[29]

If an individual falls within the provisions of the Act does that help? Some criticise the Disability Discrimination Act as paternalistic and creating a stigma for those covered by the protection offered. Indeed, the House of Commons Science and Technology Committee decided on balance that the Disability Discrimination legislation would not be an appropriate means of protection against discrimination following the results of genetic testing partly owing to the uncertainty of the genetic test results.[30] The Committee suggested that there might also be a psychological disadvantage to the employee if he or she were treated as disabled. On the other hand, it could be argued that the Act does not go far enough to prevent disability discrimination. The emphasis on the medical condition, rather than on problems of social organisation, reinforces the negative attitude towards disability.[31] A more positive approach would be to consider social arrangements and seek to enhance opportunities for employee development. In the context of genetic testing the employer should adjust workplace conditions rather than simply dismiss the employee.

48

This may itself be a question of mutual trust and confidence between the employer and employee, opening up the possibility of protection for the employee under the contract of employment and the obligations therein.

The contract of employment

The common law duty of mutual trust and respect was alluded to above in the context of unfair dismissal protection. In that context it appears that the employer has a duty of respect for his or her employees when imposing instructions on them. Does this duty require any more of the employer in terms of the conditions he or she creates for the workplace such as the requirement of employees to undergo medical examination? In the case of *Bliss v South East Thames Regional Health Authority*[32] the plaintiff, a consultant surgeon, was required by the employers to submit himself to psychiatric examination and was treated as having suspended himself from duty when he refused to submit to the examination. The Court of Appeal held that the employers had acted in breach of their obligations under the contract of employment since they had imposed the medical examination requirement without reasonable cause. As the Court observed, employers have no general power to require employees to submit to an examination. In this case, there was an implied term. The employers could require the employee to undergo a medical examination if they had reasonable ground for believing that he might be suffering from physical or mental disability which might cause harm to patients or adversely affect the quality of their treatment. Such power was only exerciseable after following a specified procedure. Having effectively breached the procedure by acting against the advice of a special committee, the Court of Appeal held the employer to be in breach of the duty of mutual trust and confidence and that the employer had therefore committed a repudiatory breach going to the root of the contract. The private intentions of the employer were considered irrelevant and what mattered for the purpose of judgment was what the employer did, which, according to Lord Justice Dillon, was "by any objective standard outrageous".[33]

From this case it might be argued that an employee could be protected by the insertion of an express clause into the contract of employment. A specified procedure could preclude the arbitrary exercise of an implied right by the employer that imposes a burden on the employee. The risk to third parties was clearly a relevant factor in *Bliss* but that was not enough to override the protection by a procedure of the employee against the employer's intrusive actions. The court considered the employers' actions to be calculated to destroy or seriously damage the relationship of trust and confidence between the employer and employee.

The case also illustrates that, even where an employer has what might be regarded as paternalistic motives, these will not be enough to justify a breach of the duty of mutual trust and confidence. In effect, the combination of an express and an implied term in the contract acted against the employers' paternalistic notions and favoured the wishes of the employee. This could act as a mechanism for developing the protection available to employees should genetic tests become more widespread. An employer might, by the express terms of the contract of employment, be required to follow a specified procedure before requiring the employee to undergo a genetic test. Through the procedure account may be taken of the potential harm to third parties of a failure to undergo a genetic test or a desire to continue working after a genetic test has displayed certain risks to the employee or to others. In an event, should the employer fail to follow the procedure then the employees' wishes would be paramount.

However, the combination of express and implied terms was considered from a converse perspective in the case of *Johnstone v Bloomsbury Health Authority*.[34] In that case the majority of the Court of Appeal held that the implied duty of safety would limit an express term in the contract which required the employee to be available for a further 48 hours in addition to his normal 40 hour week. Browne-Wilkinson VC, for example, saw no incompatibility between the plaintiff being under a duty to be available for 48 hours' overtime and the defendants having the right, subject to their ordinary duty not to injure the plaintiff, to call on him to work up to 48 hours overtime on average.[35] Browne-Wilkinson's view was that the implied term does not contradict the express term of the contract. The plaintiff argued that the public policy issue of safety to patients, as well as doctors, was a relevant factor. However, the Court was reluctant to accept this claim. Indeed, Stuart Smith LJ stated that "the courts should be wary of extending the scope of the doctrine beyond the well recognised categories."[36] What *Johnstone* highlights is that express terms may favour the wishes of the employer but that these may be limited by implied duties such as the duty of safety. Taking together both Bliss and Johnstone it seems clear that there is potential in the contract of employment to protect the employee against the abuse of genetic tests by the employer. Such protection may come from both express and implied terms, both types of term potentially being compromised by a combination of the two.

The requirement of mutual trust and respect might demand that the employer and the employee take account of each other's views. Only in clear cut cases such as immediate danger and a lack of capacity to alter the terms and conditions of employment should an employer then require an employee to undergo genetic testing. This position would respect the employee's health and would entail exchange of relevant information between the parties.

In addition to the specific employment law implications of genetic

50

testing, an employer who uses genetic tests may face legal actions in relation to violations of privacy and the infliction of psychological harm. These possibilities will be considered in turn.

Privacy

Genetic testing raises issues of privacy of health information. In general, privacy laws recognise and protect individual autonomy by stipulating that aspects of an individual's life should be free from state intrusion, thus enabling individuals to maintain control over certain areas of their lives. The intimacy of genetic information renders it suitable for specific privacy laws, for example, laws to prevent third parties such as an employer from demanding or obtaining without permission an individual's genetic information. Further, the recognition of privacy of genetic information would curtail the inappropriate use of genetic information in the workplace.

In the UK, there is no general right to privacy[37], however privacy interests may be indirectly protected by the common law for example by defamation and the law of confidence. Specific legislation has, however, been enacted to enable individuals to control or to be aware of the use and content of personal information. For example, the Data Protection Act 1998 controls the application and use of both computerised and manual personal data. More specifically, the Access to Medical Reports Act 1988 governs reports made for employment and insurance purposes and requires the permission of the subject of the report. In addition, this Act entitles the individual to the opportunity to see the medical report before it is forwarded to an employer. The 1998 Act also stipulates certain conditions for processing information and introduces individual obligations to be told about processing and to obtain information regarding processing data.[38] Importantly, the Act has introduced the concept of "sensitive data" into UK data protection law. This is data that encompasses information pertaining to an individual's physical or mental health or condition.[39] Schedule 3 stipulates the relevant conditions for the processing of sensitive data. These include that the individual has given explicit consent to processing[40] and that the processing is necessary in order to perform a right or obligation[41] and that the information has been made public by the data subject.[42]

However, the 1998 Act recognises that certain situations necessitate specific criteria in relation to data processing and storage. Amid such exceptions are health and social work. Section 29 enables the Secretary of State to make an order to exempt and modify provisions in relation to personal data concerning the physical or mental health or condition of a data subject. The disclosure of genetic information may fall within these exceptions. Despite these improvements to the law there are still criticisms to be made. For example, in the context of health care an individual may be precluded from accessing his or her medical information if, in the opinion of a health professional, disclosure would cause physical or mental harm to the individual or would breach the confidence of third parties.[43] Furthermore, in relation to employment it is possible that, in practice, the employee may feel an obligation by virtue of his or her subordinate position to enable the employer to see the information despite personal reservations. It should also be remembered that individuals require to be aware of the information about them in order to control effectively how that information might be used.[44]

The House of Commons Science and Technology Committee, in the report, *Human Genetics: The Science and its Consequences*[45] stated that, given the fallible nature of genetic testing, there is at present, no genetic diagnosis sufficiently robust to make a case for insisting that it be revealed to an employer. Thus, the Committee recommended that legislation to protect the privacy of genetic information should be so drafted as to forbid employers from testing for employee genetic traits. However, the Committee recognized that an exception would be justified in order to protect the public from direct and substantial risk as long as such testing is limited to specific conditions relevant to the particular employment.[46]

Protection from psychological harm

To what extent could an employer who forces the employee to undergo a genetic test without consent be liable for the civil actions of assault and battery or negligence? The requirement to obtain the consent[47] of an individual prior to an invasive procedure has meant that individuals who were not aware of the social ramifications of genetic information may claim that their consent was not informed and therefore invalid. In the medical context, this scenario may lead to an action in negligence against the physician for failure to obtain consent to the invasive procedure.[48] However, it is unlikely that consent would be vitiated by the failure to disclose all relevant information pertaining to the condition.

It is also possible that the results of a genetic test would have a psychological effect upon an individual or his or her family. For example, if the gene for a debilitating condition is detected in one individual it would be feasible that other family members may have a high probability of having the gene. The extent of the trauma caused by this discovery, if severe, may give rise to a claim for nervous shock against the employer who instigated the genetic test. However, following the Court of Appeal's decision in *AB v Tameside & Glossop Health Authority*,[49] it is doubtful that a civil claim for negligence would succeed where the information disseminated is accurate. On the other hand, provision of false information – a high risk with genetic test results – may well lead to a successful claim.[50]

52

As a result of these potential civil actions, it is necessary for an employer to consider the manner in which genetic tests are conducted and how the results of such tests are to be disseminated. It is submitted that individual genetic counselling prior to and after genetic testing would need to be an integral aspect of genetic testing, thus protecting the individual from psychological trauma and precluding him or her from instigating such civil actions.

POLICY CONSIDERATIONS AND RECOMMENDATIONS FOR FUTURE REGULATION

We appreciate that the use of genetic tests in employment is not a development that is likely to attract legislative intervention in the near term. However, to be prepared for such a development is to be forewarned. Thus, in anticipation of the development of genetic tests, various bodies in the UK and elsewhere have, to date, considered the manner in which the law should respond to workplace genetic testing. In the UK, both the Nuffield Council on Bioethics and the House of Commons Science and Technology Committee have recommended *inter alia* how genetic tests should be used for the purposes of employment. Both bodies recommended that genetic tests should only occur where first there is strong evidence of a clear connection between the working environment and the development of the condition that the test seeks to detect.[51] In addition, both bodies advocate that the genetic condition to be tested must be one which seriously endangers the health of the employee, dangers which cannot be eliminated or significantly reduced by the employer taking reasonable measures to modify environmental risks.[52] The Nuffield Council went further in recommending that tests could be contemplated where an affected employee is likely to present a serious danger to third parties.[53] Another body in the UK, the Trades Union Congress, recommends that genetic testing may be used if it is the only mechanism able to ensure the safety and health of employees and colleagues and the fitness of an employee to work.[54]

In an international context, other jurisdictions have proposed or enacted specific legislation to control and to regulate the use of genetic information and such provisions may prove to be of comparative interest for the UK.[55] In addition, the European Convention on Human Rights and Biomedicine,[56] though to date not ratified by the UK, has been ratified by many European countries and could be interpreted to cover the use of genetic information for the purposes of employment.[57] For example, Article 11 of the Convention provides that discrimination on the basis of genetic heritage is to be prohibited. In addition, Article 12 states that genetic tests that both identify the subject as a carrier of a gene or detect a genetic predisposition or disease susceptibility may only be performed for health purposes or scientific research purposes provided that appropriate genetic counselling is provided. According to McGleenan, the combination of Articles 11 and 12 means that workplace genetic testing will only be deemed to be acceptable if it is to benefit the health of the individual employee and not to prohibit the use of tests to exclude from the workplace those with a predispostion to a genetic condition.[58] However, as the BMA states, the accompanying explanation to the Convention stipulates that genetic testing that offers no health benefit to the individual such as by employers or insurance companies is an interference with the right to privacy. Such interference with this right may only be permitted if it is necessary on the grounds of the interests of third parties or the wider public good.[59] Thus, it is difficult to gauge at the moment the manner in which the Convention could be interpreted in relation to genetic information in the workplace. However, given that the purpose of the Convention is to 'protect the dignity and identity of all human beings and guarantee everyone, without discrimination, respect for their integrity and other rights and fundamental freedoms with regard to the application of biology and medicine', it is likely that the interpretation of the Convention would be sympathetic to the employee.[60]

It is apparent from the above that it is necessary to strike a balance between the employer's economic and business needs and the employee's personal health needs, and that occasionally this balance may tip in favour of either interest. As genetic testing will have ramifications for the individual that transcend the employment relationship, it is imperative that as far as possible the employee's interests should be acknowledged. The most appropriate way of doing this would be to give the employee the option of whether or not to undergo a genetic test. In addition, the employee should at the very least be able to participate in any decisions that are made with regard to genetic testing. For example, by the employee receiving full and adequate information about the nature of the condition and its relevance for decision-making in the employment context. To accept completely the employee's wishes, both in relation to the decision to have a genetic test and the consequential use of testing may however be problematic due to practical considerations and the wider public good. For example, an employee may have to forfeit decision-making powers in order to maintain an efficient working practice in the workplace. In addition, where the individual employee could affect health and safety of others such as his or her colleagues, then the implications of refusal of testing or the disclosure of test results would have to be taken into account.

It can be seen that ultimately when genetic tests are used during employment the exercise and protection of employee autonomy is at stake. This is in contrast to the manner in which personal health information is usually assessed and used, for it is usually the norm for health decisions to be implemented within a framework that respects and

54

attempts to advance as far as possible the autonomy of the individual. This recognition of autonomy, is illustrated in patient participation in decision-making and the expectation of information concerning health-care decisions. However, the exercise of autonomy may be restricted due to expediency, such as the need to take into account the position of others. Outwith the health-care relationship autonomy may not be regarded as an intrinsic factor that pervades all decision-making, due in part to the need to take into account the position of third parties. That is, if the exercise of autonomy were to affect third parties to their detriment then the interests of the individual employee and the third parties would have to be weighed, the result potentially being that the employee's autonomy is restricted. It would, of course, be necessary to establish the criteria by which the different interests may be weighed. For example, safety might be regarded as more important than efficiency.

A further balance has to be struck between allowing the exercise of individual autonomy and protecting employees from falling to the temptation of accepting what they know to be dangerous work. Employers should also have a defence against action taken by employees who had exercised their right to refuse genetic screening and subsequently developed a work related illness to which they were particularly susceptible. Provided of course that it is a risk which the employer could not have eliminated by more general measures or which could not have been exposed except by use of genetic tests.

At present, the relatively infrequent use of genetic testing in the UK arguably does not justify the enactment of specific legislation. We propose that the first step ought to be the establishment of an independent body to supervise and advise upon the use of genetic tests in the workplace. This body would provide information about the use and proposals for use of genetic testing in employment, information that is currently proving difficult to obtain. Ultimately, the involvement of an independent body may curtail the sporadic utilisation of genetic tests and may lead to common genetic testing practices throughout the UK.

Protective procedures are necessary for the benefit of both the employer and the employee to cover the use of genetic tests. Such procedures should balance the economic concerns of both parties: the cost to the employer of adapting workplace practices in the light of greater awareness of links between workplace practices and the onset of genetic diseases and the economic needs of the employees who seek work. The inequality of bargaining positions between the employer and the employee should be recognised, particularly in an era in which collective bargaining has become less effective. It is also necessary to avoid the introduction of legislation that creates stigma, particularly with the involvement of such sensitive personal issues as genetics that may affect parties beyond the employer and the employee.

Procedures before genetic tests that might safely be relied upon by employers should include the following basic steps as a minimum: first, the employer should seek to alter work processes in order to eliminate the risks of genetic problems. The necessary considerations for deciding if such alterations are viable should not just be economic but should include social and psychological factors. Secondly, pre-test counselling should be given so that employees are provided with an explanation of the scientific evidence of links between genetic conditions and workplace practices. Employees should be made aware of the risks involved in such practices as well as the consequences of undergoing or not undergoing a test. Thirdly, independent testers should conduct the tests and workers should have the opportunity to be accompanied by a friend. Fourthly, post-test counselling should be offered by which test results are disseminated sensitively with full explanations of the results and their implications. Employees should be given options where tests show problems with their continued involvement in the relevant working practices. Such choices might include exclusion from working on particular tasks (but without detrimental effect on earning capacity or career development), retraining for other suitable posts, and the offer of other suitable posts if possible. If employees have to leave because this is the only viable option then employment references for subsequent employment application purposes should take into account the privacy concerns of the employee recognised in the data protection legislation.

While these proposals do not provide a complete solution to the potential problems of workplace genetic testing we hope that this article provides at least a starting point for a debate that is certain to develop.

NOTES

1. For a discussion of the implications of genetic testing see, The Advisory Committee on Genetic Testing, *Code of Practice and Guidance On Human Genetic Testing Services Supplied Direct To The Public*, (1997, Department of Health, London).
2. The Trades Union Congress suggests that there is only anecdotal evidence of such practices and that there is no conclusive evidence of testing being conducted in unionised organisations: Note of a telephone conversation between the authors and the Trades Union Congress, February 1998. The Health and Safety Executive suggests that evidence is small and that only HM Forces have adopted genetic testing as a definite programme to screen for the sickle cell gene among aviators: Health and Safety Commission, Occupational Health Advisory Committee, Report of the Working Group on Genetic Screening and Monitoring, 1996, at p. 7.
3. See Rowe, Russell-Einhorn and Weinstein, "New Issues in Testing, the Workforce: Genetic Diseases" (1987), August, *Labor Law Journal*, 518–523, at 519.
4. See Ann Lusas Diamond, "Genetic Testing in Employment Situations" (1983) *The Journal of Legal Medicine* 231, at pp. 234–240.
5. The establishment of a Working Group on Genetic Testing and Employment by the Human Genetics Advisory Commission in July 1998 lends support to this prediction.
6. For example consider the genetic condition alpha-l-antitrypsin. Approximately one person in every 3000 has the gene for this condition. Those with the condition will

56

develop lung disease later in life, exposure to pollutants will increase the onset of the condition.

7. Diamond, above note 3, at p. 244.
8. The qualifying period of employment in order to claim unfair dismissal remains at two years: s. 108 Employment Rights Act 1996. This position seems not to have changed following the decision of the House of Lords in *R v Secretary of State for Employment ex parte Seymour-Smith and Perez* [1997] IRLR 315.
9. Ss. 98(2)(a) and 98(3)(a).
10. *Lynock v Cereal Packaging Ltd* [1988] IRLR 510.
11. Deakin and Morris at p. 489.
12. *Appleyard v FM Smith (Hull) Ltd* [1972] IRLR 19.
13. Employment Rights Act 1996 s. 98(2)(d).
14. Employment Rights Act s, 98(1)(b).
15. See John Bowers and Andrew Clarke, "Unfair Dismissal and Management Prerogative: a Study of 'Some other substantial reason' " (1981) 10 *ILJ* 34.
16. Collins, above note 28, at pp. 37–40.
17. This test is set out in s. 98(4) of the Employment Rights Act 1996: The determination of whether a dismissal is fair or unfair – (a) depends on whether in the circumstances the employer acted reasonably or unreasonably in treating it as a sufficient reason for dismissing the employee, and (b) shall be determined in accordance with equity and the substantial merits of the case.
18. Deakin and Morris, above note 21, at p. 427. Wedderburn also makes the point that this approach tends to lower standards and therefore reduces levels of protection: see Lord Wedderburn, *The Worker and the Law*, (1986, Penguin, Harmondsworth).
19. The common law implied duty of cooperation has been illustrated by the case of *Cresswell v Board of Inland Revenue* [1984] IRLR 190. However, the employer now has a clearly established duty of respect for the employee in the instructions he or she imposes: see *United Bank Ltd v Akhtar* [1989] IRLR 507.
20. SI 1994, No. 2865.
21. John Sanchez, "Genetic Testing: The Genesis Of A New Era in Employee Protection", *Western State University Law Review*, 11, 165, 119–220, at p. 200. See also Office of Technology Assessment, "Role of Genetic Testing in the Prevention of Occupational Disease" (1983).
22. Mary Basset-Stanford, "Genetic Testing in Employment", *Suffolk University Law Review*, Vol xv, 1187–1217 at p. 1192. The ILO refers to comparable figures thus leading us to suggest that in the UK the incidence of these genetic conditions will be similar: See, International Labour Office, "Conditions of Work Digest – Workers' Privacy – Part III: Testing in the Workplace" (1993).
23. Race Relations Act 1976, s. 1; Sex Discrimination Act 1975, s. 1.
24. Race Relations Act 1976, s. 4; Sex Discrimination Act 1975, s. 6.
25. [1982] IRLR 418 at 421, per Everleigh L.J.
26. As acknowledged in *King v Great Britain-China Centre* [1991] IRLR 513.
27. [1981] IRLR 13.
28. Richard Townshend-Smith, (1981) 11 *ILJ*, 132 at pp. 133–4.
29. See also Health and Safety Commission, Occupational Health Advisory Committee, *Report of the Working Group on Genetic Screening and Monitoring*, (1996) at para 12.
30. House of Commons Science and Technology Committee, Session 1994–5, '*Human Genetics: The Science And Its Consequences*', Vol 1, at para 221.
31. For a discussion of the Disability Discrimination Act see Sue Maynard Campbell, "Disability Rights" in Aileen McColgan, (Ed) *The Future of Labour Law*, (Pinter, London, 1996) 221.
32. [1985] IRLR 308.
33. *Ibid*, at p. 315.

34. [1992] 1 QB 333.
35. *Ibid*, at pp. 350–351.
36. *Ibid*, at p. 346.
37. The Human Rights Act has arguably created a basis for the development of such a right. Alternatively, as a common law jurisdiction we might witness the courts developing such a right to be protected by the common law. See this argument expressed by the Lord Chancellor in House of Lords Debates, 24 November 1997, Col 783–787.
38. Section 7.
39. Section 2(e).
40. Schedule 3(1).
41. Schedule 3(2).
42. Schedule 3(5).
43. Section 7.
44. The Act addresses this by involving the data subject in the control, access and use of any data. For example, section 7(a)–(d) grants the right to be informed of processing, to be given a description of data, to be informed of to whom the information is to be disclosed, the source of the data and the logic involved in any decisions made using the data. In addition, Schedules 1, 2 and 3 stipulate principles which underpin the interpretation of the Act. These principles aim to protect the individual from abuse of unauthorised use of his or her personal or sensitive data.
45. House of Commons Science and Technology Committee, Session 1994–95, *Human Genetics: The Science And Its Consequences*, Volume 1.
46. Science and Technology Committee above note 90, at para 226.
47. The phrase "informed consent" is subject to debate in relation to its interpretation. In the UK it is more usual to refer to "consent".
48. For a comprehensive account of informed consent see Faden and Beauchamp, *A History and Theory of Informed Consent*, (1986 Oxford University Press, Oxford).
49. [1997] 8 Med LR 91.
50. *Allin v City & Hackney Health Authority* [1996] 7 Med LR 167. For a discussion of this case and the case in the above note see Nicholas J Mullaney, "Liability for Careless Communication of Traumatic Information" [1998] 114 *LQR* 380.
51. Nuffield Council on Bioethics, "*Genetic Screening Ethical Issues*" December (1993), p. 90. para 10.13; Science and Technology Committee see above note 90 at para 233.
52. *Ibid.*
53. *Ibid.*
54. Telephone Conversation February 1998.
55. See T. McGleenan. 'Genetic Testing and Screening: The Developing European Jurisprudence', (1999), *Human Reproduction and Genetic Ethics*, Vol. 5, No 1, 11–19.
56. Council of Europe Convention for the Protection of Human Rights and Dignity of the Human Being with regard to the application of Biology and Medicine: Convention on Human Rights and Biomedicine 164, European Treaty Series, Oviedo, (1997).
57. For the impact of the Convention upon the UK see T. McGleenan, above note 55 at pp 13.
58. T. McGleenan, above note 55, at pp. 12.
59. BMA, *Human Genetics: Choice and Responsibility*, (1998, Oxford University Press, Oxford).
60. Article 1, Convention for the Protection of Human Rights and Dignity of the Human Being with Regard to the Application of Biology and Medicine.

[12]

Pharmacogenetics: Ethical Issues and Policy Options*

Allen Buchanan, Andrea Califano, Jeffrey Kahn, Elizabeth McPherson, John Robertson, and Baruch Brody

ABSTRACT. Pharmacogenetics offers the prospect of an era of safer and more effective drugs, as well as more individualized use of drug therapies. Before the benefits of pharmacogenetics can be realized, the ethical issues that arise in research and clinical application of pharmacogenetic technologies must be addressed. The ethical issues raised by pharmacogenetics can be addressed under six headings: (1) regulatory oversight, (2) confidentiality and privacy, (3) informed consent, (4) availability of drugs, (5) access, and (6) clinicians' changing responsibilities in the era of pharmacogenetic medicine. We analyze each of these categories of ethical issues and provide policy approaches for addressing them.

THE EMERGENCE OF PHARMACOGENETICS may mark a new era in medicine in which drug therapies can be selected for individual genotypes, increasing the safe and efficacious use of pharmaceutical agents (Weinstein 2000; Sander 2000; Roses 2000a; Beaudet 1999; Collins 1999). While offering many benefits, pharmacogenetics also raises potential ethical and policy issues. The aim of this analysis is to identify the chief ethical issues in pharmacogenetics and the policy options for addressing them, so that the clinical community can be prepared to deal with these issues as they emerge.

Pharmacogenetics is the study of the effects of genotypic variations on drug-response, including safety and efficacy, and drug-drug interaction. These variations, including single nucleotide polymorphisms (SNPs) and microsatellites, can be in genes themselves or in other parts of the genome. The variations are not tissue-specific and constitute a static, global

*From the Consortium on Pharmacogenetics, funded by an unrestricted grant from GlaxoSmithKline, First Genetic Trust, and IBM.

property of the individual. The second source of differences in drug response is variations in the expression of individual genes in the cells of particular tissues. This information is dynamic, changing in response to endogenous and exogenous stimuli. The study of this second source of variations in drug response is pharmacogenomics. Both types of variations have been shown to correlate strongly with drug response (Weinstein et al. 1997; Califano, Stolovitzky, and Tu 2000). Once such correlations are known, a *pharmacogenetic* or *pharmogenomic* test, assays to determine an individual's probable response to a drug or group of drugs, can be devised.

Because pharmacogenetics has the potential to be a large-scale methodology that will have important effects on drug research and clinical practice, a study of the ethical and policy issues is needed at this time. For the most part, our analysis will apply to pharmacogenomics as well.

POTENTIAL BENEFITS OF PHARMACOGENETICS

A number of potential benefits for this methodology have been claimed, including a better understanding of drug response mechanisms, safer and more effective use of drugs, a more streamlined drug development and approval process, and the salvage of drugs that have been removed from the market for safety reasons. If these benefits are realized, clinicians and patients will have access to safer and more effective drugs. Some of these benefits are rather speculative, but three in particular seem likely to be significant in the near term, given the realities of the drug development and utilization process.

Understanding the Genetic Bases of Drug Response Mechanisms

By identifying links between drug response phenotypes and genetic variations, it is possible to pinpoint critical aspects of drug metabolism or activity. For instance, individuals with one or more specific mutations in some members of the Cytochrome P450 family of enzymes tend to be poor or rapid drug metabolizers over a wide range of drugs (Murphy et al. 2000). Health care consumers, physicians, and payers would all find this information valuable.

Enhanced Post-Market Drug Surveillance

Significant numbers of adverse reactions sometimes appear only after a drug has been approved and in use for some time. By genotyping those who suffer adverse reactions, and restricting the use of the drug to those who do not have this genotype, drug safety could be improved. In addition, given the large investment already made in the development of the drug and provided that there is still sufficient time left on the patent, the drug maker would have an incentive to sponsor such a pharmacogenetic study. For example, Abacavir has proven to be an effective treatment for HIV, but in clinical trials hypersensitivity reactions have been reported in 5 percent of adults and pediatric patients receiving the drug. A study of DNA samples from patients exhibiting hypersensitivity is being conducted to determine whether this adverse reaction is correlated with an identifiable genetic variation, in the hope that a pharmacogenetic test could be developed to determine which HIV patients should not receive the drug (Sykes 2000).

Expediting Drug Development Through Smaller Clinical Trials

Pharmacogenetic tests could be used in Phase II drug trials to identify genotypes of individuals who respond to the drug. Phase III trials would then include only individuals who have these genotypes. The drug would then be labeled only for individuals who have the included genotypes. Because Phase III trials would use fewer subjects and the results would be available sooner, drug companies could have strong incentives to conduct a pharmacogenetic study as part of Phase II.

LIMITS

There are three *scientific* constraints on the scope and scale of pharmacogenetics. The first is the range of variability of genotypic variation regarding drug response. Will there turn out to be enough variation to make genotype testing worthwhile, yet not so much variation that an impractically large number of drugs and tests would be needed? At present, knowledge of the extent of variation is insufficient to answer this crucial question.

Second, in many, and perhaps most, cases, more than one gene influences drug response and the manifestation of side effects. This may be due to the fact that several genes are required to produce the set of proteins that are responsible for the actions of the drug. Where many genes are involved, pharmacogenetic testing may be more complex and less definitive.

Third, drug response depends not only on genotype but upon a number of other factors as well, including environment (understood broadly to cover a range of factors from workplace chemical agents to food, alcohol, and tobacco consumption), the individual's state of health (liver, kidney, and endocrine functions, immune system, and so forth), and compliance with drug treatment. For all three of these reasons, pharmacogenetic tests will yield probabilistic results rather than definitive predictions. So even where a pharmacogenetic test is available, sophisticated clinical judgment will be needed to determine (1) whether to use the pharmacogenetic test or instead to rely on phenotypic tests or drug dose adjustment based on clinical studies or on trial and error in the individual case, and (2) if the pharmacogenetic test is administered, how much weight to accord to its result, vis-à-vis other factors, in making a drug treatment decision.

The pace, direction, and extent of pharmacogenetic development will also be shaped by various *social* factors, the most important of which are likely to be (1) the nature of the existing health care insurance and delivery system, and in particular the sorts of cost-containment strategies that are pursued; (2) public attitudes toward genetic research and testing; (3) the distribution and extent of knowledge about the potential benefits and costs of pharmacogenetics among payers, providers, and consumers of health care; (4) the character of regulation and oversight practices regarding pharmacogenetic research and clinical use; and (5) the willingness of industry to invest in the technologies needed to realize the potential of the methodology.

Some combination of these factors may explain the puzzling fact that, despite the long-standing knowledge that Cytochrome P450 plays a major role in the metabolism of a wide range of drugs, pharmacogenetic tests for the relevant genetic variations have not been adopted in clinical practice.

The scientific and social factors are interrelated as well. For example, how much industry invests may depend in part upon predictions about how likely it is that cost-conscious insurers will view pharmacogenetic testing as a significant cost-saver rather than as a costly procedure of uncertain benefit. Conversely, consumer demand for pharmacogenetic tests, possibly stimulated by direct marketing and the availability of information about its potential benefits on the internet, may serve to counteract insurers' initial tendency to think that routine use of these tests is too costly. Public perception can also shape regulation: negative attitudes toward anything regarded as a genetic test (and hence as a possible source of insurance or employment discrimination) may encourage regulators to impose rigorous protections of confidentiality that will inhibit the development of the methodology.

ETHICAL ISSUES: TWO MISAPPREHENSIONS

As with other applications of genetic knowledge, the challenge of pharmacogenetics is to take genetic variations among people into account to provide better health care, while treating people with equal respect. The first step toward accomplishing this is to avoid certain misapprehensions that may distort public deliberation. Otherwise, concern about ethical issues may result in ill-conceived constraints that unnecessarily limit the benefits pharmacogenetics could deliver. Two misapprehensions are particularly important to avoid: (1) genetic exceptionalism, the assumption that all things genetic inherently involve exceptionally serious ethical concerns and therefore require novel ethical principles or special regulatory responses, and (2) overbroad genetic generalization, the assumption that all "genetic tests" raise the same ethical issues.

Genetic exceptionalism is widespread. Consider, for example, the fact that legislation is being adopted in many states to prohibit "genetic discrimination" by insurers, while at the same time little is said about the ethical issues raised by the fact that insurance companies also routinely refuse coverage or charge higher rates for those who have had cancer or have untreatable hypertension (Hall and Rich 2000). If the former raises ethical issues, why not the latter as well?

It is also a mistake to lump all genotype assays together under the uninformative heading "genetic tests"—and then to make assumptions about ethical concerns that apply only to some (Roses 2000a & b; 1997). For example, a test for the Huntington's disease gene mutations may carry a much higher risk of stigma and discrimination than one for hereditary hemochromatosis (a condition that can be effectively treated) or a pharmacogenetic test to determine whether an individual is a high-responder for a particular drug.

SIX ETHICAL ISSUES

Once these misapprehensions are dispelled, the ethical issues can be addressed under six headings: (1) regulatory oversight, (2) confidentiality and privacy, (3) informed consent, (4) availability of drugs, (5) access, and (6) clinicians' changing responsibilities in the era of pharmacogenetic medicine.

Oversight of Clinical Introduction of Pharmacogenetic Tests

If pharmacogenetic tests are to play a major role in medicine, it is essential that there be safe and effective means for determining a patient's likely response to a drug before it is prescribed. So the first ethical issue is whether current oversight mechanisms are sufficient.

Pharmacogenetic tests and their use in prescribing drugs will fall within the Food and Drug Administration's (FDA's) regulatory jurisdiction on two grounds. If sold as test kits, they are "diagnostics" that must have FDA approval before being sold in interstate commerce. Makers of pharmacogenetic test kits will have to submit to the FDA evidence of the tests' analytic validity, and clinical validity and utility, so that users will be assured that they are safe and effective for the intended purpose. However, currently these protections may not apply to "home brews," that is, to those many laboratories that offer genetic testing services using their own reagents, rather than utilizing commercial testing kits. The Secretary's Advisory Committee on Genetic Testing (2001) and the FDA are now formulating a suggested process and criteria for reviewing and approving genetic tests, including pharmacogenetic tests.

Pharmacogenetic tests also may fall within the FDA's jurisdiction when they are an integral part of a New Drug Application (NDA). As noted above, a pharmaceutical company may investigate the use of a drug to block a disease process in a pharmacogenetically-defined subgroup of the population with that disease and would be obligated to provide the pharmacogenetic data in support of the desired claim—i.e., that the drug is indicated for use in a given genetic subpopulation. These data are subject to review and inspection by FDA as part of the NDA review. The FDA may then approve the drug only for persons who have the requisite drug response profile, thus requiring a pharmacogenetic test prior to the prescription of the drug. Our analysis suggests that the current framework for FDA oversight of pharmacogenetic tests is sufficient when they are part of commercial testing kits. However, if there is continued lack of oversight for "home brew" testing it may become an issue that needs to be addressed if and when pharmacogenetic tests become widely offered by a range of laboratories.

Protection of Privacy and Confidentiality

Appropriate protection for privacy and confidentiality is crucial because pharmacogenetic test results can carry several types of secondary information that represent a risk of psychosocial harm. (1) Pharmacogenetic test results may be linkable to genetic disease or to genetic predisposition to diseases for which the individual is currently asymptomatic, because the genotype that influences drug response may also play a role in predisposition to disease. (2) Genotype-based information about drug response may have implications for disease progression and hence for prognosis. The same genotype that is correlated with a negative drug response test may also be correlated with a more rapid progression of the disease or a worse outcome. (3) The information that an individual is a nonresponder or an unsafe responder to a particular drug or class of drugs might itself have adverse insurance and/or employment implications. (4) Information about genotypic factors in drug response, like other types of genetic information, may have implications for relatives, not just for the individual tested—e.g., relatives may have a greater likelihood of responding in a similar manner to a particular drug. These concerns about secondary information may be more serious in pharmacogenetic testing than in pharmacogenomics, because pharmacogenetic information may be linked more often to an individual's stable genotype rather than to the expression of his genes in a particular tissue.

It is important to note that concerns about secondary information are not unique to pharmacogenetic tests or to genetic tests generally. For example, a positive test for HIV status has implications for individuals other than the individual tested, namely, sexual partners. Of course the value of privacy and confidentiality is not absolute, since there will be circumstances in which protection of the public's health may trump privacy or confidentiality. For example, reporting the status of nonresponders to drug therapy for an infectious disease might be required in the future, just as reporting of infectious diseases is now required. Therefore there is no reason to assume that novel mechanisms for reducing the risk of breaches of confidentiality will be needed in the case of pharmacogenetics, even though the task of protecting privacy and confidentiality is complicated by the fact that pharmacogenetic research and clinical use both require integration of genotypic information with individuals' medical records. Regulations and practices concerning confidentiality of pharmacogenetic test results, therefore, must be one element in a larger infrastructure for collecting, storing, exchanging, and accessing a complex array of information (NBAC 1999).

Policy designed to reduce the risk of psychosocial harm from secondary information should distinguish between (1) the role of "firewalls"—

barriers to the *dissemination* of sensitive information—and (2) the role of insurance and employment regulation to prevent certain damaging *uses* of information. Since neither firewalls nor legislation prohibiting the use of pharmacogenetic information is likely to be capable of dealing with the full range of issues, policy should seek to achieve an optimal balance between these two types of safeguards.

One example of the firewall strategy is a policy of including only the medicine response result in a laboratory report of pharmacogenetic test results, but no secondary information—including information about the genotypic features that are responsible for the drug response result. However, as was noted above, in some cases the drug response result itself, apart from any information about its genetic basis, may have damaging implications. For example, where there is no other effective treatment for a serious disease, indicating that the individual is a nonresponder for the only drug available for a serious condition would imply that he has an untreatable serious chronic disease and therefore will represent high medical costs. Given that insurers and sometimes employers have access to medical records, a firewall strategy cannot completely protect the individual against discrimination based on an unfavorable pharmacogenetic test result. To reduce the risk of discrimination based on information that cannot be contained by firewalls, regulations prohibiting insurers (and employers) from using such information to the detriment of the patient may be part of the solution.

Another type of firewall strategy, designed to allow for recurring access to biological samples and other relevant data, is the use of "trusted intermediaries." Such public or private entities would serve as secure repositories for biological samples and information about the individuals from whom the sample was taken. Those seeking access to samples or information for clinical or research uses could obtain them only from the intermediary, on terms designed to protect privacy and confidentiality. However, because this firewall strategy concentrates a very large amount of potentially sensitive information in the hands of the intermediary, any breach of confidentiality might be serious. Regulation may be needed to reduce this risk.

Legally, most states protect to some extent against unauthorized disclosure of medical information. Those protections almost certainly would apply to the results of pharmacogenetic tests contained in medical or hospital records. However, the scope of legal protection varies widely, and many holders of medical information may not be directly subject to legal restrictions. Recently issued federal privacy regulations under the Health Insurance Protection and Portability Act (HIPPA) of 1996 would greatly extend legal protection against unauthorized uses of medical and pharmacogenetic information (DHHS 2000).

Informed Consent

Participation in pharmacogenetic research, as with nearly all research on human subjects, requires informed consent. Subjects must be informed of the need to obtain DNA for research, the protections of confidentiality and privacy that will be in place (including how the sample and test results will be stored and their privacy and confidentiality maintained), whether additional research may be done in the future, the risks and benefits of providing DNA for participation in the trial, and whether research results will be disclosed to the subject. There is no consensus, however, about: (1) whether research results should be disclosed to subjects; (2) if so, when and under what conditions, and by whom; (3) what information, counseling or other resources ought to be provided; and (4) whether it should be included in subjects' medical records even though it may not meet mandated clinical laboratory standards for quality and accuracy.

Pharmacogenetic research will involve new and largely unpredictable uses for biological samples and related information over time. Obtaining consent for each new use would be costly and cumbersome, and would require repeated intrusions into the individual's privacy. A more promising approach would be to combine prospective blanket authorization for a wide range of future "minimal risk" uses of samples and related information with the opportunity for "dynamic" consent for especially sensitive uses that may arise in the future. One key policy question is what role IRBs should play in determining when special consent is needed for research uses of genotypic information.

Once pharmacogenetic tests have moved from research to clinical practice, a crucial issue is whether informed consent to the collection of DNA samples and to performance of pharmacogenetic tests is always ethically required. The answer will depend both upon how serious the risk of psychosocial harm from secondary information is and upon what is meant by "informed consent." In situations where there is significant risk, this risk must be disclosed to the patient and explicit consent, either oral or written, must be obtained. In low-risk situations, to avoid genetic exceptionalism, pharmacogenetic tests should be treated like other routine laboratory tests, in which minimal explanation and patient assent suffice.

Test results by themselves are not high or low risk. Whether a test result carries high risk will depend both upon the character of the information conveyed in the result and whether there are adequate safeguards to prevent misuse of information. Risk-reducing measures might include (1) the availability of counseling when appropriate before and/or after testing; (2) increased patient knowledge, both about genetics generally and about pharmacogenetics in particular; (3) physician understanding of the importance of keeping test results confidential; (4) coding technologies that erect firewalls between the pharmacogenetic test result and other information to which these parties have access; and (5) legislation to prohibit insurers and employers from discriminating on the basis of secondary information. Of particular importance in test design will be the choice of markers that carry minimal secondary information. For example, if drug response is associated with more than one marker, the choice of which marker to use in the pharmacogenetic test should reflect a concern to minimize the risk of damaging secondary information. If such measures are in place, then minimal explanation and patient assent may suffice.

In the clinical context, there exists no counterpart to IRBs that could serve to review later testing of a DNA sample. However, the consent that is obtained for the original pharmacogenetic testing might be worded broadly enough to encompass later testing that becomes necessary as the clinical situation evolves.

Influence on Research Agenda and Availability of Drugs

The increase in knowledge of how genotypic variations affect drug response may identify genotypic subgroups that carry a higher than average likelihood of nonresponse or side effects. This new knowledge may have two quite different effects on the research agenda and, eventually, on the availability of drugs. If the group is sufficiently large, pharmaceutical companies may find it economically attractive to invest in research and development of a drug to meet the needs of that population. For example, those who have the ALOX5 promoter genotype do not respond to some of the drugs used in conventional asthma treatment (Drazen et al. 1999); but given the large number of people with asthma, pharmaceutical companies may find it profitable to try to develop an alternative drug for this minority of asthma sufferers. Alternatively, if the group of nonresponders is too small, pharmaceutical companies might not find it economically attractive to try to develop an alternative drug for that population. Persons who are in such an "orphan" genotypic subgroup will lack access to effective medications, not because they lack health insurance, but because drugs have not been developed for them. However, individuals who are identified by pharmacogenetic tests as being poor responders for a drug may in fact benefit, because they can then avoid the costs and the possible side effects of drugs that would be unlikely to be efficacious for them.

There are two main policy options for ameliorating the problem of "orphan" groups of nonresponders, which might be pursued either singly or in combination. The first option is to rely on the existing orphan drug law, which encourages pharmaceutical companies to develop drugs for relatively rare diseases by granting a seven year period of market exclusivity, tax credits, and in some cases funds for orphan drug research and clinical trials (21 U.S.C. 33360aa- 360ee (1984); 26 U.S.C. 345C (supp. II 1994); 42 U.S.C. 236 (1994)). This law could be amended or could serve as the model for a new law designed to give pharmaceutical companies incentives to develop drugs for genotypic groups or diseases that would otherwise be neglected. The second option is to shape NIH and other public and private grant priorities so as to subsidize research to develop drugs for orphan genotypic subgroups or new orphan diseases. Before any such policy change should be contemplated, however, it would be necessary to ascertain whether the increasing integration of pharmacogenetics into drug research design does in fact unduly restrict research agendas and then try to ascertain the least costly, most effective policy response to the problem.

Access to Clinical Application

At present, it is unclear whether, or to what extent, pharmacogenetic tests will be included in standard insurance benefit packages or in Medicaid or Medicare coverage. Much will depend upon whether payers believe that pharmacogenetic testing will be cost-effective, in the long-run, even when the costs of integrating the methodology into the health care delivery system are taken into account. (These include the costs of learning about, administering, interpreting, and responding to pharmacogenetic testing). If payers see testing as a way of rationally limiting the use of drugs to those for whom they are likely to be effective and of reducing the incidence of costly adverse reactions, and hence as a source of cost-savings, they may be more willing to bear the costs of developing and implementing it.

But even for those who have insurance coverage for pharmacogenetic tests, the use of these tests may, paradoxically, result in restrictions on access to drugs from which the individual might benefit. If pharmacogenetic tests are used to set the standard of care regarding the use of drugs, then that standard is likely to reflect a group rather than an individual patient perspective. Especially in non-life-threatening situations, the standard may restrict the use of certain drugs to those patients who have test results that show they have a "high response" or "safe-response" genotype. Although this approach may be rational from a group perspective, it may be equally rationale for the individual to want to have access to the drug despite the test results. Despite the risk and/or the probability of a low response, the individual patient may quite rationally see the drug as his best option. This conflict between the individual and group perspective is not unique to pharmacogenetics, of course. However, the integration of pharmacogenetics into clinical practice, especially in an era of increasing cost-consciousness, may make the problem more acute.

This is one example of a more general problem that theorists of distributive justice have identified in a number of areas of health care: on the one hand, it would be unreasonable for an individual to expect that society will always provide him with whatever care would be maximally beneficial to him, regardless of cost; on the other hand, a strictly utilitarian social allocation policy that simply aims to maximize aggregate benefit is unfair to individuals (Brock 1983; Daniels, Light, and Caplan 1996). Existing pharmacoeconomic methodology does not permit consideration of fairness in the allocation of drugs and hence is insensitive to the individual's claim that he should be allowed access to uses of a drug for which there is a pharmacogenetic test that shows he is unlikely to receive benefit or is at high risk of experiencing a serious side effect (Bootman, Townsend, and McGhan 1996).

Clinicians' Responsibilities to Use and Respond to Pharmacogenetic Tests

Assuming progress in the science of pharmacogenetic testing, the benefits for patients and the health care system will depend on how rapidly and easily such testing is assimilated into medical practice. Physicians will have to learn the role that pharmacogenetic tests play in choosing therapy, and respond accordingly. The pace at which pharmacogenetic tests are integrated into medical practice may differ across diseases and between specialists and generalists.

Physicians have been ignorant of genetics in the past, and have often made errors in interpreting genetic tests when they have used them (Giardiello et al.). Without major changes in the teaching of genetics in undergraduate, graduate, and continuing medical education, physicians are unlikely to quickly become knowledgeable about genetic testing and how to use it in their clinical decision making. Pharmacists may also have a role to play in educating both physicians and patients about the use of genotypic tests for drug selection.

Ethically, physicians have a duty to become knowledgeable about pharmacogenetic testing and use it in their clinical decision making when reasonable practitioners in similar circumstances would do so. Two questions must be addressed: When is there an obligation to offer pharmacogenetic tests? When must pharmacogenetic test results be followed and when may they be ignored?

A physician would have a duty to offer a pharmacogenetic test when the benefits outweigh the risks and costs of doing so, and when reasonable medical practitioners in those circumstances agree. That situation is likely to exist when there is reliable evidence of a test's positive predictive value and its utility in determining whether to prescribe a drug, in what dosage, or in what combinations with other drugs. This information could exist in the medical literature, in clinical practice patterns, or in the drug's labeling. If the positive predictive value of the test is high enough, there may be a duty to offer the test.

A second question of duty arises once a test is done and a clinical decision has to be made on the basis of it. In most cases, if the pharmacogenetic test is clinically valid and has a high positive predictive value, the results will strongly indicate what clinical decision will be made, and no problem will arise. If the pharmacogenetic test has a more limited predictive power, the clinical decision will be more complicated, as it is in many other nongenetic cases in which diagnostic tests yield only probabilistic results. A physician will have to exercise clinical judgment in deciding whether to give a drug where a patient's genotype indicates that she has, say, a 30 percent chance of experiencing side effects. As long as the judgment is within the range of acceptable practice, and the patient has been informed of the tradeoffs and consented to the decision, no breach of duty will have occurred.

CONCLUSION

Pharmacogenetics has the potential to revolutionize clinical practice and to transform drug research. Potential benefits include a better understanding of the basic mechanisms of drug response, provision of evidence-based information on the use of the drug in the individual patient, development of drugs that can be used more safely and more effectively, and a more efficient drug development process. However, this powerful new methodology raises ethical issues, both for research and clinical applications, that will require careful consideration and wise policy responses.

REFERENCES

Beaudet, Arthur L. 1999. ASGH Presidential Address. Making Genomic Medicine a Reality. *American Journal of Human Genetics*. 64: 1–13.

Bootman, J. Lyle; Townsend, Raymond J.; and McGhan, William F. 1996. *Principles of Pharmacoeconomics*. 2d ed. Cincinnati, OH: Harvey Whitney Books Co.

Brock, Dan W. 1983. Distribution of Health Care and Individual Liberty. In *Securing Access to Health Care, Volume Two: Appendices (Sociocultural and Philosophical Studies)*, President's Commission for the Study of Ethical Problems in Medicine and Biomedical and Behavioral Research, pp. 239–63. Washington, DC: U.S. Government Printing Office.

Califano, Andrea; Stolovitzky, Gustavo; and Tu, Yuhai 2000. Analysis of Gene Expression Microarrays for Phenotype Classification. In *Proceedings of the Eighth International Conference on Intelligent Systems for Molecular Biology*, ed. Philip Bourne; Michael Gribskov; Russ Altman; et al., pp. 75–85. Menlo Park, CA: AAAI Press.

Collins, Francis S. 1999. Shattuck Lecture—Medical and Societal Consequences of the Human Genome Project. *New England Journal of Medicine* 341: 28–37.

Daniels, Norman; Light, Donald W.; and Caplan, Ronald L., eds. 1996. *Benchmarks of Fairness for Health Care Reform*. New York: Oxford University Press.

DHHS. Department of Health and Human Services. 2000. Standards for Privacy of Individually Identifiable Health Information. *Federal Register* 65 (28 December): 82462.

Drazen, Jeffrey M.; Yandava, Chandri N.; Dubé, Louise; et al. 1999. Pharmacogenetic Association Between ALOX5 Promoter Genotype and the Response to Anti-Asthma Treatment. *Nature Genetics*. 22: 168–70.

Giardiello, Francis M.; Brensinger, Jill D.; Petersen, Gloria M.; et al. 1997. The Use and Interpretation of Commercial APC Gene Testing for Familial Adenomatous Polyposis. *New England Journal of Medicine* 336: 823–27.

Hall, Mark A., and Rich, Stephen S. 2000. Patients' Fear of Genetic Discrimination by Health Insurers: The Impact of Legal Protections. *Genetics in Medicine* 2: 214–21.

Murphy, M. Paul; Beaman, Margaret E.; Clark, L. Scott; et al. 2000. Prospective CYP2D6 Genotyping as an Exclusion Criterion for Enrollment of a Phase III Clinical Trial. *Pharmacogenetics* 10: 583–90.

NBAC. National Bioethics Advisory Commission. 1999. Research Involving Human Biological Materials: Ethical Issues and Policy Guidance. Rockville, MD: NBAC.

Roses, Allen D. 1997. Genetic Testing for Alzheimer Disease. Practical and Ethical Issues. *Archives of Neurology* 54: 1226–29.

———. 2000a. Pharmacogenetics and Future Drug Development and Delivery. *Lancet* 355: 1358–61.

———. 2000b. Pharmacogenetics and the Practice of Medicine. *Nature* 405: 857–65.

Sander, Chris. 2000. Genomic Medicine and the Future of Health Care. *Science* 287: 1977–78.

Secretary's Advisory Committee on Genetic Testing. 2001. Enhancing the Oversight of Genetic Tests: Recommendations of the SACGT. (Available at *http://www4.od.nih.gov/oba/sacgt.htm*. Accessed 30 April 2001.)

Sykes, Richard. 2000. *New Medicines, the Practice of Medicine, and Public Policy: A Strategic View of How Innovation in the Development of Medicine Will Impact on the Practice of Medicine and the Delivery of Healthcare During the Next Two Decades*. Norwich: Stationery Office.

Weinstein, John N. 2000. Pharmacogenomics—Teaching Old Drugs New Tricks. *New England Journal of Medicine* 343: 1408–9.

———; Myers, Timothy G.; O'Connor, Patrick M.; et al. 1997. An Information-Intensive Approach to the Molecular Pharmacology of Cancer. *Science* 275: 343–49.

Part II
Gene Therapy/Testing/Cloning

[13]

Beware! Preimplantation genetic diagnosis may solve some old problems but it also raises new ones

Heather Draper and Ruth Chadwick *University of Birmingham and University of Central Lancashire, respectively*

Abstract

Preimplantation genetic diagnosis (PIGD) goes some way to meeting the clinical, psychological and ethical problems of antenatal testing. We should guard, however, against the assumption that PIGD is the answer to all our problems. It also presents some new problems and leaves some old problems untouched. This paper will provide an overview of how PIGD meets some of the old problems but will concentrate on two new challenges for ethics (and, indeed, law). First we look at whether we should always suppose that it is wrong for a clinician to implant a genetically abnormal zygote. The second concern is particularly important in the UK. The Human Fertilisation and Embryology Act (1990) gives clinicians a statutory obligation to consider the interests of the future children they help to create using in vitro fertilisation (IVF) techniques. Does this mean that because PIGD is based on IVF techniques the balance of power for determining the best interests of the future child shifts from the mother to the clinician?

(*Journal of Medical Ethics* 1999;25:114–120)

Keywords: preimplantation genetic diagnosis; reproductive control; new genetics; best interests of future children

Preimplantation genetic testing

Preimplantation genetic diagnosis (PIGD) is the result of combining our increasing knowledge about the human genome with techniques employed in assisted conception. It is currently employed when a couple have had one affected child and/or one or more termination of pregnancy (TOP) following conventional antenatal testing. Superovulation, egg collection and in vitro fertilisation (IVF) are followed by the genetic testing of each of the resulting embryos. Non-affected embryos are then implanted using the same protocols covering IVF for infertility. Preimplantation genetic diagnosis enables couples to found their family with greater confidence that they will neither give birth to an(other) affected child nor subject themselves to a series of terminations of pregnancy.

Advantages of preimplantation diagnosis

The most obvious advantage of PIGD is that it enables couples to have an unaffected child without also having to have a series of terminations of pregnancy. These terminations can be particularly stressful for a woman (and her partner) because each aborted fetus is potentially a wanted child. Moreover, the knowledge that each pregnancy is conditional upon a clear test result also constrains the usual joy with which news of a new pregnancy is greeted by those "trying for a baby". The potential of PIGD to reduce the physical and psychological wear and tear on the woman and her partner is one advantage. Another is that the failure to implant an embryo is viewed by many as morally preferable to the killing of a more fully developed fetus. This view is attractive to those who believe that the embryo gradually acquires greater moral status as it develops both physically and towards viability - hence the various cut-off points in its development which have been used to demark appropriate behaviour (the primitive streak for embryo research or the increased ability to exist independently which seems to mark the cut-off point for termination of pregnancy etc).[1] Preimplantation genetic diagnosis is also attractive to those who wish to draw a distinction between actively destroying life (termination of pregnancy) and a failure to save life (the decision not to implant an embryo). Indeed, we speak of "allowing embryos to perish" rather than "embryo killing".

There remains, however, a sizeable contingent for whom PIGD does not resolve any ethical problems because for them, morally significant human life begins at conception. Obviously those holding this view may be as opposed to PIGD as they are to IVF, unless they are prepared to endorse "embryo euthanasia". According to this view, the human embryo has a moral status akin to that of the fetus, infant or fully developed human. This position need not be incompatible with the view that what makes life worth living is its quality.

Thus, it could both be held that the embryo has moral significance and that it is in the best interests of the embryo that it is not implanted, provided that its life is not worth living. Such a view might also justify termination of a pregnancy (feticide) or non-voluntary euthanasia (of both small infants and incompetent adults).

Another perceived advantage of preimplantation diagnosis is that it gives greater choice to couples because it gives them the scope to make a decision about which of the embryos to implant, whereas TOP simply presents the choice of whether or not to terminate a pregnancy. This choice is, as we will show, somewhat of a mixed blessing. On the positive side it enables couples to maximise the advantages for their future child but within the limits of what nature provides (ie the number of embryos they have to choose between). So, whilst not actually manipulating the genes of their offspring, they are able to choose not to implant, for instance, not just those embryos which will be directly affected by some genetic disorder, but also any embryo which carries recessive genes for some disorder. Thus, parents are able to protect their future children not just from the direct burdens of genetic disorder, but also the worries which they have themselves experienced to ensure that their own children will not suffer. In time couples may even be able to select according to other genetic traits just as they can now select the sex of their future progeny.

Of course, even as these perceived advantages are listed, old problems begin to emerge. It is one thing to accept, as some people do, that no injustice is done to the embryo if it is not implanted. It is quite another to endorse the policies and justifications which motivate the decision not to implant. For instance, what counts as a life not worth living? Is sex selection permissible? What makes a particular genetic formation a disorder rather than a difference? Is it permissible to maximise advantage as well as to minimise disadvantage? Should we be eradicating difference or intolerance to difference? These larger questions cut across the debate even if we are prepared to accept that no one of moral significance is affected by our decision not to implant any one of the embryos in question. We do not intend to dwell on these old problems, but intend instead to explore some of the new problems which PIGD has introduced, concerning whether it is always wrong to implant an affected embryo; and the extent to which PIGD involves a transfer of power away from women. There is also a third problem, concerning the greater potential for eugenic implications in PIGD because of the possibility that embryos might be selected on the basis of carrier status for recessive conditions, but this will not be discussed here for reasons of space.

First new problem: is it always wrong to implant an affected embryo?

It would be easy to assume that the purpose of PIGD is to ensure that only unaffected children are born, and that since the rationale for desiring this end is to avoid the life of suffering which affected children are perceived to have, in the absence of available and effective treatment for the condition in question, it must be wrong to deliberately or knowingly implant an affected embryo. It is one thing, it could be argued, for couples to refuse to undertake PIGD, but quite another for them to request that an affected embryo be implanted. The assumption that it is wrong knowingly to implant an affected embryo seems to be a shift away from the rhetoric of choice which dominates when the context is counselling concerning termination of pregnancy. Yet if it is so obviously wrong to implant an affected embryo, why is it not equally obviously wrong to continue to term under similar circumstances? One answer is that embryos have less moral worth than fetuses. Another is that termination is wrapped up with the autonomy of women whereas the ex-utero embryo is the responsibility of the clinician. We will return to this point later. The assumption that it is obviously wrong to implant affected embryos also ignores at least two other issues in the wider debate about reproductive technologies. The first is the extent to which parents can expect - where possible - to be genetically related to their offspring. The second is the extent to which we are prepared to entertain debate about what makes a life worth living, what constitutes a disadvantaged life, and under what circumstances, if any, an individual could be said to be harmed by being brought into existence.

Let's put this into the context of some examples.

1. Simon and Claire are both in their mid-40s. They have been referred for PIGD because ten years ago they had a child with Fragile X who died six months ago. Since the birth of this child, Claire has had two TOPs following positive antenatal tests for Fragile X. Following the death of their first child, they are more than ever determined to extend their family, but acknowledge that time is running out for them. Claire is superovulated but even so only five gametes are collected. Only two of these become fertilised when mixed with Simon's sperm. Both embryos are affected by Fragile X. Simon and Claire decide that they are

getting too old to wait any longer for a child and ask the clinician to implant the embryos. As an alternative, he offers them a place on his IVF waiting list, arguing that they should try IVF with donor gametes. They refuse because they want a child which is genetically related to both of them. He offers further PIGD, they refuse again, concerned that next time they may not even manage to produce a single embryo and that there is no guarantee that even if they do, it will be unaffected by Fragile X. They prefer instead to take the chance that this second child - if a pregnancy is established - will be less badly affected than was their first.

2. Judith and Paul are carriers for cystic fibrosis. They have both had experience of living with sufferers and want to avoid having an affected child. Following PIGD they are informed that they have six viable embryos, but one is affected with Down's syndrome. They ask for four of the embryos to be frozen for later use and for two of them to be implanted. They are not in the least concerned about Down's syndrome, saying that they believe that it is possible to have Down's and still have a good quality of life. They want the affected embryo to be implanted/frozen without discrimination and at random.

3. Philip and Linda are both deaf. Linda is infertile and the couple have been accepted for IVF. Once on the programme they were offered PIGD by a well-meaning clinician who assumed that they would not want any of their children to be deaf. He is shocked when they steadfastly insist that out of their nine embryos the one with congenital deafness be implanted first - along with any one of the other unaffected embryos. The remaining embryos should be frozen for later use. They justify their decision by arguing that their quality of life is better than that of the hearing. As far as they are concerned, giving preference to the affected embryo is giving preference to the one which will have the best quality of life. They are very concerned that any hearing child they have will be an "outsider" - part neither of the deaf nor of the hearing community at least for the first five or so years of his/her life.

CASE NUMBER 1: SIMON AND CLAIRE

This case challenges us to determine whether it is permissible for a couple to put the desire for a child which is genetically related to them above the interests of the child, once born; and also, whether it is acceptable to offset the risk that they will not manage to have another successful attempt against the risk that the child may have a very disadvantaged life. To address these issues, we have to look at reproductive freedoms and parental responsibility.

There has been systematic and unresolved discussion over what makes an individual a parent, even more debate about what makes a woman a mother. This discussion has been provoked by what is perceived as the division of motherhood into, for instance, the social, the bearer and the genetic. Although most people agree that it is not a necessary criterion of parenting that one has a genetic relationship with a child, it is usually considered a sufficient qualification. The strength of this assumption has been revealed in a variety of contexts: the "father" distressed because a woman is going to abort "his" child; the reluctance of some to donate genetic material or frozen embryos; mothers who only reluctantly handed over children for adoption; men who discover that unbeknown to them, they have a child in the world whom they have never met. Likewise, as many of those experiencing the discomfort and expense of infertility treatment attest, the desire to have a genetically related child is an enormously strong one. From this perspective, it is not difficult to have sympathy with what Simon and Claire are proposing to do.

Their claim does, however, highlight the weakness of considering the conceptual issue of parenting in isolation from the responsibilities which flow from parenting. Simon and Claire are asserting a right to be parents of a particular kind - genetic parents. Whether such a right exists and if so what its content might be is contested.[2] Yet if such a right exists, it stands to protect the goods which flow from parenting.[3] These goods are arguably of two kinds - the fundamental and the incidental. Incidental goods include goods which one obtains as a known side effect of having children; for instance, securing council housing, producing a football team, or having company and security in one's old age. Fundamental goods, on the other hand, are the moral goods of parenting, the goods which parenting represent and which, without parenting, might never be achieved. These goods include the care and nurture of a child, the sense of affection and community found in loving families and the moral good of being responsible for the wellbeing of another. It is these kinds of goods which the concept of a right to parent seems to protect.

The question which Simon and Claire's case raises is whether being genetically related to the child is an incidental or fundamental good of parenting. Or, put another way, does the right to found a family include the right to be genetically related to one's child? On balance, the answer to this question is no. Being genetically related to one's child seems more appropriately located as an incidental good of parenting because it lacks sufficient weight to count as a fundamental good. This way of looking at the dilemma faced by Simon and Claire yields a quite different result from that which concentrates solely on future persons. In conclusion then, it would seem that Simon and Claire cannot insist upon having either of the embryos implanted on the grounds that they have a right to a child genetically related to them.

CASES 2 AND 3 JUDITH AND PAUL, PHILIP AND LINDA

In some respects both of these cases raise the same issue - that there is no definitive view either about what makes a life worth living, or about what constitutes quality of life. These are not new issues. The new dimension in PIGD is the possibility of actively choosing to have an affected child; in Judith and Paul's case because they do not see Down's as a condition incompatible with quality of life; and in Philip and Linda's case, because they think that being deaf is positively life-enhancing.

Of course, making an assessment of an individual's quality of life is a moral decision because other morally important decisions will be based upon the answer reached. In this respect, one has to look at the context in which the decision is being made to gain a full appreciation of the morality of the decision itself. So, for instance, if a quality-of-life decision is being used to allocate scarce resources, we want to locate this decision in the context of debate about justice. The context for our two couples is that they are claiming the parental authority to make decisions about the welfare of their children. The context in which their decision has to be discussed is that of whether they are making a responsible or irresponsible *parental* judgment. The moral basis for respecting the judgment of parents is the presumption that parents have the best interests of their children at heart, that they of all people can be trusted to do what is best for their children. This is the basis upon which parental judgments command moral authority. Sometimes parents do not seem to exercise good judgment when it comes to protecting their children's best interests. When this happens, we may consider them to be bad or even unfit parents, or sometimes no kind of parent at all! These kinds of judgments reinforce the view expressed above that it is almost impossible to look at what being a parent means without also considering parental responsibility. In this sense, not only might it be argued that the term parent is actually a moral term (as opposed to a term related to bearing, genetics or social roles), but also that it is perhaps inappropriate to separate our understanding of good parents from our understanding of what it means to be a parent.[3] Whilst attractive, this leaves us in danger of denying that parents can make mistakes without losing, as it were, the right to parent. We have, therefore, to accept that parents can make *misguided* judgments as well as *bad* judgments. Also, because there is an acceptably wide understanding of what constitutes a child's best interests, different parents may arrive at different decisions for their children without being either misguided or irresponsible in their judgment.

Now we need to apply these observations to the decisions made by Judith and Paul, Linda and Philip. We will then be in a better position to determine whether it is wrong for them to want to have affected embryos implanted.

JUDITH AND PAUL

To assess the strength of Judith and Paul's claim, we may actually need to know more about their views about the status of the embryo. If they are making an embryo euthanasia decision (ie attempting to make a decision about the future quality of life for what they consider to be an existing child) and if they believe that Down's is compatible with a quality of life worth having, then we will have to accept that they are making a responsible parental judgment - whether or not it is the one which we would be prepared to make for our own child.

But what if they are making a more political statement - perhaps one about the wrongness of a policy of testing for Down's? This is a much more difficult issue to resolve because the couple are much more vulnerable to the criticism of harming the future child, particularly if they believe that the embryo itself has no moral status. Judith and Paul cast as proponents of embryo euthanasia are balancing the harm of ceasing to exist against the harm of existing with Down's. Judith and Paul cast as a couple who believe that the embryo does not have moral worth (because it does not yet have a life at all) seem rather to be choosing to create a person with Down's. We need to be aware that there are different points coming together here. First there is the issue unique to PIGD - that of the active choice to implant a particular embryo. Second is the old dilemma of the extent to which parents can allow their own political views to affect the decisions they take about reproduction, rather than being motivated by considerations of the interests of their future children.

It might be argued that since Down's syndrome is not deemed to be incompatible with a life worth living, Judith and Paul do not wrong a child by bringing it into existence with Down's. But however much value is attached to the lives of those who have Down's, it nevertheless remains the case that is it not good to have the disabilities which those who have Down's have to suffer. It is one thing to say that those who have Down's are valuable and quite another to say that the choice between choosing to implant an affected or an unaffected embryo has no moral significance. Whether a condition is so severe as to make life not worth living is one question. Whether it is permissible to implant an embryo with a condition that will result in disability, but a disability that is not so severe that it makes life not worth living, is another. If the answer is not clear then there is a case for leaving it to parental choice. There remains an issue, however, about the sorts of considerations that are relevant to that choice.

What if, however, they argue that for them the moral significance in having the affected embryo implanted lies precisely in making a political statement about the value of those with Down's, a symbol of their belief that a life with Down's is nevertheless a valuable life? We cannot do justice to the wider issue here, so we will simply describe it rather than discussing it at length. In its broadest sense this question asks us to decide which takes precedence, our obligations as citizens or our obligations as parents.

The extent to which we can impose the effects of our political beliefs on our children is raised in many different contexts. The conscientious objector in an oppressive regime knows that her objections place not just herself but her children at risk of physical harm and death. Those who believe in public education for children have to decide whether or not to send their children to private schools when their children have special needs or happen to live in the catchment areas of substandard state schools. We might be happy to conclude that making these decisions is one of those areas where, once again, the boundary for responsible parental decision making is quite a wide one. Nevertheless, we also conclude that once parents cite their moral authority as parents as the reason to allow them to make such decisions for their children, the interests of the children do have to figure to a significant degree in the calculations, whatever the final decision which is reached.

LINDA AND PHILIP

Linda and Philip also want to implant an affected embryo, but in their case what they believe themselves to be doing is maximising the advantages for their future child. This decision is a difficult one to challenge. By offering a choice between possible future persons, PIGD provides not only the opportunity to avoid avoidable harms, but also the possibility of maximising advantage or enhancing quality of life.

Why are Linda and Philip able to argue that deafness is life-enhancing? One answer is that they are themselves deaf and are therefore in a good position to judge, whereas we who can hear are not. Equally, however, it could be argued that Linda and Philip because of their deafness are not in a position to appreciate what they are missing. In one sense this argument is intractable; it is impossible to claim that one form of life is better than another unless one has experienced both and even then one's preference could be said to be subjective. One could, of course, refuse to give their claim any credibility at all by arguing that it is harming a child to engineer that she is born without the capacity to enjoy all of her senses or do all that it is possible for humans to do.[4] This is surely the kind of judgment to which we appeal when we observe that whatever value attaches to the life of the person with Down's, nevertheless having Down's is a disadvantage. But Linda and Philip's claim is more difficult to dismiss than this. Few if any of us claim that the pressure of intellectual capacity is so great that we wish we'd been born without it. We might claim that the pressures of responsibilities we have as a result are so great that we'd like to give them up in favour of a more simple life, but in considering the possibilities for this more simple life, irreversible brain damage rarely, if ever, features. Many of us have, however, found noise to be obtrusive. Too much uninvited noise is itself a recognised medical condition (tinnitus) but too much intellectual capacity is not. Some of us regularly resort to ear-plugs in order to sleep or work in environments we cannot control. Thus whilst we do not literally deafen ourselves, we certainly are prepared to trade hearing for peace and quiet.

The argument, however, may be less in terms of the quality of sound-related experience or its lack, but as indicated above, in terms of a sense of belonging to a community, a language and a culture. It is this that marks out deafness from other conditions. Linda and Philip are not simply making a political claim that people like them should not be discriminated against through programmes to screen out deaf future persons, although there are arguments for the view that screening for deafness harms the deaf community. They are motivated by the genuine belief that they are acting in their child's best interests. Perhaps the best that can be argued against them is that they should allow their child to decide for herself whether she thinks that their quality of life is better than hers. This choice is removed from her if she is born irreversibly deaf. But even this solution does not address the point which they make about being an outsider.

This case highlights one of the disadvantages of giving parental decision making moral authority, namely that it can act as a trump card in those areas where there is no right answer, but where the decision which parents want to make, to choose a state in which a child is born with a hearing system which does not work, is one that most rational bystanders would not take. On the other hand, as has been suggested above, in cases where it is not the case that the child would be so disadvantaged that its life would not be worth living, it may be argued that parental choice is the best option, particularly where motivated by the child's best interests. It is precisely this that may be undermined by PIGD, however, and this is where we turn to our second problem.

Second new problem: does PIGD shift reproductive power from women to physicians?

To date, quality-of-life decisions have largely rested with women because it was impossible to make decisions over the life or death of fetuses without effecting these decisions on women's bodies. Once pregnant, a woman is free to decide whether or not to seek antenatal screening, and irrespective of this decision she is free (within legal boundaries) to determine whether or not the child is a wanted one, irrespective of any advice given to the contrary by her clinician.[5] Although women cannot be compelled to participate in PIGD, once they have parted with their gametes and once the resulting embryos are tested, it is possible for them to lose control over what happens next. Clinicians participating in IVF have a clear statutory obligation (and some would argue that this reflects an absolute moral obligation) to consider the interests of the future child. Just as the clinician cannot compel a woman to give up her gametes, or have a TOP, or be implanted, she cannot compel him to implant embryos against his wishes. Does this mean that in all the cases we discussed above, we were wrong to assume that the decisions were the parents to make at all? The clinician, it seems, has the final say in whether or not to implant.[6]

Or does he? There are several points to be made here. The first concerns the extent to which the clinician needs to rely solely on his own judgment. He can of course take account of the arguments of the parents - or indeed, his own ethics committee. However, the net effect of this could be to split hairs, since although he takes account of the parents' views he still has the final power to decide whether or not to be bound by their decision. Also, the 1990 Human Fertilisation and Embryology Act was formulated on an assumption that the clinician would arrive at this judgment *before* deciding whether or not to treat. When disagreements arise in PIGD, treatment (if this is the correct term) is already well underway. Any dispute concerns whether or not to implant *existing* embryos. In deciding not to implant on the basis of disagreement about the interests of the future child, the clinician not only retains his considerable power over reproductive freedom, he also gains power over what could be described as the property of the couple - the embryo - which is the product of their own gametes, which arguably are their property.

But, is it fair to assume that if gametes are property embryos are also property? If we consider that the embryo has an independent moral status, the claim about ownership diminishes because we readily accept that humans cannot be owned in any sense. If, however, we are working on the assumption that the embryo has no independent moral status it does seem reasonable to suppose that it can be owned and therefore belongs to the couple jointly (let's leave aside totally the issue of what to do if they disagree among themselves). This puts the clinician in an impossible position. He cannot implant because he does not think that it is in the interests of the child to do so, equally he cannot simply freeze the embryo since the purpose of this would be to afford the couple the chance of successfully finding a clinician who will agree to implant and this may result in a child whose interests he believes are best served by not coming into existence. Both freezing and implanting are using his skills to bring about a child. Of course, there is a precedent for suggesting that he should freeze and refer, namely the practice of clinicians who have a conscientious objection to abortion but who are expected to offer to refer patients to someone who will perform an abortion for them. The moral inconsistency of this position is obvious and needs no further explanation.

It seems to us highly probable that this tension is likely to cause problems in the future. The couple will, not unreasonably, assume that the decision in the case of PIGD will be theirs to make - because in antenatal screening it is. Likewise, the clinician cannot be blamed for assuming that he has the final word since in infertility treatment he does. Moreover, it seems likely that any "contract" which the couple make with their clinician as part of their consent prior to PIGD will be impossible to enforce - particularly in a case like that of

Simon and Claire, where the embryos in question became endowed with the additional status of being the ONLY chance for them to have a genetically related child. It remains to be seen, then, whether PIGD should be marketed as affording greater autonomy and reproductive freedom to couples when, as things stand, they are effectively putting the decision in the hands of the treating clinician.

Acknowledgement

Both authors would like to acknowledge the support of the Directorate General XII of the Commission of the European Communities under its Biomedicine and Health Research Programme (Biomed II). The authors gratefully acknowledge the stimulus and support provided by the commission.

Heather Draper, BA, MA, PhD, is Lecturer in Biomedical Ethics at the Centre for Biomedical Ethics, University of Birmingham. Ruth Chadwick, BPhil, MA, DPhil, LLB, FRSA, is Head of the Centre for Professional Ethics, and Professor of Moral Philosophy, University of Central Lancashire.

References and notes

1 For instance in the amendment to the Abortion Act 1967 found in the Human Fertilisation and Embryology Act 1990.
2 See for example, Chadwick RF, ed. *Ethics, reproduction and genetic control* [2nd ed]. London: Routledge, 1992.
3 For the complete development of this argument see Draper H. Assisted conception techniques, parent selection and the interests of children to adequate parents. *Bioetica* 1997;5:391-9.
4 For a discussion of this issue see Chadwick R, Levitt MA. The end of deafness? Deaf people, deaf genes and deaf ethics. *Deaf Worlds* 1997;13:2-9.
5 Granted, conditions such as extent of knowledge, subtle coercion, tests made available and other circumstances may erode autonomy.
6 Steinberg DL. *Bodies in glass: genetics, eugenics, embryo ethics.* Manchester: Manchester University Press, 1997. As Deborah Lynn Steinberg has argued, while preimplantation diagnosis may be debated in the gender-neutral language of genetic risks, it "obscures social inequalities between practitioners and female patients and between patients and their male partners, while implicitly reinforcing and relying on these power imbalances".

[14]

Predictive Genetic Testing for Conditions that Present in Childhood

Lainie Friedman Ross

ABSTRACT. There is a general consensus in the medical and medical ethics communities against predictive genetic testing of children for late onset conditions, but minimal consideration is given to predictive testing of asymptomatic children for disorders that present later in childhood when presymptomatic treatment cannot influence the course of the disease. In this paper, I examine the question of whether it is ethical to perform predictive testing and screening of newborns and young children for conditions that present later in childhood. I consider the risks and benefits of (1) predictive testing of children from high-risk families; (2) predictive population screening for conditions that are untreatable; and (3) predictive population screening for conditions in which the efficacy of presymptomatic treatment is equivocal. I conclude in favor of parental discretion for predictive genetic testing, but against state-sponsored predictive screening for conditions that do not fulfill public health screening criteria.

THERE IS A GENERAL CONSENSUS in the medical and medical ethics communities against predictive genetic testing of children for late onset conditions (IOM 1994; Working Party 1994; ASHG/ACMG 1995; Hoffmann and Wulfsberg 1995; Clayton 1997; AAP 2001). The overriding belief is that "predictive testing for an adult onset disorder should generally not be undertaken if the child is healthy and there are no medical interventions established as useful that can be offered in the event of a positive test result" (ASHG/ACMG 1995, p. 785). Even those who support some parental discretion to procure (at least some) predictive genetic testing of their children focus on predictive testing for adult onset conditions (Pelias and Blanton 1996; Cohen 1998; Robertson and Savulescu 2001).[1] Lacking from the consensus statements is a discussion about predictive testing of asymptomatic children for disorders that present later in childhood when presymptomatic treatment cannot influence the course of the disease.[2]

Predictive genetic testing for childhood onset conditions differs from predictive genetic testing for late-onset conditions because it does not involve concerns regarding the child's right, as an adult, to make the decision for him- or herself, nor the right to confidentiality with regard to the decision and the information (Clarke and Flinter 1996). These concerns are not raised in predictive testing for childhood onset disorders because the diseases most likely will manifest before the children have the autonomy to decide for themselves whether to undergo testing, and while the parents still have health care decision-making authority.[3] However, predictive testing both for childhood and for adult onset conditions does raise concerns regarding the psychosocial implication of being an individual "at risk" (Davison, Macintyre, and Smith 1994). Such testing also raises the question of what the moral basis for parental authority in health care decision making is: whether parents must focus on the children's medical best interest (Blustein 1982; Buchanan and Brock 1989) or whether they can make decisions that balance the children's interests and needs with the needs and interests of other family members (Ross 1998; Schoeman 1985).

In this paper, I examine the question of whether it is ethical to perform predictive testing and screening of newborns and young children for conditions that present later in childhood.[4] I consider the risks and benefits of (1) predictive testing of children from high-risk families; and (2) predictive population screening—both for conditions that are untreatable and for conditions in which the efficacy of presymptomatic treatment is equivocal. I then consider whether such testing can be ethically justified.

PREDICTIVE TESTING OF HIGH-RISK CHILDREN

Why do some families want to predictively test their children for childhood onset conditions? Consider the condition familial adenomatous polyposi (FAP). FAP is a condition in which individuals acquire multiple intestinal adenomas beginning in adolescence that will become malignant. Gene testing is recommended around ages 10 to 12 years to determine whether the child has the gene and needs annual colonoscopies (Laxova 1999, pp. 11–14). In families with a history of the disease, clarifying a child's risk status at a young age has potential benefits to both the child and the parents. With respect to the child, if the test is negative, the parents can rear the child free from this particular threat; and if the test is positive, the parents can share the information at "teaching moments" over time, thereby allowing the child to incorporate the information into his or her self-concept (IOM 1994; ASHG/ACMG 1995). With respect to the parents, if the test is negative, parental guilt can be assuaged; if the test is positive, parents can prepare emotionally and financially. Regardless of the results, the test eliminates the anxiety of not knowing. Data show that ambiguity can be more stressful than either a positive or a negative result (Broadstock, Michie, and Marteau 2000). In addition, the knowledge can inform parental reproductive and other life-planning decisions, such as distance from a tertiary care facility and choice of health insurance (IOM 1994; ASHG/ACMG 1995; Cohen 1998; Robertson and Savulescu 2001).

There is a small but growing empirical literature that examines predictive testing of young children in families with a familial cancer syndrome. The studies reported in the literature have focused on FAP and multiple endocrine neoplasia type 2A (MEN 2A). Both MEN 2A and FAP have virtually complete penetrance by young adulthood (Grosfeld et al. 1997, p. 64; Laxova 1999, p. 14). The recommendations for FAP are annual colonoscopies beginning in early adolescence for children who test positive and prophylactic colectomy once the individual has multiple adenomas. In MEN 2A, medullary thyroid cancer can present in childhood and prophylactic thyroidectomy is recommended. Although there is debate about when the surgery should be done given the increased risk of surgical complications in younger children, screening is recommended around age five years (Grosfeld et al. 1997, p. 64). Genetic testing of infants or children younger than five years is not clinically indicated for either condition.

Michie and colleagues (1996) report on a single case study of a family that sought FAP testing of their children who were aged two and four years. The family agreed to undergo extensive psychological interviewing and they offered seven reasons for wanting to test: (1) the technology is available to give accurate results; (2) to avoid the worry of uncertainty; (3) to be able to support children by providing information; (4) to avoid resentment from their children later in life; (5) to help prepare the child and themselves for the future; (6) to encourage vigilance about preventative screening; and (7) because it is their parental prerogative (Michie et al. 1996, p. 314).

Codori and colleagues (1996) examine the short-term psychological effect of genetic testing for FAP on children and their parents. They sampled 41 children between the ages of 6 and 16 such that at least some of the children were tested prior to clinical screening guidelines. They found that depression scores for both the children and the parents remained in the normal limits regardless of the results.

Despite the nonharmful results of these small case studies, the consensus position is not to provide predictive genetic cancer testing in children. Many support the "rule of earliest onset" proposed by Kodish (1999, p. 393): "[G]enetic testing should be permitted at an age no earlier than the age of first possible onset of cancer." Although Kodish is arguing why it is acceptable to test children for FAP but not for hereditary nonpolyposi colon cancer, a colon cancer that does not present until adulthood, his rule suggests that he would be against testing infants for FAP. Disappointingly, he does not address this issue directly. Clarke does. He specifically states that FAP testing should wait until the child is between 8 and 10 years so that the child can be involved in the decision-making process (Clarke 1997b, p. 174). However, it is not clear whether the child could refuse, given that it is not clear that the parents can refuse. Wertz, Fanos, and Reilly (1994, p. 879) note that when health benefits may accrue, parents cannot refuse testing without placing themselves at risk for being reported for child neglect. So testing a child for FAP at age 2 is not permitted, but failing to test the child at age 12 is child neglect. Parental autonomy is given a very small window of discretion.

Parents, however, may not agree with Kodish's rule. Patenaude and colleagues (1996) interviewed mothers of pediatric oncology patients about hypothetical predisposition testing for themselves and their healthy children, and 51 percent of mothers would test themselves and 42 percent would test healthy children, even with no medical benefit. If there were an established medical benefit, 86 percent of mothers would test themselves and 91 percent would test their healthy children (Patenaude et al. 1996, p. 417). Malkin and colleagues (1996) asked parents of all newly diagnosed childhood cancer patients with Beckwith-Wiedemann syndrome and Li Fraumeni syndrome—two familial genetic syndromes that predispose to childhood cancers—whether they would have elected genetic testing on their children and compared them with parents whose child was diagnosed with a nongenetic noncancer disease. Overall greater than 80 percent of parents believed that predictive genetic testing should be made available to all cancer patients (Malkin et al 1996). Other studies show that many clinicians are willing to accede to similar parental requests (Wertz and Reilly 1997; Working Party 1994, p. 796, table 4).

One argument against such testing is the concern that parents will seek clinical interventions before they are necessary, thereby putting the child at unnecessary medical risk—e.g., parents might request colonoscopies for a two-year-old child known to have the gene for FAP. There are no data to support this (Michie and Marteau 1998), and, in fact, one study found that parents of children with the gene for MEN 2A showed restraint regarding surgery, preferring annual testing instead (Grosfeld et al. 2000, p. 320).

A second objection to predictive testing is that the main reasons for testing are often parent-regarding—e.g., to relieve parental ambiguity or to allow the parents to prepare emotionally—whereas testing children should serve the children's needs and interests not those of their parents. Although morally the parents should have some child-regarding reasons for wanting the information, in actuality, parental well-being has a significant impact on children regardless of whether the child's well-being or best interest is at the core of the parent's desire to know. That is, an exclusive focus on the child's interest is too narrow (Ross 1998, pp. 20–34), and clinicians should respect parental decisions that take into account familial psychosocial factors. It is useful for parents to be able to plan realistically for their child; it is problematic when the information is used to limit a child's opportunities. There are no data to show that predictive testing for childhood onset conditions is likely to cause the latter (Michie 1996), and good reason to believe that it may be quite useful.

In summary, whether the psychosocial risks of predictive testing for childhood onset conditions outweigh the psychosocial benefits is unknown and probably will be found to vary depending on individual familial circumstances. Empirical data will be useful for families and clinicians, although each family may require an individualized benefit/risk assessment. What is known, however, is that many parents want predictive information. They believe that the benefits of such knowledge outweigh the risks and harms, even when health benefits will not accrue, and that predictive testing should be a parental prerogative (Working Party 1994). This is not to deny that the information can have detrimental implications for the child, the parents, and the family; only to argue that it is information that reasonable families may want.

PREDICTIVE POPULATION SCREENING PROGRAMS

Currently, most newborn screening in the U.S. focuses on conditions for which early treatment reduces morbidity or mortality—e.g., phenylketonuria (PKU), an inborn error of metabolism that leads to severe mental retardation unless appropriate dietary measures are undertaken. Newborn screening for such diagnostic and therapeutic purposes[5] differs significantly from predictive newborn screening for untreatable conditions that present later in childhood—e.g. Duchenne muscular dystrophy (DMD)—or for predictive newborn screening for conditions in which the efficacy of presymptomatic treatment is equivocal—e.g., cystic fibrosis (CF). Newborn screening for diagnostic purposes focuses on the known clinical benefits of testing and early treatment, and its benefits are considered so great that such screening is often mandatory.[6] In contrast, the benefits of predictive screening are equivocal, and it often is offered only in the research setting.

Predictive Screening for Untreatable Conditions That Present in Childhood

Concerns regarding predictive screening are heightened when the condition is fatal and no treatment can change the course or outcome of the disease. Consider, for example, DMD, a progressive neurological degenerative condition in boys that begins around four to six years of age and that is fatal in the third decade of life. Currently, no treatments or preventions exist that can change the course of illness. The Institute of Medicine (IOM 1994, p. 10) concludes that "[c]hildhood testing is not appropriate for . . . untreatable childhood diseases . . . ." As such, IOM would be against predictive newborn screening for DMD, although not against diagnostic testing once symptoms develop.

The problem with such a stance is that there are potential benefits to predictive genetic testing for DMD both for the child and for the family. First, the parents could use the information for lifestyle decisions that will benefit their son —e.g., buying a ranch home rather than a house with stairs. Second, they could use the information for their own reproductive planning, which could benefit the child indirectly—e.g., a decision to have only one child with handicaps may give the child greater family resources. Third, it may benefit the child directly given the recurrent problem of delayed diagnosis of DMD and its impact on family psychosocial well-being (Mohamed, Appelton, and Nicolaides 2000).

This is not to deny that there are also risks. Some parents whose children test positive may treat the child as ill even before symptoms develop, and/or they may become overprotective (vulnerable child syndrome). Presymptomatic testing may result in increased stress and stigma for the family. It may make it difficult for the family to procure or change insurance. But there are scant data regarding whether the risks outweigh the benefits generally, and whether the risks will outweigh the benefits for any particular family cannot be known *a priori*.

Bradley, Parsons, and Clarke (1993) implemented a program in 1990 that offered voluntary newborn screening of male infants for DMD on the presumption that early diagnosis benefits both the child and the family. Greater than 94 percent of families consented to screening. By 1993, they had diagnosed nine boys, and eight of the nine parents were pleased with the program, although as they acknowledge, the boys are still presymptomatic (Bradley, Parsons, and Clarke 1993). In a 1996 update, Parsons, Bradley and Clarke (1996) describe diagnosing 15 boys, and 13 of the families were positive about the program. Continued follow up is critical because the current data reflect a period in which only some of the children have become symptomatic.

It is important to distinguish predictive population screening for DMD with predictive testing in families with a known predisposition. In the latter case, parents may know their son is at risk because of other affected first degree relatives and a positive carrier test in the child's mother. These families deliberately may have chosen not to undergo prenatal testing because of the risk it would pose to a wanted child, regardless of disease status. The difference in these cases is that the families already may have an understanding of the disease and they are aware of their risk status. In contrast, in predictive population screening, individuals and couples may not know what DMD is or be aware of its genetic inheritance. To these individuals, the decision to test may be a test to confirm a healthy child, and they may not be prepared for the results that follow, even if they consented to testing (Clarke 1997a).

Predictive Population Screening Programs for Childhood Conditions in Which the Clinical Efficacy of Presymptomatic Treatment Is Equivocal

Contrast population screening for DMD with newborn screening for CF, an inherited disease of the exocrine glands primarily affecting the pulmonary and gastrointestinal systems, which can present in the newborn as meconium ileus—a gastrointestinal obstruction that requires surgery—or can present later in childhood as failure to thrive or recurrent pulmonary infections.

Newborn screening for CF became possible in 1979 after the development of the immunoreactive trypsinogen assay (IRT). Until then, CF was confirmed or ruled out by sweat test. Colorado implemented a CF screening program in 1982. A review of the first five years of testing found that testing by IRT could be done efficiently and diagnoses could be confirmed via sweat test, but the researchers acknowledged that "[i]t is still unclear whether mass screening of newborns for cystic fibrosis should be adopted, since an overall benefit of early diagnosis has not been demonstrated" (Hammond et al. 1991, p. 774). In Wisconsin, CF testing was begun experimentally as part of a randomized clinical trial. In 1994, although no clinical benefit had been firmly established, Wisconsin implemented CF screening as part of the routine mandatory newborn screening program (Wilfond and Thomson 2000).

The rationale for the development of newborn screening programs for CF were threefold. First, there was the belief that early identification would change the overall prognosis of children by improving their long-term nutritional status and pulmonary function. Second, newborn screening would avoid delay in diagnosis and the anxiety created by such a delay. Third, it was hypothesized that early knowledge might be useful for parents in their reproductive planning.

Does early screening achieve these goals? There is some current evidence of nutritional benefits of newborn screening for CF (Farrell et al. 1997; Farrell et al. 2001), although not all find these data adequate justification for a newborn screening program (Murray et al. 1999; Grosse et al. 2001). There is also some evidence of pulmonary benefit (Dankert-Roelse and te Meerman 1995), but not all studies have been able to reproduce these results (Murray et al. 1999). Finally, there is some evidence that newborn testing reduces the costs of delayed and missed diagnoses (Wilcken, Towns, and Mellis 1983; Phelan 1995). One of the problems with such studies is that a significant proportion of infants in both the screened and unscreened groups will identify themselves soon after birth, either because they are siblings of known CF patients or because they present with meconium ileus. As such, an U.K. Health Technology assessment report concluded in 1999 that "there is some circumstantial evidence favouring a benefit" but "[t]he ability of screening to alter long-term prognosis has not been conclusively proven" (Murray et al. 1999, p. iv).

Newborn screening for CF can identify at risk individuals and couples, although there are data to show that it has variable impact on reproductive decisions within these families (Mischler et al. 1998; Dudding et al. 2000). More importantly, however, is whether this is an appropriate primary goal for newborn CF screening. Any benefit-cost analysis includes assumptions about the value of a disabled person's life and about what counts as a good versus a burden (IOM 1994, pp. 152–54, 304–5). Thus, whether the medical and reproductive benefits of newborn screening for CF outweigh the costs is unclear, and the calculation may hinge on the psychosocial benefits and risks.

The psychosocial benefits of newborn screening for CF include the ability of the family to make lifestyle decisions that will ensure access to a tertiary care hospital. This must be balanced against the potential psychosocial risks including adverse impact on the family's ability to bond with a labeled child (Al-Jader et al. 1990) and on the child's own developing self-image (Boland and Thompson 1990). Such screening also may increase parental stress (Baroni, Anderson, and Mischler 1997).

Another risk of general population screening for CF is the risk of a false positive test. Data show that may parents do not understand a false positive test result (Sorenson et al. 1984; Tluczek et al. 1992), even if it is accompanied by counseling (Mischler et al. 1998). In the Wisconsin newborn screening study for CF, approximately 5 percent of parents whose infants had false-positive IRT results still believed their children might have CF when questioned a year later (Tluczek et al. 1992). In the newer CF screening programs that use IRT/DNA, the risk is further complicated because the screening also identifies carriers, and at least with other genetic conditions, data have found that families do not always understand the difference between those who carry the trait versus those who have the disease (Hampton et al. 1974). Thus, the risks of CF screening of newborns are considerable, and, given the ambiguity in medical benefit, the benefit-risk calculation is ambiguous at present.

WHEN IS PREDICTIVE TESTING FOR CHILDHOOD ONSET CONDITIONS ETHICALLY JUSTIFIED?

Predictive Testing in High-Risk Families

Elsewhere, I have argued that parents should have presumptive medical decision-making authority regarding health care decisions for their children unless there is evidence to show that their decisions are abusive or neglectful (Ross 1998). Lacking such evidence, I believe that physicians should respect parental autonomy regarding whether to test a child predictively for conditions that present in childhood.

Two caveats are in order. First, when possible, predictive testing should be done as part of a research protocol to collect data on the short- and long-term benefits and risks of testing (Michie and Marteau 1998). This is important both to ensure that such information does not lead to abuse or neglect and to collect information that will be useful for future families considering such testing. Second, given the potential adverse psychosocial impact, I would argue that physicians do not have a moral obligation to offer or to encourage predictive testing, even to high-risk families. This recommendation is consistent with Wertz, Fanos, and Reilly (1994) and Hoffman and Wulfsberg (1995).

Critics may object that a policy not to offer predictive testing is paternalistic and does not respect parental autonomy because parents cannot act autonomously if they do not have information to decide for themselves whether their child should undergo predictive testing. I believe this criticism is incorrect for three reasons. First, autonomy should not be interpreted as the right to infinite and indiscriminate choice, nor should respect for autonomy be interpreted as an obligation to respect random choices. Rather, respect for autonomy must be understood to entail not only a negative obligation of noninterference with another's choice, but also a positive obligation of physicians to help promote the conditions that allow patients to make informed voluntary decisions (Halpern 2001). Unfortunately, information about *all* possible screening tests may undermine and not enhance autonomy. Respect for autonomy means providing individuals with information that will help them make reasoned decisions, and not to undermine their decision-making capacity by providing them with so much information that they will be paralyzed by the sheer magnitude of choice.

Second, the issue at hand is not about respect for individual autonomy, but respect for parental autonomy. Parental autonomy is not absolute, but rather must be constrained by the child's developing and partially realized personhood (Ross 1998). Parents have a responsibility to make health care decisions for their child, and this requires information about their child's actual health needs. They also have obligations to promote their child's overall development (Ross 1998). Providing a parent with information about genetic tests that locate markers of potential disease is not clinically useful and can lead to inappropriate labeling and stigma, thereby adversely affecting the child's development.

Third, a policy of selective provision of medical information to parents is not paternalistic but a professional duty. Physicians should not view their role as a mere source of unfiltered information, but as moral agents responsible for the quality and quantity of medical information provided to their patients and families.

If parents do request predictive testing, I would argue that physicians should counsel directively against such testing. I would argue also that physicians should test only under ideal conditions of informed consent. Parents should undergo pretest counseling in which they explore what the test results can and cannot determine and what a positive and negative test result would mean for them. Physicians should inform families that the vast majority of experts do not favor predictive testing of children for diseases that present later in childhood. Nevertheless, if the parents still request genetic testing for childhood onset conditions, I believe that physicians should respect the parents' decision. Finally, I would recommend that test results only be given in person to ensure post-test counseling is provided.

Unlike Wertz, Fanos, and Reilly (1994), then, I believe that informed parental requests to test their children for conditions that present in childhood should always be respected. Critics may argue that I am too tolerant of parental idiosyncrasies and that there will be misinformed parents or parental decisions based on "bad" reasons. They will point out that some parents will want assurances regarding dozens of conditions and that it will encourage those with a vested interest in promoting testing (from biotech firms to pharmaceutical companies) to attempt to foster exaggerated anxiety about disease and potential disease so as to encourage unnecessary requests for testing of healthy individuals. This is a danger of a market economy for health care, but one that is not insurmountable. Physicians have an obligation to serve as the child's advocate and to educate parents about the dangers of ambiguous test results, about labeling and stigma, and about the dangers of acting upon increased susceptibility results when preventive treatments have not been shown to be safe or effective. Insurance companies can make such testing undesirable by restricting coverage for predictive genetic testing to high-risk families or symptomatic individuals. But in the end, testing requires a personal cost-benefit analysis, and parents are best situated to calculate it.

Predictive Population Screening for Conditions That Present in Childhood and Are Either Untreatable or for Which the Efficacy of Presymptomatic Treatment Is Equivocal

The development of tandem mass spectrometry and other technologies will make it possible to screen for many metabolic and genetic conditions quickly and cheaply. The differences between predictive testing in high-risk families and predictive screening in the general population are threefold. First, the decision about whether to undergo predictive testing in high-risk families often is based on actual experience with the disease. Testing in high-risk families either confirms or refutes a known risk. In contrast, predictive screening often gives information to unsuspecting families who may have consented because "the pressure of the hospital setting, the parents' physician and emotional condition immediately after birth, and the cultural belief that 'medical testing is good for you' will lead most parents to consent" (Wertz 2002, p. 107).

Second, the meaning and probability of a positive test differ dramatically. Not only is a positive test much more likely in a high-risk family, but if a gene test is positive, it is more likely to express (Frank et al. 2002). In addition, positive test results have been associated with more negative effects than anticipated in population-based screening and less negative effects than anticipated in high-risk screening programs (Marteau 1995). There are also more problems in comprehension in population-based screening (Michie and Marteau 1998).

The third difference is who suggests testing. In families with known genetic conditions, the parents may request testing either with or without consultation from their primary care physician. Population screening, on the other hand, is recommended by the state or expert panels that advise the state. Because state-sponsored programs often become mandatory—e.g., immunizations and newborn screening programs—and thereby restrict individual liberties, state-sponsored programs should provide a clear positive benefit to the individual and/or the society. State-sponsored programs that meet the public health criteria enumerated by Wilson and Jungner (1968) do just that. These criteria include: (1) that the condition should be an important health problem; (2) that there should be a suitable screening test; (3) that there should be an accepted treatment for patients detected; and (4) that early detection should have proven medical benefits.

If population screening is limited to conditions that fulfill these public health screening criteria, then neither predictive population screening for untreatable conditions that present in childhood—e.g., DMD—nor screening for childhood conditions in which the efficacy of presymptomatic treatment is equivocal—e.g., CF—meets these criteria. As such, the state cannot justify routine predictive screening, and predictive testing should be provided only upon request—usually to high-risk families, although it may be requested by low-risk anxious parents—with extensive counseling as detailed above. When offered in combination with an experimental treatment, predictive population screening programs should be limited to research protocols, and the protocols should be stopped if the treatments are not found to be clinically beneficial to the child.[7] Unfortunately, the Wisconsin program suggests that it may be very difficult to halt a program once the infrastructure is in place (Wilfond and Thomson 2000).

One of the main dangers with expanding newborn screening programs to include predictive screening programs, even within well-defined research protocols, is that they may be confused with the more traditional diagnostic newborn screening programs in which treatments provide significant clinical benefits to the child. Parents may consent to the experimental predictive screening test because they perceive it to be an extension of diagnostic screening without a real appreciation of the different risk-benefit calculations. Alternatively, parents may extrapolate that the unclear risk-benefit calculations of experimental predictive screening protocols also apply to the diagnostic screening programs and refuse both. Thus, consent for predictive screening must be clearly separated from consent for diagnostic screening.[8] Safeguards must be in place to ensure that predictive research screening programs are not confusing parents about the known clinically significant benefits of the diagnostic screening programs. If a significant number of parents is found to be confused, these research programs would need to be restructured—e.g., by offering the experimental screening tests at the two-week office visit rather than in the nursery—or abandoned.

CONCLUSION

The ethics of predictive testing of children for childhood onset conditions depends on whether the psychosocial benefits outweigh the psychosocial risks. This is a calculation that depends upon personal beliefs and values and will differ among families. There are scant empirical data regarding how families have made these calculations and the long-term implications of their decisions.

To claim that parents are the appropriate decision makers for deciding whether to procure predictive genetic testing for diseases that present in childhood does not imply unconditional support, nor does it imply that one should encourage such testing. In addition, respect for parental autonomy does not extend to respect for state-sponsored predictive population screening programs. Rather, I support constrained parental access to predictive testing and reject state-sponsored predictive population screening programs that do not provide clear medical benefit to the child, except in the research context.

Although I support respecting parental requests for predictive testing of childhood onset conditions, I have recommended that predictive testing be coupled with extensive counseling to make certain that parents understand the benefits and risks of such testing. I also have recommended that physicians directively counsel parents against predictive testing. But in the end, the question is ultimately about who knows what is best for the child and the family. Although health care providers can be clear about the medical risks and benefits, it is only within the context of a particular family that a calculation of the psychosocial and reproductive risks and benefits for the child and the family will determine whether predictive testing is appropriate for a particular child in a particular family for a particular condition at any particular time.

I thank two anonymous reviewers who made critical suggestions to this paper. Work on this project was supported by a Harris Foundation Grant, Ethical Analysis and Public Policy Recommendations Regarding the Genetic Testing of Children.

NOTES

1. The one exception is Robertson and Savulescu (2001) who argue in favor of a parental right to predictive testing of children for all conditions, both early- and late-onset. Pelias and Blanton (1996) also discuss predictive testing for early-onset conditions, but only examine the issue of whether parents can refuse such testing.
2. Actually, the issue is raised by the Working Party (1994), but the statement is quite vague and self-contradictory (Dalby 1995). The issue is also raised by the Institute of Medicine (IOM 1994, p. 10), which summarily rejects predictive testing for untreatable diseases in children as well as genetic disorders for which no effective curative or preventive treatment exist.
3. Some do use considerations of autonomy and confidentiality to argue against parental authority to test young children for diseases that present in adolescence (Wertz, Fanos, and Reilly 1994, p. 879).
4. Throughout this paper, I use the term "testing" to refer to testing of a particular child. "Screening" refers to testing of an entire population.
5. Although newborn screening for genetic conditions in which early treatment reduces morbidity and morality—e.g. PKU—are technically predictive genetic screening tests in the sense that the diagnosis is being made before

symptoms develop, I refer to these tests as "diagnostic screening" tests because they are no different than other medical diagnostic tests that are being used to determine a clinical course of action. In fact, most newborn diagnostic screening programs include diagnostic screening for medical conditions that are nongenetic—e.g., hypothyroidism. I do not mean to imply that such tests do not have costs as well as benefits (Clayton 1992), but I use the term diagnostic because of how they are used clinically to influence the course of treatment. In contrast, I use the phrase "predictive screening" to refer to those screening tests that provide information about genetic conditions in which presymptomatic treatment is inefficacious or at best equivocal in changing the course or outcome of the genetic condition.

6. Whether diagnostic newborn screening should be mandatory, or whether only the offer of it should be mandatory, is quite controversial and is beyond the scope of this paper (see AAP 1996; 2001; Clayton 1992; IOM 1994).
7. I am arguing that experimental predictive screening protocols should be stopped if they are not clinically beneficial to the child, regardless of whether they offer valuable reproductive information to the parents. Population screening programs should not be designed to use the child solely (or even primarily) as a means for discovering parental reproductive risks. The latter information can be procured by preconception or prenatal testing. This position is consistent with the recommendations of the IOM (1994, p. 6) and the American Academy of Pediatrics Committee of Bioethics (AAP 2001, p. 1453). This is not to suggest that parents should not be informed of carrier findings when discovered in newborn screening, only that such information should be a secondary benefit of newborn screening with the primary focus being the child's medical needs.
8. In most states, newborn screening is mandatory, and informed consent is not required. Although ideally all parents are informed about the nature of the study, the risks and benefits, and the meaning of a positive test, this is not the case in practice. The AAP Committee on Bioethics thought that it would be easier to integrate experimental tests if all programs routinely sought informed consent for newborn screening (AAP 2001). I suggest, in contrast, that requiring consent only for the experimental predictive screening initiatives might be confusing, further conflating the two programs. The addition of experimental predictive newborn screening programs makes consent for diagnostic screening even more critical: not to simplify integration, but rather to ensure that parents understand the difference between the diagnostic newborn screening tests and the experimental predictive screening tests.

REFERENCES

Al-Jadar, L. N.; Goodchild, M. C.; Ryley, H. C.; and Harper, P. S. 1990. Attitudes of Parents of Cystic Fibrosis Children Towards Neonatal Screening and Antenatal Diagnosis. *Clinical Genetics* 38: 460–65.

AAP. American Academy of Pediatrics. Committee on Bioethics. 2001. Ethical Issues with Genetic Testing in Pediatrics. *Pediatrics* 107: 1451–55.

———. Committee on Genetics. 1996. Newborn Screening Fact Sheets. *Pediatrics* 98: 473–501.

ASHG/ACMG. American Society of Human Genetics/American College of Medical Genetics. 1995. Points to Consider: Ethical, Legal, and Psychosocial Implications of Genetic Testing in Children and Adolescents. *American Journal of Human Genetics* 57: 1233–41.

Baroni, Mary A.; Anderson, Yvonne E.; and Mischler, Elaine. 1997. Cystic Fibrosis Newborn Screening: Impact of Early Screening Results on Parenting Stress. *Pediatric Nursing* 23: 143–51.

Blustein, Jeffrey. 1982. *Parents and Children: The Ethics of the Family*. New York: Oxford University Press.

Boland, Carol and Thompson, Norman. 1990. Effects of Newborn Screening on Reported Maternal Behavior. *Archives of Diseases of Children* 65: 1240–44.

Bradley, D. M.; Parsons, E. P.; and Clarke, A. J. 1993. Experience with Screening Newborns for Duchenne Muscular Dystrophy in Wales. *BMJ* 306: 357–60.

Broadstock, Marita; Michie, Susan; and Marteau, Theresa. 2000. Psychological Consequences of Predictive Genetic Testing: A Systematic Review. *European Journal of Human Genetics* 8: 731–38.

Buchanan, Allen E., and Brock, Dan W. 1989. *Deciding for Others: The Ethics of Surrogate Decision Making*. New York: Cambridge University Press.

Clarke, Angus J. 1997a. Newborn Screening. In *Genetics, Society and Clinical Practice*, ed. Peter S. Harper and Angus J. Clarke, pp. 107–17. Oxford: Bios Scientific Publishers.

———. 1997b. Parents' Responses to Predictive Genetic Testing in Their Children. *Journal of Medical Genetics* 34: 174–75.

———, and Flinter, Frances. 1996. The Genetic Testing of Children: A Clinical Perspective. In *The Troubled Helix: Social and Psychological Implications of the New Human Genetics*, ed. Theresa Marteau and Martin Richards, pp. 164–76. Cambridge: Cambridge University Press.

Clayton, Ellen Wright. 1992. Screening and Treatment of Newborns. *Houston Law Review* 29: 85–148.

———. 1997. Genetic Testing in Children. *Journal of Medicine and Philosophy* 22: 233–51.

Codori, Ann-Marie; Petersen, Gloria M.; Boyd, Patricia A.; et al. 1996. Genetic Testing for Cancer in Children: Short-Term Psychological Effect. *Archives of Pediatrics and Adolescent Medicine* 150: 1131–38.

Cohen, Cynthia B. 1998. Wrestling with the Future: Should We Test Children for Adult-Onset Genetic Conditions. *Kennedy Institute of Ethics Journal* 8: 111–30.

Dalby, Shirley. 1995. GIG [Genetic Interest Group] Response to the UK Clinical Genetics Society Report "The Genetic Testing of Children." *Journal of Medical Genetics* 32: 490–92.

Dankert-Roelse, Jeannette E., and te Meerman, Gerard J. 1995. Long Term Prognosis of Patients with Cystic Fibrosis in Relation to Early Detection by Neonatal Screening and Treatment in a Cystic Fibrosis Center. *Thorax* 50: 712–18.

Davison, Charlie; Macintyre, Sally; and Smith, George Davey. 1994. The Potential Social Impact of Predictive Genetic Testing for Susceptibility to Common Chronic Disease: A Review and Proposed Research Agenda. *Sociology of Health and Illness* 16: 340–71.

Dudding, Tracy; Wilcken, Bridget; Burgess, Bronwyn; et al. 2000. Reproductive Decisions after Neonatal Screening Identifies Cystic Fibrosis. *Archives of Disease in Childhood—Fetal and Neonatal Edition* 82: F124–F127.

Farrell, Philip M.; Kosorok, Michael R.; Laxova, Anita; et al. 1997. Nutritional Benefits of Neonatal Screening for Cystic Fibrosis. *New England Journal of Medicine* 337: 963–69.

Farrell, Philip M.; Kosorok, Michael R.; Rock, Michael J.; et al. 2001. Early Diagnosis of Cystic Fibrosis Through Neonatal Screening Prevents Severe Malnutrition and Improves Long-Term Growth. *Pediatrics* 107: 1–13.

Frank, Thomas S.; Deffenbaugh, Amie M.; Reid, Julia E.; et al. 2002. Clinical Characteristics of Individuals with Germline Mutations in BRCA1 and BRCA2: Analysis of 10,000 Individuals. *Journal of Clinical Oncology* 20: 1480–90.

Grosfeld, Frans J. M.; Beemer, Frits A.; Lips, Cornelis J. M.; et al. 2000. Parents' Responses to Disclosure of Genetic Test Results of Their Children. *American Journal of Medical Genetics* 94: 316–23.

Grosfeld, F. J. M.; Lips, C. J. M.; Beemer, F. A.; et al. 1997. Psychological Risks of Genetically Testing Children for a Hereditary Cancer Syndrome. *Patient Education and Counseling* 32: 63–67.

Grosse, Scott D.; Khoury, Muin J.; Hannon, W. Harry; and Boyle, Coleen A. 2001. Early Diagnosis of Cystic Fibrosis. *Pediatrics* 107: 1492.

Halpern, Jodi. 2001. *From Detached Concern to Empathy: Humanizing Medical Practice*. New York: Oxford University Press.

Hammond, Keith B.; Abman, Steven H.; Sokol, Ronald J.; and Accurso, Frank J. 1991. Efficacy of Statewide Neonatal Screening for Cystic Fibrosis by Assay of Trypsinogen Concentrations. *New England Journal of Medicine* 325: 769–74.

Hampton, Mary L.; Anderson, James; Lavizzo, Blanche S.; and Bergman, Abraham B. 1974. Sickle Cell 'Nondisease': A Potentially Serious Public Health Problem. *American Journal of Diseases of Children* 128: 58–61.

Hoffmann, Diane E., and Wulfsberg, Eric A. 1995. Testing Children for Genetic Predispositions: Is it in Their Best Interest? *Journal of Law, Medicine and Ethics* 23: 331–44.

IOM. Institute of Medicine. 1994. *Assessing Genetic Risks: Implications for Health and Social Policy*. Washington, DC: National Academy Press.

Kodish, Eric D. 1999. Testing Children for Cancer Genes: The Rule of Earliest Onset. *Journal of Pediatrics* 135: 390–95.

Laxova, Renata. 1999. Testing for Cancer Susceptibility Genes in Children. *Advances in Pediatrics* 46: 1–40.

Malkin, D.; Austalie, K.; Shuman, C.; et al. 1996. Parental Attitudes to Genetic Counseling and Predictive Testing for Childhood Cancer. *American Journal of Human Genetics* 59 (4): A7.

Marteau, Theresa M. 1995. Toward an Understanding of the Psychological Consequences of Screening. In *Psychosocial Effects of Screening for Disease Prevention and Detection*, ed. Robert T. Croyle, pp. 185–99. New York: Oxford University Press.

Michie, Susan. 1996. Predictive Genetic Testing in Children: Paternalism or Empiricism. In *The Troubled Helix: Social and Psychological Implications of the New Genetics*, ed. Theresa Marteau and Martin Richards, pp. 177–83. Cambridge: Cambridge University Press.

———, and Marteau, Theresa M. 1998. Predictive Genetic Testing in Children: the Need for Psychological Research. In *The Genetic Testing of Children*, ed. Angus J. Clarke, 169–81. Oxford: Bios Scientific Publishers.

Michie, Susan; McDonald, Valerie; Bobrow, Martin; et al. 1996. Parents' Responses to Predictive Genetic Testing in Their Children: Report of a Single Case Study. *Journal of Medical Genetics* 33: 313–18.

Mischler, Elaine H.; Wilfond, Benjamin S.; Fost, Norman; et al. 1998. Cystic Fibrosis Newborn Screening: Impact on Reproductive Behavior and Implications for Genetic Counseling. *Pediatrics* 102: 44–52.

Mohamed, Khalid; Appleton, Richard; and Nicolaides, Paola. 2000. Delayed Diagnosis of Duchenne Muscular Dystrophy. *European Journal of Paediatric Neurology* 4: 219–23.

Murray, J.; Cuckle, H.; Taylor, G.; et al.; eds. 1999. Screening for Cystic Fibrosis. *Health Technology Assessment* 3 (8).

Parsons, Evelyn; Bradley, Don; and Clarke, Angus. 1996. Disclosure of Duchenne Muscular Dystrophy after Newborn Screening. *Archives of Disease in Childhood* 74: 550–53.

Patenaude, Andrea Farkus; Basili, Laura; Fairclough, Diane L.; and Li, Frederick P. 1996. Attitudes of 47 Mothers of Pediatric Oncology Patients Toward Genetic Testing for Cancer Predisposition. *Journal of Clinical Oncology* 14: 415–21.

Pelias, Mary Z., and Blanton, Susan H. 1996. Genetic Testing in Children and Adolescents: Parental Authority, the Rights of Children, and Duties of Geneticists. *University of Chicago Law School Roundtable* 3: 525–43.

Phelan, P. D. 1995. Neonatal Screening for Cystic Fibrosis. *Thorax* 50: 705–6.

Robertson, Stephen, and Savulescu, Julian. 2001. Is There a Case in Favour of Predictive Genetic Testing in Young Children? *Bioethics* 15: 26–49.

Ross, Lainie Friedman. 1998. *Children, Families, and Health Care Decision Making*. Oxford: Oxford University Press.

Schoeman, Ferdinand. 1985. Parental Discretion and Children's Rights: Background and Implications for Medical Decision-Making. *Journal of Medicine and Philosophy* 10: 45–62.

Sorenson, James R.; Levy, Harvey L.; Mangione, Thomas W.; and Sepe, Stephen J. 1984. Parental Response to Repeat Testing of Infants with "False-Positive" Results in a Newborn Screening Program. *Pediatrics* 73: 183–87.

Tluczek, Audrey; Mischler, Elaine H.; Farrell, Philip M.; et al. 1992. Parents' Knowledge of Neonatal Screening and Response to False-Positive Cystic Fibrosis Testing. *Developmental and Behavioral Pediatrics* 13: 181–86.

Wertz, Dorothy. 2002. Testing Children and Adolescents. In *A Companion to Genethics*, ed. Justine Burley and John Harris, pp. 92–113. Oxford: Blackwell Publishers.

———, and Reilly, Philip R. 1997. Laboratory Policies and Practices for the Genetic Testing of Children: A Survey of the Helix Network. *American Journal of Human Genetics* 61: 1163–68.

Wertz, Dorothy C.; Fanos, Joanna H.; and Reilly, Philip R. 1994. Genetic Testing for Children and Adolescents. Who Decides? *JAMA* 272: 875–81.

Wilcken, B.; Towns, S. J.; and Mellis, C. M. 1983. Diagnostic Delay in Cystic Fibrosis: Lessons from Newborn Screen. *Archives of Disease in Childhood* 58: 863–66.

Wilfond, Benjamin S., and Thomson, Elizabeth J. 2000. Models of Public Health Genetic Policy Development. In *Genetics and Public Health in the 21st Cen-*

tury, ed. Muin J. Khoury, Wylie Burke, and Elizabeth J. Thomson, pp. 61–81. New York: Oxford University Press.

Wilson, J. M. G., and Jungner, F. 1968. Principles and Practice of Screening for Disease. *Public Health Papers*, no. 34. Geneva: WHO.

Working Party of the Clinical Genetics Society (UK). 1994. The Genetic Testing of Children. *Journal of Medical Genetics* 31: 785–97.

[15]

IS THERE A CASE IN FAVOUR OF PREDICTIVE GENETIC TESTING IN YOUNG CHILDREN?

STEPHEN ROBERTSON AND JULIAN SAVULESCU

ABSTRACT

Genetic testing technology has brought the ability to predict the onset of diseases many years before symptoms appear and the use of such predictive testing is now widespread. The medical fraternity has met the application of this practice to children with caution. The justification for their predominantly prohibitive stance has revolved around the lack of a readily identifiable medical benefit in the face of potential psychological harms to the child. We argue that predictive testing can have important psychosocial benefits and that the interests of the child have been construed too narrowly. Proponents of a prohibitive stance also argue that testing in childhood breaches the child's future right to make the same decision as an autonomous adult and to maintain this information as confidential. We argue that predictive genetic testing of children is not necessarily a violation of the child's future autonomy. Indeed, in some cases, such testing may facilitate the development of autonomy in the maturing child. We argue that parents are generally best placed to judge what is in their own child's overall interests, and that a parental request for testing after appropriate genetic counselling should be respected unless there is clear evidence that the child will be harmed in an overall sense as a result of testing.

INTRODUCTION

There have been rapid advances in the understanding of heritable diseases over the last decade. We now have a comprehensive understanding of the relationship between molecularly defined genetic defects and their manifestations. Diagnosis using molecular genetic methodology has become commonplace. In some instances, it is possible to determine if a predisposition to develop disease exists prior to the onset of symptoms. The predictive nature of such genetic tests sets them apart from most other forms of medical investigation.

Predictive testing for genetic disorders in adults has become widespread, including testing for some forms of bowel, breast and ovarian cancer, Huntington Disease and other degenerative neurological conditions. Professional guidelines have been established for the delivery of such programmes.[1,2,3] These guidelines emphasise the importance of informed consent based on a thorough examination of the psychological, social and medical implications of a test result for each individual being tested, prior to testing. The importance given to medical[4,5] or psychosocial[6] considerations varies according to whether 'beneficial interventions' can be employed on the basis of the test result.[7] Respect for autonomy is generally the overriding principle in deciding whether to have such a test. There has also been a recognition that psychosocial benefits can be compelling grounds to seek predictive genetic testing in adults.

Professional guidelines concerning use of predictive genetic testing in children are more guarded.[8,9] We seek to examine the consistency of the ethical justifications underpinning current practice towards children and propose a more liberal framework for predictive genetic testing in children.

Clinical scenarios

Case 1

Denise and Mark have two children, Julie, aged 10, and Jane, aged 4 years old. Mark's father and grandfather both died of Huntington Disease, a dominantly inherited degenerative

[1] Ethical Issues Policy Statement on Huntington Disease Molecular Genetics Predictive Test. *Journal of Medical Genetics* 1990; 27: 34–38.

[2] National Council for Human Genome Research. Statement on the Use of DNA Testing for the Presymptomatic Identification of Cancer Risk. *Journal of the American Medical Association* 1994; 27,1: 785.

[3] American Society of Clinical Oncology. Statement of the American Society of Clinical Oncology: Genetic Testing for Cancer Susceptibility. *Journal of Clinical Oncology* 1996; 14: 1730–6.

[4] *Ibid.*

[5] F.M. Giardiello. Genetic Testing in Hereditary Colorectal Cancer. *Journal of the American Medical Association* 1997; 278: 1278–81.

[6] *Op. cit.*, Note 1.

[7] P.S. Harper, A. Clarke. Should we Test Children for 'Adult' Genetic Diseases? *Lancet* 1990; 305: 1205–6.

[8] *Op. cit.*, Note 1.

[9] *Op. cit.*, Note 7.

brain disease for which there is no medical treatment. Mark and Denise decided not to have prenatal testing. Six years ago, Mark's behaviour began to change in subtle ways. Latterly abnormal movements and features of dementia have become more obvious. Recently a clinical diagnosis of Huntington Disease has been made and confirmed by a molecular genetic test. He is now hospitalised in a psychiatric institution. Julie asks her mother if she will get the same disease as her father and grandfather. Denise tells her that there is a one in two chance, the same as tossing a coin, of her getting the disease later in life. Worried, Julie looks on the internet for more information about Huntington Disease. She tells her mother that there is a test that will tell her whether she will get the disease. Denise presents to her local doctor seeking a genetic test for her children. She feels the anxiety borne from the uncertainty over whether they have the gene or not has become emotionally disabling for them and it is probably better to know now, so that together they can prepare for the future.

Case 2

Derek has had Charcot Marie Tooth disease, a disorder of the nerves of the legs and to a lesser extent the arms, since his teenage years. For Derek, the disease makes walking difficult, and running impossible. His grip strength is also affected. Each child of an individual with Charcot Marie Tooth has a 50% chance of inheriting the gene. Derek's apparently healthy 12 year old son, Colin, is about to enter secondary school. He is academically below average and would prefer to pursue a technical education at a college in a hill suburb across town. Derek and his wife feel a predictive gene test for Colin would aid them in deciding whether or not he should pursue an education and career path that would require robust neuromuscular health. If he does have the Charcot Marie Tooth disease mutation, they will seriously consider trying to send him to an expensive private school to maximise his opportunities to have a more sedentary career. This would require one of Colin's parents taking on another job.

Case 3

Kylie is 8 years old and Jason is 15 years old. Their father died at the age of 35 of bowel cancer. A genetic test revealed he had familial adenomatous polyposis coli (FAP), a dominantly inherited disease that carries with it the inevitability of colorectal cancer if it is left untreated in adult life. Surveillance for malignant change by colonoscopy usually begins in gene carriers from age 12–15 years. Removal of the colon is usually recommended in early adult life. Jason requests genetic testing and is found to have the mutant gene. Prior to his colonoscopy, he has to have a clean out of his bowel at home, which causes him unpleasant diarrhoea. Kylie asks her mother, Narelle, if she will have to have a test too. Her mother says that there is a 1 in 2 chance that she will. Kylie is confused and anxious. When blood tests are being done on Kylie for an intercurrent infection, Narelle suggests that 'we get the FAP test done too'. Her doctor's hesitation to acquiesce to the request is met with anger. 'Her father died of it,' says Narelle, 'she sees her brother going for a colonoscopy and she is worried. Why put her through another set of blood tests later?'

Should predictive testing be permissible in any of these circumstances?

DEFINITIONS

In the following commentary, the term genetic test is used to mean DNA tests and chromosome examinations. The term excludes tests aimed to detect the phenotypic manifestations of genetic disease. We define a predictive genetic test as one in which: (i) an individual has no evidence of pathology (and thus no symptoms or signs of disease) at the time of testing; and, (ii) the result of the test indicates a high chance of developing pathology in the future, either in the person tested or his/her offspring. Depending on the genetic disease the onset of symptoms could be during childhood or adulthood. No firm age is stated for defining childhood in the guidelines. For this discussion, we will use the term 'children' to refer to children who are incompetent in terms of their decision making abilities.

Different principles apply if the child is a competent adolescent. Dickenson[10] has promoted a framework where the competence of the maturing adolescent must be acknowledged and such individuals engaged as active participants in the decision making process. In reply Harper[11] was quick to emphasise that this approach has already been

[10] D.L. Dickenson. Can Children and Young People Consent to be Tested for Adult Onset Genetic Disorders? *British Medical Journal* 1999; 318: 1063–66.

[11] P.S. Harper, R. Glew, R. Harper. Response to Requests for Genetic Testing is not Based on Age Alone. *British Medical Journal* 1999; 319: 578.

adopted at the clinical level, at least in the United Kingdom. There the emerging consensus seems to include weighing the benefits and harms of testing and the competence of the minor on a case by case basis.[12,13] There is an evolving recognition that even children prior to adolescence may possess cognitive abilities that mean they can enter dialogue and participate in decision making to varying degrees. The degree to which a decision to undergo genetic testing should be that of the child's should be commensurate with his or her ability to understand the issues involved and engage in dialogue. In this paper we address the more contentious instance when young children cannot do this.

GUIDELINES ON PREDICTIVE TESTING

Harper and Clarke were among the first to address the ethical issues in predictive testing in childhood. They stated that the interests of the child and 'the preservation of the possibility of future choice' should be paramount, and that medical justifications should prevail in decision-making. They did not countenance early childhood testing of adult onset conditions for which there is no medical treatment.

The Clinical Genetics Society (CGS) in the United Kingdom[14] and the American Society of Human Genetics (ASHG)[15] have published guidelines that address predictive genetic testing in children. The Genetic Interest Group (GIG), a consumer umbrella group in the United Kingdom, has responded with a critique of the CGS statement.[16] All endorse testing when there is a beneficial medical intervention which can be employed. Recommendations vary, however, when testing is for a disease which may begin in adulthood or in childhood, for an asymptomatic carrier state, or primarily for the benefit of third parties.

[12] Advisory Committee on Genetic Testing. Report on Genetic Testing for Late Onset Disorders. London: Department of Health.

[13] J. Binedell, J.R. Solden, J. Scourfield, P.S. Harper. Huntington's Disease Predictive Testing: The Case for an Assessment Approach to Requests From Adolescents. *Journal of Medical Genetics* 1996; 33: 912–5.

[14] Working Party of the Clinical Genetics Society (UK). The Genetic Testing of Children. *Journal of Medical Genetics* 1994; 31: 785–97.

[15] ASHG/ACMG report. Points to Consider: Ethical, Legal and Psychological Implications of Genetic Testing in Children and Adolescents. *American Journal of Human Genetics* 1995; 57: 1233–1241.

[16] S. Dalby. GIG response to the UK Clinical Genetics Society report 'The Genetic Testing of Children' *Journal of Medical Genetics* 1995; 32: 490–491.

(a) Predictive testing for a disease that begins in adulthood

All three position statements strongly advise against testing for a disease in which neither surveillance, pre-emptive nor definitive medical treatment is available in childhood (e.g. case 1). The CGS group stated 'clear cut and exceptional' circumstances may validate a decision to test but the preservation of the autonomy and confidentiality of the future adult justifies not testing in childhood. An implication in all three arguments is that, strict medical considerations aside, there can be few if any purely psychosocial indications to perform such testing in early childhood. The argument against childhood testing has been made particularly strongly in the case of Huntington disease.[17] It is asserted that the burden of certainty – of knowing that one will get a genetic disease later in life[18] – may result in severe psychological harm and even suicide – what may be termed the 'unbearability of certainty'. The magnitude of the elevation (if any) of the risk of serious psychological disturbance is not clear even in the largest, most recent studies.[19,20] What is known is that, in Huntington disease, in the 10–15% of at risk adult individuals who opt for testing, studies show exactly the opposite – that the state of uncertainty is most frequently cited as more burdensome than knowing.[21] While those who come forward for testing may be psychologically more robust than those who do not, and while there are acknowledged difficulties in extrapolating these studies to children,[22] this study provides some empirical evidence of the value of knowledge to well-

[17] M. Bloch, M.R. Hayden. Opinion: Predicitve Genetic Testing for Huntington Disease in Childhood: Challenges and Implications. *American Journal of Human Genetics* 1990; 46: 1–4.

[18] S. Kessler. Psychiatric Implications of Presymtomatic Testing for Huntington's Disease. *American Journal of Orthopsychiatry* 1987; 57: 212–219.

[19] E.W. Almqvist, M. Bloch, R. Brinkman, D. Craufurd, M.R. Hayden. A Worldwide Assessment of the Frequency of Suicide, Suicide Attempts, or Psychiatric Hospitalisation After Predictive Testing for Huntington Disease. *American Journal of Human Genetics* 1999; 64: 1293–304.

[20] T.D. Bird. Outrageous Fortune: The Risk of Suicide in Genetic Testing for Huntington Disease. *American Journal of Human Genetics* 1999; 64: 1289–92.

[21] S. Wiggins, P. Whyte, M. Huggins, S. Adam, J. Theilman, M. Bloch, S.B. Sheps, M.T. Schechter, M.R. Hayden. The Psychological Consequences of Predictive Testing for Huntington Disease. *New England Journal of Medicine* 1992; 327: 1401–5.

[22] S. Michie. Predictive Genetic Testing in Children: Paternalism or Empiricism? In: T.M. Marteau, M.P.M. Richards, eds. *The Troubled Helix: Social and Psychological Implications of the New Human Genetics*. 1996. Cambridge. Cambridge University Press: 177–83.

being. It is simply not known if the other 85% of at risk individuals would be better or worse off if they had the test.

The 'unbearability of certainty claim' carries less weight when applied to genetic predispositions that do not confer a certainty of developing disease later in life. For example, genes have been identified which predispose to breast cancer (BRCA genes), but carriage of a mutant gene does not mean that breast cancer will inevitably develop if one lives long enough. Despite the uniformity of opinion amongst professionals that this information presents serious psychological risk, a sizeable proportion of families with conditions such as bipolar disorder[23] would favour childhood testing, if they were allowed. The availability of testing, though, may modify their desires as it has in Huntington Disease. In Huntington Disease, three out of four adults with parents who had the disease cited they would come forward to be tested prior to the availability of the test[24] but now that the test has been introduced, the proportion requesting testing is much less than that initial figure.

Little direct evidence exists to support the 'unbearability of certainty claim'. The paediatric oncology literature suggests that children cope better with concrete and frank information if it is available, as opposed to a policy of non-disclosure or suspended uncertainty.[25] Evidence from predictive testing for Huntington Disease in adults suggests that testing (and receiving either a negative or positive result) is associated with less depression and more well-being than that experienced prior to the test.[26]

(b) Predictive genetic testing for a disease that begins later in childhood/ adolescence.

This category of conditions includes disorders such as Charcot-Marie-Tooth Disease (case 2 above) where the onset is often (but not always) in childhood but there is no effective pre-

[23] LB. Smith, B. Sapers, V.I. Reus, N.B. Freimer. Attitudes Towards Bipolar Disorder and Predictive Genetic Testing Among Patients and Providers. *Journal of Medical Genetics* 1996; 33: 544–9.

[24] S. Kessler, T. Field, L. Worth, H. Mosbarger. Attitudes of Persons at Risk for Huntington Disease Toward Predictive Testing. *American Journal of Medical Genetics* 1987; 26: 259–70.

[25] L. Slavin, J. O'Malley, G. Koocher, D.J. Foster. Communication of the Cancer Diagnosis to Pediatric Patients: Impact on Long-term Adjustment. *American Journal of Psychiatry* 1982; 139: 179–83.

[26] *Op. cit.*, Note 21.

symptomatic medical treatment available. Here there is the possibility that testing will remove the child's choice to decide for herself to undergo genetic testing later as an asymptomatic adult, if the disease does not begin in childhood. The case of FAP is similar although medical intervention (colonoscopic surveillance) is available from late childhood/early adolescence.

Both the CGS and AHSG statements are more liberal in this situation. They advocate finding a balance between the psychosocial benefits and harms of testing for such diseases at a time remote from the expected onset of symptoms or medical surveillance. These risks and benefits are presented in Table 1.

The CGS document, pointing to the lack of prospective studies examining the basis of these putative harms and benefits, favours a position of non-maleficence – *primum non nocere* – and therefore argues that decisions not to test should be the rule rather than the exception.[27] Implicit in this argument is the assumption that omitting to test is not harming the individual. In the ASHG and CGS position statements, a medical justification for testing (the existence of effective surveillance and treatment in childhood) is the prime consideration. Consequently, predictive testing for FAP is allowed, although not normally for an 8 year old child, as in Case 3, where it is usually deferred until the time when surveillance begins (around 12 years of age). Psychosocial benefits carry little weight, presumably precluding testing in cases 2 and 3 above, although the GIG and others have criticised this.[28,29] In response, Clarke[30] stated that 'psychosocial evaluations may well be critical for certain medically 'borderline' cases.' This stance still gives primacy to medical benefit (early diagnosis and treatment) and relegates any potential psychosocial benefit of knowing one's genetic status to a secondary concern. The service provider remains the arbiter of who qualifies for testing. This assumption of power by the provider has also been criticised by Sharpe[31] and others.[32] The ASHG is more equivocal, stating 'if the balance of benefits and harms is uncertain, the provider should respect the decision of

[27] *Op. cit.*, Note 14.
[28] *Op. cit.*, Note 16.
[29] *Op. cit.*, Note 22.
[30] A. Clarke. The Genetic Testing of Children. *Journal of Medical Genetics* 1996; 32: 492.
[31] N.F. Sharpe. Presymptomatic Testing for Huntington Disease: Is There a Duty to Test Those Under the Age of Eighteen Years? *American Journal of Medical Genetics* 1993; 46: 250–3.
[32] *Op. cit.*, Note 22.

34 STEPHEN ROBERTSON AND JULIAN SAVULESCU

Table 1 Putative psychological harms and benefits to predictive genetic testing

Potential harms
- Development of a perception that the child is 'ill' with negative parental attitudes towards the child
- Low self-esteem on the part of the child
- Serious psychological maladjustment, even perhaps depression and suicide
- Parental guilt
- Social discrimination, including future employment and insurance discrimination

Potential benefits
- Minimises the possibility of serious psychological maladjustment later in life induced by late discovery of status
- Decreased parental and child anxiety
- Decreased uncertainty about the future
- More realistic life choices
- Elimination of risk
- More openness about genetic conditions within the family and society in general

competent adolescents and their families'.[33] Autonomous adolescents should be involved in the decision to test at any rate, so this concession adds little. It fails to give sufficient power to parents to be involved in decision-making about their own young children. There seems to be an assumption that 'children need to be protected from their parents,' an assumption to which consumer groups have taken exception.[34] The GIG in their response to the CGS guidelines[35] argued that parents, aided by appropriate counselling, should be the decision makers. Indeed, Clarke and Flinter themselves admit that parents are normally well equipped to grasp the ethical and psychological issues at stake.[36]

[33] *Op. cit.*, Note 15.
[34] *Op. cit.*, Note 16.
[35] *Ibid.*
[36] A. Clarke, F. Flinter. The Genetic Testing of Children: A Clinical Perspective. In: T.M. Marteau, M.P.M. Richards, eds. *The Troubled Helix: Social and Psychological Implications of the New Human Genetics.* 1996. Cambridge. Cambridge University Press: 164–76.

(c) Testing for an asymptomatic carrier state

A carrier state by definition does not affect the health of the carrier. Testing for a carrier state is not usually considered under the heading of predictive genetic testing. However, it warrants consideration here because the arguments presented against it are similar to the prohibitive arguments against predictive genetic testing in childhood in general.

Carrier tests include testing for the presence of chromosomal translocations or single mutant genes for recessive diseases such as cystic fibrosis. The medical implications of these states are almost exclusively reproductive. Putative psychosocial harms include misunderstanding the medically innocent nature of the carrier state.[37] This could suggest the counselling/educational process was inadequate or that irrational beliefs were held by the person tested. The existence of a small number of people who, after voluntarily agreeing to a carrier detection test, suffer because of false beliefs that are refractory to counselling should not necessarily preclude the provision of testing to an informed and rational majority.

Justifications for childhood testing for carrier status include avoiding the need to organise assessment upon maturity prior to reproduction, providing more time for forethought and discussion of future reproductive choices and choosing the optimal time during childhood to educate the individual child on their specific status. Clarke and Flinter[38] have provocatively suggested that modern families lack sufficient resources or memory to reliably relate carrier status discovered in childhood to an adolescent or adult. Others have responded[39] claiming that genetic services should not assume this responsibility but potentiate the process within the family.

The GIG concurred with the ASHG and CGS positions that carrier testing should ideally be performed at maturity. The GIG did qualify this: testing raises arguments that are centred on parenting style, and therefore the locus of control should primarily rest with the parents as the individuals with the greatest insight into the individual child's psyche.[40] Evidence aimed at

[37] S. Zeesman, C.L. Clow, L. Cartier, C.R. Scriver. A Private View of Heterozygousity: Eight Year Follow-up Study on Carriers of the Tay-Sachs Gene Detected by High School Screening in Montreal. *American Journal of Medical Genetics* 1984; 18: 769–78.

[38] *Op. cit.*, Note 36.

[39] *Op. cit.*, Note 22.

[40] *Op. cit.*, Note 16.

assessing the potential for harm inherent in carrier testing minors is beginning to emerge. A small retrospective study has suggested negligible psychological harm in carrier testing minors for Duchenne muscular dystrophy and Haemophilia A.[41]

EMPIRICAL RESEARCH

Unfortunately, there is little direct research on the psychological impact of predictive testing in children. The application of conclusions from studies in adults to children carries with it many limitations.[42] Michie et al. reported one family in which a 2-year-old and a four-year-old were tested for the presence of a mutant FAP gene and one was found to be positive.[43] No psychosocial disturbance was detected within the family at short term follow up. Clearly, larger studies need to be performed. Clarke[44] rightly questioned the ability of Michie's case study to address issues related to autonomy and confidentiality. We address these arguments below.

Is current accepted clinical practice consistent with the current prohibitive stance regarding predictive testing of children?

i. Family History

Children are routinely told information about diseases which run in the family. For example, in the case of the X-linked disorder Haemophilia A, all daughters of a father with haemophilia are obligate asymptomatic carriers of the disease gene. They may be informed of this fact when they are children.

While constructing a family tree and extracting information from it to draw conclusions about a child's carrier status are permissible, carrier testing of children is prohibited by the UK and US guidelines. In the case of daughters of a known female carrier of Duchenne muscular dystrophy (another X-linked disorder), genetic testing would not be allowed to determine

[41] O. Järvinen, A-M. Aalto, A-E. Lehesjoki, M. Lindlöf, I. Söderling, A. Uutela, H. Kääriäinen. Carrier Testing for Two X-linked Diseases in a Family Based Setting: Retrospective Long Term Psychosocial Evaluation. *Journal of Medical Genetics* 1999; 36: 615–620.

[42] *Op. cit.*, Note 22.

[43] S. Michie, V. McDonald, M. Bobrow, C. McKeown, T. Marteau. Parent's Responses to Predictive Genetic Testing in Their Children; Report of a Single Case Study. *Journal of Medical Genetics* 1996; 33: 313–318.

[44] A. Clarke. Parents' Responses to Predictive Genetic Testing in Children. *Journal of Medical Genetics* 1997; 34: 174.

whether the girls are carriers. Why is disclosure of facts derived from family history acceptable, but results from genetic testing not acceptable? It seems arbitrary to draw a moral distinction between information derived from analysis of a pedigree and the same information derived from a blood or saliva test. Both involve deliberate actions (in one case, taking a history, in the other, a blood or saliva sample).

The case of Denise and her daughter Julie (Case 1) indicates the problems with disclosing family history but not allowing genetic testing. Information regarding family history may raise anxiety, while the possible route to alleviating that anxiety is closed by those best placed to relieve it. As the case of Julie illustrates, people may choose their own ways of finding out more, and it may be impossible to prevent parents and children getting access to genetic testing in the future. A prohibitive stance taken by clinical geneticists may drive testing into an arena where the information given is inaccurate and counselling resources insufficient.

ii. Predictive Phenotypic Tests

Consider another hypothetical case. Mary's brother and uncle have Lowe syndrome which is an X-linked condition characterised by mental retardation, kidney failure and cataracts. A good carrier test for Lowe syndrome is the demonstration in the otherwise normal female of fine speckled cataracts in the lens of the eye. These cataracts do not lead to visual disturbance but can be present early on in life. Would it be unethical for an ophthalmologist to look into Mary's eyes to ascertain her carrier status? If not, why would the performance of a genetic test to the same end be unethical? The current genetic guidelines would prohibit the use of a gene test to achieve the same aim but they do not address the use of phenotypic tests in this manner. There may be an assumption that genetic tests are in some way different from phenotypic assessments, a contention that has been criticised.[45]

Ethical Principles Grounding a Prohibitive Approach to Predictive Testing of Children

The CGS and ASHG guidelines cite violation of three ethical principles as the justification for not doing predictive genetic tests on young children.

[45] S. Holm. 'There is Nothing Special About Genetic Information'. In: *Acquisition, Access, and Control.* 1999. A. Thompson, R.F. Chadwick (eds). Plenum Press.

1. Failure to respect the child's future autonomy.
2. Breach of confidentiality.
3. Harm to the child.

1. Failure to respect the child's future autonomy

At the top of the list of reasons not to test, according to Clarke and Flinter, is a breach of autonomy.[46] How might childhood testing breach a later adult's autonomy? There are at least three ways.

1. Reduced Options

This is what Clarke and Flinter have in mind when they say 'testing in childhood removes the individual's right to make their own decisions about testing as an autonomous adult',[47] which reflects similar remarks in the CGS statement.

Feinberg[48] equated liberty with having as many *open options* as possible:

> 'I have an open option with respect to a given act X when I am permitted to do X and I am also permitted to do **not**-X (that is to omit doing X) so that it is up to me entirely whether I do X or not. If I am permitted to do X but not permitted to do **not**-X, I am not in any usual sense at liberty to do X, for if X is the only thing I am permitted to do, it follows that I am compelled to do X, and compulsion, of course is the plain opposite of liberty.'

There are two major objections to the claim that childhood testing is bad because it restricts the child's future options. The first is that more choice is not necessarily better.[49,50] Is my autonomy enhanced by being offered the option of donating a kidney to my brother? Maybe not. I might prefer that I had never been given this choice. I might be worse off either if I do or do not donate the kidney, than I would have been if I had never been given this choice. It may be rational to prefer not to have such a choice.

[46] *Op. cit.*, Note 36.

[47] *Ibid.*

[48] J. Feinberg. The Interest in Liberty on the Scales. In: J. Feinberg. *Rights, Justice, and the Bounds of Liberty: Essays in Social Philosophy.* 1980. Princeton. Princeton University Press: 36.

[49] G. Dworkin. *The Theory and Practice of Autonomy.* 1988. Cambridge. Cambridge University Press: 12.

[50] G. Dworkin. Is More Choice Better Than Less? *Midwest Studies in Philosophy* 1982; 7: 47–61.

Secondly, the formulation that childhood testing results in terms of reduced options is incorrect. The child who is not tested is denied an option of growing up and adapting to the knowledge of their genetic status during their formative years. Thus the choice is not between two courses of action, one which simply has more choice for the later adult, but between two mutually exclusive futures:

i. A future in which the child grows up with information about her future but has no choice about whether to have this information.
ii. A future in which the child grows up in uncertainty or ignorance, but has the choice herself of whether to have the information in the future.

An intermediate course, advocated by the CGS Working Party, is a future in which a child is aware of the risk of a heritable disorder and has the choice to find out her status in the future.[51] The putative benefits of this approach are to enhance self-esteem and to explicitly promote the autonomy of the child being tested.

There are, then, at least three possibilities:

1. Growing up in complete ignorance of genetic risk.
2. Growing up in uncertainty of one's genetic risk, though with information that one can have testing later.
3. Growing up in the presence of maximum available information about one's own risk of disease later in life, but with no choice over whether to have testing later or not.

Many parents may perceive that (2) is superior to (3). Our overall argument is that parents should be allowed to choose the option they believe is best for their children. Our argument is that (3) may not infringe future autonomy any more than (2). Being a parent necessarily involves making choices between mutually exclusive futures for one's child. Foreclosing some options is *not necessarily* a significant violation of the child's future autonomy. When parents decide to send their child to a private school hoping to provide a better education, they necessarily prevent the child realising all the friendships she would have formed at her local state-funded school. She will grow up with different goods. Likewise, the child who is raised in the presence of maximum information about genetic risk grows up formulating her goals and values, her life plans, in the presence of a

[51] *Op. cit.*, Note 14.

maximally realistic picture of herself and her relationship to the world – an important good in many people's eyes.

If parents provide a reasonable amount of knowledge and number of skills to the child in order for the child to have a reasonable range of choices and to be able to effectively deliberate about these, then they have created the conditions for an autonomous life, rather than limiting it. Parents can significantly infringe future autonomy when they limit the child's capacities which are necessary for autonomous action (e.g. by causing psychological or physical disability) or by severely restricting the range of options open to that child.

Clarke and Flinter[52] make a distinction between parental rights and duties in making such decisions. While we agree that any appeal to an innate right of parents to know their child's genetic makeup lacks ethical validity, our argument is consistent with a parental duty to facilitate a future that maximises autonomy for their child. Genetic testing may, in some circumstances, be part of fulfilling this duty.

2. Own Choice

Clarke and Flinter's remarks suggest another way in which childhood testing might be an infringement of the later adult's autonomy. Parents make the decision. It is not the child's own choice. There are two responses to the claim that this is a significant breach of autonomy:

(a) Learning to become autonomous requires actually making important decisions for oneself about oneself, and one such decision might be whether to know some fact about one's genetic make up. Indeed, the CGS statement acknowledged the importance of slowly imparting information to children as they mature but stopped short of suggesting definitive information should be imparted until the child is mature enough to take that decision for themselves.[53] Testing could be viewed by some as a necessary step *towards* the attainment of autonomy.

(b) Parents make many decisions on behalf of their children, in what they predict to be their child's interests. However, children retain control over many aspects of their own life, such as deciding which subjects to take at school, their peer group, and how to spend their leisure time. Through making these choices,

[52] *Op. cit.*, Note 36.
[53] *Op. cit.*, Note 14.

children become autonomous individuals. Undeniably parents have an influential role but ultimately the child who grows up knowing the truth about herself has different choices, not fewer choices. Again testing facilitates the attainment of autonomy by parents exercising their proxy powers to choose on their child's behalf. Choices that stem from that new knowledge will then be her *own* to make and her life will still be hers to direct.

3. Later Adult Regrets About Being Tested

The third way in which testing children may infringe their later autonomy is when an adult later comes to regret having been tested. When someone says, 'I wish you hadn't done that to me', it seems that there has been a paradigmatic violation of that person's autonomy.

Consider the possible sequel to case 2 as an illustration of this point:

Colin's parents decide to have him tested for CMT. Colin agrees and he is found to have the mutant gene. His parents send him to the non-technical college which Colin reluctantly agrees is probably the wisest choice. He gets a job as a clerk in the postal service. There, he meets his wife, Susan. It eventually becomes clear that Colin's disease is much milder than his father's was and he begins to resent being tested as a minor. He regrets not following his dream of obtaining an apprenticeship in automotive mechanics.

Is Colin worse off than he would have been if he had not been tested? Even if Colin has not been able to become a mechanic, there is much that is worthwhile in his life that he would not have had if he had chosen the path of a mechanic. Colin at 30 has a particular set of projects, friendships, and relations. Had he not been tested and gone to technical college, these would have been very different.

Does Colin's resentment of having been tested imply that his parents should not have tested him? His parents requested testing in good faith since it represented the wisest course of action to promote his overall interests. Humans being fallible, there is always a chance that such decisions are wrong. The fact that Colin was harmed is not sufficient to show that his parents acted wrongly. In order for them to be blameworthy, it is necessary to show that, based on the facts available at the time, it was more likely that Colin would be harmed than have benefited.

42 STEPHEN ROBERTSON AND JULIAN SAVULESCU

Good evidence to suggest predictive genetic testing in children is harmful is conspicuously lacking.

The Value of Genetic Knowledge

Our main argument is therefore that the objection to childhood testing based on a violation of future autonomy fails. Just as authors have listed potential harms of genetic testing, there are conceivable potential benefits as well. Although untested, to add balance to this debate, we set out some ways in which genetic knowledge could be of value to children.

Genetic knowledge may promote autonomy

Far from frustrating future autonomy, genetic testing of children may promote the development of autonomous decision-making. Autonomy is self-government or self-determination. Being autonomous involves freely and actively making one's own evaluative choices about how one's life should go. Evaluative choice requires holding true beliefs.

Colin in Case 2 either does or does not have the mutated gene for Charcot Marie Tooth disease. If he is trying to imagine what it would be like to work in various professions, including that of a motor mechanic or clerk, it is relevant for him to know his future physical capabilities. What is problematic about the case of Colin is not the genetic knowledge but rather the use to which it is put. Colin *reluctantly* agrees to give up his dream which suggests an element of coercion. However, if testing was done, and parents used that information to engage Colin in a discussion about what was in his interests, and he decided freely to be a mechanic and pursued that course, then there would have been nothing objectionable.

Autonomy requires some concept of self and self-knowledge. Autonomy, at least in the Millian sense, is related to forming one's own conception of the good life for oneself, and acting on it. But to decide what the good life is for oneself, it is necessary to know what kind of entity one is. A fundamentally important fact about ourselves is how long we will live and how robust our health will be. To take the extreme case, it might make a great difference to our actions, if we learnt we were to live one more day or 40 more years. Predictive genetic testing offers such options for adults today.

Genetic knowledge may promote well-being

Knowledge of one's lifespan and disease susceptibility is important in prudential terms. If one were going to live to 75, it might be prudent to invest time in long term projects, defer child bearing until one's career was established, put money aside for retirement, and so on. However, if one were only to live until 45 decisions made in these areas may be very different. It may not be best for either oneself or one's children to have children in one's early 40s.

If this kind of information is of value to adults, why is it not also of value to children? Are children unable to understand and make use of this information? Experience from oncology shows that children's capacity to understand information about their death and themselves is often underestimated. In some cases, it is parents and doctors who have difficulty with imparting information to children, not children receiving it. Could it be *our* psychological hang-ups and inadequacies, and not the child's immaturity, which prevent meaningful discussion of these important issues? In some ways this information could be less psychologically challenging than it is for adults. Children do not generally have established life plans or expectations of the future, especially the distant future. Adults by contrast have established plans and expectations which early death often thwarts. Moreover, it is a psychological fact that human beings discount harms that occur in the distant future.[54] The harm appears less bad the further in the future it is. This would suggest that the earlier such information is discovered, the less psychologically damaging it would be.

The point of this section is to speculate on the potential value of genetic knowledge to children. While genetic knowledge may be of value to children, the point of this paper is not primarily to promote genetic testing in childhood. Our purpose has been to critically evaluate the basis for objections, in particular, the claim that such testing is a violation of autonomy, the application of a moratorium on testing for that reason, and to stimulate debate. While testing may be of benefit, there are also issues, discussed below, concerning psychological harm to the child which need further research.

2. Breach of confidentiality

Both the CGS and ASHG guidelines cite a possible breach of confidentiality as a reason for not testing children because their parents will be aware of the results.

[54] D. Parfit. *Reasons and Persons.* 1984. Part II. Clarendon Press. Oxford.

44 STEPHEN ROBERTSON AND JULIAN SAVULESCU

Disclosure of such information to parents is hardly a significant breach of confidentiality in itself. A request to test must have the child's best interests at heart. If the parents merely wish the test to allay their own guilt, anxiety or curiosity, then clearly this requirement is breached. It is only then, in our view, that a gratuitous request to test constitutes an invasion of the child's privacy.

Parents are privy to all sorts of sensitive and personal information about their children. Family relations are built around trust. What is more important is what parents do with that information. To be sure, counselling needs to make clear to parents its sensitive nature but it is paternalistic to claim that parents cannot be entrusted with such confidences if they see a utility in it for the child.

3. Maleficence: harm to the child

In the absence of any medical benefit from predictive genetic testing, do the potential psychological harms preclude its practice? Current guidelines suggest so. 'First, do no harm' is the operative maxim.

There are two major problems:

1. This stance gives little weight to psychosocial benefits of genetic knowledge.
2. There is no empirical evidence supporting the claim that testing results in psychological harm.

Given the potential benefits of genetic knowledge, we, like others[55,56] feel that research is urgently needed to establish whether there are serious harms associated with genetic testing.

In other areas of medical practice, there are two prime justifications for interfering with parental choice. The first is when the parental choice is clearly significantly harmful to the child (e.g. child abuse, mandating life-saving medical treatment contrary to parental religious belief). There is no evidence, only

[55] S. Michie. Predictive Genetic Testing in Children: Paternalism or Empiricism? In: T.M. Marteau, M.P.M. Richards, eds. *The Troubled Helix: Social and Psychological Implications of the New Human Genetics.* 1996. Cambridge. Cambridge University Press: 177–83.

[56] A. Clarke, F. Flinter. The genetic testing of children: a clinical perspective. In: *The troubled helix: social and psychological implications of the new human genetics.* T.M. Marteau, M.P.M. Richards, (eds). 1996. Cambridge. Cambridge University Press: 164–76.

suspicions, to suggest that predictive testing causes harm of this order. However, convictions that testing will result in harm are not only based on empirical evidence. We take seriously the concerns that children may not be able to understand the full significance of information about the distant future. The negative effects of death and disease in their immediate family may distort their understanding of their own genetic predisposition so much that testing in these circumstances may be best delayed until such an age that a mature perspective can be obtained.

Secondly, a request to test must be likely to be of benefit to the child. This means that testing must not be performed gratuitously, or the child used as a means to an end. The determination of whether the test is 'likely' to be beneficial must take seriously arguments presented by parents. In the next section we address this tension between two potentially conflicting assessments, by the parent(s) and the provider, of the potential for benefit and harm in the face of a request for a predictive genetic test in a child.

Why parents must be given a greater responsibility: privileged access

In competent adults, great weight is accorded to the value judgements of individuals about how they should live their own lives, or their personal autonomy. There are two reasons for this:

i. Privileged access. Individuals know best their social relations, abilities, economic situation, and so on. Each individual is best placed to know what is in her overall interests. John Stuart Mill described this as the privileged access each individual has into his own circumstances.[57]
ii. The subject of the decision. An individual who makes a decision about how she should live her life (which does not significantly affect others) is the one most affected by that decision. She will be the one to experience the consequences. She is both object and subject of her choice. Since she is primarily affected, and not others, she should have a greater say in the decision.

Similar justifications extend to giving parents the power to decide what is in their children's overall interests and how they should live their lives, rather than doctors, the judiciary or others. These

[57] J.S. Mill. *Utilitarianism, On Liberty and Consideration on Representative Government*. 1910. Everyman Library. London. J.M. Dent and Sons: 133.

justifications ground a concept of respect for parental autonomy which is similar to and based on the same justifications as the idea of reproductive autonomy.[58,59,60,61]

1. Privileged access. Parents know best their child's social relations, abilities, economic situation, and so on. The GIG consumer group drew attention to privileged access parents have into their child's circumstances. Narelle knows best Kylie's disposition, and her worries, and what she already knows, and whether she will cope better with the certainty that she has, like her brother and father, FAP, rather than the uncertainty of not knowing. With the aid of informative and sensitive genetic counselling, she is best placed to make the decision for her daughter. One specific source of harm that many cite may arise from the practice of predictive genetic testing is that of employment and insurance discrimination. In the context of the arguments set out here we believe that these reservations do not differ in principle from any of the other potential harms in considering a predictive genetic test on a child (Table 1). Although in reality legislative controls may curb its impact, the prospect of employment and insurance discrimination must be weighed carefully by the decision makers. It may be that they have not considered it before requesting the test for their child and the counsellor may have to place the issue before them for deliberation. Again though, it should be the parent's judgement that carries the greater weight in deciding about the relative impact of such benefits and harms for their child in undergoing a test.
ii. A subject of the decision. While clearly parents are not affected to the same degree as the child, they are affected by decisions about their child far more than doctors and others. Parents will have to live with the consequences of their decision for the rest of their lives. Their interests are intimately connected to the child's. This interconnectedness is seen by many as a compelling reason not to test children. The objection here is

[58] R. Dworkin. Life's Dominion: An Argument about Abortion and Euthanasia. 1993. London. HarperCollins.

[59] J. Harris. Rights and Reproductive Choice. In: *The Future of Reproduction.* 1998. J. Harris and S. Holm (eds). Oxford. Clarendon Press.

[60] J.A. Robertson. *Children of Choice: Freedom and the New Reproductive Technologies.* 1994. Princeton. Princeton University Press.

[61] C. Strong. *Ethics in Reproductive and Perinatal Medicine.* 1997. New Haven. Yale University Press: 26.

that we should not test children when the benefit accrues to the parents. We agree when this is the primary motivation. In the presence of a reasoned request for a predictive test where there is a perceived benefit to the child as the prime motivation, the existence of a secondary benefit to the parents (perhaps in alleviating their anxiety) should not preclude testing from proceeding. This approach should not be confused with one that fails to engage the child in dialogue at all. In the counselling process the child's feelings and opinion's should be explored in an age-appropriate manner. This approach has been used successfully in other areas of paediatric medical practice and Alderson in particular has emphasised that a child's ability to reason and comprehend is often underestimated.[62] Parents retain the decision making power under this model but the child's cooperation and perspective are still considered important elements in the process.

iii. For these reasons, parents (after appropriate dialogue with their doctors and others)[63] should generally be the arbiters of what is in their child's current interests, and whether they should have predictive genetic testing. What we have rejected is the claim that when it is uncertain whether benefits outweigh risks, professionals can refuse testing on grounds of uncertainty, *invalidly* invoking the principle *primum non nocere.* Health professionals need to take seriously the value judgements of parents of the child's overall interests. The GIG alluded to this when it stated that the CGS statement 'gave little credence to psychosocial issues as indications to perform testing and overemphasised the strictly medical justifications to test'[64] (although the GIG ultimately supported CGS guidelines on predictive testing for adult-onset conditions).

Perhaps the hardest case is that of adult onset neurodegenerative disorders like Huntington Disease, as illustrated in Case 1. Denise might argue that it is better for her to tell her children whether they have the gene, at the right time. She may prefer a culture of openness in their family where Huntington disease is an accepted part of their family. She might say,

> 'This way, they will have more time to adapt to the knowledge. Children don't have plans for the distant future. It won't affect

[62] P. Alderson. *Children's Consent to Surgery.* 1993. London. Open University Press.

[63] J. Savulescu. Liberal Rationalism and Medical Decision-making. *Bioethics* 1997; 11: 115–129.

[64] *Op. cit.*, Note 16.

them in the same way as it would if they found out when they were 30. Besides, if they have Huntington Disease, that is a part of their life. Why shouldn't they know the boundaries of their own lives? The sooner they know themselves, the better they can make decisions about themselves.'

Before acceding to such a request, it is important to evaluate the level of insight that Julie might have with regard to being diagnosed an HD gene carrier. There is an important distinction between being provided with information from a genetic test, and *understanding* that information and putting it in the context of an overall life. If children cannot put this information into an appropriate context (e.g. the child falsely believes that she will get Huntington Disease before this Christmas), then the information creates harm and confusion, and should not be provided. Perhaps children can understand more than we credit, and much depends on how the information is presented. The issue is open.

In many cases, genetic testing yields probabilistic information (e.g. it is not certain when the neurological manifestations of Charcot-Marie-Tooth will commence or how severe they will be; a child with a BRCA1 gene mutation has a 50–70% chance of developing breast cancer in later life).[65] While conveying probabilistic information raises special problems[66,67] it is likely that children can deal with probabilistic information just as they are provided with risk information derived from their family history (e.g. that they are at 50% chance of developing Huntington Disease).

The drive to request a test may be derived from parental anxiety rather than rational concern for child's interests – a commonly encountered scenario in the context of requests for predictive testing in children in our experience. Health professionals still need to evaluate whether testing is being requested with the child's best interests as the paramount concern.

Ultimately, the decision of whether or not to test must be sensitive to the particularities of the case, the reasons provided by parents, and the individual clinician's judgement. A blanket prohibition cannot be justified.

[65] D. Ford, D. Easton, D. Bishop, S. Narod, D. Goldgar, the Breast Cancer Linkage Consortium. Risks of Cancer in BRCA1–mutation Carriers. *Lancet* 1994; 343: 692–5.

[66] D. Kahneman, P. Slovic, A. Tversky (eds.) *Judgement Under Uncertainty: Heuristics and Biases.* 1982. Cambridge. Cambridge University Press.

[67] A. Tversky, D. Kahneman. The Framing of Decisions and the Psychology of Choice. *Science* 1981; 211: 453–8.

CONCLUSIONS

Are the CGS and ASHG guidelines paternalistic and regressive, a return to the 'doctor knows best' practice of 20 years ago which accorded so little weight to individual and parental values? We think not, although the balance of power and the acknowledgement of parental autonomy are not given sufficient weight.

Predictive genetic testing in incompetent children should only be performed in the best interests of the child. We have argued that current guidelines construe benefit too narrowly. We have argued that psychosocial benefits must be given greater weight. Most importantly, we have argued that the claim that testing violates the child's future autonomy is mistaken. More empirical research is needed into the psychological effect of providing genetic information to young children. However, in the absence of such research, current guidelines give too much weight to medical evaluations of the child's best interests in determining whether testing is in the child's interests. We have argued that parents have privileged access into the child's and family's circumstances which clinicians lack. Decision-making should proceed on a case by case basis. When there is no justified grounds for believing that testing would harm the child, parents should be allowed to decide whether to employ these tests which are available to older competent children and adults.

Victorian Clinical Genetics Service
Melbourne, Australia

Centre for the Study of Health and Society
University of Melbourne

[16]

Inheritable Genetic Modification
Inheritable Genetic Modification
Inheritable Genetic Modification
and a Brave New World:

Did Huxley Have It Wrong?

by MARK S. FRANKEL

What makes inheritable genetic modification attractive is not its ability to treat disease, but its capacity, someday, to enhance human traits beyond what mere good health requires. But, these discoveries will not be imposed on us by government, as Huxley thought. If they take over our lives, it will be because they were sold to us on the open market, as commodities we cannot do without.

In the fall of 2000, the American Association for the Advancement of Science called on science to slow down. In a report it issued on inheritable genetic modification, AAAS took the position that no genetic modifications affecting the germ line, whether intentional or inadvertent, should be undertaken until the technology's safety, efficacy, and social implications had been subject to widespread public discussion. Further, said AAAS, there should be no work on inheritable genetic modification until a system of public oversight was in place that exercised authority over research in both the public and private sectors.[1]

Yet only six months after the report was released, a fertility clinic reported "the first case of human germline modification resulting in normal healthy children."[2] The work was done through the transfer of ooplasm, which surrounds the nucleus of the egg and is essential for it to thrive, from donor eggs into the eggs of women who have experienced recurring implantation failure—fertilization occurs, but the resulting embryo will not implant in their uterus. An inadvertent consequence of this procedure was that mitochondrial DNA found in the ooplasm of the donated material was introduced into the recipient eggs.

The clinic reported that the technique had "led to the birth of 30 babies worldwide." The clinic also reported that both the donated mitochrondrial DNA and that of the birth mother were found in all the cells of those babies born by this method—a modification of the children's genome, since they inherited mitochondrial DNA from two mothers. Presumably, they will pass this inherited DNA on to their offspring. The report was met with ethical disapproval in some quarters of the United States,[3] and the British reminded us that the procedure would be illegal in the United Kingdom.[4]

In his 1932 book, *Brave New World*, Aldous Huxley led us to believe that when it came to our genes and reproductive futures, our worst nightmare was

Mark S. Frankel, "Inheritable Genetic Modification and a Brave New World: Did Huxley Have It Wrong?," *Hastings Center Report* 33, no. 2 (2003): 31-36.

government involvement in procreative activities and a society that devalued individual decisionmaking. But as we begin the twenty-first century, the greater danger, I believe, is a highly individualized marketplace fueled by an entrepreneurial spirit and the free choice of large numbers of parents that could lead us down a path, albeit incrementally, toward a society that abandons the lottery of evolution in favor of intentional genetic modification. The discoveries of genetics will not be imposed on us. Rather, they will be sold to us by the market as something we cannot live without.

State of the Art

By inheritable genetic modification, or IGM, I mean interventions capable of modifying genes that are transmitted to offspring and to generations beyond. IGM includes interventions made early enough in embryonic development to have global effects—that is, to affect all of one's cells—as well as any interventions targeted at the reproductive cells—sperm and ova—or their precursor stem cells. It encompasses modifications both of nuclear and of extra-nuclear genomes, and modifications that are inadvertent side effects of other, deliberate genetic interventions (of, for example, somatic cell gene transfer). The only criterion is that the modification be inheritable.

What can one say about the promise and the risks of IGM? Some light might be shed on this question by work on somatic cell gene transfer, which is designed to treat or eliminate disease through genetic intervention only in the person receiving treatment. There are different technical approaches to such gene transfer. In "gene augmentation or addition," new genetic material is inserted into the body in order to take over the function of a faulty gene. The offending genes are not removed; instead, their adverse effects are masked by the new material. In "gene correction," a normal gene segment is swapped for the segment with the defect. Yet another approach is "gene repair," in which a normal fragment of DNA is introduced into a cell and the cell's own DNA repair machinery permanently corrects the faulty DNA sequence.[5] None of these interventions is easy to do; nor is it clear that any of them will do what we want them to do—restore health to the person treated. Despite considerable investment by the government and by the private sector, the reported successes have been very rare.[6]

Somatic cell gene transfer came under fire following the death in 1999 of Jesse Gelsinger, who was enrolled in a clinical trial that involved the insertion of genetic material.[7] A review of the "latest developments in somatic gene transfer technology" concluded that "it is not easy to assess exactly all these risk factors. . . . A number of complex issues must be addressed to evaluate the probability of having adverse effects in patients related to the treatment, and to establish the extent of the possible harm that patients may sustain."[8]

A very recent incident has highlighted other possible harms to patients. In October 2002, researchers reported that a child treated successfully with gene therapy in France for an ordinarily fatal immune deficiency disease had developed a form of leukemia, and that there was persuasive evidence that the treatment was a major contributing factor.[9] In January 2003, another child was discovered to have developed leukemia.

IGM poses even greater uncertainties related to risk because the interventions would be passed on to the progeny of those treated. Thus we need compelling evidence that the procedures are safe. But as the AAAS report concluded, that evidentiary standard cannot yet be met.[10] How do we assess whether, for example, in future generations a gene necessary for healthy development will be accidentally turned off, or a gene that contributes to a certain type of cancer turned on? In a provocative article published a year ago, one scientist claims that genomic researchers have underestimated, if not ignored, a phenomenon known as "alternative splicing," which involves the formation of new combinations of gene sequences containing the building blocks of life—proteins—each of them different from the unspliced original and each capable of being inherited. This raises questions about the predictability of the inheritability of traits based solely on DNA, leaving the author to warn that "any artificially altered genetic system . . . must sooner or later give rise to unintended, potentially disastrous, consequences."[11] Yet how would one design and execute a protocol that enabled us to make such a determination, since many of the effects from IGM may not show up for generations? Other, even more compelling questions relate to how society would assess the potential benefits and risks of a successful germ line intervention in the context of various social and ethical considerations.

Although we have much to learn about applying IGM techniques to humans, researchers have established proof of principle in animals, where foreign genes introduced into mice have been transmitted and expressed for at least three generations.[12] Moreover, recent advances in stem cell and cloning research, which were reported after completion of the AAAS study, will likely provide more options for doing IGM.[13] As knowledge of human genetics grows in the years ahead, the technical obstacles may fall by the wayside sooner than we expect. To what uses, then, might IGM be put? One would be to target IGM toward the alleviation or elimination of genetic diseases. The other would be to enhance human traits.

The Promise of Health

In principle, IGM would have the benefit of preventing the inheritance of genetic diseases in families rather than treating it every time it appears, generation after generation. And by targeting either germ cells or the embryo, IGM could intervene before a condition occurs—before it causes irreversible damage. Some possible health-related uses: transferring ooplasmic mitochondrial DNA to avoid potentially lethal diseases and problems with repeated miscarriages that are caused by faulty mitochondrial DNA; treating sperm or sperm stem cells to help men overcome infertility caused by a genetic mutation; and treating gametes or early-stage embryos to allow couples in which both partners share a recessively inherited disorder to have a genetically-related child free of the disease. All of these cases, it should be pointed out, are relatively rare, and all of them require some form of assisted reproduction.

There already exist, however, other, better-tested techniques to avoid passing on mutant genes. These include genetic screening and counseling, prenatal diagnosis and abortion, egg or sperm donation, and adoption either of a child or an embryo. Another technique, known as pre-implantation genetic diagnosis, combines in vitro fertilization (IVF) with pre-implantation diagnosis and selection of embryos. Individual cells are removed from an embryo, fertilized outside the body, and tested for the presence of genetic mutations (to the extent that tests are available). Embryos without known mutations are then implanted in the woman via IVF. This approach could have wide application, although it would not work in those cases where both parents have identical versions of the same mutant gene.

With these techniques available, and in light of the enormous difficulties associated with determining risks, why bother with IGM? The answer lies in its possible use for genetic enhancement.

Pursuing Enhancement

It is this prospect, I believe, that generates the most excitement over IGM, and the most uneasiness. I will argue that enhancement applications more than medical uses will determine the scope, direction, pace, and acceptance of IGM in the United States. And it will be the market and free choice, not government, that pushes it along. But I am getting ahead of myself.

By "genetic enhancement" I mean improving human traits that without intervention would be within the range of what is commonly regarded as normal, or improving them beyond what is needed to maintain or restore good health. Examples could include increasing height, improving intelligence, altering behavior, or changing eye color, all of which have been shown to have some underlying genetic connection. IGM offers the promise that genes associated with characteristics found to be undesirable (or less desirable) could be replaced by those linked to desired traits.

There are promises and perils associated with genetic enhancement. On the one hand, we know that some people are born more "genetically fit" than others, giving them certain advantages. The promise of enhancing the capabilities of those who are genetically less fortunate is an exciting and noble prospect for some people. And by increasing, for example, the intelligence of individuals, all of society may gain from the knowledge they generate.

On the other hand, genetic enhancement could also lead us to devalue various social and environmental factors that influence human development in concert with genes. There might be less appreciation for productive social interaction in a classroom, for example, or for the hard work traditionally required to become a successful professional. These conventional methods of enhancement may have some intrinsic value that could never be duplicated by a genetic intervention. In fact, a preoccupation with genetic enhancement may place too much emphasis on the genes and ultimately prevent us from solving problems that are really embedded in the structure of our society.

In principle, IGM would have the benefit of preventing the inheritance of genetic diseases in families rather than treating it every time it appears, generation after generation. Yet, there already exist other, better-tested techniques to avoid passing on mutant genes. In light of the enormous difficulties associated with determining risks, why bother with IGM?

Another complication is that the technology developed for therapeutic purposes will be the same as that used for enhancement. So while we might approve of IGM for medical treatment, its availability will likely promote creeping enhancement applications as well. For example, scientists are now testing the use of gene transfer to strengthen the muscles of children with muscular dystrophy, but the same technique could be used to increase an athlete's strength and endurance. An even more revealing example concerns the use of Human Growth Hormone, or HGH, which is genetically engineered to supplement natural growth hormone. While originally approved to treat children deficient in natural growth hormone, it could be used to make normal children taller, and indeed one newspaper has reported that some parents have requested HGH for children within the normal height range for their age because they want to improve their chances in competitive basketball.[14]

Further complicating matters is that distinguishing between treatment and enhancement may get increasingly difficult. The line where one ends and the other begins may become blurred as our experience with IGM expands. Hence, interventions which now give us pause may become more acceptable in the future. Surveys have shown that 40 to 45 percent of Americans approve of using genetic technologies to bolster their children's physical and mental traits.[15] I suspect that as more people get used to the idea, it will become even more appealing. Americans already avail themselves of cosmetic surgery to make them look better, drugs to make them more alert, and herbs to promote sexual performance. And we expect and praise parents for doing all that they can to enhance their children's well-being; it is "the natural expression of parental affection."[16] For many Americans, IGM will merely be seen as a logical extension of what is commonplace throughout America today, and it will be increasingly difficult to draw a clear line between the use of genetics for therapeutic purposes and its use for other ends.

Yet enhancement by genetics is also qualitatively different from enhancement by other means. Existing methods of enhancement, from pharmacology to advanced music lessons, are aimed at the current generation of adults and children. They are not biologically intrusive in a manner that will significantly shape our evolutionary course. Inheritable genetic enhancement would have long-term effects on persons yet to be born. Thus we have little, if any, precedent for this way of using IGM. We would be venturing into unknown territory, but without any sense of where the boundaries should lie, much less with an understanding of what it means to cross such boundaries.

To Market, To Market

There are good reasons to believe that market forces are more likely than other factors to determine the path we take on IGM. Science in general is increasingly valued for its commercial promise, and some recent developments reinforce the specific connection between genetics and the market. In the case of somatic cell gene transfer research, the focus of the field has shifted from rare genetic disorders, now viewed as offering limited profits, to more common ailments that promise greater financial gain. As one observer of the field noted, "The whole concept of gene therapy for genetic diseases doesn't fit the business model."[17] Another foresees movement in the direction of "the most profitable human conditions because there is . . . far more money to be made in curing baldness and wrinkles than there ever will be in cancer of HIV/AIDS."[18]

To some extent, that is what is occurring now. For example, the president of Anticancer, Inc., a San Diego company working on a genetic cure for baldness, has publicly stated that FDA approval will be sought for marketing the product for hair regrowth in cancer patients who become permanently bald due to chemotherapy, but that once such approval is granted, the product will be marketed to all those experiencing baldness.[19] The experience so far with somatic cell gene transfer suggests that, if left to its own, IGM is likely to follow the push and pull of the market, as applications for treating disease are either limited or more efficiently handled by other means, as Americans become more accustomed to the notion of enhancement, and as businesses offer commercially attractive products.

Genetics and Reproduction

Advances in reproductive technology have given thousands of infertile couples the chance to have a child. But merging these advances with those in genetics promises to extend choices beyond whether and when to have a child to what sort of child to have. Greater knowledge of genetics now enables people considering pregnancy to use new reproductive technologies to increase the possibility that their child has a certain genetic makeup.

Until very recently, that typically meant avoiding a genetically related disease that could be passed on. For example, in 2001 doctors reported on the use of PGD to identify human embryos that lacked a specific cancer-causing gene mutation. They implanted only embryos without the disease gene into a woman's uterus, and she gave birth to a baby free of the cancer syndrome that runs in the father's family.[20] It was also announced in early 2002 that the same procedure had been used to help a woman who has the gene for early-onset Alzheimer's disease have a baby free of that mutation.[21]

But the technology has now gone well beyond that limited purpose. Two years ago in the United States, IVF and PGD were used to produce a child whose genetic make-up made possible a life-saving treatment for an older sister. Several embryos were created and tested to ensure that they were genetically similar to the older sister yet lacked mutant gene that had given the sister a fatal disease. The result was a healthy baby boy, from whom umbilical cord cells were successfully used to treat his sister.[22] Within days of the story appearing in the popular press, authorities in England received more than a dozen requests for approval to undergo the same procedure. In response, the British government subsequently approved use of a pre-implantation technique that detects a range of genetic abnormalities,[23] and earlier this year a British couple gave birth to a baby born with a desired genetic characteristic—cells capable of saving her older sibling from leukemia.[24]

Who among us would not want to avoid passing on mutant genes to future children or help existing children overcome serious diseases? What these examples show is a willingness on the part of people to select future children for specific genetic traits. A technique that was originally created to help people screen *out* certain characteristics has crept in a different direction, now enabling people to screen *in* desirable characteristics. For now, the latter has been for the noble purpose of saving the lives of exiting children. But where does one draw the line? Should we screen out for short stature? Or screen in for eye color?

Poised to help meet, if not fuel, parents' desire to make their offspring healthier, and perhaps prettier, smarter, and more athletic as well, is a cottage industry that has sprung up in the past two decades in assisted reproductive technology. What began as an effort by fertility clinics to help infertile couples have a child is now a growth industry offering a range of services no longer confined to the infertile.

Some infertility clinics are offering couples the opportunity for "family balancing" via techniques that can virtually assure parents a child of a particular sex.[25] The idea is to use PGD to distinguish between embryos with the Y chromosome and those lacking it, and then implant the "right" embryo or embryos. In October 1999, in a statement declaring that the use of solely for sex selection "should be discouraged,"[26] the American Society for Reproductive Medicine said, "Those who argue that offering parental choices of sex selection is taking a major step toward 'designing' offspring present concerns that are not unreasonable in a highly technologic culture."[27] Web sites have sprung up where persons can market their gametes and couples can assess the height, weight, hair and eye color, education, and musical and athletic abilities of potential suppliers of eggs or sperm.[28] Some egg donors "are becoming shrewd businesswomen, asking top dollar for their high IQs or good looks."[29] And all this is occurring in an industry where there are strong economic interests in expanding services under the banner of enhancing consumer choice, and where choice is reinforced by the very high value that our society places on reproductive freedom.[30] It will be difficult to overcome the resistance of consumers to any effort to restrict their access to such technologies.

The market is now poised to take advantage of the increasing power of genetic technology, producing financial profit for some and giving parents the "best product" for their money. In the vernacular of the marketplace, people are not parents; they are consumers, and distinctions long recognized between reproduction and production begin to fade.

There should be no backdoors, whether due to gaps in public policy or an aggressive marketplace, through which inheritable genetic modification inches its way into our lives. These technologies are highly seductive, and we could easily get used to them without fully considering their consequences.

Whither Policy?

For society, deciding the fate of IGM will be among the most profound decisions it ever faces. Former Senator Daniel Moynihan once remarked that while it was important in the development of civilization for someone to have invented the wheel, it was equally important that soon thereafter someone invent the brake. We must balance our scientific efforts with a better understanding of where they are leading. Not all social values are well served by the push and pull of commerce. Individual decisions regarding the use of these genetic technologies may be personally beneficial, but they may not lead us toward a socially desirable outcome. In the meantime, the larger moral and social climate can be changed in ways that make applying the brakes difficult.

An editorial in the *St. Louis Post-Dispatch* articulated the problem nicely:

> private entities tend to be profit driven—which should be the last consideration in how we alter the human race. . . . [T]he . . . critical question is whether government should set rules for both public and private genetic manipulation of the species. If there continues to be no public oversight of the private entities, there will be nothing to stop fertility clinics from offering whatever genetic manipulation becomes possible and marketable.[31]

So what do we do? The 2000 AAAS report recommended several steps that remain important. Most important among them was that no IGM, whether involving intentional or "reasonably foreseeable" inadvertent transmission, should go forward at this time. In a subsequent article appearing in *Science* following the report by the fertility clinic of mitochondria DNA transmission, my coauthor and I stressed the urgency of moving more quickly to put in place a system of public oversight with authority over IGM efforts in both the public and private sectors.[32] Accompanying this oversight mechanism should be a national public dialogue on whether and, if it is deemed acceptable, how such research and its applications should proceed.

These proposals are premised not on a belief that IGM should never be tried, but that it must pass the test of public discourse, undergo rigorous assessment of its potential impacts, and receive explicit public approval. There should be no backdoors, whether due to gaps in public policy or an aggressive marketplace, through which IGM inches its way into our lives. These technologies are highly seductive, and we could easily get used to them without fully considering their consequences.

In the book *Future Shock*, Alvin Toffler wrote that the future arrives too soon and in the wrong order. If the future occurred in the right order, we could understand changes before they happened, rather than after, and could better prepare. That is not, of course, the way it works. But the future is not fixed, either. The question with regard to IGM is whether we will shape it or be shaped by it.

Acknowledgement

An earlier version of this article was presented at "Germ-Line Interventions and Human Research Ethics," a conference held on 5 April 2002, at Washington University in St. Louis. Complete conference proceedings are available at law.wustl.edu/centeris/apr5agenda_ns.html. The conference was sponsored by the Center for Interdisciplinary Studies, Washington University Law School and was coordinated by Rebecca Dresser, professor of law and of ethics in medicine at Washington University.

References

1. M.S. Frankel and A.R. Chapman, *Human Inheritable Genetic Modifications* (Washington, D.C.: American Association for the Advancement of Science, 2000).

2. J.A. Barritt et al., "Mitochondria in Human Offspring Derived from Ooplasmic Transplantation: Brief Communication," *Human Reproduction* 16 (2001): 513-516, at 513.

3. E. Parens and E. Juengst, "Inadvertently Crossing the Germ Line," *Science* 292 (2001): 397.

4. David Whitehouse, "Genetically Altered Babies Born," BBC News, May 4, 2001; http://news.bbc.co.uk/hi/english/sci/tech/newsid_1312000/1312708.stm.

5. T. Friedmann, Approaches to Gene Transfer to the Mammalian Germ Line, in *Designing Our Descendants: The Promises and Perils of Genetic Modifications*, ed. A.R. Chapman and M.S. Frankel (Baltimore, MD: The Johns Hopkins University Press, 2003).

6. Ibid.

7. J. Brainard, "Citing Patient Deaths, Key Senator Urges Better Oversight of Gene-Therapy Trials," *The Chronicle of Higher Education* (3 February 2000); http://www.chronicle.com/daily/2000/02/2000020301n.htm.

8. G. Romano et al., "Latest Developments in Gene Transfer Technology: Achievements, Perspectives, and Controversies over Therapeutic Applications," *Stem Cells* 18 (January 2000); http://stemcells.alphamedpress.org/cgi/content/full/18/1/1.

9. E. Marshall, "What to Do When Clear Success Comes with an Unclear Risk? *Science* 298 (2002): 510-11.

10. Frankel and Chapman, Human Inheritable Genetic Modifications," p. 7.

11. B. Commoner, "Unraveling the DNA Myth," *Harper's Magazine*, (February 2002): 39-47, at 47.

12. M. Nagano et al., "Transgenic Mice Produced by Retroviral Transduction of Male Germ-Line Stem Cells," *Proceedings of the National Academy of Sciences* 98 (2001): 13090-13095.

13. Friedmann, "Approaches to Gene Transfer to the Mammalian Germ Line."

14. R. Rubin, "Giving Growth a Synthetic Hand, Use of Hormone Sparks Debate," *Dallas Morning News*, 7 July 1986: A1.

15. R. Weiss, "Cosmetic Gene Therapy's Thorny Traits," *The Washington Post*, 12 October 1997.

16. R. Cole-Turner, "Do Means Matter? Evaluating Technologies of Human Enhancement," *Report from the Institute for Philosophy & Public Policy* 18 (Fall 1998); http://www.puaf.umd.edu/ippp/fall98/do_means_matter.htm.

17. Quoted in A. Pollack, "Gene Therapy's Focus Shifts from Rare Illnesses," *The New York Times*, 4 August 1998.

18. M.S. Langan, "Prohibit Unethical 'Enhancement' Gene Therapy," statement delivered at NIH Gene Policy Conference, Bethesda, Md., 11 September 1997.

19. Weiss, "Cosmetic Gene Therapy's Thorney Traits."

20. Y. Verlinsky et al., "Preimplantation Diagnosis for P53 Tumor Suppressor Gene Mutations," Reproductive Biomedicine Online 2, (March/April 2001); http://www.rbmonline.com/4DCGI/Article/Detail?38%091%09=%2055%09.

21. Y. Verlinsky et al., "Preimplantation Diagnosis for Early-Onset Alzheimer Disease Caused by V717L Mutation," *JAMA* 287 (2002): 1018-1021.

22. L. Belkin, "The Made-To-Order Savior," *New York Times Magazine*, 1 July 2001: 36 passim.

23. A. Ferriman, "UK Approves Preimplantation Genetic Screening Technique," *British Medical Journal* 323 (2001): 125.

24. J. Meek, "Baby with Selected Gene Born in Britain," Guardian Unlimited, 16 February 2002; http://www.guardian.co.uk/Archive/Article/0,4273,4357446,00.html.

25. L. Belkin, "Getting the Girl," *New York Times Magazine*, 25 July 1999: 26 passim; also see http://www.centerforhumanreprod.com/treatment_assisted.htm.

26. Ethics Committee of the American Society of Reproductive Medicine, "Sex Selection and Preimplantation Genetic Diagnosis," *Fertility and Sterility* 72 (1999): 595-98, at 598.

27. Ibid., p. 597.

28. For one such site, see http://www.conceptualoptions.com/ExtraordinaryDonors.htm.

29. C. Cleary, "Altruism, Money Motivate Egg Donors: More Young Women Aiding Infertile Couples," *The Seattle Times*, 4 November 2002.

30. O.D. Jones, "Reproductive Autonomy and Evolutionary Biology: A Regulatory Framework for Trait-Selection Technologies," *American Journal of Law and Medicine* XIX (1993): 187-231.

31. Editorial, "Altering Man: Who Decides," *St. Louis Post-Dispatch*, 21 September 2000.

32. M.S. Frankel and A.R. Chapman, "Facing Inheritable Genetic Modification," *Science* 292 (2001): 1303.

[17]

Gene Therapies and the Pursuit of a Better Human

SARA GOERING

As a philosopher interested in biomedical ethics, I find recent advances in genetic technologies both fascinating and frightening. Future technologies for genetic therapies and elimination of clearly deleterious genes offer us the ability to get rid of the cause of much human suffering, seemingly at its physiological root. But memories of past eugenics programs gone horribly awry (whether we speak of Hitler's program, California sterilization laws and practices of the 1920s, or even contemporary practices, such as attempts to work out deals that exchange sterilization for early prison release)[1] must make cautious our initial optimism for these generally well-intentioned programs. Most often the scientist proceeds in research with the best of intentions, but that does not make all scientific investigation worth pursuing.

Surely no one would dispute the claim that the aim of bettering humanity and/or our own children is morally acceptable. Indeed, most of us see as ideal a world in which every parent works toward improving the lot of his or her child, or the lot of all of our children. But while no one denies the importance of this quite general goal, we are still left with difficult issues about *how* we ought to proceed in addressing that goal. When we try to dodge diseases or disadvantages through genetic intervention, are we solving problems or just moving them to a different level? I want to briefly address two quite general questions in regard to this topic. First, what *means* should we take in trying to better our children? Second, how are we to decide what really counts as "bettering" them? I do not claim to solve these difficult issues here, but only to consider some ways we might approach the problems.

Consider the various ways in which we try to better our children. First and foremost, we try to educate them. That is, we provide public education for all children with the intention of teaching them not only how to read, write, and do arithmetic, but also how to function in this society and how to make important decisions in their lives. No one questions this means to bettering our children, unless they find fault with particular styles of teaching (for instance, they see the teacher as indoctrinating a particular view rather than presenting basic facts and promoting the child's own rational critical faculties in assessing those facts). Generally, as long as the education exercises the child's mind and allows the child some autonomy in regard to how he or she will understand the material, this means of bettering is considered morally acceptable and even required.

In addition to education, we find it morally acceptable to better children by giving them appropriate medical care. If a child needs surgery or a painful treatment to survive or to thrive, we allow ourselves room to do what is "best for the child" even if that may involve unavoidable pain that the child is not able to consent to. Furthermore, we often go beyond merely treating clearly defined diseases and actually allow enhancement of our children in the medical setting. We regularly give our children vitamin supplements, vaccinations, and dental enhancements (e.g., braces), and we generally do not blink at such interventions.[2] Thus we have no clear moral concern with imposing medical treatments paternalistically, so long as we are fairly certain that we are promoting the child's best interests.

There are no doubt numerous other ways that we try to make things better for our children and future generations. Many of them are indirect: we try to balance the national budget (or at least keep it under control); we try to pass legislation that will save the environment and preserve a fair quality of living for future generations; and we put federal and state money into exploratory research that is unlikely to produce immediate results but may lead to improvements in future lives. So whether we are trying to enhance children themselves or their environment, we are generally quite at ease with working toward improvements.

But how do genetic therapies fit into this classification? If we are speaking of somatic cell therapies (performed on the body cells of a fetus, infant, or adult, so that the genetic changes will not be passed on to the next generation), then we at least have an identifiable being who may be benefited by the changes; but if we speak of germline genetic therapies, performed pre-embryonically, then what is in question is who will come into existence, and we find ourselves in the Parfitian paradox of future generations.[3] Should we be compelled to try to make things better for unidentifiable future persons? What could the compelling reason be? Without wading through the vast literature debating this topic, I think it is safe to say that most of us are at least willing to admit that duties of beneficence and intuitions about morality suggest that we do care about these future people (whether or not we are required to do so by rules of justice). But in trying to work things out for future people (or even for young children who are not considered competent to decide for themselves), we must decide what kind of interventions are morally legitimate and most likely to be truly beneficial.

If we think that a genetic therapy will benefit a fetus or child, should we perform that therapy? The initial response might appear to be a resounding yes. This sort of treatment appears to get to the root of the problem and eliminates the need for any suffering from the disorder. Some common arguments posed against genetic interventions generally have been rather soundly disposed of in the relevant literature, e.g., arguments from playing God, from messing with nature, from the inevitability of slippery slopes.[4] But in a society that tends to overvalue the quick fix solution, we might do well to exercise some caution even here.

First, we might get unexpected results. If we perform genetic therapies to remove or change a clearly deleterious gene, then we might find that other important physical or psychological traits were also controlled by that gene (or by its influence on another gene or its expression). The commonest example here is the link between sickle cell trait and resistance to malaria. Another such link commonly discussed in the literature is that between creativity and various forms of mental illness.[5] Although we may be interested in relieving the suffering caused by the expression of certain genes, we are not yet certain what else we may be removing or changing inadvertently. Although these are certainly reasonable worries, even with such possibilities, treating painful and restrictive genetic disorders (e.g., Tay-Sachs or cystic fibrosis) might be worth the risk, so long as traditional rules regarding informed consent for clinical trials are respected.[6]

Second, we might be losing something valuable if we are able genetically to engineer around our problems. Erik Parens, for instance, suggests that part of what we value about humanness is our fragility, and the capricious nature of our lives, which necessitate our taking care of one another in times of need.[7] If we are able to use genetic engineering to get rid of this fragility (or at least to change the kind of fragility or to make people less willing to feel sympathetic to one who is fragile), then we may inadvertently destroy something very valuable. This is not to say that we must keep people suffering so that we can be caring creatures, but only that we might lose part of what makes us really appreciate our lives.[8] Parens has tongue-in-cheek suggested that we might all vote to make it so that no one had to experience adolescence as we know it (it is painful to go through and it is painful to be around those who are going through it), but he also notes that most of us place value on the process of working through such a time, and that effort is part of what makes us appreciate our adult lives.[9] Thus he counsels caution in our eagerness for genetically eliminating *anything* that appears to cause pain or discomfort, and a deeper analysis of what it is that we receive from the experience of living with disadvantages and diseases.

This brings me to my second issue: How are we to decide what is to count as "bettering" children? Few of us would dispute the claim that eliminating Tay-Sachs disease or Lesch-Nyan syndrome or cystic fibrosis would count as an improvement for future generations. A future in which no one has to suffer from these debilitating diseases seems undeniably worth pursuing. On the other hand, disability rights advocates are quick to point out practical problems with holding this view without devaluing existing persons with those diseases.[10] Even if we can conceptually distinguish between the value of individuals with disabilities and the relative value of bringing such individuals into existence given other options,[11] in practice, public attitudes toward such individuals are likely to be prejudiced and will likely affect public financial support of the disabled.[12,13]

Even if we could reach agreement about the value of genetically intervening for clear cases of debilitating disorders, there are some physiologically or genetically based conditions that offer disadvantages to children in our society that might not so clearly be candidates for intervention. What about cases in which the real cause of the disadvantage is located in unjustified societal prejudices or values? For instance, children who are shorter than average (and grow into shorter than average adults) have a smaller statistical chance for success in classes and in athletics (and ultimately in the job market) because of the biased perception of them based on their inferior height.[14] Physicians who offer growth hormone treatments treat the physiological symptoms of shortness as a way of solving the social problem for the child. But the community is then allowed to continue its arbitrary preference for taller people. In this case, society is at fault for creating the disadvantage—solving the real problem seems to require addressing societal values, *not* just engineering a way around the problem. This may seem obvious, since when height is the feature in question, there is no absolute advantage to be had, but only a relative advantage. There is no inherent value to being six feet tall, but it is advantageous if you are *taller* than others (within limits). But the same can be said of other features that do not rely on relative advantage, such as societal standards of beauty (having a symmetric face, hair in appropriate places and not in inappropriate places, etc.).[15] Some interventions that appear to be beneficial for the recipient may not be real benefits after all if they leave the root of the disadvantages unaddressed.

Another difficult case is deafness. Most hearing people consider deafness to be a defect, a physiological problem that deserves medical attention if and when it is available. But at least a segment of the Deaf community values their physiology as different but equal to that of hearing people, and they may argue that the only reason that deafness confers any disadvantage in society (when it does), is because of unfair societal discrimination and the fact that society is set up for the benefit of the hearing. Indeed, a 1994 publication of the Denver Ear Institute notes that many deaf people consider deafness "a birthright to a distinctive and rewarding way of life."[16] The Deaf community is rich and complex in terms of language, art, and social association.[17] Deaf community advocates suggest that the Deaf are more appropriately considered a cultural and linguistic minority (on par, for instance, with Hispanic-Americans) than a disabled group.[18,19] The availability of cochlear implants for deaf children has sparked the debate about the future of the Deaf community, and it will surely only be enflamed by the possibility of genetically engineering to avoid some forms of deafness. Segments of the dwarf community have made similar claims about the value of their genetic condition and problems with therapies that try to "rectify" it.[20] Is the elimination of deafness or dwarfism a benefit, or is it a systematic destruction of special minority communities?

We might also ask ourselves about the case of homosexuality. Simon LeVay's announcement in 1991 of a statistical difference in the sizes of a particular hypothalamic nucleus between heterosexual males and homosexual males evoked a loud public debate about the relevance of genetic or biology-based explanations for homosexual lifestyles.[21] If there is a genetic basis for homosexuality, then we must ask ourselves what to do about it. A homophobic parent might aspire to have this "defect" fixed, even though homosexuality itself does not bring about disadvantages; rather, homosexuals are often unfairly discriminated against by a society that arbitrarily devalues their lifestyle. Should we try to engineer a solution to this supposed "problem," or should we work to educate people that the only problem is within their unreasonable biases?

In general, then, it appears that we may not want to genetically treat (or eliminate) just any condition or trait that confers disadvantage to our children, especially when the disadvantages are not a direct result of the trait; rather, we need to find a reasonable decisionmaking process that will help to delineate what traits are acceptable candidates for genetic therapy or genetic engineering. The dominant paradigm in the literature for drawing this line has been a distinction between treatment and enhancement, based on standard medical practices. If we use the standard model of medical practice that relies on a principle of beneficence and is tied to a "normal" human capabilities model, then it seems justifiable (or perhaps even obligatory[22]) for us to treat defects or diseases. But nothing follows about the permissibility of using genetic intervention for the purpose of mere enhancement. This distinction is intuitively appealing to doctors, genetic scientists, and the general public, including many medical ethicists,[23] but it has been rejected more recently by ethicists on the basis of its vagueness.

According to the medical model, the basis for pursuing any genetic therapies is the relief of pain and suffering. W. French Anderson argues that genetic diseases that "produce significant suffering and premature death"[24] (p. 690) ought to be the first candidates for genetic therapies, and then, if we succeed with those cases, we might be justified in extending genetic treatments to other diseases. He claims, however, that we should not undertake any genetic engineering for the purpose of enhancement. He offers two reasons why we should not engage in any enhancement engineering. First, he thinks that sort of engineering is "medically hazardous" because we are less sure about what "adding" a gene could do to the complex system compared to fixing an existing gene. Second, he believes that it is "morally precarious" because we don't have a clear way of determining what genes should be provided, who should receive them, and how to prevent discrimination against those who don't receive them.[25] His proposal involves the claim that it is problematical for us to determine the details of enhancement engineering, but not problematical, and in fact defensible, for us to employ treatment of disease as a clear category for use of genetic engineering. This proposal is sound only if there is indeed no clear line between acceptable improvements and problematic ones, and if there is a clear line between treatment and enhancement. But is such a view defensible?

How are we to define "disease," if that is what we are allowed to treat? If we look more closely at the concepts of health and disease, we discover that the label "disease" is not metaphysically pure. Indeed, while we often assume that a disease is an objectively identifiable state, in fact, the identification of something as a disease is dependent at least in part on evaluative judgments of the physician or general society.[26] Some commentators assume that the physiological conditions to be included under the label "disease" are ones that are identified either as abnormal or as dysfunctional relative to species norms. On this view, to be diseased is merely to exemplify a certain abnormal physiological state, whether or not it is disadvantageous or painful to the person. Anyone in any society could be objectively labeled as diseased simply by reference to his or her exemplification of the relevant physiological state. But this view does not stand up to difficult cases, as numerous commentators have pointed out. Disease is not simply a physiological state, but a physiological state that bears significantly on the functioning of the individual in his or her society. Colorblindness is not considered a significant disorder or disease in the United States, but in some places in Africa "in which the capacity to distinguish a great variety of shades of green is needed to function at a minimal level for survival" it is highly problematic.[27] As Harlan Lane notes, alcohol use, tobacco addiction, large body weights, the need for eyeglasses, and hookworm infestation are all considered diseases in certain societies and not in others. The American medical system has recently medicalized many conditions that were not previously considered appropriate for "treatment," including "contraception, fertility, pregnancy, childbirth, child development, hyperactivity in children, reading difficulty, learning problems, drug addiction, criminality, child abuse, physical disability, exercise, hygiene, sleeplessness, diet, breast and nose size, wrinkles, baldness, obesity, and shortness."[28] We view certain physical states as diseases because of our judgments about what is dysfunctional, and those judgments depend on our values and social norms, resources, and standard medical practices. These may differ across societies and across time periods, as well as within societies and time periods. Consider, for example, the fact that masturbation and homosexuality were both once identified as diseases.[29] When we realize that the definition of disease is norm based, then we find that "the intuitively attractive reply that 'If we stick to curing disease and promoting health, all will be well' begins to lose its attraction."[30] What counts as a disease depends on physiology but also on what the particular society values. Consequently, the line between treatment of disease and enhancement-directed engineering seems itself to rely on a rather fuzzy distinction.

There is also the further problem of defining when something is a disease in a society that agrees that a particular condition is not valuable. For instance, no one desires atherosclerosis and the heart attacks and strokes that often follow it. Evidence suggests that there is a gene that determines the body's ability to regulate blood cholesterol levels by production of low density lipoprotein (LDL) receptors on body cells. Inserting additional LDL receptor genes in otherwise normal individuals might reduce the probability for their developing atherosclerosis. But atherosclerosis is not so rigidly tied to genetic production of LDL receptors. Treatment could also take the form of reducing consumption of low density fat in the regular diet. Is genetic elimination of this "disposition" for heart disease rightly considered treatment or enhancement? Have we treated a condition that significantly contributes to much morbidity and mortality, or have we simply enhanced our systems so that we can be gluttonous, so that we can unrestrainedly eat according to our heart's desire rather than our body's needs? These difficult cases illustrate the difficulty of attempting to distinguish the treatment of disorder from creating enhancements. Given the problems with trying to force fit the genetic debate into these problematic categories, we ought to look for a new way to conceptualize the debate. What we care about is improving the lives of our children and future generations, but we are not certain what ought to clearly count as justifiable *genetic* improvement.

One possibility that I would like to propose involves using a famous sort of thought experiment in philosophy, proposed by John Rawls for devising a fair distributive justice scheme.[31] From behind the veil of ignorance (which obscures each individual's detailed knowledge of his or her own position in society), Rawls has rational creatures attempt to figure out what basic rules of justice would be fair for all society. Because no one is certain if he or she will be at the lowest rung of the social ladder or at the highest one, Rawls believes that people in this "original position" would opt for rules of justice that require equal basic liberties and a "maximin" policy that requires any changes in distribution of goods to benefit the worst off in society as well as the ones who have reason to cause the change to occur.

What if we were to put ourselves behind the veil of ignorance in respect to our children's genetic makeup? That is, what if we tried to determine what traits we would desire for them, and what traits we would prefer for them not to have if we did not know the details of our society (that is, if we did not know the particular patterns of racial/sexual/gender discrimination that we find in our own society, or we did not know what society we would find ourselves in)? The veil of ignorance, then, is a way to conceal from us the particular biases that our society has for traits that are otherwise not genuinely physically desirable. When we put on this veil of ignorance, we assume that we do not know which society we will be living in—we do not know physical or social details about the majority class for instance. We then try to determine what physical traits would lead to clear advantages or disadvantages in *any* society. This test allows us to decide for our children and future generations what sorts of traits should *not* be genetically manipulated.

It seems that things like race and sexual preferences would be quickly eliminated as genetic engineering candidates, as well as cultural standards of beauty (including particular features as well as height) because they are only valued by particular societies. That is, if you live in a predominantly white society, having white skin would tend to confer advantages, but this would not be so in a predominantly black society. If there is no reason to prefer a particular trait from behind the veil of ignorance, then perhaps we should rule it out as a candidate for genetic engineering. Other things might be clear candidates for genetic intervention, because they would be disabilities (or bring disadvantages) for anyone in any society (e.g., conditions like Tay-Sachs or aminodeaminase deficiency).

Deafness would be more difficult—whether you need hearing depends on the social structure of society, and we can imagine a majority deaf society designed for the benefit of the deaf. Such societies are not completely imaginary. Although the deaf were not a majority, the history of Martha's Vineyard can be used as an illustration of what such a society might be like.[32] But there is an asymmetry. If you are deaf in a world in which the norm is nondeaf, then you are likely to experience significant disadvantages, both in terms of social goods and physical safety, because most communication, transportation, and warning systems are designed for the hearing. However, if you are nondeaf in a world where deaf is the norm, then you are not so clearly disadvantaged. (This claim is not without contention. You might be more easily distracted, and there is the possibility of experiencing some kind of schizophrenic symptoms, since you might respond to stimuli not perceived by others. This sort of experience is not, however, borne out by the early childhood stories of hearing children raised in deaf families.[33]) There is an asymmetric pattern of disability or disadvantage. The deaf individual might experience disadvantage in a hearing world, but the same might not be true (or at least not to the same extent) for the hearing person in a deaf world.

On the other hand, try height: the disability is symmetrical there. If you are tall in a short world, you hit your head often and cannot fit in cars or through doorways; if you are short in a tall world, you cannot reach the pedals or the countertops, etc. The feature of asymmetrical disadvantage might help to pick out factors that are worth changing genetically (or worth considering as candidates for genetic change). Perhaps changing the thing that "veiled" rational people agree would *always* be a detriment is permissible (or even required), as is changing what could not be a harm in *any* society (i.e., what involves asymmetrical conditions), but changing what finds its value only in the particular society should not be allowed.

While this suggestion is clearly susceptible to many of the criticisms of Rawls's work,[34] it may at least help to figure out a way to start the difficult process of distinguishing between legitimate genetic intervention and discriminatory or arbitrary intervention. Furthermore, given limitations on our imaginations (whether we are behind the veil of ignorance or not), I would certainly propose that those who make policies on genetic therapies should represent a wide variety of physical abilities and conditions (so that we do not hastily presume, for instance, that deafness is a clear defect without first consulting with those who are deaf).[35] Thus we must surely bring a fair representative sample to the table to consider these possible interventions.[36] As Susan Wendell eloquently illustrates:

> The desire for perfection and control of the body, or for the elimination of differences that are feared, poorly understood, and widely considered to be marks of inferiority, easily masquerades as the compassionate desire to prevent or stop suffering. It is not only a matter of being deceived by others, but all too often a matter of deceiving ourselves. It is easy to make the leaps from imagining that I would not want to live in certain circumstances to believing that no one would want to live in those circumstances, to deciding to prevent people from being born into those circumstances, to supporting proposals "mercifully" to kill people living in those circumstances—all without ever consulting anyone who knows life in those circumstances from experience.[37]

To ensure that such experience is taken into account, the ideal decisionmaking procedure would bring together a number of differently abled individuals who would first openly discuss the benefits and harms, delights and difficulties of living with various physical conditions (as a way to inform the rest of the group about conditions they may only understand superficially or peripherally). Then, each representative would perform the thought experiment suggested by the veil-of-ignorance strategy. While I do not claim that this strategy would produce *unanimous* agreement on appropriate candidates for genetic engineering, I do believe that it is likely to bring us closer to agreement on what traits we should *not* be genetically engineering. Furthermore, it should help us to uncover some of our societal biases regarding genetic and/or physical traits and to stretch our imaginations regarding how we might address these biases through nongenetic means. If we work to imagine societies where being short or deaf or homosexual is not a disadvantage (as it unfortunately is in our present society), then we may be able to apply that thinking to social structures in our present society.

One interesting and fairly worrisome implication of this sort of analysis is that it does seem to leave open the possibility of tampering with intelligence.[38] Considering intelligence from behind the veil of ignorance results in an asymmetrical finding, like that of having hearing. Presumably, improving intellectual capabilities would be desirable in any society, and having more of it in a society that is less intelligent would not result in any significant disadvantage.[39] This implication of my analysis is a matter for concern and undoubtedly requires further investigation. One possible answer might be that we should be able to try to better our intellectual capabilities, but that doing so genetically is not really feasible. That is, we may all applaud programs like "Hooked on Phonics" if they do indeed provide a more efficient and effective means of teaching children language skills, but genetic alterations may never be able to produce that sort of change. A genetic intervention might increase the speed at which we can pick up or process information, but perhaps nothing more can be affected genetically. A child cannot be born *with* knowledge, but only with an improved ability to gain knowledge. But if we *could* improve intelligence via genetic enhancement, we might be tempted to do so. Although intelligence might offer relative advantages (having more of it than another offers an advantage), it also seems to offer absolute advantages.[40] Even if we did desire to enhance intelligence and found that it was an acceptable candidate for genetic engineering, it might not be our first priority for funding, since other medical and/or social needs are more pressing. Still, this sort of difficulty deserves further consideration.

In conclusion, genetic enhancement is not clearly an evil deserving of outright prohibition. Rather, we need to make careful and reasoned decisions about what genetic and/or physical changes would truly constitute human improvement, and what changes would only serve to reproduce our societal biases. My decisionmaking proposal is a brief suggestion that requires further thought and consideration, but one that appears to take into account both what we are interested in preserving and what we are interested in changing. My main hope is that it will provoke further discussion.

Two final points require attention. First, leaving decisions about what traits we ought to genetically change to the market (as suggested, for instance, by Nozick[41]) is likely to be disastrous. Not only will it most likely increase the disparity between the "haves" and the "have nots," but even if we could ensure equal access to such therapies, we might move ourselves toward a highly homogenous society. Within a society, many people have similar values and desires (e.g., polls show that most American women would prefer to have a boy child first, followed by a girl[42]) and would probably prefer stereotyped masculine and feminine traits in their children as well. We need to consider what it is that we value about diversity, and how we can best promote that. We may be unintentionally drawn to a world where any difference is considered a disease.[43]

Second, we cannot be drawn back into an assumption of genetic determinism. Nature/nurture debates periodically come back into vogue, but we must continue to recognize that phenotypes and behavioral patterns are not wholly determined by genetic data, but by the interaction of the genetic information with the environment (both internal and external). Despite solid arguments against deterministic views,[44] the amount of time and attention given to books such as *The Bell Curve* ought not only to warn us of our ability to ignore the environmental influences when we so desire, but also of our need to take care to ensure that science is done properly. We should not expect genetic engineering to solve all of our problems, nor should we reduce anyone's identity and value to their genetics.

Notes

1. For a historical review of eugenic practices, see Kevles D. Eugenics and the Human Genome Project: is the past prologue? In: Murphy T, Lappé M, eds. *Justice and the Human Genome Project*. Berkeley: University of California Press, 1994:14–29.
2. Many of these noncontroversial interventions are discussed by Glenn McGee in his recent book *The Perfect Baby*. Lanham, Maryland: Rowman & Littlefield, 1997, esp. Chapter 7.
3. This problem was made famous by Derek Parfit in *Reasons and Persons*. Oxford: Oxford University Press, 1984.
4. For example, these issues are dealt with in Resnick D. Debunking the slippery slope argument against human germ-line gene therapy. *Journal of Medicine and Philosophy* 1994;19:23–40; Munson R, Davis L. Germ-line gene therapy and the medical imperative. *Kennedy Institute of Ethics*

Gene Therapies and the Pursuit of a Better Human

Journal 1992;2(2):137–58; Boone K. Bad axioms in genetic engineering. *Hastings Center Report* 1988;18(4):9–13; Glover J. *What Sort of People Should There Be?* Harmondsworth, England: Penguin, 1984 (see also note 2, McGee 1997:ch. 4).

5. E.g., Beeman CA. *Just This Side of Madness: Creativity and the Drive to Create.* Conway AK: UCA Press, 1990. Beeman argues that while psychotic illnesses are not correctly construed as tied to creativity, the creative drive that exists to some degree in all humans is intensified in people who also inherit any of a range of affective disorders, especially manic depression or bipolar disorder. She believes that mental illness is a hindrance to creativity, but because those who inherit affective disorders often have an increased creative drive, they are often more artistic and creative until the illness overcomes them. She concludes that we ought to look to the social context to cure psychiatric disorders, rather than eliminating them genetically (which might eliminate the intense creative drive as well).
6. For a discussion of these rules, see McCormick R. Proxy consent in the experimentation situation. *Perspectives in Biology and Medicine* 1974;18(1):2–20.
7. Parens E. The goodness of fragility: on the prospect of genetic technologies aimed at the enhancement of human capacities. *Kennedy Institute of Ethics Journal* 1995;5(2):141–53.
8. As Ann Davis suggested to me, this argument bears striking similarities to philosophy of religion replies to the problem of evil, e.g., John Hick's soul-building theodicy, "soul-making and suffering." See Hick J. Evil and the god of love. In: Adams MM, Adams RM, eds. *The Problem of Evil.* New York: Oxford University Press, 1990: 6–88.
9. See note 7, Parens 1995.
10. Susan Wendell makes this point in her book *The Rejected Body: Feminist Philosophical Reflections on Disability*. New York: Routledge, 1996. She notes, "It sends a message to children and adults with disabilities, especially people who have genetic or prenatal disabilities, that 'We do not want any more like you.' Knowing that your society is doing everything possible to prevent people with bodies like yours from being born is bound to make you feel as though you are not valued and do not really belong" (p. 153).
11. See, for example, the argument in Glover J. *Ethics of New Reproductive Technologies*. Dekalb, Ill.: Northern Illinois Press, 1989:ch. 12, 13; and note 4, McGee, 1997. McGee claims "It is not insult to these people to suggest that serious genetic defects should be prevented—unless we equate the patient with the disease. Many patients would not resist the idea that the world would be a better place if they did not have to suffer and die from the disease. We can over-romanticize the courage of sufferers" (p. 59).
12. Murphy T. The genome project and the meaning of difference. In: Murphy T, Lappé M, eds. *Justice and the Human Genome Project*. Berkeley: University of California Press, 1994:1–13.
13. See note 10, Wendell 1996: 153–6.
14. See, e.g., Jackson L, Ervin K. Height stereotypes of women and men: the liabilities of shortness for both sexes. *Journal of Social Psychology* 1992;132(4):433–45; and Allen D, Fost N. Growth hormone therapy for short stature: panacea or Pandora's box? *Journal of Pediatrics* 1990;117(1):17–9.
15. See Landy D, Sigall H. Beauty is talent: task evaluation as a function of the performer's physical attractiveness. *Journal of Personality and Social Psychology* 1974;29(3):299–304; and Adams G. Physical attractiveness research: toward a developmental social psychology of beauty. *Human Development* 1977;20:217–39.
16. Boddie C. Debate heats up in the deaf community. *Denver Ear Institute Highlights* 1994;13(3):1–5.
17. For an interesting discussion of the deaf community's identity as a political and social group, as well as a distinctive culture, see Sacks O. *Seeing Voices*. Berkeley, Calif.: University of California Press, 1989.
18. Lane H. *The Mask of Benevolence: Disabling the Deaf Community*. New York: Alfred A. Knopf, 1992.
19. See Silvers A. "Defective" agents: equality, difference and the tyranny of the normal. *Journal of Social Philosophy*, 25th Anniversary Special Issue 1994;25:154–75. Silvers notes that "Deafness is not a natural disadvantage in interpersonal communication. Signing members of the deaf community communicate with one another as effectively as do hearing persons who speak to each other. That the majority of Americans know speech rather than sign may be thought of as simply another in a long list of practices imposed by the dominant group to suit its members while suppressing a minority whose practices are otherwise" (p. 164).
20. Berreby D. Up with people: dwarves meet identity politics. *New Republic* 1996;214(18):14–9.
21. LeVay S. A difference in hypothalamic structure between heterosexual and homosexual men. *Science* 1991;253:1034.

22. See Buchanan A. Equal opportunity and genetic intervention. *Social Philosophy & Policy* 1995;12(2): 105–35.
23. This kind of distinction is supported, for example, by Fletcher JC. Moral problems and ethical issues in prospective human gene therapy. *Virginia Law Review* 1983;69:515–46; Daniels N. The genome project, individual differences, and just health care. In: Murphy T, Lappé M, eds. *Justice and the Human Genome Project*. Berkeley, Calif.: University of California Press, 1994:110–32; and to a certain extent, Munson R, Davis L. Germ-line gene therapy and the medical imperative. *Kennedy Institute of Ethics Journal* 1192;2(2):137–158. W. French Anderson was one of its original proponents, and now accepts it with some exceptions (e.g., enhancement that may be considered preventive medicine, such as increasing LDL receptors to prevent atherosclerosis later in life). See Anderson WF. Human gene therapy: why draw a line? *Journal of Medicine and Philosophy* 1989;14:681–93. He is followed by many American and European scientists. Most of the proponents of the treatment-enhancement distinction recognize the difficulty of holding this line (due to its vague grounding), but conclude that it is the best alternative available.
24. See note 23, Anderson 1989.
25. See note 23, Anderson 1989:687. He argues that we can allow treatment for serious genetic diseases that everyone can agree are significant, but recognizes that we have no clear way of distinguishing "a serious disease from a minor disease from a cultural discomfort."
26. This popular debate of the 1970s pitted objectivists (e.g., Christopher Boorse, who believed that diseases and their diagnosis contained no judgments of value) against normativists (e.g., Tristram Englehardt or Thomas Szasz, who believed that the concept of disease and also diagnosis of diseases are necessary partially evaluative and thus relative to the prevailing views of the culture or society). See Boorse C. On the distinction between disease and illness. In: Caplan AL, Englehardt HT, McCartney JJ, eds. *Concepts of Health and Disease*. Addison-Wesley, 1981:545–60; Englehardt HT, Jr., The concepts of health and disease. In: Caplan AL, Englehardt HT, McCartney eds. *Concepts of Health and Disease*. Addison-Wesley, 1981:31–46; and Szasz T. *The Myth of Mental Illness*. New York: Harper, 1961.
27. See Cohen C."Give me children or I shall die!": new reproductive technologies and harm to children. *Hastings Center Report* 1996; 26(2):19–27.
28. See note 18, Lane 1992.
29. See note 26, Englehardt, 1981; and Green R. Homosexuality as a mental illness. In: Caplan AL, Englehardt HT, McCartney JJ, eds. *Concepts of Health and Disease*. Chicago: Addison-Wesley, 1981: 333–51.
30. Kitcher P. *The Lives to Come: The Genetic Revolution and Human Possibilities*. New York: Simon & Schuster, 1996:212.
31. Rawls J. *A Theory of Justice*. Cambridge, Mass.: Harvard University Press, 1971.
32. See Groce NE. *Everyone Here Spoke Sign Language: Hereditary Deafness on Martha's Vineyard*. Cambridge, Mass.: Harvard University Press, 1985.
33. These stories are contained in Carol Padden and Tom Humphries' book, *Deaf in America: Voices from a Culture*. Cambridge, Mass.: Harvard University Press, 1988.
34. For example, that it only works within a particular kind of liberal society; that it presumes that rational creatures can successfully abstract away from their individual realities when behind the veil of ignorance; that it is not sufficiently attentive to what exists in this society, etc. These are important objections, and I will no doubt have to assume some general conception of the human form (to avoid bizarre but possible genetic traits that might be beneficial in some imaginable societies, and not a harm in ours—like gills, or philosophers designed to sit for long periods of time reading and drinking port, with no ill results for health). (The latter suggestion was proposed in Englehardt HT, Jr.: Germ-line genetic engineering and moral diversity: moral controversies in a post-christian world. *Social Philosophy & Policy* 1996; 13(2):47–62), but I think that this strategy is still a useful one for assessing which physical traits are valued in our society for arbitrary or socially biased reasons, and which traits are genuinely physically beneficial.
35. It is important to recognize the particular limitations that able-bodied or "normal" people seem to have in regard to imagining themselves with disabilities. Anita Silvers comments, "the prospect of being so impaired seems to paralyze the normal imagination," resulting in a great number of people who report that they would "rather be dead than confined to a wheelchair." See note 19, Silvers 1994:158–9.
36. Here I draw from Susan Moller Okin's defenses of Rawls in Okin SM. *Justice, Gender and the Family*. New York: Basic Books, 1989. Again, this suggestion is clearly prone to objections raised in response to Okin's work, regarding how to decide on a fair representative sample, etc.

Gene Therapies and the Pursuit of a Better Human

37. See note 10, Wendell 1996:156.
38. Not everyone agrees that it does leave this possibility open. See, for example, Ledley F. Distinguishing genetics and eugenics on the basis of fairness. *Journal of Medical Ethics* 1994; 20:157–64.
39. Except perhaps for the frustration of listening to illogical arguments—nothing that any teacher of an introduction to philosophy has not survived, or even found rather humorous.
40. Despite the frequent use of the phrase "ignorance is bliss," increases in intelligence through standard means in education are certainly not rejected outright, and are not considered mere attempts to seek relative advantage in the sense of "keeping up with the Joneses."
41. Nozick R. *Anarchy, State and Utopia*. New York: Basic Books, 1974.
42. See, e.g., Pebley A, Westhoff C. Women's sex preferences in the United States: 1970–1975. *Demography* 1982;19(2):177–89; and Coombs LC. Preferences for sex of children among U.S. couples. *Family Planning Perspectives* 1977;9(6):259–65.
43. Murphy T. The genome project and the meaning of difference. In: Murphy T, Lappé M, eds. *Justice and the Human Genome Project*. Berkeley, Calif.: University of California Press, 1994:1–13.
44. The best I have seen are offered by Philip Kitcher in Kitcher P. *The Lives to Come: The Genetic Revolution and Human Possibilities*. New York: Simon & Schuster, 1996; and in Lewontin R. *Biology as Ideology: The Doctrine of DNA*. New York: HarperCollins, 1992.

[18]

Protecting the Endangered Human: Toward an International Treaty Prohibiting Cloning and Inheritable Alterations

George J. Annas, Lori B. Andrews and Rosario M. Isasi†

I. INTRODUCTION

We humans tend to worry first about our own happiness, then about our families, then about our communities. In times of great stress, such as war or natural disaster, we may focus temporarily on our country but we rarely think about Earth as a whole or the human species as a whole. This narrow perspective, perhaps best exemplified by the American consumer, has led to the environmental degradation of our planet, a grossly widening gap in living standards between rich and poor people and nations and a scientific research agenda that focuses almost exclusively on the needs and desires of the wealthy few. Reversing the worldwide trends toward market-based atomization and increasing indifference to the suffering of others will require a human rights focus, forged by the development of what Vaclav Havel has termed a "species consciousness."[1]

In this Article we discuss human cloning and inheritable genetic alterations from the human species perspective, and suggest language for a proposed international "Convention of the Preservation of the Human Species" that would outlaw all efforts

---

† George J. Annas is the Edward R. Utley Professor and Chair of the Health Law Department, Boston University School of Public Health, Professor of SocioMedical Sciences and Community Medicine, Boston University School of Medicine, Professor of Law, Boston University School of Law, and cofounder of Global Lawyers and Physicians. A.B., Harvard College (1967); J.D., Harvard Law School (1970), M.P.H., Harvard School of Public Health (1972).

Lori B. Andrews is Distinguished Professor of Law, Chicago-Kent College of Law, and Director, Institute for Science, Law, and Technology, Illinois Institute of Technology. B.A., Yale College (1975); J.D., Yale Law School (1978).

Rosario M. Isasi is the Health Law and Bioethics Fellow, Health Law Department, Boston University School of Public Health. B.A., Pontificia Universidad Catolica del Peru (1987); J.D., Pontificia Universidad Catolica del Peru (1992); M.P.H., Boston University School of Public Health (2002).

1 "We still don't know how to put morality ahead of politics, science and economy. We are still incapable of understanding that the only genuine backbone of all our actions, if they are to be moral, is responsibility—responsibility to something higher than my family, my country, my company, my success." Vaclav Havel, *Excerpts from Czech Chief's Address to Congress*, N.Y. TIMES, Feb. 22, 1990, at A14. *See also* AMARTYA SEN, DEVELOPMENT AS FREEDOM (1999); George J. Annas, *Mapping the Human Genome and the Meaning of Monster Mythology*, 39 EMORY L.J. 629, 661-64 (1990).

to initiate a pregnancy by using either intentionally modified genetic material or human replication cloning, such as through somatic cell nuclear transfer. We summarize international legal action in these areas over the past five years, relate these actions to arguments for and against a treaty and conclude with an action plan.

II. HUMAN RIGHTS AND THE HUMAN SPECIES

The development of the atomic bomb not only presented to the world for the first time the prospect of total annihilation, but also, paradoxically, led to a renewed emphasis on the "nuclear family," complete with its personal bomb shelter. The conclusion of World War II (with the dropping of the only two atomic bombs ever used in war) led to the recognition that world wars were now suicidal to the entire species and to the formation of the United Nations with the primary goal of preventing such wars.[2] Prevention, of course, must be based on the recognition that all humans are fundamentally the same, rather than on an emphasis on our differences. In the aftermath of the Cuban missile crisis, the closest the world has ever come to nuclear war, President John F. Kennedy, in an address to the former Soviet Union, underscored the necessity for recognizing similarities for our survival:

> [L]et us not be blind to our differences, but let us also direct attention to our common interests and the means by which those differences can be resolved . . . . For, in the final analysis, our most basic common link is that we all inhabit this small planet. We all breathe the same air. We all cherish our children's future. And we are all mortal.[3]

That we are all fundamentally the same, all human, all with the same dignity and rights, is at the core of the most important document to come out of World War II, the Universal Declaration of Human Rights, and the two treaties that followed it (together known as the "International Bill of Rights").[4] The recognition of universal human rights, based on human dignity and equality as well as the principle of nondiscrimination, is fundamental to the development of a species consciousness. As Daniel Lev of Human Rights Watch/Asia said in 1993, shortly before the Vienna Human Rights Conference:

> Whatever else may separate them, human beings belong to a single biological species, the simplest and most fundamental commonality before which the significance of human differences quickly fades. . . . We are all capable, in exactly the same ways, of feeling pain, hunger,

---

[2] *See* THE CHARTER OF THE UNITED NATIONS: A COMMENTARY 49 (Bruno Simma et al. eds., 1995).

[3] Commencement Address at American University in Washington, PUB. PAPERS 459, 462 (June 10, 1963). President George W. Bush echoed Kennedy's words almost forty years later:

> All fathers and mothers, in all societies, want their children to be educated and live free from poverty and violence. No people on earth yearn to be oppressed or aspire to servitude or eagerly await the midnight knock of the secret police. . . . America will lead by defending liberty and justice because they are right and true and unchanging for all people everywhere.
>
> No nation owns these aspirations and no nation is exempt from them. We have no intention of imposing our culture, but America will always stand firm for the non-negotiable demands of human dignity: the rule of law; limits on the power of the state; respect for women; private property; free speech; equal justice; and religious tolerance.

Address Before a Joint Session of the Congress on the State of the Union, 38 WEEKLY COMP. PRES. DOC. 133, 138 (Jan. 29, 2002).

[4] *See* HENRY J. STEINER & PHILIP ALSTON, INTERNATIONAL HUMAN RIGHTS IN CONTEXT: LAW, POLITICS, MORALS 137-41 (2d ed. 2000).

> and a hundred kinds of deprivation. Consequently, people nowhere routinely concede that those with enough power to do so ought to be able to kill, torture, imprison, and generally abuse others. . . . The idea of universal human rights shares the recognition of one common humanity, and provides a minimum solution to deal with its miseries.[5]

Membership in the human species is central to the meaning and enforcement of human rights, and respect for basic human rights is essential for the survival of the human species. The development of the concept of "crimes against humanity" was a milestone for universalizing human rights in that it recognized that there were certain actions, such as slavery and genocide, that implicated the welfare of the entire species and therefore merited universal condemnation.[6] Nuclear weapons were immediately seen as a technology that required international control, as extreme genetic manipulations like cloning and inheritable genetic alterations have come to be seen today. In fact, cloning and inheritable genetic alterations can be seen as crimes against humanity of a unique sort: they are techniques that can alter the essence of humanity itself (and thus threaten to change the foundation of human rights) by taking human evolution into our own hands and directing it toward the development of a new species, sometimes termed the "posthuman."[7] It may be that species-altering techniques, like cloning and inheritable genetic modifications, could provide benefits to the human species in extraordinary circumstances. For example, asexual genetic replication could potentially save humans from extinction if all humans were rendered sterile by some catastrophic event. But no such necessity currently exists or is on the horizon.

As a baseline, if we take human rights and democracy seriously, a decision to alter a fundamental characteristic in the definition of "human" should not be made by any individual or corporation without wide discussion among all members of the affected population. No individual scientist or corporation has the moral warrant to redesign humans (any more than any individual scientist or corporation has the moral warrant to design a new, lethal virus or bacteria that could kill large numbers of humans). Altering the human species is an issue that directly concerns all of us, and should only be decided democratically, by a body that is representative of everyone on the planet.[8] It is the most important decision we will ever make.

The environmental movement has adopted the precautionary principle to help stem the tide of environmental alterations that are detrimental to humans. One version of this principle holds that "when an activity raises threats of harm to human health or the environment . . . the proponent of that activity, rather than the public, should bear the burden of proof [that the activity is more likely to be beneficial than harmful]."[9] The only way to shift the burden of proof is to outlaw potentially lethal

---

5 Quoted in MARY ANN GLENDON, A WORLD MADE NEW: ELEANOR ROOSEVELT AND THE UNIVERSAL DECLARATION OF HUMAN RIGHTS 223 (2001).

6 *See generally* M. CHERIF BASSIOUNI, CRIMES AGAINST HUMANITY IN INTERNATIONAL CRIMINAL LAW (1992) (exploring the history and evolution of "crimes against humanity"). *See also* George J. Annas, *The Man on the Moon, Immortality, and Other Millennial Myths: The Prospects and Perils of Human Genetic Engineering*, 49 EMORY L.J. 753, 778-80 (2000) (discussing the possibility of species-alteration becoming a new category of "crimes against humanity").

7 *See, e.g.*, FRANCIS FUKUYAMA, OUR POSTHUMAN FUTURE: CONSEQUENCES OF THE BIOTECHNOLOGY REVOLUTION (2002). Of course, these actions have not yet been recognized as crimes against humanity or any other type of international crime, and this is one reason why some still see these activities as legitimate.

8 Obviously, the only current candidate is the United Nations.

9 PROTECTING PUBLIC HEALTH AND THE ENVIRONMENT: IMPLEMENTING THE PRECAUTIONARY PRINCIPLE 354 (Carolyn Raffensperger & Joel A. Tickner eds., 1999).

activities, thus requiring proponents to change the law before proceeding. This can be done nation by nation, but can only be effective (because scientists and laboratories can move from country to country) by an internationally-enforceable ban. The actual text of a treaty banning human replicative cloning and inheritable modifications will be the subject of debate. We suggest the following language, obviously subject to future negotiations as well as added details, as a basis for going forward:

Convention on the Preservation of the Human Species

The Parties to this Convention,

Noting that the Charter of the United Nations affirms human rights, based on the dignity and worth of the human person and on equal rights of all persons;

Noting that the Universal Declaration of Human Rights affirms the right of every person not to be discriminated against;

Realizing that human dignity and human rights derive from our common humanity;

Noting the increased power of genetic science, which opens up vast prospects for improving health, but also has the power to diminish humanity fundamentally by producing a child through human cloning or by intentionally producing an inheritable genetic change;

Concerned that human cloning, which for the first time would produce children with predetermined genotypes, rather than novel genotypes, might cause these children to be deprived of their human rights;

Concerned that by altering fundamental human characteristics to the extent of possibly producing a new human species or subspecies, genetic science will cause the resulting persons to be treated unequally or deprived of their human rights;

Recognizing the history of abuses of human rights in the name of genetic science;

Believing that no individual, nation or corporation has the moral or legal warrant to engage in species-altering procedures, including cloning and genetic alteration of reproductive cells or embryos for the creation of a child;

Believing that the creation of a new species or subspecies of humans could easily lead to genocide or slavery; and

Stressing the need for global cooperation to prevent the misuse of genetic science in ways that undermine human dignity and human rights;

Have agreed on the following:

Article 1

Parties shall take all reasonable action, including the adoption of criminal laws, to prohibit anyone from initiating or attempting to initiate a human pregnancy or other form of gestation using embryos or reproductive cells which have undergone intentional inheritable genetic modification.

Article 2

Parties shall take all reasonable action, including the adoption of criminal laws, to prohibit anyone from utilizing somatic cell nuclear transfer or any other cloning technique for the purpose of initiating or attempting to initiate a human pregnancy or other form of gestation.

Article 3
Parties shall implement systems of national oversight through legislation, executive order, decree or other mechanism to regulate facilities engaged in assisted human reproduction or otherwise using human gametes or embryos for experimental or clinical purposes to ensure that such facilities meet informed consent, safety, and ethical standards.

Article 4
A Conference of the Parties and a Secretariat shall be established to oversee implementation of this Convention.

Article 5
Reservations to this Convention are not permitted.

Article 6
For the purpose of this Convention, the term "somatic cell nuclear transfer" shall mean transferring the nucleus of a human somatic cell into an ovum or oocyte. "Somatic cell" shall mean any cell of a human embryo, fetus, child or adult, other than a reproductive cell. "Embryo" shall include a fertilized egg, zygote (including a blastomere and blastocyst) and preembryo. "Reproductive cell" shall mean a human gamete and its precursors.[10]

---

[10] This proposed Convention is the product of many people, including the participants at a September 21-22, 2001 conference at Boston University on "Beyond Cloning: Protecting Humanity from Species-Altering Procedures." The treaty language was the subject of a roundtable that concluded the conference. The authors, together with others, most especially Patricia Baird and Alexander Morgan Capron, had drafted language to be considered at the conference, and revised it after the conference based on the discussion that occurred there and comments on the draft by others.

The original draft also included the following codicil to encourage individual countries to examine broader issues as well:

ISSUES FOR NATIONS TO CONSIDER IN FURTHERANCE OF THE CONVENTION ON THE PRESERVATION OF THE HUMAN SPECIES

In the course of discussions about the Convention on the Preservation of the Human Species, countries may desire to expand the provisions to deal in greater detail with other matters. Perhaps there will be a desire to add a moratorium on the creation of cloned human embryos for research. It may also be thought useful to include provisions that deal more comprehensively with assisted reproduction and life-science patents. Such provisions could take into consideration the following issues:

Assisted Human Reproduction

Potential Regulation:
The regulation of the practice of assisted human reproduction could include such provisions as requirements of a license for any healthcare professional who or healthcare facility that:

- Facilitates assisted human reproduction, e.g., via donor insemination or in vitro fertilization;
- Undertakes research or treatment using an in vitro embryo;
- Collects, stores, transfers, destroys, imports or exports sperm, ova or in vitro embryos for reproduction or research purposes; or
- Undertakes genetic screening on an *ex utero* embryo.

The regulation of the practice of assisted human reproduction could also include provisions to ensure:

- Free and informed consent of prospective parents and gamete donors as a prerequisite to the use of the techniques;
- Quality assurance and proficiency testing for labs;
- Reporting to a governmental entity the outcomes (including births per attempt and data about morbidity and mortality of the resulting children for the first five years) and

Perhaps the most difficult challenge in implementing this treaty is setting up the monitoring and enforcement mechanisms. Article Four would have to address these in detail. Although the specifics are beyond the scope of this article, some general comments are needed. Monitoring and compliance bodies must be broadly representative, possess authority to oversee activities related to human cloning and

---

disclosure to the public of this information;

- Non-misleading advertising (to the extent that advertising is permitted at all); and
- Confidentiality of individually identifiable health information.

Other prohibitions beyond those on human cloning and germline intervention might include bans on:

- Extracorporeal gestation of a human being;
- Transfer of a human embryo into an animal;
- Creation of embryos solely for research purposes; or
- Transplanting reproductive material (including gametes, ovaries or testes) from animals into humans.

Gene Patents

The purpose of the patent system is to encourage innovation and the development of products by providing the holder of a patent with a twenty-year monopoly over the use of an invention. Patenting genes runs counter to this purpose because gene patents are stifling innovation and impeding access to genetic diagnostic and treatment technologies. Many researchers who are searching for genes that predispose individuals to diseases are reluctant to share information and tissue samples with other researchers because they want to discover the gene themselves and to reap the financial rewards of discovery. These rewards can be high. For example, one particular gene patent in the United States is worth $1.5 billion annually.

Once a gene is discovered, the patent holder can prevent any doctor or laboratory from even checking a person's body to see if he or she has a mutation of the gene. Alternatively, the patent holder can collect a very high royalty from the doctors and laboratories that examine the gene. The patent holder can even stop any use of the patented gene. One patent holder, for example, will not permit the use of its gene in prenatal screening because of the controversy surrounding abortion. Another patent holder, a major European pharmaceutical company, will not allow anyone to use its patented gene to develop a test which shows which patients will benefit from one of the company's drugs and which will not. Another biotechnology company has a patent on the genetic sequence of a particular infectious disease and is stopping another company from instituting inexpensive public health screening to determine if people are infected.

On the other hand, patent holders themselves may encourage premature adoption of genetic diagnostic tests and unsafe efforts at gene transfer experiments to benefit the patent holder rather than patients or research subjects. Moreover, special issues are raised in the case of patenting human tissue, including the ethical and legal propriety of ownership of one person's genetic information by another.

Potential regulation:

No patents shall be granted on human genes, parts of human genes or unaltered products of human genes, nor on the genes of bacteria, viruses or other infectious agents that cause disease in humans.

Work on a national regulatory scheme for the new reproductive technologies will, of course, be most relevant to countries that have an in-vitro fertilization (IVF) industry. We also believe that the best existing guidance for approaching such regulation is contained in the final report of Canada's Royal Commission on New Reproductive Technologies. 1 Patricia Baird, PROCEED WITH CARE: FINAL REPORT OF THE ROYAL COMMISSION ON NEW REPRODUCTIVE TECHNOLOGIES, 564-76 (1993). Also, to the extent that a country wants to proceed with research cloning (e.g., for the purpose of making stem cells or studying embryonic growth), regulation of the infertility industry will be needed to prevent a cloned embryo from being implanted in a woman's uterus. Such regulation could include, for example, the prohibition of freezing cloned embryos, and the prohibition of any physician or embryologist involved in IVF from making or possessing a cloned embryo.

human genetic modification, and be able to enforce bans by announcing and denouncing potential violators. Moreover, we believe the commission (and the countries themselves) should support, through the Convention and through their national criminal laws, the establishment of two new international crimes: initiation of a pregnancy to create a human clone and initiation of a pregnancy using a genetically-altered embryo. [11]

III. AN INTERNATIONAL CONVENTION: WHY NOW?

Five years after the announcement of the cloning of Dolly the sheep it is time to ask not *if* cloning and inheritable alterations should be regulated, but *how*. Had a five-year moratorium for further thought and discussion been placed on cloning humans, as the National Bioethics Advisory Commission (NBAC) recommended in 1997, for example, the time would now have expired.[12] What new have we learned in the last five years?

First, virtually every scientist in the world with an opinion believes it is unsafe to attempt a human pregnancy with a cloned embryo.[13] This is, for example, the unanimous conclusion of a 2002 report from the U.S. National Academy of Sciences, which recommended that human "reproductive" cloning be outlawed in the United States following a study that included the viewpoints of the only two scientists in the world who publicly advocate human cloning today.[14] Although scientists seldom like

---

11 While we believe these crimes should be subject to the jurisdiction of the International Criminal Court, this may not be possible in the near future, and it is more important to establish them as international crimes than to broaden the definition of "crimes against humanity" as it applies to the International Criminal Court at this time.

12 NAT'L BIOETHICS ADVISORY COMM'N, CLONING HUMAN BEINGS 109 (1997), *available at* http://bioethics.georgetown.edu/nbac/pubs.html.

13 For example, during the 1998 debate on cloning in the U.S. Senate, more than sixteen scientific and medical organizations, including the American Society of Reproductive Medicine and the Federation of American Societies for Experimental Biology (which includes more than sixteen scientific and medical organizations), believed that there should be a moratorium on the creation of humans by cloning. *See* 144 CONG. REC. S434-38 (1998); 144 CONG. REC. S661 (1998). None of these organizations has since changed their position. *See, e.g.*, Press Release, AM. SOC'Y REPRODUCTIVE MED., *ASRM Statement on Attempts at Human Cloning* (Apr. 5, 2002) ("[W]e caution policy makers not to be rushed into approving over-reaching legislation that will criminalize valid scientific and medical research and the therapies they might lead to.") *available at* http://www.asrm.org/Media/Press/cloningstatement4-02; Letter from Carl B. Feldbaum, President, Biotechnology Ind. Org., to President George W. Bush (Feb. 1, 2002) ("The current moratorium on cloning humans should remain until our nation has had time to fully explore the impact of such cloning."), *available at* http://www.bio.org/bioethics/cloning_letter_bush.html. *See also* Rudolf Jaenisch & Ian Wilmut, *Don't Clone Humans!*, 291 SCIENCE 2552 (2001) ("We believe attempts to clone human beings at a time when the scientific issues of nuclear cloning have not been clarified are dangerous and irresponsible."); Editorial, *Reasons to be Cloned*, 414 NATURE 567 (2001) ("[T]he health risks to mother and child inherent in [cloning] . . . demand that it be banned.").

14 NAT'L RESEARCH COUNCIL, SCIENTIFIC AND MEDICAL ASPECTS OF HUMAN REPRODUCTIVE CLONING 1 (2002).

> Human reproductive cloning . . . is dangerous and likely to fail. The panel therefore unanimously supports the proposal that there should be a legally enforceable ban on the practice of human reproductive cloning. . . . The scientific and medical considerations related to this ban should be reviewed within 5 years. The ban should be reconsidered only if at least two conditions are met: (1) a new scientific and medical review indicates that the procedures are likely to be safe and effective and (2) a broad national dialogue on the societal, religious, and ethical issues suggests that a reconsideration of the ban is warranted.

Id. at ES-1 to ES-2. *See also* NAT'L RESEARCH COUNCIL, STEM CELLS AND THE FUTURE OF REGENERATIVE MEDICINE (2001).

to predict the future without overwhelming data to support them, many believe that human cloning or inheritable genetic alternations at the embryo level will never be safe because they will always be inherently unpredictable in their effects on the children and their offspring. As Stewart Newman has noted, for example, it is unlikely that a human created from the union of "two damaged cells" (an enucleated egg and a nucleus removed from a somatic cell) could ever be healthy.[15] Of course, adding genetic modification to the somatic cell's nucleus just adds another series of events that could go wrong, because genes seldom have a single function, but will usually interact in complex and unpredictable ways with other genes.[16] It is worth underlining that the dangers are not just physical, but also psychological. Whether cloned children could ever overcome the psychological problems associated with their origins is unknown and perhaps unknowable.[17] In short, the safety issues, which inherently make attempts to clone or genetically alter a human being unethical human experiments, provide sufficient scientific justification for the treaty alone.

If and when safety can be assured, assuming this will ever be possible, two primary arguments have been set forth in favor of proceeding with cloning (and its first cousin, inheritable genetic alterations). First, cloning is a type of human reproduction that can help infertile couples have genetically-related children. Second, cloning is a part of human "progress" that could lead to a new type of genetic immortality, therefore, to prevent it is to be anti-scientific.

The infertility argument is made by physiologist Panos Zavos and his former Italian colleague, infertility specialist Severino Antinori. They argue that the inability of a sterile male to have a genetically-related child is such a human tragedy that it justifies human cloning.[18] This view not only ignores the rights and interests

---

15 Stuart A. Newman, Speech at the "Beyond Cloning" Conference, Boston University (Sept. 21, 2001).

16 *See, e.g.*, Jon W. Gordon, *Genetic Enhancement in Humans*, 283 SCIENCE 2023, 2023 (1999).

17 Hans Jonas, for example, argued that it is a crime against the clone by depriving the cloned child of his or her "existential right to certain subjective terms of [ ] being." HANS JONAS, PHILOSOPHICAL ESSAYS: FROM ANCIENT CREED TO TECHNOLOGICAL MAN 160 (1974). Jonas believes that a clone will not have a "right to ignorance" or the "right . . . to a unique genotype." *Id.* Instead, a clone knows:

> [A]ltogether too much about himself and is known . . . altogether too well by others. Both facts are paralyzing for the spontaneity of becoming himself . . . . [T]he clone is antecedently robbed of the freedom which only under the protection of ignorance can thrive: and to rob a human-to-be of that freedom deliberately is an inexplicable crime that must not be committed even once.

Id. at 161. Human reproductive cloning poses both physical and psychological risks to children who might be conceived using this technique. In animals, cloning currently only results in a successful pregnancy three to five percent of the time. And, even in those rare instances, many of the resulting offspring suffer—one-third die shortly before or right after birth. Other cloned animals seem perfectly healthy at first and then suffer heart and blood vessel problems, underdeveloped lungs, diabetes, immune system deficiencies and severe growth abnormalities. The mothers who gestate clones are also at risk, due to the often abnormally large size of the offspring produced—some cattle clones for example, are born up to twice the normal weight expected for calves.

18 Tim Adams, *Interview: The Clone Arranger*, THE OBSERVER, Dec. 2, 2001, at 3 (comments of Severino Antinori).

> Male infertility grows . . . . My invention of ICSI has helped. I have helped men whose sperm are misformed or too slow. I have helped men whose sperm does not come out from their testes! And the next step [cloning] is to help men who—traumatico!—have lost their ability to produce any sperm at all. Through war or accident or cancer. I will help only stable, loving couples. Some doctors say this is a step too far, but those same doctors have said that about all the other steps too. Very few doctors are pioneers! Very few have both the knowledge and the, the, the . . . courage.

Id. See also ROBERT WINSTON, *The Promise of Cloning for Human Medicine*, 314 BRIT. MED. J. 913

of women and children (even if only males are to be cloned, eggs must be procured from a woman, the embryos must be gestated by a woman and the child is the subject of the experiment), but also contains a highly-contested assertion: that asexual genetic replication or duplication should be seen as "human reproduction."[19] In fact, humans are a sexually reproducing species and have never reproduced or replicated themselves asexually.

Asexual replication may or may not be categorized by future courts as a form of human reproduction, but there are strong arguments against it. First, asexual reproduction changes a fundamental characteristic of what it means to be human (i.e., a sexually reproducing species) by making sexual reproduction involving the genetic mixture of male and female gametes optional. Second, the "child" of an asexual replication is also the twin brother of the male "parent," a relationship that has never existed before in human society. The first clone, for example, will be the first human being with a single genetic "parent" (unless the biological grandparents are taken to be the actual "parents" of the clone).[20] Third, the genetic replica of a genetically sterile man would be sterile himself and could only "reproduce" by cloning. This

---

(1997) (advocating for the use of cloning to help infertile men have genetically-related children). Zavos and Antinori dissolved their partnership in 2002. David Brown, *Human Clone's Birth Predicted*, WASH. POST, May 16, 2002, at A8.

19 *See generally*, Michael H. Shapiro, *I Want a Girl (Boy) Just Like the Girl (Boy) that Married Dear Old Dad (Mom): Cloning Lives*, 9 S. CAL. INTERDISC. L.J. 1 (arguing, in part, that cloning should be considered reproduction, for the essence of reproduction is the creation of a new person).

A variety of personal desires may interest people in creating a child through cloning or germline genetic engineering. The NBAC report suggests it would be "understandable, or even, as some have argued, desirable" to create a cloned child from one adult if both members of the couple have a lethal recessive gene; from a dying infant if his father is dead and the mother wants an offspring from her late husband; or from a terminally ill child to create a bone marrow donor. CLONING HUMAN BEINGS, *supra* note 12, *at* 78-80. Some of the experts testifying before the NBAC suggested that cloning should be appropriate in exceptional circumstances. Rabbi Elliot Dorff opined that it would be "legitimate from a moral and a Jewish point of view" to clone a second child to act as a bone marrow donor so long as the "parents" raise that second child as they would any other. *Id.* at 55. Rabbi Moshe Tendler raised the scenario of a person who was the last in his genetic line and whose family was wiped out in the Holocaust. "I would certainly clone him," said Tendler. *Id.* For other Jewish perspectives supporting cloning, see Peter Hirschberg, *Be Fruitful and Multiply and Multiply and Multiply*, JERUSALEM REP., Apr. 16, 1998, at 33. In contrast, the Catholic viewpoint is that cloning "is entirely unsuitable for human procreation even for exceptional circumstances." CLONING HUMAN BEINGS, *supra* note 12, at 55.

20 Before a U.S. Senate Committee, which also heard from Ian Wilmut shortly after he had announced the birth of Dolly, one of us made the argument that a human clone would be the first human being with one genetic parent. *Testimony on Scientific Discoveries and Cloning: Challenges for Public Policy, Before the Sen. Subcomm. on Public Health and Safety, Sen. Comm. on Labor and Human Resources*, 105th Cong. 25 (1997) (statement of George J. Annas), *available at* http://www.bumc.bu.edu/www/sph/lw/pvl/Clonetest.htm. Population geneticist Richard Lewontin challenged this assertion, writing:

> A child by cloning has a full set of chromosomes like anyone else, half of which were derived from a mother and half from a father. It happens that these chromosomes were passed through another individual, the cloning donor, on the way to the child. The donor is certainly not the child's "parent" in any biological sense, but simply an earlier offspring of the original parents.

R.C. Lewontin, *Confusion over Cloning*, N.Y. REV. BOOKS, Oct. 23, 1997, at 20.

It should be noted that Lewontin's position takes genetic reductionism to its extreme: people become no more than containers of their parent's genes, and their parents have the "right" to treat them not as individual human beings, but rather like embryos—entities that they can "split" or "replicate" without consideration of the child's choice or welfare. Children, even adult children, under this view have no say as to whether or not they are replicated because it is their "parents," not them, who are "reproducing." This radical redefinition of reproduction and the denial to children of the choice to procreate or not turns out to be an even stronger argument against cloning children than its biological novelty. GEORGE J. ANNAS, SOME CHOICE: LAW, MEDICINE & THE MARKET 13 (1998).

means either that infertility is not a major problem (because if it were, it would be unethical for a physician to intentionally create a child with this problem), or that the desire of existing adults should take precedent over the welfare of children. We find neither conclusion persuasive, and this is probably why, although some ethicists believe that cloning could be considered a form of human reproduction, infertility specialists have not joined Antinori's call for human cloning as a treatment for infertile males. In fact the organization that represents infertility specialists in the United States, and is generally opposed to the regulation of the infertility industry, the American Society of Reproductive Medicine, has nonetheless consistently opposed human cloning.[21]

There are, nonetheless, legal commentators who believe that human cloning should be classified as a form of human reproduction, and protected as such, at least if it is the only way for an individual to have a "genetically-related child." The strongest proponent of this view is probably John Robertson,[22] although Ronald Dworkin[23] shares his enthusiasm as well. Suffice it to say here that it is very unclear that human reproduction or procreation of a kind protected by principles of autonomy and self-fulfillment can be found in a "right to have a genetically-related child." It cannot be just the genetic tie that is important in human reproduction, because if it were, this could be accomplished by having one's twin brother have a child with one's wife[24]—the genetic tie would be identical, yet few, if any, would argue that this method of reproduction should satisfy the twin's right to have a "genetically-related child." Genes are important, but there is more to human reproduction, as protected by the U.S. Constitution, than simple genetic replication.

The second major argument in favor of human cloning is that it can lead to a form of immortality. This is the premise of the Raelian cult that has chartered its own corporation, Clonaid, to engage in human cloning. The leader of the cult, who calls himself Rael (formerly Claude Vorilhon, the editor of a French motor sport magazine), believes that all humans were created in the laboratories of the planet Elohim and that the Elohims have instructed Rael and his followers to develop cloning on Earth to provide earthlings with a form of immortality.[25] The Raelians, of course, can believe whatever they want to; but just as human sacrifice is illegal, experiments that pose a significant danger to women and children can also be outlawed,[26] and the religious beliefs of this cult do not provide a sufficient justification to refrain from outlawing cloning.

Just as two primary arguments in favor of cloning and inheritable genetic alterations have emerged over the past five years, so have two basic arguments about the future regulation of these technologies. The first, exemplified by Lee Silver, is that these technologies, while not necessarily desirable, are unstoppable because the market combined with parental desire will drive scientists and physicians to offer these services to demanding couples.[27] Similar to the way parents now seek early educational enrichment for their children, he believes that parents of the future will

---

21 *See, e.g.*, ETHICS COMM., AM. SOC'Y REPRODUCTIVE MED., *Human Somatic Cell Nuclear Transfer (Cloning)*, 74 FERTILITY & STERILITY 873, 873-76 (2000).

22 John A. Robertson, *Two Models of Human Cloning*, 27 HOFSTRA L. REV. 609 (1999).

23 RONALD DWORKIN, SOVEREIGN VIRTUE: THE THEORY AND PRACTICE OF EQUALITY 437-42 (1997).

24 Leon Kass made this point in another context. LEON R. KASS, TOWARD A MORE NATURAL SCIENCE: BIOLOGY AND HUMAN AFFAIRS 110-111 (1985).

25 *See* RAEL, THE TRUE FACE OF GOD (Int'l Raelian Movement 1998).

26 *See* JAY KATZ, EXPERIMENTATION WITH HUMAN BEINGS (Russell Sage Found. 1972).

27 LEE SILVER, REMAKING EDEN: CLONING AND BEYOND IN A BRAVE NEW WORLD 123 (1997).

seek early genetic enhancement to give them a competitive advantage in life. Silver thinks this will ultimately lead to the creation of two separate species or subspecies, the GenRich and "the naturals."[28]

A related "do nothing" argument is that regulation may not be needed because the technologies will not be widely used. The thought is that humans may muddle through, either because the science of human genetic alterations may never prove possible, or because it will be used by only a handful of humans because most will instinctively reject it. Colin Tudge, a proponent of this argument, also accepts Silver's argument that the market is powerful and often determinative, but nonetheless believes that the three fundamental principles of all religions—personal humility, respect for fellow humans and reverence for the universe as a whole—could lead the vast majority of humans to reject cloning and genetic alterations.[29] In his words:

> The new technologies, taken to extremes, threaten the idea of humanity. We now need to ask as a matter of urgency who we really are and what we really value about ourselves. It could all be changed after all—we ourselves could be changed—perhaps simply by commercial forces that we have allowed to drift beyond our control. If that is not serious, it is hard to see what is.[30]

We agree with Tudge that the issues are serious. We think that they are too serious to be left to religions or human instinct, or even to individual national legislation, to address.

In this regard, we find a second approach, that of a democratically-formed regulatory scheme more reasonable. Indeed, in our view the widespread condemnation of human replicative cloning by governments around the world means that cloning provides a unique opportunity for the world to begin to work together to take some control over the biotechnology that threatens our very existence.[31]

The primary arguments against cloning and inheritable genetic alterations, which we believe make an international treaty the appropriate action, have been summarized in detail elsewhere. In general, the arguments are that these interventions would require massive dangerous and unethical human experimentation,[32] that cloning would inevitably be bad for the resulting children by restricting their right to an "open future,"[33] that cloning would lead to a new eugenics movement for "designer children" (because if an individual could select the entire genome of their future child, it would seem impossible to prohibit individuals from choosing one or more specific genetic characteristics of their future children),[34] and that it would likely lead to the creation of a new species or subspecies of humans,

---

28 *Id.* at 4.

29 COLIN TUDGE, THE IMPACT OF THE GENE: FROM MENDEL'S PEAS TO DESIGNER BABIES 4 (2000).

30 *Id.* at 342. *See also* IAN WILMUT ET. AL., THE SECOND CREATION: DOLLY AND THE AGE OF BIOLOGICAL CONTROL 267-98 (2000) (discussing the implications of cloning for humankind).

31 *See* the appendix to this Article for current national laws on human cloning and inheritable modifications.

32 *See generally* THE NAZI DOCTORS AND THE NUREMBERG CODE 3 (George J. Annas & Michael A. Grodin eds., 1992) (exploring the "history, context, and implications of the Doctor's Trial at Nuremberg and the impact of the Nuremberg Code on subsequent codes of research ethics and international human rights").

33 JONAS, *supra* note 17, at 161-62.

34 It is in this sense that children become "manufactured" products. *See* KASS, *supra* note 24, at 71-73.

sometimes called the "posthuman."[35] In the context of the species, the last argument has gotten the least attention, and so it is worth exploring.

Specifically, the argument is that cloning will inevitably lead to attempts to modify the somatic cell nucleus not to create genetic duplicates of existing people, but "better" children.[36] If this attempt fails, that is the end of it. If it succeeds, however, something like the scenario envisioned by Silver and others such as Nancy Kress,[37] will unfold: a new species or subspecies of humans will emerge. The new species, or "posthuman," will likely view the old "normal" humans as inferior, even savages, and fit for slavery or slaughter. The normals, on the other hand, may see the posthumans as a threat and if they can, may engage in a preemptive strike by killing the posthumans before they themselves are killed or enslaved by them. It is ultimately this predictable potential for genocide that makes species-altering experiments potential weapons of mass destruction, and makes the unaccountable genetic engineer a potential bioterrorist. It is also why cloning and genetic modification is of species-wide concern and why an international treaty to address it is appropriate.[38] Such a treaty is necessary because existing laws on cloning and inheritable genetic alterations, although often well-intentioned, have serious limitations.

IV. INTERNATIONAL RESTRICTIONS ON CLONING AND GENETIC MODIFICATIONS

Despite the fact that no children have been born as a result of these species-altering interventions, policymakers around the world have expressed concerns about the use of these technologies. Some countries' lawmakers have enacted bans on these proposed experimental technologies, while others have assumed that existing laws apply to the techniques. However, both categories of laws have shortcomings.

A. MORATORIA

Some countries have approached species-altering procedures with caution, instituting moratoria in order to consider the wide range of impacts of the technologies. Israel, for example, has stated that the purpose of such moratoria is to have time "to examine the moral, legal, social and scientific aspects of such types of

---

35 *See* Annas, *supra* note 6, at 776-780; Fukuyama, *supra* note 7, at 22; Francis Fukuyama, *Natural Rights and Human History*, NAT'L INTEREST, Summer 2001, at 19, 30. For arguments favoring inheritable genetic modifications, see, for example, ENGINEERING THE HUMAN GERMLINE (Gregory Stock & John Campbell eds., 2000); GREGORY STOCK, REDESIGNING HUMANS (2002) (arguing, among other things, that it is inherent in our human nature to want to change our human nature, and that an international treaty would be unenforceable because every nation would have an economic incentive to defect and capture the market for inheritable modifications).

36 *See, e.g.*, WILMUT ET AL., *supra* note 30, at 5-6 (discussing how the post-Dolly experiments were designed to use cloning techniques to make "better animals," which was always Ian Wilmut's and Keith Campbell's plan for cloning technology). *See also* Angelika E. Shnieke et al., *Human Factor IX Trans-genic Sheep Produced by Transfer of Nuclei from Transplanted Fetal Fibroblasts*, 278 SCIENCE 2130 (1997).

37 See, for example, Nancy Kress's *Beggars* series: BEGGARS IN SPAIN (1993); BEGGARS AND CHOOSERS (1994); and BEGGAR'S RIDE (1996).

38 *See* Annas, *supra* note 6, at 778-81. An alternative scenario, that sees equal access to genetic "improvement" by all seems like pie in the sky to us in a world where fewer than ten percent of the population has access to contemporary medical care, and even in the world's richest country, more than forty millior people lack health insurance. We do not think it is reasonable to even discuss equal access to genetic alterations until all members of the species have access to current medical technologies as a matter of right.

intervention and their implication on human dignity."[39] In 1998, Israel adopted a five-year moratorium on cloning a human being, defining cloning as "the creation of an entire human being, who is genetically identical to another person or fetus, alive or dead."[40] The same law banned interventions to create a child through the use of reproductive cells that have undergone a permanent intentional genetic modification.[41] Some countries are using that same time period to consider the wealth of issues involved in species-altering procedures.[42] Others, though, have already determined that such technologies are inimical to human values and human dignity.[43]

B. LIMITATIONS IN HUMAN CLONING BANS

Some countries have attempted to ban human cloning, but have used language that inadvertently creates ambiguities. In other countries, policymakers may believe that their laws ban human cloning, but that may not be the case. Japan, for example, explicitly and clearly bans cloning.[44] Germany bans attempts to bring to birth a human embryo having the same genetic information as another embryo.[45] Spain, Victoria, Australia and Western Australia prohibit cloning and other procedures that bring about the birth of an identical human being.[46] But because cloning includes mitochondrial DNA from the woman whose egg is used, the clone will not have a completely identical genome (unless a woman clones herself and uses her own egg) and thus the practice of cloning may not be adequately banned.

Some countries ban embryo research,[47] but cloning through the Dolly technique of somatic cell nucleus transfer (SCNT) may not be viewed as embryo research. The SCNT technique utilizes an experimental procedure involving an egg to create an embryo.[48] Once the embryo has been created, no experimental technique is necessary. The resulting embryo can be implanted into a woman using the same standard clinical technique as is used in the in vitro fertilization (IVF) process.

British lawmakers thought they had a ban in place to prevent human cloning. The British have created a regulatory structure for IVF and related technology under the Human Fertilisation and Embryology Act of 1990 (HFEA).[49] The statute requires

---

39 Prohibition of Genetic Intervention Law No. 5759 (1998).

40 *Id.* § 3(1).

41 *Id.* § 3(2).

42 *See* Ania Lichtarowicz, *Scientist Warns on Human Cloning*, BBC NEWS, *at* http://news.bbc.co.uk/hi/English/ world/Europe/newsid_1719000/1719195.stm (Dec. 21, 2001) (noting that Spain and Belgium are still considering different types of legislation for adoption).

43 *Britain to Ban Human Cloning*, CNN.COM, *at* http://www.cnn.com/2001/WORLD/europe/UK/04/19/cloning. legislation/index.html (Apr. 19, 2001). *See also* Human Reproductive Cloning Act 2001, U.K. Stat. 2001, ch. 23, Enactment Clause (Eng.). (stating that the law "prohibit[s] the placing in a woman of a human embryo which has been created otherwise than by fertilisation").

44 *Ministry Bans Cloning Technology for Humans*, DAILY YOMIURI, July 29, 1998, at 2.

45 Gesetz zum Schutz von Embryonen (Embryonenschutzgesetz), v.13.12.1990 (BGBI. I S.2747). [Federal Embryo Protection Law].

46 Manipulacion Gentica y Reproduccion [Genetic Manipulation and Reproduction]; Victoria Infertility Treatment Act, 2000; Human Reproductive Technology Act, 1991, § 7(1)(d)(i) (W. Austl.).

47 *See, e.g.*, *The Logical Next Step? An International Perspective on the Issues of Human Cloning and Genetic Technology*, 4 ILSA J. INT'L & COMP.L. 697, 721-25 (1998).

48 *See, e.g.*, Valerie S. Rup, *Human Somatic Cell Nuclear Transfer Cloning, the Race to Regulate, and the Constitutionality of the Proposed Regulations*, 76 U. DET. MERCY L. REV. 1135, 1138-39 (1999); Christine Willgoos, Note, *FDA Regulation: An Answer to the Questions of Human Cloning and Germline Gene Therapy*, 27 AM. J.L. & MED. 101, 103 (2001).

49 Human Fertilisation and Embryology Act, 1990, ch. 37, Enactment Clause (Eng.). *See generally* Ruth Deech, *The Legal Regulation of Infertility Treatment in Britain*, *in* CROSSCURRENTS:

that activities falling within the act, such as the creation, storage, handling and use of human embryos outside of the body, must only be undertaken in licensed facilities.[50] Only activities enumerated in the Act, or approved by the Human Fertilisation and Embryology Authority, may be undertaken.[51] Certain activities, such as placing a human embryo in an animal, are completely prohibited. The British Act defines an "embryo" as a "live human embryo where fertilisation is complete" or "an egg in the process of fertilisation."[52] British lawmakers assumed human reproductive cloning was prohibited under the Act because it was not listed as an allowable activity with human embryos.[53]

In November 2000, the Pro-Life Alliance brought suit claiming that embryos created through cloning are not covered by HFEA. On November 15, 2001, the British High Court of Justice, Queen's Bench Division, Administrative Court, ruled in favor of the Pro-Life Alliance. The judge said, "With some reluctance, since it would leave organisms produced by CNR [cell nuclear replacement] outside the statutory and licensing framework, I have come to the conclusion that to insert these words would involve an impermissible rewriting and extension of the definition."[54] In response, Parliament passed new legislation just two weeks after the ruling making it an offense, punishable by up to ten years in prison, for a person to place "in a woman a human embryo which has been created otherwise than by fertilisation."[55] Ultimately, a higher court ruled that a human embryo created by cloning was in fact covered by HFEA.[56]

C. BANS ON INHERITABLE GENETIC INTERVENTIONS

Internationally, the bans on inheritable or germline genetic interventions are general enough to reach a wide range of technologies. These laws reflect a profound understanding of the need to avoid the social pressures to engineer a "better" race, as occurred in the Nazi era. German law understandably forbids germline intervention.[57] Victoria, Australia, in its Infertility Treatment Act of 1995, has comprehensive language prohibiting germline genetic alterations.[58] The law prohibits altering the genetic constitution of gametes[59] or altering the genetic, pro-nuclear or nuclear constitution of a zygote.[60] A Western Australia law prohibits the alteration of the genetic structure of an egg in the process of fertilization or an embryo.[61] In Norway, a 1994 law provides that the "human genome may only be altered by means of somatic gene therapy for the purpose of treating serious disease

---

FAMILY LAW AND POLICY IN THE U.S. AND ENGLAND 165-86 (Sanford Katz et al., eds, 2000).

50 Human Fertilisation and Embryology Act. ch. 27, §§ 3, 12.

51 *Id.* ch. 37, § 41.

52 *Id.* ch. 37, Enactment Clause.

53 The Act also had a ban, predating Dolly, on the replacement of the nucleus of a human embryo cell with that of any person or embryo, but that prohibition does not cover somatic cell nucleus transfer into a human egg.

54 Pro-Life Alliance v. Sec'y State for Health, CO/4095/2000 (Q.B. 2001), *available at* 2001 WL 1347031.

55 Human Reproductive Cloning Act, 2001, U.K. Stat. 2001 ch 23 § 1.

56 R (Quintavalle) v. Sec'y of State for Health, 2 WLR 550 (C.A. 2002), *reprinted at Cell Nuclear Replacement Organism is "Embryo,"* THE TIMES (London), Jan. 25, 2002.

57 Federal Embryo Protection Law, 1990 (Eng.)

58 Victoria Infertility Treatment Act, 1995.

59 Federal Embryo Protection Law, 1990 (Eng.), at Part 5, § 39(1).

60 Federal Embryo Protection Law, 1990 (Eng.), at Part 5, § 39(2).

61 Human Reproductive Technology, 1991, § 7(1)(j) (Austl.).

or preventing serious disease from occurring."[62] Sweden prohibits research that attempts to modify the embryo.[63] France, too, prohibits such interventions.[64] Costa Rica bans any manipulation or alternation of an embryo's genetic code.[65]

V. THE LEGAL STATUS OF HUMAN CLONING AND GERMLINE GENETIC INTERVENTION IN THE UNITED STATES

In 1997 President Clinton issued an executive order banning the use of federal funds for human cloning.[66] However, such a ban has little effect on private fertility clinics. For twenty years, the federal government has refused to provide funds for research on IVF, but that has not stopped the hundreds of privately-financed IVF clinics from creating tens of thousands of babies. The ban on federal funding of embryo research and human cloning does not, of course, apply to scientists who wish to undertake either activity with private funds.

A. THE APPLICATION OF U.S. LAWS BANNING EMBRYO RESEARCH TO HUMAN CLONING AND INHERITABLE GENETIC INTERVENTION

Existing American laws banning embryo research, dating back in some states to the mid-1970s, could potentially be used to prohibit certain species-altering technologies at the experimental stage.[67] Eleven states have laws regulating research and/or experimentation on conceptuses, embryos, fetuses or unborn children that use broad enough language to apply to early embryos.[68] It should be noted, however, that these bans would not apply once the techniques are no longer considered to be research and instead are thought of as standard practice.

Several arguments could be made to suggest that most of the embryo research statutes should be construed narrowly so as not to apply to cloning. Eight of the eleven states prohibit some form of research on some product of conception, referred

---

62 The Act Relating to the Application of Biotechnology in Medicine, ch. 7.

63 Law No. 115 of March 14, 1991, Act Concerning Measures for the Purposes of Research or Treatment in Connection with Fertilized Human Oocytes (1993).

64 Law No. 94-654 of July 29, 1994, on the Donation and Use of Elements and Products of the Human Body, Medically Assisted Procreation, and Prenatal Diagnosis.

65 Decree No. 24029-S: A Regulation on Assisted Reproduction, Feb. 3, 1995.

66 *See* Memorandum on the Prohibition on Federal Funding for Cloning of Human Beings, 33 WEEKLY COMP. PRES. DOC. 281 (Mar. 4, 1997); *see also Transcript of Clinton Remarks on Cloning*, U.S. NEWSWIRE, Mar. 4, 1997, *available at* 1997 WL 5711155.

67 Yet despite the risks, only six states—California, Iowa, Louisiana, Michigan, Rhode Island and Virginia—have passed legislation that prohibits human reproductive cloning. CAL. HEALTH & SAFETY CODE ANN. § 24185 (West 2002); IOWA CODE § 707B, CSB 218 (S.F. 2118) (2002); LA. REV. STAT. 40:1299.36.2 (West 2002); MICH. COMP. LAWS ANN. § 750.430a (West 2001); R.I. GEN. LAWS § 23-16.4 (2001); VA. CODE ANN. §§ 32.1-162.21, 162.22 (Michie 2002). In addition, Missouri prohibits the use of any state funds to bring about the birth of a child via cloning techniques. MO. ANN. STAT. § 1.217 (West 2002). The U.S. House of Representatives in July 2001 voted to ban human cloning. *See* The Human Cloning Prohibition Act of 2001, H.R. 2505, 107th Cong. (2001). *See also* Sheryl Gay Stolberg, *House Backs Ban on Human Cloning for any Objective*, N.Y. TIMES, Aug. 1, 2001, at A1. At the time of this writing, the U.S. Senate was scheduled to consider this issue in 2002.

68 FLA. STAT. ANN. § 390.0111(6) (West 2002); LA. REV. STAT. ANN. § 9:121-129 (West 2002); ME. REV. STAT. ANN. tit. 22, § 1593 (2002); MASS. GEN. LAWS ANN. ch. 112, § 12J West (2002); MICH. COMP. LAWS ANN. § 333.2685-.2692 (West 2002); MINN. STAT. § 145.421 (2001); N.D. CENT. CODE § 14-02.2-01 (2001); N.H. REV. STAT. ANN. § 168-B:15 (2002); 18 PA. CONS. STAT. § 3216 (2001); R.I. GEN. LAWS § 11-54-1 (2001). A South Dakota law bans research that destroys an embryo, when such research has not been undertaken to preserve the life and health of the particular embryo. S.D. CODIFIED LAWS § 34-14-18 (Michie 2001).

to in the statutes as a conceptus,[69] embryo,[70] fetus[71] or unborn child.[72] With cloning, an argument could be made that the experimentation is being done on an egg, not the product of conception, and thus these statutes should not apply.[73] By the time the egg is re-nucleated, the experiment or research has already been completed and the resulting embryo could be implanted under standard practices, as with IVF.

Moreover, two of the eleven states define the object of protection—the conceptus (Minnesota) or unborn child (Pennsylvania)—as the product of fertilization. If transfer of nucleic material is not considered fertilization (as was the case in the initial court decision in England), then these laws would not apply. In addition, at least eight of the states banning embryo research are sufficiently general that they might be struck down as unconstitutionally vague.[74]

Under New Hampshire's embryo research law, a researched-upon pre-embryo may not be transferred to a uterine cavity.[75] Thus, if a re-nucleated oocyte is considered to be a pre-embryo and if cloning is considered to be research, it would be impermissible in New Hampshire to implant the resulting conceptus to create a child. Possibly as a result of the deficiencies in the embryo research laws, three of the states with embryo research bans have new laws banning cloning.[76]

The embryo research bans could potentially affect the practice of inheritable genetic alterations. Under these laws, research attempts to insert genes into embryos would be prohibited if undertaken strictly to gain scientific knowledge. If the genes were added in an attempt to "cure" a particular embryo that was destined to go to term, however, it is likely to be permissible in most states. Maine might still ban it, because it prohibits "any form of experimentation."[77] But several of the other embryo research bans explicitly allow procedures for the purpose of providing a health benefit to the fetus or embryo, and therefore might not affect gene alterations.[78] In some states, the embryo research bans might forbid the use of evolving or insufficiently-tested therapies if such therapies were not necessary to the preservation of the life of the fetus. However, these laws or related laws generally require the protection and preservation of viable fetuses. Therefore, it seems unlikely that the embryo research laws in these states would be invoked to enforce the

69 MINN. STAT. ANN. § 145.421.

70 MICH. COMP. LAWS ANN. § 333.2685.

71 FLA. STAT. ANN. § 390.0111(6); ME. REV. STAT. ANN. tit. 22, § 1593; MASS. GEN. LAWS ANN. ch. 112, § 12J; MICH. COMP. LAWS ANN. § 333.2685-.2692; N.D. CENT. CODE § 14-02.2-01; R.I. GEN. LAWS § 11-54-1.

72 18 PA. CONS. STAT. § 3216.

73 *See* Ronald M. Green, *The Ethical* Considerations, 286 SCIENTIFIC AM. 4850, 4850 (Jan. 2002) (arguing that when Advanced Cell Technology created what the company called the "world's first human cloned embryo," all it had really done was create an "activated egg"). The company's president, Michael West, had previously argued that the company's work did not violate the Massachusetts Federal Research statute, and we believe he is correct in this argument.

74 Four states' fetal research bans—those of Arizona, Illinois, Louisiana, and Utah—have already been struck down on those grounds. Forbes v. Napolitano, 236 F.3d 1009 (9th Cir. 2000); Margaret S. v. Edwards, 794 F.2d 994, 998-99 (5th Cir. 1996); Jane L. v. Bangerter, 61 F.3d 1493, 1499-1502 (10th Cir. 1995); Lifchez v. Hartigan, 735 F. Supp. 1361, 1363-66 (N.D. Ill. 1990).

75 N.H. REV. STAT. ANN. § 168-B:15(II) (2002).

76 LA. REV. STAT. ANN. § 40:1299.36.2 (West 2002); MICH. COMP. LAWS ANN. §§ 333.16275, 750.430(a) (West 2001); R.I. GEN. LAWS § 23-16.4-2 (2001).

77 ME. REV. STAT. ANN. tit. 22, § 1593 (2002).

78 *See, e.g.*, FLA. STAT. ANN. § 390.0111(6) (West 2002); MASS. GEN. LAWS ANN. ch. 112, § 12J (West 2002); MICH. COMP. LAWS ANN. § 333.2685-.2692 (West 2002); MINN. STAT. § 145.421 (2001); N.D. CENT. CODE § 14-02.2-01 (2001); 18 PA. CONS. STAT. § 3216 (2001); R.I. GEN. LAWS § 11-54-1.

withholding of gene therapy as a form of treatment if doctors argued that the procedure held out some actual promise of a health benefit to the embryo and prospective child.

The New Hampshire and Louisiana laws have unique twists. New Hampshire's law might ban creating a child with inheritable alterations because it prohibits the transfer of any embryo donated for research to a uterus.[79] Louisiana's law has the opposite effect, prohibiting farming or culturing embryos solely for research purposes,[80] but apparently allowing research as long as the embryo is implanted.

B. NATIONAL IMPLICATIONS OF BANNING CLONING AND INHERITABLE ALTERATIONS

We do not believe there is any constitutional prohibition that would limit the legal authority of the federal government to enter into an international treaty banning human cloning and inheritable genetic alterations, although this question has been the subject of wide discussion in the legal literature.[81]

---

79 N.H. REV. STAT. ANN. § 168-B:15(II).

80 LA. REV. STAT. ANN. § 9:122 (West 2002).

81 The right to make decisions about whether or not to bear children is constitutionally protected under the constitutional right to privacy. *See, e.g.*, Eisenstadt v. Baird, 405 U.S. 438 (1972); Griswold v. Connecticut, 381 U.S. 479 (1965). The constitutional right to liberty also affords such protection. *See* Planned Parenthood of S.E. Pa. v. Casey, 505 U.S. 833, 857 (1992). The U.S. Supreme Court in 1992 reaffirmed the "recognized protection accorded to liberty relating to intimate relationships, the family, and decisions about whether to bear and beget a child." *Id.* at 857. Early decisions held that the right to privacy protected married couples' ability to make procreative decisions, but later decisions focused on individuals' rights as well. The U.S. Supreme Court has stated, "If the right of privacy means anything, it is the right of the individual, married or single, to be free from unwarranted governmental intrusion into matters so fundamentally affecting a person as the decision whether to bear or beget a child." *Eisenstadt*, 405 U.S. at 453.

A federal district court has indicated that the right to make procreative decisions encompasses the right of an infertile couple to undergo medically-assisted reproduction, including IVF and the use of a donated embryo. Lifchez v. Hartigan, 735 F. Supp. 1361, 1367-69. (N.D. Ill. 1990). *Lifchez* held that a ban on research on conceptuses was unconstitutional because it impermissibly infringed upon a woman's fundamental right to privacy. *Id.* at 1363. Although the Illinois statute banning embryo and fetal research at issue in the case permitted IVF, it did not allow embryo donation, embryo freezing or experimental prenatal diagnostic procedures. *Id.* at 1365-70. The court stated, "It takes no great leap of logic to see that within the cluster of constitutionally protected choices that includes the right to have access to contraceptives, there must be included within that cluster the right to submit to a medical procedure that may bring about, rather than prevent, pregnancy." *Id.* at 1377. The court also held that the statute was impermissibly vague because of its failure to define "experiment" or "therapeutic." *Id.* at 1376.

Some commentators argue that the Constitution similarly protects the right to create a child through cloning. *See* John Robertson, Views on Cloning: Possible Benefits, Address Before the National Bioethics Advisory Commission (Mar. 14, 1997), *available at* http://bioethics.georgetown.edu /nbac/transcripts/index.html. This seems to be a reversal of Robertson's earlier position that cloning "may deviate too far from prevailing conception of what is valuable about reproduction to count as a protected reproductive experience. At some point attempts to control the entire genome of a new person pass beyond the central experiences of identity and meaning that make reproduction a valued experience." JOHN ROBERTSON, CHILDREN OF CHOICE: FREEDOM AND THE NEW REPRODUCTIVE TECHNOLOGIES 169 (1994).

However, cloning is sufficiently different from normal reproduction and the types of assisted reproduction protected by the *Lifchez* case that constitutional protections should not apply. In even the most high-tech reproductive technologies available, a mix of genes occurs to create an individual with a genotype that has never before existed. In the case of twins, two such individuals are created. Their futures are open and the distinction between themselves and their parents is acknowledged. In the case of cloning, however, the genotype already exists. Even though it is clear that the individual will develop into a person with different traits because of different social, environmental and generational influences, there is evidence that the fact that he or she posses an existing genotype will affect how the

The United States itself currently has no federal law on either cloning or inheritable genetic modification, even though both President George W. Bush and former President Bill Clinton are in favor of outlawing human "reproductive" cloning.[82] In August 2001, the House of Representatives voted to outlaw both

---

resulting clone is treated by himself, his family and social institutions.

In that sense, cloning is sufficiently distinct from traditional reproduction or alternative reproduction to not be considered constitutionally protected. It is not a process of genetic mix, but of genetic duplication. It is not reproduction, but a sort of recycling, where a single individual's genome is made into someone else. This change in kind in the fundamental way in which humans can "reproduce" represents such a challenge to human dignity and the potential devaluation of human life (even comparing the "original" to the "copy" in terms of which is to be more valued) that even the search for an analogy has come up empty handed. *Testimony on Scientific Discoveries and Cloning: Challenges for Public Policy, Before the Sen. Subcomm. on Public Health and Safety, Sen. Comm. on Labor and Human Resources*, 105th Cong. 25 (1997) (statement of George J. Annas), *available at* http://www.bumc.bu.edu/www/sph/lw/pvl/Clonetest.htm. Gilbert Meilaender, in testifying before NBAC, pointed out the social importance of children's genetic independence from their parents: "They replicate neither their father nor their mother. That is a reminder of the independence that we must eventually grant to them and for which it is our duty to prepare them." CLONING HUMAN BEINGS, *supra* note 12, at 81.

Even if a constitutional right to clone were to be recognized, any legislation which would infringe unduly upon this fundamental right would be permissible if it furthered a compelling interest in the least restrictive manner possible in order to survive this standard of review. *See Lifchez*, 735 F. Supp. at 1377. Along those lines, the NBAC raised concerns about physical and psychological risks to the offspring, as well as about "a degradation of the quality of parenting, and family if parents are tempted to seek excessive control over their children's characteristics, to value children according to how well they meet every detailed parental expectation, and to undermine the acceptance and openness that typify loving families." CLONING HUMAN BEINGS, *supra* note 12, at 77. The NBAC also noted how cloning might undermine important social values, such as opening the door to a form of eugenics, or by tempting some to manipulate others as if they were objects instead of persons, and exceeding the moral boundaries of the human condition. *Id.*

The potential physical and psychological risks of cloning an entire individual are sufficiently compelling to justify banning the procedure. The notion of replicating existing humans seems to fundamentally conflict with our legal system, which emphatically protects individuality and uniqueness. Banning procreation through nuclear transplantation is justifiable in light of common law and constitutional protection of the sanctity of the individual and personal privacy. Francis C. Pizzulli, Note, *Asexual Reproduction and Genetic Engineering: A Constitutional Assessment of the Technology of Cloning*, 47 S. CAL. L. REV. 476, 502 (1974).

In the United States, couples' constitutional arguments regarding a privacy right or liberty right to use inheritable genetic interventions would appear to be stronger than those regarding access to cloning. In decisions construing the Americans with Disabilities Act, including one before the U.S. Supreme Court, individuals with AIDS were judged to be disabled because their disease was seen as interfering with a major life function—reproduction. Bragdon v. Abbott, 524 U.S. 624, 631 (1998). The argument seems to be that "normal" reproduction involves the creation of children without diseases.

Couples who both have sickle cell anemia or some other recessive genetic disorder might argue that a ban on germline interventions deprives them of reproductive liberty because it is the only way they can have healthy children. (There are several fallacies in that argument. The children born may have other diseases. And the genetic modification intervention itself might harm the children or be ineffective.) Forbidding the use of the techniques, it would be argued, forces them to go childless.

The couple might bolster their argument with a reference to another aspect of the *Lifchez* holding. The court in that case also held the ban on embryo research unconstitutional because it forbade parents from using experimental diagnostic techniques to learn the genetic status of their fetus. *See Lifchez*, 735 F. Supp. at 1366-67. The court reasoned that, if the woman has a constitutional right to abort, she has a right to genetic information upon which to make the decision. *See id.* Using an expansive interpretation, *Lifchez* could be understood as saying that it was understandable that couples would choose to have only children of a certain genetic makeup and that such a decision was constitutionally protected. However, even if there were a constitutional right to use inheritable genetic interventions, such interventions could be banned if they posed compelling physical, psychological or social risks. To be constitutional, the ban would also need to be narrowly focused to operate in the least restrictive manner possible.

82 *See* KAISER FAMILY FOUND., *Lawmakers Vow to Introduce Cloning Restrictions, Bush Signals He Will "Work to Pass" Ban*, KAISER DAILY REPROD. HEALTH REP., Mar. 29, 2001, *available*

research and reproductive cloning, and this proposal, known as the Weldon bill, has reached the Senate.[83] When the Senate takes up the issue, it will have to decide whether to agree with the Weldon bill (in which case it will be signed by the President and become law), or to try to craft a bill that outlaws reproductive cloning, but permits research cloning, as recommended by the National Academy of Sciences. In this case, the Senate bill will be sent to a conference committee where, unless the politics of the issue changes radically, it will likely die.[84] Of course, unless the United States passes legislation outlawing reproductive cloning, it cannot take any meaningful leadership role in the international treaty area on this issue.

VI. PROMULGATING AN INTERNATIONAL TREATY

The adoption of an international treaty is the most appropriate approach to prohibit species-altering interventions. A rogue doctor or scientist who wishes to offer the procedure can easily move across borders if a particular nation bans the procedure. When the American physicist Richard Seed announced he intended to clone human beings and U.S. lawmakers threatened to clamp down on the procedure, he responded that he would open up a clinic in Mexico[85] or join a Japanese-based project.[86] Restrictions on European biotechnology companies have stimulated some to move to Africa.[87]

Various international declarations and laws already oppose human cloning or

---

at http://report.kff.org/archive/repro/2001/3/ kr010329.2.htm.

83 As of June, 2002, the Senate had three bills to consider. S. 790, introduced by Senator Brownback of Kansas, is substantially the same as the Weldon bill passed by the House of Representatives in August, 2001. Human Cloning Prohibition Act, S. 790, 107th Cong. (2001). It would ban both the creation of human embryos by cloning as well as attempts to create a human child by cloning. *Id.* at § 3. S. 1758, introduced by Senator Dianne Feinstein of California bans attempts at human cloning, defined as "asexual reproduction by implanting or attempting to implant the product of nuclear transplantation into a uterus." Human Cloning Prohibition Act, S. 1758, 107th Cong. (2001). It also specifically permits certain activities, including "nuclear transplantation to produce human stem cells." *Id.* at § 4. It was slightly modified on May 1 and reintroduced as S. 2439 with the endorsement of Senator Orin Hatch. 148 Cong. Rec. S36,633 (2002) Finally, S. 1893, introduced in late January, 2002 by Senator Harkin of Iowa would simply ban cloning as defined as "asexual human reproduction by implanting or attempting to implant the product of nuclear transplantation [defined as "introducing the nuclear material of a human somatic cell into a fertilized or unfertilized oocyte from which the nucleus has been or will be removed or inactivated"] into a woman's uterus or a substitute for a woman's uterus." S. 1893, 107th Cong. § 498C (2002). For more details, see George J. Annas, *Cloning and the U.S. Congress*, 346 NEW ENG. J. MED. 1599 (2002).

84 The outstanding question is whether abortion politics will permit members of Congress to outlaw so-called reproductive cloning (which they all agree should be done) without also outlawing research cloning (a prohibition included in the Weldon and Brownback bills because some supporters object to any creation of human embryos in the laboratory, and others believe that once created by cloning, it is inevitable that a cloned human embryo will be introduced into a woman's uterus and eventually result in the birth of a cloned child). The slippery slope from research to reproductive cloning is real, of course, but could be made much less likely by adding restrictions to what physicians involved in infertility treatment could do (e.g., no creation or use of cloned embryos by infertility specialists). Three further steps would virtually eliminate the danger: creation of a federal oversight panel that would have to approve any research projects involving the creation of cloned embryos; outlawing the purchase and sale of human eggs (as is done now for organs and tissues for transplant); and outlawing the freezing or storage of cloned human embryos, eliminating the potential for stockpiling human embryos, and making it almost impossible for a research embryo to be used for reproduction in practice. George J. Annas, *Cell Division*, BOSTON GLOBE, Apr. 21, 2002, E1.

85 Gene Weingarten, *Strange Egg*, WASH. POST, Jan. 25, 1998, at F1.

86 *Radical Scientist to Help Open Cloning Clinics in Japan*, JAPAN SCI. SCAN, Dec. 7, 1998, *available at* 1998 WL 8029927.

87 Thomas Hirenee Atenga, *Africa: Biotech Firms Have Their Eyes on Africa, Euro MPS Say*, INT'L PRESS SERV., Oct. 14, 1998.

inheritable genetic interventions, either directly or indirectly. As summarized above,[88] many of these existing legal documents have shortcomings. Some are mere moratoria, set to expire in 2003. Some are limited in the type of species-altering technologies they ban, covering only cloning and not inheritable genetic interventions, or even just applying to cloning via a limited range of techniques.

Some of the existing laws have also been outpaced by technology and do not comprehensively ban all forms of reproductive cloning and inheritable interventions. Others are ambiguous as to what they cover. In some cases, potentially relevant laws were adopted more than two decades ago to deal with a different set of technologies and concerns; it is unclear whether their expansive prohibitions will be applied to the newer technologies of reproductive cloning and inheritable interventions. Moreover, many of the existing declarations and laws do not include sanctions. Thus, there is a need for an international treaty to encourage participating nations to clarify what is prohibited and have them commit to effective criminal penalties for breaches.

The treaty we propose takes a strong human rights perspective. This approach comports with international human rights traditions because it conceptualizes medical research issues as human rights matters.[89] It also comports with people's concerns about cloning and inheritable intervention. For example, in a survey of 2,700 Japanese doctors and academics, ninety-four percent of respondents found cloning to be ethically unacceptable, primarily because it insulted human dignity.[90] In Portugal, the National Ethics Council for Human Sciences deemed human cloning unacceptable due to concerns about human dignity and about the equilibrium of the human race and social life.[91] Human rights language is also evident in calls for a prohibition on cloning, such as one by the Council of Europe's Parliamentary Assembly, emphasizing that "every individual has a right to his own genetic identity."[92]

Concerns about human cloning run sufficiently deep that even those who would make money on the procedure have come out against it. Ian Wilmut, the scientist whose team cloned Dolly the sheep, might benefit financially if humans were cloned because his group holds a patent on a cloning process. But he has testified around the world against human cloning.[93] Similarly, BIO (the Biotechnology Industry

---

88 *See supra* notes 39 to 56 and accompanying text.

89 BRIT. MED. ASS'N, THE MEDICAL PROFESSION AND HUMAN RIGHTS 205-40 (2001), *see also* sources cited *supra* note 32.

90 *Most Doctors, Academics Oppose Human Cloning*, JAPAN ECON. NEWSWIRE, Nov. 7, 1998.

91 Conselho Nacional de Etica para as Ciencias de Vida [National Council on Ethics for the Life Sciences], Opinion on Embryo Research and the Ethical Implications of Cloning, No. 21/CNEV/97 (1997).

92 Resolution on Human Cloning, European Parliament, Jan. 15, 1998, O.J. (C 34) 164 (1998).

93 *See* Ian Wilmut, *Cloning for Medicine*, SCIENTIFIC AM., Dec. 1998, at 58:

> None of the suggested uses of cloning for making copies of existing people is ethically acceptable to my way of thinking, because they are not in the interests of the resulting child. It should go without saying that I strongly oppose allowing cloned human embryos to develop so that they can be tissue donors.

Id. Wilmut has testified around the world against human cloning. *See, e.g.*, Christine Corcos et al., *Double-Take: A Second Look at Cloning, Science Fiction and Law*, 59 LA. L. REV. 1041, 1051 (denouncing cloning human beings at a talk at Princeton University); *Creator of Dolly Stresses Benefits of Further Research on Cloning*, DAILY YOMIURI, June 7, 1997, *available at* 19997 WL 1211052 (advocating a worldwide prohibition against human cloning); *Cult in the First Bid to Clone Human*, EXPRESS, Oct. 11, 2000, *available at* 2000 WL 24217743 (responding to a British couple's plan to clone their deceased daughter, stating that "it is absolutely criminal to try this [cloning] in a human."); Curt Suplee, *Top Scientists Warn Against Cloning Panic; Recreating Humans Would Be Unethical Experts Say*, WASH. POST, Mar. 13, 1997, at A03 (testifying against human cloning before the NBAC).

Organization, a U.S. trade association of biotechnology companies) opposes human reproductive cloning.[94]

Numerous entities have called for an enforceable international ban on species-altering interventions. The World Health Organization (WHO) at its fifty-first World Health Assembly reaffirmed that "cloning for replication of human beings is ethically unacceptable and contrary to human dignity and integrity."[95] WHO urges member states to "foster continued and informed debate on these issues and to take appropriate steps, including legal and juridical measures, to prohibit cloning for the purpose of replicating human individuals."[96]

The European Union's Council of Europe adopted the Council of Europe Protocol, prohibiting the cloning of human beings. Twenty-nine countries have signed the treaty.[97] Similarly, the European Parliament has adopted a Resolution on Human Cloning. The Resolution indicates that people have a fundamental human right to their own genetic identity.[98] It states that human cloning is "unethical, morally repugnant, contrary to respect for the person and 'a grave violation of fundamental human rights which cannot under any circumstances be justified or accepted.'"[99] The Resolution calls for member states to enact binding national legislation banning cloning and also urges the United Nations to secure an international ban on cloning.[100]

UNESCO's Universal Declaration on the Human Genome and Human Rights specifically addresses cloning.[101] Like the treaty we propose, the Declaration is based on "universal principles of human rights."[102] The Declaration specifically refers to UNESCO's constitution, which underscores "the democratic principles of the dignity, equality and mutual respect of men."[103] Article 11 of the Declaration states, "Practices which are contrary to human dignity, such as reproductive cloning of human beings, shall not be permitted."[104] However, the Declaration does not have an enforcement mechanism. Rather, it calls upon nations and international

---

94 Press Release, BIOTECHNOLOGY IND. ORG., *BIO Reiterates Unequivocal Opposition to Reproductive Cloning; Support for Therapeutic Applications*, Nov. 25, 2001, *available at* http://www.bio.org/newsroom/news.asp. *See also* Frances Bishop, *11th Annual Bio Conference: Ethical Issues in Genetics Create Challenges for Biotech Industry*, 8 BIOWORLD TODAY 112 (1997), *available at* 1997 WL 11130296.

95 Press Release, W.H.O., WORLD HEALTH ASSEMBLY, *World Health Assembly States its Position on Cloning Human Reproduction* (May 14, 1997), *available at* http://www.who.int/archives/inf-pr-1997/en/97wha9.html.

96 W.H.O., WORLD HEALTH ASSEMBLY, 51st Sess., *Ethical, Scientific and Social Implications of Cloning in Human Health*, WHA51.10, (1998).

97 1) Croatia, 2) Cyprus, 3) Czech Republic, 4) Denmark, 5) Estonia, 6) Finland, 7) France, 8) Georgia, 9) Greece, 10) Hungary, 11) Iceland, 12) Italy, 13) Latvia, 14) Lithuania, 15) Luxembourg, 16) Moldova, 17) Netherlands, 18) Norway, 19) Poland, 20) Portugal, 21) Romania, 22) San Marino, 23) Slovenia, 24) Slovenia, 25) Spain, 26) Sweden, 27) Switzerland, 28) the former Yugoslav Republic of Macedonia and 29) Turkey. *See also* ADDITIONAL PROTOCOL (EXPLANATORY REPORT) TO THE CONVENTION ON HUMAN RIGHTS AND BIOMEDICINE, Jan. 12, 1998, *available at* http://conventions.coe.int/Treaty/en/Reports/Html/168.htm.

98 Resolution on Human Cloning, Eur. Parliament, Jan. 15, 1998 O.J. (C34) 164 (1998).

99 *Id.*

100 *Id.*

101 U.N.E.S.C.O., UNIVERSAL DECLARATION ON THE HUMAN GENOME AND HUMAN RIGHTS, 29th Sess., 29 C/Res. 16 (1997), *available at* http://www.unesco.org/human_rights/hrbc.htm. The declaration was adopted by the General Assembly in 1999. G.A. Res. 152, U.N. GAOR, 53rd Sess., U.N. Doc. A/53/152 (1999).

102 UNIVERSAL DECLARATION ON THE HUMAN GENOME AND HUMAN RIGHTS, Introduction.

103 *Id.*

104 *Id.* art. 11.

organizations to enact national and international policies to prohibit cloning and to identify and prohibit those genetic practices that are contrary to human dignity.[105]

A. The Process for Creating an International Treaty

On August 7, 2001, France and Germany urged the U.N. Secretary-General to add an International Convention against reproductive cloning of human beings to its agenda.[106] The French-German initiative is focused on banning only reproductive cloning apparently because there is an international consensus on this issue. It is worth noting that the laws of both countries ban research cloning and other forms of inheritable genetic interventions as well, and that political leaders in both countries have spoken out publicly in favor of imposing a broader ban. Nonetheless, both of these countries seem content to pursue a two-step process at the United Nations: securing as soon as possible a ban on reproductive cloning, and leaving negotiations on other issues, including inheritable genetic alterations and research cloning, for a second round of international negotiations. This may in fact be the only practicable way to proceed.

In November 2001, the Legal Committee of the United Nations added its support to a ban on reproductive cloning,[107] and the first meeting on the treaty was held in February 2002. There was virtual unanimous support for the treaty among the approximately 80 countries that attended, although the United States took the position that it would only support the treaty if it also outlawed research cloning. The issues of reproductive cloning discussed in this Article can, of course, be separated from those involved in research cloning, and will likely have to be if a treaty on reproductive cloning is to be adopted.[108] Whether the U.S. position will change remains to be seen. It seems likely that this U.N. treaty process is the only way a cloning treaty is likely to be achieved.[109] The treaty we propose is an attempt to provide language that could be used in both stages of a two stage process, or in one process if they are combined. It is drafted in a way to reflect the broad social concerns against species-altering technologies and to close loopholes in existing legal documents and declarations. Like the U.N. Legal Committee, we believe that the time is ripe for a flat-out international ban on human cloning. We also advocate a similar ban on inheritable genetic interventions.

---

105 *Id.*

106 *Request for the Inclusion of a Supplementary Item in the Agenda of the 56th Session, International Convention Against the Reproductive Cloning of Human Beings*, U.N. GAOR, 56th Sess., U.N. Doc. A/56/192 (2001).

107 *United Nations Calls for a Treaty to Ban Human Cloning*, BIRMINGHAM POST, Nov. 21, 2001, at 8.

108 *See supra* note 84. The next meeting of the ad hoc committee is scheduled at the United Nations for September 23-27, 2002. It is anticipated that the mandate to guide subsequent treaty negotiations will be adopted at this meeting, and that treaty language may be agreed upon a year or so later.

109 Stephen P. Marks, *Tying Prometheus Down: The International Law of Human Genetic Manipulation*, 3 CHI. J. INT'L L. (forthcoming 2002). The other U.N. treaty method is known as a framework convention, which is used when countries agree that a particular field needs to be regulated (such as the environment), but do not yet agree on the specifics of how the regulation should work. Because there is basic international agreement on human reproductive cloning, the framework convention is inappropriate. *See, e.g.*, Daniel Bodansky, *The United Nations Framework Convention on Climate Change: A Commentary*, 18 YALE J. INT'L L. 451, 494 (1993). *See also* ANTHONY AUST, MODERN TREATY LAW AND PRACTICE 97 (2000); Donald M. Goldberg, *Negotiating the Framework Convention on Climate Change*, 4 TOURO J. TRANSNAT'L L. 149 (1993); Lee A. Kimball, *The Biodiversity Convention: How to Make it Work*, 28 VAND. J. TRANSNAT'L L. 763 (1995).

VII. CONCLUSION

Biotechnology, especially human cloning and inheritable genetic alteration, has the potential to permit us to design our children and to literally change the characteristics of the human species. The movement toward a posthuman world can be characterized as "progress" and enhancement of individual freedom in the area of procreation; but it also can be characterized as a movement down a slippery slope to a neo-eugenics that will result in the creation of one or more subspecies or superspecies of humans. The first vision sees science as our guide and ultimate goal. The second is more firmly based on our human history as it has consistently emphasized differences, and used those differences to justify genocidal actions. It is the prospect of "genetic genocide" that calls for treating cloning and genetic engineering as potential weapons of mass destruction, and the unaccountable genetic researcher as a potential bioterrorist.

The greatest accomplishment of humans has not been our science, but our development of human rights and democracy. Science cannot tell us what we should do, or even what our goals are, therefore, humans must give direction to science. In the area of genetics, this calls for international action to control the techniques that could lead us to commit species suicide. We humans clearly recognized the risk in splitting the atom and developing nuclear weapons; and most humans recognize the risk in using human genes to modify ourselves. Because the risk is to the entire species, it requires a species response. Many countries have already enacted bans, moratoria and strict regulations on various species-altering technologies. The challenge, however, is global, and action on the international level is required to be effective.

We believe that the action called for today is the ratification of an international convention for the preservation of the human species that outlaws human cloning and inheritable genetic alterations. This ban would not only be important in itself; but it would also mark the first time the world worked together to control a biotechnology. Cloning and inheritable genetic alterations are not bioweapons per se, but they could prove just as destructive to the human species if left to the market and individual wants and desires.

We think an international consensus to ban these technologies already exists, and that countries, non-governmental organizations and individual citizens should actively support the treaty process, as they did with the recent Convention on the Prohibition of the Use, Stockpiling, Production and Transfer of Anti-Personnel Mines and their Destruction (Land Mine Treaty).[110]

Cloning may not seem as important as landmines, as no clone has yet been born and thus no children have been harmed by this technique. Nonetheless, cloning has the potential to harm all children, both directly by physically and mentally harming them, and indirectly by devaluing all children—treating them as products of their parents' genetic specifications. Likewise, inheritable genetic alteration carries the prospect of developing a new species of humans that could turn into either destroyers or victims of the human species. Opposition to cloning and inheritable genetic alteration is "conservative" in the strict sense of the word: it seeks to conserve the human species. But it is also liberal in the strict sense of the word: it seeks to preserve democracy, freedom and universal human rights for all members of the human species.

---

110 Convention on the Prohibition of the Use, Stockpiling, Production and Transfer of Anti-Personnel Mines and on their Destruction, G.A. res. 47/39, 47 U.N. GAOR Supp. (No. 49) at 54, U.N. Doc. A/47/49 (1992).

174 AMERICAN JOURNAL OF LAW & MEDICINE VOL. 28 NOS. 2&3 2002

APPENDIX

LEGISLATION IN FORCE RELATED TO HUMAN SPECIES PROTECTION

ASIA

Japan
The "Law concerning Regulation Relating to Human Cloning Techniques and Other Similar Techniques," Nov. 2000, in effect since June 2001. English version, *available at* http://www.mext.go.jp/english/shinkou/index.htm.

The Japanese law prohibits the transfer of embryos created by techniques of human cloning, and those created by xenotransplantation. However, it allows the application of these techniques and other similar ones for research purposes as long as the embryo created is not allowed to be transplanted in a human or an animal. It also imposes criminal sanctions.

Guidelines to the "Law concerning Regulation Relating to Human Cloning Techniques and Other Similar Techniques," Dec. 4, 2001, Minister of Education and Science, *available at* http://www.mext.go.jp/a_menu/shinkou/seimei/2001/hai3/17_shishin.pdf (in Japanese).

Commentaries to the Guidelines mentioned above by the Ministry of Education and Science, *available at* http://www.mext.go.jp/a_menu/shinkou/seimei/2001/hai3/20_shishin.pdf (in Japanese).

EUROPE

Council of Europe
Additional Protocol (Explanatory Report) to the Convention on Human Rights and Biomedicine, 12 January 1998., *available at* http://conventions.coe.int/Treaty/EN/Treaties/Html/168.htm.

Article 1.
Any intervention seeking to create a human being genetically identical to another human being, whether living or dead is prohibited. For the purpose of this article, the term human being 'genetically identical' to another human beings means a human being sharing with another the same nuclear gene set.

Convention for the Protection of Human Rights and Dignity of the Human Being with regard to the application of Biology and Medicine - Convention on Human Rights and Biomedicine; Oviedo, Apr. 4, 1997, *available at* http://conventions.coe.int/Treaty/EN/Treaties/Html/164.htm.

Article 13. Interventions on the human genome.
An intervention seeking to modify the human genome may only be undertaken for preventive, diagnostic or therapeutic purposes and only if its aim is not to introduce any modification in the genome of any descendants.

Austria
Federal Law of 1992 (Serial 275) Regulating Medically Assisted Procreation (The Reproductive Medicine Law), and Amending the General Civil Code, The Marriage

Law and the Rules of Jurisdiction (1993), *available at* http://www.bmbwk.gv.at/ (in German).

The law does not explicitly prohibit the cloning of human beings, but it limits research on human embryos (defined as "developable cells"). Its central principle is that reproductive medicine is acceptable only within a stable heterosexual relationship for the purpose of reproduction. The law provides that embryos can be used only for implantation in the woman who has donated the oocytes and cannot be used for other purposes. The donation of embryos or gametes is explicitly prohibited.

Denmark

Act No. 460 on Medically Assisted Procreation in connection with medical treatment, diagnosis and research, June 10, 1997, in force Oct. 1, 1997; and Act No. 503 on a Scientific, Ethical Committee System and the Handling of Biomedical Research Projects.

According to §28 (Act No. 460) research with the following aims is forbidden:

a) research where the aim is to develop human reproductive cloning;

b) research where the aim is to facilitate the creation of a human identity by melting together genetically unidentical embryos or parts of embryos before the implantation in the woman womb.

Provision §4 states that it is forbidden to implant identical unfertilized or fertilized ova in one or more women. The Act establishes penalties of fine and imprisonment for the doctor and the authorized health persons that violate its provisions.

Finland

Medical Research Act (No. 488/1999).

The legislation applies to embryo research. Section 15 explicitly prohibits any research which has the objective of modifying the germ line, but makes an exception for research done for the purposes of curing or preventing serious hereditary disease.

France

Law No. 94-653 of July 29 1994, on Respect for the Human Body, *available at* http://www.cnrs.fr/SDV/loirespectcorps.html (in French).

The law prohibits the invasion into the integrity of the human species, eugenic behaviors intended to organize selection of human beings, and conversion of genetic characteristics leading to any change in descendants of humans (except for studies aiming at prevention and treatment of hereditary diseases) (articles 16-4). The act also amends the Penal Code, prescribing penalties of imprisonment and fine for the implementation of eugenic activities on human beings (Article 511-1).

Law No. 94-654 of July 29, 1994, on the Donation and Use of Elements and Products of the Human Body, Medically Assisted Procreation, and Prenatal Diagnosis, *available at* http://www.cnrs.fr/SDV/loidocorps.html (in French).

The bioethics legislation and its amendments (Law No. 94-653 and Law No. 94-654) specifically prohibit human cloning, creation of hybrids and chimeras, germline gene therapy, the creation of embryos purely for research purposes, and eugenic experiments.

Germany
Gesetz zum Schutz von Embryonen (Embryonenschutzgesetz), v.13.12.1990 (BGBI. I S.2747) [Federal Embryo Protection Law], *available at* http://www.bmgesundheit.de/rechts/genfpm/embryo/embryo.htm (in German).

This special criminal law prohibits human reproductive cloning and prescribes criminal penalties (imprisonment or fine) against violations. Regarding germ cells, the act prohibits any artificial changes in the genetic characteristics of human cells and prohibits the use of such altered cells for fertilization.

Hungary
Law No. 154 (1997) on Genetic Research. The law bans germline engineering.

Iceland
Artificial Fertilisation Act No. 55/1996, Regulation No. 568 on Artificial Fertilisation Act English version, *available at* http://brunnur.stjr.is/interpro/htr/htr.nsf/pages/lawsandregs0002.

Art 12(d): "It is prohibited to perform cloning."

Norway
Law No. 56 of August 5 1994 on the Medical Use of Biotechnology, (1995) 46 (1), *available at* http://www.stortinget.no/english/index.html.

Article 3-1 prohibits "research on fertilized eggs."

Ministry of Health and Social Affairs, The Act relating to the Application of Biotechnology in Medicine, 1994.

Prohibits germline therapy and prescribes criminal sanctions for its violation.

Russia
Law on Reproductive Human Cloning, Apr. 19, 2002.

The law establishes a moratorium on human reproductive cloning and the importation of cloned embryos for five years.

Spain
Law No. 35/1988, Nov. 22, 1988, on techniques of assisted reproduction, *available at* http://www.geocities.com/Eureka/9068/SANIDAD/reproduc.html (in Spanish), modified by Organic Law No. 10/995 of 23 November 1995, *available at* http://www.webcom.com/kruzes/legislac09.htm (in Spanish).

The law No. 35/1988 establishes in sections §13.3(d) and 15.2(b) that any therapeutic intervention, investigation or research activity in pre-embryos in vitro, pre-embryos, embryos and fetuses in utero, will be authorized only if such intervention or activity does not alter its genetic make-up (in so far as it does not contain any anomaly), or if it is not aimed to individual or race selection.

The Organic Law introduced in section II of the Penal Code a Title V: Offenses relating to genetic engineering, prescribing criminal and civil sanctions for its violation.

The Spanish Penal Code (Article 16 1,2) prohibits bringing about the birth of identical human beings as a result of cloning or other procedures aimed at the selection of humans.

Sweden
Law No.115 of Mar. 14, 1991, Act concerning measures for the purposes of Research or Treatment in connection with Fertilized Human Oocytes (1993). This statute and the *in vitro* Fertilization law of 1988 govern embryo research. Any research, which seeks to genetically modify the embryo, is prohibited.

United Kingdom
Human Reproductive Cloning Act 2001, UK Stat. 2001 c23 &1, in force Dec. 4, 2001.

Makes it an offense for a person to place "in a woman a human embryo which has been created otherwise than by fertilisation."

MIDDLE EAST

Israel
Prohibition of Genetic Intervention Law No.5759 (cloning on Human Beings and Genetic Modifications for Reproductive Cells) (1998), *available at* http://www.knesset.gov.il/index.html.

The law introduces a five-year moratorium on human reproductive cloning and germline engineering and prescribes criminal sanctions for its violation.

NORTH AMERICA

Costa Rica
Decree No.24029-S: A Regulation on Assisted Reproduction, Feb. 3, 1995, *available at* http://www.netsalud.sa.cr/ms/decretos/dec5.htm (in Spanish).

Article 11.
Any manipulation or alteration of an embryo's genetic code is prohibited, as well as any kind of experimentation with embryos.

OCEANIA

Australia
Victoria Infertility Treatment Act 2000 - No.37/200 (Amendment of the Act No.37/1997, 63/1995), *available at* http://www.dms.dpc.vic.gov.au.

The State of Victoria explicitly bans human reproductive cloning and germline engineering, and prescribes criminal sanctions (fines and imprisonment) for its violation.

178 AMERICAN JOURNAL OF LAW & MEDICINE VOL. 28 NOS. 2&3 2002

SOUTH AMERICA

Argentina
Decree No.200/97 of Mar. 7, 1997: A Prohibition on Human Cloning Research, *available at* http://infoleg.mecon.gov.ar/txtnorma/42213.htm (in Spanish).

Article 1: Cloning experiments regarding human beings are prohibited.

Brazil
Law No.8974 (1995) on genetically modified organisms, *available at* http://www.mct.gov.br/legis/leis/8974_95.htm (in Portuguese).

Art. 8 of the law prohibits the genetic manipulation of the germline as well as the intervention on the human genetic material in vivo, with the exception of the treatment of genetic defects.

Peru
Law No.26842, General Health Law, July 9, 1997, *available at* http://www.congreso.gob.pe.

Art. 7
"[T]he fertilization of a human ovum with an intent other than procreation is prohibited, as well as human cloning." (Unofficial translation).

Law No. 27636, Criminal Code: Genetic Manipulation.
The genetic manipulation with the purpose of cloning a human being is prohibited. The law establishes criminal sanctions of imprisonment for its violation.

[19]

"Goodbye Dolly?" The ethics of human cloning

John Harris *The Institute of Medicine, Law and Bioethics, University of Manchester*

Abstract

The ethical implications of human clones have been much alluded to, but have seldom been examined with any rigour. This paper examines the possible uses and abuses of human cloning and draws out the principal ethical dimensions, both of what might be done and its meaning. The paper examines some of the major public and official responses to cloning by authorities such as President Clinton, the World Health Organisation, the European parliament, UNESCO, and others and reveals their inadequacies as foundations for a coherent public policy on human cloning. The paper ends by defending a conception of reproductive rights or "procreative autonomy" which shows human cloning to be not inconsistent with human rights and dignity.

The recent announcement of a birth[1] in the press heralds an event probably unparalleled for two millennia and has highlighted the impact of the genetic revolution on our lives and personal choices. More importantly perhaps, it raises questions about the legitimacy of the sorts of control individuals and society purport to exercise over something, which while it must sound portentous, is nothing less than human destiny. This birth, that of "Dolly", the cloned sheep, is also illustrative of the responsibilities of science and scientists to the communities in which they live and which they serve, and of the public anxiety that sensational scientific achievements sometimes provokes.

The ethical implications of human clones have been much alluded to, but have seldom been examined with any rigour. Here I will examine the possible uses and abuses of human cloning and draw out the principal ethical dimensions, both of what might be done and its meaning, and of public and official responses.

There are two rather different techniques available for cloning individuals. One is by nuclear substitution, the technique used to create Dolly, and the other is by cell mass division or "embryo splitting". We'll start with cell mass division because this is the only technique for cloning that has, as yet, been used in humans.

Key words

Cloning; Dolly; human dignity; procreative autonomy, human rights.

Cell mass division

Although the technique of cloning embryos by cell mass division has, for some time been used extensively in animal models, it was used as a way of multiplying human embryos for the first time in October 1993 when Jerry Hall and Robert Stillman[2] at George Washington Medical Centre cloned human embryos by splitting early two- to eight-cell embryos into single embryo cells. Among other uses, cloning by cell mass division or embryo splitting could be used to provide a "twin" embryo for biopsy, permitting an embryo undamaged by invasive procedures to be available for implantation following the result of the biopsy on its twin, or to increase the number of embryos available for implantation in the treatment of infertility.[3] To what extent is such a practice unethical?

Individuals, multiples and genetic variation

Cloning does not produce identical copies of the same individual person. It can only produce identical copies of the same genotype. Our experience of identical twins demonstrates that each is a separate individual with his or her own character, preferences and so on. Although there is some evidence of striking similarities with respect to these factors in twins, there is no question but that each twin is a distinct individual, as independent and as free as is anyone else. To clone Bill Clinton is not to create multiple Presidents of the United States. Artificial clones do not raise any difficulties not raised by the phenomenon of "natural" twins. We do not feel apprehensive when natural twins are born, why should we when twins are deliberately created?

If the objection to cloning is to the creation of identical individuals separated in time, (because the twin embryos might be implanted in different cycles, perhaps even years apart), it is a weak one at best.

We should remember that such twins will be "identical" in the sense that they will each have the same genotype, but they will never (unlike some but by no means all natural monozygotic twins) be identical in the more familiar sense of looking identical at the same moment in time. If we think of expected similarities in character, tastes and so on, then the same is true. The further separated in time, the less likely they are to have similarities of *character* (the more different the environment, the more different environmental influence on individuality).

The significant ethical issue here is whether it would be morally defensible, by outlawing the creation of clones by cell mass division, to deny a woman the chance to have the child she desperately seeks. If this procedure would enable a woman to create a sufficient number of embryos to give her a reasonable chance of successfully implanting one or two of them, then the objections to it would have to be weighty indeed. If pre-implantation testing by cell biopsy might damage the embryo to be implanted, would it be defensible to prefer this to testing a clone, if technology permits such a clone to be created without damage, by separating a cell or two from the embryonic cell mass? If we assume each procedure to have been perfected and to be equally safe, we must ask what the ethical difference would be between taking a cell for cell biopsy and destroying it thereafter, and taking a cell to create a clone, and then destroying the clone? The answer can only be that destroying the cloned embryo would constitute a waste of human potential. But this same potential is wasted whenever an embryo is not implanted.

Nuclear substitution: the birth of Dolly

This technique involves (crudely described) deleting the nucleus of an egg cell and substituting the nucleus taken from the cell of another individual. This can be done using cells from an adult. The first viable offspring produced from fetal and adult mammalian cells was reported from an Edinburgh-based group in *Nature* on February 27, 1997.[4] The event caused an international sensation and was widely reported in the world press. President Clinton of the United States called for an investigation into the ethics of such procedures and announced a moratorium on public spending on human cloning; the British Nobel Prize winner, Joseph Rotblat, described it as science out of control, creating "a means of mass destruction",[5] and the German newspaper *Die Welt*, evoked the Third Reich, commenting: "The cloning of human beings would fit precisely into Adolph Hitler's world view".[6]

More sober commentators were similarly panicked into instant reaction. Dr Hiroshi Nakajima, Director General of the World Health Organisation said: "WHO considers the use of cloning for the replication of human individuals to be ethically unacceptable as it would violate some of the basic principles which govern medically assisted procreation. These include respect for the dignity of the human being and protection of the security of human genetic material".[7] The World Health Organisation followed up the line taken by Nakajima with a resolution of the Fiftieth World Health Assembly which saw fit to affirm "that the use of cloning for the replication of human individuals is ethically unacceptable and contrary to human integrity and morality".[8] Federico Mayor of UNESCO, equally quick off the mark, commented: "Human beings must not be cloned under any circumstances. Moreover, UNESCO's International Bioethics Committee (IBC), which has been reflecting on the ethics of scientific progress, has maintained that the human genome must be preserved as common heritage of humanity".[9]

The European parliament rushed through a resolution on cloning, the preamble of which asserted, (paragraph B):

"[T]he cloning of human beings . . . , cannot under any circumstances be justified or tolerated by any society, because it is a serious violation of fundamental human rights and is contrary to the principle of equality of human beings as it permits a eugenic and racist selection of the human race, it offends against human dignity and it requires experimentation on humans," And which went on to claim that, (clause 1) "each individual has a right to his or her own genetic identity and that human cloning is, and must continue to be, prohibited".[10]

These statements are, perhaps un-surprisingly, thin on argument and rationale; they appear to have been plucked from the air to justify an instant reaction. There are vague references to "human rights" or "basic principles" with little or no attempt to explain what these principles are, or to indicate how they might apply to cloning. The WHO statement, for example, refers to the basic principles which govern human reproduction and singles out "respect for the dignity of the human being" and "protection of the security of genetic material". How, we are entitled to ask, is the security of genetic material compromised? Is it less secure when inserted with precision by scientists, or when spread around with the characteristic negligence of the average human male?[11]

Human dignity

Appeals to human dignity, on the other hand, while universally attractive, are comprehensively vague and deserve separate attention. A first question to ask when the idea of human dignity is invoked is: whose dignity is attacked and how? Is it the duplication of a large part of the genome that is supposed to constitute the attack on human dignity? If so we might legitimately ask whether and how the dignity of a natural twin is threatened by the existence of her sister? The notion of human dignity is often also linked to Kantian ethics. A typical example, and one that attempts to provide some basis for objections to cloning based on human dignity, was Axel Kahn's invocation of this principle in his commentary on cloning in *Nature*.[12]

"The creation of human clones solely for spare cell lines would, from a philosophical point of view, be in obvious contradiction to the principle expressed by Emmanuel Kant: that of human dignity. This principle demands that an individual – and I would extend this to read human life – should never be thought of as a means, but always also as an end. Creating human life for the sole purpose of preparing therapeutic material would clearly not be for the dignity of the life created."

The Kantian principle, crudely invoked as it usually is without any qualification or gloss, is seldom helpful in medical or bio-science contexts. As formulated by Kahn, for example, it would outlaw blood transfusions The beneficiary of blood donation, neither knowing of, nor usually caring about, the anonymous donor uses the blood (and its' donor) simply as a means to her own ends. It would also outlaw abortions to protect the life or health of the mother.

Instrumentalization

This idea of using individuals as a means to the purposes of others is sometimes termed "instrumentalization". Applying this idea coherently or consistently is not easy! If someone wants to have children in order to continue their genetic line do they act instrumentally? Where, as is standard practice in *in vitro* fertilisation (IVF), spare embryos are created, are these embryos created instrumentally? If not how do they differ from embryos created by embryo splitting for use in assisted reproduction?[13]

Kahn responded in the journal *Nature* to these objections.[14] He reminds us, rightly, that Kant's famous principle states: "respect for human dignity requires that an individual is *never* used . . . *exclusively* as a means" and suggests that I have ignored the crucial use of the term "exclusively". I did not of course, and I'm happy with Kahn's reformulation of the principle. It is not that Kant's principle does not have powerful intuitive force, but that it is so vague and so open to selective interpretation and its scope for application is consequently so limited, that its utility as one of the "fundamental principles of modern bioethical thought", as Kahn describes it, is virtually zero.

Kahn himself rightly points out that debates concerning the moral status of the human embryo are debates about whether embryos fall within the *scope* of Kant's or indeed any other moral principles concerning persons; so the principle itself is not illuminating in this context. Applied to the creation of individuals which are, or will become autonomous, it has limited application. True the Kantian principle rules out slavery, but so do a range of other principles based on autonomy and rights. If you are interested in the ethics of creating people then, so long as existence is in the created individual's own best interests, and the individual will have the capacity for autonomy like any other, then the motives for which the individual was created are either morally irrelevant or subordinate to other moral considerations. So that even where, for example, a child is engendered exclusively to provide "a son and heir" (as so often in so many cultures) it is unclear how or whether Kant' principle applies. Either other motives are also attributed to the parent to square parental purposes with Kant, or the child's eventual autonomy, and its clear and substantial interest in or benefit from existence, take precedence over the comparatively trivial issue of parental motives. Either way the "fundamental principle of modern bioethical thought" is unhelpful and debates about whether or not an individual has been used *exclusively* as a means are sterile and usually irresolvable.

We noted earlier the possibility of using embryo splitting to allow genetic and other screening by embryo biopsy. One embryo could be tested and then destroyed to ascertain the health and genetic status of the remaining clones. Again, an objection often voiced to this is that it would violate the Kantian principle, and that "one twin would be destroyed for the sake of another".

This is a bizarre and misleading objection both to using cell mass division to create clones for screening purposes, and to creating clones by nuclear substitution to generate spare cell lines. It is surely ethically dubious to object to one embryo being sacrificed for the sake of another, but not to object to it being sacrificed for nothing. In *in vitro* fertilisation, for example, it is, in the United Kingdom, currently regarded as good practice to store spare embryos for future use by the mother or for disposal at her direction, either to other women who require donor embryos, or for research, or simply to be destroyed. It cannot be morally worse to use an embryo to provide information about its sibling, than to use it for more abstract research or simply to destroy it. If it is permissible to use early embryos for research or to destroy them, their use in genetic and other health testing is surely also permissible. The same would surely go for their use in creating cell lines for therapeutic purposes.

It is better to do good

A moral principle, that has at least as much intuitive force as that recommended by Kant, is that it is better to do some good than to do no good. It cannot, from the ethical point of view, be better or more moral to waste human material that could be used for therapeutic purposes, than to use it to do good. And I cannot but think that if it is right to *use* embryos for research or therapy then it is also right to *produce* them for such purposes.[15] Kant's prohibition does after all refer principally to use. Of course some will think that the embryo is a full member of the moral community with all the rights and protections possessed by Kant himself. While this is a tenable position, it is not one held by any society which permits abortion, post-coital contraception, or research with human embryos.

The UNESCO approach to cloning is scarcely more coherent than that of WHO; how does cloning affect "the preservation of the human genome as common heritage of humanity"? Does this mean that the human genome must be "preserved intact", that is without variation, or does it mean simply that it must not be "reproduced a-sexually"? Cloning cannot be said to impact on the variability of the human genome, it merely repeats one infinitely small part of it, a part that is repeated at a natural rate of about 3·5 per thousand births.[16]

Genetic variability

So many of the fears expressed about cloning, and indeed about genetic engineering more generally, invoke the idea of the effect on the gene pool or upon genetic variability or assert the sanctity of the human genome as a common resource or heritage. It is very difficult to understand what is allegedly at stake here. The issue of genetic variation need not detain us long. The numbers of twins produced by cloning will always be so small compared to the human gene pool in totality, that the effect on the variation of the human gene pool will be vanishingly small. We can say with confidence that the human genome and the human population were not threatened at the start of the present millennium in the year AD one, and yet the world population was then perhaps one per cent of what it is today. Natural species are usually said to be endangered when the population falls to about one thousand breeding individuals; by these standards fears for humankind and its genome may be said to have been somewhat exaggerated.[17]

The resolution of the European parliament goes into slightly more detail; having repeated the, now mandatory, waft in the direction of fundamental human rights and human dignity, it actually produces an argument. It suggests that cloning violates the principal of equality, "as it permits a eugenic and racist selection of the human race". Well, so does prenatal, and pre-implantation screening, not to mention egg donation, sperm donation, surrogacy, abortion and human preference in choice of sexual partner. The fact that a technique could be abused does not constitute an argument against the technique, unless there is no prospect of preventing the abuse or wrongful use. To ban cloning on the grounds that it might be used for racist purposes is tantamount to saying that sexual intercourse should be prohibited because it permits the possibility of rape.

Genetic identity

The second principle appealed to by the European parliament states, that "each individual has a right to his or her own genetic identity". Leaving aside the inevitable contribution of mitochondrial DNA,[18] we have seen that, as in the case of natural identical twins, genetic identity is not an essential component of personal identity[19] nor is it necessary for "individuality". Moreover, unless genetic identity is required either for personal identity, or for individuality, it is not clear why there should be a right to such a thing. But if there is, what are we to do about the rights of identical twins?

Suppose there came into being a life-threatening (or even disabling) condition that affected pregnant women and that there was an effective treatment, the only side effect of which was that it caused the embryo to divide, resulting in twins. Would the existence of the supposed right conjured up by the European parliament mean that the therapy should be outlawed? Suppose that an effective vaccine for HIV was developed which had the effect of doubling the natural twinning rate; would this be a violation of fundamental human rights? Are we to foreclose the possible benefits to be derived from human cloning on so flimsy a basis? We should recall that the natural occurrence of monozygotic (identical) twins is one in 270 pregnancies. This means that in the United Kingdom, with a population of about 58 million, over 200 thousand such pregnancies have occurred. How are we to regard human rights violations on such a grand scale?

A right to parents

The apparently overwhelming imperative to identify some right that is violated by human cloning sometimes expresses itself in the assertion of "a right to have two parents" or as "the right to be the product of the mixture of the genes of two individuals". These are on the face of it highly artificial and problematic rights – where have they sprung from, save from a desperate attempt to conjure some rights that have been violated by cloning? However, let's take them seriously for a moment and grant that they have some force. Are they necessarily violated by the nuclear transfer technique?

If the right to have two parents is understood to be the right to have two social parents, then it is of course only violated by cloning if the family identified as the one to rear the resulting child is a one-parent family. This is not of course necessarily any more likely a result of cloning, than of the use of any of the other new reproductive technologies (or indeed of sexual reproduction). Moreover if there is such a right, it is widely violated, creating countless "victims", and there is no significant evidence of any enduring harm from the violation of this supposed right. Indeed war widows throughout the world would find its assertion highly offensive.

If, on the other hand we interpret a right to two parents as the right to be the product of the mixture of the genes of two individuals, then the supposition that this right is violated when the nucleus of the cell of one individual is inserted into the de-nucleated egg of another, is false in the way this claim is usually understood. There is at least one sense in which a right expressed in this form might be violated by cloning, but not in any way which has force as an objection. Firstly it is false to think that the clone is the genetic child of the nucleus donor. It is not. The clone is the twin brother or sister of the nucleus donor and the genetic offspring of the nucleus donor's own parents. Thus this type of cloned individual is, and always must be, the genetic child of two separate genotypes, of two genetically different individuals, however often it is cloned or re-cloned.

Two parents good, three parents better

However, the supposed right to be the product of two separate individuals is perhaps violated by cloning in a novel way. The de-nucleated egg contains mitochondrial DNA – genes from the female whose egg it is. The inevitable presence of the mitochondrial genome of the egg donor, means that the genetic inheritance of clones is in fact richer than that of other individuals, richer in the sense of being more variously derived.[20] This can be important if the nucleus donor is subject to mitochondrial diseases inherited from his or her mother and wants a child genetically related to her that will be free of these diseases. How this affects alleged rights to particular combinations of "parents" is more difficult to imagine, and perhaps underlines the confused nature of such claims.

What good is cloning?

One major reason for developing cloning in animals is said to be[4] to permit the study of genetic diseases and indeed genetic development more generally. Whether or not there would be major advantages in human cloning by nuclear substitution is not yet clear. Certainly it would enable some infertile people to have children genetically related to them, it offers the prospect, as we have noted, of preventing some diseases caused by mitochondrial DNA, and could help "carriers" of X-linked and autosomal recessive disorders to have their own genetic children without risk of passing on the disease. It is also possible that cloning could be used for the creation of "spare parts" by for example, growing stem cells for particular cell types from non-diseased parts of an adult.

Any attempt to use this technique in the United Kingdom, is widely thought to be illegal. Whether it would in fact be illegal might turn on whether it is plausible to regard such cloning as the product of "fertilisation". Apparently only fertilised embryos are covered by the *Human Fertilisation and Embryology Act 1990.*[21] The technique used in Edinburgh which involves deleting the nucleus of an unfertilised egg and then substituting a cell nucleus from an existing individual, by-passes what is normally considered to be fertilisation completely and may therefore turn out not to be covered by existing legislation. On the other hand, if as seems logical, we consider "fertilisation" as the moment when all forty-six chromosomes are present and the zygote is formed the problem does not arise.

The unease caused by Dolly's birth may be due to the fact that it was just such a technique that informed the plot of the film "The Boys from Brazil" in which Hitler's genotype was cloned to produce a fuehrer for the future. The prospect of limitless numbers of clones of Hitler is rightly disturbing. However, the numbers of clones that could be produced of any one genotype will, for the foreseeable future, be limited not by the number of copies that could be made of one genotype (using serial nuclear transfer techniques 470 copies of a single nuclear gene in cattle have been reported),[22] but by the availability of host human mothers.[23] Mass production in any democracy could therefore scarcely be envisaged. Moreover, the futility of any such attempt is obvious. Hitler's genotype might conceivably produce a "gonadically challenged" individual of limited stature, but reliability in producing an evil and vicious megalomaniac is far more problematic, for reasons already noted in our consideration of cloning by cell mass division.

Dolly collapses the divide between germ and somatic cells

There are some interesting implications of cloning by nuclear substitution (which have been clear since frogs were cloned by this method in the 1950s) which have not apparently been noticed.[24] There is currently a world-wide moratorium on manipulation of the human germ line, while therapeutic somatic line interventions are, in principal, permitted.[13] However, inserting the mature nucleus of an adult cell into a de-nucleated egg turns the cells thus formed into germ line cells. This has three important effects. First, it effectively eradicates the firm divide between germ line and somatic line nuclei because each adult cell nucleus is in principle "translatable" into a germ line cell nucleus by transferring its nucleus and creating a clone. Secondly, it permits somatic line modifications to human cells to become germ line modifications. Suppose you permanently insert a normal copy of the adenosine deaminase gene into the bone marrow cells of an individual suffering from severe combined immuno-deficiency (which affects the so called "bubble boy" who has to live in a protective bubble of clean air) with obvious beneficial therapeutic effects. This is a somatic line modification. If you then cloned a permanently genetically modified bone marrow cell from this individual, the modified genome would be passed to the clone and become part of his or her genome, transmissible to her offspring indefinitely through the germ line. Thus a benefit that would have perished with the original recipient and not been passed on for the protection of her children, can be conferred on subsequent generations by cloning.[25] The third effect is that it shows the oft asserted moral divide between germ line and somatic line therapy to be even more ludicrous than was previously supposed.[15]

Immortality?

Of course some vainglorious individuals might wish to have offspring not simply with their genes but with a matching genotype. However, there is no way that they could make such an individual a duplicate of themselves. So many years later the environmental influences would be radically different, and since every choice, however insignificant, causes a life-path to branch with unpredictable consequences, the holy grail of duplication would be doomed to remain a fruitless quest. We can conclude that people who would clone themselves would probably be foolish and ill-advised, but would they be immoral and would their attempts harm society or their children significantly?

Whether we should legislate to prevent people reproducing, not 23 but all 46 chromosomes, seems more problematic for reasons we have already examined, but we might have reason to be uncomfortable about the likely standards and effects of child-rearing by those who would clone themselves. Their attempts to mould their child in their own image would be likely to be more pronounced than the average. Whether they would likely be worse than so many people's attempts to duplicate race, religion and culture, which are widely accepted as respectable in the contemporary world, might well depend on the character and constitution of the genotype donor. Where identical twins occur naturally we might think of it as "horizontal twinning", where twins are created by nuclear substitution we have a sort of "vertical twinning". Although horizontal twins would be closer to one another in every way, we do not seem much disturbed by their natural occurrence. Why we should be disturbed either by artificial horizontal twinning or by vertical twinning (where differences between the twins would be greater) is entirely unclear.

Suppose a woman's only chance of having "her own" genetic child was by cloning herself; what are the strong arguments that should compel her to accept that it would be wrong to use nuclear substitution? We must assume that this cloning technique is safe, and that initial fears that individuals produced using nuclear substitution might age more rapidly have proved groundless.[26] We usually grant the so called "genetic imperative" as an important part of the right to found a family, of procreative autonomy.[27] The desire of people to have "their own" genetic children is widely accepted, and if we grant the legitimacy of genetic aspirations in so many cases, and the use of so many technologies to meet these aspirations,[28] we need appropriately serious and weighty reasons to deny them here.

It is perhaps salutary to remember that there is no necessary connection between phenomena, attitudes or actions that make us uneasy, or even those that disgust us, and those phenomena, attitudes, and actions that there are good reasons for judging unethical. Nor does it follow that those things we are confident *are* unethical must be prohibited by legislation or regulation.

We have looked at some of the objections to human cloning and found them less than plausible, we should now turn to one powerful argument that has recently been advanced in favour of a tolerant attitude to varieties of human reproduction.

Procreative autonomy

We have examined the arguments for and against permitting the cloning of human individuals. At the heart of these questions is the issue of whether or not people have rights to control their reproductive destiny and, so far as they can do so without violating the rights of others or threatening society, to choose their own procreative path. We have seen that it has been claimed that cloning violates principles of human dignity. We will conclude by briefly examining an approach which suggests rather that failing to permit cloning might violate principles of dignity.

The American philosopher and legal theorist, Ronald Dworkin has outlined the arguments for a right to what he calls "procreative autonomy" and has defined this right as "a right to control their own role in procreation unless the state has a compelling reason for denying them that control".[29] Arguably, freedom to clone one's own genes might also be defended as a dimension of procreative autonomy because so many people and agencies have been attracted by the idea of the special nature of genes and have linked the procreative imperative to the genetic imperative.

"The right of procreative autonomy follows from any competent interpretation of the due process clause and of the Supreme Court's past decisions applying it. . . . The First Amendment prohibits government from establishing any religion, and it guarantees all citizens free exercise of their own religion. The Fourteenth Amendment, which incorporates the First Amendment, imposes the same prohibition and same responsibility on states. These provisions also guarantee the right of procreative autonomy."[30]

The point is that the sorts of freedoms which freedom of religion guarantees, freedom to choose one's own way of life and live according to one's most deeply held beliefs are also at the heart of procreative choices. And Dworkin concludes:

"that no one may be prevented from influencing the shared moral environment, through his own private choices, tastes, opinions, and example, just because these tastes or opinions disgust those who have the power to shut him up or lock him up."[31]

Thus it may be that we should be prepared to accept both some degree of offence and some social disadvantages as a price we should be willing to pay in order to protect freedom of choice in matters of procreation and perhaps this applies to cloning as much as to more straightforward or usual procreative preferences.[32]

The nub of the argument is complex and abstract but it is worth stating at some length. I cannot improve upon Dworkin's formulation of it.

"The right of procreative autonomy has an important place . . . in Western political culture more generally. The most important feature of that culture is a belief in individual human dignity: that people have the moral right – and the moral responsibility – to confront the most fundamental questions about the meaning and value of their own lives for themselves, answering to their own consciences and convictions. . . . The principle of procreative autonomy, in a broad sense, is embedded in any genuinely democratic culture."[33]

In so far as decisions to reproduce in particular ways or even using particular technologies constitute decisions concerning central issues of value, then arguably the freedom to make them is guaranteed by the constitution (written or not) of any democratic society, unless the state has a compelling reason for denying its citizens that control. To establish such a compelling reason the state (or indeed a federation or union of states, such as the European Union for example) would have to show that more was at stake than the fact that a majority found the ideas disturbing or even disgusting.

As yet, in the case of human cloning, such compelling reasons have not been produced. Suggestions have been made, but have not been sustained, that human dignity may be compromised by the techniques of cloning. Dworkin's arguments suggest that human dignity and indeed democratic constitutions may be compromised by attempts to limit procreative autonomy, at least where greater values cannot be shown to be thereby threatened.

In the absence of compelling arguments against human cloning, we can bid Dolly a cautious "hello". We surely have sufficient reasons to permit experiments on human embryos to proceed, provided, as with any such experiments, the embryos are destroyed at an early stage.[34] While we wait to see whether the technique will ever be established as safe, we should consider the best ways to regulate its uptake until we are in a position to know what will emerge both by way of benefits and in terms of burdens.

Acknowledgements

This paper was presented to the *UNDP/WHO/World Bank Special Programme of Research, Development and Research Training in Human Reproduction* Scientific and Ethical Review Group Meeting, Geneva 25th April 1997 and to a hearing on cloning held by the *European parliament* in Brussels, 7th May 1997. I am grateful to participants at these events for many stimulating insights. I must also thank Justine Burley, Christopher Graham and Pedro Lowenstein for many constructive comments.

John Harris is Sir David Alliance Professor of Bioethics and Research Director of The Centre For Social Ethics and Policy, University of Manchester and a Director of The Institute of Medicine, Law and Bioethics.

References and notes

1 The arguments concerning human dignity are developed in my Cloning and human dignity in *The Cambridge Quarterly of Healthcare Ethics* [in press].The issues raised by cloning were discussed in a special issue of the *Kennedy Institute of Ethics Journal* 1994; **4,3** and in my *Wonderwoman and Superman: the ethics of human biotechnology*. Oxford University Press, Oxford 1992, especially ch 1.
2 Human embryo cloning reported. *Science* 1993; **262:** 652–3.
3 Where few eggs can be successfully recovered or where only one embryo has been successfully fertilised, this method can multiply the embryos available for implantation to increase the chances of successful infertility treatment.
4 Wilmut I *et al.* Viable offspring derived from fetal and adult mammalian cells. *Nature* 1997; **385:** 810–13.
5 Arlidge J. *The Guardian* 1997 Feb 26: 6.
6 Radford T. *The Guardian* 1997 Feb 28: 1.
7 WHO press release (WHO/20 1997 Mar 11).
8 WHO document (WHA50.37 1997 May 14). Despite the findings of a meeting of the Scientific and Ethical Review Group (see **Acknowledgements**) which recommended that "the next step should be a thorough exploration and fuller discussion of the [issues]".
9 UNESCO press release No 97-29 1997 Feb 28.

360 *"Goodbye Dolly?" The ethics of human cloning*

10 The European parliament. Resolution on cloning. Motion dated March 11 1997. Passed March 13 1997.
11 Perhaps the sin of Onan was to compromise the security of his genetic material?
12 Kahn A. Clone mammals . . . clone man. *Nature* 1997; **386:** 119.
13 For use of the term and the idea of "instrumentalization" see: *Opinion of the group of advisers on the ethical implications of biotechnology to the European Commission No 9.* 1997 28 May. Rapporteur, Dr Anne McClaren.
14 Kahn A. Cloning, dignity and ethical revisionism. *Nature* 1997; **388:** 320. Harris J. Is cloning an attack on human dignity? *Nature* 1997; **387:** 754.
15 See my *Wonderwoman and Superman: the ethics of human biotechnology.* Oxford University press, Oxford 1992: ch 2.
16 It is unlikely that "artificial" cloning would ever approach such a rate on a global scale and we could, of course, use regulative mechanisms to prevent this without banning the process entirely. I take this figure of the rate of natural twinning from Moore KL and Persaud TVN. *The developing human* [5th ed]. Philadelphia: WB Saunders, 1993. The rate mentioned is 1 per 270 pregnancies.
17 Of course if *all* people were compulsorily sterilised and reproduced only by cloning, genetic variation would become fixed at current levels. This would halt the evolutionary process. How bad or good *this* would be could only be known if the course of future evolution and its effects could be accurately predicted.
18 Mitochondrial DNA individualises the genotype even of clones to some extent.
19 Although of course there would be implications for criminal justice since clones could not be differentiated by so called "genetic fingerprinting" techniques.
20 Unless of course the nucleus donor is also the egg donor.
21 Margaret Brazier alerted me to this possibility.
22 Apparently Alan Trounson's group in Melbourne Australia have recorded this result. *The Herald Sun* 1997 Mar 13.
23 What mad dictators might achieve is another matter; but such individuals are, almost by definition, impervious to moral argument and can therefore, for present purposes, be ignored.
24 Except by Pedro Lowenstein, who pointed them out to me.
25 These possibilities were pointed out to me by Pedro Lowenstein who is currently working on the implications for human gene therapy.
26 Science and technology: *The Economist* 1997 Mar 1: 101–4.
27 *Universal Declaration of Human Rights* (article 16). *European Convention on Human Rights* (article 12). These are vague protections and do not mention any particular ways of founding families.
28 These include the use of reproductive technologies such as surrogacy and Intra Cytoplasmic Sperm Injection (ICSI).
29 Dworkin R. *Life's dominion.* London: Harper Collins, 1993: 148.
30 See reference 28: 160.
31 Dworkin *R. Freedom's law.* Oxford: Oxford University Press,1996: 237–8.
32 Ronald Dworkin has produced an elegant account of the way the price we should be willing to pay for freedom might or might not be traded off against the costs. See his *Taking rights seriously*, London: Duckworth, 1977: ch 10. And his *A matter of principle,* Cambridge, Mass: Harvard University Press, 1985: ch 17.
33 See reference 27: 166–7.
34 The blanket objection to experimentation on humans suggested by the European parliament resolution would dramatically change current practice on the use of spare or experimental human embryos.

[20]

Cloning and Infertility

CARSON STRONG

Although there are important moral arguments against cloning[1] human beings, it has been suggested that there might be exceptional cases in which cloning humans would be ethically permissible.[2-3] One type of supposed exceptional case involves infertile couples who want to have children by cloning. This paper explores whether cloning would be ethically permissible in infertility cases and the separate question of whether we should have a policy allowing cloning in such cases. One caveat should be stated at the beginning, however. After the cloning of a sheep in Scotland, scientists pointed out that using the same technique to clone humans would, at present, involve substantial risks of producing children with birth defects.[4-6] This concern over safety gives compelling support to the view that it would be wrong to attempt human cloning now. Thus, we do not reach the debate about exceptional cases unless the issue of safety can be set aside. I ask the reader to consider the possibility that in the future humans could be cloned without a significantly elevated risk of birth defects from the cloning process itself. The remainder of this paper assumes, for sake of argument, that cloning technology has advanced to that point. Given this assumption, would cloning in the infertility cases be ethically permissible, and should it be legally permitted?

An example of the type of case in question is a scenario in which the woman cannot produce ova and the man cannot produce sperm capable of fertilizing ova.[7] Like many couples, they wish to have a child genetically related to at least one of them. One approach would use sperm and ova donated by family members, but suppose that no family donors are available in this case. Let us assume, in other words, that cloning is the only way they could have a child genetically related to one of them. Imagine the couple asking their infertility doctor to help them have a child by cloning. This would involve replacing the nucleus of a donated ovum with that of a cell taken from either member of the couple. Suppose that an ovum donor is available who is willing to participate in this process. The infertile couple would decide whether to duplicate genetically the woman or the man. They could try to have a girl or a boy or possibly a child of each sex—fraternal twins. They could use cloning again to have subsequent children: perhaps one the opposite sex of a first child; or another the same sex; or twins again, among other possibilities.

Many would consider cloning ethically unjustifiable in such cases. Following the birth of Dolly the sheep clone, the response from ethicists, politicians, and journalists was overwhelmingly against cloning human beings.[8-12] In fact, few issues in bioethics seem to have reached the high level of consensus found in our society's opposition to human cloning. This opposition rests on a number of concerns, religious and secular. I will focus on the secular arguments, which include at least the following main ones. First, the persons produced would lack genetic uniqueness, and this might be psychologically harmful to them. Second, this reproductive method transforms babymaking into a process similar to manufacturing. Children would become products made according to specification; this would objectify children and adversely affect parental attitudes toward children and other aspects of parent–child relationships. Third, additional abuses might occur if this technology were obtained by totalitarian regimes or other unscrupulous persons.

My main thesis is that the ethics of cloning is not as clear-cut as many seem to think. Specifically, when the arguments against cloning are applied to infertility cases like the one described above, they are not as strong as they might initially appear. Such cases can reasonably move us away from the view that cloning humans is always wrong. Moreover, the arguments for legally prohibiting cloning in such cases are not strong enough to support such restrictions.

Whether cloning in the above case is ethically justifiable rests on the following question: Which is weightier, infertile couples' reproductive freedom to use cloning or the arguments against cloning humans? To address this question, we need to examine closely both the importance of the freedom of infertile couples to utilize cloning and the arguments against its use.

Freedom of Infertile Couples to Use Cloning

Let us begin by asking why the freedom of infertile couples to use cloning should be valued. Because a main reason to use cloning in the above case is to have children who are genetically related to at least one member of the couple, we need to ask whether reasons can be given to value the having of genetically related children. It is worth noting that studies[13-18] have identified a number of reasons people actually give for having genetically related offspring, some of which seem selfish and confused. For example, some people desire genetic children as a way to demonstrate their virility or femininity. The views on which these reasons seem to be based—that virility is central to the worth of a man, and that women must have babies to prove their femininity—are unwarranted. They stereotype sex-roles and overlook ways self-esteem can be enhanced other than by having genetic offspring. Another example involves desiring a genetic child in order to "save" a shaky marriage. This reason fails to address the sources of the marital problems, and the added stress of raising the child might make the marital relationship even more difficult. Some commentators seem to think that the desire for genetic offspring is always unreasonable, as in these examples.[19] Rather than make this assumption, we should consider whether there are defensible reasons that could be given for desiring genetic offspring.

I have explored this question elsewhere, focusing on a category of procreation commonly referred to as "having a child of one's own," sometimes stated simply as "having a child" or "having children."[20] Although these expressions can be interpreted in several ways, I use them to refer to begetting, by sexual intercourse, a child whom one rears or helps rear. This, of course, is the common type of procreation, in which parents raise children genetically their own. My strategy was to try to understand why having genetic offspring might be meaningful to people in this ordinary scenario, and then use this understanding to address assisted reproductive technologies. For the common type of procreation, I identified six reasons people might give for valuing the having of genetic offspring. Briefly, they are as follows: having a child involves participation in the creation of a person; it can be an affirmation of a couple's mutual love and acceptance of each other; it can contribute to sexual intimacy; it provides a link to future persons; it involves experiences of pregnancy and childbirth; and it leads to experiences associated with child rearing. However, the above infertility case differs in several ways from the ordinary type of procreation, including the fact that children would not be created by sexual intercourse. Because of these differences, not all the reasons that might be given in justifying the desire for genetic offspring in the common scenario would be strong reasons in the cloning situation. Nevertheless, I believe that at least the following two reasons would be significant.

Participation in the Creation of a Person

When one 'has a child of one's own,' as defined above, a normal outcome is the creation of an individual with self-consciousness. Philosophers have regarded the phenomenon of self-consciousness with wonder, noting that it raises perplexing questions: What is the relationship between body and mind? How can the physical matter of the brain give rise to consciousness? It is ironic that although we have difficulty giving satisfactory answers to these questions, we can create self-consciousness with relative ease. Each of us who begets or gestates a child who becomes self-conscious participates in the creation of a person. One might say that in having children we participate in the mystery of the creation of self-consciousness. For this reason, among others, some might regard creating a person as an important event, perhaps one with metaphysical or spiritual dimensions (p. 114).[21] Perhaps not all who have children think about procreation in these terms, but this is a reason that can be given to help justify the desire for genetic offspring.

Similarly, the infertile couple might reasonably value the use of cloning because it would enable them to participate in the creation of a person. The member of the couple whose chromosomes are used would participate by providing the genetic material for the new person. Regardless of whose chromosomes are used, if the woman is capable of gestating, she could participate by gestating and giving birth to the child.

It might be objected that the infertile couple could participate in the creation of a person by using donor gametes or preembryos, in the sense that they would authorize the steps taken in an attempt to create a person. Also, if the woman were the gestational mother and used donor gametes or preembryos, then she would participate biologically in the creation of the person. In reply, although these would constitute types of participation, a more direct involvement would occur if one member of the couple contributed genetically to the creation of the child. From the body of one of them would come the makeup of the new person. Cloning would be the only way that the man, in fact, could participate biologically in the creation of the person. This more direct involvement would increase the degree to which the couple participates in the creation of a person, and for some this greater participation might be especially meaningful.

Carson Strong

Affirmation of Mutual Love

In the ordinary type of procreation, intentionally having children can be an affirmation of a couple's mutual love and acceptance of each other. It can be a deep expression of acceptance to say to another, in effect, "I want a child to come forth from your body and mine." In such a context there might be an anticipation that the bond between the couple will grow stronger because of common children to whom each has a biological relationship. To intentionally seek the strengthening of their personal bond in this manner can be a further affirmation of mutual love and acceptance.

In the infertility case in question, if cloning is used, then the child would not receive genes from both parents. Nevertheless, a similar affirmation of mutual love is possible if the woman is capable of being the gestational mother and the man's genes are used. In that situation, it remains true that the child comes forth from their two bodies. Assuming mutual love, the woman bears a child having the genes of the man who loves her and is loved by her. Alternatively, suppose that the woman's genes are used. The man then can become the social father of a child having the genes of the woman who loves him and is loved by him; to seek to become social parents in this manner can also be an affirmation of mutual acceptance.

It might be objected that having children by donor gametes or preembryos—or even adopting children—can also be an affirmation of mutual acceptance, for each member of the couple selects the other to be a social parent of their children. These types of affirmation could enrich the couple's relationship with each other. In response, although these would indeed be forms of affirmation, they can be viewed as different from the affirmation involved in trying to have genetically related children. Intentionally to create a child having the partner's genes might be regarded by some as a special type of affirmation, one that would enrich the couple's relationship in its own distinctive way. For some couples, this type of affirmation might have special significance.

In stating these two reasons, I do not mean to imply that one *ought* to desire genetic offspring, much less that one ought to desire cloning if necessary to have genetically related children. Rather, the point is that the desire for genetic offspring—and hence the desire for cloning in the situation being considered—could be supported by reasons that deserve consideration. Although not everyone in the infertile couple's situation would want to pursue cloning, some might. These reasons also help explain why *freedom* to use cloning to have biological children might be considered valuable; namely, because some couples might value either the opportunity to participate directly in the creation of a person or an affirmation of mutual love that can be associated with that endeavor, or both.

Arguments against Cloning

Let us turn to the considerations against cloning, beginning with the arguments based on lack of genetic uniqueness. What exactly are the adverse effects envisioned for persons who are genetically identical to others? Perhaps the most obvious concerns involve the possibility of being one of *many* genetically identical persons—perhaps one of hundreds of clones, or thousands, or even more. One argument is that the clones would be psychologically harmed, in that they would feel insignificant and have low self-esteem. If I know that I am one of many who physically are exact duplicates, then I might easily believe that there is little or nothing special about me. Apart from what the clones would feel, objections to multiple cloning can also be made from a deontological perspective; it would be an affront to their human dignity to be one of so many genetic replicas. There would also be a serious violation of personal autonomy if the clones were under the control of those who produced them.

Lack of genetic uniqueness can raise concerns even if there are not many clones. Imagine, for example, being the single clone of a person 30 years older. There might be a tendency for the older person's life to be regarded as a standard to be met or exceeded by the younger one. If the clone feels pressured to accept that standard, this might be a significant impediment to freedom in directing one's own life. In addition, knowing that one is a clone might be psychologically harmful in this situation too. For example, self-esteem could be diminished; perhaps the child would regard herself as nothing more than a copy of someone who has already traveled the path ahead.

Although these arguments initially appear persuasive, we need to consider the conclusions reached when we apply them to the infertility case. To begin, the arguments based on large cohorts of clones would not be relevant. The couple might create only one or two children using cloning; thus, their use of cloning can be distinguished from scenarios in which large numbers of clones are produced.

The argument that the parent's life might be regarded as a standard would be relevant, but a response can be given. For one thing, parents' lives often are held up as standards, even in the absence of cloning. This can be either good or bad for the child, depending on how it is handled; it has the potential to promote as well as inhibit development of the child's talents, abilities, and autonomy. Similarly, a clone's being given a role model or standard is not necessarily bad. It depends on how the standard is used and regarded by those directly involved. If it is used by parents in a loving and nurturing manner, it can help children develop their autonomy, rather than inhibit it.

It might be objected that when the child is genetically identical to the parent, there will be a tendency for the parent to be less forgiving when the child fails to meet expectations. If so, the parental standard might tend to thwart rather than promote the child's developing autonomy. In reply, it should be noted that this concern is rather speculative. We do not know the extent, if any, to which the child's being genetically identical would tend to promote a domineering attitude on the part of the parent. Moreover, there is a way to address this concern other than forbidding the infertile couple to use cloning; in particular, the couple could be counseled about the possible psychological dimensions of parenthood through cloning. Psychological counseling already is widely accepted in preparing infertile couples for various noncoital reproductive methods, such as donor insemination and surrogate motherhood. Similarly, it would be appropriate to offer psychological counseling if cloning is made available to infertile couples. The aims of such counseling could include raising awareness about, and thereby attempting to prevent or reduce problems associated with, a possible tendency of parents to be too demanding.

With regard to the claim that cloning even a single child would be psychologically harmful, it can be replied that such claims misuse the concept of 'harm.' Specifically, there is a serious problem with the claim that it would *harm* a child to bring her into being in circumstances where she would experience adverse psychological effects from being a clone. To explain this problem, we need to consider what it means to be harmed.

Harming versus Wronging

I shall draw upon Joel Feinberg's detailed and helpful discussions of what is involved in being harmed.[22-23] A key point is that persons are harmed only if they are caused to be worse off than they otherwise would have been. As Feinberg expresses it, one harms another only if the victim's personal interest is in a worse condition than it would have been had the perpetrator not acted as he did.[24] The claim that cloning harms the children who are brought into being, therefore, amounts to saying that the children are *worse off than they would have been if they had not been created*. Many readers will see problems with such an assertion. Some will say that it fails to make sense because it attempts to compare nonexistence with something that exists, and therefore it is neither true nor false. Others will maintain that it is *false*. Whether incoherent or false, it should be rejected. Because I have addressed the relative merits of these two criticisms elsewhere, I will not repeat that discussion, except to say that the better explanation of what is wrong with the statement seems to be that it is false.[25-26] To see its falsity, let us consider what a life would have to be like in order reasonably to say that it is worse for the person living it than nonexistence. I suggest that a life would have to be so filled with pain and suffering that these negative experiences greatly overshadow any pleasurable or other positive experiences the individual might have. If a neonate were born with a painful, debilitating, and fatal genetic disease, for example, we could reasonably make such a statement. However, the gap between such a neonate and a child cloned by an infertile couple is exceedingly great. Even if the cloned child experienced adverse psychological states associated with being a clone, that would not amount to a life filled with pain and suffering. Thus, the concept of harm is not appropriate for describing the cloned individual's condition.

I have applied the usual concept of harm to situations involving cloning. However, there is an objection that should be considered. It might be asserted that applying this concept of harm to this type of situation leads to implausible conclusions. Specifically, it seems to imply that it is ethically justifiable knowingly to create a child who will suffer disadvantages—even serious ones—as long as those disadvantages are not so severe that the life will be worse than nonexistence. To illustrate, consider a hypothetical situation in which a person with cystic fibrosis asks to be cloned, and a cure for cystic fibrosis is not yet available. Suppose that a physician knows about the cystic fibrosis but nevertheless provides the cloning, and the child later suffers the adverse effects of cystic fibrosis. It seems wrong for the physician to carry out the cloning in this example. According to the usual concept of harm, however, the child in this scenario is not harmed by being cloned, given that her life is better than nonexistence. If we cannot say that the child is harmed, then how can we account for our view that the cloning is unethical?

In reply, we can account for cloning being unethical in such a case without inventing a new concept of harm. Although the child is not harmed by being cloned, we can say that she is *wronged*. In particular, we can say that a certain right of the child is violated. It has been suggested that there is a type of birthright according to which people have a right to be born free of serious impediments to their well being.[27–30] It is important not to misunderstand this right; it is not a 'right to be born' but rather a right possessed by all persons who *are* born. Also, it is not what one might call a 'right against nature'; if a child is born with serious handicaps through the fault of no one, then the right in question is not violated. The right would only be violated if someone negligently or intentionally created a child with the requisite handicaps. I suggest that we can account for the wrongness of cloning in the cystic fibrosis example by positing such a birthright, a right that one's circumstances of birth be free of impediments that would seriously impair one's ability to develop in a healthy manner and to realize a normal potential. Negligently or intentionally creating a child who faces such severe impediments is a violation of that right, which I shall refer to as a right to a decent minimum opportunity for development. The handicaps imposed on the child with cystic fibrosis are not so severe that we could reasonably say that her life is worse than nonexistence, but they are severe enough to impair seriously her ability to develop.

An objection can be made against using the concept 'being wronged' to describe what is unethical about cloning in the cystic fibrosis example. Specifically, if the child were to claim that she was wronged by those who created her, it would commit her to the judgment that their duty had been to refrain from doing what they did; but if they had refrained, it would have led to her never being born, an even worse result from her point of view. Thus, it is argued, the child cannot reasonably claim that her creators should have acted otherwise, given that her life with cystic fibrosis is preferable to nonexistence. If she cannot make this claim, then she has no genuine grievance against them and cannot claim that they wronged her.[31–32]

However, this objection is mistaken. Perhaps we can see this more clearly by considering a similar type of situation. Instead of acts that cause a person to come into being, let us consider actions that cause an existing person to continue to live—that is, *life-saving* actions. Both types of acts have the result that a person exists who otherwise would not be in existence. Consider a patient who has suffered substantial bleeding because of a ruptured ulcer. Suppose that blood transfusions are necessary to save her life but she refuses transfusions on religious grounds and is considered mentally competent. Imagine that a physician provides transfusions despite her refusal, with the result that the patient's life is saved. Let us suppose, also, that the patient later states that she is better off having been kept alive. According to the objection in question, the patient cannot reasonably claim that the physician should have refrained from treating, given that her continued life is better than nonexistence. If she cannot make this claim, then she cannot claim that the physician wronged her.

But this conclusion is incorrect. Clearly, the patient can validly claim that she was wronged; her rights to informed consent and self-determination were violated. The benefits caused by the physician's act do not alter the fact that these rights were infringed. If a beneficial outcome removed all wrongness of the act, that would mean that paternalism, when successfully carried out, is ethically justifiable. This would be inconsistent with the view that competent patients have a right to refuse life-saving treatment. Similarly, the fact that the cloned child with cystic fibrosis has a life that is better than nonexistence does not mean that the child was not wronged. More generally, a child who is intentionally or negligently brought into being in circumstances where she lacks a decent minimum opportunity for development is wronged, even if her life is better than nonexistence, just as a competent patient who is forced to receive life-saving treatment is wronged, even though her subsequent life is preferable to nonexistence.[33]

Assuming there is a right to a decent minimum opportunity for development, the question arises concerning how serious the impediments must be in order for the right to be violated. No doubt, there is room for disagreement concerning this issue, and a sharp line probably cannot be drawn. Nevertheless, a basic concept can be stated: the impediments must be severe, not minor. As examples, I would suggest that creating a clone who has cystic fibrosis or Down's syndrome would violate the right in question, but creating a clone with, say, nearsightedness would not in itself constitute violation of the right.

Can we say, in the infertility case, that the cloned child would experience psychological problems of such magnitude that the right to a decent minimum opportunity for development would be violated? I suggest that the answer would depend largely on the approach taken by the parents in raising the child. Consider a hypothetical world in which the parents of cloned children always undermine their self-esteem. Then we might reasonably say that being a clone is associated with such serious impediments that the act of cloning violates the child's right to a decent minimum opportunity for development. Consider another hypothetical world in which many but not all parents of cloned children undermine their self-esteem. Then we might reasonably say that cloning puts a child at risk of experiencing obstacles severe enough to constitute a violation of the right in question. Depending on the level of risk involved, we might decide that cloning is wrong because the risk is unacceptably high. In yet another hypothetical world, the parents of cloned children are no more likely to undermine their self-esteem than the parents of other children. In that world it would not be reasonable to say that cloning in itself violates the right in question. If we were to allow infertile couples to use cloning, then which of these hypothetical worlds would the real world be most like? The fact is, we do not know. And this is the problem with the argument that cloned children would experience severe psychological obstacles to well-being; it is based on empirical assumptions concerning how cloned children would generally be treated by parents and others, and we lack evidence supporting those assumptions.

The Argument from Parent–Child Relationships

Let us consider the argument that children would be objectified and parent-child relationships generally would be adversely affected. This argument arises from reflection on what it would be like if there were a widespread practice of controlling the characteristics of our offspring. This practice might involve the insertion and deletion of genes in human preembryos, as well as cloning and other laboratory techniques not yet envisioned. In some cases, such manipulations might have a therapeutic goal; perhaps disease-causing genes would be replaced by normal ones. Objections to such therapeutic manipulations have been based mainly on concerns about whether they can be performed safely. But other forms of genetic control might have the much different goal of enhancing offspring nondisease characteristics, such as height, intelligence, and body build, and it is especially this type of control that raises concerns about undesirable changes in the attitudes and expectations of parents toward their children.[34-35] The specific concerns can be expressed by a number of questions: If a child failed to manifest the qualities she was designed to have, would the parents be less inclined to accept the child's weaknesses? Would children be regarded more as objects and less as persons? Would less tolerance for imperfection result in less compassion toward the handicapped? Would children who recognize their own shortcomings blame their parents for failing to design them better? Would such feelings sometimes disrupt family relationships? Would knowledge of being designed make a child feel more controlled by parents? Would this result, for example, in greater adolescent rebelliousness? These and other questions suggest a number of ways in which disharmony could enter into parent–child relationships.

However, a reply can be made. Although these are important concerns, their bearing on infertility cases is tangential. Because cloning does not involve inserting and deleting genes, concerns over whether these particular manipulations can be done safely are not directly applicable. Also, cloning in the infertility cases does not involve efforts to enhance the child's genetic makeup. Thus, the concerns expressed above that are specific to enhancement do not directly apply, either; these include the concern that modifying children in order to enhance their characteristics objectifies them. The claim that cloning in the infertility case objectifies the child is weakened by the fact that the purpose is not to design the child but to have a genetically related child, in the only way that is possible. In addition, the number of cases like the one being considered—those in which there is both male and female infertility and use of family donors has been ruled out—would be relatively small. Thus, it is difficult to argue that cloning restricted to such cases would result in widespread changes in parent–child relationships. For these reasons, the arguments in question do not succeed in showing that cloning in the infertility cases would be wrong.

The Argument Based on Abuses

The third argument is that abuses might occur if the technology of cloning is used by unscrupulous persons. A paradigm of such envisioned abuse is found in Aldous Huxley's *Brave New World*, in which multiple genetically identical persons are produced and conditioned to fill defined social roles.[36] Other variations of such abuse could be imagined, in which cloning plays a role in the systematic control of persons by determining their genetic makeup and upbringing.

In reply, it seems clear that we can distinguish such abuse from the infertility case in question; the couple's trying to have a child would not constitute such abuse or even remotely approach it. Perhaps it will be objected that permitting cloning in the infertility cases would facilitate development of the technology, thereby making it more likely that unscrupulous persons could use it. In reply, this objection assumes that human preembryos would not be created by cloning as part of research. In the future it might become useful to create such preembryos, not for the purpose of transferring them to a woman's uterus, but for studies in any of a number of scientific areas, perhaps including preembryo development, cell differentiation, immunologic properties of cells, or the creation of cell lines using stem cells. Some of these scientific areas—or others not mentioned here—might be considered important enough in the future to justify creating human preembryos by cloning for purposes of laboratory studies.[37]

Thus, the technology of cloning humans might go forward, even if use of that technology to produce babies is proscribed.

Although the main arguments against cloning are not persuasive when applied to the infertility cases, at least two additional arguments can be presented that focus more directly on those cases. First, consider a son who is, as we might put it, genetically identical to his mother's husband. Some might be concerned by the somewhat oedipal nature of this situation. Would knowing that one's mother is sexually attracted to someone genetically identical to oneself cause special psychological problems for the child? Would the ill effects be great enough to make life worse than nonexistence? Would the right to a decent minimum opportunity for development be violated? Second, consider a father whose daughter is genetically identical to his wife. Would there tend to be a higher incidence of sexual abuse in this type of situation? These questions raise legitimate concerns, but because the answers are highly speculative at present, these concerns do not constitute definitive arguments against the particular use of cloning in question. However, they suggest possible topics for inclusion in preimplantation counseling, if such counseling is provided.

In summary, some main reasons have been identified supporting freedom to use cloning in the type of infertility case being considered. Also, each of the main arguments against cloning has been shown to involve substantial difficulties when applied to the infertility cases. It seems reasonable to conclude that the arguments against cloning are not compelling enough, either singly or collectively, to support the conclusion that cloning is wrong in such cases.

Cloning versus Collaborative Reproduction

It might be objected that it would be ethically preferable for the couple to have children who are not genetically related to them. Adoption would be a possibility, or donor gametes or preembryos could be used. If the woman is capable of gestating using donor gametes or preembryos, then at least she could be biologically related to the child. In reply, some infertile couples prefer not to adopt, even if that is the only way they can have children to raise. Moreover, if the couple tried to adopt, there would be a significant chance that their attempt would be unsuccessful because of the difficulties involved in adoption.

The claim that using donated gametes or preembryos would be ethically preferable to cloning overlooks the problems associated with third-party collaborative reproduction. Such arrangements raise a number of difficult issues because of the separation of genetic and social parenthood: What should the children be told about their origins? When and how should any informing take place? What if the child later wants to meet the genetic parents?

For example, when there is male-factor infertility, the man often prefers secrecy concerning his inability to beget. As a result, couples often choose not to reveal the fact of gamete donation to the child and others. This creates the problem of there being a significant deception at the center of the family relationship. Maintaining this deception can take its toll, including a substantial emotional burden on the couple. Also, such deception is at odds with the values of honesty and trust that should bind families together. On the other hand, children who are told the truth about their origins might develop strong desires to meet their genetic parents. If these wishes are frustrated, the result might be substantial emotional distress for the child. Thus, depending on how these various issues are handled, adverse psychological consequences are possible for the child and family. It would be reasonable for the couple to prefer not to encounter these problems, and cloning would provide a way to avoid them. The existence of these problems calls into question the claim that third-party collaborative reproduction would be less ethically problematic for the couple than cloning.

It might be objected that cloning would also raise difficult issues for the couple. For example, should the fact of cloning be revealed to the child, and if so, when and how? If the child is not told, then she will not suffer any psychological ill effects arising from knowledge of being a clone. But if such children were never told, then they would be deprived of important information about their background. Should family and friends be told about the cloning? Or should they be led to believe that the child simply "looks like" one of the parents?[38] There might be disagreements between husband and wife over whom, when, and how to tell. In reply, although these too are difficult issues, their existence seems to indicate that the two approaches to reproduction are at a standoff in this regard; both raise issues that carry the potential for interpersonal conflict within the family. Because of this parity, it is not obvious that third-party collaborative reproduction is ethically preferable to cloning.

I have been discussing a situation in which cloning is the only way for a couple to have a child genetically related to one of them. However, the arguments I have stated in support of cloning are applicable to other infertility cases in which the alternative is third-party collaboration. Suppose the woman cannot produce ova, but donor ova and the husband's sperm could be used. The couple might nevertheless prefer cloning in order to avoid the complications associated with third-party gamete donation discussed above.[39] Alternatively, suppose the man cannot produce sperm but the woman's ova could be used with donor insemination. A preference for cloning might be based, not solely on male ego, but also on a desire to avoid the problems associated with third-party reproduction. Moreover, the three main arguments against cloning do not fare better when applied to these types of infertility cases. Again, cloning does not harm the child, nor is it clear that the right to a decent minimum opportunity for development would be violated. The argument that parent-child relationships generally would be adversely affected continues to be unpersuasive because enhancement is not involved and, although we now are dealing with a larger class of infertile couples, the number still is relatively small compared to the general population. These scenarios also would be distinguishable from *Brave New World* abuses. Thus, the arguments against cloning do not constitute compelling reasons to override the freedom of infertile couples to use cloning in these cases either.

Cloning as a Bridge to Future Remedies for Infertility

Research in gene therapy is resulting in the discovery of ways to insert genes into human cells. This is increasing the plausibility of the view that in the future it might be possible to insert, and perhaps delete, a variety of chosen genes. Such modifications could be performed on cells prior to using them for cloning. By means of such techniques, a child created with cloning technology could have genes from both parents. Starting with a cell from one parent, one might change hair color, skin complexion, and eye color, using genes from the other parent. Perhaps genetic defects causing infertility and susceptibilities to other diseases would also be corrected. The child then would be genetically unique. Thus, the argument against cloning based on lack of genetic uniqueness would no longer be applicable. Prohibiting cloning in the future might prevent us from helping infertile couples in these ways.

Such modifications would not necessarily include changes that constitute 'enhancement,' such as higher intelligence, better body build, or greater height; the goal could be to make the child genetically different from either parent, rather than to produce a 'superior' child. In that event, objections to genetically enhancing our offspring would not be applicable. Moreover, this particular use of genetic technology—to help an infertile couple have a child—would also be distinguishable from *Brave New World* abuses.

This type of reproduction would not be cloning, strictly speaking. The term "cloning" in both scientific and lay usage, implies the production of a genetically identical copy. In fact, we lack a common term to refer specifically to the type of reproduction being envisioned.

It might be objected that we could not ethically create children in this manner because developing the technology would involve experimenting on unconsenting subjects. It might be claimed, for example, that it would be unethical to try to alter genetically a child's hair color because some unintended adverse genetic modification might occur. It might be argued that altered hair color is not a significant enough benefit to justify the risks involved. In reply, it is conceivable that our technology might advance to the point where the risks involved in such an attempt would be low. Moreover, if being a clone exposes one to the risk of adverse psychological effects, as opponents of cloning maintain, then alterations that prevent one from being a clone might have benefits significant enough to outweigh the risks. Therefore, I do not believe we can reasonably claim that such genetic modifications could never ethically be done. If such modifications could be performed safely, this type of reproduction might be ethically *preferable* to reproduction using donor gametes or preembryos because the problems associated with third-party collaboration would be avoided.

Should Exceptional Cases Be Permitted?

I have argued that cloning humans could sometimes be ethically defensible in cases of infertility. It might be objected that cloning should not be permitted even in those cases in which it is ethically justifiable. One argument supporting this objection begins by claiming that a general prohibition of human cloning is warranted, based on the reasons against cloning discussed above. Legal restrictions are needed to prevent the creation of cohorts of multiple clones, as well as other clear abuses of cloning technology. Restrictions also are needed to prevent a widespread practice of cloning, thereby avoiding the feared ill effects on parent-child relationships. Moreover, it is claimed that practical problems involved in attempting to enforce a general policy against human cloning while permitting exceptions provide grounds for not allowing the exceptions.[40] In particular, it would be difficult for authorities to gather the evidence needed to distinguish allowable from nonallowable cases. For example, fertile couples might have children by cloning, yet claim that they are infertile. Prosecution could not reasonably proceed unless evidence ruling out infertility were obtained. Assuming that an infertility doctor assisted the couple in the cloning, often that doctor's testimony and records would be crucial evidence. However, such confidential records are protected in the absence of a court order, and establishing 'reasonable cause' for such court orders might be difficult in many cases. Moreover, the couple could refuse to release the records voluntarily, claiming (perhaps disingenuously) that the information is too personal and sensitive. The legal protection of medical records behind which such couples could hide is itself important and based on constitutional guarantees against unreasonable invasions of privacy. It can be argued that relaxing such protections in order to distinguish permitted from nonpermitted cloning would be too intrusive of reproductive privacy. Thus, if we permit cloning in the infertility cases, in effect we open the door for anyone to use cloning and get away with it.

In reply, it is possible to have widespread compliance with a law even though there are difficulties in detecting violations. If we were to make cloning except in infertility cases illegal not only for the couples using it but also for the physicians carrying it out, I believe that we would see widespread compliance. Most physicians will choose to avoid illegal activity, even if authorities would face difficulties in detecting violations. This tendency could be reinforced if the penalties for being convicted of the violation are high, even though the likelihood of detection is relatively low. Although a few cases of cloning might occur outside the allowed exceptions, it is doubtful that we would see the sort of widespread practice that the policy in question would attempt to prevent. Reproductive privacy can be protected while having a generally effective policy that prohibits cloning except in cases of infertility.

Another objection is that it would be inconsistent to permit cloning for infertile couples but not fertile ones. If reproductive freedom is important enough to permit infertile couples to use cloning, then why shouldn't all couples be allowed to use it, if that is their desire? According to this argument, if permitting the infertile to clone children commits us logically to allowing everyone to do so, then we should not allow the infertile to clone.

A reply to this objection can be based on the fundamental reasons we give in explaining why procreative freedom is worthy of protection. I identified six reasons that can be given in helping explain why freedom to have children is important. It is worth noting that the goals and values reflected in those six reasons can be achieved by fertile couples without resorting to cloning. By having a child through sexual intercourse, they can: participate in the creation of a person; affirm their mutual love through procreation; deepen their sexual intimacy; obtain a link to future persons; and have experiences associated with pregnancy, childbirth, and child rearing. Therefore, in prohibiting those who are fertile from using cloning, we do not deprive them of the ability to pursue what is valuable about having children. However, when we forbid the infertile to use cloning, we force them to choose either not to realize any of those valued goals or to pursue collaborative reproduction with its associated difficulties. This is the morally relevant difference, I would suggest, that can justify differing policies for fertile and infertile couples. In conclusion, there do not appear to be good reasons to disallow exceptions to cloning for infertile couples.

Carson Strong

Notes

1. In this paper, the term "cloning" refers to creating a child by transferring the nucleus of a somatic cell into an enucleated egg cell. This method should be distinguished from blastomere separation, which involves the division of a preembryo when its cells are totipotent. Although both produce individuals with identical chromosomes, the two methods have different ethical implications. For example, cloning by nuclear transfer involves the possibility of creating numerous duplicates of the original individual, but in blastomere separation only a few copies can be produced, as explained in Cohen J, Tomkin G. The science, fiction, and reality of embryo cloning. *Kennedy Institute of Ethics Journal* 1994;4:193–203.
2. National Bioethics Advisory Commission. *Cloning Human Beings: Report and Recommendations of the National Bioethics Advisory Commission.* Rockville, Maryland: National Bioethics Advisory Commission, 1997:79–81. World Wide Web: http://www.nih.gov/nbac/nbac.htm.
3. Winston R. The promise of cloning for human medicine: not a moral threat but an exciting challenge. *British Medical Journal* 1997;314:913–4.
4. See note 2, National Bioethics Advisory Commission 1997:ii-iii,13,23–4.
5. The researchers who produced Dolly used nuclei from three sources: late embryos, fetal cell cultures, and cell cultures derived from the mammary gland of an adult sheep. Of 277 preembryos created using mammary cells, only one developed into a live lamb. Sixty-two percent of fetuses from all three sources failed to survive until birth, compared to an estimated 6% fetal loss rate after natural mating. This high rate of fetal loss suggests an increased incidence of genetic anomalies. For data on the total number of preembryos and live births, see Wilmut I, Schnieke AE, McWhir J, Kind AJ, Campbell KHS. Viable offspring derived from fetal and adult mammalian cells. *Nature* 1997;385:810–3.
6. Stewart C. Nuclear transplantation: an udder way of making lambs. *Nature* 1997;385:769,771.
7. This situation could result from various medical conditions: the woman's ovaries might have been surgically removed, or she might suffer from premature ovarian failure—the inability of ovaries to produce ova; the man could have testes that produce no sperm, or perhaps a small number of sperm are produced but attempts to perform intracytoplasmic sperm injection (ICSI) using donor ova have been unsuccessful, among other possibilities. For a discussion of the uses and limitations of ICSI, see Silber SJ. What forms of male infertility are there left to cure? *Human Reproduction* 1995;10:503–4.
8. Recer P. Clone fear may slow research. *AP Online* 1997; Mar 5:19:20EST.
9. Nash JM. The age of cloning. *Time* 1997;Mar 10:60–61,64–5.
10. Butler D, Wadman M. Calls for cloning ban sell science short. *Nature* 1997;386:8–9.
11. Masood E. Cloning technique "reveals legal loophole." *Nature* 1997;Feb 27. World Wide Web: http://www.nature.com.
12. Harris J. Is cloning an attack on human dignity? *Nature* 1997;387:754.
13. Pohlman E, assisted by Pohlman JM. *The Psychology of Birth Planning.* Cambridge, Massachusetts: Schenkman, 1969:48–81.
14. Pohlman E. Motivations in wanting conceptions. In: Peck E, Senderowitz J, eds. *Pronatalism: The Myth of Mom and Apple Pie.* New York: Crowell, 1974:159–90.
15. Veevers JE. The social meanings of parenthood. *Psychiatry* 1973;36:291–310.
16. Arnold F. *The Value of Children: A Cross-National Study.* Honolulu: East-West Population Institute, 1975.
17. Laucks EC. *The Meaning of Children: Attitudes and Opinions of a Selected Group of U.S. University Graduates.* Boulder, Colorado: Westview, 1981.
18. Gould RE. The wrong reasons to have children. In: Peck E, Senderowitz J, eds. *Pronatalism: The Myth of Mom and Apple Pie.* New York: Crowell, 1974:193–8.
19. Kahn A. Clone mammals . . . clone man? *Nature* 1997;386:119. Kahn states, "the debate has in the past perhaps paid insufficient attention to the current strong social trend towards a fanatical desire for individuals not simply to have children but to ensure that these children also carry their genes."
20. Strong C. *Ethics in Reproductive and Perinatal Medicine: A New Framework.* New Haven: Yale University Press, 1997:13–22.
21. Ellin J. Sterilization, privacy, and the value of reproduction. In: Davis JW, Hoffmaster B, Shorter S, eds. *Contemporary Issues in Biomedical Ethics.* Clifton, New Jersey: Humana Press, 1978:109–25.
22. Feinberg J. *Harm to Others.* New York: Oxford University Press, 1984:31–64.
23. Feinberg J. Wrongful life and the counterfactual element in harming. *Social Philosophy and Policy* 1987;4:145–78.

Cloning and Infertility

24. See note 23, Feinberg 1987:149. Feinberg's discussion is more extensive; this is only one of six conditions he identifies as necessary and sufficient for harming, pp. 150–53.
25. See note 20, Strong 1997:90–92.
26. Also see note 23, Feinberg 1987:158–9.
27. Bayles MD. Harm to the unconceived. *Philosophy and Public Affairs* 1976;5:292–304.
28. Also see note 22, Feinberg 1984:99.
29. Steinbock B, McClamrock R. When is birth unfair to the child? *Hastings Center Report* 1994;24(6):15–21.
30. Also see note 20, Strong 1997:92–5.
31. This type of objection is put forward by Feinberg. See note 23, Feinberg 1987:168.
32. A similar objection is stated by Brock DW. The non-identity problem and genetic harms—the case of wrongful handicaps. *Bioethics* 1995;9:269–75.
33. For this response to the objection in question see note 20, Strong 1997:93–4.
34. Botkin JR. Prenatal screening: professional standards and the limits of parental choice. *Obstetrics and Gynecology* 1990;75:875–80.
35. Strong C. Tomorrow's prenatal genetic testing: should we test for 'minor' diseases? *Archives of Family Medicine* 1993;2:1187–93.
36. The method of genetic duplication that Huxley described did not involve replacement of the nucleus of an egg cell. In his fictional account, the fertilized egg was described as "budding" when "Bokanovsky's Process" was applied to it; the result could be as many as "ninety-six identical twins."
37. See note 3, Winston 1997.
38. We can imagine friends saying, for example, "He's the spitting image of his father," but not realizing, because of the age difference, that they are genetically identical.
39. Cloning in such a case would involve an ovum donor, but the chromosomes would be removed from the ovum. Although mitochondrial DNA in the ovum would be inherited by the child, the ovum donor would not be a "genetic mother" in the ordinary sense of that term. Thus, although there would be third-party collaboration, it would not involve the difficulties typically associated with third-party genetic parentage.
40. This type of objection is suggested in *Cloning Human Beings*. See note 2, National Bioethics Advisory Commission 1997:81–2.

[21]

Should we clone human beings? Cloning as a source of tissue for transplantation

Julian Savulescu *The Murdoch Institute, Royal Children's Hospital, Melbourne, Australia*

Abstract

The most publicly justifiable application of human cloning, if there is one at all, is to provide self-compatible cells or tissues for medical use, especially transplantation. Some have argued that this raises no new ethical issues above those raised by any form of embryo[1] experimentation. I argue that this research is less morally problematic than other embryo research. Indeed, it is not merely morally permissible but morally required that we employ cloning to produce embryos or fetuses for the sake of providing cells, tissues or even organs for therapy, followed by abortion of the embryo or fetus.

(*Journal of Medical Ethics* 1999;25:87–95)

Keywords: Cloning; transplantation; autonomy; embryonic stem cells; fetal tissue; embryo experimentation; abortion; potential

Introduction

When news broke in 1997 that Ian Wilmut and his colleagues had successfully cloned an adult sheep, there was an ill-informed wave of public, professional and bureaucratic fear and rejection of the new technique. Almost universally, human cloning was condemned.[2-6] Germany, Denmark and Spain have legislation banning cloning; Norway, Slovakia, Sweden and Switzerland have legislation implicitly banning cloning.[7] Some states in Australia, such as Victoria, ban cloning. There are two bills before congress in the US which would comprehensively ban it.[8 9] There is no explicit or implicit ban on cloning in England, Greece, Ireland or the Netherlands, though in England the Human Embryology and Fertilisation Authority, which issues licences for the use of embryos, has indicated that it would not issue any licence for research into "reproductive cloning". This is understood to be cloning to produce a fetus or live birth. Research into cloning in the first 14 days of life might be possible in England.[7]

There have been several arguments given against human reproductive cloning:

1. It is liable to abuse.
2. It violates a person's right to individuality, autonomy, selfhood, etc.
3. It violates a person's right to genetic individuality (whatever that is—identical twins cannot have such a right).
4. It allows eugenic selection.
5. It uses people as a means.
6. Clones are worse off in terms of wellbeing, especially psychological wellbeing.
7. There are safety concerns, especially an increased risk of serious genetic malformation, cancer or shortened lifespan.

There are, however, a number of arguments in favour of human reproductive cloning. These include:

1. General liberty justifications.
2. Freedom to make personal reproductive choices.
3. Freedom of scientific enquiry.
4. Achieving a sense of immortality.
5. Eugenic selection (with or without gene therapy/enhancement).
6. Social utility - cloning socially important people.
7. Treatment of infertility (with or without gene therapy/enhancement).
8. Replacement of a loved dead relative (with or without gene therapy/enhancement).
9. "Insurance" - freeze a split embryo in case something happens to the first: as a source of tissue or as replacement for the first.
10. Source of human cells or tissue.
11. Research into stem cell differentiation to provide an understanding of aging and oncogenesis.
12. Cloning to prevent a genetic disease.

The arguments against cloning have been critically examined elsewhere and I will not repeat them here.[10 11] Few people have given arguments in favour of it. Exceptions include arguments in favour of 7-12,[12] with some commentators favouring only 10-11[13 14] or 11-12.[15] Justifications 10-12 (and possibly 7) all regard cloning as a way of treating or avoiding disease. These have emerged as arguably the strongest justifications for cloning. This paper examines 10 and to some extent 11.

Human cloning as a source of cells or tissue

Cloning is the production of an identical or near-identical genetic copy.[16] Cloning can occur by fission or fusion. Fission is the division of a cell mass into two equal and identical parts, and the development of each into a separate but genetically identical or near-identical individual. This occurs in nature as identical twins.

Cloning by fusion involves taking the nucleus from one cell and transferring it to an egg which has had its nucleus removed. Placing the nucleus in the egg reprogrammes the DNA in the nucleus to replicate the whole individual from which the nucleus was derived: nuclear transfer. It differs from fission in that the offspring has only one genetic parent, whose genome is nearly identical to that of the offspring. In fission, the offspring, like the offspring of normal sexual reproduction, inherits half of its genetic material from each of two parents. Henceforth, by "cloning", I mean cloning by fusion.

Human cloning could be used in several ways to produce cells, tissues or organs for the treatment of human disease.

HUMAN CLONING AS A SOURCE OF MULTIPOTENT STEM CELLS

In this paper I will differentiate between totipotent and multipotent stem cells. Stem cells are cells which are early in developmental lineage and have the ability to differentiate into several different mature cell types. Totipotent stem cells are very immature stem cells with the potential to develop into any of the mature cell types in the adult (liver, lung, skin, blood, etc). Multipotential stem cells are more mature stem cells with the potential to develop into different mature forms of a particular cell lineage, for example, bone marrow stem cells can form either white or red blood cells, but they cannot form liver cells.

Multipotential stem cells can be used as

a. a vector for gene therapy.
b. cells for transplantation, especially in bone marrow.

Attempts have been made to use embryonic stem cells from other animals as vectors for gene therapy and as universal transplantation cells in humans. Problems include limited differentiation and rejection. Somatic cells are differentiated cells of the body, and not sex cells which give rise to sperm and eggs. Cloning of somatic cells from a person who is intended as the recipient of cell therapy would provide a source of multipotential stem cells that are not rejected. These could also be vectors for gene therapy. A gene could be inserted into a somatic cell from the patient, followed by selection, nuclear transfer and the culture of the appropriate clonal population of cells in vitro. These cells could then be returned to the patient as a source of new tissue (for example bone marrow in the case of leukaemia) or as tissue without genetic abnormality (in the case of inherited genetic disease). The major experimental issues which would need to be addressed are developing clonal stability during cell amplification and ensuring differentiation into the cell type needed.[13] It should be noted that this procedure does not necessarily involve the production of a multicellular embryo, nor its implantation in vivo or artificially. (Indeed, cross-species cloning—fusing human cells with cow eggs—produces embryos which will not develop into fetuses, let alone viable offspring.[17])

A related procedure would produce totipotent stem cells which could differentiate into multipotent cells of a particular line or function, or even into a specific tissue. This is much closer to reproductive cloning. Embryonic stem cells from mice have been directed to differentiate into vascular endothelium, myocardial and skeletal tissue, haemopoietic precursors and neurons.[18] However, it is not known whether the differentiation of human totipotent stem cells can be controlled in vitro. Unlike the previous application, the production of organs could involve reproductive cloning (the production of a totipotent cell which forms a blastomere), but then differentiates into a tissue after some days. Initially, however, all early embryonic cells are identical. Producing totipotent stem cells in this way is equivalent to the creation of an early embryo.

PRODUCTION OF EMBRYO/FETUS/CHILD/ADULT AS A SOURCE OF TISSUE

An embryo, fetus, child or adult could be produced by cloning, and solid organs or differentiated tissue could be extracted from it.

Cloning as source of organs, tissue and cells for transplantation

THE NEED FOR MORE ORGANS AND TISSUES

Jeffrey Platts reports: "So great is the demand that as few as 5% of the organs needed in the United States ever become available".[19] According to David K C Cooper, this is getting worse: "The discrepancy between the number of potential recipients and donor organs is increasing by approximately 10-15% annually".[20] Increasing procurement of cadaveric organs may not be the solution. Anthony Dorling and colleagues write:

"A study from Seattle, USA, in 1992 identified an annual maximum of only 7,000 brain dead donors in the USA. Assuming 100% consent and suitability, these 14,000 potential kidney grafts would still not match the numbers of new patients commencing dialysis each year. The clear implication is that an alternative source of organs is needed."[21]

Not only is there a shortage of tissue or organs for those with organ failure, but there remain serious problems with the compatibility of tissue or organs used, requiring immunosuppressive therapy with serious side effects. Using cloned tissue would have enormous theoretical advantages, as it could be abundant and there is near perfect immunocompatibility.[22]

There are several ways human cloning could be used to address the shortfall of organs and tissues, and each raises different ethical concerns.

1. PRODUCTION OF TISSUE OR CELLS ONLY BY CONTROLLING DIFFERENTIATION

I will now give an argument to support the use of cloning to produce cells or tissues through control of cellular differentiation.

The fate of one's own tissue

Individuals have a strong interest or right in determining the fate of their own body parts, including their own cells and tissues, at least when this affects the length and quality of their own life. A right might be defended in terms of autonomy or property rights in body parts.

This right extends (under some circumstances) both to the proliferation of cells and to their transmutation into other cell types (which I will call the Principle of Tissue Transmutation).

Defending the Principle of Tissue Transmutation

Consider the following hypothetical example:

Lucas I Lucas is a 22-year-old man with leukaemia. The only effective treatment will be a bone marrow transplant. There is no compatible donor. However, there is a drug which selects a healthy bone marrow cell and causes it to multiply. A doctor would be negligent if he or she did not employ such a drug for the treatment of Lucas's leukaemia. Indeed, there is a moral imperative to develop such drugs if we can and use them. Colony-stimulating factors, which cause blood cells to multiply, are already used in the treatment of leukaemia, and with stored marrow from those in remission in leukaemia before use for reconstitution during relapse.

Lucas II In this version of the example, the drug causes Lucas's healthy skin cells to turn into healthy bone marrow stem cells. There is no relevant moral difference between Lucas I and II. We should develop such drugs and doctors would be negligent if they did not use them.

If this is right, there is nothing problematic about cloning to produce cells or tissues for transplantation by controlling differentiation. All we would be doing is taking, say, a skin cell and turning on and off some components of the total genetic complement to cause the cell to divide as a bone marrow cell. We are causing a differentiated cell (skin cell) to turn directly into a multipotent stem cell (bone marrow stem cell).

Are there any objections? The major objection is one of practicality. It is going to be very difficult to cause a skin cell to turn **directly** into a bone marrow cell. There are also safety considerations. Because we are taking a cell which has already undergone many cell divisions during terminal differentiation to give a mature cell such as a skin cell, and accumulated mutations, there is a theoretical concern about an increased likelihood of malignancy in that clonal population. However, the donor cell in these cases is the same age as the recipient (exactly), and a shorter life span would not be expected. There may also be an advantage in some diseases, such as leukemia, to having a degree of incompatibility between donor and recipient bone marrow so as to enable the donor cells to recognise and destroy malignant recipient cells. This would not apply to non-malignant diseases in which bone marrow transplant is employed, such as the leukodystrophies. Most importantly, all these concerns need to be addressed by further research.

Lucas IIA In practice, it is most likely that skin cells will not be able to be turned directly into bone marrow cells: there will need to be a stage of totipotency in between. The most likely way of producing cells to treat Lucas II is via the cloning route, where a skin cell nucleus is passed through an oocyte to give a totipotent cell. The production of a totipotent stem cell is the production of an embryo.

Production of an embryo as a source of cells or tissues

There are two ways in which an embryo could be a source of cells and tissues. Firstly, the early embryonic cells could be made to differentiate into cells of one tissue type, for example, bone marrow. Secondly, differentiated cells or tissues from an older embryo could be extracted and used directly.

Are these permissible?

In England, the Royal Society[15] has given limited support to cloning for the purposes of treating human disease. The Human Genetics Advisory Commission (HGAC) defines this as "therapeutic cloning," differentiating it from "reproductive cloning",[7] Both bodies claim that embryo experimentation in the first 14 days is permitted by English law, and question whether cloning in this period would raise any new ethical issues.

Cloning in this circumstance raises few ethical issues. What is produced, at least in the first few days of division after a totipotent cell has been produced from an adult skin cell, is just a skin cell from that person with an altered gene expression profile (some genes turned on and some turned off). In one way, it is just an existing skin cell behaving differently from normal skin cells, perhaps more like a malignant skin cell. The significant processes are ones of *cellular multiplication* and later, *cellular differentiation*.

If this is true, why stop at research at 14 days? Consider the third version of the Lucas case:

Lucas III The same as Lucas IIA, but in this case, Lucas also needs a kidney transplant. Therefore, in addition to the skin cell developing blood stem cells (via the embryo), the process is adjusted so that a kidney is produced.

The production of another tissue type or organ does not raise any new relevant ethical consideration. Indeed, if Lucas did not need the kidney, it could be used for someone else who required a kidney (if, of course, in vitro maturation techniques had been developed to the extent that a functioning organ of sufficient size could be produced).

Consider now:

Lucas IV In addition to the blood cells, all the tissue of a normal human embryo is produced, organised in the anatomical arrangement of an embryo. This (in principle) might or might not involve development in a womb. For simplicity, let us assume that this occurs in vitro (though this is impossible at present).

Is there any morally relevant difference from the previous versions? It is not relevant that many different tissues are produced rather than one. Nor is the size of these tissues or their arrangement morally relevant. If there is a difference, it must be that a special kind of tissue has been produced, or that some special relationship develops between existing tissues, and that a morally significant entity then exists. When does this special point in embryonic development occur?

The most plausible point is some point during the development of the brain. There are two main candidates:

1. when tissue differentiates and the first identifiable brain structures come into existence as the neural plate around day 19.[23]
2. when the brain supports some morally significant function: consciousness or self-consciousness or rational self-consciousness. The earliest of these, consciousness, does not occur until well into fetal development.

On the first view, utilisation of cloning techniques in the first two weeks to study cellular differentiation is justifiable. The most defensible view, I believe, is that our continued existence only becomes morally relevant when we become self-conscious. (Of course, if a fetus can feel pain at some earlier point, but is not self-conscious, its existence is morally relevant in a different way: we ought not to inflict unnecessary pain on it, though it may be permissible to end its life painlessly.) On this view, we should use the drug to cause Lucas IV's skin cells to transmutate and remove bone marrow from these. What is going on in Lucas IV is no different, morally speaking, from cloning. If this is right, it is justifiable to extract differentiated tissues from young fetuses which have been cloned.

Conception and potentiality

The other usual point in development which is taken to be morally significant is conception.[24] However, in the case of cloning, there is no conception. There is just a process of turning some switches in an already existing cell. Proponents of the persons-begin-to-exist-at-conception view might reply that cloning is like conception. An individual begins to exist at the point of nuclear transfer. But why should we accept this? Conception seems quite different. Conception involves the unification of two different entities, the sperm and the egg, to form a new entity, the totipotent stem cell. In the case of cloning, there is identity between the cell before and after nuclear transfer—it is the same cell. Something new and important does happen to the entity when it undergoes nuclear transfer, just as something new and important happens when a cell with a malignant potential becomes malignant. But it is the same cell.

POTENTIALITY

In response, one might claim that after nuclear transfer the cell undergoes a radically and morally significant change: it acquires the potential to be a person. On this view, the cell immediately prior to nuclear transfer does not have the potential to be a human being but after nuclear transfer, it has the potential to be a human being. This has a jarring ring to it. What happens when a skin cell turns into a totipotent stem cell is that a few of its genetic switches are turned on and others turned off. To say it doesn't have the potential to be a human being until its nucleus is placed in the egg cytoplasm is like saying my car does not have the potential to get me from Melbourne to Sydney unless the key is turned in the ignition. (Rather, we should say that it has the potential but that that potential may not be *realised* if the key is not turned in the ignition.) Or it is like saying that a stick of dynamite *acquires* the potential to cause an explosion when placed in the vicinity of a lighted match. Of course, a stick of dynamite has the potential to cause an explosion, and various conditions, including placing it in the vicinity of a lighted match, are sufficient to realise this potential. In general terms, X has the potential to be Y, if X would be a Y if conditions c, d, e, . . . obtained. Nuclear transfer is like a number of other conditions (such as adequate placental blood flow) which must obtain if a skin cell is to become a person.

There may be another difference between a mature skin cell and a fertilised egg. Totipotent cells directly give rise to human beings but mature skin cells do not. The latter must go through a further stage of totipotency first. And it may be that that change is significant enough to say that the skin cell does not itself have the potential to create a human being. However, something with the potential to cause A may not lead directly to A. Killing the president may not lead directly to a world war. However, it may lead to political destabilisation which will cause a world war. Killing the president does then have the potential to cause a world war.

At bottom, these issues may be semantic, and depend on how we choose to define "potential." What matters morally is whether skin cells *can* become human beings with the application of technology, and whether they *should*. That is an important moral feature of nuclear transfer. Nuclear transfer is a technical intervention which it is necessary to employ if a skin cell is to become a person, just as microsurgical transfer of an embryo formed in vitro is necessary if the embryo is to become a person.

I cannot see any intrinsic morally significant difference between a mature skin cell, the totipotent stem cell derived from it, and a fertilised egg. They are all cells which could give rise to a person if certain conditions obtained. (Thus, to claim that experimentation on cloned embryos is acceptable, but the same experimentation on non-cloned embryos is not acceptable, because the former are not embryos but totipotent stem cells, is sophistry.)

Looking at cloning this way exposes new difficulties for those who appeal to the potential of embryos to become persons and the moral significance of conception as a basis for opposition to abortion. If all our cells could be persons, then we cannot appeal to the fact that an embryo could be a person to justify the special treatment we give it. Cloning forces us to abandon the old arguments supporting special treatment of fertilised eggs.

PRODUCTION OF A FETUS

If one believes that the morally significant event in development is something related to consciousness, then extracting tissue or organs from a cloned fetus up until that point at which the morally relevant event occurs is acceptable. Indeed, in law, a legal persona does not come into existence until birth. At least in Australia and England, abortion is permissible throughout fetal development.

PRODUCTION OF A CHILD OR ADULT AS A SOURCE OF CELLS OR TISSUES

Like the production of a self-conscious fetus, the production of a cloned child or adult is liable to all the usual cloning objections, together with the severe limitations on the ways in which tissue can be taken from donors for transplantation.

Many writers support cloning for the purposes of studying cellular differentiation because they argue that cloning does not raise serious new issues above those raised by embryo experimentation.[15] Such support for cloning is too limited. On one view, there is no relevant difference between early embryo research and later embryo/early fetal research. Indeed, the latter stand more chance of providing viable tissue for transplantation, at least in the near future. While producing a cloned live child as a source of tissue for transplantation would raise new and important issues, producing embryos and early fetuses as a source of tissue for transplantation may be morally obligatory.

Consistency

Is this a significant deviation from existing practice?

1. FETAL TISSUE TRANSPLANTATION

In fact, fetal tissue has been widely used in medicine. Human fetal thymus transplantation is standard therapy for thymic aplasia or Di George's syndrome. It has also been used in conjunction with fetal liver for the treatment of subacute combined immunodeficiency.

Human fetal liver and umbilical cord blood have been used as a source of haematopoietic cells in the treatment of acute leukaemia and aplastic anaemia. Liver has also been used for radiation accidents and storage disorders. The main problem has been immune rejection.[25]

One woman with aplastic anaemia received fetal liver from her own 22-week fetus subsequent to elective abortion over 20 years ago.[26]

Fetal brain tissue from aborted fetuses has been used as source of tissue for the treatment of Parkinson's disease. Neural grafts show long term survival and function in patients with Parkinson's disease, though significant problems remain.[27 28]

Fetal tissue holds promise as treatment for Huntington's disease,[29 30] spinal cord injuries,[31] demyelinating disorders,[27] retinal degeneration in retinitis pigmentosa,[32 33] hippocampal lesions associated with temporal lobe epilepsy, cerebral ischaemia, stroke and head injury,[34] and beta thalassemia in utero using fetal liver.[35] Fetal pancreas has also been used in the treatment of diabetes.

Fetal tissue banks

Indeed, in the US and England, fetal tissue banks exist to distribute fetal tissues from abortion clinics for the purposes of medical research and treatment. In the US, the Central Laboratory for Human Embryology in Washington, the National Diseases Research Interchange, and the International Institute for the Advancement of Medicine and the National Abortion Federation, all distribute fetal tissue.

In the UK, the Medical Research Council's fetal tissue bank was established in 1957 and disperses about 5,000 tissues a year.

2. CONCEPTION OF A NON-CLONED CHILD AS A SOURCE OF BONE MARROW: AYALA CASE

Not only has fetal tissue been used for the treatment of human disease, but human individuals have been deliberately conceived as a source of tissue for transplantation. In the widely discussed Ayala case, a 17-year-old girl, Anissa, had leukaemia. No donor had been found in two years. Her father had his vasectomy reversed with the intention of having another child to serve as a bone marrow donor. There was a one in four chance the child would be compatible with Anissa. The child, Marissa, was born and was a compatible donor and a successful transplant was performed.[36]

A report four years later noted: "Marissa is now a healthy four-year-old, and, by all accounts, as loved and cherished a child as her parents said she would be. The marrow transplant was a success, and Anissa is now a married, leukaemia-free, bank clerk."[37]

Assisted reproduction (IVF) has been used to produce children to serve as bone marrow donors.[38] It is worth noting that had cloning been available, there would have been a 100% chance of perfect tissue compatibility and a live child need not have been produced.

Objections

While there are some precedents for the proposal to use cloning to produce tissue for transplantation, what is distinctive about this proposal is that human tissue will be: (i) cloned and (ii) deliberately created with abortion in mind. This raises new objections.

ABORTION IS WRONG

Burtchaell, a Catholic theologian, in considering the ethics of fetal tissue research, claims that abortion is morally wrong and that fetal tissue cannot be used for research because no one can give informed consent for its use and to use it would be complicity in wrongful killing.[39] He claims that mothers cannot consent: "The flaw in this claim [that mothers can consent] is that the tissue is from within her body but is the body of another, with distinct genotype, blood, gender, etc." Claims such as those of Burtchaell are more problematic in the case of cloning. If the embryo were cloned from the mother, it would be of the same genotype as her, and, arguably, one of her tissues. Now at some point a cloned tissue is no longer just a tissue from its clone: it exists as an individual in its own right and at some point has interests as other individuals do. But the latter point occurs, I believe, when the cloned individual becomes self-conscious. The presence or absence of a distinct genotype is irrelevant. We are not justified in treating an identical twin differently from a non-identical twin because the latter has a distinct genotype.

In a society that permits abortion on demand, sometimes for little or no reason, it is hard to see how women can justifiably be prevented from aborting a fetus for the purpose of saving someone's life. And surely it is more respectful of the fetus, if the fetus is an object of respect, that its body parts be used for good rather than for no good purpose at all.

IT IS WORSE TO BE A CLONE

Some have argued that it is worse to be a clone.[40] This may be plausible in the sense that a person suffers in virtue of being a clone - living in the shadow of its "parent", feeling less like an individual, treated as a means and not an end, etc. Thus cloning in the Ayala case would raise some new (but I do not believe overwhelming) issues which need consideration. But cloning followed by abortion does not. I can't make any sense of the claim that it is worse to be a cloned cell or tissue. These are not the things we ascribe these kinds of interests to. Cloning is bad when it is bad for a person. Likewise, arguments regarding "instrumentalisation" apply to persons, and not to tissues and cells.

CREATING LIFE WITH THE INTENTION OF ENDING IT TO PROVIDE TISSUE

Using cloning to produce embryos or fetuses as a source of tissue would involve deliberately creating life for the purposes of destroying it. It involves intentionally killing the fetus. This differs from abortion where women do not intend to become pregnant for the purpose of having an abortion.

Is it wrong deliberately to conceive a fetus for the sake of providing tissue? Most of the guidelines on the use of fetal tissue aim to stop women having children just to provide tissues.[41] The reason behind this is some background belief that abortion is itself wrong. These guidelines aim to avoid moral taint objections that we cannot benefit from wrong-doing. More importantly, there is a concern that promoting some good outcome from abortion would encourage abortion. However, in this case, abortion would not be encouraged because this is abortion in a very special context: it is abortion of a *cloned* fetus for medical purposes.

But is it wrong deliberately to use abortion to bring about some good outcome?

In some countries (for example those in the former Eastern bloc), abortion is or was the main available form of birth control. A woman who had intercourse knowing that she might fall pregnant, in which case she would have an abortion, would not necessarily be acting wrongly in such a country, if the alternative was celibacy. When the only way to achieve some worthwhile end - sexual expression - is through abortion, it seems justifiable.

The question is: is the use of cloned fetal tissue the best way of increasing the pool of transplantable tissues and organs?[42]

AN OBJECTION TO THE PRINCIPLE OF TISSUE TRANSMUTATION

Another objection to the proposal is that we do not have the right to determine the fate of all our cells. For example, we are limited in what we can do with our sex cells. However, we should only be constrained in using our own cells when that use puts others at risk. This is not so in transmutation until another individual with moral interests comes into existence.

SURROGACY CONCERNS

At least at present, later embryonic and fetal development can only occur inside a woman's uterus, so some of the proposals here would require a surrogate. I have assumed that any surrogate would be freely consenting. Concerns with surrogacy have been addressed elsewhere,[43] though cloning for this purpose would raise some different concerns. There would be no surrogacy concerns if the donor cell were derived from the mother (she would be carrying one of her own cells), from the mother's child (she would be carrying her child again) or if an artificial womb were ever developed.

SHOULD WE GIVE GREATER IMPORTANCE TO SOMATIC CELLS?

I have claimed that the totipotent cells of the early embryo, and indeed the embryo, do not have greater moral significance than adult skin cells (or indeed lung or colon or any nucleated cells). I have used this observation to downgrade the importance we attach to embryonic cells. However, it might be argued that we should upgrade the importance which we attach to somatic cells.

This is a *reductio ad absurdum* of the position which gives importance to the embryo, and indeed which gives weight to anatomical structure rather than function. If we should show special respect to all cells, surgeons should be attempting to excise the very minimum tissue (down to the last cell) necessary during operations. We should be doing research into preventing the neuronal loss which occurs normally during childhood. The desquamation of a skin cell should be as monumental, according to those who believe that abortion is killing persons, as the loss of a whole person. These claims are, I think, all absurd.

YUK FACTOR

Many people would find it shocking for a fetus to be created and then destroyed as a source of organs. But many people found artificial insemination abhorrent, IVF shocking and the use of animal organs revolting. Watching an abortion is horrible. However, the fact that people find something repulsive does not settle whether it is wrong. The achievement in applied ethics, if there is one, of the last 50 years has been to get people to rise above their gut feelings and examine the reasons for a practice.

PERMISSIVE AND OBSTRUCTIVE ETHICS

Many people believe that ethicists should be merely moral watch-dogs, barking when they see something going wrong. However, ethics may also be permissive. Thus ethics may require that we stop interfering, as was the case in the treatment of homosexuals. Ethics should not only be obstructive but constructive. To delay unnecessarily a good piece of research which will result in a life-saving drug is to be responsible for some people's deaths. It is to act wrongly. This debate about cloning illustrates a possible permissive and constructive role for ethics.

Conclusion

The most justified use of human cloning is arguably to produce stem cells for the treatment of disease. I have argued that it is not only reasonable to produce embryos as a source of multipotent stem cells, but that it is morally required to produce embryos and early fetuses as a source of tissue for transplantation. This argument hinges on:

1. The claim that the moral status of the cloned embryo and early fetus is no different from that of the somatic cell from which they are derived.
2. The claim that there is no morally relevant difference between the fetus and the embryo until some critical point in brain development and function.
3. The fact that the practice is consistent with existing practices of fetal tissue transplantation and conceiving humans as a source of tissue for transplantation (the Ayala case).
4. An argument from beneficence. This practice would achieve much good.
5. An argument from autonomy. This was the principle of tissue transmutation: that we should be able to determine the fate of our own cells, including whether they change into other cell types.

This proposal avoids all the usual objections to cloning. The major concerns are practicality and safety. This requires further study.

The HGAC and The Royal Society have broached the possibility of producing clones for up to 14 days: "therapeutic cloning". Those bodies believe that it is acceptable to produce and destroy an embryo but not a fetus. Women abort fetuses up to 20 weeks and later. We could make it mandatory that women have abortions earlier (with rapid pregnancy testing). However, we do not. Moreover, while the decision for most women to have an abortion is a momentous and considered one, in practice, we allow women to abort fetuses regardless of their reasons, indeed occasionally for no or bad reasons. If a woman could abort a fetus because she wanted a child with a certain horoscope sign, surely a woman should be able to abort a fetus to save a person's life.

I have been discussing cloning for the purposes of saving people's lives or drastically improving their quality. While we beat our breasts about human dignity and the rights of cells of different sorts, people are dying of leukaemia and kidney disease. If a woman wants to carry a clone of her or someone else's child to save a life, it may not be society's place to interfere.

The recent development of human totipotent stem cell lines from embryonic tissue[44 45] means that we are closer to understanding cellular development and differentiation, generating the hope that we may be able to produce tissue for transplantation directly from totipotent stem cells without going to the stage of producing a mature embryo or fetus. But that is still some way off, and at present requires deriving the cell lines from embryonic tissue. The use of nuclear transfer may still be the best way to produce highly *compatible* tissue, even coupled with this technology.

We could address the shortage of tissue for transplantation now. We could routinely employ embryo-splitting during IVF and create embryo banks as a source of fetal tissue. Indeed, rather than destroying millions of spare embryos, we could use them as a source of human tissue. As opposed to using nuclear transfer as a source of tissue, such proposals could not be instituted with the consent of a person who both needs the tissue and is the source of the tissue. That is, we could not appeal to the Principle of Tissue Transmutation to justify these proposals, though they may be justifiable on other grounds.

Acknowledgement

Many thanks to Jeff McMahan, Bob Williamson, David McCarthy, Edgar Dahl, Peter Singer, Lynn Gillam and Ainsley Newson.

Funding

The Murdoch Institute and The Cooperative Research Centre for Discovery of Genes for Common Human Diseases.

Julian Savulescu, MB, BS, BMedSci, PhD, is Director of the Ethics Unit, The Murdoch Institute, Royal Children's Hospital, Melbourne, Australia and Associate Professor at the Centre for the Study of Health And Society, University of Melbourne.

References and notes

1 I use the word "embryo" to cover all stages of development from the single-cell stage to the stage at which there is signifi-

cant organogenesis (about the eighth week), after which I will refer to the growing human as a "fetus". I do not differentiate between a pre-embryo and an embryo.
2 National Bioethics Advisory Commission. *Cloning human beings.* Maryland: National Bioethics Advisory Commission, 1997.
3 World Health Organisation. *Proposed international guidelines on ethical issues in medical genetics and genetics services.* Geneva: WHO, 1998.
4 The European Parliament. *Resolution on cloning.* Motion dated March 11 1997. Passed March 13 1997. The European Parliament, 1997.
5 UNESCO. *Declaration on the human genome and human rights,* adopted on 11 November 1997(13), article 11. UNESCO, 1997.
6 Butler D. Europe brings first ban. *Nature* 1998;**391**:219.
7 HGAC. HGAC Papers. *Cloning issues in reproduction, science and medicine.* Issued for comment January 1998 at http://www.dti.gov.uk/hgac/papers/papers_c.htm.
8 Human Cloning Prohibition Act of 1998, 105th Congress, 2nd Session. S 1599
9 Human Cloning Prohibition Act of 1998, 105th Congress, 1st Session. HR 923.
10 Harris J. Goodbye Dolly? The ethics of human cloning. *Journal of Medical Ethics* 1997;**23**:353-60.
11 Singer P. *forthcoming.*
12 Childress JF. The challenges of public ethics: reflections on the NBAC's report. *Hastings Center Report* 1997; **27, 5**.
13 Trounson A. Cloning: potential benefits for human medicine. *Medical Journal of Australia* 1997;**167**:568-9.
14 Kassirer JP, Rosenthal NA. Should human cloning research be off limits? *New England Journal of Medicine* 1998;**338**:905-6.
15 Council of the Royal Society. *Whither cloning?* London: The Royal Society: The UK Academy of Science, 1998: 1-8.
16 Every cell division results in genetic differences between progeny which are not completely repaired by DNA repair mechanism. Though the vast majority of these differences have no functional implications, they are differences none the less and cellular division does not normally produce an exactly identical copy of DNA (Pannos Iannou, personal communication).
17 Cohen P. Organs without donors. *New Scientist* 1998 Jul 11: 4.
18 Weiss MJ, Orkin SH. In vitro differentiation of murine embryonic stem cells: new approaches to old problems. *Journal of Clinical Investigation* 1996;**97**:591-5.
19 Platts JL. New directions for organ transplantation. *Nature* 1998;**392**:11-7.
20 Cooper DKC. Xenotransplantation—state of the art. *Frontiers of Bioscience* 1996;**1**:248-65.
21 Dorling A *et al.* Clinical xenotransplantation of solid organs. *Lancet* 1997;**349**:867-71.
22 It is not clear whether different patterns of X inactivation or the different mitochondrial complement of cloned cells will affect immunocompatibility.
23 Prior to this point, cells have the potential to develop into several tissue types and there is a lack of coordinated functional activity. For example, Michael Lockwood has argued that early on it is not determined which cells will become the fetus and which the placenta (Lockwood M. Human identity and the primitive streak. *Hastings Center Report* 1995;Jan-Feb:45). It is not clear until much later which cells will become the brain and which the liver. Indeed, early on, the primitive embryo has the capacity to divide into identical twins. Lockwood's argument is that it is only when the definite precursor of a person's brain is formed that a discrete and determined individual can be said to exist.
24 Even an embryo formed by conception in the usual way may not become a person—a significant proportion are spontaneously aborted.
25 Vawter DE, Kearney W, Gervaise KG, Caplan AL, Garry D, Tauer C. *The use of human fetal tissue: scientific, ethical and policy concerns.* Minnesota:University of Minnesota, 1990: 38-41.
26 Keleman E. Recovery from chronic idiopathic bone marrow aplasia of a young mother after intravenous injection of unprocessed cells from the liver (and yolk sac) of her 22mm CR-length embryo. A preliminary report. *Scandinavian Journal of Haematology* 1973;**10**:305-8.
27 Lindvall O. Neural transplantation. *Cell Transplantation* 1995;**4**, 4:393-400.
28 Freeman TB. From transplants to gene therapy for Parkinson's disease. *Experimental Neurology* 1997;**144,1**:47-50.
29 Shannon KM, Kordower JH. Neural transplantation for Huntington disease: experimental rationale and recommendations for clinical trials. *Cell Transplantation* 1996;**5,2**:339-52.
30 Freeman TB, Sanberg PR, Isacson O. Development of the human striatum: implications for fetal striatal transplantation in the treatment of Huntington disease. *Cell Transplantation* 1995;**4,6**:539-45.
31 Zompa EA, Cain LD, Everhart AW *et al.* Transplant therapy: recovery of function after spinal cord injury. *Journal of Neurotrauma* 1997;**14,8**:479-506.
32 del Cerro M, Lazar ES, Diloreto D Jr. The first decade of continuous progress in retinal transplantation. *Microscopy Research and Technique* 1997;**36,20**;130-41.
33 Litchfield TM, Whitely SJ, Lund RD. Transplantation of retinal pigment epithelial, photoreceptor and other cells as treatment for retinal degeneration. *Experimental Eye Research* 1997;**64,5**:655-66.
34 Shetty AK, Turner DA. Development of fetal hippocampal grafts in intact and lesioned hippocampus. *Progress in Neurobiology* 1996; **50,5-6**:597-653.
35 Touraine JL. In utero transplantation of fetal liver stem cells into human fetuses *Journal of Hematotherapy* 1996; **5,2**:195-9.
36 Rachels J. When philosophers shoot from the hip. *Bioethics* 1991;**5**:66-71.
37 Note in *Hastings Center Report* 1994; May/June: 2.
38 Lamperd R. Race for Life. *Sun Herald* 1998 Jul 2:1.
39 Burthchaell JT. University policy on experimental use of aborted fetal tissue. *IRB: A Review of Human Subjects Research* 1988;**10**:7-11.
40 Holm S. A life in the shadow: one reason why we should not clone humans. *Cambridge Health Care Quarterly* 1998;7:160-2.
41 See reference 25: the whole book.
42 Some argue that opt-out systems of organ procurement, umbilical cord blood and xenotransplantation offer viable alternatives. I would take an organ derived from one of my own cells, if concerns about aging and increased mutation were adequately addressed.
43 Oakley J. Altruistic surrogacy and informed consent. *Bioethics* 1992;**6,4**:269-87.
44 Thomson JA, Itskovitz-Eldor J, Shapiro SS, Waknitz MA, Swiergiel JJ *et al.* Embryonic stem cell lines derived from human blastocysts. *Science* 1998;**282**:1145-7.
45 Shamblott MJ, Axelman J, Wang S, Bugg EM, Littlefield JW *et al.* Derivation of pluripotent stem cells from cultured human primordial germ cells. *Proceedings of the National Academy of Sciences*1998;95:13,726-31.

[22]

GOING TO THE ROOTS OF THE STEM CELL CONTROVERSY

SØREN HOLM

ABSTRACT

The purpose of this paper is to describe the scientific background to the current ethical and legislative debates about the generation and use of human stem cells, and to give an overview of the ethical issues underlying these debates.

The ethical issues discussed are 1) stem cells and the status of the embryo, 2) women as the sources of ova for stem cell production, 3) the use of ova from other species, 4) slippery slopes towards reproductive cloning, 5) the public presentation of stem cell research and 6) the evaluation of scientific uncertainty and its implications for public policy.

INTRODUCTION

The ability to produce and culture human embryonic stem cells has raised hopes for a range of new cell based therapies, but has at the same time created intense national and international debate.

The purpose of this paper is to describe the scientific background to the current ethical and legislative debates about the generation and use of human stem cells, and to give an overview of the ethical issues that are central to these debates. Because the paper is intended to be reasonably comprehensive the presentation and analysis of each individual argument must necessarily be rather brief.[1]

[1] One major topic has been left out of this paper because of space constraints. That is the question of intellectual and actual property rights in human stem cell lines and the techniques by which they are produced. This is a huge topic on its own, actualising all the issues of ownership of the human body, body parts and human genetic material.

494 SØREN HOLM

THE SCIENTIFIC BACKGROUND TO THE STEM CELL CONTROVERSY

Three partially independent scientific developments underlie the current debates about stem cell research. These are 1) the discovery of methods to derive and culture human embryonic stem cells, 2) the discovery of nuclear replacement techniques, and 3) the discovery of new and previously unsuspected potentialities of stem cells in the adult human body.

A stem cell is a non-differentiated cell that can divide and multiply in its undifferentiated state, but which can also give rise to more specialised differentiated cells. It has been known for a long time that adult human tissues contain stem cells that can replenish cells lost through normal wear and tear or through trauma or disease. This fact has been utilised as a basis for a number of different treatments including bone marrow and skin transplants.

It has also been known that cells from the inner cell mass of the early embryo are stem cells (since we know that they must necessarily be able to become every cell in the body during the development from embryo to adult individual), but no method existed by which these embryonic stem cells could be grown in culture in the laboratory in a way that preserved their stem cell character.

In 1998 researchers at the University of Wisconsin published a method for deriving and culturing human embryonic stem cells indefinitely.[2] This development made it possible to create stable human stem cell lines and generate (in principle) unlimited quantities of any particular embryonic stem cell, and thereby the possibility to 1) standardise research into human stem cells, and 2) create reproducible stem cell therapies.

Almost at the same time as the Wisconsin group developed the method for culturing human embryonic stem cells, a group at the Roslin Institute in Scotland developed methods for the cloning of adult mammals using nuclear replacement techniques.[3] The techniques basically work by removing a cell from an adult animal, and then taking the cell nucleus from the adult cell and placing it in an ovum from which the original nucleus

[2] J.A. Thomson, J. Itskovitz-Eldor, S.S. Shapiro, M.A. Waknitz, J.J. Swiergiel, V.S. Marshall, J.M. Jones. Embryonic stem cell lines derived from human blastocysts. *Science* 1998; 282: 1145–1147.

[3] I. Wilmut, A.E. Schnieke, J. McWhir, A.J. Kind, K.H. Campbell. Viable offspring derived from foetal and adult mammalian cells. *Nature* 1997; 385: 810–813.

has been removed. This procedure reprogrammes the adult nucleus to an embryonic state and creates a cell that is more than 99% genetically identical with the original adult cell from which the nucleus was taken.[4] It is, however, not the ability to reproduce a fully-grown mammal by nuclear replacement that is of main interest to the stem cell debate. It is the combination of nuclear replacement techniques and embryonic stem cell culture. When these two techniques are combined it becomes possible to produce embryonic stem cells that are almost genetically identical to any given adult human being.

Research into the potentialities of the remaining stem cells in the adult human body has also progressed apace in recent years. Stem cells have been found in a number of tissues in which it was previously 'common knowledge' that they did not exist (e.g. neuronal stem cells in the brain),[5] many kinds of adult stem cells[6] have been cultured, and adult stem cells have been shown to be capable of transdifferentiation into different kinds of cells than the cells of the tissues in which they originated.[7] These discoveries have opened the possibility that adult stem cells may be used in a range of stem cell therapies far beyond what was thought possible.[8]

At present there are thus three main research programmes that are pursued in stem cell research: 1) research on adult stem cells, 2) research on embryonic stem cells from embryos

[4] The mitochondria in this cell come from the ovum, and contain their own genetic material. It is thus only if both nucleus and ovum come from the same woman that 100% genetic identity is achieved.

[5] C.B. Johanson, S. Momma, D.L. Clarke, M. Risling, U. Lendahl, J. Friesen. Identification of a neural stem cell in the adult mammalian central nervous system. *Cell* 1999; 96: 25–34.

[6] In this paper 'adult stem cell' will be used for any stem cell derived from a human being after birth.

[7] D.L. Clarke, C.B. Johansson, J. Wilbertz, B. Veress, E. Nilsson, H. Karlstrom, U. Lendahl, J. Friesen. Generalized potential of adult neural stem cells. *Science* 2000; 288: 1559–1561; P.A. Zuk, M. Zhu, H. Mizono, J. Huang, J.W. Futrell, A.J. Katz, P. Benhaim, H.P. Lorenz, M.H. Hedrick. Multilineage cells from human adipose tissue: implications for cell-based therapies. *Tissue Engineering* 2001; 7: 211–228.

[8] Two recent papers cast some doubt on these possibilites for transdifferentiation, but their validity and relevance is contested. N. Terada, T. Hamazaki, M. Oka, M. Hoki, D.M. Mastalerz, Y. Nakano, E.M. Meyer, L. Morel, B.E. Petersen, E.W. Scott. Bone marrow cells adopt the phenotype of other cells by spontaneous cell fusion. *Nature* 2002; 416: 542–545; Q-L. Ling, J. Nichols, E.P. Evans, A.G. Smith. Changing potency by spontaneous fusion. *Nature* 2002; 416: 545–548. N. Dewitt, J. Knight. Biologists question adult stem-cell versatility. *Nature* 2002; 416: 354.

produced through IVF techniques, and 3) research on embryonic stem cells produced through nuclear replacement techniques.[9]

All three research programmes are directed at 1) increasing our knowledge about basic cell biology, 2) creating new therapies through stem cell culture and control of cell differentiation, and 3) producing commercially viable stem cell products either by the direct patenting of stem cell lines, or by combining stem cell technology with genetic engineering or other patentable interventions.

As we will see below, much of the discussion on stem cells is concerned with the ethical issues raised by each of these programmes, and with whether or not these ethical issues should influence decisions about regulation and/or funding of the research programmes.

THE EXPECTED BENEFITS FROM STEM CELL RESEARCH

Stem cell research is undoubtedly going to increase our knowledge about basic cell biology considerably, but this is not the benefit of stem cell research that excites most people. The really exciting thing about stem cell research is in the therapeutic potential of stem cells.

If we can develop methods to grow human stem cells in unlimited quantities, and if we can further learn how to control their differentiation, then a whole range of therapeutic possibilities becomes (theoretically) available.[10] The most immediate therapeutic gains are likely to be in the area of cell therapy. Many diseases are caused by, or accompanied by, loss of specific cell types. The lost cell types could be produced in the laboratory and later implanted to cure or alleviate the disease.

[9] The term 'research programme' is here used in the sense given to it by Lakatos, i.e. a group of concrete research endeavours kept together by a common core of relatively stable assumptions about the goals of research, the proper research methodologies and the most fruitful research topics. What distinguishes the three stem cell research programmes from each other is primarily different beliefs about what kind of stem cell is going to be the basis for the most progressive (i.e. productive in terms of scientific and commercial results) research. I. Lakatos. 1974. Falsification and the methodology of scientific research programmes. In *Criticism and the Growth of Knowledge*. I. Lakatos and A. Musgrave, eds. Cambridge. Cambridge University Press: 91–196.

[10] R.P. Lanza, J.P. Cibelli, M.D. West. Prospects for the use of nuclear transfer in human transplantation. *Nature Biotechnology* 1999; 17: 1171–1174; E. Fuchs, J.A. Segre. Stem Cells: A New Lease of Life. *Cell* 2000; 100: 143–155.

Further into the future it may become possible to grow whole organs from stem cells and use these for transplantation, removing the need for organ donation; and even further into the future we may be able to use stem cells for rejuvenating therapies leading to an increased life-span.

The therapeutic potential of stem cells spans such a wide range of diseases and conditions that it will constitute a major medical breakthrough if only even a small percentage of the most likely uses (e.g. in the area of cell therapy) become a reality. Even if stem cell therapy turned out only to be effective in myocardial infarction it would still alleviate huge amounts of human suffering.

These very large, and very likely benefits of stem cell research indicate that prohibition of certain kinds of stem cell research needs strong justification. The ethical and regulatory debates have therefore concentrated on whether such justification can be found.

THE ETHICAL ISSUES

Stem cells and embryos

One of the main ethical issues discussed concerning stem cell research originates in the fact that embryonic stem cells have to be generated from embryos that are destroyed in the process. This means that stem cell research again raises the question of whether there are any ethical limits concerning the destruction of human embryos for research or therapeutic purposes, as well as the more fundamental question of the moral status of the human embryo. If human embryos have any moral status we need a good justification to destroy them, and the greater their moral status the more important or weighty the justification has to be.[11]

The question of the moral status of the embryo was not resolved during the abortion debate nor during the debates about various forms of assisted reproductive technologies. It is unlikely to be resolved during the current debates about stem cells, since no really new arguments seem to be forthcoming.[12]

[11] R.M. Doerflinger. The ethics of funding embryonic stem cell research: a Catholic viewpoint. *Kennedy Institute of Ethics Journal* 1999; 9: 137–150.

[12] L.H. Harris. Ethics and politics of embryo and stem cell research: Reinscribing the abortion debate. *Women's Health Issues* 2000; 10: 146–151; D.C. Wertz. Embryo and stem cell research in the USA: a political history. *TRENDS in Molecular Medicine* 2002; 8: 143–146.

If one looks at the legislation about abortion and assisted reproductive technologies it is evident that no jurisdiction has legislation which is compatible with the view that human embryos are just things with no moral status, and that no jurisdiction has legislation compatible with the view that embryos have the same moral status as born human beings. Most legislations implicitly or explicitly adopt some kind of middle position, although it is often unclear to what extent this represents a considered view or whether it is the result of a political compromise.

The important question with regard to regulation or legislation therefore becomes how the use of embryos for stem cell research and therapy can be fitted into a legislative structure that either relies on a view that embryos have some moral value, or is a direct result of political compromise. Giving some moral status to embryos does not automatically rule out embryonic stem cell research, since it can be argued that the likely benefits in terms of reduction of human suffering and death in many cases outweigh the sacrifice of a (small?) number of human embryos.[13]

All of the ethical questions concerning the use of embryos would be by-passed if it became technically possible to produce cells equivalent to embryonic stem cells, without the creation of embryos. This could, for instance, be the case if other methods for re-programming nuclei from adult cells became available.

PPL Therapeutics PLC has claimed to have done this using bovine cells and is working towards doing it with human cells, but very few details have been released because of commercial concerns.[14]

The spare embryo

In arguments about the use of embryos for stem cell research the distinction between embryos produced for research and spare embryos left over after IVF and other forms of assisted reproduction has also been invoked. It has been argued that the use of spare embryos is less problematic than the use of embryos produced for research, and that at present the use of specifically produced embryos for stem cell research should not be allowed.[15] No new arguments to support or refute this

[13] G. McGee, A. Caplan. The ethics and politics of small sacrifices in stem cell research. *Kennedy Institute of Ethics Journal* 1999; 9: 151–158.

[14] PPL Therapeutics PLC. 2001. *Interim Report 2001.* Edinburgh. PPL Therapeutics PLC.

[15] See for instance the report from the American National Bioethics Advisory Commission. National Bioethics Advisory Commission. 1999. *Ethical*

distinction have, however, been forthcoming in the stem cell debate.[16]

Women and the need for ova

If stem cells are to be produced from embryos that are not 'spare' after IVF, the ova for this production must come from women.[17] In the initial research phase the number of ova needed will be relatively small, but for stem cell therapy the number may become very large. If, for instance, a specific therapy is based on nuclear replacement from the intended recipient in order to ensure perfect immunological compatibility, at least one ovum will be needed for each patient (and probably more since the techniques for nuclear replacement are unlikely to become 100% effective any time soon).

This raises general problems concerning how we can ensure that the ova are obtained without coercion or exploitation of the ova donors, sellers or providers, but also more specific questions about how a new practice of non-reproduction related ova procurement would influence the status of women in society.

At an even more general level there is a connection to the debate about the rights and wrongs of the commodification of human body parts.[18]

Issues in Human Stem Cell Research. Rockville. NBAC. A number of jurisdictions have legislation concerning assisted reproductive technologies that allow research on spare embryos, but prohibit the creation of embryos for research purposes.

[16] On the cogency of the distinction see S. Holm. The spare embryo – A red herring in the embryo experimentation debate. *Health Care Analysis* 1993; 1: 63–66.

[17] Unless it is possible to use ova obtained from aborted foetuses, dead women, or ovaries removed as part of surgical interventions. The first two of these alternative sources of ova may in themselves raise ethical issues but these are beyond the scope of this paper.

[18] L.S. Cahill. Genetics, Commodification, and Social Justice in the Globalization Era. *Kennedy Institute of Ethics Journal* 2001; 11: 221–238; S. Holland. Contested Commodities at Both Ends of Life: Buying and Selling Gametes, Embryos, and Body Tissues. *Kennedy Institute of Ethics Journal* 2001; 11: 263–284; L. Andrews, D. Nelkin. 2001. *Body Bazaar: The Market for Human Tissue in the Biotechnology Age.* New York. Crown Publishers; M.J. Radin. 1996. *Contested Commodities.* Cambridge, MA. Harvard University Press; R. Macklin. 1996. What is Wrong with Commodification? In *New Ways of Making Babies: The Case of Egg Donation.* C.B. Cohen, ed. Bloomington. Indiana University Press: 106–121.

Stem cells produced using ova from other species

One way of solving the problem of shortage of ova, and the potential ethical problems in using women as donors of ova for these purposes, is to use ova from other species (e.g. bovines) in the creation of stem cells by means of nuclear replacement techniques.

It is, as yet, unknown whether the use of ova from other species is technically possible, and if possible whether the stem cells produced would be functionally and immunologically equivalent to stem cells produced using human ova. The technique has been patented by the American firm Advanced Cell Technology, but there is still doubt in the scientific community whether it actually works.[19]

The additional ethical problems created by this different source of ova can, however, be argued to be small as long as the resulting embryos are only used for stem cell production and not for reproductive purposes.[20]

On some lines of argument the ethical problems may actually be less than if human ova are used, since it could be argued that the embryos produced are not really human embryos. If the moral status of human embryos is based in their being human, then the moral status of these 'less than human' embryos could be argued to be less important.

Slippery slopes towards reproductive cloning

A classical slippery slope argument has been prominent in the specific debate about whether the creation of stem cells by means of cell nuclear replacement techniques should be allowed. Opponents of this technique have claimed that allowing this would put us on a slippery slope towards reproductive cloning. The slope that is imagined is of a technical nature. If all the

[19] Advanced Cell Technology. *Advanced Cell Technology Announces Use of Nuclear Replacement Technology for Successful Generation of Human Embryonic Stem Cells.* Press Release November 12, 1998. Available at *http://www.advancedcell.com/pr_11-12-1998.html* E. Marshall. Claim of human-cow embryo greeted with scepticism. *Science* 1998; 282: 1390–1391.

[20] There are two lines of argument seeing major ethical problems in the use of non-human ova. The first sees the technique in itself as a transgression of an important boundary line between human and animal. The second points to a possible slippery slope from the use of this technique for the production of stem cells, to a use for reproductive purposes.

technical problems in the first steps of cell nuclear replacement techniques are solved succesfully then it becomes both easier and more tempting (because certain risks have been reduced) to try to use nuclear replacement techniques for reproductive cloning.

This is clearly not a problem if reproductive cloning does not raise any serious ethical problems because in that case there is no slope, slippery or not.[21]

If reproductive cloning is ethically problematic the question then becomes how to respond to the existence of the slope. The slope has to be taken seriously by politicians as a policy problem. Whatever the analysis of bioethicists as to the cogency of the belief that reproductive cloning is a serious ethical problem, there is no doubt that this belief is shared by many people and by many politicians.

The political reaction to the perceived slippery slope depends on whether it is seen as a possible threat to the positive development of stem cell research (as it is perceived by the government in the UK and a number of other European countries), or whether it is seen as a possible tool to justify the prohibition of stem cell research by nuclear replacement as part of a more comprehensive ban on all kinds of human cloning (as it is perceived by the government in the US).[22]

If the slope is seen as a possible threat to the acceptance of stem cell research the logical response is to legally prohibit human reproductive cloning, and to try to convince the public that such a prohibition will be effective.[23] Whether legal prohibition can be effective given the possibilities for international reproductive tourism to more permissive jurisdictions is, however, questionable.[24]

[21] The literature on the ethics of reproductive cloning is extensive. A range of views can be found in a thematic issue of the *Journal of Medical Ethics* 1999; 25(2), and in a thematic issue of the *Cambridge Quarterly of Health Care Ethics* 1998; 7(2).

[22] E. Check. Call for cloning ban splits UN. *Nature* 2002; 416: 3.

[23] This is the approach chosen by the governments of the UK, Denmark and the Netherlands among others. For an overview of European policies in this area see: L. Matthiessen. 2001. *Survey on opinions from National Ethics Committees or similar bodies, public debate and national legislation in relation to human embryonic stem cell research and use.* Bruxelles. European Commission Research Directorate-General.

[24] P.G. Wood. To what extent can the law control human cloning? *Medicine, Science & the Law,* 1999; 39: 5–10.

The presentation of stem cell research – Promising too much too early?

The public presentation of the benefits of stem cell research has often been characterised by the promise of huge and immediate benefits. Like with many other scientific breakthroughs the public has been promised real benefits within 5–10 years, i.e. in this case significant stem cell therapies in routine clinical use.[25] Several years have now elapsed of the 5–10 years and the promised therapies are still not anywhere close to routine clinical use.[26] There are similarities to the initial enthusiastic presentation of gene therapy in the late 1980s and the later problems encountered, and some reason to fear that stem cell therapies will have an equally long trajectory between theoretical possibility and clinical practice. It is likely that many of the current sufferers from some of the conditions for which stem cell therapies have been promised will be long dead before the therapies actually arrive.[27]

It is clearly ethically problematic to raise false expectations in seriously ill people, and even more problematic if this is partly done from self-interest (e.g. to promote one's own research in the media). But the problem may go deeper because the optimistic predictions and the targeting of these predictions on certain groups of diseases also have a function in the political arena where public policy is decided. When gene therapy was initially promoted, and the public and political resistance overcome, gene therapy was promoted as a treatment for the unfortunate people suffering from genetic disorders. Gene therapy was put forward as their only hope of cure and alleviation. Today we do know however, that most gene therapy projects are not directed towards genetic disease, but towards the treatment of common diseases (partly for commercial reasons). The groups that were used as symbolic 'battering rams' to gain political and public acceptance of the gene therapy, have not yet benefited significantly from gene therapy, and many of the people having rarer forms of genetic disorders are unlikely ever to benefit.

[25] Anon. Taking stock of spin science. *Nature Biotechnology* 1998; 16: 1291.

[26] Given the time needed for basic research, clinical research and regulatory approval it is unlikely that any therapy using biological materials, and based on a truly novel therapeutic approach could move from initial discovery to clinical use in 5–10 years. See also R. Lovell-Badge. The future for stem cell research. *Nature* 2001; 14: 88–91.

[27] B. Albert. *Presentation to the All-Party Disablement Group* – July 25th 2000. Unpublished manuscript.

Scientific uncertainty, ethical unease and the formulation of public policy

At the current point in time it is not known which (if any) of the three main lines of research described above is going to be most successful in terms of a) generating scientific knowledge about cell biology, and b) generating new stem cell based therapies for common diseases. That each is, at least at the moment, seen as a viable approach with regard to therapy is attested by the fact that many biotech firms have been founded aiming at exploiting each of the approaches.[28]

The question is important because it has been argued that there is no need to permit more ethically contentious ways of generating stem cells, if the same benefits can be realised using less contentious stem cells, either adult stem cells or stem cells from aborted foetuses.[29]

What factors could we use to decide whether one line of research is more promising than another?[30] One possibility is to think about what characteristics a stem cell should have in order to be therapeutically useful and then try to decide which of the research programmes is most likely to be able to lead to the production of such cells, and if more than one can produce the required cells, which one will progress fastest to the goal.[31] We do know (some of) the characteristics that the therapeutically optimal stem cell should display:

1. No immunological rejection
2. Immediate availability

[28] N. Axelsen. 2001. Commercial interests in stem cells. In *Nordic Committee on Bioethics. The Ethical Issues in Stem Cell Research.* Copenhagen. Nordic Council of Ministers: 79–80.

[29] J.R. Meyer. Human embryonic stem cells and respect for life. *Journal of Medical Ethics* 2000; 26: 166–170; V. Branick, M.T. Lysaught. Stem cell research: licit or complicit? Is a medical breakthrough based on embryonic and foetal tissue compatible with Catholic teaching? *Health Progress* 1999; 80: 37–42. This kind of reasoning also seems to underlie the National Bioethics Advisory Committee report *op. cit.* note 15, although it draws the line of contentiousness between the spare embryo and the embryo produced for research.

[30] Most of this debate has centred on the therapeutic uses of stem cells. With regard to the 'pure' scientific production of knowledge about cell biology it seems clear that each of the research programmes will produce at least some unique bits of knowledge, and that each of them must therefore be pursued if complete scientific knowledge is the goal.

[31] A difference in speed of development between two research programmes is important, even if they will both eventually lead to the same goal, since any delay in implementation of stem cell therapies entail costs in term of human suffering.

3. Availability in large numbers
4. Controlled differentiation to desired cells
5. Controlled integration into existing tissues and biological niches leading to normal function
6. No other biological risks

From a theoretical point of view embryonic stem cells created by nuclear replacement should be able to fulfil most of these requirements. We know that they can become all types of cells, and we know that they are immunologically perfectly compatible. We are, however, not yet able to control their differentiation into all desired cell types, and there may be situations of acute organ or cell failure where we do not have the necessary time to grow a sufficient number of cells to initiate therapy in time.

Embryonic stem cells derived in other ways have the disadvantage of not being immunologically perfectly compatible, but they do, on the other hand, offer the advantage of being potentially immediately available from a stem cell bank in the necessary quantities. Adult stem cells are immunologically compatible, but it is still uncertain whether we can derive all types of cells from adult stem cells, and they may also not be available in sufficient quantities in acute cases.

No type of stem cell therefore fulfils all the criteria for a therapeutically optimal stem cell. How should we evaluate this evidence in order to decide what research programmes to pursue?

At approximately the same time, the American National Bioethics Advisory Commission and a British government expert group reviewed the evidence and came to two rather different conclusions. The National Bioethics Advisory Commission concluded that:

> Currently, we believe that cadaveric fetal tissue and embryos remaining after infertility treatments provide an adequate supply of research resources for federal research projects involving human embryos. Therefore, embryos created specifically for research purposes are not needed at the current time in order to conduct important research in this area.
>
> [...]
>
> We conclude that at this time, because other sources are likely to provide the cells needed for the preliminary stages of research, federal funding should not be provided to derive ES cells from SCNT. Nevertheless, the medical utility and

> scientific progress of this line of research should be monitored closely.[32]

Whereas the British Chief Medical Officer's Expert Group concluded that:

> For some people, particularly those suffering from the diseases likely to benefit from the treatments that could be developed, the fact that research to create embryos by cell nuclear replacement is a necessary step to understanding how to reprogramme adult cells to produce compatible tissue provides sufficient ethical justification for allowing the research to proceed.[33]

What was a fact for one group of experts was clearly not a fact for the other. What is at play here is a different evaluation of the available scientific evidence, but possibly also a different approach to the decision of whether a line of research should be deemed 'necessary'. Is a particular line of research only necessary if it is the only way to get the knowledge we need for stem cell therapies, or is it necessary if scientific progress will otherwise be slowed down and will be much more costly, but will eventually lead to stem cell therapies any way even if this particular line of research is not pursued?[34]

The policy-maker is thus left with a very difficult problem. If we knew that adult stem cell research could deliver therapies for all the conditions where stem cell therapy seems to be a possibility, then there would be a straight forward policy argument for choosing only to support this ethically uncontentious research programme. If the same goal can be obtained in two ways, and if one of them is less contentious than the other it makes good political sense to choose the uncontentious one.[35] If on the other hand there was unequivocal certainty that research using embryonic stem cells was necessary for the development of stem cell therapies for one or more important diseases, a relatively

[32] National Bioethics Advisory Commission, *op.cit,* note 15, pp. 71–72.

[33] Chief Medical Officer's Expert Group. 2001. *Stem Cell Research: Medical Progress with Responsibility – A Report from the Chief Medical Officer's Expert Group Reviewing the Potential of Developments in Stem Cell Research and Cell Nuclear Replacement to Benefit Human Health.* London. Department of Health. p. 40.

[34] S. Holm. 2001. European and American ethical debates about stem cells – common underlying themes and some significant differences. In *Nordic Committee on Bioethics. The Ethical Issues in Stem Cell Research.* Copenhagen. Nordic Council of Ministers: 35–45.

[35] This might be the proper policy response even if it would lead to some delay in the development of treatments.

strong consequentialist argument would offer itself based on a moral imperative to reduce human suffering, and this could be combined with appeals to consistency in those jurisdictions that already allow some kinds of embryo research.

Because there is scientific uncertainty each of these two lines of argument is, however, considerably weakened because an opponent can always point to uncertainty about the underlying empirical premises concerning whether embryonic stem cell research is necessary or not.

CONCLUSION

It should by now be evident that many of the most discussed ethical issues in connection with stem cell research are minor variants of issues that have been discussed in reproductive ethics since the beginning of modern bioethics in the late 1960s and early 1970s. Many arguments in the stem cell debate, for instance, merely re-iterate arguments for or against giving moral status to embryos, or arguments concerning the validity of the distinction between 'spare' embryos and embryos produced specifically for research. The underlying points of contention in these recycled arguments have not been resolved during the abortion debate, or during the debates about assisted reproductive technologies, and they are unlikely to be resolved now. Each side has arguments that it sees as compelling, but which the other side rejects utterly. It is probably this re-ignition of old debates that has added to the heat of the stem cell debates, because neither side can give ground without fearing a knock on effect on the political accommodations or compromises reached in the abortion and the assisted reproduction areas.

If we take all of these already well known debates into account it seems that there is a rough hierarchy of contentiousness ordering the different ways of producing human stem cells according to how many issues each raise. This would look something like the following (with the most contentious first):

Embryonic stem cells created by nuclear replacement
Embryonic stem cells from embryos created for research
Embryonic stem cells from spare embryos
Adult stem cells

This proposed hierarchy is not very illuminating for ethical analysis, but it may well influence public policy.

There are, however, also a few issues raised by the stem cell debate that are not as well worn. The most interesting of these are the questions surrounding how public policy should be formed in an area where there is 1) agreement about the value of the goal of a particular kind of research (i.e. the creation of effective stem cell therapies), 2) genuine scientific uncertainty about exactly what line of research is most likely to achieve this goal, and 3) disagreement about the ethical evaluation of some of these lines of research but not about others. This question is perhaps more a question of political or legal philosophy than a question of ethics, but it is nevertheless an issue that should be of interest to those bioethicists who want their elegant analyses transformed into public policy.

Søren Holm
Institute of Medicine, Law and Bioethics
University of Manchester
Manchester M13 9PT
UK
Soren.holm@man.ac.uk

[23]

Stem Cells, Sex, and Procreation

JOHN HARRIS

Sex is not the answer to everything, though young men think it is,[1] but it may be the answer to the intractable debate over the ethics of human embryonic stem cell research. In this paper, I advance one ethical principle that, as yet, has not received the attention its platitudinous character would seem to merit.[2] If found acceptable, this principle would permit the beneficial use of any embryonic or fetal tissue that would, by default, be lost or destroyed. More important, I make two appeals to consistency, or to parity of reasoning, that I believe show that no one who either has used or intends to use sexual reproduction as their means of procreation,[3] nor indeed anyone who has unprotected heterosexual intercourse, nor anyone who finds in vitro fertilization (IVF) acceptable, nor anyone who believes that abortion is ever permissible can consistently object on principle[4] to human embryo research nor to the use of embryonic stem cells for research or therapy.

This paper has four parts. I begin by simply reviewing the range of ethical issues raised by human embryo stem cell (HESC) research or therapy. I then examine why human stem cells are so important. Third, I review the current state of social and regulatory policy on stem cells, and finally, I say some positive things about the ethics of HESC research and therapy.

What Are the Ethical Issues?

The ethical aspects of human stem cell research raise a wide variety of controversial and important issues. Many of these issues have to do with the different sources of stem cells. In principle, stem cells can be obtained from adults, from umbilical cord blood, from fetal tissue, and from embryonic tissue. Clearly, there are widely differing views as to the ethics of sourcing stem cells in these four different ways. For the moment, there is general consensus that embryos are the best source of stem cells for therapeutic purposes, but this may, of course, change as the science develops. Then there is the question of whether embryos or fetuses may be deliberately produced to be sources of stem

---

This paper draws on, but also answers objections to, my earlier paper "The ethical use of human embryonic stem cells in research and therapy" published in: Burley JC, Harris J, eds. *A Companion to Genethics: Philosophy and the Genetic Revolution*. Oxford: Blackwell; 2001. I am indebted to my colleague Louise Irving for the data on social policy in Europe and to Julian Savulescu for many stimulating conversations and exchanges. Work on this paper was supported by a project grant from the European Commission for EUROSTEM under its Quality of Life and Management of Living Resources Programme, 2002.

cells, whether they are also intended to survive stem cell harvesting and grow into healthy adults.

The European Group on Ethics, which advises the European Parliament, is one of the few to have highlighted the women's rights issues that arise here. In particular, we should bear in mind that women as the most proximate sources of embryonic and fetal material and hence also of cord blood may be under special pressures and indeed risks if these are to be the sources of stem cells. There are issues of free and informed consent, both of donors and recipients; the responsibility of accurate risk-benefit assessment; and the fact that particular attention needs to be paid to appropriate ethical standards in the conduct of research on human subjects. There are issues concerning the anonymity of the donors and security and safety of cell banks and of the confidentiality and privacy of the genetic information as well as the tissue they contain. Finally, there are issues of commerce and remuneration for those taking part and of the transport and security of human tissue and genetic material and information across frontiers both within the European Union and worldwide. All of these issues are important, but most of them have received extensive discussion over the past few years. For this reason, I shall not look in detail at these issues.

Before considering the ethics of such use in detail, we need to understand the possible therapeutic and research uses of stem cells and, equally, the imperatives for research and therapy.

Why Embryonic Stem Cells?

Embryonic stem cells were first grown in culture as recently as February 1998 by James A. Thomson of the University of Wisconsin. In November of that year, Thomson announced in *Science* that such human embryo stem cells formed a wide variety of recognizable tissues when transplanted into mice.[5] As Roger A. Pedersen noted recently:

> Research on embryonic stem cells will ultimately lead to techniques for generating cells that can be employed in therapies, not just for heart attacks, but for many conditions in which tissue is damaged.
>
> If it were possible to control the differentiation of human embryonic stem cells in culture the resulting cells could potentially help repair damage caused by congestive heart failure, Parkinson's disease, diabetes, and other afflictions. They could prove especially valuable for treating conditions affecting the heart and the islets of the pancreas, which retain few or no stem cells in an adult and so cannot renew themselves naturally.[6]

Stem cells then might eventually enable us not only to grow tailor-made human organs that, using cloning technology of the type that produced Dolly the sheep, could be made individually compatible with their designated recipients. In addition to tailor-made organs or parts of organs, such as heart valves, it may be possible to use human embryonic stem cells to colonize damaged parts of the body, including the brain, and to promote the repair and regrowth of damaged tissue. These possibilities have long been theoretically understood, but it is only now with the isolation of human embryonic stem cells that their benefits are being seriously considered.

Stem Cells, Sex, and Procreation

Now that we have noted some of the research and therapeutic possibilities, it is important to remind ourselves of the moral reasons we have to pursue stem cell research. "Research" always sounds like such an abstract and even vainglorious objective when set against passionate feelings of fear or distaste. We need to remind ourselves of the human benefits that stem from research and the human costs of not pursuing research.

Stem Cells for Therapy

It is difficult to estimate how many people might benefit from the products of stem cell research should it be permitted and prove fruitful. Most sources agree that the most proximate use of HESC therapy would be for Parkinson's disease. Parkinson's disease is a common neurological disease, the prevalence of which increases with age. The overall prevalence (per 100 population in persons 65 years of age and older) is 1.8.[7] Parkinson's disease has a disastrous effect on the quality of life. Another source speculates that "the true prevalence of idiopathic Parkinson's disease in London may be around 200 per 100,000."[8] In the United Kingdom, around 120,000 individuals have Parkinson's disease,[9] and it is estimated that Parkinson's disease affects between one and one-and-a-half million Americans.[10] Untold human misery and suffering could be prevented if Parkinson's disease became treatable. If Roger Pedersen's hopes for stem cell therapy are realized and treatments become available for congestive heart failure, diabetes, and other afflictions and if, as many believe, tailor-made transplant organs will eventually be possible, then literally millions of people worldwide will be treated using stem cell therapy.

When a potential new therapy holds out promise of dramatic cures we should, of course, be cautious, if only to dampen false hopes of an early treatment. Equally, however, for the sake of all those awaiting therapy, we should pursue the research that might lead to therapy with all vigor. To fail to do so would be to deny people who might benefit from the possibility of therapy.

Immortality

Finally, I want to note the possibility of therapies that would extend life, perhaps even to the point at which humans might become in some sense "immortal."[11] This, albeit futuristic, dimension of stem cell research raises important issues that are worth serious consideration. Many scientists[12] now believe that death is not inevitable, that the process whereby cells seem to be programmed to age and die is a contingent "accident" of human development that can in principle and perhaps in fact be reversed, and that part of that reversal may flow from the regenerative power of stem cells.[13] I have discussed immortality at length elsewhere,[14] but, before turning to the ethics of stem cell research and therapy, I wish to note one important possible consequence of life-extending procedures.

Human Evolution and Species Protection

HESC research in general and the immortalizing properties of such research in particular raise another acute question. If we become substantially longer lived and healthier, and certainly if we transformed ourselves from "mortals" into "immortals," we would have changed our fundamental nature. One of the common defining characteristics of a human being is our mortality. Indeed, in English we are "mortals," or persons; not "immortals" or gods, demigods, or devils. Is there then any moral reason to stay as we are simply because it is "as we are"? Is there something sacrosanct about the human life form? Do we have moral reasons against further evolution whether it is "natural" Darwinian evolution or evolution determined by conscious choice?

One choice that may confront us concerns whether to attempt treatments that might enhance human functioning, so-called enhancement therapies. For example, it may be that, because of their regenerative capacities, stem cells inserted into the brain to repair damage might in a normal brain have the effect of enhancing brain function. Again, it would be difficult if the therapies are proved safe in the case of brain-damaged patients to resist requests for their use as enhancement therapies. What after all could be unethical about improving brain function? We don't consider it unethical to choose schools on the basis of their (admittedly doubtful) claims to achieve this; why would a more efficient method seem problematic?[15]

Marx famously said, "The purpose of philosophy is not to understand the world but to change it." Perhaps the purpose of genetics, and indeed of life sciences more generally, is not to understand humanity but to change it. We should not, of course, attempt to change human nature for the worse, and we must be very sure that in making any modifications we would in fact be changing it for the better and that we can do so safely, without unwanted side effects. However, if we could change the genome of human beings, say by adding a new manufactured and synthetic gene sequence that would protect us from most major diseases and allow us to live on average 25% longer with a healthy life throughout our allotted time, I for one, would want to benefit from this and I have not been able to find an argument against so doing that is even worth citing for rebuttal. In the West, human beings now do live on average 25% longer than we did 100 years ago, but this is usually cited as an unmitigated advantage of "progress." It is not widely regretted, there is no wailing and gnashing of teeth; why would regrets or fears be appropriate if a further health gain could be obtained only by species modification or "directed" evolution? The point is sometimes made that, so long as humans continued to be able to procreate after any modifications, which changed our nature, we would still be, in the biological sense, members of the same species. But, the point is not whether we remain members of the same species in some narrow biological sense but whether we have changed our nature and perhaps with it our conception of normal species functioning.

Stem Cell Research and Social Policy[16]

The United Kingdom's Welcome for Stem Cell Research

On 22 January 2001, the United Kingdom became the first country, certainly in Europe, to approve HESC research, albeit with what the government described as "adequate safeguards." The United Kingdom government had set up an "expert group" under the Chief Medical Officer (CMO's Expert Group), and this group finally reported in June 2000. In August 2000, the government published its response broadly welcoming the report and accepting all of its major recommendations.[17] These recommendations were the subject of a free vote in both houses of the U.K. Parliament, and this vote was overwhelmingly for approval of stem cell research and so-called therapeutic cloning. The CMO's Expert Group relied for such argument mainly on the consistency of such research with embryo research already permitted and well established in the United Kingdom under the Human Fertilization and Embryology Act 1990 and the regulation of research under that Act by the Human Fertilization and Embryology Authority. Basically, under that act, research on embryos is permitted to investigate problems of infertility and other limited purposes. Now the list of permitted purposes is extended to include HESC research.

The U.K. government's policy on stem cells suffered a reverse when a legal action brought by the Pro-Life Alliance succeeded, on 15 October 2001, in getting a declaration that cloning by cell nuclear substitution was outside the terms of the Human Fertilization and Embryology Act 1990. This was because that Act had foolishly and erroneously defined an embryo as "the product of fertilization," which of course embryos produced by "the Dolly method" are not, unless, because they use a cell nucleus produced by fertilization when the original organism was conceived, the relevant act of fertilization can be displaced a generation. However, the Pro-Life Alliance lost its case on appeal, and emergency legislation rushed through the U.K. Parliament has banned reproductive cloning, the government repeating the unsupported[18] claim that human reproductive cloning was "ethically unacceptable."[19]

Before addressing head-on the ethics of stem cells research as I see it, it is important to place stem cell research in a European and world perspective. There are few comprehensive legal or regulatory frameworks for stem cell research throughout the European Union. Many countries are without any legislation, and where laws are in place, they range from an absolute prohibition on embryo research[20] to the permissibility of the creation of embryos for research purposes.[21] This diversity of opinion is a reflection of existing cultural and religious differences. The strength of feeling in some countries regarding embryo research makes even compromise positions difficult to achieve. Governments have to balance strongly held beliefs regarding the moral status of the embryo and fears of instrumentalization against the promise of remarkable advances in the treatment of disease. There are conflicting duties between state responsibility for the health of their populations and the protection of their moral sensibilities.

The Position of European Union Countries

In most E.U. countries, there is a parallel between the permissibility of embryo research and the permissibility of abortion. Ireland is the only E.U. country whose constitution affirms the right to life of the unborn, where this right is equal to that of the mother,[22] although it is unclear whether this constitutional right applies from fertilization or implantation. Despite the constitutional wording, abortion is legitimate if the life of the mother is in immediate danger. Rape, incest, or fetal abnormalities are no justification, however. There is a tension between this attitude and the European Court of Justice decision that abortion constitutes a medical service within the meaning of the European Treaties and that any limitation on the provision of such services by a Member

State was a matter for the European Union rather than Irish law.[23] Ireland had to negotiate special provisions in the Maastricht treaty to maintain its antiabortion measures. Many applicant countries with bans or restrictions on abortions, such as Poland, Slovakia, Lithuania, Hungary, Slovenia, the Czech Republic, and Malta, may have to do the same.

Belgium and the Netherlands conduct embryo research without a framework of legislation. Portugal, where abortion is illegal except in cases of rape or serious medical reasons, and banned regardless after the twelfth week, has no legislation but no research. It is banned in Austria, Germany, and even France, but the latter allows "the study of embryos without prejudicing their integrity"[24] and preimplantation diagnosis. The Spanish constitution offers protection only to the in vitro viable embryo; the criteria for viability leave out spare embryos.[25] Embryo research is permissible under specified conditions in Finland, Spain, and Sweden. The most liberal research conditions are to be found in the United Kingdom, where even the creation of embryos for research purposes has been legal since the 1990 Human Fertilisation and Embryology Act came into force. The legal situation in nine European countries is either under review or being revised or amended. For those countries, and the ones with no legislation at all, the situation may be guided by international regulations.

The United States' Position

The United States seems to share some of Germany's hypocrisy and indecision on this issue. Ten states have passed laws regulating or restricting research on human embryos, fetuses, or unborn children, and at the federal level funding is prohibited to support any research in which embryos are destroyed. As I discuss later, however, this federal prohibition is ominously restrictive and would seem to condemn a number of other practices as well.[26]

International Guidelines

International guidelines provide little clarity specifically on human embryo research. Apart from the wide international agreement on the prohibition of human reproductive cloning, agreements at the European level have left the permissibility of particular research to the discretion of each member state. There are few guidelines, but if research is authorized by a member state, then respect for human dignity requires an appropriate regulatory framework and the provision of guarantees "against risks of arbitrary experimentation and the instrumentalisation of embryos."[27] Both Italy and Greece rely on the Council of Europe's Convention of Human Rights and Biomedicine, Article 18, which stipulates only two conditions: a prohibition on producing embryos for research purposes and the adoption of rules designed to ensure adequate protection of embryos.[28] To date, only three countries have ratified this convention. An added protocol prohibiting human cloning took effect in 2002. Human cloning was also banned by the Charter of Fundamental Rights of the European Union in December 2000, as are eugenic practices but, surprisingly for such a recent statement, it does not comment explicitly on embryo research.[29] The European Parliament has stated its opposition both to therapeutic cloning and to the creation of spare embryos. Subsequently, the European Group of Ethics in Science and New Technologies to the European Commission, while advocating

Stem Cells, Sex, and Procreation

the allocation of a community budget to research on spare embryos from IVF treatment, confirmed the position that it considered the creation of embryos from donated gametes for research purposes ethically unacceptable and deemed therapeutic cloning as premature.[30] Those countries that have commissioned an exploration of the issues from their National Ethics Committees or similar bodies provide an insight into the problems of achieving consistency of legislation.

Consistency of Legislation

There are many problems regarding the consistency of legislation throughout the E.U. countries. There often exist a constitutional right to freedom of research and a responsibility to ensure the health of their citizens. Again, Germany is an interesting illustration of the paradoxes that stem cell research has generated in Europe. Abortion is technically illegal in Germany, but women are not penalized, provided they receive counseling at a state-approved center, which may then issue them a certificate.[31] So, there is a situation where abortion is permissible for a variety of reasons, where the abortion pill RU-486 is available,[32] but where research on embryos is prohibited.

Germany also provides a constitutional right to freedom of research at the individual and institutional levels and a constitutional duty for the state to protect the life and health of its citizens.[33] This was a consideration for many states who debated the ethics of this research. Germany, like the United States, opted for a compromise position when the Federal Parliament voted to permit limited import of embryonic stem cell lines created before 30 January 2002 while maintaining a ban on their derivation within German laboratories.[34] For France, where embryo research is also prohibited, the ethics committee struggled with the fact that prohibition had halted HESC research when the therapeutic possibilities make it very desirable. The law is currently under review there, and supporters of HESC research pointed out that a "duty of solidarity" with individual suffering prohibits any attempt to stop research.[35] There were pragmatic considerations in the acknowledgment that this research will continue elsewhere and if it produces the results it promises it was considered that French researchers would have no choice but to pursue it anyway.[36] The dilemma now is whether to legislate directly and have safeguards that reflect the sensibilities of French society. The French also raise the concern that improved technical skills in IVF will lead to a decrease in the number of spare embryos. Their ethics committee recommends that the question of oocyte extraction and culture will need to be dealt with explicitly by law to prevent any risk of creating a market situation that would put psychological pressure on women.[37]

Benefiting from Evil?

Nations whose constitution (or, for that matter, democratic will) provides for freedom of research and imposes an obligation on the state to protect the lives and health of citizens and that have outlawed HESC research may face an agonizing dilemma should this research produce a therapeutic success. They will have to decide whether to make the resulting therapy available to their citizens, thereby risking the charge of exploiting and benefiting from the wickedness of others, or face the unhappy prospect of watching their citizens die while those of other countries receive treatment. Of course, in reality, this will not quite be the dilemma because many citizens will seek treatment abroad, but the poor, as ever, are most likely to suffer from such a policy.[38]

The Italian National Bioethics Committee was split on the permissibility of creating embryos for research, a split grounded in the status of the embryo. Some members thought even the use of cryogenically frozen embryos, of which there is a considerable surplus, was not ethically justifiable, as respect for human beings prevents instrumental use of these embryos.[39] Those who were in favor of research mentioned the additional consideration of the autonomy of women and couples in deciding to donate their eggs and decide on the fate of their nonimplanted embryos.[40] Despite the legal and constitutional issues and the concerns of pragmatism and consistency, the status of the embryo is a continuous sticking point in the attempt to guide social policy. The religious positions all comment on the point that they believe life begins, and this gives an insight into the debate.

The Ethics of Stem Cell Research

Stem cell research is of ethical significance for three major reasons:

1. It will for the foreseeable future involve the use and sacrifice of human embryos.
2. Because of the regenerative properties of stem cells, stem cell therapy may always be more than therapeutic—it may involve the enhancement of human functioning and indeed the extension of the human lifespan.
3. So-called therapeutic cloning—the use of cell nuclear replacement to make the stem cells clones of the genome of their intended recipient—involves the creation of cloned pluripotent and possibly totipotent cells, which some people find objectionable.

Elsewhere I have discussed in detail the ethics of genetic enhancement[41] and the ethics of cloning,[42] and I noted above the immortalizing potential of stem cell research. In this essay, I concentrate on objections to the use of embryos and fetuses as sources of stem cells.

Given that, currently, the most promising source of stem cells for research and therapeutic purposes is either aborted fetuses or preimplantation embryos, their recovery and use for current practical purposes seems to turn crucially on the moral status of the embryo and the fetus. A number of recent indications are showing promise for the recovery and use of adult stem cells. It was reported recently that Catherine Verfaillie and her group at the University of Minnesota had successfully isolated adult stem cells from bone marrow and that these seemed to have pluripotent properties (i.e., capable of development in many ways but not in all, and not capable of becoming a new separate creature) like most HES cells.[43] Simultaneously, *Nature Online* published a paper from Ron McKay at NIH showing the promise of embryo derived cells in the treatment of Parkinson's disease.[44]

This indicates the importance of pursuing both lines of research in parallel. The dangers of abjuring embryo research in the hope that adult stem cells will be found to do the job adequately is highly dangerous and problematic for a number of reasons. The first is that we do not yet know whether adult cells will prove as good as embryonic cells for therapeutic purposes. At the moment there is simply much more accumulated data and much more therapeutic promise from human embryonic stem cells. The second is that it might turn out that adult cells will be good for some therapeutic purposes and human embryonic stem cells for others. Third, we already know that it is possible to modify or replace virtually any gene in human embryonic stem cells; whether this will also be true of adult stem cells has yet to be established. Finally, it would be an irresponsible gamble with human lives to back one source of cells rather than another and to make people wait and possibly die while what is still the less favored source of stem cells is further developed. This means that the ethics of HESC research is still a vital and pressing problem and cannot for the foreseeable future be bypassed by concentration on adult stem cells.

Stem Cells from Early Embryos

It is possible to remove cells from early preimplantation embryos without damage to the original embryo. This may be one solution to the problem of obtaining embryonic stem cells. However, if the cells removed are totipotent (i.e., capable of becoming literally any part of the creature, including the whole creature), and if moreover they are capable of deciding until the cell mass achieves sufficient cells for autonomy (i.e., the ability to implant successfully and continue to grow to maturity),[45] then they are in effect separate zygotes, they are themselves "embryos," and so they must be protected to whatever extent embryos are protected. If however, such cells are merely pluripotent, then they could not be regarded as embryos and the use of them would, presumably, not offend those who regard the embryo as sacrosanct. Unfortunately, it is not at present possible to tell in advance whether a particular cell is totipotent or simply pluripotent. This can only be discovered for sure retrospectively by observing the cells capabilities.

I will now set out one ethical principle that I believe must be added to the central principles cited in guiding our approach to HESC research and raise two issues of the consistency of attitudes and judgments about stem cell research with other practices and treatments used and considered acceptable (albeit with qualifications) not only in the European Union but indeed in the world at large. The two issues of consistency are:

1. consistency of stem cell research with what is regarded as acceptable and ethical with respect to normal sexual reproduction
2. consistency with attitudes to and moral beliefs about abortion and assisted reproduction.

The ethical principle that I believe we all share and that applies to the use of embryos in stem cell research is the Principle of Waste Avoidance.

The Principle of Waste Avoidance

This widely shared principle assumes that it is right to benefit people, if we can, and wrong to harm them, and it states that, faced with the opportunity to use resources for a beneficial purpose when the alternative is that those resources are wasted, we have powerful moral reasons to avoid waste and do good instead. I will start with consideration of the first requirement of consistency.

Lessons from sexual reproduction. Let us start with the free and completely unfettered liberty to establish a pregnancy by sexual reproduction without any "medical" assistance. What are people and societies who accept this free and unfettered liberty committing themselves to? What has a God who has ordained natural procreation committed herself to?

We now know that for every successful pregnancy that results in a live birth many, perhaps as many as five,[46] early embryos will be lost or will "miscarry" (although these are not perhaps "miscarriages" as the term is normally used because this sort of very early embryo loss is almost always entirely unnoticed). Many of these embryos will be lost because of genetic abnormalities, but some would have been viable. Many people believe that the fact that perhaps a large proportion of these embryos are not viable somehow makes their sacrifice irrelevant. But those who believe that the embryo is morally important do not usually believe that this importance applies only to healthy embryos. Those who accept the moral importance of the embryo would be no more justified in discounting the lives of unhealthy embryos than those who accept the moral importance of adult humans would be in discounting the lives of the sick or of persons with disability.

How are we to think of the decision to attempt to have a child in the light of these facts? One obvious and inescapable conclusion is that God and/or nature has ordained that "spare" embryos be produced for almost every pregnancy and that most of these will have to die in order that a sibling embryo can come to birth. Thus, the sacrifice of embryos seems to be an inescapable and inevitable part of the process of procreation. It may not be intentional sacrifice, and it may not attend every pregnancy, but the loss of many embryos is the inevitable consequence of the vast majority (perhaps all) pregnancies. For everyone who knows the facts, it is a conscious, knowing, and therefore deliberate sacrifice; and for everyone, regardless of "guilty" knowledge, it is part of the true description of what they do in having or attempting to have children.

We may conclude that the production of spare embryos, some of which will be sacrificed, is not unique to assisted reproduction technologies (ART); it is an inevitable (and presumably acceptable, or at least tolerable?) part of all reproduction.

Both natural procreation and ART involve a process in which embryos, additional to those that will actually become children, are created only to die. I will continue to call these "spare" embryos in each case. If either of these processes is justified it is because the objective of producing a live healthy child is judged worth this particular cost. The intentions of the actors, appealed to in the frequently deployed but fallacious doctrine of double effect,[47] are not relevant here. What matters is what the agents knowingly and voluntarily bring about. That this is true can be seen by considering the following example.

Suppose we discovered that the use of mobile telephones within 50 meters of a pregnant woman resulted in a high probability, near certainty, of early miscarriage. No one would suggest that, once this is known, it would be legitimate to continue use of mobile telephones in such circumstances on the grounds that phone owners did not intend to cause miscarriages. Any claim by phone users that they were merely intent on causing a public nuisance or, less probably, making telephonic communication with another person and therefore not responsible for the miscarriages would be rightly dismissed. It might, of course, be the case that we would decide that mobile communications were so important that the price of early miscarriage and the consequent sacrifice of embryos was a price well worth paying for the freedom to use mobile telephones. And this is, presumably, what we feel about the importance of establishing pregnancies and having children. Mobile telephone users, of course, usually have an alternative method of communication available, but let us suppose they do not.

This example shows the incoherence of the so-called doctrine of double effect. The motives or primary purposes of the phone user are clearly irrelevant to the issue of their responsibility for the consequences of their actions. They are responsible for what they knowingly bring about. The only remaining question is whether, given the moral importance of what they are trying to achieve (phoning their friends), the consequent miscarriages are a price it is morally justifiable to exact to achieve that end. Here the answer is clearly "no." Sometimes proponents of the doctrine of double effect attempt to make proportionality central to the argument. It is not, so it is claimed, the fact that causing miscarriage is not the primary or first intention or effect that matters but the fact that miscarriage is a serious wrong compared with the benefit of using a mobile telephone. However, this is to miss the point of the doctrine of double effect. Proportionality cannot be the issue because the doctrine of double effect was designed to exculpate people from the wrong of intending a forbidden act. The proportionality of the various outcomes cannot speak to the issue of primary or second effects. Only the true account of what the agents wanted to achieve or were "trying to do," of what the main intention or purpose actually was or is, can do that.

However, when we pose the same question about the moral acceptability of sacrificing embryos in pursuance of establishing a successful pregnancy, the answer seems different. My point is that the same issues arise when considering the use of embryos to obtain embryonic stem cells. Given the possible therapeutic uses I have reviewed, it would be difficult, I suggest, to regard such uses as other than morally highly significant. Given that decisions to attempt to have children using sexual reproduction as the method (or even decisions to have unprotected intercourse) inevitably create embryos that must die, those who believe having children or even running the risk of conception is legitimate cannot consistently object to the creation of embryos for comparably important moral reasons. The only remaining question is whether the use of human embryonic stem cells for therapies designed to save lives and ameliorate suffering are purposes of moral importance comparable to those of attempting to have (or risking the conception of) children by sexual reproduction.

The conscious voluntary production of embryos for research, not as the by-product of attempts (assisted or not) at reproduction, is a marginally different case, although some will think the differences important. However, if the analysis so far is correct, then this case is analogous in that it involves the production and destruction of embryos for an important moral purpose. All that remains is to decide what sorts of moral objectives are comparable in importance to that of producing a child. Although some would defend such a position,[48] it would seem more than a little perverse to imagine that saving an existing life could rank lower in moral importance to creating a new life. Assisted reproduction is, for example, given relatively low priority in the provision of healthcare services. Equally, saving a life that will exist in the future seems morally comparable to creating a future life. In either case, the moral quality and importance of the actions and decisions involved and of their consequences seem comparable.

Instrumentalization. It is important to note that prolife advocates or Catholics are necessarily acting instrumentally when they attempt to procreate. They are treating the 1-4 embryos that must be sacrificed in natural reproduction as a conscious (though not intended) means to have a live birth. This is something Catholics certainly and probably most others who hold a "prolife" position should not do.

However, the issue is not whether Catholics or those who take a prolife position may or may not be permitted to create embryos, which certainly or highly probably will die prematurely, and whether this constitutes reckless endangerment of embryos or even the unjustifiable killing of embryos. Rather, the facts of life, the facts of natural reproduction, show that the creation and destruction of embryos is something that all those who indulge in unprotected intercourse and certainly all those who have children are engaged in. It is not something that only those who use assisted reproduction or those who accept experimentation on embryos are "guilty" of. It is a practice in which we are all, if not willing, at least consenting participants, and it shows that a certain reverence for or preciousness about embryos is misplaced.

Embryo-sparing ART. It might be said that there is a difference—those who engage in assisted reproduction engage in the destruction of embryos at a greater rate than need be. Those who engage in sex are not engaged in the destruction of embryos at a greater rate than is required for the outcome they seek. It would be interesting to know whether, if a creating a single embryo by IVF became a reliable technique, prolife supporters would feel obliged to use this method rather than sexual reproduction because of its embryo-sparing advantages. It looks as though there would indeed be a strong moral obligation to abandon natural procreation and use only embryo-sparing ART.

Consider a fictional IVF scenario. A woman has two fertilized eggs and is told it is certain that if she implants both only one will survive but that if she implants only one it will not survive. Would she be wrong to implant two embryos to ensure a successful singleton pregnancy? This example is, of course, fictional only in terms of the degree of certainty supposed. It is good practice in IVF to implant two or three embryos in the hope of achieving the successful birth of one child. Thus, in normal IVF as in normal sexual reproduction, the creation and "sacrifice" of embryos in pursuit of a live child is not only accepted as necessary but is part of the chosen means for achieving the objective. Most people would, I believe, judge this to be permissible, and indeed it is what often happens in successful IVF pregnancies, where up to three embryos are implanted in the hope of one live birth. Even in Germany, where stem cell research using embryos is currently banned and where legal protections for the embryo are enshrined in the constitution, IVF is permitted, and it is usual to implant three embryos in the hope and expectation of achieving no more than a single live birth.

Even if we could accurately predict in advance which embryos would survive and which would not, the ethics would not change. Suppose that for some biological reason there was a condition that required that, for one embryo to implant, it was necessary to introduce a companion embryo that would not, and we could tell in advance which would be which. It is difficult to imagine how or why this fact would alter the ethics of the procedure; it would remain the case that one must die in order that the other would survive. If people in this condition wanted ART, would we judge it unethical to provide it to them but not to "normal" IVF candidates when the "costs" were the same in each case—namely, the loss of one embryo in pursuit of a healthy birth?

It might be objected that the parallel with sexual reproduction is like saying that, because we know that road traffic causes thousands of deaths per year, to drive a car is to accept that the sacrifice of thousands of lives in almost every country, for example, is a price worth paying for the institution of motor transport. This might seem a telling analogy showing that we do not willingly accept the inevitable consequences of what we do. There are, of course, many disanalogous features of the purported *reductio ad absurdum* comparison with road deaths. The vast majority of drivers will go all their lives without having an injury-causing incident, let alone a fatality, and the probability of any individual causing a death once exacerbating factors such as alcohol use and reckless fatigue are taken into account is vanishingly small by any standards and insignificant when compared with the high risk of production of embryos in unprotected sex between fertile partners. However, suppose an individual knew that, despite a long driving career without accidents, today is the day that either they will surely be involved in a fatal accident and cause someone's death or that the probability of this happening is very high indeed. Would it be conceivable that it might be permissible, let alone ethical, to drive today? And yet that is the situation with normal sexual intercourse, at least for those who regard the embryo as protected.

The natural is not connected to the moral. It is important to be clear about the form of this argument. I am not, of course, suggesting that because something happens in nature it must be morally permissible for humans to choose to do it. I am not suggesting that, because embryos are produced only to die in natural procreation, that the killing of embryos must be morally sound. I am saying, rather, that if something happens in nature *and* we find it acceptable in nature given all the circumstances of the case, then if the circumstances are relevantly similar it will for the same reasons be morally permissible to achieve the same result as a consequence of deliberate human choice. I am saying that we do as a matter of fact and of sound moral judgment accept the sacrifice of embryos in natural reproduction, because although we might rather not have to sacrifice embryos to achieve a live healthy birth, we judge it to be defensible to continue natural reproduction in the light of the balance between the moral costs and the benefits. And if we make this calculation in the case of normal sexual reproduction we should, for the same reasons, make a similar judgment in the case of the sacrifice of embryos in stem cell research.

To take a different but analogous case: if we say that God and/or nature "approves" of cloning by cell division because of the high rate of natural monozygotic twinning[49] and that therefore the duplication of the human genome is not per se unethical we are not saying that cloning by cell division is ethically unproblematic because it occurs naturally. The point of the analogy is rather that, because the birth of natural identical twins is generally not considered regrettable, we are reminding ourselves that there is nothing here to regret. Indeed, it is the occasion for unmitigated joy or at least moral neutrality. We should, therefore, unless we can find a difference, feel the same about choosing deliberately to create twins by duplication of the human genome.[50] If we then object to cloning by a different method, cell nuclear transfer objections must obviously be to features that arise uniquely in cell nuclear transfer and cannot simply be to such features as duplication of the human genome. Our acceptance of the natural does not, of course, apply to naturally occurring premature death; here we do think there is something to regret, even if it is natural and inevitable.

Instrumentalization revisited. Another possible concern involves a version of the instrumentalization objection that demands that embryos not be produced only to be used for the benefit of others but that, as in sexual reproduction, they should all have some chance of benefiting from a full normal lifespan.[51] In normal sexual reproduction, embryos must be created only to die so that a sibling embryo can come to birth. But, arguably, it is in each embryo's interest that reproduction continues because it is the embryo's only chance to be the one that survives. Embryos (if they had rationality) would have a rational motive to participate (albeit passively) in sexual reproduction. By contrast, so it might be claimed, embryos produced specifically for research would not rationally choose to participate for they stand to gain nothing. All research embryos will die, and none have a chance of survival. If this argument is persuasive against the production of research embryos, it is easily answered by ensuring that the embryos produced for research have to some appropriate extent a real chance of survival. One would simply have to produce more embryos than are required for research, randomize allocation to research, and ensure that the remainder are implanted with a chance to become persons. To ensure that it would be in every embryo's interest to be "a research embryo," all research protocols permitting the production of research embryos would have to produce extra embryos for implantation. To take a figure at random but one that, as it happens, mirrors natural reproduction and gives a real chance of survival to all embryos, we could ensure that for every, say, 100 embryos needed for research another 10 would be produced for implantation. The 100 embryos would be randomized 90 for research, 10 for implantation, and all would have a chance of survival and an interest in the maintenance of a process that gave them this chance.

The third case concerns spare embryos that become available for research as a result of an ART program in which they have been produced and to which they are now superfluous because their "mother" has now declined for whatever reason to accept more embryos for implantation and has refused consent for their implantation into others. Here it might be suggested that these embryos are also like the research embryos just considered. However, this is not the case. These embryos have had their chance of implantation, but unfortunately for them, they have missed out. The fact that now they are irredeemably surplus to requirements for implantation does not show that they always were. These embryos have had their chance of life, their "motive" for participating in the program is as strong as in sexual reproduction or randomized research embryos.

Born to die. The force of the sexual reproduction analogy may seem vulnerable to the following claim.[52] It can be said that, just as parents are responsible for the deaths of the embryos inevitably produced as a consequence of unprotected intercourse, so also and to the same extent are they responsible for the deaths of the children they actually produce when these children eventually die of old age. This is because in each case the parents have produced a life, which will end at a particular point and that point is in each case out of the parents' control. So, if parents are responsible for the deaths of the embryos lost as a result of unprotected intercourse, they are also responsible for the deaths of their children lost in old age. In neither case, however, have the parents been the proximate cause of death, but they have caused the life and death cycle. This objection, like the objection from the acceptability of motorized road transport, purports to constitute a *reductio ad absurdum*.

This is a puzzling objection. As I have argued, people accept the necessity of and the justification for producing surplus embryos because they wish to have a baby. Those who judge the embryo to have moral importance comparable to adults or children will have to justify their instrumentalization of the embryos that are sacrificed to this end.

On the other hand, those who think that dying of old age or being given a worthwhile life is a good will see nothing to justify. The parents are responsible for that life, to be sure, but they are morally justified in that responsibility, and in that the life for which they are responsible has been or is reasonably likely to be a worthwhile life, then, unless they have also arranged the death, their responsibilities have been exercised in a way that is both morally and socially appropriate.

The life of their child was in this case neither created nor ended to be a means to the interests of others. It is a good life, the creation of which requires no justification and the end of which was neither caused by the parents nor was its timing predictable by them. They therefore have no excuses to make. By contrast, the lives of the embryos that must die early are, if those lives are morally important at all, not lives the ending of which is a reasonable price to pay for the life lived.

The United States condemns human reproduction! Shocked by the idea of any activity that threatens the embryo, the U.S. government has adopted the revolutionary strategy of attempting to condemn human reproduction and, for good measure, has included all unprotected intercourse in the condemnation and to ban all federal support for such activities.

How have our cousins in the United States arrived at this daring and groundbreaking social policy? In the United States, current federal law prohibits the use of federal funds for "the creation of a human embryo" explicitly for research purposes or, more crucially, for "research in which a human embryo or embryos are destroyed, discarded or knowingly subjected to the risk of injury or death."[53] Such law is presumably animated by concern about the morally problematic nature of such actions and also by the idea that federal support in the form (among others) of "tax dollars" should not be given to activities that a significant number of people find offensive or objectionable. As I have noted, normal sexual reproduction inevitably involves a process in which a human embryo or embryos are destroyed or discarded. It is also incontrovertibly an activity in which a human embryo or embryos are "knowingly subjected to the risk of injury or death," at least for anyone who knows the facts of life. The perpetuation of this position seems likely, as President George W. Bush had made an election promise never to provide federal support for research that involves living human embryos. Those who can read his lips may have less confidence that this promise will be kept.

Consistency with attitudes to and moral beliefs about abortion and assisted reproduction. In most countries of the European Union and indeed in most countries of the world, abortion is permissible under some circumstances. Usually, permissibility is considered greater at very early stages of pregnancy, permissibility waning with embryonic and fetal development. The most commonly accepted ground for abortion (where it is acceptable) is to protect the life and the health of the mother. Sometimes the idea of protection of the life and health of the mother is very broadly and liberally interpreted, as it is in the United Kingdom; sometimes the requirement is very strict, demanding real and present danger to the life and health of the mother (Northern Ireland, for example). Given that the therapies initially posited for stem cell research—the treatment of Parkinson's disease and the development of tailor-made transplant organs—are all for serious diseases that threaten life and dramatically compromise health, it is difficult to see how those who think the sacrifice of early embryos for these purposes is or could be justified could find principled objections to the use of embryos in other lifesaving therapies.[54]

The same is, of course, true, as I have already noted of ART. All IVF involves the creation of spare embryos, and all IVF now practiced is built on research done on many thousands of embryos. Most countries and most religions accept IVF and its benefits and in doing so accept that spare embryos will be produced only to die. Even Germany, which has, as I noted, an Embryo Protection Act, accepts the practice of implanting up to three embryos in the hope and expectation that at least one will survive. The acceptance of the practice of IVF is necessarily an acceptance that embryos may be created and destroyed for a suitably important moral purpose.

The Principle of Waste Avoidance. As I stated previously, this widely shared principle states that it is right to benefit people if we can and wrong to harm them, and that, faced with the opportunity to use resources for a beneficial purpose when the alternative is that those resources are wasted, we have powerful moral reasons to avoid waste and do good instead.

That it is surely better to do something good than to do nothing good should be reemphasized. It is difficult to find arguments in support of the idea that it could be better (more ethical) to allow embryonic or fetal material to go to waste than to use it for some good purpose. It must, logically, be better to do something good than to do nothing good; it must be better to make good use of something than to allow it to be wasted. It must surely be *more* ethical to help people than to help no one. This principle—that, other things being equal, it is better to do some good than no good—implies that tissue and cells from aborted fetuses should be available for beneficial purposes in the same way that it is ethical to use organs and tissue from cadavers in transplantation.

It does not follow, though, that it is ethical to create embryos specifically for the purposes of deriving stem cells from them. However, as I discussed, there may be problems in objecting to creating embryos for this purpose from people who do not object to the sacrifice of embryos in pursuit of another supposedly beneficial objective—namely, the creation of a new human being. Only those who think that it is more important to create new humans than to save existing ones will be attracted to the idea that sexual reproduction is permissible whereas the creation of embryos for therapy is not.

Notes

1. And we were young. With apologies to A. E. Houseman, *Parta quies*.
2. Whereas other equally platitudinous but also totally incoherent principles, like the appeal to human dignity, so often occupy center stage.
3. Nor anyone who even believes that unprotected sexual intercourse is permissible.
4. Of course, there may be specific research projects that are unethical.
5. Thomson JA, Itskovitz-Elder J, Shapiro SS, Waknitz MA, Swiergiel VS, Jones JJ. Embryonic stem cell lines derived from human blastocysts. *Science* 1998;282(Nov 6):1145–47.
6. Pedersen R. Embryonic stem cells for medicine. *Scientific American*. 1999;April:69.
7. De Rijk MC, Launer LJ, Berger K, Breteler MMB, Dartigues J-F, Baldereschi M et al. Prevalence of Parkinson's disease in Europe. *Neurology* 2000;54(11;S5):21–3.
8. Schrag A, Ben-Shlomo Y, Quinn NP. Cross sectional prevalence survey of idiopathic Parkinson's disease and Parkinsonism in London. *BMJ* 2000;321(7252):21–2.
9. Parkinson's Disease Society. Who gets Parkinson's? Available at: http://www.parkinsons.org.uk/docs/viewdoc.asp?ID=105&catid=29&nodeid=224.
10. Parkinson's Disease Foundation. Parkinson's disease: an overview. Available at: http://www.pdf.org/aboutPD/.
11. Harris J. Intimations of immortality. *Science* 2000;288(5463):59. Harris J. Intimations of immortality: the ethics and justice of life extending therapies. In: Freeman M, ed. *Current Legal Problems*. Oxford: Oxford University Press; 2002.
12. And perhaps most religionists too, but that is a different kind of immortality.
13. Kirkwood T. *Time of Our Lives*. London: Weidenfeld and Nicolson; 1999; Kirkwood T. *The End of Age*. London: Profile Books; 2001.
14. See note 11 Harris 2000, Harris 2002.
15. For more on the ethics of genetic enhancement, see: Harris J. *Clones, Genes, and Immortality*. New York: Oxford University Press; 1998.
16. This section was researched and substantially drafted by my colleague Louise Irving. I am grateful to her here as elsewhere for help and advice.
17. Stem Cell Research: Medical Progress with Responsibility Department of Health June 2000. Government Response to the Recommendations made in the Chief Medical Officer's Expert Group Report "Stem Cell Research: Medical Progress with Responsibility." Presented to Parliament by the Secretary of State for Health by Command of Her Majesty, August 2000. The Stationary Office CM4833.
18. And also insupportable claim unless the immorality consists solely in the fact that as yet cloning by cell nuclear transfer is untested and probably unsafe in humans. For a detailed attempt to demonstrate that there are no good moral arguments against cloning, see: Harris J. Goodbye Dolly: the ethics of human cloning. *Journal of Medical Ethics* 1997;23(6):353–60; Harris J. Cloning and human dignity. *Cambridge Quarterly of Healthcare Ethics* 1998;7(2):163–8; Harris J. Genes, clones, and human rights. In: Burley JC, ed. *The Genetic Revolution and Human Rights: The Amnesty Lectures 1998*. New York: Oxford University Press; 1999:61–95.
19. See note 18, Harris 1999.
20. This is the case in France, Germany, Ireland, Lithuania, and Switzerland.

John Harris

21. This has been legal in the United Kingdom since the 1990 Human Fertilisation and Embryology Authority Act came into force.
22. The Eighth Amendment of the Constitution Act 1983 reads, "Acknowledged the right to life of the unborn, with due regard to the equal right to life of the mother," 7 October 1983, in *Constitution of Ireland*. Available at: http://www.taoiseach.gov.ie/viewitem.asp?id=297&lang=ENG.
23. Frydrych M. Abortion not considered in enlargement rules. Available at http://www.euobserver.com/index.phtml?aid-4646.
24. The Opinion of the European Group on Ethics in Science and New Technologies to the European Commission, No. 15. *Ethical Aspects of Stem Cell Research and Use* 2000 Nov 14:11.
25. Spanish regulation is through The Assisted Reproduction Techniques Act of 1988, taken from: European Commission Research Directorate-General. The Spanish Survey on Human Embryonic Stem Cells. In: Matthiessen L, ed. *Survey on Opinions from National Ethics Committees or Similar Bodies Public Debate and National Legislation in Relation to Human Embryonic Stem Cell Research and Use*. Brussels, Belgium: European Commission Research Directorate-General; 2001.
26. See: Lanza RP, Caplan AL, Silver LM, Cibelli JB, West MD, Green RM. The ethical validity of using nuclear transfer in human transplantation. *JAMA* 2000;284(24):3175.
27. See note 24, European Commission 2000:12.
28. The Convention of the Protection of Human Rights and Dignity of the Human Being with Regard to the Application of Biology and Medicine: Convention on Human Rights and Biomedicine states: "(1) where the law allows research on embryos in vitro, it shall ensure adequate protection of the embryo; and (2) the creation of embryos for research purposes is prohibited."
29. The Charter of Fundamental Rights of the European Union. Available at: http://ue.eu.int/df/docs/en/CharteEN.pdf.
30. The European Group on Ethics in Science and New Technologies to the European Commission. Available at: http://europa.eu.int/comm/european_group_ethics/index_en.htm.
31. German bishop capitulates on abortion. *BBC News* [online] 2002 Mar 8. Available at: http://news.bbc.co.uk/2/hi/europe/1862662.stm.
32. *Marantha Christian Journal* [online]. Available at: http://www.mcjonline.com.
33. The European Commission Research Directorate-General. Statement of the Central Ethics Committee on Stem Cell Research 23.11.01 (nonofficial translation) annexe. In: Matthiessen L, ed. *Survey on Opinions from National Ethics Committees or Similar Bodies Public Debate and National Legislation in Relation to Human Embryonic Stem Cell Research and Use*. Brussels, Belgium: European Commission Research Directorate-General; 2001.
34. Schiermeier Q. German Parliament backs stem-cell research. *Nature* 7 Feb 2002;415:566. See also: Heinemann T, Honnefelder L. Principles of ethical decision making regarding embryonic stem cell research. *Bioethics* 2002;16:530–43.
35. European Commission Research Directorate-General. Opinion on the preliminary draft revision of the laws on bioethics of the Comité Consultatif National d'Ethique pour les Sciences (CCNE). In: Matthiessen L, ed. *Survey on Opinions from National Ethics Committees or Similar Bodies Public Debate and National Legislation in Relation to Human Embryonic Stem Cell Research and Use*. Brussels, Belgium: European Commission Research Directorate-General; 2001.
36. See note 36, European Commission Research Directorate-General 2001.
37. See note 36, European Commission Research Directorate-General 2001.
38. de Beaufort I, English V. Between pragmatism and principles? on the morality of using the results of research that a country considers immoral. In: Gunning J, ed. *Assisted Conception: Research, Ethics, and Law*. Aldershot, UK: Dartmouth; 2000. See also: Green RM. Benefiting from evil: an incipient moral problem in human stem cell research and therapy. *Bioethics* 2002;16:544–56.
39. European Commission Research Directorate-General. Italian National Bioethics document. In: Matthiessen L, eds. *Survey on Opinions from National Ethics Committees or Similar Bodies Public Debate and National Legislation in Relation to Human Embryonic Stem Cell Research and Use*. Brussels, Belgium: European Commission Research Directorate-General; 2001.
40. See note 40, European Commission Research Directorate-General 2001.
41. See: Harris J. *Wonderwoman and Superman: The Ethics of Human Biotechnology*. New York: Oxford University Press; 1992; see also note 15, Harris 1998.
42. See note 19, Harris 1997, Harris 1998, Harris 1999.
43. *International Herald Tribune* 2002 Jun 22–23:1. In a paper presented at FENS Forum Workshop, Paris, 13 Jul 2002, Austin Smith emphasized the importance of pursuing research on all sources of stem cells simultaneously.

Stem Cells, Sex, and Procreation

44. McKay R. Stem cells—hype and hope. *Nature* 2000;406:361-4.
45. Pedersen R. Embryonic steps towards stem cell medicine. Paper presented at the EUROSTEM conference Regulation and Legislation under Conditions of Scientific Uncertainty. 2002 Mar 6-9; Bilbao, Spain. The conference was supported by a Project Grant from The European Commission Directorate-General for Research, "Quality of Life."
46. Robert Winston gave the figure of five embryos for every live birth some years ago in a personal communication. Anecdotal evidence to me from a number of sources confirms this high figure, but the literature is rather more conservative, making more probable a figure of three embryos lost for every live birth. See: Boklage CE. Survival probability of human conceptions from fertilization to term. *International Journal of Fertility* 1990;35(2)75-94. See also: Leridon H. *Human Fertility: The Basic Components*. Chicago: University of Chicago Press; 1977. Again, in a recent personal communication, Henri Leridon confirmed that a figure of three lost embryos for every live birth is a reasonable conservative figure.
47. For a conclusive refutation of that doctrine, see: Harris J. *Violence and Responsibility*. London: Routledge and Kegan Paul; 1980. For a more recent discussion of these broad issues, see: Kamm FH. The doctrine of triple effect and why a rational agent need not intend the means to his end. *Proceedings of the Aristotelian Society* (Supplementary volume) 2000;74:21-39; and Harris J. The moral difference between throwing a trolley at a person and a throwing a person at a trolley: a reply to Francis Kamm. *Proceedings of the Aristotelian Society* (Supplementary volume) 2000;S74:40-57.
48. Some hedonistic utilitarians, for example.
49. Human monozygotic twinning occurs in roughly one per 270 births (three per 1,000). I take this figure of the rate of natural twinning from: Moore KL, Persaud TVN. *The Developing Human*, 5th ed. Philadelphia: W. B. Saunders; 1993.
50. Or indeed cloning by cell nuclear substitution, but that is another story. For the full story, see note 18, Harris 1997.
51. The possible objection was put to me by Julian Savulescu—the response to it with all its defects is mine.
52. A point made to me by Louis G Aldrich at the Third International Conference of Bioethics, National Central University, Shungli, Taiwan. 2002 Jun 24-29.
53. U.S. Public Law 105-277, sect. 511, 1998 Oct 21, slip copy. H.R. 4328.
54. See Harris J. Should we experiment on embryos? In: Lee R, Morgan D, eds. *Birthrights: Law and Ethics at the Beginnings of Life*. London: Routledge; 1988:85-95.

[24]

Stem Cells, Superman, and the Report of the Select Committee

Roger Brownsword*

Introduction

In 1998, the Wisconsin-based biologist James Thomson announced that his research group had made a major breakthrough by successfully isolating human embryonic stem cells (hES cells). On this side of the Atlantic, too, there have been important developments, both scientific and regulatory. Most recently, in February 2002, the Human Fertilisation and Embryology Authority announced that, pursuant to the Human Fertilisation and Embryology (Research Purposes) Regulations 2001,[1] it had issued the first licences in the United Kingdom for basic hES cell research to be undertaken. Why have these, and similar, announcements prompted a sense of great excitement and anticipation but also, and in just about equal measure, a sense of tremendous anxiety and deep concern? Attempting to answer such questions, a House of Lords Select Committee has recently published the most detailed review yet in the United Kingdom of the science and ethics of stem cell research and the regulatory issues to which it gives rise.[2]

First, why the excitement? Stem cells are special because they are a potential source of new cells – for example, blood stem cells will replenish blood cells, neural stem cells will replenish brain cells, skin stem cells will replace skin, and so on. ES cells, however, are special because they are still in a so-called pluripotent state. As pluripotent stem cells, hES cells have the potential to develop into any one of the 200 or so human cell types and, in natural development, they will duly differentiate in this way. Those stem cells that have already developed to perform one or more specialised functions are called 'adult' stem cells (regardless of whether they are located in an adult person); and, whilst adult stem cells might have a number of possible differentiated functions – and, thus, are multipotent – the othodox view is that their function in the human body is relatively stable. In other words, whilst say blood stem cells might be able to regenerate blood cells, they are not thought to have the capacity to regenerate brain cells; and, similarly, neural stem cells are not thought to have the capacity to regenerate blood cells.

Having isolated, purified and cultured hES cells, the next step is to seek to understand and simulate the mechanisms by which they differentiate to become specialised adult stem cells. The importance of such an advance in our knowledge

---

* Professor of Law, University of Sheffield and one of the two specialist advisers to the House of Lords Select Committee on Stem Cell Research. The views expressed in this paper are my own. In no sense am I speaking for the Committee collectively or for any of its members personally. Drafts of this paper were presented at the University of Hull and at the ALT annual conference; I am grateful for comments made by participants.

1 SI 2001 No. 188.

2 *Stem Cell Research* (London: HMSO, February 2002) HL Paper 83(i) (Report) and 83(ii) (Evidence).

is explained by Thomson, the pioneering hES cell researcher, in the following way:

> As developmental biologists become more accomplished at directing hES cells to specific cell types, the differentiated derivatives of the cells should have an important role in developing new therapies. Large, purified populations of hES cell-derived cells, such as heart muscle cells or neurons, could be used to screen for new drugs. Purified, normal human cells would allow accurate screening of candidate drugs, greatly reduce the need for animal testing during the early screening process, and accelerate drug discovery. Differentiated derivatives of hES cells also could be used to test for possible toxic side effects of drugs identified by other methods, and hES cells would be particularly useful for identifying compounds that interfere with normal development.
>
> Finally, differentiated derivatives of hES cells could be applied to transplantation therapies for treatment of a range of human diseases. ... Numerous diseases might be treated by this approach, including heart disease, juvenile-onset diabetes, Parkinson's disease, and leukemia.[3]

Interestingly, in contrast with the stem cell debate in the United Kingdom, Thomson identifies the drug development implications first, before going on to highlight the possibility of cell-based transplantation therapies for serious diseases.[4]

If cell-based therapies were to be delivered in future health care regimes, the current thinking is that this might be sourced in one of two ways. One way would be to draw on cell lines held in a national cell bank; it is unlikely that the bank would hold lines that would be perfect matches for patients but they would aim to hold at least some lines that would be close enough to serve. The other way would be to use the patient's own tissue, and then generate a custom-made embryo and ES cells using cell nuclear replacement (CNR) technology.[5] Such bespoke cells should be perfectly compatible with the patient and, in many popularised and schematic accounts, this is seen as the answer to degenerative diseases. For example, a patient suffering from Alzheimer's disease supplies some tissue, say, skin cells; they are reprogrammed using CNR; ES cells are derived from the CNR-embryo and cultured to differentiate into millions of brain cells which are then transplanted into the patient.

So much for the promise and the excitement, but why the concern arising from developments in hES cell research and its regulation? Briefly, even though we seem to be on the cusp of a medical revolution, the price that we pay for deriving hES cells from an embryo is the destruction of that embryo which, at this point, will have been developing for some 5–6 days. Thus, the procedure is not quite comparable to, say, pre-implantation diagnosis (PGD) where a cell can be subtracted from the developing embryo without jeopardising the viability of that embryo – although, of course, where PGD reveals an abnormality, the embryo will be discarded and duly destroyed. The vast majority of the embryos terminated by stem cell researchers will be donated by couples who no longer require them for IVF purposes. To be sure, they would have been destroyed anyway; but they are now destroyed in a research (or, at a later stage, therapeutic) programme rather

3 J.A. Thomson, 'Human Embryonic Stem Cells' in S. Holland, K. Lebacqz and L. Zoloth (eds), *The Human Embryonic Stem Cell Debate* (Cambridge, MA: MIT Press, 2001) 15, 21–22.

4 We might also note the reluctance in Warnock (note 10 below) to make human embryos available for routine testing of drugs (in para 12.5).

5 Cell nuclear replacement (CNR), or somatic cell nuclear transfer (SNT) involves introducing the nucleus from an adult cell into an enucleated egg. The engineered egg, retaining its own mitochondrial DNA but otherwise drawing its DNA from the inserted nucleus, is then electronically stimulated to develop into a CNR-embryo.

than as surplus to reproductive requirements. Where CNR is used, an embryo that has been specially created for research (or, at a later stage, therapeutic) purposes is once again destroyed at the 5–6 day stage. Clearly, then, if we believe that the human embryo has a special value or dignity, these procedures for disaggregating embryos will require some very considerable justification; and the utilisation of CNR, which is associated with 'cloning', will serve only to aggravate these concerns.

Although Europe is a patchwork of rapidly changing regulatory provisions in relation to research on human embryos,[6] the concerns just sketched are widely held and they are articulated most prominently in Article 18 of the Convention on Human Rights and Biomedicine.[7] According to Article 18(2), the *creation* of human embryos for research is categorically prohibited; and, Article 18(1) provides that, where the law permits research on human embryos (created for some other purpose as in an IVF programme), then 'it shall ensure adequate protection of the embryo'. In the absence of any guidance as to the meaning of 'adequate protection', we might translate the Convention as requiring that human embryos should be treated with respect.[8]

It is in this broader European and international setting that we should consider how we came to be where we are; and how the Select Committee's Report[9] fits into this picture. We can also consider where we are going; and whether the United Kingdom can defend its permissive but explicit regulatory position against its critics. This comment will look, first, at the principles underlying the regulatory position in the United Kingdom coupled with the extension to the approved purposes for research on human embryos; and then the four principal objections to permitting hES cell research will be reviewed before offering some thoughts about the likely focal points for future ethical and regulatory debates.

Research on human embryos: the principles underlying the regulatory position in the United Kingdom

The regulatory story begins in the United Kingdom with the Committee of Inquiry into Human Fertilisation and Embryology (the Warnock Report).[10] Controversially, Warnock took the view that research on human embryos was morally permissible; and some members of the Committee (in fact, just a majority) were prepared, even then, to sanction the creation of embryos for research. However, as a counterweight, it was emphasised that the human embryo has a special status and should not regarded as simply a ball of cells.

Despite considerable Parliamentary unease about licensing research on human embryos, the Government took forward the main thrust of Warnock's recommen-

6 See D. Beyleveld and S.D. Pattinson, 'Embryo Research in the UK: Is Harmonistation in the EU Needed or Possible?' in M.B. Friele (ed), *Embryo Experimentation in Europe* (Bad Neuenahr-Ahrweiler: Europäische Akademie, 2001) 58; and Report (n 2 above) Chapter 7.

7 This is sometimes referred to as 'the Bioethics Convention'. Its full title is the Convention for the Protection of Human Rights and Dignity of the Human Being with Regard to the Application of Biology and Medicine: Convention on Human Rights and Biomedicine (Oviedo, April 4, 1997).

8 Within the EU, see, too, the *Opinion on Ethical Aspects of Human Stem Cell Research and Use* adopted by the European Group on Ethics in Science and New Technologies (Paris, November 14, 2000) (revised edition, January 2001).

9 See n 2 above.

10 Cmnd 9314 (1984).

dations by enacting the Human Fertilisation and Embryology Act 1990.[11] The legislation is complex; it was never easy to interpret; and, against a moving scientific background, the task of interpretation and application does not become any easier. Nevertheless, the governing policy and the key regulatory principles underlying the legislation are reasonably clear. The governing policy is that, if embryos are to be available for research, it is better to regulate such research openly and effectively. This demands that the legislation declares quite explicitly (for the guidance of scientists and funders as well as for the re-assurance of the public) what is lawful and what is not; and it calls for a dedicated regulatory body (the HFEA) to license and monitor such research. So far as research on embryos is concerned, the three central principles are:

- that the HFEA should license such work only if it is necessary (the necessity principle);[12]
- that, if the HFEA is satisfied that such research is necessary (because it cannot be done in any other way), then a licence should be granted only if the particular activity is judged to be necessary or desirable in relation to one of the approved statutory purposes (this principle involves a sense of proportionality and good purpose);[13] and
- that, in no circumstances, should research on embryos run beyond 14 days or the appearance of the primitive streak.[14]

In the 1990 legislation, five purposes, are listed as approved. These are:

(a) promoting advances in the treatment of infertility,
(b) increasing knowledge about the causes of congenital disease,
(c) increasing knowledge about the causes of miscarriages,
(d) developing more effective techniques for contraception, [or]
(e) developing methods for detecting the presence of gene or chromosome abnormalities in embryos before implantation.

The Act also provides that research may be licensed 'for such other purposes as may be specified in regulations'.[15] However, this enabling provision is limited: the said 'other purposes' must be designed to 'increase knowledge about the creation and development of embryos, or about disease, or enable such knowledge to be applied'.[16]

11 During the period between Warnock and the 1990 Act, attempts were made to prohibit all embryo research, including Enoch Powell's Unborn Children (Protection) Bill in 1985.
12 Sched 2, para 3(6).
13 Sched 2, para 3(2). Compare the Opinion of the European Group on Ethics in Science and New Technologies, n 8 above para 2.7:
In the opinion of the Group, in such a highly sensitive matter, **the proportionality principle and a precautionary approach** must be applied: it is not sufficient to consider the legitimacy of the pursued aim of alleviating human sufferings, it is also essential to consider the means employed. In particular, the hopes of regenerative medicine are still very speculative and debated among scientists. Calling for prudence, the Group considers that, at present, **the creation of embryos by somatic cell nuclear transfer for research on stem cell therapy would be premature**, since there is a wide field of research to be carried out with alternative sources of human stem cells (from spare embryos, foetal tissues and adult stem cells). [Phrases in bold as per emphasis of original].
14 ss 3(3)(a) and 3(4).
15 Sched 2, para 3(2).
16 Sched 2, para 3(3).

Extending the purposes

Under the 1990 provisions, three applications to carry out stem cell research on human embryos were in fact licensed; but, such licences were exceptional and, of course, they could be granted only where the research activity was judged to be necessary or desirable in relation to one of the five original approved purposes. Given that these purposes are mainly concerned with research that is aimed at improving (or disimproving) the chances of successful reproduction, they are far too narrow to support the range of ends now contemplated by stem cell researchers. It follows that the licences recently announced by the HFEA could not have been issued under the original terms of the 1990 legislation. Accordingly, following a consultation exercise jointly conducted by the HFEA and Human Genetics Advisory Commission,[17] and an important report by an expert group chaired by the Chief Medical Officer,[18] the Government decided to extend the purposes in order to open the way for stem cell research to deliver on its apparent potential.

There were two ways in which the Government might have sought to extend the purposes: (i) by issuing Regulations drawing on the enabling provisions in the 1990 Act[19] or (ii) by drafting a bespoke statutory amendment to the 1990 Act.[20] Not altogether surprisingly, the Government chose the first of these options, putting the Human Fertilisation and Embryology (Research Purposes) Regulations 2001[21] through in the period December 2000 to January 2001. The new Regulations virtually 'copy out' the terms of the enabling provisions adding the following three new purposes to the original five:

(a) increasing knowledge about the development of embryos,
(b) increasing knowledge about serious disease, or
(c) enabling any such knowledge to be applied in developing treatments for serious disease.

Coming to these Regulations without any explanation, one would be unlikely to appreciate that they were designed to facilitate stem cell research, nor would it be apparent that licences for embryonic research using CNR might be approved under these purposes. Parliamentarians, however, were fully apprised as to the significance of these Regulations. Thus, introducing the debate on the Regulations in the House of Lords, Lord Hunt said:

> The principle and law that were established in the 1990 Act are clear. Embryo research may be allowed now, but only for conditions such as infertility, contraception and congenital disease, including cystic fibrosis and haemophilia. The question before the House today is whether those purposes should be extended to include serious diseases such as Parkinson's disease, Alzheimer's disease, cancer and diabetes – a provision anticipated and included as a regulation-making power in the 1990 Act.

17 *Cloning Issues in Reproduction, Science and Medicine* (December 1998).
18 *Stem Cell Research: Medical Progress with Responsibility* (London: Department of Health, June 2000).
19 See n 15 above and n 16 above.
20 Effectively, as the Government subsequently chose to do with the Human Reproductive Cloning Act 2001.
21 SI 2001 No 188. When asked to justify the use of Regulations rather than primary legislation, Yvette Cooper (the Parliamentary Under-Secretary of State for Health) simply responded that the 1990 Act gave the Government the power to extend the approved purposes by Regulation, see Hansard (HC) November 17, 2000, Col 1176.

> That question arises because the late 1990s saw developments in cell nuclear replacement technology in animals and because of the announcement in the US of the extraction of stem cells from a human embryo.[22]

A fierce, full, lengthy, and extremely high quality debate ensued in the Upper House.[23] On one side, Lord Alton and his supporters proposed that the Regulations should be rejected until a Select Committee had had the opportunity to examine the implications of what was now being brought forward. This amendment was defeated and, in the event, a compromise proposed by Lord Walton was accepted. Under this amendment, the Regulations were approved but a Select Committee was set up to consider their implications. Although the HFEA did *not* formally undertake to put any 'new purpose' applications on hold until after the Select Committee had reported, in fact, no licences under the Regulations were granted prior to the publication of the Report.

While the Select Committee was deliberating, the Pro-Life Alliance launched a much publicised judicial review challenging the HFEA's claimed jurisdiction to regulate CNR research.[24] Shortly after the Court of Appeal rejected this challenge, the Select Committee reported and, broadly speaking, endorsed the spirit and intent of the new purposes. Almost at once the HFEA began to issue licences under the new purposes and funding bodies announced that money for hES cell research would now be released.[25]

Was the Select Committee right to take this view? Is the extended regulatory framework fit for the purpose and proof against reasonable objection?

The Select Committee and the four principal objections

The debate in the House of Lords foreshadowed the principal objections to the Regulations that were to be rehearsed in the voluminous evidence taken by the Select Committee. Essentially, there are four recurring objections: one scientific, one ethical, one practical, and one legal.

The scientific objection

During the House of Lords' debate on the Regulations, it was forcibly suggested that new research was coming through to challenge the ruling view that adult stem cells have a relatively limited and specialised range. And, as the Select Committee sat, evidence was regularly adduced to add plausibility to the possibility that adult stem cells might be much more plastic than is generally assumed.[26] If the claims for the potency of adult stem cells are made out, the objection runs, surely it would be better to direct research into this channel rather than into hES cells. After all, if a patient's own adult stem cells are the starting point for that person's therapy, then

22 Hansard (HL) January 22, 2001, Cols 16–17.

23 The debate on the Regulations ran for some seven hours and involved over 40 speakers.

24 See nn 50 and 52 below.

25 J. Meek, 'Millions in grants for embryo stem cell research' *The Guardian*, February 28, 2002, 2. To underline the point made in the text: there was *no* understanding that the HFEA would suspend applications under the Regulations until after the Committee had reported (although the HFEA did undertake not to issue licences for CNR until after the Pro-Life Alliance case was resolved). The fact that the HFEA met on the very same day that the Select Committee reported was purely a coincidence.

26 See, in particular, Dr Elizabeth Allan's Memorandum, Evidence (n 2 above) 306.

there should be no risk of rejection when they or their derivatives are re-introduced into the patient.

The Select Committee could have dealt with this issue quite shortly. It could have simply said, for example, that the weight of scientific evidence presented to the Committee overwhelmingly supported a 'dual track' complementary approach to stem cell research, with work being conducted on both adult and hES cells.[27] Moreover, the Committee had evidence from a number of leading international *adult* stem cell researchers, each of whom subscribed to the dual track approach.[28] Equally, the Select Committee could have kept the issue at arm's length by pointing out that the necessity principle, which is at the heart of the legal framework, means that the HFEA will have to keep this issue under review anyway – in a sense, the operative question is not whether the Select Committee judges that research on hES cells is a poor investment of scientific resource, it is whether the HFEA is persuaded that the research can be done in no other way. Taking this approach, the Committee could have made the telling point that, if the proponents of one track adult stem cell work are right, it will soon become clear that research on embryos is unnecessary and the HFEA will decline to issue licences for such hES cell research. However, the Committee eschews such short responses, dedicating a whole chapter to a detailed and extremely careful consideration of the relative advantages and limitations of research-leading-to-therapy focused on, respectively, adult and embryonic stem cells. One of the most interesting points made here is that, even if research is focusing on adult stem cells, 'ES cells provide the only realistic means at present of studying the mechanisms and control of the processes of differentiation and dedifferentiation'.[29] In other words, if therapies derived from adult stem cells involve some reverse engineering (dedifferentiation) before being redirected (differentiated), ES cells provide a crucial point of comparison along the way. At the end of this discussion, the Committee concludes that, whilst recent research on adult stem cells looks promising and should be strongly encouraged by funding bodies and Government, the dual track approach is essential if maximum medical benefit is to be obtained.[30]

The ethical objection

For a Select Committee of non-scientists, chaired by the Bishop of Oxford, one of the leading issues was bound to be whether the destruction of embryos for research purposes could possibly be consistent with the principle of respect for the embryo[31] or (*per* Warnock[32]) recognition of its special status. The Committee's central conclusion on this issue is expressed in the following terms:

> Whilst respecting the deeply held views of those who regard any research involving the destruction of a human embryo as wrong and having weighed the ethical arguments carefully, the Committee is not persuaded, especially in the context of current law and social attitudes, that all research on human embryos should be prohibited.[33]

27 Report, para 3.16.
28 Report, para 3.20.
29 Report, para 3.17.
30 Report, para 3.22.
31 For an attempt to give this principle some meaning, see K. Lebacqz, 'On the Elusive Nature of Respect' in S. Holland, K. Lebacqz and L. Zoloth (eds), *The Human Embryonic Stem Cell Debate* (Cambridge, MA: MIT Press, 2001) 149.
32 See n 10 above.
33 Report, para 4.21.

Unpacking this conclusion, the Committee is accepting not simply that this is a pluralistic society but that moralists hold very different views about the status of the embryo and, concomitantly, whether it has rights or protectable interests or whether it represents a protectable good, and so on. As with the abortion debate, where proponents of choice can respect that rival proponents of life are arguing from a moral viewpoint, and vice versa, there can be mutual respect between moralists who debate the legitimacy of conducting research on human embryos.[34] The differences, however, are irreconcilable and a public position has to be taken. Faced with this, the Committee declines to re-open Warnock and rests not only on Warnock itself but on the subsequent accretion of permissive regulation and social acceptance thereof.[35]

Although it would have been startling, indeed, if the Select Committee had concluded that, current law and social attitudes notwithstanding, research on human embryos can in no circumstances be justified, this is the view of some. Moreover, those who hold this view will accuse the Committee of having evaded the issue by favouring consistency with present practice over a fundamental re-evaluation of that practice.

To some extent, the Committee responds to the deepest concerns of those who are fundamentally opposed to embryo research by endorsing the current 14 day limit,[36] by underlining the sensitivity that is required where human tissue is handled,[37] and by arguing that embryos, whether standard or CNR, 'should not be created specifically for research purposes unless there is a demonstrable and exceptional need which cannot be met by the use of surplus embryos'.[38] The Report also makes a pair of mutually re-inforcing recommendations that have, as yet, attracted little comment but which are very significant extensions of the necessity principle (and, thus, aim to minimise the use of embryos in research). These recommendations are: (i) that, where the HFEA grants licences for hES cell research, it should impose a condition requiring that any ES cell line generated in the course of the research should be deposited in a national cell bank; coupled with (ii) that, before granting any new licence for hES cell line research, the HFEA should be satisfied that no suitable cell lines are already available in the cell bank.[39]

Now, much of the opposition in Europe to research on human embryos, including the opposition reflected in Article 18 of the Convention on Human Rights and Biomedicine, is predicated on a fundamental principle of respect for

34 cf the strategy in R. Dworkin, *Life's Dominion* (London: Harper Collins, 1993).

35 Once one accepts that there will be surplus embryos from IVF programmes, and that these embryos will have to be destroyed, it is relatively easy to go along with their commitment to research. This kind of pragmatism is forthrightly put by E. Jackson in *Regulating Reproduction* (Oxford: Hart Publishing, 2001) at 230: 'If the disposal of spare embryos is inevitable, it is difficult to see why washing an embryo down the drain would be morally preferable to using it in order to carry out valuable research.'
See, too, the views put to the Select Committee on behalf of the Office of the Chief Rabbi, by Dayan Chanoch Ehrentreu, Evidence (Examination of Witnesses July 9, 2001) 83 *et seq.*

36 Report, para 4.22.

37 Report, para 4.25.

38 Report, paras 4.28 (standard embryos) and 5.14 (CNR embryos). Compare Lebacqz, n 31 above, at 160:

Researchers show respect towards autonomous persons by engaging in careful practices of informed consent. They show respect toward sentient beings by limiting pain and fear. They can show respect toward early embryonic tissue by engaging in careful practices of research ethics that involve weighing the necessity of using *this* tissue, limiting the way it is to be handled and even spoken about, and honoring its potential to become a human person by choosing life over death where possible.

39 Report, para 8.29.

human dignity. The Select Committee notes that human dignity lies at the root of much international concern about new science and medicine but finds the idea elusive and short on practical guidance in the context of setting limits to permissible research on human embryos.[40] Interestingly, whilst the Committee fully supports the ban on reproductive human cloning in the Human Reproductive Cloning Act 2001, it does not rest on human dignity – rather it is the risk of physical abnormality coupled with the ambiguity of a cloned child's relationship with its parents that troubles the Committee.[41]

Undoubtedly, human dignity has little to contribute to our ethical advancement if it simply acts as an emphatic way of expressing our concern about some new possibility in science or medicine. Indeed, for many bioethicists, the rediscovery of the value (or virtue) of human dignity is a cause for concern. Even if this is not an attempt to reverse the secularisation of the discipline, it is, so it is claimed, a retrograde step. For instance, according to the bio-ethicist Helga Kuhse:

> [T]he notion of human dignity plays a very dubious role in contemporary bioethical discourse. It is a slippery and inherently speciesist notion, it has a tendency to stifle argument and debate and encourages the drawing of moral boundaries in the wrong places. Even if the notion could have some use as a short-hand version to express principles such as 'respect for persons', or 'respect for autonomy', it might, given its history and the undoubtedly long-lasting connotations accompanying it, be better if it were for once and for all purged from bioethical discourse.[42]

Many (including the Select Committee) would agree with Kuhse that human dignity is a slippery idea. However, it is also an extremely powerful idea and the thought that we might do better if we purged it from bioethical discourse is not merely wishful thinking, it is almost certainly misconceived. What we need is a defensible conception of human dignity as the basis on which human rights and human virtue are founded.

In our recent book, *Human Dignity in Bioethics and Biolaw*,[43] Deryck Beyleveld and I have attempted to map the discourse of human dignity – which runs right through from the dignity of human conception to death with dignity[44] – and develop a rationally defensible conception of this idea. Put shortly, the map is dominated by two ideologies. On one side, human dignity is proclaimed as the foundation of human rights which leads to dignity being argued in support of

40 Report, paras 7.3–7.7.

41 Report, para 5.21 and Appendix 6.

42 H. Kuhse, 'Is There a Tension Between Autonomy and Dignity?', in P. Kemp, J. Rendtorff and N. Mattson Johansen (eds), *Bioethics and Biolaw Volume II: Four Ethical Principles* (Copenhagen: Rhodos International Science and Art Publishers and Centre for Ethics and Law, 2000), 61, at p 74.

43 D. Beyleveld and R. Brownsword, *Human Dignity in Biothics and Biolaw* (Oxford: OUP, 2001).

44 Most recently, two important 'death with dignity' cases have been heard in the English courts: namely, *The Queen on the Application of Mrs Dianne Pretty* v *DPP and Secretary of State for the Home Department* [2001] UKHL 61, and *Ms B* v *An NHS Hospital* [2002] EWHC 429 (Fam). Whereas *Pretty* (in which a subsequent appeal to the ECHR was unsuccessful: see *Case of Pretty* v *The United Kingdom*, application no. 2346/02, Judgment of April 29, 2002) is an assisted suicide case, that of *Ms B* is one of withdrawal of treatment at the self-conscious, explicit and repeated request of the patient. The position taken by the House of Lords in the *Pretty* case, rejecting the application, parallels that taken by the majority of the Canadian Supreme Court in *Re Rodriguez and A-G of British Columbia* (1993) 107 DLR 4th 342. Although the House's cautious approach to relaxation of the law involves a restriction on individual autonomy, it is, it is submitted, not a misreading of human dignity as such. Rather, the ruling against Mrs Pretty rests on placing the interests of those who are perceived to be vulnerable above the interests of those whose autonomy could (if only the Courts permitted) be exercised. By contrast, the decision in *Ms B* (in favour of the applicant) is fully in line with respect for human dignity and Dame Elizabeth Butler-Sloss, P's analysis of the issue as one involving, not the right to die, but the right to choose, is exactly right.

individual autonomy (the conception of 'human dignity as empowerment' as we term it). On the other side, a coalition of 'new dignitarians' (including Catholics, communitarians and Kantians) contend that certain practices or actions are intrinsically wrong as compromising human dignity (the conception of 'human dignity as constraint' as we term it). Whilst we think that a rationally defensible conception of human dignity will start with the idea of agents having purposive capacities and, concomitantly, rights against one another, there is more to our position than a simple endorsement of human dignity as empowerment. Our view is that, immanent within an autonomy-centred conception of human dignity, there is a dimension of responsibility that must be drawn out – which dimension begins to take shape once individual autonomy is placed in the context of human finitude and vulnerability in which rational agents strive to co-exist. Such a theory of human dignity sometimes yields very clear answers; sometimes it does not; but one thing that dignity enjoins is a degree of humility in the face of uncertainty. We may strive to be rational but we are not omniscient – and we should try to build in some practical safeguards against the possibility that we might be wrong about our moral theory or its applications. So far as human embryos are concerned, they do not yet seem to have the dignity-relevant capacities that would qualify them as rights-holders; but, if we are mistaken about this, we would do them a terrible wrong by terminating them. The calculation that has to be made is the possibility that we might be making just such a mistake against the possibility that we might be able to assist ostensible rights-holders whose basic well-being is under threat. It is not an easy calculation and, given such uncertainty, the regulatory regime in the United Kingdom seems to be on the right track in counselling precaution via the necessity principle.[45]

The practical (slippery slope) objection

Slippery slopes are only a cause for concern if they create uncertainty or if they carry us towards end states that we judge to be problematic. The momentum of the slope can be relevant to the way in which we address issues of principle or it can be important for its practical consequences; and, whether the bearing is on principle or practice in the first instance, the one can have implications for the other. Although there can be a tendency to appeal rather quickly to slippery slope fears, the objection is far from rhetorical or trivial – witness, for example, the way in which the Select Committee builds on current acceptance of the reproductive revolution to respond to ethical objections to embryo research.[46]

At the time that the Select Committee commenced its work, the most serious practical objection to the Regulations arose from the possibility that the HFEA might (almost certainly would, in fact) draw on the new purposes to license CNR in order to create customised embryos.[47] Such research, if successful, would improve our understanding of what happens when an adult nucleus is placed in an enucleated egg and apparently returns to an embryonic state as well as aiding the development of so-called therapeutic cloning techniques and therapies delivered by this route. However, in its early stages, CNR is common to both non-reproductive

45 Compare the views of the European Group, n 13 above.
46 And see Jackson, n 35 above.
47 Since the 1990 Act has been in force, a mere 118 embryos have been specially created for research (as against more than 50,000 donated from IVF programmes) (see Report, 4.26). It is quite likely that, with the new purposes now in play, considerably more embryos, including CNR-embryos, will be specially created for research.

(therapeutic) and reproductive cloning. In practice, it would be difficult to control the spread of this knowledge and prevent it falling into the hands of scientists who might be tempted to implant CNR embryos and go on to attempt human reproductive cloning. During the Summer recess in 2001, this concern was significantly inflamed by the various declarations of intent issued by the Italian embryologist, Professor Severino Antinori, who is seemingly determined to be the first person to reproductively clone a human being.[48]

At the same time that Professor Antinori was in the headlines, a judicial review initiated by Bruno Quintavalle, for the Pro-Life Alliance, was waiting to be heard. Essentially, the purpose of the Pro-Life Alliance's application was to question the regulatory jurisdiction of the HFEA in relation to CNR. Put simply, the argument was that, because an 'embryo' as defined by the 1990 legislation explicitly contemplates production by a process involving fertilisation, a CNR-embryo (having been generated by electronic stimulation of an engineered egg rather than by sperm fertilisation of an egg) cannot be an embryo for the purposes of the Act.[49] If this argument succeeded, as it did before Mr Justice Crane,[50] then CNR work fell outwith the jurisdiction of the HFEA and, in effect, was entirely unregulated. Professor Antinori, encouraged (paradoxically) by the Alliance's initial success, issued further declarations of intent involving his operation coming to England. The Government, on the back foot and perceiving a cloning crisis, rushed through the Human Reproductive Cloning Act 2001.[51] Subsequently, the High Court decision was to be reversed by the Court of Appeal[52] and the Pro-Life Alliance seemed set to abandon the case. There is more to be said about this,[53] but the question here is whether the restoration of the HFEA's jurisdiction over CNR coupled with the passing of the emergency legislation puts an end to the practical objection.

So far as CNR research within the United Kingdom is concerned, the legislation now draws a line at the point of implantation by prohibiting the placing in a woman of 'a human embryo which has been created otherwise than by fertilisation'.[54] Until the Court of Appeal reversed Mr Justice Crane's ruling, it would not have been unlawful to do CNR work preparatory to implantation; but, after the reversal, research involving CNR embryos will only be lawful where it is authorised by a licence given by the HFEA – and, so long as the 2001 Act remains in place, it is inconceivable that the HFEA would license research directed at human reproductive cloning. This does not quite cover all eventualities. For example, if reproduction became possible by utilising artificial wombs, there might be a way round the legislation. More importantly, as the activities of Professor Antinori highlight, practical concerns about human reproductive cloning arise within a

48 See, eg, J. Borger, 'Maverick scientists promise a human clone' *The Guardian*, August 8, 2001, p 1.

49 s 1(1)(a) of the 1990 Act defines an embryo as 'a live human embryo where fertilisation is complete'. Further, s 1(1)(b) provides that references to an embryo 'include an egg in the process of fertilisation'. The argument that '[n]uclear substitution constitutes propagation not fertilisation' was prefigured by M. Brazier in 'Regulating the Reproductive Business?' (1999) 7 *Medical Law Review* 166, at 189. In line with the decision to be handed down by Mr Justice Crane, Brazier concluded that 'nuclear substitution into an egg cell is unregulated in the United Kingdom today.' (*ibid*).

50 *R* v *Secretary of State for Health, ex parte Bruno Quintavalle (on behalf of Pro-Life Alliance)* 15 November, 2001.

51 The Bill was introduced a mere six days after the High Court decision and it became law two weeks later.

52 *R (Quintavalle)* v *Secretary of State for Health* [2002] EWCA, 18 January 2002.

53 See below at 581.

54 s 1(1) of the Human Reproductive Cloning Act 2001. The Act, which comprises just two sections, does not define 'fertilisation' but, presumably, it means human sperm fertilisation of a human egg.

global context. Closing the door within one legal system is one thing; closing it worldwide is another matter altogether. On this point, the Select Committee takes a realistic view:

> Securing international agreement is never easy, whatever the field of activity, and there would no doubt be formidable practical difficulties in negotiating a ban (or moratorium) even on such an apparently straightforward issue [as human reproductive cloning], on which there is known to be widespread agreement.[55]

The Committee concludes by recommending that 'the Government should take an active part in any move to negotiate an international ban on human reproductive cloning'.[56] Bearing in mind that the United Kingdom has declined to sign up to the European Convention on Human Rights and Biomedicine (partly because of the restrictions in Article 18), it has to be said that the Government would have some difficulty in claiming to be in the moral vanguard for the purposes of securing an international ban to match the domestic prohibition that is now in place in the United Kingdom.

If regulation, whether local or global, can only go so far, what else can one say about the practical objection? First, the technology is very primitive (even if Dolly is a story of success, by and large, the story of reproductive cloning in animals is one of massive failure). Secondly, the demand for human eggs far outstrips supply using present techniques for harvesting eggs. Thirdly, peer pressure within the scientific community will make life relatively difficult for would-be human reproductive cloners. And, fourthly, funding for such research will not be forthcoming from the principal backers of CNR-based work. Having said this, one suspects nevertheless that somewhere, some day, the 'worst' (assuming that this is the worst) will happen: a cloned child will be born.[57]

The legal objection

The fourth objection, hinted at in the House of Lords debate and pleaded initially in the Pro-Life Alliance case, was that the Regulations were *ultra vires*. Given that the Regulations reproduce virtually verbatim the enabling provisions in the 1990 Act, that the *ultra vires* point was dropped in the judicial review, and that the Court of Appeal decided against whatever was left of the Pro-Life Alliance's case, it might be thought that there is nothing more to be said against the Regulations from a legal perspective. Nevertheless, the Select Committee looked in some detail at how well the Regulations map onto stem-cell science and its potential development.[58]

55 Report, para 7.17.
56 Report, para 7.22.
57 We might recall the remarks of the United States Supreme Court in the landmark case of *Diamond* v *Chakrabarty* 447 US 303 (1980) at 317:

> The grant or denial of patents on micro-organisms is not likely to put an end to genetic research or its attendant risks. The large amount of research that has already occurred when no researcher had sure knowledge that patent protection would be available suggests that legislative or judicial fiat as to patentability will not deter the scientific mind from probing into the unknown any more than Canute could command the tides.

To similar effect, see the remarks by Yvette Cooper, Hansard (HC) November 17, 2000, col 1230. And, indeed, Professor Antinori has very recently claimed that one of the women in his programme is eight weeks pregnant with a cloned embryo: see James Meek, 'Woman "expecting first cloned baby"' *The Guardian*, April 6, 2002, p 1.
58 Report, paras 8.7–8.15.

First, the Regulations refer in the second and the third of the new purposes to 'serious disease'. Recalling the proportionality principle in the framework legislation, this might seem to be a commendable limitation – indeed, a fine example of self-restraint because the enabling provisions actually authorise additional purposes including increasing knowledge about 'disease' without any such qualification.[59] However, by introducing the restriction to *serious* disease, the Regulations immediately raise the question of (i) where the line lies between serious and non-serious and (ii) from whose perspective the disease is serious. There is also room for argument about the meaning of 'disease'. In their coverage of the Report, many newspapers highlighted the plight of the Superman filmstar, Christopher Reeve, who was paralysed in a riding accident. However, where the problem is spinal injury following an accident, would one characterise this as a 'disease'? Can we assume that 'disease' also covers disabilities and disorders? Leading the debate on the Regulations in the House of Commons, Yvette Cooper (the Parliamentary Under-Secretary of State for Health) had no doubt that 'serious disease' encompassed spinal injury:

> In response to the concerns raised, we have revised the regulations to ensure that the research relates to serious disease. Ultimately, it will be for the HFEA or the courts to interpret the term 'serious disease'. There can be no question but that it includes spinal injury, Parkinson's disease and cancer, the treatment of which could benefit from the research. We are not talking about a cure for the common cold.[60]

Such confidence notwithstanding, and sharing the concerns of some parliamentarians about possible interpretive difficulties, the Select Committee invites the Department of Health jointly with the HFEA to consider drawing up indicative guidance as to what qualifies as a serious disease for the purposes of the Regulations.[61]

Yet, it might be thought, this is surely unnecessary. Following the guidance in *Pepper* v *Hart*,[62] reference to Hansard would clear up most of the doubts about the meaning of 'serious disease'. After all, the Government seemed entirely clear in its own mind about the scope of the Regulations. Unfortunately, this is not really the point. Rather, the point is that the meaning of 'disease' (ignore the 'serious' qualifier) cannot be more inclusive than intended in the 1990 parent legislation. Provided that the Government's use of 'disease' in 2001 accords with the legislative understanding of that concept in 1990, all is well; otherwise it is not. It follows that the Select Committee's recommendation that there should be some clarification with regard to the interpretation of 'serious disease' is a shade more complex than it appears at first blush. For, on closer analysis, it is apparent that there are two rather different matters that invite clarification, namely: (i) the distinction between 'serious' disease and 'non-serious' disease (for which purpose reference to Hansard might be of some considerable assistance); and (ii) the meaning of 'disease' whether serious or non-serious (for which purpose reference to Hansard at the time of the passage of the Regulations is not strictly decisive).

59 As first drafted, the Regulations referred to 'human disease and disorders and their treatment', see Hansard (HC) November 17, 2000, Col 1178. However, the Regulations were then revised to ensure that research licensed was concerned with serious disease, see Hansard (HC) December 15, 2000, Cols 878 and 935.

60 Hansard (HC) December 15, 2000, Col 879. And, shortly after, at Cols 879–880: 'I have made it clear that the Government believe that spinal cord injury, heart disease, muscular dystrophy and osteoporosis are serious diseases...'.

61 Report, para 8.8.

62 [1993] AC 593.

Secondly, the Report rightly points out that many of the early applications to the HFEA for new purpose licences will be for basic research. For example, researchers must start by developing sound techniques for extracting and purifying hES cells. But, do the terms of the Regulations clearly cover such research purposes? The Select Committee suggests that, if there is any interim doubt about the application of the Regulations to basic hES cell research, it could be covered in one of two ways. One way would be to follow the legal advice already obtained by the HFEA to the effect that, where an application is directed at understanding how human stem cells behave and differentiate, such research 'may be appropriately described' as being concerned with increasing knowledge about the development of the embryo (ie under new purpose (a)).[63] However, if a court took a cautious literal approach – possibly backed by a principle of minimum derogation from basic values, such as the value of respect for human embryos – this argument might come unstuck. Moreover, an application to develop sound techniques for extracting and purifying hES cells is probably a more difficult test case for the Regulations than an application that is directed at understanding how human stem cells behave and differentiate. Another way of handling the point, the Committee suggests, is to take an altogether more purposive approach. As we have just underlined in connection with the meaning of 'disease', because the new purposes purport to be authorised by the 1990 legislation, it is the spirit and intent of that earlier legislation that is vital. Thus, the fact that the Government undoubtedly intended the Regulations to extend to basic hES cell research is not decisive.[64] Quite simply the question is: assuming that stem cell research using human embryos was not foreseen in 1990, does research of this kind (including basic research) fit with the purposes of that legislation? In the Pro-Life Alliance case, the Alliance relied on the guidance given by Lord Wilberforce in *Royal College of Nursing of the United Kingdom* v *Department of Health and Social Security*.[65] What Lord Wilberforce said in that case was that purposive interpretation could take on new developments 'if they fall within the same genus of facts as those to which the expressed policy has been formulated ... [or] if there can be detected a clear purpose in the legislation which can only be fulfilled if the extension is made'.[66] However, he emphasised that this was no licence for gap-filling and second-guessing what Parliament would have provided had they contemplated the new situation. Furthermore, in the RCN case, Lord Wilberforce (albeit in the minority) ruled that the extension sought by the Department involved a radical reconstruction of the legislation and was properly a matter for Parliament rather than the Courts. Of course, it was the restrictive side of Lord Wilberforce's guidance and its application in the case that appealed to the Pro-Life Alliance. However, in the event, the Court of Appeal turned the argument on its head and applied the permissive dimension of Lord Wilberforce's guidance to rule that CNR-embryos count as embryos. We will, it seems, not now know whether the House of Lords would have seen this as an unacceptable act of radical reconstruction but the important point is the general one: namely, that the line between acceptable purposive adjustment and unacceptable gap-filling is a fine one. Sensibly, the Select Committee, deeming it prudent to put the matter beyond any possible doubt,

63 See Report, para 8.11.

64 See Hansard (HC) December 15, 2000, Col 879, where Yvette Cooper asserts that, under the Regulations, research into the development of stem cells includes 'basic research'.

65 [1981] AC 800.

66 [1981] AC 800, 822.

recommends that 'when the Government bring forward legislation they should consider making express provision for such basic research as is necessary as a precursor for the development of cell-based therapies'.[67]

Thirdly, there is reason to think that, pursuant to the new purposes, licences will be sought to *create* more embryos for research than hitherto. Insofar as these specially created embryos are, so to speak, standard human embryos (donated *ab initio* rather than as surplus to IVF requirement), there should be no particular legal difficulty. However, if we are dealing with the production of customised embryos (such as CNR embryos), and if the real research focus is on the process by which such embryos are created (as with improving the process of CNR itself and embryo stimulation), then there might be a difficulty. The enabling provisions in the 1990 Act, it will be recalled, authorise the addition of further purposes provided that they are designed to 'increase knowledge about the *creation* and development of embryos, or about disease, or enable such knowledge to be applied'.[68] For some reason, however, the new purposes added by the Regulations do not include 'increas[ing] knowledge about the *creation* ... of embryos'[69] and, in the circumstances, it might be difficult to argue that such a purpose should be read in by implication. It would be ironic if, after the struggle to bring CNR research within the ambit of the HFEA's regulatory control, it was found that the Regulations had omitted to extend the licensing powers to this strand of research.

Fourthly, we can confidently assume that the 1990 Act and, in consequence, the Regulations cover at least two classes of 'embryos', those produced by fertilisation and those produced by CNR. What do we make, however, of embryos produced by other means? What would we make, for example, of eggs or even stem cells that are induced to become embryonic, or of CNR embryos generated from enucleated *animal* eggs, or of embryos that have used an egg engineered to overcome mitochondrial defects (so-called oocyte nucleus transfer)?[70] How far can we continue to stretch the scope of the legislative framework as science devises ever more ways of producing organisms that seem to have embryonic functions and potential?

Future debates and issues for the future

Debate about the ethics of stem-cell research will surely continue to reverberate around the world as one legal system after another, as well as regional and international legal communities, seek to clarify their regulatory positions. Within Europe, whereas the United Kingdom, Sweden and the Netherlands are generally regarded as being at the permissive end of the spectrum, Ireland and Germany are towards the restrictive end. Beyond Europe, in Australia, the House of

67 Report, para 8.15.
68 Sched 2, para 3(3), emphasis added.
69 Emphasis added.
70 See Report, paras 5.15–5.20. The Committee concludes that, if CNR is permitted in some circumstances then so, too, should oocyte nucleus transfer be allowed for research purposes. If this goes beyond research, the question of whether the procedure involves germ-line gene manipulation and, if so, whether this is problematic remains to be answered.

There are three related matters here: (1) research into mitochondrial disease/disorder; (2) research into possible treatments for mitochondrial disorders; and (3) actually treating mitochondrial disorders. The Government's view is that (1) is permitted under the 1990 Act and (2) under the 2001 Regulations, see Hansard (HC) December 15, 2000, Col 937. As I understand it, the Select Committee is agreeing that (2) is appropriate.

Representatives Standing Committee on Legal and Constitutional Affairs recently took a relatively restrictive view by recommending that, whilst extraction of stem cells from spare embryos should be permitted, there should be bans on both the creation of embryos for research and on human reproductive cloning coupled with a three year moratorium on the use of CNR.[71] Similarly, in Japan – at any rate, so the Select Committee was informed – while there is an explicit legislative ban on human reproductive cloning, national guidelines are being drafted with a view to authorising the extraction of stem cells from spare embryos; and it has been recently reported that there is growing pressure in China for the prohibition of human reproductive cloning and tight controls on CNR.[72] In the United States, President Bush's agonising last Summer about the use of federal funds for stem cell research has been well-publicised. There, the position is that federal funds can be applied to support research on hES cells but only in relation to cell lines derived before August 9, 2001. Where federal funds are not involved, research on hES cells is subject to State regulation which varies from extremely restrictive to extremely permissive.

Where legal systems have taken a restrictive view, the debate will focus on whether there should be some relaxation of the restriction, or some kind of compromise, in order to find acceptable ways of reaping the benefits of stem-cell research. So, for example, at the end of January 2002, the German *Bundestag* called for legislation to permit the importation of stem cell lines established before January 2, 2002. As in the United States, this move was designed to assuage concerns about government support for the creation and destruction of embryos for research but, at the same time, to facilitate hES cell research.[73] In those legal systems towards the permissive end of the spectrum (especially in the United Kingdom, where hES cell research has been given the green light), the debate will move on to review a range of questions concerning the increased use of customised embryos, as well as issues concerning informed consent and commercialisation (including patenting and the possibility that drugs companies might profit from access to cell lines for testing – a matter arguably underplayed in the Select Committee's Report).[74] And, everywhere, whenever a new step is proposed, arguments about human dignity will abound.

In the domestic debates, three perspectives (already dominant in current thinking about the regulation of biotechnology) can be expected to continue to lead the way. These are:

71 *Human Cloning: Scientific, Ethical and Regulatory Aspects of Human Cloning and Stem Cell Research* (Canberra, Australia, August 2001).

72 See J. Gittings, 'Experts call for curbs on human cloning in China' *The Guardian*, April 16, 2002, p 15.

73 See Report para 7.9 for the debate in Germany and paras 7.11–7.12 for the position in the United States.

On December 20, 2001, the European Life Sciences Group issued a press release expressing agreement that the EU should support stem cell research but that human reproductive cloning should be banned. The Group also agreed that, whilst the human embryo has a special moral status, the use of spare embryos for hES cell research is acceptable provided that it is carefully regulated, peer reviewed, scientifically sound, directed towards substantial goals and ethically controlled. The similarity between this agreement and the position taken by the Select Committee is obvious.

74 See Report para 6.3. For parallel concerns about informed consent and commercialisation, but in relation to the gene bank for medical research now being constructed in the UK, see J. Meikle, 'Biggest gene bank seeks 500,000 volunteers' *The Guardian*, April 17, 2002, p 9.

- utilitarian cost/benefit thinking
- the human rights perspective (grounded on human dignity and emphasising the importance of individual autonomy and choice)
- various rights-restricting appeals to human dignity (as articulated by the 'new dignitarian' alliance).

For much of the second half of the Twentieth Century, the principal tension in bioethics and biolaw was between utilitarian and human rights perspectives, with the latter at least seeking to qualify the former (Beauchamp and Childress' seminal approach to medical law and ethics reflects this kind of accommodation).[75] However, the most striking development of recent years is the reassertion of the importance of respect for human dignity. For instance, whilst the European Convention on Human Rights (likewise the Human Rights Act 1998) makes no mention at all of human dignity, the Council's Convention on Human Rights and Biomedicine advances respect for human dignity as its cornerstone principle. Human dignity is also the key value in the UNESCO Universal Declaration on Human Rights and the Human Genome;[76] and it is noteworthy that the Charter of Fundamental Rights of the European Union[77] opens with human dignity in Article 1.[78]

Where utilitarian thinking is operated in an impartial way, there should be no bias towards either benefit or cost. However, where promising developments such as those associated with stem-cell research are debated, there is a temptation to talk up the benefits while playing down concerns about risk and safety – not that this is a charge that can be levelled at the Select Committe.[79] Against utilitarianism, human rights thinking prioritises the importance of individual interests. Where embryo research is concerned, there can be some convergence with utilitarian outcomes because rights theorists will tend to privilege the choices of rights-holders (to donate embryos) against the questionable moral status of the embryo. However, in relation to informed consent requirements, utilitarians and rights theorists are likely to disagree, the former favouring convenience where procuring consent is problematic. For example, in *Kingdom of the Netherlands* v *European Parliament and Council of the European Union*,[80] the Netherlands brought an action under Article 230 EC (ex Article 173 EC) seeking annulment of Directive 98/44/EC on the Legal Protection of Biotechnological Inventions.[81] One of the grounds invoked for the annulment of the Directive was that it breaches fundamental rights, specifically by failing to provide for the free and informed consent of the donor of human biological material prior to an application to patent inventions developed from or using such material. In fact, it would have been very easy to read the Directive as requiring patent examiners to satisfy themselves that

75 T.L. Beauchamp and J.F. Childress, *Principles of Biomedical Ethics* (New York: Oxford University Press, 1979).
76 This Declaration was adopted at the 29th Session of the General Conference on November 11, 1997; and, on December 9, 1998, the UN General Assembly adopted Resolution A/RES/53/152 endorsing the Declaration.
77 As declared at the Nice European Council on December 7, 2000.
78 Generally, on the new dignity-based bioethics, see n 43 above, ch 2.
79 But, for a good example, see the reconsidered decision by the examiners in the Harvard Onco-mouse case OJ EPO 10/1992, 590. For an extended commentary, see D. Beyleveld and R. Brownsword, *Mice, Morality and Patents – The Oncomouse Application and Article 53(a) of the European Patent Convention* (London: Common Law Institute of Intellectual Property, 1993).
80 Case C-377/98, [2000] ECR I-6229.
81 [1988] OJ L 213/13.

proper informed consents were in place. Astonishingly, however, the ECJ (inspired by utilitarian thinking) ruled that such requirements, although involving matters of fundamental right, fall outwith patent law. And, one of the principal reasons for so holding, is that rigorous scrutiny of informed consents within the patent system would be a burden for the system and an inconvenience for researchers.

Against both utilitarian and (dignity-based) human rights thinking, it is the new (restrictive) dignitarian thinking that is perhaps most interesting. Typically, dignitarians refuse to contemplate research on embryos. As Lord Alton put it in the debate on the Regulations:

> One does not have to believe in the sanctity of human life, or that life begins at fertilisation, to be concerned about the general commodification of life. Every generation is tempted by the seductive and tantalising prospect of universal happiness as a trump over all other values and principles, but human dignity must always be defended against the abuse of scientific techniques.[82]

Similarly, the Select Committee heard it said by many witnesses that embryo research simply should not be permitted. In relation to consent issues the *Relaxin Opposition*[83] at the European Patent Office is instructive. This case, it will be recalled, is one of the milestones in accepting that inventions relating to copies of human genes are, in principle, patentable. There, it was unclear whether the pregnant women not only agreed that their tissue could be taken for research purposes but also understood that the research product might be patentable (and commercially exploitable). From a human rights perspective, such uncertainty might be troubling. However, the opponents of the patent, arguing from altogether a different perspective, objected quite simply that the patent compromised human dignity. This proved too simple for the Opposition Division who rejected the challenge without ever really being attuned to the dignitarian perspective pleaded.[84]

Gathering together these perspectives, what should we make of the possibility that cell lines in the cell bank might have commercial value? In the view of the Select Committee, it is not a good policy to introduce commercialisation upstream from the point at which the inventive research work takes place. Thus:

> It has been suggested that those who donate an embryo for stem cell research might subsequently expect a share in any benefits accruing from commercial exploitation of research on stem cell lines derived from it. In our view it would be undesirable for legislation to permit such claims: any commercial benefits will have come about as a result of the research and subsequent development rather than any intrinsic quality of a particular embryo donated. However, it makes it even more important that potential donors should fully understand the implications if embryos they are donating may be used for the production of stem cell lines, and in particular that the material donated may be used for a purpose other than the immediate one.[85]

Accordingly, the Committee recommends that the implications of stem cell line immortality should be fully explained to embryo donors and that, in order to avoid

82 n 22 above, at Col 28. Dignity, however, can be appealed to in more than one way. Baroness Greengross, for example, gave it a telling utilitarian application (*ibid* at Col 86):

> Dignity is important and it extends to many people who live with the results of horrific accidents, burns and acute illnesses, as well as degenerative diseases. They want to be free from the pain and suffering brought about by the diseases which have attacked them. Their dignity is important, too, and it is often hard to retain when in the grip of many of those conditions.

83 Howard Florey/Relaxin [1995] EPOR 541.

84 See R. Brownsword, 'The *Relaxin Opposition* Revisited' (2001) 9 *Jahrbuch für Recht und Ethik* 3.

85 Report, para 8.32.

restrictions on the use of cell lines, donation of embryos for hES research must be free of any specific constraints.[86] Such recommendations will go a long way towards meeting the concerns of rights theorists, for whom the principal issue will be that embryo donors act on the basis of free and fully informed consent (although the all-or-nothing nature of the donation might be seen as problematic). For utilitarians, unless the terms of donation put off too many potential donors, the Committee's recommendation should be welcome – after all, having obtained a blanket consent, there is then none of the practical inconvenience associated with going back for consents for secondary purposes and the like.[87] Alas, for the new dignitarians, although opposed to any form of commodification of human tissue, the Committee's recommendations for information and consent can never retrieve things once embryos are to be committed for research: it is the interests of embryos that must be prioritised, not those of donors, and nor those of the general public. No doubt, dignitarians will also note with some concern that the Committee is not radically opposed to the commodification of the embryo; for, it is not as though the Committee sees non-commodification as an integral aspect of respect for the embryo and its derivatives. Rather, the Committee's position seems to involve an acceptance that research on embryos will lead to commercial exploitation downstream, the only question then being one of who is to be permitted to profit from such commodification.[88]

Alongside the ethical debate, there will also be an ongoing regulatory debate. In the United Kingdom, the principal focus for this debate will be about the demands being put upon the HFEA[89] and the ageing Human Fertilisation and Embryology Act 1990.[90] If there is to be a major re-enactment of this legislative framework, scientists and funders are liable to put research on hold until the new framework is settled (which will, no doubt, evoke further utilitarian arguments about lost opportunities and benefits foregone). If there is to be no re-enactment, the question is whether we can make do with patching (whether via Regulations or soft law) coupled with a willingness on the part of the courts to apply a robust purposive approach in order to keep the law in touch with new developments. Elsewhere, the regulatory challenge will be considered for the first time, with attempts being made to find the right mix of hard law and soft law, and the right blend of facilitation and restriction.[91] Internationally, we can expect the regulatory debate to be rather different. It will be less concerned with striking the right balance between certainty and flexibility or with resolving profound ethical questions than with finding effective ways of enforcing whatever international consensus can be achieved.

86 Report, para 8.33.

87 Compare Article 22 of the European Convention on Human Rights and Biomedicine, which requires compliance with 'appropriate information and consent procedures' where it is proposed to store or use body parts for a purpose other than that for which they were originally removed.

88 This, of course, was the question famously litigated in *Moore* v *Regents of the University of California* (1990) 793 P 2d 479.

89 See Report para 8.5 where the Committee recommends that 'the Government should keep the funding of the HFEA under review and ensure that it is commensurate with its increased responsibilities'.

90 cf Brazier, n 49 above, for an insightful analysis of the Act, now situated in a context in which the reproductive revolution has generated a *global* fertility *business*.

91 See eg, the debate initiated in New Zealand by the Independent Biotechnology Advisory Council (IBAC) in *Cloning and Stem Cell Research* (2001) (available at <www.ibac.org.nz>).

Conclusion

Good science has a lot in common with a healthy policy debate about science. In both cases, judgements must be provisional; there needs to be a willingness to keep the arguments and the evidence under review. Whatever one makes of the particular position taken by the Select Committee – which, although relatively permissive, does draw the line at human reproductive cloning, does seek to minimise the use of human embryos for research, and does prefer the use of surplus embryos destined for destruction rather than the creation of new embryos – the spirit of the Report surely makes good sense. By supporting the Regulations, the Report opens the way for new and apparently extremely promising lines of research, but it does so without in any sense seeking to close the door on debate about the necessity, desirability, or acceptability of using human embryos.

In an area of fast moving science, there is a sense in which the law, too, is provisional. Quite clearly, there are at least two major questions that now need to be considered. First, although the present 1990 framework does well to make the necessity principle focal, it is rapidly losing touch with the science that it is intended to regulate. The only thing that is going to slow down embryology is if we pause to give the legislation an opportunity to catch up. Can we afford to do this – or, as lawyers might prefer to ask, can we afford not to? Secondly, what is the right frame of reference for the regulation of stem cell research? Where we are dealing with hES cell work, it naturally falls within the remit of the HFEA. However, as the Select Committee recognises, we need to monitor progress in relation to adult stem cell research; and there are also important possibilities arising from stem cells recovered from fetuses and blood cord. Moreover, with the development of a cell bank, one wonders whether the regulatory centre of gravity should be embryos or stem cells. And, as and when therapeutic applications of hES cell research are ready for clinical trial, is this the time to involve the Gene Therapy Advisory Committee?[92] This is not to suggest that the regulatory architecture has to have classical lines but it needs to be workable, efficient and effective. How best to achieve this is another large item of unfinished business to add to the agenda.[93]

92 cf Report para 8.23.

93 Please note, at the time that this piece was drafted, my assumption was that the Pro-Life Alliance case would not proceed beyond the Court of Appeal. However, it seems that the possibility of an appeal is still being considered.

Name Index